D1307570

THE WORLD OF BIOLOGY

THIRD EDITION

P. WILLIAM DAVIS
ELDRA PEARL SOLOMON

HILLSBOROUGH COMMUNITY COLLEGE

SAUNDERS COLLEGE PUBLISHING

Philadelphia   New York   Chicago
San Francisco   Montreal   Toronto
London   Sydney   Tokyo   Mexico City
Rio de Janeiro   Madrid

Address orders to:
383 Madison Avenue
New York, NY   10017

Address editorial correspondence to:
West Washington Square
Philadelphia, PA   19105

Text Typeface: Auriga
Compositor: York Graphic Services
Acquisitions Editor: Ed Murphy
Developmental Editors: Lynne Gery and Don Reisman
Project Editor: Carol Field
Copy Editor: Sally Atwater
Art Director: Carol Bleistine
Text Design: Emily Harste
Cover Design: Lawrence R. Didona
Text Artwork: J & R Technical Services
Production Manager: Tim Frelick
Assistant Production Manager: JoAnn Melody

Cover credit: © Nicholas Foster/THE IMAGE BANK

Library of Congress Cataloging in Publication Data

Solomon, Eldra Pearl.
  The world of biology.

  Rev. ed. of: The world of biology/P. William
Davis, Eldra Pearl Solomon. 2nd ed. c1979.
Bibliography: p.
Includes index.

  1. Biology.   I. Davis, P. William.   II. Davis,
P. William. World of biology.   III. Title.
QH308.2.S66   1985   574   85-11961
ISBN 0-03-059997-0

ISBN 0-03-059997-0.

6   071   98765432

CBS COLLEGE PUBLISHING
Saunders College Publishing
Holt, Rinehart and Winston
The Dryden Press

For Freda, Nathan, and Seth,
Mical, Amy, and Belicia

. . . and for their generation

PREFACE

Today, as never before, we have the knowledge to appreciate the marvelous complexity and precise function of life both on the cellular and ecosystem levels. One of our principal goals in preparing this book has been to contribute a sense of the excitement of modern biological science to the student's education.

AUDIENCE

THE WORLD OF BIOLOGY has been designed as a textbook for an introductory course in general biology. The book has been tailored to meet the needs of students who have only a minimal foundation in the physical sciences. The student population served by this textbook is very broad, encompassing not only some who might have a professional reason for taking a biology course but even more those whose majors lie in other fields entirely.

PHILOSOPHY AND APPROACH

What part might a biology course play in a student's total educational experience? Surely such issues as the nature and possible societal impact of genetic engineering have their place, as do problems resulting from the shortsighted human exploitation of the ecosphere. But the study of molecular genetics and ecology is also intrinsically justified. A college graduate should be familiar with the names and accomplishments not only of Julius Caesar, Charlemagne, and Cromwell, but of Pasteur, Miescher, and Crick as well.

All of the basic biologic principles are here, and wherever possible they are discussed in their social and environmental context. An important goal of this book is to provide the student with the tools needed to function as a biologically literate citizen. To this end, the emphasis in this book is on the interrelationships of living organisms—on their dependence upon one another and upon the environment. In addition to the three chapters devoted to ecology, including one on human ecology, interactions of the organisms that share our planet are stressed throughout the book. The mechanisms by which organisms have become adapted to their ecologic niches, and just how they pass on their genetic recipes for survival to future generations are also emphasized. Evolution and behavior receive three chapters, and genetics another five.

As compared to former editions of THE WORLD OF BIOLOGY, greater attention is given to the survey of the kingdoms of organisms. In particular, the coverage of the plant kingdom has been greatly augmented. The green fellow citizens of our world of life are accorded four substantial chapters covering systematics, structure, reproduction, and growth and development. In the chapters devoted to animal life processes, biological principles are discussed both comparatively and in terms of human body function. Thus, the integument and flight muscles of insects are given attention along with vertebrate skin and muscle.

READABILITY

Special care has been taken to make the reading level appropriate for today's college student. New technical terms are set in boldface and carefully defined. Each concept is thoroughly explained and, wherever appropriate, related to topics familiar and important to the reader. Students will find the writing style easy to read and enjoyable.

ORGANIZATION

THE WORLD OF BIOLOGY is organized into seven parts, each containing several chapters.

Part I The Organization of Life

Chapter 1 introduces the student to the basic characteristics and organization of living things, and provides a general overview of the subject matter of biology. An introduction to the methods of science is given in Chapter 1, and throughout the book examples of experimental work are discussed to give the student insight into the processes by which scientific knowledge is acquired. Chapters 2 and 3 lay the foundations in chemistry needed for an understanding of biological processes. These chapters have been designed to be biologically relevant. The structure and functions of cells and their membranes are discussed in Chapters 4 and 5, with special attention to recent advances in cellular biology.

Part II Life and the Flow of Energy

Chapters 6, 7, and 8 focus on energy and the energetic relationships of photosynthesis and cellular respiration.

Part III The Continuity of Life: Cell Division and Genetics

Chapter 9 integrates the behavior of chromosomes both in mitosis and meiosis. These topics are presented here because an understanding of chromosome behavior is an essential foundation for understanding genetic mechanisms discussed in Chapter 10. In Chapters 11 through 13 the student is introduced to the world of molecular genetics. Recombinant DNA, recombinant RNA, and other aspects of genetic engineering are discussed.

Part IV The Diversity of Life

In this unit we introduce basic concepts of taxonomy and survey the five kingdoms. Chapter 14 focuses on microbial life including the viruses, monera, protista, and fungi. Chapter 15 is devoted to the plant kingdom and Chapter 16 to the animal kingdom. Interrelationships of organisms are stressed throughout the discussion of diverse life forms.

Part V Plant Structure and Function

Chapter 17 discusses the basic structure and function of the vascular plant body. Chapter 18 summarizes reproduction, particularly sexual reproduction, in the more complex plants. Plant growth and development are described in Chapter 19.

Part VI Animal Structure and Function

Each system of the complex animal body is comparatively discussed, with emphasis upon human structure and function. Chapter 20 reviews animal tissues and introduces the student to the detailed material that follows. Chapter 21 unites three systems—integumentary, skeletal, and muscle—in one chapter on the grounds that the three typically function together. Neural control is discussed in Chapters 22 and 23, with Chapter 22 devoted to the neuron and neural transmission. Circulation is the subject of Chapter 24, and Chapter 25 is focused on immunity, reflecting the great research emphasis currently devoted to this area. Chapter 26 summarizes gas exchange processes.

Chapters 27 and 28 cover the related areas of food processing and nutrition, including some discussion of world hunger and malnutrition. Chapter 29 focuses on body fluid homeostasis with emphasis on the vertebrate kidney. Endocrine regulation, the subject of Chapter 30, is introduced here after the student has studied most of the other body systems so that the complex integrative functions of the endocrine system can be appreciated. Reproduction is the topic of Chapter 31, and development, an often slighted subject, is the focus of Chapter 32.

Part VII Evolution, Behavior, and Ecology

Chapters 33 and 34 explore organic evolution mainly from the standpoint of genetics and population dynamics. The origin of living things and the evidence for evolution are discussed. Behavior is discussed in Chapter 35, mainly from a social and ecological perspective. Chapters 36 through 38 conclude the book with a discussion of ecology both as an academic discipline and as the basis of necessary planning for the future of the human species.

LEARNING AIDS

To help the student succeed in mastering the principles of biology, a variety of pedagogic devices have been employed. A **chapter outline** at the beginning of each chapter shows the student how the material is organized and divides the material into manageable units. **Learning objectives** at the beginning of each chapter indicate exactly what the student must be able to do to demonstrate mastery of the material in the chapter.

Numerous **tables,** some of them illustrated, organize and summarize material throughout the text. **Focus boxes** present enrichment material or discuss topics such as glycolysis and the citric acid cycle in greater depth than presented in the text.

Illustrations have been carefully designed and selected to support the textual material. Conceptual diagrams, photographs, photomicrographs including many electron micrographs, and medical art are included to help the student accurately visualize concepts presented in the text. Full color is used throughout, increasing the teaching value of the illustrations and adding visual appeal to the text.

A **summary** in outline form at the end of each chapter helps the student to review the main ideas presented in the chapter. An objective **post-test,** included at the end of each chapter, permits the student to evaluate mastery of the concepts presented in the chapter. Answers to post-test questions are given at the back of the book. **Review questions** give the student the opportunity to check understanding of concepts, to apply them, and synthesize some of the material presented.

A glossary of terms is included in the combined **Glossary/Index.** An **appendix** of common prefixes, suffixes, and word roots is included.

SUPPLEMENTS

The **Study Guide** that accompanies the text has been designed around the learning objectives. Each chapter includes a list of key concepts, a scientific vocabulary matching test, questions testing mastery of each learning objective, a test evaluating overall mastery of the chapter, and a set of comprehensive questions. Diagrams for the student to label are included. Answers are provided to all questions and tests.

The **Instructor's Resource Manual** includes suggestions for course organization, lists of sources for audiovisual materials, and references. For each chapter there is an overview as well as suggestions for enrichment, including readings, films and filmstrips, topics for class discussion, and essay questions usable in tests or homework assignments. Also, the resource manual coordinates all the elements of the package on a chapter by chapter basis.

The **Test Bank** includes both chapter and unit tests. Answers are provided for all questions. Tests are presented in such a way that the instructor can duplicate them directly from the printed page. A **computerized test bank** is also available for the Apple II series and the IBM PC series.

A set of 105 **overhead transparencies** includes two- and four-color illustrations chosen and designed for optimum utility in the classroom and laboratory. There is also a set of 100 **35-mm slides,** all in full color. A **Laboratory Manual** and an accompanying **Instructor's Manual** are available. Also, a set of 40 **overhead transparencies of electron micrographs** is available.

ACKNOWLEDGMENTS

We thank the very talented and dedicated editorial staff of Saunders College Publishing for their help and support throughout this project. Our Developmental Editors Don Reisman and Lynne Gery worked along with us as we wrote, reorganized, and rewrote. They provided valuable suggestions for improving both content and clarity. Developmental Assistant Amy Leary helped us develop the striking four-color illustration program that graces the text. We are much indebted to Carol Field, our Project Editor, who stayed with us through mountains of manuscript, art boards, and halftones. Art Director Carol Bleistine and Manager of Editing, Design, and Production Tim Frelick contributed their considerable expertise in producing the final product. We thank our Publisher, Don Jackson, for his continued confidence and support.

We are grateful to Phala Pesano who has worked with us since we began preparing the first edition of THE WORLD OF BIOLOGY more than twelve years ago. Her skills at word processing, preparing the index/glossary, and rendering preliminary art, along with her wonderful optimism, have helped prevent panic through many deadline crunches. We also thank our families for their steadfast support and encouragement.

Much valuable input has been provided by the many instructors and students who have taken time

to share with us their responses to the second edition of THE WORLD OF BIOLOGY. We thank them and ask here for comments and suggestions from those who use this new edition. You can reach us through our editors at Saunders College Publishing.

REVIEWERS

We want to express our thanks to the many professors and researchers who have read the manuscript during various stages of its preparation and provided valuable input for improving it. Their suggestions have contributed greatly to our final product.

Melvin L. Beck, Memphis State University
Michael C. Bell, Richland University
Brenda C. Blackwelder, Central Piedmont
    Community College
Richard Boohar, University of Nebraska, Lincoln
E.F. Carell, University of Pittsburgh
John Chisler, Glenville State College
Earl Creutzburg, Parkland College
Donald Deeds, Drury College
Stephen J. Dina, St. Louis University
Alison Duxbury, Seattle Central Community College
Lynn Ebersole, Northern Kentucky University
Joseph Faryniarz, Mattatuck Community College
Robert J. Ferl, University of Florida
Richard Freiburg, MacMurray College
David Fromson, California State University,
    Fullerton
B.L. Frye, University of Texas
Paul Goldstein, University of North Carolina at
    Charlotte
Adair B. Gould, University of Delaware
Paul H. Gurn, Mattatuck Community College
Jim Hall, Central Piedmont Community College
William Hartig (deceased), St. John's University
Mary D. Healy, Springfield College
John J. Heise, Georgia Institute of Technology
Curt Huffman, Danville Area College
Charles Joungwirth, Roger Williams College
Arnold J. Karpoff, University of Louisville
Paul H. Monson, University of Minnesota
William O'Dell, University of Nebraska, Omaha
Patricia M. O'Mahoney-Damon, University of
    Southern Maine
R. Harvard Riches, Pittsburg State University
Herbert C. Robbins, Dallas Baptist College
Carol D. Schofield, Sacred Heart University
David M. Senseman, The University of Texas at San
    Antonio
R. Bruce Sundrud, Harrisburg Area Community
    College
Robert Tolbert, Moorhead State University
Richard R. Tolman, Brigham Young University
P. Kelly Williams, The University of Dayton

CONTENTS OVERVIEW

CONTENTS

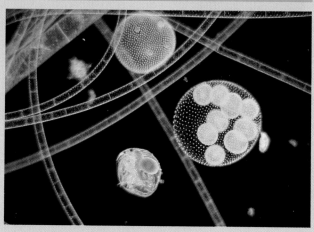

Volvox and *Spirogyra* (Tom Adams)

Female mantis eating and mating with male
(E.S. Ross)

PART III THE CONTINUITY OF LIFE: CELL DIVISION AND GENETICS, 153

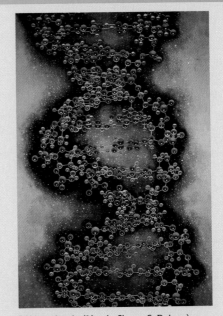

DNA molecule (Merck, Sharp, & Dohme)

PART IV THE DIVERSITY OF LIFE, 255

"89" butterfly (P.R. Ehrlich, Stanford Univ./BPS)

PART V PLANT STRUCTURE AND FUNCTION, 349

Leopard lily (R. Humbert/BPS)

PART VI ANIMAL STRUCTURE AND FUNCTION, 409

Skink (E.D. Brodie, Jr., Adelphi Univ./BPS)

PART VII EVOLUTION, BEHAVIOR, AND ECOLOGY, 683

Hoh River, Washington (L. Egede-Nissen/BPS)

PART I
THE ORGANIZATION
OF LIFE

Planet Earth seen from Apollo II, about 98,000 nautical miles away. (NASA)

Chapter 1
LIFE, SCIENCE, AND SOCIETY

Outline

Learning Objectives

After you have studied this chapter you should be able to:

1. Define biology and discuss its applications to human life and society.
2. List and briefly describe the characteristics of living systems.
3. Define metabolism and homeostasis and give specific examples of these life processes.
4. Define adaptation and describe its function in promoting perpetuation of a species.
5. List in sequence and briefly describe each of the levels of biological organization.
6. Describe the roles and interdependence of producers, consumers, and decomposers.
7. Identify the five kingdoms of living organisms and give examples of each group.
8. Design an experiment to test a given hypothesis by using the procedure and terminology of the scientific method.
9. Outline the ethical dimensions of science and give examples of ethical problems that may arise in the course of a scientific investigation.

The world of biology encompasses all of the living things that inhabit our planet, from the minute virus to the giant redwood tree. This diverse science extends to the interrelationships of living organisms, their origins, and their interaction with the environment (Fig. 1–1). **Biology,** then, is the study of life.

This book stresses the interdependence of living things and examines our own interactions with other organisms and with the environment. Early human beings were a harmonious part of the biological world, for their activities had little impact upon the environment. As human society has become increasingly technological, however, our activities have exerted a significant and often damaging effect upon our planet. The expanding human population, coupled with increasing consumption of natural resources, has transformed the earth. Chemicals from industries and modern agricultural practices have spread throughout the soil, water, and atmosphere and threaten to disrupt the delicate network of life upon earth's surface. As a result, we face many critical problems. Of all the sciences biology is perhaps of the greatest interest to us, for a knowledge of the principles of biology may be the key to human survival on planet Earth.

Understanding the principles of biology can also help us deal more intelligently with a wide range of routine concerns. Health care, nutrition, dieting, smoking, the use of drugs, and the care of domestic plants and animals are a few of the topics that may be of immediate interest. But the study of biology should enable us to extend beyond these personal interests by expanding our awareness of the millions of diverse life forms that share our planet, and our appreciation for the exquisite precision and complexity of living processes and systems.

WHAT IS LIFE?

Can you define the word *life?* Most likely not as easily as you could define a straight line in geometry. Even biologists sometimes have difficulty in defining such terms as *life, living,* and *alive.* The diversity of the living things that inhabit our planet precludes lumping them together with a simple definition. There are, however, certain characteristics and activities that a human being has in common with an earthworm, a tree, or even a single-celled ameba. Taken together, these features constitute life. All living things—more formally referred to as **living systems** or **organisms**—grow, carry on self-regulated metabolism, move, respond to stimuli, reproduce, and adapt to environmental changes. In this section we explore these characteristics in some detail.

(a)

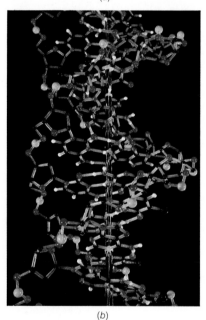
(b)

Figure 1–1 Modern biology examines the world of life in all its details and interactions. (*a*) An organism's ability to reproduce itself is essential to the continuity of life. The "secret" behind the hatching of these hognose snakes may be found in (*b*), a model of DNA, the basic hereditary material of life. The colored balls represent the different atoms that make up DNA: black = carbon, red = oxygen, white = hydrogen, blue = nitrogen, yellow = phosphorus. (*c*) An angelfish guarding a section of coral reef. Biologists are concerned with the physical characteristics of the fish, how its body functions, its behavior, and its interaction with other living things in its environment. (*a*, Animals Animals, Zig Leszczynski. *b*, Phil Degginger. *c*, Charles Seaborn.)

(c)

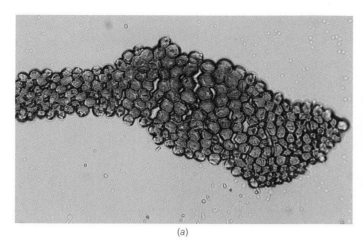

(a)

(b)

Figure 1–2 Multicellular and single-celled aquatic plants. (a) Single-celled organisms are both smaller and much less differentiated than multicellular organisms. The colonial green alga, *Chlamydomonas*, carries on all of its life functions within its one cell. Many individual algae are grouped together here forming a colony. The water lily (b), living in the same environment, is a much more complex plant. Its body parts contain specialized cells that carry on certain tasks. The leaves contain cells that carry on photosynthesis; the flower, reproduction; and the roots, absorption of nutrients. (a, E.R. Degginger.)

Cellular Structure

Each kind of organism is recognized by its characteristic appearance and structure. Living things are not homogeneous but are, instead, made of different parts, each with special functions. Their bodies are characterized by a specific complex organization.

All living things (except the viruses) are composed of cells, the basic structural and functional unit of life. The cell is the simplest unit of living matter that can carry on all the activities necessary for life. Some of the simplest organisms, such as bacteria, consist of a single cell. In contrast, the body of a human or an oak tree is made of billions of cells (Fig. 1–2). In such complex organisms, the processes of the entire organism depend upon the coordinated functions of the constituent cells.

Growth

Some nonliving things appear to grow. A snowball rolling down a hill becomes larger as snow gathers around it, and when a supersaturated solution cools, crystals may aggregate to form an enlarging mass of rock (Fig. 1–3). These inanimate objects increase in size by adding on pre-existing materials externally. In contrast, living systems grow by taking in raw materials from the environment and refashioning them into their own specific types of substances. Biologic growth, then, usually proceeds from the inside out.

Metabolism

To grow and maintain itself, an organism must be able to convert food materials into living cells. Such a conversion requires the expenditure of energy—energy that is also obtained, at least by animals and decomposing organisms, from the food they consume. Complicated chemical reactions liberate this energy, whereas other complex chemical reactions convert food materials into new tissue. Still other reactions maintain the routine operations of cell, skeleton, muscle, and nerve that keep the "engines of life" turning over.

Metabolism is the sum total of all these chemical events taking place within an organism. It includes the chemical reactions essential to nutrition, growth, and repair of the system, and the conversion of energy into forms useful to the cells (Fig. 1–4). Metabolic reactions are constantly occurring in every living system. When they cease, the organism dies.

(a)

(b)

Figure 1–3 Biological growth involves the refashioning of raw materials in accordance with the organism's internal organization. (a) Crystals of copper sulfate can grow in a supersaturated solution, but their internal structure is undifferentiated. (b) This sword plant, living in a volcanic crater in Hawaii, uses the energy of sunlight and a supply of carbon dioxide, water, and minerals to maintain its cells and grow. (b, Charles Seaborn.)

Homeostasis

Living systems need the appropriate machinery for carrying on metabolic activities. But this alone is insufficient. Metabolic activities must also be self-regulated so as to maintain a balanced state within the organism. The organism must "know" when to synthesize what, or when more nutrients or extra energy is required. On the other hand, it must not produce too much of any specific substance. When enough of a product has been made, the synthesizing mechanisms must be turned off. The organism should also be capable of adjusting its metabolism in response to changes in its external environment. The automatic tendency to maintain a constant internal environment is called **homeostasis,** and the mechanisms designed to accomplish this task are known as homeostatic mechanisms.

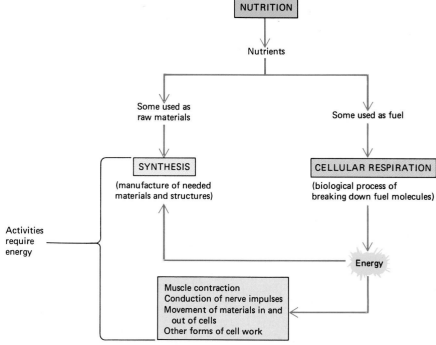

Figure 1–4 Relationships of some metabolic activities. Some of the nutrients provided by proper nutrition are used to synthesize needed materials and cell parts; other nutrients are used as fuel for cellular respiration, a process that captures energy stored in food. The energy is needed for synthesis and other forms of cellular work. Cellular respiration also requires oxygen, which is provided by the process of gas exchange.

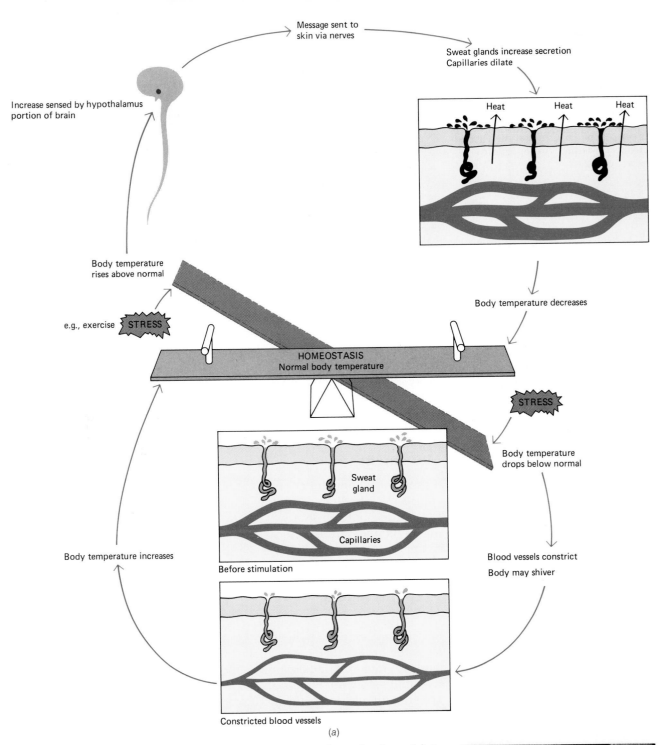

Message sent to
skin via nerves

Sweat glands increase secretion
Capillaries dilate

Heat   Heat   Heat

Increase sensed by hypothalamus
portion of brain

Body temperature decreases

Body temperature
rises above normal

e.g., exercise STRESS

HOMEOSTASIS
Normal body temperature

STRESS

Body temperature
drops below normal

Sweat
gland

Capillaries

Body temperature increases

Before stimulation

Blood vessels constrict
Body may shiver

Constricted blood vessels

(a)

Figure 1–5 Regulation of body temperature by homeostatic mechanisms. (*a*) An increase in body temperature above the normal range stimulates special cells in the brain to send messages to sweat glands and capillaries in the skin. Increased circulation of blood in the skin and increased sweating are mechanisms that help the body get rid of excess heat. When body temperature falls below the normal range, blood vessels in the skin constrict so that less heat is carried to the body surface. Shivering, in which muscle contractions generate heat, may also occur. (*b*) The sunning behavior of this marine iguana, *Amblyrhynchus cristatus,* a native of the Galapagos Islands, is homeostatic. The animal positions itself to maximize the heat it receives from the sun, thus increasing its body temperature. (*b*, J.N.A. Lott, McMaster University/BPS.)

(b)

Regulation of body temperature is a good example of a homeostatic mechanism (Fig. 1–5). When body temperature rises above its normal 37° Celsius (that is, 98.6° Fahrenheit), the increase is sensed by a special

Figure 1–6 A frog-eating bat, emerging with its prey. Not all movement in the biological world is as dramatic or as difficult to photograph as what you see here. This photograph was taken at night using a high-speed strobe. (Merlin Tuttle, Photo Researchers, Inc.)

"thermostat" in the hypothalamus of the brain, which sends impulses via nerves to the sweat glands in the skin. Sweat production then increases, and as sweat evaporates from the skin, body temperature is lowered. At the same time, capillaries (tiny blood vessels) in the skin dilate (expand), permitting the blood to carry body heat to the surface more efficiently. The heat radiates from the surface. When the body temperature decreases again, these changes are reversed. If body temperature falls even lower, messages from the hypothalamus cause blood vessels in the skin to constrict. Heat may also be generated by the muscular contractions associated with shivering.

Movement

Movement, although not necessarily locomotion, is characteristic of life (Fig. 1–6). The living material of cells is itself in continuous motion. A tree cannot pull up its roots and walk away, but it moves as it grows, opens its buds, transports its food, and responds to stimuli in the environment. Complex animals, such as human beings, possess groups of muscles that make complicated, purposeful movements possible.

Responsiveness

Figure 1–7 A few plants, such as the Venus's flytrap, can respond to the touch of an insect by trapping it. Here a leaf of the Venus's flytrap is shown attracting and capturing a lacewing. The leaves of this plant have a scent that attracts insects. Trigger hairs on the leaf surface detect the presence of an insect, and the leaf, hinged along its midrib, folds. The edges come together and hairs interlock, preventing the escape of the prey. The leaf then secretes enzymes that kill and digest the insect. (E.R. Degginger.)

Living things actively respond both to changes (stimuli) in the external environment and to changes within themselves. Suppose you were to hit an automobile with a heavy stick. The only consequence would be a dent (unless another living organism, such as the owner, were nearby!). Even the simplest living thing is able to preserve its own integrity by detecting stimuli and reacting to them. The cell itself is **irritable**—that is, sensitive to changes. When in need of nutrients a one-celled organism, such as an ameba, reacts positively to food in its aqueous surroundings by flowing toward it and engulfing it. Plants also respond to stimuli (Fig. 1–7). Most of us have at some time observed how plants grow toward light, roots tend to grow toward water, and vines wrap around solid objects.

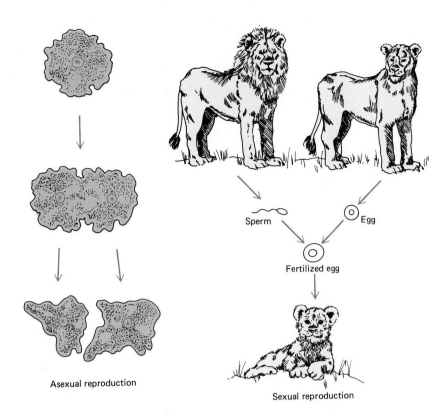

Figure 1–8 Approaches to reproduction. (*a*) In asexual reproduction one individual gives rise to two or more offspring—all identical to the parent. In sexual reproduction two parents each contribute a sex cell; these join to give rise to the offspring, which is a combination of both parents. (*b*) Asexual reproduction in *Micrasterias,* a unicellular green alga. (*c*) A pair of tropical flies, mating. (*b*, Biophoto Associates, Photo Researchers, Inc.; *c*, L.E. Gilbert, University of Texas at Austin/ BPS.)

Sperm

Egg

Fertilized egg

Asexual reproduction

Sexual reproduction

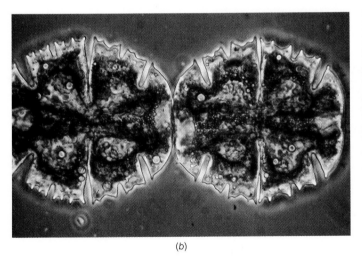

(*b*)

(*c*)

Reproduction

Although the life spans of various organisms range from minutes to centuries, they are always limited. The death of an individual, or even a generation of individuals, does not mark the end of a species,[1] however. Perpetuation of a species is provided for by the process of reproduction, and a biologically successful individual reproduces before it dies (Fig. 1–8).

In simple organisms, such as the ameba, reproduction may be **asexual,** in which a single parent buds or divides, giving rise to offspring. When an ameba has grown to a certain size, it reproduces by dividing in half to form two new amebas. Before it divides, an ameba makes a dupli-

---

[1] A species is a group of organisms that share a common origin and that freely interbreed in their natural environment to produce fertile offspring. For example, dogs belong to one species, horses to another.

Figure 1–9 This female bengal tiger is normally colored, but her offspring is white (albinistic) due to an inherited inability to produce normal pigments. The father, not shown, is also normally pigmented. The parents of this white tiger are both genetic carriers for the white trait. The white animal is genetically pure for this trait. Sexual reproduction makes it possible for offspring to differ genetically, and therefore in appearance, from their parents. Only about 50 white tigers are known to exist in the world, all descendants of one white male captured in India in 1951. (Photographed by the authors at Busch Gardens, Florida.)

cate copy of its hereditary material, or **genes,** and one complete set of genes is distributed into each new cell. Except for size, each new ameba is identical to the parent cell. Unless eaten or destroyed by environmental conditions, such as pollution, an ameba does not die; rather, it becomes a part of the new generation.

In human beings and other complex organisms, **sexual reproduction** is carried out by certain specialized cells. Usually two types of reproductive cells fuse to form a fertilized egg. Generally, though not always, the sexual process involves two individuals, male and female.

Sexual reproduction is biologically important because it increases variation in a species. Each offspring is not a duplicate of a single parent but is, instead, the product of various genes contributed by both the mother and the father (Fig. 1–9). Variety is the raw material for the vital processes of evolution and adaptation.

Adaptation

Adaptations are traits that enhance an organism's ability to survive in a particular environment. Whether structural, physiological, behavioral, or a combination of these, adaptations may enable an organism to more effectively maintain homeostasis, grow, or reproduce. For example, the long necks of giraffes are adaptations for reaching leaves of trees (Fig. 1–10). Woodpeckers have structural adaptations—powerful neck muscles, beaks fitted for chiseling, and long chisel-like tongues—that enable them to secure insects from tree trunks. Cactus plants have adaptations needed for living in dry areas. Every biologically successful organism may be viewed as an impressive collection of coordinated adaptations.

The process of adaptation involves changes in populations rather than in individual organisms. Many adaptations occur over long periods of time and involve many generations. They are the result of such processes as **mutation** (chemical change in a gene) and natural selection. Through time the ability of genes to mutate spontaneously has often been the key to survival. If every organism were exactly like every other individual of its population, any change in the environment might be disastrous to all, and the population would become extinct. Differences among individuals initiated by random mutation and enhanced by sexual reproduction provide for a differential in the ability of these individuals to cope with changes in their surroundings. Those best suited to cope with any specific change live to pass on their genetic "recipe" for survival.

One of the most interesting cases of adaptation has been documented in England since 1850. The tree trunks in a certain region of

(b)

(a)

(c)

Figure 1–10 Some diverse adaptations. (a) The long neck of the giraffe is an adaptation for reaching the leaves high on the trees. (b) The scorpion fish, which is adapted to look much like a rock, is waiting to gulp the next small organism that unwarily swims by. (c) The leaves of the tropical pitcher plant, *Nepenthes*, are modified for trapping insects. The tubular leaves hold a liquid containing digestive enzymes. An insect crawls into the pitcher and over the downward-pointing hairs that line the inside. These hairs prevent the insect from escaping, and eventually it falls into the liquid and drowns. (a,c, E.R. Degginger; b, Charles Seaborn.)

England were once white because of a plantlike organism, a lichen, that grew on them. The common peppered moths were beautifully adapted for lighting upon these white tree trunks, since their light color blended with the trunks and protected them from predaceous birds (Fig. 1–11). At that time a dark moth was an oddity. Then human beings changed the environment. They built industries that polluted the air with soot, killing the lichens and coloring the tree trunks black. The light-colored moths became easy prey to the birds, but now those that were dark blended with the dark trunks and escaped the sharp eyes of predators. In these new surroundings, the dark moths were more suited for survival. Eventually, more than 90% of the peppered moths in the industrial areas of England were dark. Interestingly, with recent efforts to control air pollution, there has been an increase in the population of the light-colored moths.

(a)

Figure 1–11 Peppered moths illustrate the advantage of protective coloration and the ability of organisms to adapt to changes in their environment. (a) In England until the mid-19th century, dark moths were very rare. Tree trunks were covered with white lichens, and dark moths were at a definite disadvantage, for predaceous birds quickly spotted and devoured them. (b) When the pollution that came with the Industrial Revolution killed the lichens on the tree trunks and covered the trunks with soot, the light-colored moths fell easy prey to birds. The dark-colored moths had the advantage and their numbers increased dramatically. With recent efforts to control pollution, this trend has begun to reverse. (Courtesy of Laurence Cook, University of Manchester.)

(b)

Focus on . . .

THE EVOLUTIONARY PERSPECTIVE

In his book *The Origin of Species,* published in 1859, Charles Darwin synthesized many new findings in geology and biology and outlined a comprehensive theory that has helped shape the nature of biological science to the present day. Darwin presented a wealth of evidence that the present forms of life on earth descended with modifications from previously existing forms—the theory of organic evolution. His book raised a storm of controversy in both religion and science, some of which still lingers. It also generated a great wave of scientific research and observation that has provided much additional evidence that evolution is responsible for the great diversity of organisms present on our planet.

Darwin observed that individual members of a species show some variation from one another. For example, there may be differences in size, body structure, and color. He also observed that many more organisms are born than survive into adulthood and reproduce. His theory of natural selection explained the mechanism of evolution—organisms with traits that best enable them to cope with pressures exerted by the environment are the ones most likely to survive and produce offspring. As a result, these traits become more widely distributed in the population. Over long periods of time, as organisms continue to change (and as the environment itself changes, bringing different selective pressures), the members of the population begin to look increasingly unlike their ancestors. The structural, physiological, and behavioral adaptations that result are the end product of evolution (see figure).

The applications of and controversies surrounding the theory of evolution will be discussed in Chapters 33 and 34. However, in this first chapter of a book that is designed to introduce you to modern biology, it is important to emphasize that although evolution is itself a subdiscipline of biology, some element of an evolutionary perspective is present in almost every specialized field within biology. Darwin's theory of evolution has proved to be one of the great unifying concepts of biology. Biologists in almost every subdiscipline try to understand the features and functions of organisms and their constituent cells and parts by considering them in light of the long, continuing process of evolution. Additionally, biologists are constantly checking for verification of the evolutionary relationships among different organisms and often reinterpret these in the light of new evidence. Biology today is more than a science of describing and naming organisms and life processes. Biologists are concerned not only with the existence of structural similarities, but also with what these similarities (and differences) tell us about how organisms are related to one another and how things have come to be as they are.

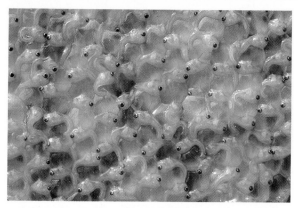

Eggs of the midshipman fish. Random events might be largely responsible for determining which of these developing organisms will reach adulthood and reproduce. However, certain desirable or undesirable traits that each organism might have will also contribute to its probability for success in its environment. Although not all organisms are as prolific as the midshipman fish, the generalization that more organisms are born than survive is true throughout the living world. (Charles Seaborn.)

A tool-using woodpecker finch (Camarhynchus pallidus). The behavior of this species from the Galapagos Islands is adapted for feeding on insects in crevices within the bark of trees. Lacking a long beak, the finch uses twigs or cactus spines as tools to dig out its prey. (Miguel Castro, Photo Researchers, Inc.)

LIFE'S ORGANIZATION

Order is the hallmark of life. Whether we study an individual organism or the world of life as a whole, its exquisite organization is evident. Several levels of biological organization are shown in Figure 1–12.

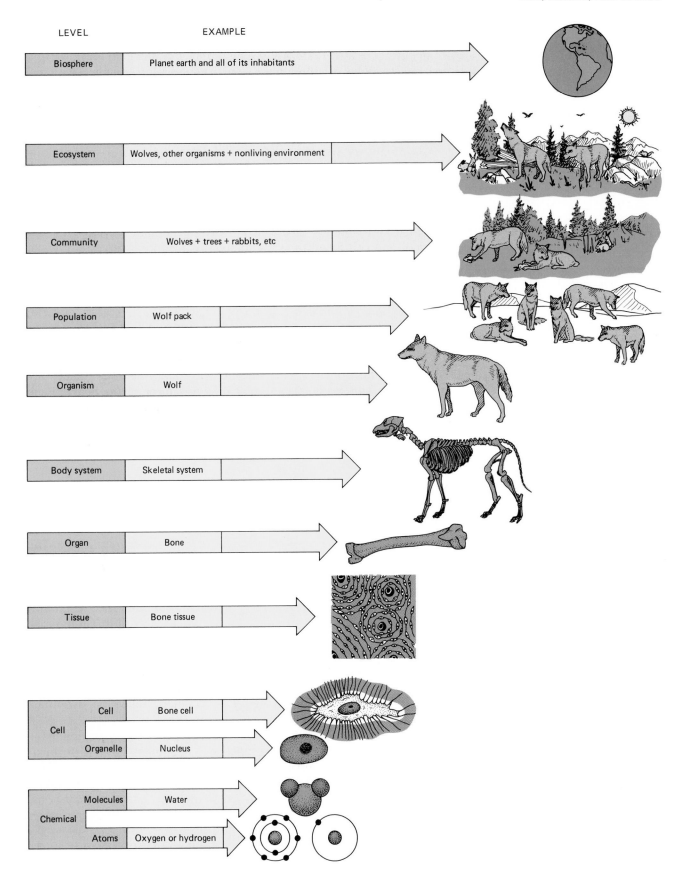

Figure 1–12 Levels of biological organization.

The Organization of the Organism

At the bottom of the organizational hierarchy, the **chemical level,** we find the basic particles of all matter, the minute and relatively simple atoms. An **atom** is the smallest amount of a chemical element (fundamental substance) that retains the characteristic properties of that element. For example, an atom of iron is the smallest possible unit of iron, and an atom of sodium is the smallest unit of sodium. As we will see in Chapter 2, even atoms consist of subatomic particles and have a very definite organization. Atoms combine chemically to form **molecules.** For example, two atoms of hydrogen combine with one atom of oxygen to form water. In turn, molecules associate with one another in prescribed ways to form the specialized structures that make up the **cellular level.**

Each cell consists of a discrete body of jelly-like **cytoplasm** surrounded by a cell membrane. Cytoplasm is a complex mixture of water and many other chemical substances. Some of these substances are organized into specialized structures called **organelles,** which perform specific functions inside the cell. Most types of cells contain a large organelle, called the **nucleus,** that holds the genes, the hereditary material containing the information governing the structure and function of the organism.

In most multicellular organisms, including human beings, cells associate to form **tissues** (e.g., muscle tissue), and various tissues are arranged into functional structures called **organs** (e.g., the heart). Each major biological function, such as circulation, is performed by a coordinated group of tissues and organs, an **organ system** (e.g., the circulatory system). Working together with far greater precision and complexity than the most complicated machine created by human beings, the organ systems make up the complex living **organism.**

Ecological Organization

Organisms interact to form several higher categories of organization. The members of a species are not uniformly or even randomly scattered over the earth. They are found in groupings called **populations** that inhabit a particular locality. Various populations of different species occupying the same area interact with one another forming **communities.** A community may be composed of hundreds or even thousands of different types of organisms. An **ecosystem** consists of the community plus its nonliving environment. An ecosystem is a community that is self-sufficient in an ecological sense. The study of how organisms of a community interact with one another and relate to their nonliving environment is called **ecology,** a word derived from the Greek *oikos,* meaning "house" or "dwelling" (Fig. 1–13).

An ecosystem generally consists of three varieties of organisms—producers, consumers, and decomposers—and must have a physical environment that is appropriate for their survival. **Producers** are algae, plants, and certain bacteria that can produce their own food from simple raw materials. These organisms are also known as autotrophs (self-nourishing). Most autotrophs carry on **photosynthesis** using sunlight as an energy source.

During photosynthesis the energy from sunlight is used to make complex food molecules from carbon dioxide and water. The light energy is transformed into chemical energy, which is stored in the food molecules produced. Oxygen, which is required not only by the plant cells but also by the cells of most other organisms, is produced as a by-product of photosynthesis.

(a)

(b)

Figure 1–13 Ecological organization. (a) A population of sable antelope at a water hole in Okavango Delta, Botswana. (b) A characteristic of ecosystems is the presence of various organisms with specialized lifestyles. In this scene from southern Africa, hyenas and vultures are scavenging the carcass of an animal that probably grazed on grasses and other plants. Though not particularly good hunters, hyenas are able to digest what other animals are likely to leave behind. The long, somewhat bald neck of vultures is especially suited to picking meat out of the interior of the carcass. (a,b, E.R. Degginger.)

$$\text{Carbon dioxide} + \text{Water} + \text{Energy (from sunlight)} \longrightarrow \text{Food} + \text{Oxygen}$$

Animals, including human beings, are **consumers** (Fig. 1–14). The consumers, like the decomposers, are **heterotrophs,** organisms that are dependent upon producers for food—and for energy and oxygen as well (Fig. 1–15). However, these organisms also contribute to the balance of the ecosystem. Like all living things (including producers), consumers and decomposers obtain energy by breaking down food molecules origi-

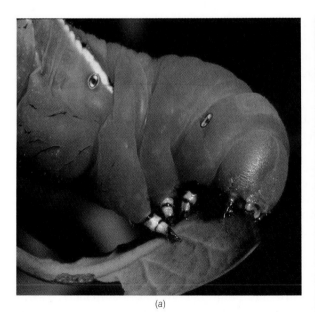

(a)

(b)

Figure 1–14 Consumers. (a) A tomato hornworm larva, *Manduca sexta*, feeding on a plant. All animals ultimately depend upon producers for their source of energy. Insects often do great damage to agricultural crops, thereby placing them in direct competition with humans for food resources. (b) An example of a consumer feeding upon another consumer. Shown here is a fishing spider feeding on a small fish. Its rapid movements and venomous sting make the spider an extremely effective hunter. (a, Peter J. Bryant, UC-Irvine/BPS; b, Robert Noonan, Photo Researchers, Inc.)

Figure 1-15 Interdependence of producers, consumers, and decomposers. Producers provide oxygen and food containing energy and nutrients for consumers. In turn, the consumers provide the carbon dioxide needed for photosynthesis by the producers. The decomposers break down wastes and dead bodies so that minerals are recycled.

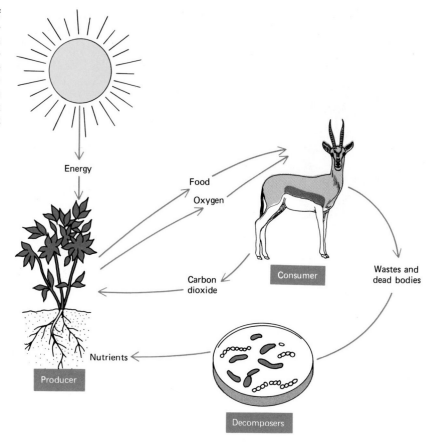

nally produced during photosynthesis. The biological process of breaking down these fuel molecules is termed **cellular respiration.** When food molecules are broken down during cellular respiration, their stored energy is made available for life processes.

$$\text{Food} + \text{Oxygen} \longrightarrow \text{Carbon dioxide} + \text{Water} + \text{Energy}$$

The life-sustaining mixture of gases in the atmosphere is a consequence of gas exchange between producers and heterotrophs by way of the non-living environment.

Decomposers—the bacteria and fungi—are an important component of an ecosystem, since they break down wastes and the bodies of dead organisms, making their constituents available for reuse. If decomposers did not exist, nutrients would become progressively locked up in the dead bodies of plants and animals, and the supply of elements required by living systems would soon be exhausted.

A familiar example of an ecosystem is a balanced aquarium. Plants or algae serve as producers, providing oxygen, food, and energy for the consumers, the fish. Bacteria and fungi function as decomposers, breaking down waste products and dead bodies so that nutrients within them can be recycled. As long as there is a continuing input of light energy for the plants, such a system may persist for months or even years. Eventually, though, it will collapse. A natural ecosystem, such as a pond, is far more stable and likely to last much longer. Stability is characteristic of such natural ecosystems, for unlike the balanced aquarium, they consist of a vast diversity of organisms, so chemicals and energy have a multitude of alternative pathways within the system. Should one type of organism die out, blocking one pathway, other organisms take its place and the system as a whole is hardly disturbed. Diversity is nature's grand tactic of survival.

Figure 1–16 Runoff of fertilizers from nearby farmland has caused a buildup of nutrients in this pond. As a consequence, there has been an explosive growth of algae on the surface of the water. Among its many effects, this layer of algae prevents sunlight from reaching plants and other algae below, causing them to die. Their decomposition reduces the concentration of oxygen in the water, which leads to fish kills. (Tom Adams.)

By biological tradition, the major natural communities, such as the Great Plains or the arctic tundra, are considered ecosystems, for they appear to be somewhat self-sufficient. But the largest ecosystem of all, and ultimately the only self-sustaining one known, is planet Earth with all its inhabitants—the **ecosphere.**[1]

SOCIETY AND THE ECOSPHERE

Our technological society exerts tremendous stress on the delicately balanced ecosphere. When we stress one aspect of an ecosystem, we often trigger a chain of events that magnifies the impact. For example, sulfur oxides emitted from fossil fuel power plants and industry are converted to sulfuric acid mists in the moist atmosphere. Such acid mist is often dispersed over wide areas by air flow patterns in the atmosphere. The acid comes back down to earth as acid rain, which damages soil and destroys vegetation (see Fig. 2–10). Acid rain also acidifies ponds and lakes, making them unsuitable environments for living organisms.

Many forms of air and water pollution destroy plants needed as a source of oxygen and food (Fig. 1–16). These producers serve as the nutritional base for animals that eat and in turn are eaten in the food webs that make up our ecosphere. And a significant number of the chemicals we dump into the environment return to haunt us within our own bodies. For example, each of us now carries strontium-90 and lead in our bones, mercury in our blood and other tissues, DDT in our fatty tissues, asbestos and other particulate matter in our lungs, and an unhealthy concentration of carbon monoxide in our blood.

In the early 1970s there was a widespread awareness of environmental problems, and public concern was responsible for passage of legislation and other actions directed at solving them. Unfortunately, just as styles of clothing change, so do intellectual fashions, and the environmental movement enjoyed only a brief heyday. The need for concern and action, however, is even greater today, for the problem, unlike the fad, shows no signs of disappearing.

---

[1] The terms ecosphere and biosphere are both used in this book. Although often used interchangeably, they have slightly different meanings. The term ecosphere emphasizes the living world as an ecosystem, whereas biosphere refers merely to the totality of life on our planet. Of course both terms encompass the same organisms.

Any chance of dealing with environmental problems depends upon our degree of commitment, and enlightened commitment demands sound knowledge. However overpopulated the earth may be, energetic, knowledgeable, and committed people are in short supply. Many aspects of biology are intertwined with the environmental predicament. The principles presented in this book should provide the foundation necessary to understand the critical relationship between society and the ecosphere. The preservation of life itself depends upon it.

THE VARIETY OF ORGANISMS

The variety of living organisms that inhabit the ecosphere challenges the imagination. If we are to study their interrelationships, we need to make some sense of this diversity. To facilitate effective communication with one another, biologists have developed a formal system of naming and classifying organisms. Although the details of this system will be discussed in Chapter 14, it will be helpful to introduce a few basic concepts here. The basic unit of classification is the species. Several closely related species may be grouped in a common genus. For example, the dog, *Canis familiaris,* and the timber wolf, *Canis lupus,* belong to the same genus. The cat, *Felis domestica,* belongs to a different genus. Note that each organism has its own scientific name; the first part of the name indicates the genus and is capitalized, and the second part of the name indicates the species and is not capitalized. Both names are italicized.

According to the system of classification used in this book every organism (except the viruses) is assigned to one of five kingdoms: Monera, Protista, Fungi, Plantae, or Animalia (Fig. 1–17). The members of kingdom Plantae, the plants, and of kingdom Animalia, the animals, are the organisms most familiar to us. The single-celled bacteria and cyanobacteria (blue-green algae) belong to kingdom Monera. They differ from all other organisms in that they lack a membrane around the nucleus, and also lack other membranous structures within the cell. Kingdom Protista is composed of single-celled animal-like organisms known as protozoa and also of several groups of algae. Kingdom Fungi consists of the true fungi and another interesting group of organisms known as slime molds. These organisms serve as decomposers, generally absorbing nutrients from dead leaves and other organic matter in the soil. Table 14–2 in Chapter 14 compares the five kingdoms.

HOW BIOLOGY IS STUDIED: SCIENTIFIC METHOD

This book is about the science of biology—the systematic study of life. Because it is a science, biology offers you knowledge about life that is different from what you can gain from casual intuition or insight from your daily experiences. What distinguishes science is its insistence on a rigorous method to examine a problem and its attempts to design experiments and tests of reason to validate its findings. The essence of the scientific method is the posing of questions and the logical, disciplined search for answers to those questions. But the questions must arise from observations and experiments, and the answers must be testable by further observation and experiment.

What makes science systematic is the attention it gives to organizing knowledge so that it is readily accessible to all those who wish to build upon its foundation. Science is not mysterious; by its sets of rules and procedures it makes itself open to all who wish to take on its challenges. Science seeks to gain precise knowledge about those aspects of the world that are accessible to its methods of inquiry. It is not a replace-

Figure 1–17 A survey of life. (*a*) This bacterium, *Spirillum volutans,* which has been magnified several hundred times in this photomicrograph, propels itself by the whiplike flagellum at each of its ends. (*b*) *Paracineta,* a protist belonging to the protozoa, attached to a filamentous alga. Although one-celled, *Paracineta* is a complex organism. Note the ring of poisonous tentacles surrounding the entrance to its "mouth." (*c*) These sulfur tufts, *Hypholoma fasciculare,* are members of the kingdom Fungi. These common mushrooms are inedible and can usually be found living on old tree stumps or decaying logs. (*d*) Red algae are simple plants, (*e*) *Bougainvillea,* a flowering plant. The blossoms are tiny and inconspicuous, but the brightly colored leaves attract insects. (*f*) Some relationships between animals are not as simple as one consumer eating another consumer. In this coral reef scene from the Philippines, a clownfish, *Amphiprion percula,* is shown living in the tentacles of a sea anemone. The tentacles of the sea anemone contain a poison used to capture small marine animals; however, the mucus coating of the clownfish prevents the anemone from stinging it. Thus the clownfish is afforded protection from its enemies—and also has the opportunity to obtain a meal that the anemone had intended for itself. (*g*) African lions, *Panthera leo,* among the fiercest of animals, are also among the most sociable. They live peaceably in prides (groups) consisting of as many as 35 lions. (*a,* Carolina Biological Supply Company; *b,* P.W. Johnson, University of Rhode Island/BPS; *c,* courtesy of Leo Frandzel; *d,* Carolina Biological Supply Company; *f,* Charles Seaborn; *g,* courtesy of Busch Gardens, Tampa.)

(a)

(b)

(c)

(d)

(e)

(f)

(g)

19

ment for philosophy, religion, or art, and being a scientist does not mean excluding oneself from participation in these other fields of human endeavor.

Science in Action

In July 1983 a research report published in the *New England Journal of Medicine* challenged traditional medical practice of treating diabetics.[1] (Diabetes mellitus is a disorder in which carbohydrates [sugars and starches] are not utilized effectively by the cells; as a result the blood sugar level tends to be abnormally high.) For many years diabetics have been instructed to avoid dietary intake of sucrose (table sugar). They gave up all types of desserts and avoided sucrose as a sweetener. This practice had been accepted so widely for so long that few researchers ever tested the concept. The research team set up an experiment to test the hypothesis that diabetics should not eat sugar and found that there was no significant difference in their blood sugar levels when sucrose was included in *moderate* amounts in *nutritionally balanced* meals.

This example of science in action illustrates several important aspects of how science works. The research team investigating use of sugar by diabetics followed the steps of the scientific method, a general sequence of steps followed by scientists in their work. They may have worked something like this.

Steps in the scientific method	*Example*
1. Recognize and state the problem.	1. Diabetics are instructed to avoid sugar in their diet. Is there evidence to support this?
2. Collect information or data on the problem.	2. Researchers read journal articles (that is, do a literature search) relating to the problem.
3. Formulate a testable hypothesis.	3. Diabetics can include moderate amounts of sugar in their diet without ill effect.
4. Make a prediction on the basis of the hypothesis.	4. If the hypothesis is correct, the pattern of increase in blood sugar level will not be significantly different for ingested sugar from that for ingested starch (or other complex carbohydrate).
5. Make observations or design and perform experiments to test the hypothesis.	5. The responses of diabetic patients and healthy subjects to meals containing different carbohydrates were tested.
6. Formulate a conclusion.	6. Because there was no significant difference in blood sugar level responses, diabetic patients can include moderate amounts of sugar in their diet.

---

[1] "Postprandial Glucose and Insulin Responses to Meals Containing Different Carbohydrates in Normal and Diabetic Subjects," John P. Bantle, et al., *The New England Journal of Medicine,* Vol. 309, No. 1, July 7, 1983.

FORMULATING A HYPOTHESIS

A **hypothesis** is a tentative explanation, or trial idea, of observations or other known facts. A good hypothesis should be:

1. Reasonably consistent with all known facts, or more consistent with them than competing hypotheses.
2. Capable of being tested; that is, it should generate definite predictions whether the results will be positive or negative. Test results must also be repeatable by independent observers.
3. Simpler than competing hypotheses.

Can you apply these characteristics to the hypothesis in the example of the use of sugar by diabetics?

SETTING UP AN EXPERIMENT

An **experiment** is a controlled situation that produces observations or results relating to a hypothesis. In a properly designed experiment there is a control. A **control** is a test performed under the same conditions as the experimental test, except that the factor being tested is varied (Fig. 1–18). The research team that tested the responses of diabetics to sugar set up a complex experiment in which 12 patients with Type I (insulin-requiring) diabetes and 10 patients with Type II (non-insulin-requiring) diabetes were fed a breakfast of common foods on five mornings. On three days the breakfasts included simple sugars—glucose, fructose, or sucrose. On two days the breakfasts included complex carbohydrates— potato starch and wheat starch. The breakfasts including the complex carbohydrates served as controls. From these breakfasts the researchers could gather data indicating what the blood level responses were when the diabetics were not challenged with simple sugars. The blood sugar level patterns resulting from the experimental breakfasts could then be compared with the controls. In addition, 10 healthy subjects were included in the experiments as an extra control to provide data regarding the responses of healthy subjects to each type of breakfast.

In any experiment it is important that sufficient numbers of subjects be tested. If the research team testing the diabetics had used only one or two subjects, the experiment would not be considered statistically significant—the subjects selected may not have been typical of most diabetics. Because a total of 22 diabetics (and 10 healthy subjects) were included in the study, the results probably reflect the responses that would be made by most diabetics. Many members of the medical profession, however, may be skeptical of the results because only 22 diabetics

Figure 1–18 Pasteur's experiments disproving the spontaneous generation of microorganisms. Nutrient broth (sugar and yeast) was placed in two types of flasks and boiled to kill any bacteria present. (*a*) As his control, Pasteur used flasks with straight necks that permitted bacteria to settle into the broth. In these flasks, the broth was soon teeming with bacteria. However, Pasteur's experimental flasks, shown in (*b*), had long S-shaped necks that did not permit bacteria to enter, even though the flask was open to the air. Bacteria did not grow in such a flask unless its neck was removed.

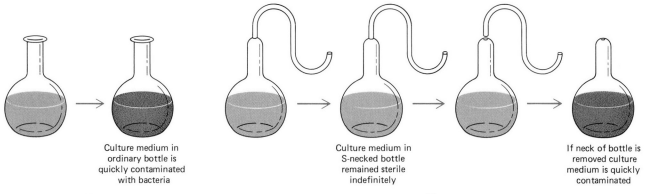

Culture medium in ordinary bottle is quickly contaminated with bacteria

Culture medium in S-necked bottle remained sterile indefinitely

If neck of bottle is removed culture medium is quickly contaminated

(*a*)

(*b*)

Focus on . . .

RECOGNIZING A PROBLEM: THE PREPARED MIND

According to legend, Isaac Newton was first led to think about gravity while idly watching the fall of ripe apples. Millions before him had observed the same phenomenon, as well as thousands of other happenings like it. But as Pasteur said, "Chance favors the prepared mind." It would seem that Newton's mind was the first that was prepared to see anything more in that apple's fall than the simple fact that it fell.

Discoveries are usually made by those who are in the habit of looking sharply at nature and whose minds are open and prepared—usually by some preliminary idea of what they are looking for, and by a subconscious collection of miscellaneous information that eventually begins to take shape. For this reason it is also helpful to have background knowledge of what others have discovered and the theories they have formulated. This is why Nobel prizes in science are more often won by scholars than by laypersons. Not that great scientific discoveries are hard to understand. Quite the contrary. Some of the most fundamental concepts are really quite simple. It is amazing how obvious the truly great hypotheses appear in retrospect, but how difficult they were to think of in the first place! Nothing is harder, it seems, than thinking a truly original, self-evident thought.

The discovery of penicillin is such an example. In 1929 the British bacteriologist Alexander Fleming noticed that one of his bacterial cultures had become invaded by a blue mold. He almost discarded it, but then he noticed that the area contaminated by the mold was surrounded by a zone where bacterial colonies did not grow well. His culture looked somewhat similar to the one shown here, in which a disc of penicillin has inhibited the growth of bacteria.

The bacteria were disease bacteria of the genus *Staphylococcus*. Anything that could kill them was interesting. Fleming saved the mold, a variety of *Penicil-*

lium, of which one kind of blue bread mold is a familiar example. It was subsequently discovered that the mold produced a substance that slowed bacterial growth—penicillin, the first antibiotic.

We may well wonder how often the same thing happened to other bacteriologists who failed to notice the key fact and just threw away their contaminated cultures. Fleming benefited from chance, yes, but his mind was prepared to observe it, and his pen to publish it.

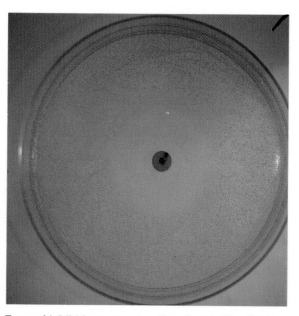

Zone of inhibition around a disc of penicillin. (Walter Dawn.)

were tested. It is likely that this experiment will be repeated, perhaps many times, using larger numbers of subjects and with many variations before physicians change the treatment plans for their diabetic patients.

How Hypotheses Become Theories or Principles

A hypothesis supported by a large body of observations and experiments becomes a **theory,** defined as a scientifically acceptable general principle offered to explain observations or events. The hypothesis that diabetics can include sucrose with their meals without ill effect cannot become a theory until it has been tested and retested many times. Theories may embody a number of related and generally accepted hypotheses. The atomic theory, for example, contains interrelated hypotheses from many areas of chemistry and physics.

If a theory has yielded true predictions with unvarying uniformity over a long period of time, and thus becomes almost universally accepted, it may be referred to as a scientific **principle.** Some time-honored scientific principles are known as **scientific laws,** for example, the law of gravity. However, the term *law* sounds so absolute that it may discourage challenge, and as a result, most scientists prefer the designation *principle*.

The Ethics of Science

Even though ethics is sometimes believed to be outside the realm of scientific investigation, science has many ethical dimensions. Ethics may be applied to the scientific method itself, to the conduct of experiments, or to the public release and use of scientific findings.

Honesty and objectivity at every stage of the scientific method are indispensable if the results are to have any value. Deliberately or subconsciously doctored data may mislead scientists for generations. Damage can also be done by withholding a portion of the results of a scientific investigation. Such a temptation is especially great among scientists who are directed to find "scientific" evidence that will support the claims made by their employers. Another area of ethical concern is the use of human subjects in medical research. For example, is it ethical to place patients in control groups when they might benefit from a treatment being tested?

Ethical concerns have also been raised regarding some types of basic research being conducted today, most notably research in genetics. We understand many of the details of the genetic code and are perhaps not far from being able to manipulate it extensively, altering and remaking living organisms. In recombinant DNA techniques, genes are transplanted from one organism into another, usually a bacterium, enabling the recipient to produce new substances. In this way bacteria can be used to produce needed therapeutic products, such as insulin, which can then be marketed commercially (Fig. 1–19). Further research in this area could lead to cures for certain genetic diseases. However, the potential misuses of such research are staggering. For example, superbacteria, resistant to antibiotics, could be developed for use in biological warfare. As members of society, we share a responsibility with scientists to prevent such abuses from occurring.

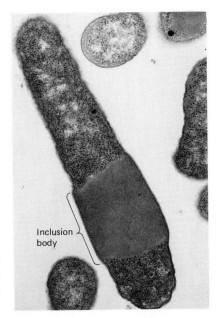

Inclusion body

Figure 1–19 A genetically engineered *Escherichia coli* bacterium (magnified ×68,000). Its hereditary material has been modified so that the bacterium produces large quantities of the hormone human insulin. Individuals with the disease diabetes mellitus have insufficient amounts of this hormone, which is necessary for the normal metabolism of sugar. Since the bacterium itself has no use for the insulin and has no way of excreting it, the hormone accumulates in an "inclusion body" within the bacterial cell. The insulin can be harvested by destroying the bacterium. (Courtesy of Dr. Daniel C. Williams and the Lilly Electron Microscope Laboratory.)

CHAPTER SUMMARY

I. Life may be defined in terms of the characteristics shared by living things.
  A. All living things (except the viruses) are composed of cells.
  B. Living things grow by taking in raw materials from the environment and refashioning them into their own specific types of substances.
  C. Metabolism is the sum of all chemical activities that take place in the organism.
    1. Metabolism includes the chemical reactions essential to nutrition, growth and repair, and conversion of energy to useful forms.
    2. Metabolic activities are self-regulated so as to maintain a steady state; the automatic tendency to maintain a steady state is called homeostasis.
  D. Although not all living things can move from one place to another (i.e., carry on locomotion), all exhibit movement.
  E. Living things respond to stimuli in their internal and external environments.
  F. Reproduction may be asexual, in which the offspring are identi-

cal to the parent, or sexual, in which the offspring reflect the characteristics of two parents.

    G. Through the process of adaptation, living systems acquire traits that enable them to survive in changing environments.

II. There is a hierarchy of biological organization.

    A. In a complex organism there is a chemical level of organization consisting of atoms and molecules, a cellular level consisting of organelles and cells, a tissue level, an organ level, and an organ system level. The organ systems working together make up the organism.

    B. The basic unit of ecological organization is the population. The various populations that inhabit an area form a community; a community and its physical environment make up an ecosystem. Planet Earth and all its inhabitants constitute the ecosphere.

        1. An ecosystem is a balanced community of producers, consumers, and decomposers living within an appropriate nonliving environment.

        2. During photosynthesis producers capture energy from sunlight and transform it to chemical energy. All living cells obtain energy by breaking down molecules originally produced during photosynthesis; this is called cellular respiration.

III. Human beings are part of a delicately balanced environmental system. Their activities are threatening the stability of that system so that their very survival is threatened. Increased, enlightened commitment to the task of solving environmental problems is needed.

IV. Each type of organism is assigned to a genus and species and classified in one of five kingdoms—Monera (bacteria and cyanobacteria), Protista (protozoa and some types of algae), Fungi, Plantae, and Animalia.

V. In their work, scientists follow a general sequence of action referred to as the scientific method.

    A. After identifying a problem and collecting data, a scientist formulates a creative hypothesis, a tentative explanation or trial idea that can be tested.

    B. The scientist makes a prediction based on the hypothesis and sets up an experiment to test the hypothesis.

        1. A properly designed experiment has a control and includes sufficient numbers of subjects (or trials) to be statistically significant.

        2. An experiment generates data, information upon which a conclusion can be based.

    C. A hypothesis supported by a large body of observations and experiments becomes a theory. A theory that is almost universally accepted becomes a scientific principle.

    D. Ethics can be applied to the scientific method, to the conduct of experiments, and to the public release and use of scientific findings.

Post-Test

1. The term *biology* literally means _____ .
2. The chemical and energy transformations that take place within an organism are referred to as its _____ .
3. The automatic tendency to maintain a steady state is called _____ .

4. A chemical change in a gene is referred to as a _____ .
5. Traits that enhance an organism's ability to survive in its environment are called _____ .
6. All living things except the viruses are composed of one or more _____ .

7. Various tissues are organized into functional structures called _____ .

8. Various populations interact with one another and with the environment to form _____ .

9. _____ is the study of how organisms of a community relate to one another and to the environment.

10. During _____ producers use the energy of sunlight to make food from carbon dioxide and water.

11. In cellular respiration food is broken down in the presence of oxygen, yielding carbon dioxide, _____ , and _____ .

12. Bacteria and fungi are ecologically classified as _____ .

13. Protozoa and some types of algae belong to the kingdom _____ .

14. Several closely related species may be assigned to a common _____ .

15. A _____ is a tentative explanation or trial idea.

16. A hypothesis supported by a large body of observations or experiments may become a _____ .

Review Questions

1. A child might argue that an automobile is alive. After all, it drinks water, guzzles gasoline, moves, and even responds to certain types of stimuli. How would you explain that a car is inanimate?

2. How does growth of a living organism differ from "growth" of a snowball as it rolls downhill?

3. Give an example of: (a) homeostasis. (b) metabolism. (c) adaptation.

4. Contrast producers, consumers, and decomposers. What would happen if all the decomposers on earth were suddenly to disappear?

5. Contrast photosynthesis with cellular respiration.

6. Organisms A and B are classified in the same species; organisms X and Y are assigned to the same genus but not to the same species. Which pair of animals would have the most characteristics in common?

7. In which kingdom would you classify a bacterium? an elephant?

8. How can the study of biology help you to function as a more enlightened citizen?

9. What is a control and why is it important? Why do you think that medical researchers often make themselves deliberately unaware of which patients are in their experimental group and which are in the control group?

10. Devise a suitably controlled experiment to test each of the following hypotheses: (a) Memory can be improved in elderly persons by including choline in the diet. (b) Heart attacks can be prevented by taking two aspirins daily. (c) Fertilizer X significantly increases crop yields.

Readings

Baker, J., and G. Allen: "Hypothesis, Prediction and Implication in Biology," Addison-Wesley, Reading, Mass., 1968. A good source for the beginning student who wishes to delve more deeply into the subject but who does not wish to become buried in a tome.

Bettex, A.: "The Discovery of Nature," Simon & Schuster, New York, 1965. A fascinating pictorial history of science. Teachers will find it a rich source of examples of the scientific method in action.

Cloud, P. "The Biosphere," *Scientific American*, September 1983, 176–189. A fascinating discussion of the relationship between microbial, animal, and plant life on earth and the physical environment.

Lederman, L.M.: "The Value of Fundamental Science," *Scientific American*, November 1984, 40–47. A discussion of how basic science research contributes greatly to technology and to the enrichment of our culture.

McMahon, T.A., and J.T. Bonner: *On Size and Life.* San Francisco: 1984. W.H. Freeman and Company. How the design of living things is influenced by their size.

Morris, I.: "Is Science Really Scientific?" *Science Journal*, Dec. 1966, p. 76. An excellent short summary of modern scientific philosophy.

Chapter 2

THE CHEMISTRY OF LIFE: ATOMS AND MOLECULES

Outline

I. Chemical elements
II. The atom
    A. Atomic structure
    B. Electron configuration
III. Chemical compounds
    A. What are formulas?
    B. Chemical equations
IV. How atoms combine: chemical bonds
    A. Covalent bonds
    B. Ionic bonds
    C. Hydrogen bonds
V. Oxidation and reduction
VI. Inorganic compounds
    A. Water
        1. Polar properties
        2. Properties as a solvent
        3. Cohesive and adhesive forces
        4. Temperature stabilization
    B. Acids and bases
    C. Salts
    D. Buffers
Focus on isotopes

Learning Objectives

After reading this chapter you should be able to:

1. Diagram the basic structure of the atom in accordance with the conventions presented in this chapter, showing the position of protons, neutrons, and electrons.
2. Identify the biologically significant elements in Table 2–1 by their chemical symbols, and summarize the main functions of each in living organisms.
3. Interpret simple chemical formulas, structural formulas, and equations.
4. Define the term *electron orbital* and relate orbitals to energy levels; relate the number of valence electrons to the chemical properties of the element.
5. Distinguish between the types of chemical bonds that join atoms to form ionic and covalent compounds, and give the characteristics of each type.
6. Contrast oxidation and reduction and explain how these processes are linked.
7. Distinguish between inorganic and organic compounds and identify the biologically important inorganic compounds.
8. Discuss the properties of water molecules and their importance to living things.
9. Compare *acids* and *bases,* use the pH scale in describing the hydrogen ion concentration in living systems, and describe how buffers help minimize changes in pH.

Despite their great diversity, the chemical composition and metabolic processes of all living things are remarkably similar. At one time it was believed that there was a unique substance in their chemical composition that distinguished living organisms from inanimate matter. The search for this substance proved fruitless. Instead, biologists came to understand that it is the specific organization and precise interaction of its components that characterize life. To understand life processes, then, one must know the basic principles of chemistry, or, to be more specific, *bio*chemistry. For those who have never enjoyed a formal chemistry course, Chapters 2 and 3 provide the basic definitions and concepts necessary for an understanding of biology. Those who have already studied chemistry may find them a useful review.

CHEMICAL ELEMENTS

All matter, living and nonliving alike, is composed of chemical **elements,** substances that cannot be broken down into simpler substances by chemical reactions. The matter of the universe is composed of 92 naturally occurring elements, ranging from hydrogen, the lightest, to uranium, the heaviest. In addition to the naturally occurring elements, about 17 elements heavier than uranium have been made by bombarding elements with subatomic particles in devices known as particle accelerators.

About 98% of an organism's mass[1] is composed of only six elements—oxygen, carbon, hydrogen, nitrogen, calcium, and phosphorus. Approximately 14 other elements are consistently present in living things, but in smaller quantities. Some of these, such as iodine and copper, are known as **trace elements** because they are present in such minute amounts. Table 2–1 lists the elements that make up a living organism and explains why each is important.

Instead of writing out the name of each element, chemists use a system of abbreviations called **chemical symbols**—usually the first one or two letters of the English or Latin name of the element. For example, O is the symbol for oxygen, C for carbon, Cl for chlorine, N for nitrogen, and Na for sodium (its Latin name is *natrium*). Chemical symbols for the elements found in living organisms are given in Table 2–1.

THE ATOM

Imagine a bit of gold being divided into smaller and smaller pieces. The smallest possible particle of gold that could be obtained would be an atom of gold. The **atom** is the smallest subdivision of an element that retains the characteristic chemical properties of that element. The subdivision of any kind of matter ultimately yields atoms. This is true no matter what physical state matter may assume—solid, liquid, or gas. Atoms are almost unimaginably small—much smaller than the tiniest particle visible under a light microscope. By special scanning electron microscopy, with magnification as much as 5 million times, researchers have been able to photograph some of the larger atoms, such as uranium.

Atomic Structure

An atom is composed of smaller components called **subatomic particles.** For our purposes we need consider only three types—protons, neutrons, and electrons. **Protons** have a positive electric charge; **neutrons** are un-

---

[1] For convenience we will consider weight and mass as equal, although this is not always true. Mass does not depend upon the force of gravity, but weight does. Thus a person on the moon has the same mass as a person on earth, but, because of the lower gravity, his or her body weight is less.

Table 2–1
ELEMENTS THAT MAKE UP THE HUMAN BODY

Name	Chemical symbol	Approximate composition by mass (%)	Importance or function
Oxygen	O	65	Required for cellular respiration; present in most organic compounds; component of water
Carbon	C	18	Backbone of organic molecules; can form four bonds with other atoms
Hydrogen	H	10	Present in most organic compounds; component of water
Nitrogen	N	3	Component of all proteins and nucleic acids
Calcium	Ca	1.5	Structural component of bones and teeth; important in muscle contraction, conduction of nerve impulses, and blood clotting
Phosphorus	P	1	Component of nucleic acids; structural component of bone; important in energy transfer
Potassium	K	0.4	Principal positive ion (cation) within cells; important in nerve function; affects muscle contraction
Sulfur	S	0.3	Component of most proteins
Sodium	Na	0.2	Principal positive ion in interstitial (tissue) fluid; important in fluid balance; essential for conduction of nerve impulses
Magnesium	Mg	0.1	Needed in blood and other body tissues
Chlorine	Cl	0.1	Principal negative ion (anion) of interstitial fluid; important in fluid balance; component of sodium chloride
Iron	Fe	Trace amount	Component of hemoglobin and myoglobin; component of certain enzymes
Iodine	I	Trace amount	Component of thyroid hormones

Other elements found in very small amounts in the body include manganese (Mn), copper (Cu), zinc (Zn), cobalt (Co), fluorine (F), molybdenum (Mo), selenium (Se), and a few others. They are called trace elements.

charged particles with about the same mass as protons. Protons and neutrons make up almost all the mass of an atom and are concentrated in the **atomic nucleus. Electrons** have a negative electrical charge and an extremely small mass (only about 1/1800 of the mass of a proton). The electrons, as we will see, spin about in the space surrounding the atomic nucleus (Fig. 2–1).

Each kind of element has a fixed number of protons in the atomic nucleus. This number, called the **atomic number,** is written as a subscript to the left of the chemical symbol. Thus $_1H$ and $_8O$ indicate that the hydrogen nucleus contains one proton and the oxygen nucleus has eight protons. It is the atomic number, the number of protons in the nucleus, that determines the chemical identity of the atom. The total number of protons plus neutrons in the nucleus is termed the **mass number** and is indicated by a superscript to the left of the chemical symbol. The common form of oxygen atom, with eight protons and eight neutrons in its nucleus, has an atomic number of 8, and a mass number of 16. It is indicated by the symbol $_8^{16}O$ (see Focus on Isotopes).

Focus on . . .

ISOTOPES

Atoms of the same element containing the same number of protons but different numbers of neutrons have different mass numbers and are called **isotopes.** The three isotopes of hydrogen, ^1_1H, ^2_1H, and ^3_1H, contain zero, one, and two neutrons, respectively. Elements usually occur in nature as a mixture of isotopes.

All isotopes of a given element have essentially the same chemical characteristics. Some isotopes with excess neutrons are unstable and tend to break down, or decay, to a more stable isotope (usually of a different element). Such isotopes are termed **radioisotopes** since they emit high-energy radiation when they decay.

Radioisotopes like ^3H (tritium) and ^{14}C have been extremely valuable research tools in biology and are useful in medicine for both diagnosis and treatment. Despite the difference in the number of neutrons, the body treats all isotopes of a given element the same chemically. The reactions of a chemical—a fat, a hormone, a drug—can be followed in the body by tagging the substance with a radioisotope, such as carbon-14 or tritium. For example, the active component in marijuana (tetrahydrocannabinol) has been tagged and administered intravenously. By measuring the amount of radioactivity in the blood and urine at successive intervals, experimenters determined that this compound remains in the blood for more than three days and products of the metabolism of this substance can be detected in the urine for more than eight days. Because radiation from radioisotopes can interfere with cell division, such isotopes have been used in the treatment of cancer (a disease characterized by rapidly dividing cells). Radioisotopes are also used to test thyroid gland function, measure the rate of red blood cell production, and study many other aspects of body chemistry.

Anthropologists use radioisotope content to date and study fossils. The skeleton of this 11th-century inhabitant of a South African Iron Age village posed an anthropological puzzle. Physically, the man's skeleton was different from those of the other villagers, suggesting that he was not a native of the area. When the skeleton was analyzed for isotopes, its carbon-12 to carbon-13 ratio was found to be similar to that of other skeletons from the same village. Since different kinds of plants incorporate different proportions of isotopes into the food produced from them, this similarity of isotope content indicates that these individuals all ate the same foods. Thus, anthropologists concluded that this man had probably spent most of his life in the village after migrating there from a distant region. (Nikolaas J. van der Merwe, American Scientist *70: 596–606, 1982.)*

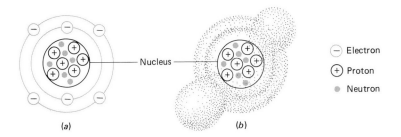

(a)　　　　　　　　(b)

— Nucleus

⊖ Electron
⊕ Proton
● Neutron

Figure 2–1 Two ways of representing an atom. (*a*) Bohr model of an atom. Although the Bohr model is not an accurate way to depict electron configuration, it is commonly used because of its simplicity and convenience. (*b*) An electron cloud. Dots represent the probability of the electron's being in that particular location at any given moment.

Electron Configuration

The way electrons are arranged around an atomic nucleus is referred to as the atom's **electron configuration.** Knowing the locations of electrons enables chemists to predict how atoms can combine to form different types of chemical compounds.

Because they are negatively charged, electrons are attracted to the positively charged protons in the atomic nucleus (opposite charges tend to attract). At the same time, electrons repel one another (like charges tend to repel one another). These considerations help determine the locations of electrons.

An atom may have several **energy levels,** or **electron shells,** where electrons are located. The lowest energy level is the one closest to the nucleus. Only two electrons can occupy this energy level. The second energy level can accommodate a maximum of eight electrons. Although the third and outer shells can each contain more than eight electrons, they are most stable when only eight are present. We may consider the first shell complete when it contains two electrons, and every other shell complete when it contains eight electrons. The atomic structures of some elements that are important in biological systems—carbon, hydrogen, oxygen, nitrogen, sodium, and chlorine—are shown in Figure 2–2.

Although the simple diagrams, called **Bohr models,** of electron configuration shown in Figure 2–2 are helpful in understanding atomic structure, they are highly oversimplified. Within energy levels, electrons occur in characteristic regions of space, termed **orbitals.** There may be several orbitals within a given energy level, and each orbital can contain at most two electrons. Electron orbitals may be represented by spherical, dumbbell-shaped, or more complex three-dimensional coordinates.

Electrons are thought to whirl around the nucleus, now close to it, now farther away. Orbitals represent the places where electrons are most

Figure 2–2 Bohr models of some biologically important atoms. (*a*) Hydrogen. (*b*) Carbon. (*c*) Oxygen. (*d*) Nitrogen. (*e*) Sodium. (*f*) Chlorine. Each circle represents an energy level, or electron shell. Electrons are represented by dots on the circles; *p*, proton; *n*, neutron.

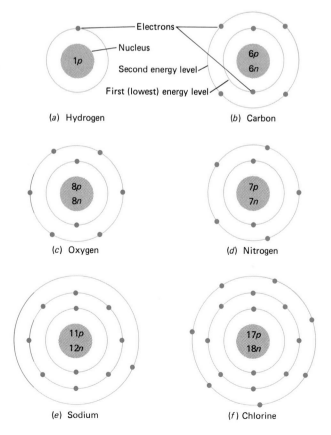

probably found. In fact, one way of illustrating an atom is to show its orbitals as electron clouds. Each occupied orbital can be represented as an electron cloud. The density of the shaded areas represents the likelihood that an electron is present at a given time (see Fig. 2–1).

The distance of the electrons from the atomic nucleus depends upon their respective energy levels. The specific orbital that an electron occupies is determined by the total energy of the system and the number of electrons available to the atom. An electron can be moved to an orbital farther from the nucleus by providing it with more energy, or an electron can give up energy and sink back to a lower energy level in an orbital nearer the central nucleus. Energy is required to move a negatively charged electron further away from the positively charged nucleus.

When energy is added to the system, an electron can jump from one level to the next, *but it cannot stop in the space in between.* To move an electron from one level to the next, the atom must absorb a discrete packet of energy known as a quantum, which contains just the right amount of energy needed for the transition—no more and no less. The term "quantum jump" is used in everyday language to indicate a sudden discontinuous move from one level to another.

CHEMICAL COMPOUNDS

Two or more atoms may combine chemically to form a **molecule.** When two atoms of oxygen combine, a molecule of oxygen is formed. Different kinds of atoms can combine, forming **chemical compounds.** A chemical compound is a substance that consists of two or more different elements combined in a fixed ratio. Water is a chemical compound consisting of two atoms of hydrogen combined with one atom of oxygen.

What Are Formulas?

A **chemical formula** is a shorthand method for describing the chemical composition of a molecule. Chemical symbols are used to indicate the types of atoms in the molecule, and subscript numbers are used to indicate the number of each type of atom present. The chemical formula for molecular oxygen, O_2, tells us that this molecule consists of two atoms of oxygen. The chemical formula for water, H_2O, indicates that each molecule consists of two atoms of hydrogen and one atom of oxygen. (Note that when a single atom of one type is present, it is not necessary to write 1; one does *not* write H_2O_1.)

Another type of formula is the **structural formula,** which shows not only the types and numbers of atoms in a molecule but also their arrangement. In any type of chemical compound the atoms are always arranged in the same way. From the chemical formula for water, H_2O, you could only guess whether the atoms were arranged H—H—O or H—O—H. The structural formula, H—O—H, settles the matter, indicating that the two hydrogen atoms are attached to the oxygen atom. In water this arrangement is the only one chemically possible. However, there are other substances that consist of the same atoms yet have different chemical properties due to alternative arrangements. Such compounds are known as **structural isomers.** The sugars glucose and fructose are examples (Fig. 3–2).

Chemical Equations

During any moment in the life of an organism, be it an earthworm or a pine tree, many complex, highly organized chemical reactions are taking place. The chemical reactions that occur between atoms and molecules—for instance, between methane (natural gas) and oxygen—can be described on paper by means of chemical equations:

$$CH_4 \ + \ 2O_2 \ \longrightarrow \ CO_2 \ + \ 2H_2O$$
$$\text{Methane} \quad \text{Oxygen} \qquad \text{Carbon dioxide} \quad \text{Water}$$

In a **chemical equation** the **reactants** (the substances that participate in the reaction) are written on the left side of the equation and the **products** (the substances formed by the reaction) are written on the right side. The arrow means **yields** and indicates the direction in which the reaction tends to proceed. The number preceding a chemical symbol or formula indicates the number of atoms or molecules reacting. Thus $2O_2$ means two molecules of oxygen and $2H_2O$ means two molecules of water. The absence of a number indicates that only one atom or molecule is present.

In some cases the reaction will proceed in the reverse direction as well as forward; at **equilibrium** a certain amount of the product continuously breaks up to form the reactants, and the rate of the forward reaction equals the rate of the reverse action. Reversible reactions are indicated by double arrows:

$$N_2 \ + \ 3H_2 \ \rightleftharpoons \ 2NH_3$$
$$\text{Nitrogen} \quad \text{Hydrogen} \qquad \text{Ammonia}$$

HOW ATOMS COMBINE: CHEMICAL BONDS

The chemical properties of an element are determined primarily by the number and arrangement of electrons in the *outermost* energy level (electron shell). In a few elements, called the **noble gases,** the outermost shell is filled. These elements are chemically inert, meaning that they will not readily combine with other elements. Two such elements are helium with two electrons (a complete inner shell) and neon with ten electrons (a complete inner shell of two and a complete second shell of eight). The electrons in the outermost energy level of an atom are referred to as **valence electrons.** When the outer shell of an atom contains fewer than eight electrons, the atom tends to lose, gain, or share electrons to achieve an outer shell of eight (zero or two in the case of the lightest elements).

The elements in a given compound are always present in a certain proportion by mass. This reflects the fact that atoms are attached to each other by chemical bonds in a precise way to form the compound. A **chemical bond** is the attractive force that holds two atoms together. Each bond represents a certain amount of potential chemical energy. The atoms of each element form a specific number of bonds with the atoms of other elements—a number dictated by the number of valence electrons. The two principal types of chemical bonds are covalent bonds and ionic bonds.

Covalent Bonds

Covalent bonds involve the sharing of electrons between atoms. The more precise definition of a **molecule** is a combination of two or more atoms joined by covalent chemical bonds. A simple example of a covalent bond is the one joining two hydrogen atoms in a molecule of hydrogen gas, H_2 (Fig. 2–3). Each atom of hydrogen has one electron, but two electrons are required to complete the first energy level. Each hydrogen atom has the same capacity to attract electrons, so one does not donate an electron to the other. Instead, the two hydrogen atoms share their single electrons so that each of the two electrons is attracted simultane-

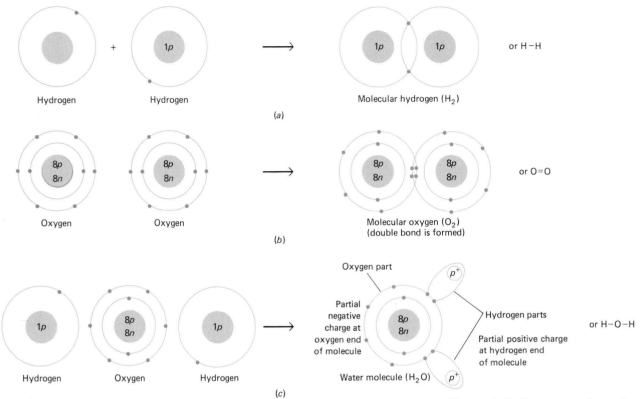

(a)

(b)

(c)

Figure 2–3 Formation of covalent compounds. (a) Two hydrogen atoms achieve stability by sharing electrons, thereby forming a molecule of hydrogen. The structural formula shown on the right is a simpler way of representing molecular hydrogen. The straight line between the hydrogen atoms represents a single covalent bond. (b) Two oxygen atoms share two pairs of electrons to form molecular oxygen. Note the double bond. (c) When two hydrogen atoms share electrons with an oxygen atom, the result is a molecule of water. Note that the electrons tend to stay closer to the nucleus of the oxygen atom than to the hydrogen nuclei. This results in a partial negative charge on the oxygen portion of the molecule, and a partial positive charge at the hydrogen end. Although the water molecule as a whole is electrically neutral, it is a polar covalent compound.

ously to the two protons in the two hydrogen nuclei. The two electrons are thus under the influence of *both* atomic nuclei, and they join the two atoms together.

The carbon atom has four electrons in its outer energy level. These four electrons are available for covalent bonding: $-\overset{\displaystyle |}{\underset{\displaystyle |}{C}}-$. When one carbon and four hydrogen atoms share electrons a molecule of methane (CH_4) is formed:

$$
\begin{array}{c}
\quad\ H \\
\quad\ | \\
H-C-H \\
\quad\ | \\
\quad\ H
\end{array}
$$

Each atom shares its outer-level electrons with the others, thereby completing the first energy level of each hydrogen atom and the energy level of the carbon atom. When one electron pair is shared between two atoms, the covalent bond is referred to as a single bond.

The nitrogen atom has five electrons in its outer shell. When a nitrogen atom shares electrons with three hydrogen atoms, a molecule of ammonia, NH_3, is formed:

$$
\begin{array}{c}
H-N-H \\
\quad\ | \\
\quad\ H
\end{array}
$$

Two oxygen atoms may achieve stability by forming covalent bonds with one another. Each oxygen atom has six electrons in its outer shell. To become stable, the two atoms share two pairs of electrons, forming molecular oxygen (Fig. 2–3). When two pairs of electrons are shared in this way, the covalent bond is referred to as a **double bond.** Some atoms form triple bonds with one another, sharing three pairs of electrons.

Table 2–2
SOME BIOLOGICALLY IMPORTANT IONS

Name	Formula	Charge
Sodium	Na^+	1 +
Potassium	K^+	1 +
Hydrogen	H^+	1 +
Magnesium	Mg^{2+}	2 +
Calcium	Ca^{2+}	2 +
Iron	Fe^{2+} or Fe^{3+}	2 + [iron(II)] or 3+ [iron(III)]
Ammonium	NH_4^+	1 +
Chloride	Cl^-	1 −
Iodide	I^-	1 −
Carbonate	CO_3^{2-}	2 −
Bicarbonate	HCO_3^-	1 −
Phosphate	PO_4^{3-}	3 −
Acetate	CH_3COO^-	1 −
Sulfate	SO_4^{2-}	2 −
Hydroxide	OH^-	1 −
Nitrate	NO_3^-	1 −
Nitrite	NO_2^-	1 −

Electronegativity is a measure of an atom's attraction for the electrons in chemical bonds. A covalent bond between atoms of different electronegativities is known as a **polar covalent bond.** The polarity of compounds is very important in understanding the structure of biological membranes and their properties. Covalent bonds have many degrees of polarity, from **nonpolar,** in which the electrons are exactly shared (as in the hydrogen molecule), to those in which the electrons are *much* closer to one atom than to the other.

Ionic Bonds

One extreme of polarity is the ionic bond, in which electrons are pulled completely from one atom to the other. The number of protons in the nucleus remains unchanged, so the loss or gain of electrons produces an atom with a net positive or negative charge. Such electrically charged atoms are termed **ions.** Atoms with one, two, or three electrons in the outer shell generally donate electrons to other atoms. These atoms, then, become positively charged because of the excess of protons in the nucleus. For example, sodium has one valence electron. It tends to donate this electron, becoming a sodium ion (Na^+) with a one plus charge. Calcium tends to lose its two valence electrons to become a calcium ion (Ca^{2+}) with a two plus charge. Positively charged ions are termed **cations** (see Table 2–2).

Atoms with five, six, or seven electrons in their outer shell tend to *gain* electrons from other atoms and become negatively charged anions (e.g., Cl^-, chloride ion). Charged particles, both anions and cations, play many important roles in biological systems, such as the transmission of nerve impulses and the contraction of muscles (Fig. 2–4).

An **ionic bond** is the force of electrical attraction between two oppositely charged ions. When held together by ionic bonds, oppositely charged ions form an ionic compound. Sodium chloride is a good example of an ionic compound. A sodium atom, with atomic number 11, has two electrons in its inner shell, eight in the second, and one in the third shell. A sodium atom cannot fill its third shell by obtaining seven electrons from other atoms because it would then have a vast excess of nega-

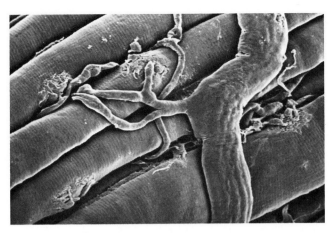

Figure 2–4 Sodium, potassium, and chlorine ions are among the ions essential in the conduction of a nerve impulse. This scanning electron micrograph shows a nerve fiber communicating with several muscle cells (approximately ×900). The nerve fiber transmits impulses to the muscle cells, stimulating them to contract. The muscle cells are rich in calcium ions, which are essential for muscle contraction. (From Desaki, J.: "Vascular Autonomic Plexuses and Skeletal Neuromuscular Junctions: A Scanning Electron Microscopic Study," *Biomedical Research Supplement*, 139–143, 1981.)

tive charge. Instead, it gives up the single electron in its third shell to some electron acceptor, leaving the second shell as the complete outer shell (Fig. 2–5). A chlorine atom, atomic number 17, has two electrons in its inner shell, eight in the second, and seven in the third shell. The chlorine atom achieves a complete outer shell not by losing the seven electrons in its third shell (for it would then have a vast positive charge) but by accepting an electron from an electron donor, such as sodium, to complete its third shell.

When an electron donor, such as sodium, meets an electron acceptor, such as chlorine, the electron may be transferred completely from the donor to the acceptor. The sodium ion now has 11 protons in its nucleus and ten electrons around the nucleus, giving it a net charge of 1^+. The chlorine ion has 17 protons in its nucleus and 18 electrons around the nucleus, so it has a net charge of 1^-. These ions attract each other as a result of their opposite charges. They are held together by this electrical attraction to form sodium chloride (NaCl), common table salt. This transfer of one or more electrons from one atom to another and the binding together of two ions of opposite charge (the ionic bond) results in the formation of an ionic compound. Ionic bonds occur between electron donors and electron acceptors (Fig. 2–5).

Whether an ionic compound is in solid form or is dissolved in water, its constituent particles (ions) do not share electrons. Because of this, the term *molecule* does not adequately explain the properties of ionic compounds such as NaCl. Chemists simply refer to them as compounds.

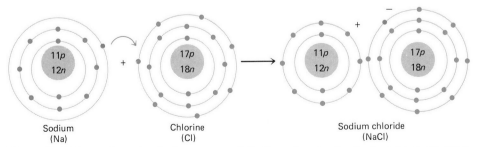

Sodium
(Na)

Chlorine
(Cl)

Sodium chloride
(NaCl)

Figure 2–5 Formation of an ionic compound. Sodium donates its single valence electron to chlorine, which has seven electrons in its outer energy level. With this additional electron, chlorine completes its outer energy level. The two atoms are now electrically charged ions. They are attracted to one another by their unlike electrical charges, forming the ionic compound sodium chloride. The force of attraction holding these ions together is called an ionic bond.

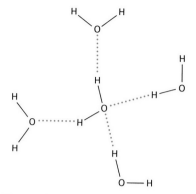

Figure 2–6 Hydrogen bonding of water molecules. Each water molecule tends to form hydrogen bonds with four neighboring water molecules. The hydrogen bonds are indicated by dotted lines. The covalent bonds between the hydrogen and oxygen atoms are represented by solid lines.

Hydrogen Bonds

Another type of bond that is important in biological systems is the **hydrogen bond** (Fig. 2–6). This type of bond is formed when a hydrogen atom is attracted to two other atoms, one of which is usually oxygen or nitrogen. Hydrogen bonds tend to form between a hydrogen atom covalently bonded to oxygen or nitrogen and some other electronegative atom, usually oxygen or nitrogen. The atoms involved may be in two parts of the same molecule or in two molecules. When hydrogen is combined with an electronegative atom, such as oxygen, it has a partial positive charge because its electron is positioned closer to the oxygen atom. Hydrogen bonds are weak and are readily formed and broken. They have a specific length and orientation, which is very important in their role in helping to determine the three-dimensional structure of large molecules, such as nucleic acids and proteins. Though relatively weak individually, the large numbers of hydrogen bonds that occur in the double helix of DNA and in the alpha helix of proteins (Chapter 4) are together very effective.

OXIDATION-REDUCTION

Rusting—the combination of iron with oxygen—is a familiar example of oxidation and reduction:

$$4Fe + 3O_2 \longrightarrow 2Fe_2O_3$$

Oxidation is a chemical process in which a substance loses electrons. In rusting, iron is changing from its metallic state to its iron(III) (Fe^{3+}) state. We say it is being oxidized:

$$4Fe \longrightarrow 4Fe^{3+} + 12e^-$$

The e^- stands for electron. At the same time, oxygen is changing from its molecular state to its charged state:

$$3O_2 + 12e^- \longrightarrow 6O^{2-}$$

Oxygen accepts the electrons removed from the iron and is said to be reduced. **Reduction** is a chemical process in which a substance gains electrons. An oxidation cannot take place without a reduction because electrons have to go somewhere; free electrons are not usually found in nature. Some substance must accept the electrons that are lost. Oxidation-reduction reactions are sometimes called **redox reactions.**

Electrons are not easy to remove from covalent compounds unless an entire atom is removed. In living cells oxidation almost always involves the removal of a hydrogen atom from a compound. Reduction often involves a gain in hydrogen atoms. As will be discussed in later chapters, redox reactions are an essential part of photosynthesis, cellular respiration, and other aspects of metabolism.

INORGANIC COMPOUNDS

Chemical compounds can be divided into two broad groups—inorganic and organic. **Inorganic compounds** are relatively small, simple substances. **Organic compounds** are usually large and complex and always contain carbon. Among the biologically important groups of inorganic compounds are water, simple acids and bases, and salts. Organisms depend upon appropriate amounts of these substances for fluid balance, acid-base balance, and many cell activities, such as transporting materials through cell membranes.

Water

Water is an essential ingredient of life. It accounts for about 80% of the weight of an average active cell. In fact, the human body is about 70% water by weight. Many organisms make their homes in lakes, rivers, or the sea, and the cells of terrestrial organisms are bathed in body fluids composed largely of water.

POLAR PROPERTIES

The physical and chemical properties of water permit life to exist on our planet. Water molecules are **polar;** that is, they bear a partial positive and a partial negative charge. The water molecules in liquid water and in ice are held together in part by hydrogen bonds (Fig. 2–6). The hydrogen atom of one water molecule, with its partial positive charge, is attracted to the oxygen atom of a neighboring water molecule, with its partial negative charge, forming a hydrogen bond. Each water molecule can form hydrogen bonds with a maximum of four neighboring water molecules.

PROPERTIES AS A SOLVENT

Because its molecules are polar, water is an excellent **solvent,** a liquid capable of dissolving many polar substances. For example, the ions of a salt, such as sodium chloride, are held together by strong ionic bonds; in fact, they form a crystal lattice. Considerable energy is required to pull the positively and negatively charged ions apart. However, when the sodium chloride is placed in water, the strong electrical attractions between the polar water molecules and Na^+ and Cl^- ions result in the formation of stable dissociated Na^+ and Cl^- ions (see Fig. 2–11).

$$\underset{\text{Sodium chloride}}{NaCl} \xrightarrow{\text{in } H_2O} \underset{\substack{\text{Sodium} \\ \text{ion}}}{Na^+} + \underset{\substack{\text{Chloride} \\ \text{ion}}}{Cl^-}$$

Because of its solvent properties and its tendency to cause the ionization of compounds in solution, water is important in facilitating chemical reactions.

COHESIVE AND ADHESIVE FORCES

Water exhibits both cohesive and adhesive forces. Water molecules have a very strong tendency to stick to each other; that is, they are **cohesive.** This is due to the hydrogen bonds among the molecules. Water molecules also stick to many other kinds of substances (i.e., those substances that have charged groups of atoms or molecules on their surfaces). These **adhesive** forces explain how water makes things wet. Water has a high degree of surface tension because of the cohesiveness of its molecules; its molecules have a much greater attraction for other water molecules than for molecules in the air. Thus water molecules at the surface crowd together, producing a strong layer as they are pulled downward by the attraction of other water molecules beneath them (Fig. 2–7).

Adhesive and cohesive forces account for the tendency, termed **capillary action,** of water to rise in very fine tubes. Capillary action plays some part in the rise of water through the stems of plants to their leaves.

TEMPERATURE STABILIZATION

Other properties are responsible for water's ability to minimize temperature changes. Water has a high specific heat. **Specific heat** is the amount of heat required to raise the temperature of 1 gram of water 1°C. The high

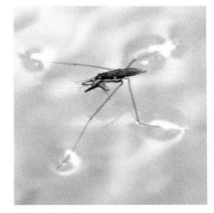

Figure 2–7 A water strider on the surface of a pond. Fine hairs at the end of its legs spread its weight over a large area, allowing the body of the animal to be supported by the surface tension of the water.

specific heat of water is due to its hydrogen bonds. For the temperature of a substance to be raised, heat energy must be added to make its molecules move faster—to increase the kinetic energy of the molecules. Before water molecules can move freely, some of the hydrogen bonds holding the water molecules together must be broken. Thus a great deal of the heat energy added to water is used up in breaking the hydrogen bonds, and only a part of this energy is available to speed the motion of the water molecules (increase the temperature of the water).

Because so much heat input (or heat loss) is required to raise (or lower) the temperature of water, the oceans and other large bodies of water have relatively constant temperatures. Thus the aquatic environment provides the multitude of organisms that inhabit it with a relatively constant environmental temperature. The water within all organisms, even those that live on land, contributes to a relatively constant internal temperature. This is important because metabolic reactions can take place only within a relatively narrow temperature range.

Because its molecules are held together by hydrogen bonds, water also has a high heat of vaporization, another property that helps stabilize temperature. More than 500 calories are required to change 1 gram of liquid water into 1 gram of water vapor. A calorie is a unit of heat energy (defined as 4.184 joules). Because of the heat of vaporization, we can rid ourselves of excess heat by the evaporation of sweat. And plants can remain cool in the midday heat by evaporating water from their surface.

Although most substances become more and more dense as the temperature decreases, water reaches its maximum density at 4°C and then begins to expand again as the temperature decreases. Hydrogen bonds become more rigid and ordered, and ice floats upon the denser cold water (Fig. 2–8). This important property of water explains why lakes and ponds freeze from the surface down rather than from the bottom up. The sheet of ice that forms at the pond surface insulates the water below from the wintry chill so that it is less likely to freeze. Further, organisms that inhabit northern lakes and ponds are able to carry on their life activities despite the frigid winter.

Acids and Bases

An **acid** is a compound that ionizes in solution to yield hydrogen ions (H^+)[1] and an anion. An acid is a proton *donor.* Acids turn blue litmus paper red and have a sour taste. Hydrochloric acid (HCl) and sulfuric acid (H_2SO_4) are examples of inorganic acids. The strength of an acid depends upon the degree to which it ionizes in water. Thus HCl is a very strong acid because most of its molecules dissociate, producing hydrogen and chloride ions.

$$\text{HCl} \xrightarrow{\text{in } H_2O} H^+ + Cl^-$$

Hydrochloric   Hydrogen ion   Chloride ion
acid

A base is defined as a proton *acceptor.* Most bases are substances that yield a hydroxide ion (OH^-) and a cation when dissolved in water. Bases turn red litmus paper blue and feel slippery to the touch. Sodium hydroxide (NaOH) and aqueous ammonia (NH_4OH) are inorganic bases.

$$\text{NaOH} \longrightarrow Na^+ + OH^-$$

Sodium hydroxide   Sodium ion   Hydroxide ion

---

[1] The H^+ immediately combines with an electronegative region of a water molecule forming a hydronium ion (H_3O^+).

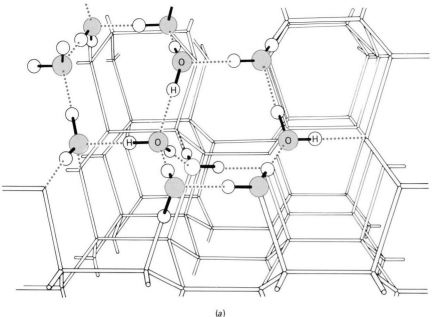

(a)

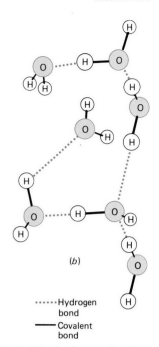

(b)

······· Hydrogen
bond

——— Covalent
bond

(c)

Figure 2–8 The hydrogen bonding in ice compared with that in liquid water. (a) Note the regular, evenly distanced hydrogen bonds in the superstructure of ice. (b) When ice melts, the hydrogen bonds occur less consistently and are of unequal length, and the crystal structure collapses. (c) Because of its unique crystalline structure, water is one of the few substances that is less dense in its solid form than in its liquid form. This property is of great biological importance, for it prevents most bodies of water from freezing to the bottom. (a,b, Redrawn with permission of Arthur Geis; c, B.F. Molnia, TERRA PHOTOGRAPHICS/BPS.)

Since the concentration of hydrogen or hydroxide ions is usually small, it is convenient to express the degree of acidity or alkalinity in a solution in terms of **pH,** formally defined as the logarithm of the reciprocal of the hydrogen ion concentration, $\log(1/[H^+])$. The pH scale is logarithmic, extending from 0, the pH of a strong acid, such as HCl, to 14, the pH of a strong base, such as NaOH (Fig. 2–9). The pH of pure water is 7. Even though water does ionize slightly, the concentrations of H^+ ions and OH^- ions are exactly equal. Solutions with a pH of less than 7 are acidic and contain more H^+ ions than OH^- ions. Solutions with a pH greater than 7 are alkaline, or basic, and contain more OH^- ions than H^+ ions. Because the scale is logarithmic, a solution with a pH of 6 has a hydrogen ion concentration that is ten times greater than a solution with a pH of 7 and is much more acidic. A pH of 5 represents another tenfold increase, and so a solution with a pH of 4 is 10×10 or 100 times more acidic than a solution with a pH of 6. The contents of most animal and plant cells are neither strongly acidic nor alkaline but are an essentially

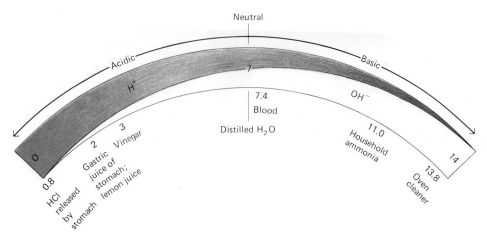

Neutral

Acidic

Basic

H⁺

OH⁻

7

7.4
Blood

Distilled H₂O

0.8
HCl
released
by
stomach

Gastric
juice of
stomach;
lemon juice

2

Vinegar

3

O

11.0
Household
ammonia

13.8
Oven
cleaner

14

Figure 2–9 The pH scale. A solution with a pH of 7 is neutral because the concentrations of H^+ and OH^- are equal. The lower the pH below 7, the more H^+ ions are present, and the more acidic the solution is. As the pH increases above 7, the concentration of H^+ ions decreases and the concentration of OH^- increases, making the solution more alkaline (basic).

neutral mixture of acidic and basic substances. Life is not compatible with any considerable change in the pH of the cell (Fig. 2–10).

Salts

When an acid and a base are mixed together, the H^+ of the acid unites with the OH^- of the base to form a molecule of water. The remainder of the acid (anion) combines with the remainder of the base (cation) to form a **salt.** This type of reaction is called acid-base neutralization. Hydrochloric acid reacts with sodium hydroxide to form water and sodium chloride:

$$HCl + NaOH \longrightarrow H_2O + NaCl$$

Hydrochloric acid   Sodium hydroxide   Water   Sodium chloride

A **salt** can be defined as a compound in which the hydrogen atom of an acid is replaced by some other cation. A salt contains a cation other than H^+ and an anion other than OH^-. Sodium chloride, NaCl, is a compound in which the hydrogen ion of HCl has been replaced by the cation, Na^+.

When a salt, an acid, or a base is dissolved in water, its constituent ions separate (Fig. 2–11). Because these charged particles can conduct an electric current, these substances are called **electrolytes.** Sugars, alco-

Figure 2–10 The effects of acid rain. Sulfur oxides emitted from fossil fuel plants and industry, and nitrogen oxides, mainly from automobile exhaust, are converted in the moist atmosphere into acids of, respectively, sulfur and nitrogen, such as sulfurous and nitrous acid. These acids are dispersed over wide areas by airflow patterns in the atmosphere. Whereas the pH of unpolluted rain averages 5.6, in some parts of the United States and Canada the pH of rain has been measured at values of 4.2 and even lower. Most fish species die at a pH of 4.5 to 5.0. Acid rain also affects vegetation. The roots of this spruce tree have withered and died. Even before that, the rest of the plant had for some time been suffering from nutrient deficiency and reduced efficiency of photosynthesis. (Townsend P. Dickinson, Photo Researchers, Inc.)

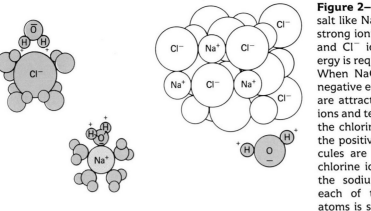

Figure 2–11 The crystal lattice of a salt like NaCl is held together by the strong ionic bonds between the Na$^+$ and Cl$^-$ ions, and considerable energy is required to pull the ions apart. When NaCl is added to water, the negative ends of the water molecules are attracted to the positive sodium ions and tend to pull them away from the chlorine ions. At the same time, the positive ends of the water molecules are attracted to the negative chlorine ions, separating them from the sodium ions. When dissolved, each of the sodium and chlorine atoms is surrounded by water molecules electrically attracted to it.

hols, and many other substances do not ionize when dissolved in water; they do not conduct an electrical current and are termed **nonelectrolytes.**

The cells and extracellular fluids (such as blood) of plants and animals contain a variety of dissolved salts. They are a source of many important mineral ions. Such ions are essential for fluid balance, acid-base balance, nerve and muscle function in animals, blood clotting, bone formation, and many other aspects of body function. Although the concentration of salts in the cells and body fluids of plants and animals is small, these salts are of great importance for normal cell function. The concentrations of the respective cations and anions are kept remarkably constant under normal conditions. Any marked change results in impaired cellular functions and ultimately in death.

Buffers

Many homeostatic mechanisms operate to maintain appropriate pH levels. For example, the pH of human blood is about 7.4 and must be maintained within very narrow limits. Should the blood become too acidic, coma and death may result. Excessive alkalinity can result in overexcitability of the nervous system and even convulsions. A **buffer** is a substance or combination of substances that resists changes in pH when acids or bases are added. The buffer either accepts or donates hydrogen ions. A buffer consists of a weak acid and a salt of that acid, or a weak base and a salt of the base.

One of the most common buffering systems and one that is important in human blood is carbonic acid and the bicarbonate ion. Bicarbonate ions are formed in the body as follows:

$$CO_2 \; + \; H_2O \; \rightleftharpoons \; H_2CO_3 \; \rightleftharpoons \; H^+ \; + \; HCO_3^-$$

Carbon dioxide　　Water　　Carbonic acid　　　　Bicarbonate ion

As indicated by the arrows, the reactions are reversible. When excess hydrogen ions are present in blood or other body fluids, bicarbonate ions combine with them to form carbonic acid, a weak acid.

$$H^+ + HCO_3^- \; \rightleftharpoons \; H_2CO_3$$

Carbonic acid

In this way a strong acid can be converted to a weak acid. The carbonic acid is unstable and quickly breaks down into carbon dioxide and water.

Buffers also work to reduce pH when excessive numbers of hydroxide ions are present. A buffer may release hydrogen ions, which combine with the hydroxide ions to form water.

$$OH^- + 2H^+ + CO_3^- \; \rightleftharpoons \; HCO_3 + H_2O$$

CHAPTER SUMMARY

I. The atom is the smallest unit of a chemical element that retains the characteristic properties of that element.
   A. An atom consists of subatomic particles, including protons, neutrons, and electrons.
   B. Electrons are found in orbitals located within energy levels. The electrons in the outermost energy level of an atom are valence electrons; they help determine the chemical properties of the atom.

II. A chemical compound consists of two or more elements combined in a fixed ratio.
   A. The composition of a compound may be described by a chemical formula, such as H_2O, or by a structural formula, such as H—O—H.
   B. The chemical reactions that occur between atoms and molecules can be described by means of chemical equations.

III. The atoms of a chemical compound are attached to one another by chemical bonds.
   A. In a covalent bond, atoms share electrons.
   B. An ionic bond is formed when one atom donates an electron to another, each atom thereby becoming charged and attracted to the other because of these electrical charges.
   C. Hydrogen bonds are weak chemical bonds formed between hydrogen atoms and an electronegative atom, usually oxygen or nitrogen. Molecules of water form hydrogen bonds with one another.

IV. When a substance is oxidized, it loses electrons; when a substance is reduced, it gains electrons.

V. Among the biologically important inorganic compounds are water, acids, bases, and salts.
   A. Water is an essential component of living things and is necessary for many vital activities, such as transport of materials.
   B. Acids ionize in solution, yielding hydrogen ions; bases usually ionize, yielding hydroxide ions. The pH scale extends from 0 to 14, with 7 considered neutral. As the pH decreases below 7, the solution is more acidic. As a solution becomes more basic, its pH increases above 7.
   C. A salt is a compound in which the hydrogen of an acid is replaced by some other cation.
   D. Buffering, which usually involves a weak acid plus a salt of that acid, helps to maintain appropriate pH.

Post-Test

1. A(n) _____ is the smallest amount of an element that retains the chemical properties of the element.

2. Isotopes are atoms of the same element that differ in their number of _____ .

3. Within energy levels electrons may be found within specific _____ .

4. What is the composition of a compound with the formula C_2H_6O?

5. Atoms with one to three valence electrons generally behave as _____ .

6. The type of bond in which atoms share electrons is a _____ bond.

7. A chemical process in which a substance gains electrons is referred to as _____ .

8. A compound that ionizes in solution to yield hydrogen ions and an anion is a(n) _____ .

9. A solution with a pH of 8 is best described as _____ .

10. A substance that resists changes in pH is a _____ .

11. The tendency for water to rise in very fine tubes is called _____ _____ ; it is due to _____ and _____ forces.

12. Water molecules in liquid water and ice are held together by _____ bonds.

Review Questions

1. Distinguish between (a) an atom and an element, (b) a molecule and a compound, and (c) an atom and an ion.
2. How do isotopes of the same element differ? What is a radioisotope?
3. How do valence electrons help determine the chemical properties of an atom?
4. Compare ionic and covalent bonds and give specific examples of each.
5. Write a chemical equation depicting the ionization of (a) sodium hydroxide (NaOH), and (b) hydrochloric acid (HCl).
6. What properties of water make it an essential component of living matter?
7. How would a solution with a pH of 5 differ from one with a pH of 9? of 7?
8. Why are buffers important in living organisms? Give a specific example of how a buffer system works.
9. Differentiate clearly among acids, bases, and salts. What are the functions of salts in living organisms?
10. Why do oxidation and reduction occur simultaneously?
11. Describe a reversible reaction that is at equilibrium.
12. What are hydrogen bonds? What is their significance?

Readings

Baker, J.J.W., and G.E. Allen. *Matter, Energy, and Life.* Reading, Mass., Addison–Wesley, 1981. A presentation of the principles of thermodynamics and their application to studies of living systems. A difficult subject clarified.

Bettelheim, F.A., and J. March. *Introduction to General,* *Organic and Biochemistry,* Philadelphia, Saunders College Publishing, 1984. A readable reference for those who would like to know more about basic chemistry.

Frieden, E. "The Chemical Elements of Life." *Scientific American.* 1972, 52–64. A discussion of the biological actions of various elements.

Chapter 3

THE CHEMISTRY OF LIFE: ORGANIC COMPOUNDS

Outline

Learning Objectives

After you have studied this chapter you should be able to:

1. Compare the major groups of organic compounds—carbohydrates, fats, proteins, and nucleic acids—with respect to their chemical composition and function.
2. Distinguish among monosaccharides, disaccharides, and polysaccharides, giving examples of each.
3. Distinguish among neutral fats, phospholipids, and steroids, giving the biologic functions of each group.
4. Describe the functions and chemical structure of proteins.
5. Describe the chemical structure of nucleotides and nucleic acids and explain the importance of these compounds in living organisms.

I n a way, the chemistry of the carbon atom is the chemistry of life itself. With four electrons in its outer energy level, the carbon atom can share electrons with other atoms, forming four covalent bonds. This ability of carbon atoms yields an immense variety of **organic compounds,** complex compounds that contain carbon. Organic compounds are the main structural components of cells and tissues, the participants in and regulators of thousands of metabolic reactions, and the fuel molecules of living systems.

In all organic compounds the main chain of atoms that makes up the principal axis of the molecule consists of carbon atoms (Fig. 3–1; Table 3–1). Thus carbon atoms share electrons with other carbon atoms to form chains of varying lengths. These chains may be unbranched or branched, or the carbon atoms may join to form rings. Adjacent carbon atoms may form single bonds, or by sharing additional pairs of electrons, they may form double (—C=C—) or triple (—C≡C—) bonds.

In this chapter we discuss some of the major groups of organic compounds that are important in living organisms, including carbohydrates, lipids, proteins, and nucleic acids. Vitamins will be discussed in Chapter 28.

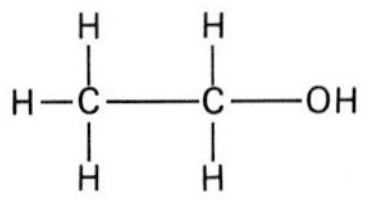

CARBOHYDRATES

Familiar to us as sugars and starches, **carbohydrates** serve as fuel molecules and are also important structural components, especially in plant cells. Carbohydrates contain carbon, hydrogen, and oxygen atoms in a ratio of approximately one carbon to two hydrogens to one oxygen $(CH_2O)_n$. The term carbohydrate, meaning "hydrate (water) of carbon," stems from the 2 to 1 ratio of hydrogen to oxygen, the same ratio found in water (H_2O). Carbohydrates may be classified as monosaccharides, disaccharides, or polysaccharides.

Monosaccharides

Monosaccharides are simple sugars that usually contain from three to six carbon atoms. **Glucose** (also called dextrose) and **fructose** are examples of monosaccharides. Each is composed of a single **hexose** unit (consists of six carbon atoms) with the formula $C_6H_{12}O_6$. Recall that such compounds as glucose and fructose, which have identical molecular formulas but different arrangements of atoms, are termed **isomers.** Because of their different arrangement of atoms, the two sugars have different chemical properties.

Glucose, often referred to as blood sugar, is the most abundant hexose in the bodies of humans and other animals. Its concentration is kept at a homeostatic level in the blood, and glucose is utilized by the cells as a fuel molecule. The **pentoses** (five-carbon sugars) ribose and deoxyribose are components of nucleotides and nucleic acids.

The "stick" formulas in Figure 3–2 give a clear but somewhat unrealistic picture of the structures of some common monosaccharides. Actually, molecules are not the simple two-dimensional structures depicted on a printed page. In fact, the properties of each compound depend in part on its three-dimensional structure. Molecules of glucose and other monosaccharides in solution exist mainly as rings (Fig. 3–3).

In some compounds certain atoms can be arranged in more than one position in space around the carbon atom to which they are bonded. Such stereoisomers differ from one another only in some geometric or three-dimensional way. There are two ring forms of glucose, for instance, that differ only in the orientation of an —OH group.

Ethyl alcohol

(a)

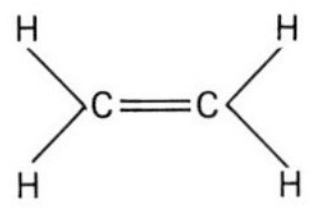

Ethylene

(b)

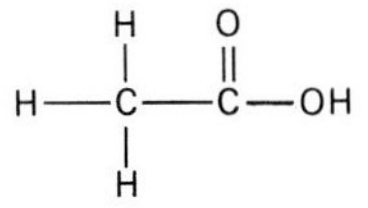

Acetic acid

(c)

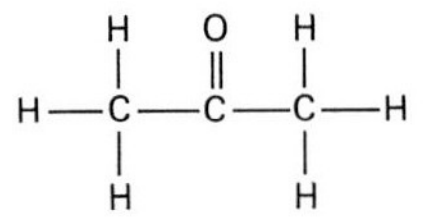

Acetone

(d)

Figure 3–1 Some simple organic compounds. Note that each carbon atom has four covalent bonds. (a) Ethyl alcohol, the type of alcohol used in alcoholic beverages. (b) Ethylene, a raw material used in making polyethylene plastic. Note the double bond. (c) Acetic acid, an organic acid found in vinegar. (d) Acetone, used as a fingernail polish remover, but also important in fat metabolism.

Table 3–1
SOME OF THE BIOLOGICALLY IMPORTANT GROUPS OF ORGANIC COMPOUNDS

Class of compound	Description	Component elements	How to recognize	Principal functions
Carbohydrates	Carbohydrates have approximately as many oxygen atoms as carbon atoms and twice as many hydrogens as carbons (when allowance is made for water loss when sugar units are linked).	C, H, O	Count the carbons, hydrogens, and oxygens.	Cellular fuel; energy storage; components of other compounds, e.g., nucleic acids.
	1. Monosaccharides, the simple sugars $(CH_2O)_n$. For the most part, these are six-carbon (hexose) molecules like glucose whose backbones may be shown as straight lines but are actually arranged in ring formations. Glucose and fructose are examples. Ribose is a pentose (five-carbon) monosaccharide.		Look for the ring shapes: hexose or pentose	
	2. Disaccharide (two sugars) e.g., sucrose, maltose, lactose.		Two sugar units.	
	3. Polysaccharide (many units) e.g., starch, glycogen, cellulose.		Many sugar subunits.	
Lipids	1. Neutral fats. Combination of glycerol with one to three fatty acids. One fatty acid—monoacylglycerol; two fatty acids—diacylglycerol; three fatty acids—triacylglycerol. If fatty acids contain double carbon-to-carbon linkages (C=C), they are unsaturated; otherwise they are saturated.	C, H, O	Look for glycerol at one end of molecule:	Energy storage; thermal insulation; support of organs.
	2. Phospholipids. Glycerol attached to one or two fatty acids and to an organic base.	C, H, O, P, N	Look for glycerol and side chain containing phosphorus and nitrogen.	Components of cell membranes.
	3. Steroids. Complex molecules containing carbon atoms arranged in four interlocking rings (three rings contain six carbon atoms each and the fourth ring contains five).		Look for four interlocking rings.	Some are hormones; other biologically important steroids are cholesterol, bile salts, and vitamin D.
Proteins	One or more polypeptides (chains of amino acids) coiled or folded in characteristic shapes.	C, H, O, N, usually S	Look for amino acid units joined by C—N bonds.	Structural components; serve as enzymes; muscle proteins.

**Table 3–1
(Continued)**

Class of compound	Description	Component elements	How to recognize	Principal functions
Nucleic acids	Backbone composed of alternating pentose and phosphate groups, from which nitrogenous bases protrude. Occurs as a double (DNA) or single (RNA) strand. DNA contains deoxyribose, guanine, cytosine, adenine, and thymine. RNA contains ribose, guanine, cytosine, adenine, and uracil.	C, H, O, N, P	Look for a pentose-phosphate backbone. DNA forms a characteristic double helix.	Storage, transmission, and expression of genetic information.

Disaccharides

A **disaccharide** (two sugars) is a carbohydrate that can be degraded into two monosaccharide units. The disaccharide **maltose** (malt sugar) consists of two chemically combined glucose units. **Sucrose,** the sugar we use to sweeten our foods, consists of a glucose unit combined with a fructose unit. **Lactose** (the sugar present in milk) is composed of one molecule of glucose and one of galactose, another hexose monosaccharide. The covalent bond that joins two monosaccharide units is called a **glycosidic bond.**

During digestion maltose is cleaved (degraded) to form two molecules of glucose:

Maltose + Water $\longrightarrow$ Glucose + Glucose

Similarly, sucrose is cleaved during digestion to form glucose and fructose:

Sucrose + Water $\longrightarrow$ Glucose + Fructose

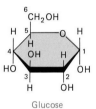

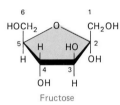

Figure 3–3 Two monosaccharides, glucose and fructose, drawn to represent their ring structures. At each angle in the ring is a carbon atom; its presence is understood by convention. Each number on the ring or on an attached group corresponds to the numbered carbons in the stick diagrams of glucose and fructose shown in Figure 3–2. The thick, tapered bonds in the lower portion of each ring indicate that the molecule is a three-dimensional structure. The thickest portion of the bond is interpreted as being the part of the molecule "nearest" the viewer.

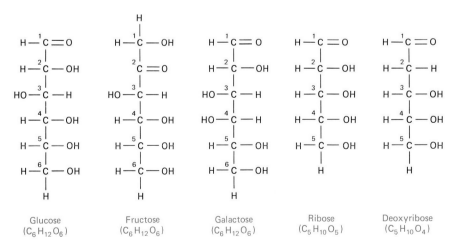

| Glucose ($C_6H_{12}O_6$) | Fructose ($C_6H_{12}O_6$) | Galactose ($C_6H_{12}O_6$) | Ribose ($C_5H_{10}O_5$) | Deoxyribose ($C_5H_{10}O_4$) |

Figure 3–2 Structural formulas of some important monosaccharides (simple sugars). The monosaccharides are represented here as straight chains, called stick formulas. Although it is convenient to show monosaccharides in this form, they are more accurately depicted as ring structures (Fig. 3–3). Note that glucose, fructose, and galactose are structural isomers—they have the same chemical formula, $C_6H_{12}O_6$, but their atoms are arranged differently.

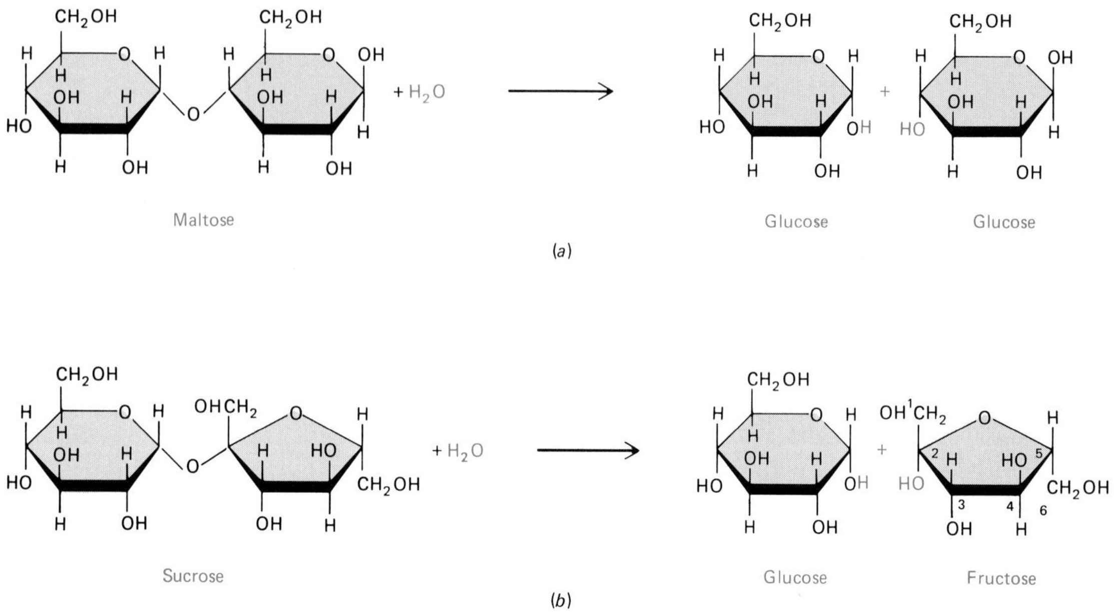

Maltose

Glucose   Glucose

(a)

Sucrose

Glucose   Fructose

(b)

Figure 3–4 A disaccharide can be cleaved to yield two monosaccharide units. (a) Maltose may be broken down (as it is during digestion) to form two molecules of glucose. This is a hydrolysis reaction that requires the addition of water. (b) Sucrose can be cleaved to yield a molecule of glucose and a molecule of fructose.

Structural formulas for the compounds in these reactions are shown in Figure 3–4. Because water is added during the cleavage of a disaccharide, this type of reaction is called a **hydrolysis** reaction.

Polysaccharides

The most abundant carbohydrates are **polysaccharides,** such as starches, glycogen, or celluloses. A polysaccharide is a single long chain, or a branched chain, consisting of repeating units of a simple sugar, usually glucose. The precise number of sugar units varies, but typically, thousands of units may be present in a single molecule of a polysaccharide. Because they are composed of different stereoisomers of glucose, or because the glucose units are arranged differently, these polysaccharides have very different properties.

Starch is the typical storage form of carbohydrate in plants, whereas **glycogen** (sometimes referred to as animal starch) is the form in which glucose is stored in animal tissues (Fig. 3–5). Glycogen is a highly branched polysaccharide, more water-soluble than plant starch. Glucose cannot be stored as such because its small, readily soluble molecules would leak out of the cells. The larger, less soluble starch and glycogen molecules do not readily pass through the cell membrane. Thus, instead of storing simple sugars, cells store the more complex polysaccharides, such as glycogen, which can readily be broken down into simple sugars.

Carbohydrates are the most abundant group of organic compounds on earth, and **cellulose** is the most abundant carbohydrate, accounting for 50% or more of all the carbon in plants. Wood is about half cellulose, and cotton is at least 90% cellulose. Plant cells are surrounded by a strong supporting cell wall consisting mainly of cellulose. Cellulose is an insoluble polysaccharide composed of many glucose molecules joined together (Fig. 3–6). The bonds joining these sugar units are different from those in starch and are not split by the enzyme that cleaves the bonds in starch.

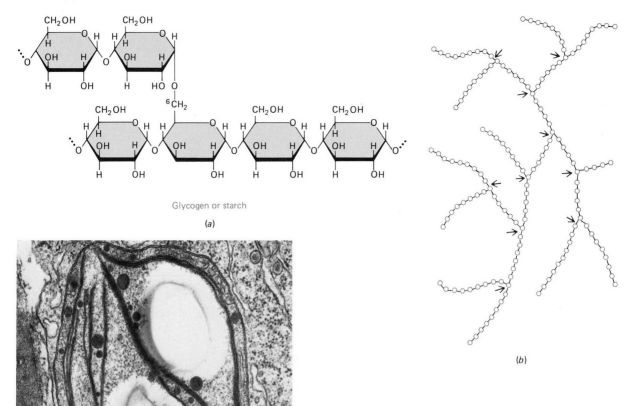

Glycogen or starch

(a)

(b)

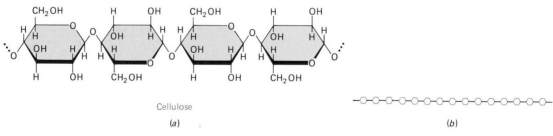

(c)

Figure 3–5 Molecular structure of glycogen or starch. (a) These molecules are branched polysaccharides composed of glucose molecules joined by glycosidic bonds. At the branch points there are bonds between carbon-6 of a glycogen in the straight chain and carbon-1 of the glucose in the branching chain. Glycogen is more highly branched than starch. (b) Diagrammatic representation of starch or glucose. The arrows represent the branch points. (c) Starch stored in a specialized organelle, a leukoplast, of an algal cell (approximately ×50,000). (c, Biophoto Associates, Photo Researchers, Inc.)

Cellulose

(a)

(b)

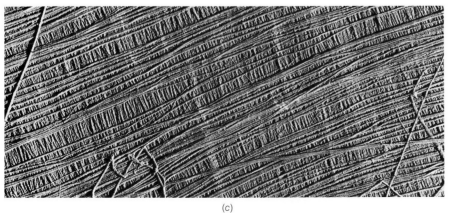

(c)

Figure 3–6 The structure of cellulose. (a) The cellulose molecule is an unbranched polysaccharide composed of approximately 10,000 glucose units joined by glycosidic bonds. (b) A more diagrammatic representation of cellulose structure. Each hexagon represents a glucose molecule bound by a glycosidic bond to the adjacent glucose molecule. (c) An electron micrograph of cellulose fibers from the cell wall of a marine alga (approximately ×24,000). (c, Omikron, Photo Researchers, Inc.)

(a)

(b)

Figure 3–7 Chitin is a polysaccharide that rivals cellulose as perhaps the most common organic compound in the ecosphere. Chitin forms the armor-like integument of such arthropods as lobsters, shrimp, and insects. (a) This fork-tailed bush katydid, *Scudderia mexicana*, has just molted and is eating its exoskeleton, thereby recycling some of its chitin and protein. (b) A ground-growing lichen. Lichens are actually compound organisms consisting of fungal cells and algae (plantlike, single-celled organisms) living together. Thus the lichen contains both chitin (in the cell walls of the fungal component) and cellulose (in the cell walls of the algal component). (a, Peter J. Bryant, University of California, Irvine/BPS.)

The amino sugars glucosamine and galactosamine are modified carbohydrates, compounds in which a hydroxide group (—OH) of a monosaccharide is replaced by an amino group (—NH$_2$; see the section on proteins). Glucosamine is the molecular unit found in chitin; galactosamine is found in cartilage. **Chitin,** a tough modified polysaccharide, is the main component of the external skeletons of insects, crayfish, and other arthropod animals (Fig. 3–7).

LIPIDS

Lipids are a heterogeneous group of compounds that have a greasy or oily consistency and are relatively insoluble in water. Like carbohydrates, lipids are composed of carbon, hydrogen, and oxygen atoms, but they have relatively less oxygen in proportion to the carbon and hydrogen than carbohydrates do. Among the groups of lipids especially significant biologically are the neutral fats, phospholipids, steroids, carotenoids (red and yellow plant pigments), and waxes. Lipids are important biological fuels; they serve as structural components of cell membranes, and some are essential hormones. Glycolipids, lipids that contain water-soluble carbohydrate components, are important in interactions among cells.

Neutral Fats

The most abundant group of lipids in living things is the neutral fats. These compounds yield more than twice as much energy per gram as do carbohydrates and are an economical form for the storage of fuel energy. Carbohydrates and proteins can be transformed by enzymes into fats and stored within the cells of adipose tissue.

A neutral fat consists of glycerol joined to one, two, or three molecules of a fatty acid. **Glycerol** is a three-carbon alcohol[1] that contains three —OH groups (Fig. 3–8). A **fatty acid** is a long, straight chain of carbon atoms with a carboxyl group (—COOH) at one end. About 30 varieties of fatty acids are commonly found in animal lipids. They typically have an even number of carbon atoms. For example, butyric acid, present in rancid butter, has four carbon atoms, and oleic acid, the most widely distributed fatty acid in nature, has 18 carbon atoms.

Saturated fatty acids contain the maximum possible hydrogen atoms, while **unsaturated fatty acids** contain carbon atoms that are doubly bonded with one another and are not fully saturated with hydrogens.

---

[1] For our purposes an alcohol may be defined as a compound in which an —OH group is bonded to a carbon atom.

Glycerol   (*a*)   Fatty acid

A triacylglycerol

$+ 3H_2O$ $\xrightarrow{\text{Enzyme}}$

Products

Oleic acid

Glycerol

Linoleic acid

Palmitic acid

(*b*)

(*c*)

Figure 3–8 Neutral fats. (*a*) Structure of glycerol and of a fatty acid. The carboxyl (—COOH) group is present in all fatty acids. The R represents the remainder of the molecule, which varies with each type of fatty acid. (*b*) Hydrolysis of a triacylglycerol yields glycerol plus three fatty acids. (*c*) Honeybees on a brood comb. The comb is composed of wax secreted by special abdominal glands of the bees. It is a compound consisting of fatty acids and alcohols, and although it is classified as a lipid, it can be digested by very few animals.

Fatty acids with several double bonds are called polyunsaturated fatty acids. Fats containing unsaturated fatty acids are the oils, and most of them are liquid at room temperature. Saturated fats are solids; butter and animal fat are examples. At least two fatty acids (linoleic and linolenic) are **essential nutrients,** which must be included in the diet.

When a glycerol molecule combines chemically with one fatty acid, a **monoacylglycerol** (sometimes called monoglyceride) is formed. When two fatty acids combine with a glycerol, a **diacylglycerol** (or diglyceride) is formed, and when three fatty acids combine with one glycerol molecule, a **triacylglycerol** (or triglyceride) is formed. In combining with glycerol, the carboxyl end of the fatty acid attaches to one of the —OH groups. This is a dehydration reaction in which H^+ and OH^-, the equivalent of water, is split off. During digestion, on the other hand, the neutral fats may be hydrolyzed to produce fatty acids and glycerol (Fig. 3–8).

Phospholipids

Phospholipids are important constituents of the cell membranes of plants and animals. A phospholipid consists of a glycerol molecule attached to one or two fatty acids, but the glycerol is also bonded to phosphorus, which is part of an organic base. Phospholipids also usually contain nitrogen (Fig. 3–9). (Note that phosphorus and nitrogen are absent from neutral fats.)

The two ends of the phospholipid molecule differ physically as well as chemically. The fatty acid portion of the molecule is **hydrophobic** (water-hating) and is not soluble in water. However, the portion composed of glycerol and the organic base is ionized and readily water-soluble. This end of the molecule is said to be **hydrophilic** (water-loving). The polarity of these lipid molecules causes them to take up a certain configuration in the presence of water, with their hydrophilic water-soluble heads facing outward toward the surrounding water. The hydrophobic tails face in the opposite direction. The cell membrane (see Fig. 5–2) is a lipid bilayer composed of two layers of phospholipid mole-

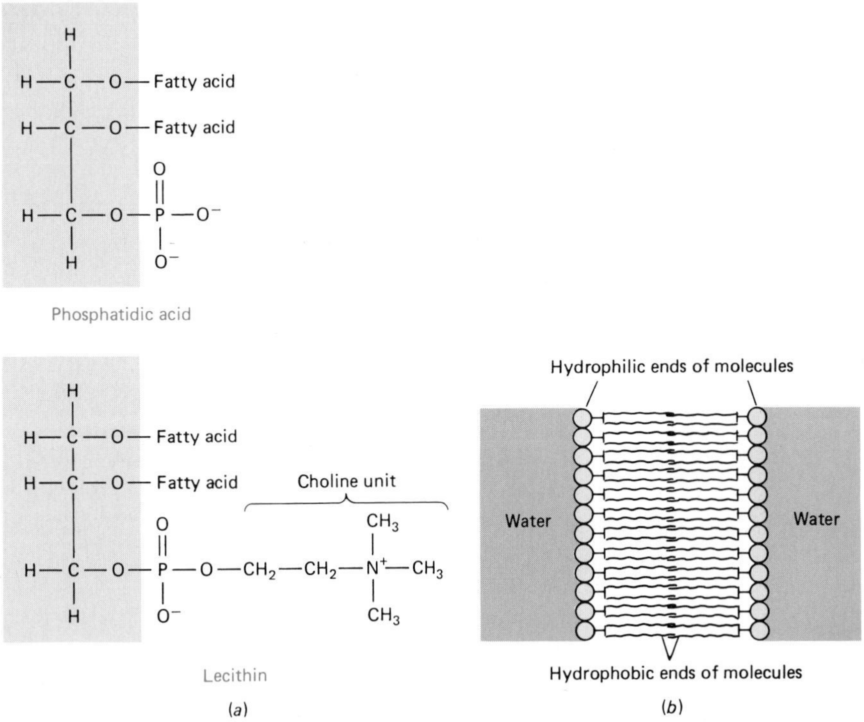

Figure 3–9 Phospholipids. *(a)* Many phospholipids are derivatives of phosphatidic acid, a compound consisting of glycerol chemically combined with two fatty acids and a phosphate group. Lecithin is a phospholipid found in cell membranes. It forms when phosphatidic acid combines with the compound choline. *(b)* A lipid bilayer, such as is found in cell membranes.

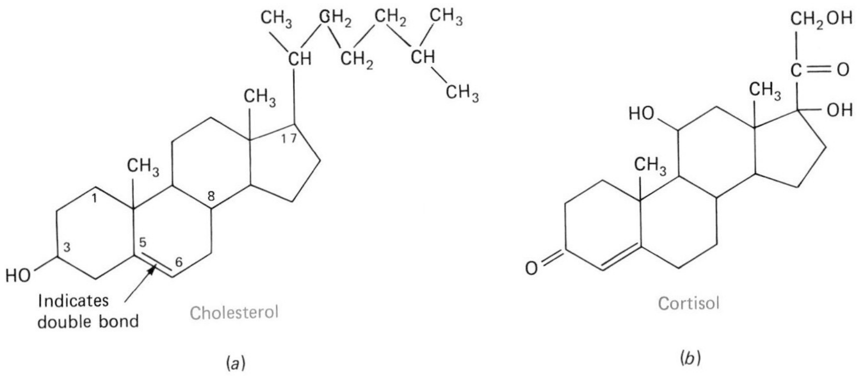

Figure 3–10 Steroids. All steroids
have the basic skeleton of four interlocking rings of carbon atoms. Note
that a carbon atom is present at each
point in each ring. Each of the first
three rings contains six carbon
atoms, and the fourth ring contains
five. For simplicity, hydrogen atoms
have not been drawn within the ring
structures.

cules with their hydrophobic tails meeting in the middle and their hydrophilic heads oriented toward the outside and inside surfaces of the cell
membrane.

Steroids

Although steroids are classified as lipids, their structure is quite different from other lipids. A **steroid** molecule contains carbon atoms arranged in four interlocking rings; three of the rings contain six carbon
atoms and the fourth contains five (Fig. 3–10). The length and structure
of the side chains that extend from these rings distinguish one steroid
from another.

Among the steroids of biological importance are cholesterol, bile
salts, the male and female sex hormones, and the hormones secreted by
the adrenal cortex. Cholesterol is a structural component of animal cell
membranes. Steroid hormones regulate certain aspects of metabolism in
a variety of animals, including vertebrates, insects, and crabs.

PROTEINS

Proteins serve as important structural components of cells and tissues,
so growth and repair, as well as maintenance of the organism, depend
upon an adequate supply of these compounds. Some proteins serve as
enzymes, catalysts that regulate the thousands of different chemical reactions that take place in a living system (see Focus on Classifying Proteins).

The protein constituents of a cell are the clue to its life-style. Each
cell type has characteristic types, distributions, and amounts of protein
that determine what the cell looks like and how it functions. A muscle
cell is different from other cell types by virtue of its large content of the
contractile proteins myosin and actin, which are largely responsible for
its appearance as well as its ability to contract. The protein hemoglobin,
found in red blood cells, is responsible for the specialized function of
oxygen transport.

Most proteins are species-specific; that is, they vary slightly in each
species, so the protein complement (as determined by the instructions in
the genes) is also mainly responsible for differences among species.
Thus the types and distributions of proteins in the cells of a dog vary
somewhat from those in the cells of a fox or a coyote. The degree of
difference in the proteins of two species is thought to depend upon evolutionary relationships. Organisms less closely related by evolution have
proteins that differ more markedly than those of closely related forms.
Some proteins differ slightly even among individuals of the same spe-

Focus on . . .

CLASSIFYING PROTEINS

There are many ways to classify proteins. Three common methods are by function, by composition, and according to their solubility.

I. Proteins grouped according to biological function

Enzymes—all enzymes are proteins
Structural proteins—proteins that are part of cells and tissues, such as collagen and elastin in connective tissues, and keratin in skin, hair, and nails
Contractile proteins—the muscle proteins actin and myosin
Hormones—insulin, growth hormone, and several other hormones are proteins
Transport proteins—hemoglobin and myoglobin both transport oxygen; a type of albumin in the blood transports fatty acids
Defense proteins—immunoglobulins (antibodies) protect the body against disease; fibrinogen in the blood is important in clotting

II. Proteins grouped according to solubility

Globular proteins—tend to be soluble in water because of their polar surface. Globular proteins are the most numerous group of proteins. They include all the enzymes, the plasma proteins, and proteins found in cell membranes

Fibrous proteins—insoluble in water; elongated to form strong fibers; function as structural and supporting proteins; include collagen, elastin, keratin, myosin, and fibrin (the protein in blood clots)

III. Proteins grouped according to composition—conjugated proteins can be classified according to the nonprotein component

Lipoproteins—contain fat and other lipids, such as the lipoproteins in the blood
Glycoproteins—contain sugars, such as immunoglobulins, that function in defense against microorganisms; many membrane proteins are glycoproteins; collagen and other proteins found in connective tissues
Nucleoproteins—bound to nucleic acids; found in chromosomes and viruses
Chromoproteins—contain a heme group; hemoglobin, myoglobin, and certain enzymes (cytochromes)
Metalloproteins—contain one or more metallic ions; some enzymes

cies, so each individual is biochemically unique. Only genetically identical organisms—identical twins or members of closely inbred strains of organisms—have identical proteins.

Amino Acid Structure

A basic knowledge of protein chemistry is essential for understanding nutrition as well as other aspects of metabolism. Proteins are composed of carbon, hydrogen, oxygen, nitrogen, and usually sulfur. Atoms of these elements are arranged into molecular subunits called **amino acids.** All of the more than 20 kinds of amino acids commonly found in proteins contain an **amino group ($-NH_2$)** and a **carboxyl group ($-COOH$)** bonded to the same carbon atom, but they differ in their side chains, abbreviated as "R" groups. **Glycine,** the simplest amino acid, has a hydrogen atom as its R group or side chain; alanine has a methyl ($-CH_3$) group (Fig. 3–11).

Amino acids combine chemically with one another by bonding the carboxyl carbon of one molecule to the amino nitrogen of another. This covalent bond linking two amino acids is referred to as a **peptide bond.** When two amino acids combine, a **dipeptide** is formed; a longer chain of amino acids is a **polypeptide.**

Each protein may contain hundreds of amino acids joined in a specific linear order. An almost infinite variety of protein molecules is possible. It should be clear that the various proteins differ from one another in the number, types, and arrangement of amino acids that they contain. The 20 types of amino acids found in biologic proteins may be thought of as letters of a protein alphabet. Each protein is a word made up of amino acid letters.

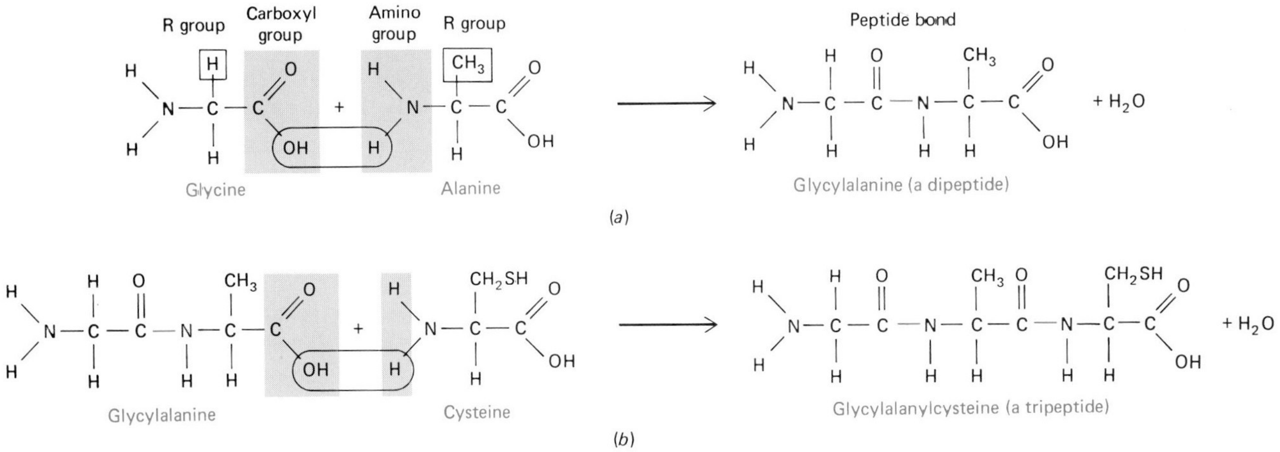

(a)

(b)

With some exceptions, plants can synthesize all their needed amino acids from simpler substances. The cells of humans and animals in general can manufacture some, but not all, of the various kinds of biologically significant amino acids if the proper raw materials are available. Those that animals cannot synthesize but must obtain in the diet are known as **essential amino acids.** Animals differ in their biosynthetic capacities; what is an essential amino acid for one species may not be essential in the diet of another.

Figure 3–11 (a) Formation of a dipeptide. Two amino acids combine chemically to form a dipeptide. Water is produced as a by-product during this reaction. (b) A third amino acid is added to the dipeptide to form a chain of three amino acids (a tripeptide or small polypeptide). The bond between two amino acids is a peptide bond.

Protein Structure—Levels of Organization

Several levels of organization can be distinguished in the protein molecule. The sequence of amino acids in a polypeptide chain constitutes its **primary structure** (Fig. 3–12). This sequence, as we shall see later, is specified by the instructions in a gene.

The **secondary structure** of protein molecules involves the coiling of the peptide chain into a helix or some other regular conformation (shape). Peptide chains ordinarily do not lie out flat or coil randomly, but undergo coiling to yield a specific three-dimensional structure. A common secondary structure in protein molecules is known as the α-helix. This involves the formation of spiral coils of the polypeptide chain. The α-helix is a very uniform geometric structure with 3.6 amino acids occupying each turn of the helix. The helical structure is determined and maintained by the formation of hydrogen bonds between amino acids in successive turns of the spiral coil.

The **tertiary structure** of a protein molecule involves the folding of the α-helix upon itself, folds that impart a specific overall structure to the protein molecule. Hydrogen bonds between one part of the peptide chain and another part help hold the folds in place. Disulfide bonds (—S—S—) between certain amino acids and other covalent bonds may also be important in maintaining the tertiary structure of many proteins. The biological activity of the protein depends in large part on the specific tertiary structure of the molecule, held together by these bonds. When a protein is heated or treated with any of a number of chemicals, the tertiary structure becomes disordered and the coiled peptide chains unfold to give a more random conformation. This unfolding is accompanied by a loss of the biological activity of the protein, for example, its ability to act as an enzyme. This change is termed **denaturation** of the protein.

Proteins composed of two or more subunits have a **quaternary structure.** This refers to the combination of two or more like or unlike peptide chain subunits, each with its own primary, secondary, and tertiary structures, to form the biologically active protein molecule. Hemoglo-

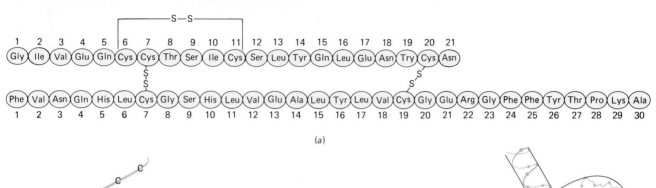

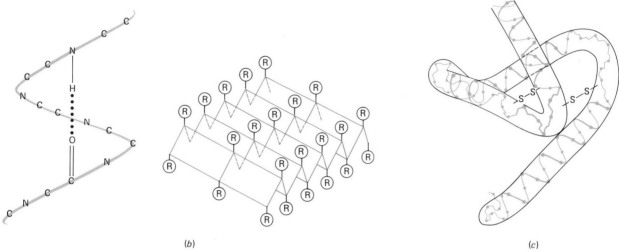

(a)

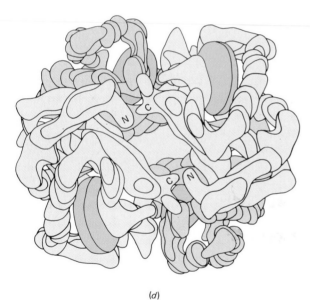

(b)

(c)

(d)

Figure 3–12 Protein structure. (a) The primary structure of the two polypeptide chains that make up the protein insulin. The primary structure is the linear sequence of amino acids. Each oval in the diagram represents an amino acid. The letters inside the ovals are symbols for the names of the amino acids. Insulin is a very small protein. (b) The secondary structure of proteins is commonly an alpha helix. The folds in the helix are held together mainly by hydrogen bonds between oxygen and hydrogen atoms. In some proteins, such as the silk protein fibroin, the backbone of the polypeptide chain is stretched out into a zigzag structure. (c) The tertiary structure results from the coiling and folding of the alpha helix (or other secondary structure) into an overall globular or other shape. Hydrogen bonds, bonds between sulfur atoms, and other attractions between atoms are among the forces that hold the parts of the molecule in the designated shape. (d) Proteins that consist of more than one polypeptide subunit assume a final quaternary shape. Hemoglobin, a globular-shaped protein containing four polypeptide subunits, is illustrated here. Its quaternary structure consists of the final shape in which the subunits combine. In hemoglobin each polypeptide encloses an iron-containing structure (shown as green discs). (e) The silk used by this garden spider to wrap its prey is an extremely strong and flexible protein. The silk fibers harden as they are spun from the glands within the spider's abdomen. (e, R.K. Burnard, Ohio State University/BPS.)

(e)

bin, the protein in red blood cells that is responsible for oxygen transport, consists of 574 amino acids arranged in four polypeptide chains— two identical α and two identical β chains. Its chemical formula is $C_{3032}H_{4816}O_{872}N_{780}S_8Fe_4$.

NUCLEIC ACIDS AND NUCLEOTIDES

Nucleic acids, like proteins, are large, complex molecules. They were first isolated by Friedrich Miescher in 1870 from the nuclei of pus cells, and their name stems from the fact that they are acidic and were first identified in nuclei. There are two classes of nucleic acids, **ribonucleic acids (RNA)** and **deoxyribonucleic acids (DNA);** the different kinds of RNA and DNA vary in some of their structural components and in their metabolic functions. DNA contains the instructions for making all the proteins needed by the organism and constitutes the genes themselves, the hereditary material of the cell. The various kinds of RNA function in the process of protein synthesis.

Nucleic acids are composed of **nucleotides,** molecular units that consist of (1) a five-carbon sugar, either ribose or deoxyribose, (2) a phosphate group, and (3) a nitrogenous base that may be either a double-ringed purine or a single-ringed pyrimidine (Fig. 3–13). DNA contains the purines adenine (A) and guanine (G), and the pyrimidines cytosine (C) and thymine (T), together with the sugar deoxyribose and phosphate. RNA contains the purines adenine and guanine and the pyrimidines cytosine and uracil (U), together with the sugar ribose and phosphate.

The molecules of nucleic acids are made of linear chains of nucleotides, each attached to the next by bonds between the sugar molecule of one and the phosphate group of the next. As we will see in our discussion of the genetic code (Chapter 11), the specific information of the nucleic acid is coded in the unique sequence of the four kinds of nucleotides present in the chain.

Besides the importance of nucleotides as subunits of nucleic acids, a number of them serve other vital functions in living cells. **Adenosine triphosphate (ATP),** composed of adenine, ribose, and three phosphates (Fig. 6–11), is of major importance as the energy currency of all cells.

Figure 3–13 A nucleic acid consists of subunits called nucleotides. Each nucleotide consists of (1) a nitrogenous base, which may be either a purine or a pyrimidine, (2) a five-carbon sugar, either ribose (in RNA) or deoxyribose (in DNA), and (3) a phosphate group. (a) The three major pyrimidine bases found in nucleotides. (b) The two major purine bases found in nucleotides. (c) A nucleotide (adenylic acid). (d) A diagrammatic representation of part of a nucleic acid molecule.

Cytosine　　Thymine　　Uracil

(a)

Adenine　　Guanine

(b)

Phosphate group

Adenine (an amine)

Ribose (a five-carbon sugar)

A nucleotide, adenosine monophosphate (AMP)

(c)

Part of a nucleic acid molecule

(d)

The two terminal phosphate groups are joined to the nucleotide by special "energy-rich" bonds, indicated by the ~P symbol. These are energy-rich bonds in the sense that much free energy is released when the bonds are hydrolyzed. The biologically useful energy of these bonds can be transferred to other molecules. Most of the chemical energy of the cell is stored in these high energy phosphate bonds of ATP, ready to be released when the phosphate group is transferred to another molecule.

A nucleotide may be converted by enzymes called cyclases to a cyclic form. ATP, for example, is converted to cyclic adenosine monophosphate (**cyclic AMP**) by the enzyme adenylate cyclase. Cyclic nucleotides play important roles in mediating the effects of hormones and in regulating various aspects of cellular function.

Cells contain several dinucleotides that are of great importance in metabolic processes. For example, nicotinamide adenine dinucleotide (NAD^+) is very important as a primary electron and hydrogen acceptor and donor in biological oxidations and reductions within cells (Chapter 6).

CHAPTER SUMMARY

I. The major groups of organic compounds are carbohydrates, lipids, proteins, and nucleic acids.

II. Carbohydrates contain carbon, hydrogen, and oxygen in a ratio of approximately 1 carbon to 2 hydrogens to 1 oxygen. Sugars, starches, and celluloses are typical carbohydrates.
   A. Monosaccharides are simple sugars such as glucose, fructose, or ribose. Glucose is an important fuel molecule in living cells.
   B. Most carbohydrates are polysaccharides, long chains of repeating units of a simple sugar.
      1. Carbohydrates are typically stored in plants as starch and in animals as glycogen.
      2. The cell walls of plant cells are composed mainly of the polysaccharide cellulose.

III. Lipids are composed of carbon, hydrogen, and oxygen but have relatively less oxygen in proportion to carbon and hydrogen than do carbohydrates. Lipids have a greasy or oily consistency and are relatively insoluble in water.
   A. The body stores fuel in the form of neutral fats. A fat consists of a molecule of glycerol combined with one to three fatty acids.
      1. Three types of neutral fats are monoacylglycerols, diacylglycerols, and triacylglycerols.
      2. Fatty acids, and therefore fats, can be saturated or unsaturated.
   B. Phospholipids are structural components of cell membranes.
   C. Steroid molecules contain carbon atoms arranged in four interlocking rings. Cholesterol, bile salts, vitamin D, and certain hormones are important steroids.

IV. Proteins are large, complex molecules made of simpler components termed amino acids that are joined by peptide bonds. They are composed of carbon, hydrogen, oxygen, nitrogen, and sulfur.
   A. Proteins are important structural components of cells and tissues. Many serve as enzymes or as hormones, and they may also be used as fuel.
   B. Four levels of organization can be distinguished in protein molecules.

1. Primary structure is the sequence of amino acids in the peptide chain.
2. Secondary structure refers to the coiling of the peptide chains in a helix or some other regular conformation.
3. Tertiary structure is the folding of the chain upon itself.
4. Quaternary structure is the spatial relationship of the combination of two or more peptide chains.

V. The nucleic acids DNA and RNA store information that governs the structure and function of the organism. Nucleic acids are composed of carbon, hydrogen, oxygen, nitrogen, and phosphorus.
   A. Nucleic acids are composed of long chains of nucleotide units, each composed of a nitrogenous base; a purine or a pyrimidine; a five-carbon sugar (ribose or deoxyribose); and a phosphate group.
   B. ATP is a nucleotide of special significance in energy metabolism. NAD is an electron and hydrogen acceptor in biological oxidations.

Post-Test

Select the most appropriate term from column B for each entry in column A.

Column A

_____ 1. a monosaccharide
_____ 2. a steroid
_____ 3. a nucleic acid
_____ 4. oleic acid

Column B

a. cellulose
b. DNA
c. glucose
d. cholesterol
e. none of the preceding

_____ 5. important constituent of cell membranes
_____ 6. subunits of proteins
_____ 7. energy currency of cell
_____ 8. component of fatty acids

a. ATP
b. glycerol
c. phospholipids
d. amino acids
e. none of the preceding

9. Peptide bonds are found linking _____ _____ .
10. The primary structure of a protein refers to the _____ .
11. _____ is an important component of the cell walls of plant cells.
12. Animals store glucose in the form of _____ .

Review Questions

1. What property of carbon makes it so important in living organisms?
2. Contrast a monosaccharide, such as glucose, with a polysaccharide, such as starch.
3. Why are each of the following biologically important? (a) steroids (b) phospholipids (c) polysaccharides (d) nucleic acids (e) amino acids
4. Draw a structural formula of a simple amino acid and identify the carboxyl and amino groups.
5. There are thousands of types of proteins. How does one protein differ chemically from another?
6. Compare proteins with nucleic acids.
7. Why are neutral fats important? What are the molecular components of a neutral fat?

Readings

Caplan, A. "Cartilage," *Scientific American*, November 1984, 84–94. This fundamental skeletal tissue has unique properties of strength and resilience which now can be explained in terms of the molecular structure of the chemical constituents of the tissue. An excellent example of the chemical basis of biology.

Sharon, N. "Carbohydrates," *Scientific American*, November 1980, 90–116. A discussion of the chains of sugar units that are the most abundant component of living cells, and of their roles in normal biological processes and disease.

(Also consult the readings for Chapter 2.)

Chapter 4
CELL STRUCTURE AND FUNCTION

Outline

I. General characteristics of cells
II. How cells are studied
III. Prokaryotic and eukaryotic cells
IV. Inside the cell
   A. Endoplasmic reticulum and ribosomes: internal transport and synthesis
   B. The Golgi complex: a packaging plant
   C. Lysosomes: breaking down harmful materials
   D. Microbodies
   E. Mitochondria: power plants of the cell
   F. Plastids: energy traps and storage sacs
   G. Microtubules and microfilaments: cell shape and movement
   H. The microtrabecular lattice: intracellular framework
   I. Centrioles
   J. Cilia and flagella
   K. Vesicles and vacuoles
   L. The cell nucleus
     1. The nuclear envelope
     2. The nucleolus
     3. Chromatin and chromosomes
Focus on viewing the cell

Learning Objectives

After reading this chapter you should be able to:

1. Explain why the cell is considered the basic unit of life and state the cell theory.
2. Describe the general characteristics of cells, for example, size, range, and shape.
3. Identify methods by which scientists study cells.
4. Contrast prokaryotic and eukaryotic cells; contrast plant and animal cells.
5. Draw and label a diagram of a prokaryotic cell, a plant cell, and an animal cell. Describe and list the functions of the principal organelles.
6. Distinguish between smooth and rough endoplasmic reticulum and describe the functional relationship between ribosomes and endoplasmic reticulum.
7. Describe how the Golgi complex packages secretions and manufactures lysosomes.
8. State the function of the mitochondria and explain why these organelles are called the power plants of the cell.
9. Compare microtubules and microfilaments and describe the microtrabecular lattice.
10. Describe the structure and function of the cell nucleus.

Although some living systems consist of only one cell and others of several billion, even the most complex organism begins life as a single cell, the fertilized egg. In most multicellular organisms, including human beings, this cell divides to form two cells, and each new cell divides again and again, eventually forming the complex tissues, organs, and systems of the developed organism. (Recall the hierarchy of biological organization described in Chapter 1; see Figure 1–12.) Like the bricks of a building, cells are the building blocks of the organism.

The cell is the smallest unit of living material capable of carrying on all the activities necessary for life. We might say that it is the smallest structure with a complete metabolism because it has all of the physical and chemical components needed for its own maintenance and growth. When provided with essential nutrients and an appropriate environment, cells can be kept alive in laboratory glassware for many years. No cell part is capable of such survival.

One of the most basic generalizations of biology is the **cell theory.** In 1838 and 1839 two German biologists, Matthias Schleiden and Theodor Schwann, proposed that all living things are made up of cells and cell products, and that the cell is the basic unit of living organisms. The cell theory was extended in 1855 by Rudolf Virchow, who stated that new cells come into existence only by the division of previously existing cells. Cells cannot arise by spontaneous generation from nonliving matter. In 1880 August Weismann pointed out the corollary of this, that all the cells living today can trace their ancestry back to ancient times (Fig. 4–1).

GENERAL CHARACTERISTICS OF CELLS

A cell consists of jelly-like cytoplasm surrounded by a cell membrane, a physical boundary that separates the cell from the outside environment. Most cells contain a **nucleus** (or **nucleoid** in monerans), which contains the DNA, the chemically coded instructions for synthesizing all the proteins needed for growth, repair, and reproduction. Many other organelles, internal structures with specialized functions, are suspended within the cytoplasm.

Almost all cells are microscopic. An "average" cell measures about 10 micrometers (1/2500 inch) in diameter. This means that if you could

Figure 4–1 Every cell comes from a pre-existing cell, according to the cell theory, and here we see the drama enacted whereby a new individual comes into existence by the union of two cells. (*a*) Shaped like a scimitar, the head of a hamster sperm is enveloped by finger-like microvilli of the egg surface into which it gradually sinks. When these two reproductive cells have united, the resulting fertilized egg, a single cell, soon divides, and does so again and again. (*b*) A cluster of cells results, each of which continues to divide while the new individual takes form. Though the organism may be composed of billions of cells, the ancestry of each cell can be traced back to the fertilized egg. (*a*, Drs. David M. Phillips and Ryuzo Yanagimachi, *Development, Growth and Differentiation* 24, 1982; *b*, Drs. Yehuda Ben-Shaul, Karen Atzt, and Dorothea Bennett.)

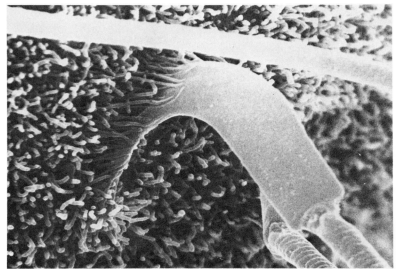

(a)

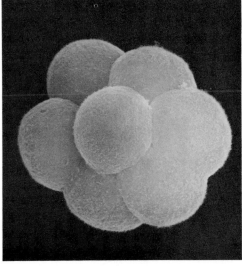

(b)

line up about 2500 typical cells end to end, the resulting cellular parade would measure less than 3 centimeters (only about 1 inch).

Although they have many features in common, cells vary widely in size and appearance (Fig. 4–2). In fact, the size and shape of a cell are related to its specific function. Epithelial cells, which are specialized to cover body surfaces, look like tiny building blocks. Nerve cells have long extensions that receive or transmit messages long distances through the body. An extension of the sciatic nerve, for example, may extend from spinal cord to foot. Although such a nerve cell may be more than a meter long, its diameter is so tiny that no part of it can be seen without the aid of a microscope. Certain white blood cells in the body resemble unicellular amebas in their ability to change shape as they flow along from one location to another. Plant cells often have large fluid-filled structures called vacuoles, and they contain chloroplasts. The largest cells are birds' eggs, which consist largely of yolk that provides nourishment for the developing bird. However, yolk is an inert material that is not really part of the metabolic structure of the cell. Neither the shell nor the white of the bird's egg is considered part of the cell because these structures consist of nonliving material secreted by the mother's reproductive tract.

Figure 4–2 The size and shape of cells are related to their functions. (a) Photomicrograph of a one-celled organism, the *Paramecium*, which is equipped with hairlike cilia that beat in a coordinated motion, allowing it to move from place to place. (b) Among the most highly specialized cells, the nerve cell may live as long as a hundred years without dividing. Its long extensions are adaptations for transmitting neural messages from one part of the organism to another. (c) Plant cells from the outer layer of a leaf. Each brick-shaped structure is a plant cell. These cells are specialized to carry on photosynthesis. The tiny green structures are the chloroplasts where photosynthesis takes place. (a, Michael Abbey, Photo Researchers, Inc.; b, Ed Reschke; c, J.F. Gennaro, Photo Researchers, Inc.)

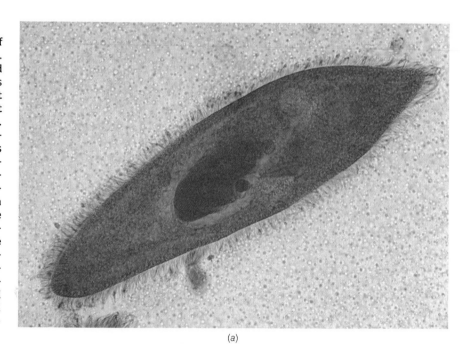

(a)

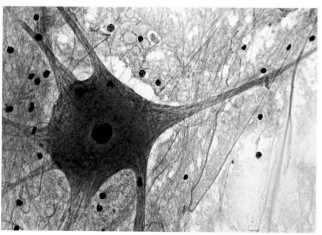

(b)

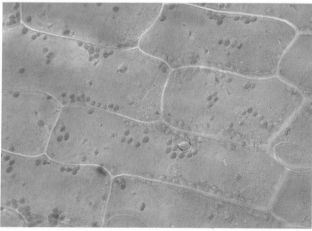

(c)

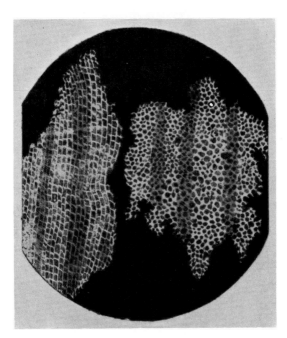

Figure 4–3 Drawing by Robert Hooke of the microscopic structure of a thin slice of cork. Hooke was the first to describe cells—his observations were based on the cell walls of these dead cork cells. (From the book *Micrographia*, published in 1665, in which Hooke described many of the objects he had viewed using the compound microscope he constructed.)

HOW CELLS ARE STUDIED

Cells are so small that one might well wonder how we know so much about what goes on inside them. The biologist's most important tool for studying the internal structure of cells has been the microscope. Anton van Leeuwenhoek (1632–1723) is credited with developing some of the earliest microscopes and with leaving written records of the structures he studied. In 1665 Robert Hooke (1635–1703) examined a slice of cork with the aid of a crude, homemade microscope. Because the tiny compartments he saw reminded him of the little rooms, or cells, of a monastery, he called them cells. What Hooke saw were the cell walls of dead cork cells (Fig. 4–3). In later observations he described cell contents, but it was not until two centuries later that scientists realized that the important part of the cell is its contents and not its outer walls or membranes. During the 1800s scientists studied various cells and observed a variety of intracellular structures, which they referred to as **organelles** (little organs) because they were thought to perform special jobs within the cell, just as our organs perform specific jobs within our bodies.

The ordinary compound light microscope, the kind used by students in most college laboratories, was responsible for the discovery of most cell structures. The **light microscope,** gradually improved since Hooke's time, uses visible light as the source of illumination (see Focus on Viewing the Cell). During the last three decades the development of the electron microscope has enabled researchers to study the fine detail (ultrastructure) of cells. The **electron microscope** floods the specimen being studied with a beam of electrons rather than with light waves.

Magnification is the ratio of the size of the image to the size of the specimen. Whereas the ordinary light microscope can magnify a structure about 1000 times, the electron microscope can magnify it 250,000 times or more. Another advantage of the electron microscope is its superior resolving power. Even more important than magnification, **resolving power** is the ability to reveal fine detail; it is expressed as the minimum distance between two points that can be distinguished as separate and distinct points, rather than as one single point. Whereas a light microscope equipped with very fine lenses can resolve objects about 500 times

VIEWING THE CELL

Viewing the Cell with the Compound Light Microscope

The light microscope uses visible light as the source of illumination (see figure). It has a resolving power of about 200 nanometers (0.2 meter). Objects can be magnified about 1000 times with good resolution using the light microscope. Beyond that magnification, structures may appear larger, but they are not clearer. To be viewed with the light microscope, specimens must be very thin. Single-celled organisms—such as amebas and some cells from multicellular organisms, like sperm cells—can be viewed in the living state. However, tissues are too thick and must be sectioned (cut) into very thin slices. They are generally preserved and then stained with special dyes.

In ordinary light microscopy the absorption of light by the specimen is important. Both the phase-contrast microscope and the interference microscope permit us to view to advantage parts of living cells that would ordinarily not absorb much light (see figures). The technique converts small differences in the way specimens refract (bend) light to much greater differences, causing the specimen to appear brighter. In the dark-field microscope only the light scattered from edges or particles in the specimen can enter the microscope lenses that produce the image. Thus the cell shows up as bright against a dark background. Fluorescence microscopy depends upon a similar technique.

Viewing the Cell with the Transmission Electron Microscope

The electron microscope uses a beam of electrons as a source of illumination instead of light. The electron beams have much shorter wavelengths than visible light. Because electrons have electrical charges, a magnetic field can be used to direct them. In the transmis-

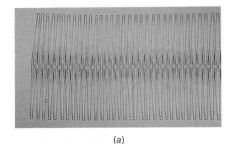

(a)

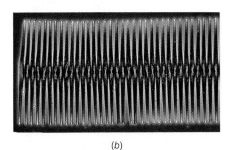

(b)

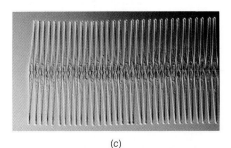

(c)

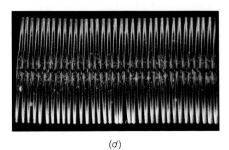

(d)

Fragilaria, a freshwater colonial diatom, as seen with four different types of light microscopy. In each case the magnification is approximately ×160. (a) Bright-field microscopy. (b) Phase contrast microscopy. (c) Nomarski interference microscopy. (d) Darkfield microscopy. (J. Robert Waaland, University of Washington/BPS.)

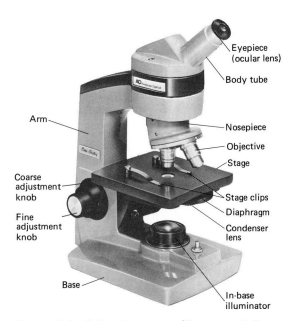

Eyepiece (ocular lens)
Body tube
Arm
Nosepiece
Objective
Stage
Coarse adjustment knob
Stage clips
Diaphragm
Fine adjustment knob
Condenser lens
Base
In-base illuminator

Parts of the light microscope. (Courtesy of the American Optical Corporation.)

sion electron microscope an electromagnetic field is used to channel electrons to a focal point. For the image formed to be viewed, it must first be projected onto a television screen or photographic plate. A photograph taken with an electron microscope is called an **electron micrograph (EM).**

In the transmission electron microscope a beam of electrons is transmitted through the specimen and falls upon a photographic plate or a fluorescent screen. Before it can be viewed, the specimen must be embedded in plastic and cut in ultrathin sections so that the beam of electrons can pass through it. This is a disadvantage because live specimens cannot be viewed. However, the transmission electron microscope has been valuable for studying details of internal cell structure.

Viewing the Cell with the Scanning Electron Microscope

In scanning electron microscopy the specimen is coated with a metal, often gold. The electron beam does not pass through the specimen. Instead, it is directed along the surface of the specimen. The contour of the specimen causes variations in the angle with which the beam strikes the various points of the specimen. This leads to variations in the intensity with which secondary electrons are emitted from the specimen's surface. A recording of the emission from the specimen provides a picture of the three-dimensional nature of the specimen. This provides information about the shape and surface of the specimen that could not be gained from transmission electron microscopy. Scanning electrons provide striking views of a specimen's surface but cannot be used to study internal structures without special preparation.

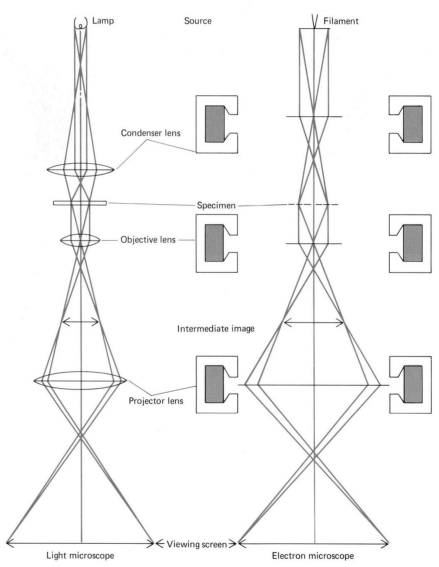

A comparison of a light microscope with an electron microscope. In each kind of instrument light rays or an electron beam are focused by the condenser lens onto the specimen. The objective lens forms a first magnified image of the specimen, which is further magnified by the projector lens onto a ground glass screen in the light microscope or onto a fluorescent screen in the electron microscope. The lenses in the electron microscope are actually magnets that bend the beam of electrons.

(Continued)

Focus on . . .

VIEWING THE CELL (Continued)

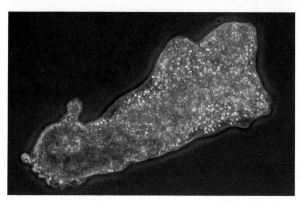

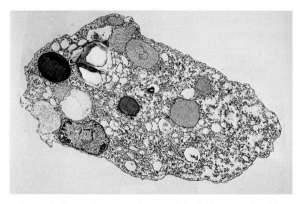

Light micrograph of Amoeba proteus, *a sarcodinian. (approximately* ×*120). (Walker England, Photo Researchers, Inc.)*

Transmission electron micrograph of Entamoeba histolytica, *a parasitic sarcodinian. The dark bodies are recently ingested red blood cells (approximately* ×*550). (Science Photo Library, Photo Researchers, Inc.)*

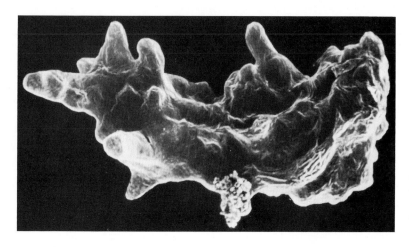

Scanning electron micrograph of Amoeba proteus, *a sarcodinian, giving a remarkably clear picture of the surface of the animal. (Courtesy of Dr. Eugene Small.)*

better than the unaided human eye, the electron microscope has a resolving power of more than 10,000 times that of the human eye (Fig. 4–4).

Cell structure has also been studied extensively by physical and chemical methods. Cells can be broken apart and then centrifuged (spun) at high speeds to separate the cellular components. Using a variety of techniques, organelles can be separated from one another on the basis of size or density. Once separated from its surrounding cellular components, a given type of organelle can be analyzed biochemically to determine its composition. Another technique used to determine the function of an organelle is to monitor its activities in a test tube under controlled conditions.

PROKARYOTIC AND EUKARYOTIC CELLS

Members of kingdom Monera—the bacteria and cyanobacteria—consist of **prokaryotic cells,** which are fundamentally different from the **eukaryotic cells** of all other organisms. Prokaryotic cells are simpler and gener-

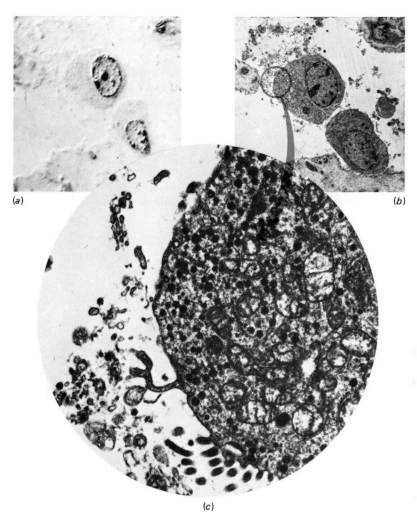

(a)

(b)

(c)

Figure 4–4 Comparison of a photograph taken with a modern light microscope with two taken using an electron microscope. (a) Lung cancer cells magnified 1800 times as seen using a light microscope. (b) The same cells, at the same magnification, seen through an electron microscope. The clearer detail is a result of the greater resolving power of the electron microscope. (c) A portion of one of the cells seen in (b) magnified about 17,000 times by the electron microscope. Note the black granules and other detail not visible at the lower magnification. (Courtesy of Zeiss. The electron micrographs were taken with Zeiss EM 9S-2 by Dr. Harry Carter.)

ally smaller than eukaryotic cells and are thought to have evolved first. The word prokaryotic means "before the nucleus," and these cells have no nuclear membrane. Prokaryotic cells do have one or more nuclear regions, sometimes referred to as **nucleoids,** in which DNA is concentrated (Fig. 4–5).

Prokaryotic cells also lack other membrane-bound organelles typical of eukaryotic cells. In some prokaryotic cells the cell membrane is folded inward to form a complex of internal membranes along which the reactions of cellular respiration are thought to take place. Prokaryotic cells that carry out photosynthesis contain the green pigment chlorophyll associated with flat, sheetlike membranes called **lamellae,** but these lamellae are not distinct organelles.

As implied by the term *eukaryotic,* which means "good nucleus," or "true nucleus," eukaryotic cells have distinct nuclei surrounded by nuclear membranes. They also have many types of membrane-bound organelles that partition the cytoplasm into compartments. Table 4–1 summarizes the types of organelles typical of eukaryotic cells. Some organelles are found only in specific kinds of cells. For example, chloroplasts, structures that trap sunlight, are found only in cells that carry on photosynthesis.

Plant cells differ from animal cells in several ways (Figs. 4–6 and 4–7): (1) although all cells are limited by cell membranes, plant cells are also surrounded by stiff cell walls of cellulose, which limits any change in position or shape; (2) plant cells contain plastids, membrane-bound structures that produce and store food material, the most familiar and

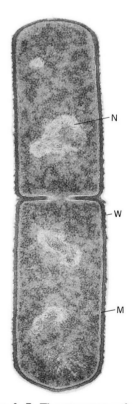

N

W

M

Figure 4–5 The structure of a prokaryotic cell. An electron micrograph of the bacterium, *Bacillus subtilis.* This cell has a prominent cell wall (W) surrounding the cell membrane (M). The nuclear regions (N) are clearly visible. (Courtesy of A. Ryter.)

Table 4–1
EUKARYOTIC CELL STRUCTURES AND THEIR FUNCTIONS

Structure	Description	Function
Cell membrane* (plasma membrane)	Lipid bilayer in which variety of proteins are distributed in mosaic pattern	Protection; regulates passage of materials in and out of cell; helps maintain cell shape; communicates with other cells
Endoplasmic reticulum (ER)	Network of internal membranes extending through cytoplasm; forms system of tubes and vesicles	Intracellular transport of materials
smooth	Lacks ribosomes on outer surfaces	Produces steroids in certain cells; lipid metabolism; detoxifies drugs
rough	Ribosomes stud outer surfaces	Manufactures and transports proteins
Ribosomes	Granules composed of RNA and protein; some attached to ER; some float free in cytoplasm	Manufacture protein
Golgi complex	Stacks of flattened membranous sacs	Packages secretions; manufactures lysosomes
Lysosomes	Membranous sacs containing digestive enzymes	Release enzymes that break down worn-out parts or unwanted materials (debris or bacteria) within cell; play role in cell death
Microbodies (e.g., peroxisomes)	Membranous sacs containing oxidative enzymes	Site of many metabolic reactions
Vesicles and vacuoles	Membranous sacs; vesicle is small vacuole	Transport and store ingested materials, cellular secretions, or wastes
Mitochondria	Sacs consisting of two membranes; inner membrane is folded to form cristae	Site of most reactions of cellular respiration; power plants of cell
Plastids	Membranous; chloroplast has thylakoids that contain chlorophyll	Chloroplasts contain chlorophyll that traps energy in photosynthesis
Microtubules	Hollow rods with walls of tubulin protein	Provide structural support; have role in cell movement; components of centrioles, cilia, and flagella
Microfilaments	Solid, rodlike structures consisting of contractile protein	Provide structural support; play role in cell movement
Microtrabecular lattice	Network of slender protein threads forming framework of cell	Links microtubules, microfilaments, and organelles into functional unit; positions organelles
Centrioles	Pair of hollow cylinders located within region called centrosome; each centriole consists of nine triple microtubules	During nuclear division, spindle forms between these structures in animal cells
Cilia	Hollow tubes made of smaller microtubules; extend outside cell	Movement of cell or of material outside cell. Ciliated cells that line respiratory tract beat path of mucus away from the lungs; not present in all cells
Flagella	Hollow tubes made of smaller microtubules; extend outside cell	Cellular locomotion; in human body, found only in sperm cells
Nucleus	Large spherical structure surrounded by nuclear membrane; contains nucleolus and chromosomes	Control center of cell; contains chromosomes
Nucleolus	Nonmembranous; rounded granular body within nucleus; consists of RNA and protein	Assembles ribosomes; may have other functions
Chromosomes	Long threadlike structures composed of DNA and proteins; when cell is not dividing, chromosomes are elongated, forming chromatin	Contain genes (hereditary units) that govern structure and activity of cell

*The cell membrane will be discussed in Chapter 5.

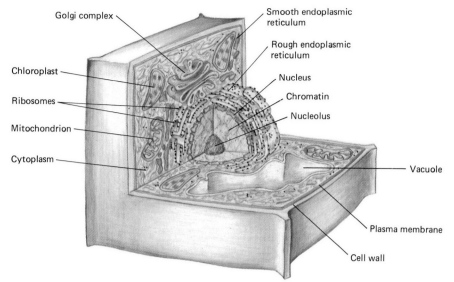

Golgi complex

Smooth endoplasmic reticulum

Rough endoplasmic reticulum

Chloroplast

Nucleus

Chromatin

Ribosomes

Nucleolus

Mitochondrion

Cytoplasm

Vacuole

Plasma membrane

Cell wall

Figure 4–6 A "typical" plant cell, as visualized by an artist. Notice the green chloroplasts, responsible for the process of photosynthesis upon which virtually all life depends directly or indirectly. The large vacuole is filled with watery fluid. The pressure of this fluid, combined with the resistance of the tough cell wall, lends turgor and stiffness to the plant tissues of which such a cell may be a part.

abundant of which are chloroplasts; (3) most plant cells have one large or several small conspicuous compartments, called vacuoles, used for transporting and storing nutrients and waste products; and (4) certain organelles—centrioles (and probably lysosomes)—are absent in the cells of complex plants.

INSIDE THE CELL

Early biologists believed that the cell consisted of a homogeneous jelly, which they referred to as protoplasm. With the electron microscope and other modern research tools, perception of the world within the cell has been greatly expanded. We now know that the cell is a highly organized, amazingly complex structure (Figs. 4–8 and 4–9). It has its own control center, internal transportation system, power plants, factories for making needed materials, packaging plants, and even a "self-destruct" system. Today the word protoplasm, if used at all, is used in a very general way. Specifically, the portion of protoplasm outside the nucleus is called **cytoplasm,** and the corresponding jelly-like material within the nucleus is termed **nucleoplasm.** Within the fluid component of the cytoplasm and nucleoplasm the various organelles are suspended.

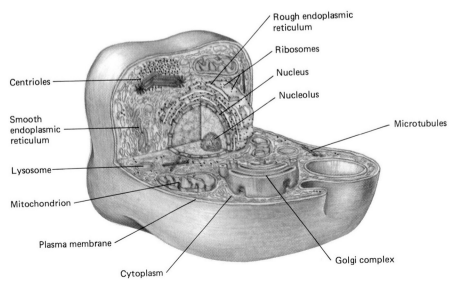

Rough endoplasmic reticulum

Ribosomes

Nucleus

Nucleolus

Centrioles

Smooth endoplasmic reticulum

Microtubules

Lysosome

Mitochondrion

Plasma membrane

Cytoplasm

Golgi complex

Figure 4–7 An artist's conception of a "typical" animal cell.

Figure 4–8 The structure of an animal cell. (*a*) Electron micrograph of a human pancreas cell, magnified ×16,000. Most of the structures of a typical animal cell are present here; however, like most of the cells of a complex, multicellular animal, this cell has certain features that permit it to carry out a specialized function. The larger, circular dark bodies within the cell are zymogen granules containing inactive enzymes. Released from their storage cells and activated, these enzymes facilitate reactions such as the breaking down of peptide bonds during the digestion of proteins. (*b*) A drawing based on the electron micrograph, emphasizing the important structures of the cell. Desmosomes, important structures in maintaining adhesion between cells, are discussed in Chapter 5. (*a*, courtesy of Dr. Susumu Ito, Harvard Medical School.)

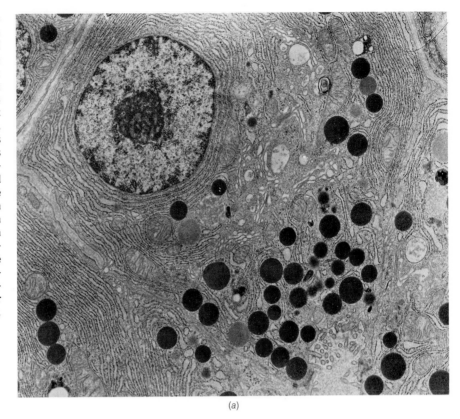

(*a*)

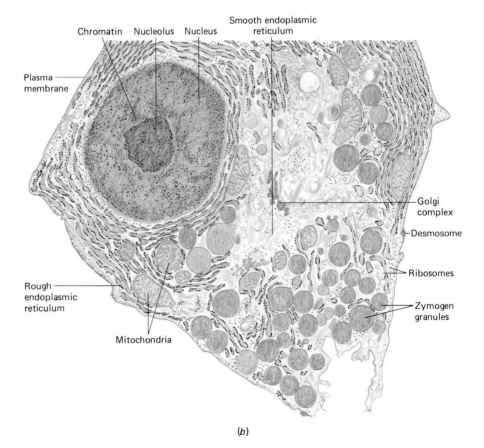

(*b*)

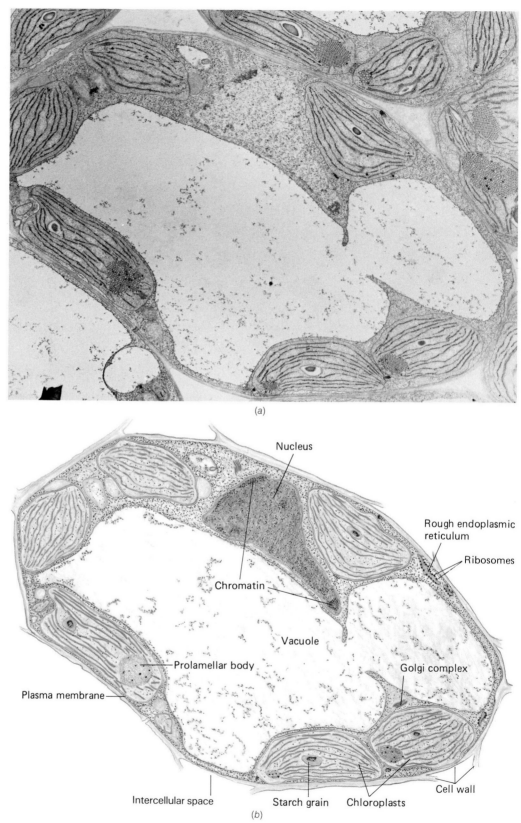

Figure 4–9 The structure of a plant cell. (a) Electron micrograph of a cell from the leaf of a young bean plant, *Phaseolus vulgaris,* magnified ×14,000. (b) A drawing highlighting the structures of the plant cell. Prolamellar bodies are membranous regions typically seen in developing chloroplasts. The structure of this cell would be different were it taken from some other section of the plant, such as the stem or root. (a, courtesy of Kenneth Miller, Brown University.)

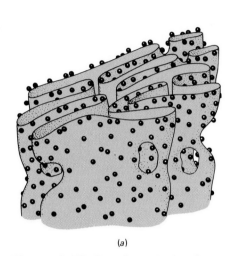

(a)

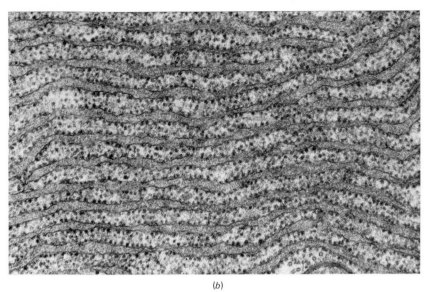

(b)

Figure 4–10 Rough endoplasmic reticulum (a) Diagram of rough ER. (b) Electron micrograph of the rough ER from a secretory cell of the sea anemone *Metridium*. This form of ER consists of parallel arrays of broad flat sacs or cisternae. The outer surface of their limiting membranes is studded with ribosomes (approximately ×70,000). (Courtesy of Dr. E. Anderson.)

Endoplasmic Reticulum and Ribosomes: Internal Transport and Synthesis

A maze of internal membranes extends throughout the cytoplasm of many eukaryotic cells, forming an extensive complex of branching tubules, the **endoplasmic reticulum (ER).** The ER appears to be continuous with the cell membrane and nuclear membrane. In electron micrographs, such as Figure 4–10, the ER may appear to be discontinuous. This is because such photographs are taken of a thin slice through the cell, but the ER is continuous in three dimensions.

The membranes of the ER divide the cytoplasm into a multitude of compartments in which different types of reactions may occur. In fact, the membranes of the ER contain a variety of enzymes and serve as a framework of surfaces on which many reactions take place. The ER also functions as a system for transporting materials from one part of the cell to another and perhaps to the outside environment as well. Expanded regions of the ER may serve as temporary storage areas for certain substances. Still another important function of the ER is synthesis of some types of compounds.

Two types of ER can be distinguished, smooth and rough. **Smooth ER,** so called because the outer surfaces of its membranes have a smooth appearance, produces steroids in certain types of cells. In other cells it functions in cellular secretion. The smooth ER of liver cells functions in lipid metabolism and also in detoxifying drugs. When an experimental animal is injected with the barbiturate phenobarbital, the amount of smooth ER in the liver cells increases over a period of several days. In addition, enzymes known to break down phenobarbital increase in concentration within the smooth ER membranes when this drug is administered over a period of time.

Rough ER has a granular appearance due to the presence of minute organelles called ribosomes that stud its outer walls. **Ribosomes** are the site of protein synthesis. Rough ER is especially extensive in cells that synthesize proteins for export from the cell. Not all ribosomes are attached to the ER. Some float freely in the cytoplasm. These assemble proteins for use inside the cell.

Ribosomes are found in all kinds of cells, from bacteria to complex plant and animal cells. Composed of RNA and protein, a ribosome consists of two subunits that are combined to form the complete protein-synthesizing ribosome. In many cells, clusters of five or six ribosomes termed **polysomes** appear to function in protein synthesis.

The Golgi Complex: A Packaging Plant

The **Golgi complex,** first described in 1898 by the Italian investigator Camillo Golgi, consists of stacks of platelike membranes. The membranes may be distended in certain regions forming vesicles filled with cell products (Fig. 4–11). In animal cells the Golgi complex is usually located at one side of the nucleus.

The Golgi complex functions mainly as a packaging plant and is most highly developed in cells that are specialized to secrete products. Proteins manufactured along the rough ER pass to the Golgi. They are released from the ER sealed off in little packets of membrane, or **vesicles.** These vesicles fuse with the older membranes of the Golgi complex. Within the Golgi complex the protein secretion may be concentrated by the actions of its membranes. During storage proteins may be modified. Often some carbohydrate component is added to form a glycoprotein. (Polysaccharides may be synthesized from simple sugars within the Golgi complex.) A vesicle containing the modified product is released from the Golgi complex. In cells that secrete substances, such as gland cells, the vesicles fuse with the cell membrane and release their contents to the exterior of the cell. Although it is especially prominent in cells specialized to secrete products, the Golgi complex also performs an im-

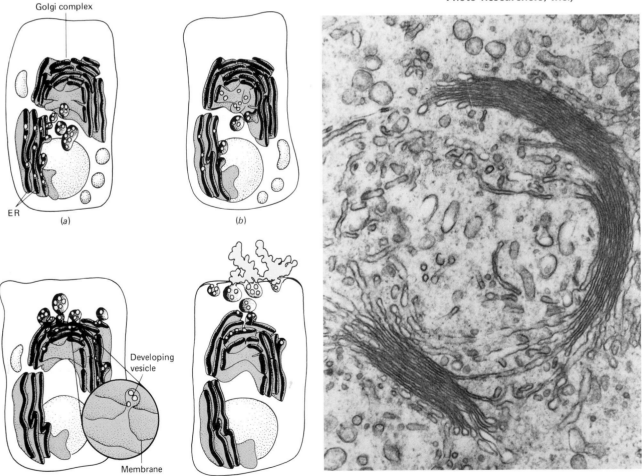

Figure 4–11 The Golgi complex. Diagrams (*a*) through (*d*) show the Golgi complex during a secretory cycle in a goblet cell. The tiny mucus droplets join to form larger drops, which are then released from the cell. (*e*) Electron micrograph of a section through the Golgi complex from a sperm cell of a ram. (*e*, Don Fawcett, Photo Researchers, Inc.)

Golgi complex

ER

(*a*)

(*b*)

Developing vesicle

Membrane

(*c*)

(*d*)

(*e*)

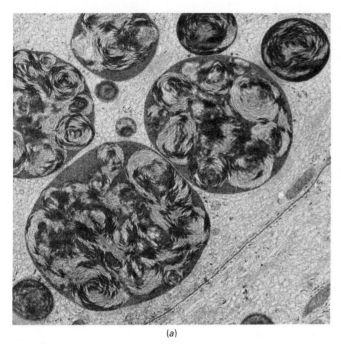

(a)

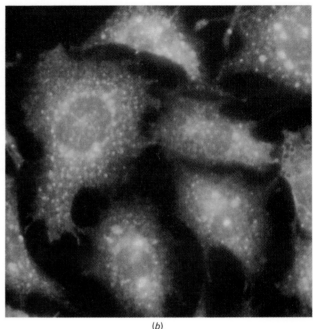

(b)

Figure 4–12 Lysosomes. (a) Electron micrograph showing different stages of lysosome formation. Primary lysosomes bud off from the Golgi complex. After a lysosome encounters material to be digested it is known as a secondary lysosome. The secondary lysosomes shown here contain various materials being digested. (b) Distribution of lysosomes in cells. (a, Don Fawcett, Photo Researchers Inc.; b, courtesy of Dr. Paul Gallup.)

portant function in nonsecreting cells: It packages intracellular digestive enzymes in little organelles called lysosomes.

Lysosomes: Breaking Down Harmful Materials

Small sacs of digestive enzymes called **lysosomes** are released from the Golgi complex and dispersed throughout the cytoplasm (Fig. 4–12). With their potent enzymes, lysosomes break down worn-out organelles and foreign matter into their component molecules. In a cell that is short of fuel, lysosomes may break down organelles so that their component molecules can be used as fuel.

When a white blood cell ingests a bacterium or some other scavenger cell ingests debris or dead cells, the foreign matter is enclosed in a vesicle composed of part of the cell membrane. One or more of the cell's lysosomes then fuse with the vesicle containing the foreign matter. The lysosome pours its powerful digestive enzymes into the vesicle, destroying the material within it. These enzymes can digest almost any type of large molecule found in cells. The lysosomal membrane itself is able to resist the digestive action of these powerful enzymes, but this is thought to require a continuous expenditure of energy.

When a cell dies, the lysosome membrane breaks down, releasing digestive enzymes into the cytoplasm, where they break down the cell itself. This "self-destruct" system accounts for the rapid deterioration of many cells following death.

Some forms of tissue damage as well as the aging process may be related to leaky lysosomes. Rheumatoid arthritis is thought to result in part from damage done to cartilage cells in the joints by enzymes that have been released from lysosomes. Cortisone-type drugs, which are used as anti-inflammatory agents, stabilize lysosome membranes so that leakage of damaging enzymes is reduced.

Microbodies

Microbodies are membrane-bound organelles that contain a variety of oxidative enzymes rather than digestive enzymes. These structures contain enzymes that promote an assortment of metabolic reactions, such as

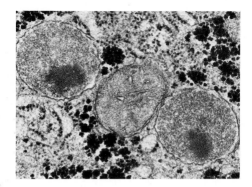

Figure 4–13 Transmission electron micrograph of a rat liver cell showing two circular peroxisomes with a section of a mitochondrion (center). Each peroxisome shows a region of crystalline material that may consist of oxidizing enzymes (approximately ×66,000). (Courtesy of Drs. Christian deDuve and Helen Shio, Rockefeller University.)

the conversion of fats to carbohydrates. During some of these reactions, hydrogen peroxide, a substance toxic to the cell, is produced. **Peroxisomes,** the type of microbody (Fig. 4–13) in which these reactions occur, also contain enzymes that split hydrogen peroxide, rendering it harmless.

Mitochondria: Power Plants of the Cell

power house of cell

Often referred to as the power plants of the cell, the **mitochondria** (singular, mitochondrion) are the sites of cellular respiration (Fig. 4–14). More than 1000 mitochondria have been counted in a single liver cell, but the number varies among cell types. In general they are more abundant in cells that are very active metabolically. Mitochondria are capable of changing size and shape, of fusing with other mitochondria to form larger structures, or of splitting to form smaller ones. They may appear as spheres, rods, sausages, or threads.

The mitochondrion is bounded by a double membrane. Both the outer and inner membranes consist of lipid bilayers in which a variety of protein molecules are embedded. The outer layer of the mitochondrial membrane is smooth, but folds of the inner layer, called **cristae,** project into the interior of the mitochondrion. Cristae serve to increase the available surface area, and some of the enzymes needed for cellular respiration are organized along these folds. Other enzymes are found in the semifluid material within the inner compartment (matrix). Each mitochondrion has a small amount of DNA, enough to code for about 15 proteins, and also contains ribosomes. Some proteins are indeed synthesized within these organelles.

Figure 4–14 The mitochondrion. (*a*) Diagram of a mitochondrion cut open to show the cristae. (*b*) Electron micrograph of a typical mitochondrion from the pancreas of a bat showing the cristae and matrix. Note the extensive rough endoplasmic reticulum at the lower right and some lysosomes at the upper right (approximately ×80,000). (Courtesy of Keith R. Porter.)

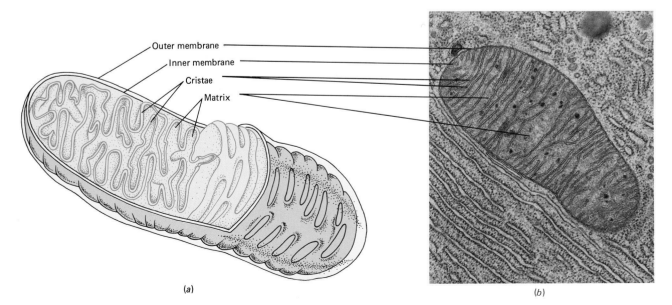

Outer membrane
Inner membrane
Cristae
Matrix

(*a*)

(*b*)

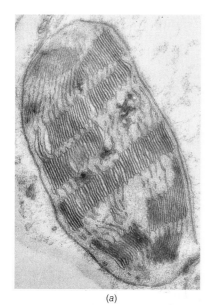

(a)

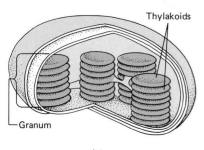

(b)

Figure 4–15 The chloroplast. (a) Electron micrograph of a chloroplast from a leaf of the tobacco plant, *Nicotiana rustica*, showing its fine structure. The thylakoids are arranged in stacks called grana (approximately ×30,000). (b) An inside view of a chloroplast showing the arrangement of the thylakoids. The membranous interconnections between thylakoids can only be suggested in this diagram. (a, courtesy of Dr. E.T. Weier.)

Plastids: Energy Traps and Storage Sacs

Structures known as plastids produce and store food materials in algae and plant cells. **Chloroplasts,** the most common type, contain the green pigment **chlorophyll,** which traps light energy for photosynthesis. Chloroplasts also contain a variety of yellow and orange pigments known as carotenoids. Although a unicellular alga may have only a single large chloroplast, cells of complex plants often possess 20 to 40 of these interesting organelles.

The chloroplasts of plant cells are typically disc-shaped structures bounded by an inner and outer membrane (Fig. 4–15). Inside the chloroplast there are parallel platelike membranes called lamellae. At various points along the lamellae the membranes are expanded to form disclike structures called **thylakoids.** Chlorophyll is present within the thylakoids, and the light-dependent reactions of photosynthesis (those involved in capturing the energy of sunlight) take place within these structures. The thylakoids are arranged in stacks called **grana** (singular, granum). Like mitochondria, chloroplasts contain DNA and ribosomes and manufacture some proteins. They are also able to grow and divide to form daughter chloroplasts. Details of chloroplast structure and function will be discussed in Chapter 7.

Plant cells also contain colorless plastids termed **leukoplasts** (see Fig. 3–5c) which serve as centers for the storage of starch and other materials. Starch is commonly deposited in leukoplasts present in storage roots and stems and in seeds. A third type of plastid, the **chromoplasts,** contains pigments that give color to flowers, autumn leaves, and ripe fruit.

Microtubules and Microfilaments: Cell Shape and Movement

Microtubules are very small, hollow cylinders composed mainly of proteins called tubulins (Fig. 4–16). Microtubules can grow by the addition of more tubulin subunits or can shorten by the disassembly of subunits. Microtubules located just beneath the cell membrane determine the shape of many cells. These structures also play a role in some types of cell movements. For example, they help move chromosomes during nuclear division. In nerve cells microtubules are important in the rapid transport of substances down the axon (the elongated extension of a nerve cell). Microtubules are the main structural components of cilia and flagella, organelles that are discussed in a later section.

Solid cytoplasmic **microfilaments** play additional roles in cell structure and movement. They are composed of the proteins actin (or actin-like substances) and myosin, both important in muscle contraction. Microfilaments are essential in cellular movement, such as the flowing of cytoplasm that enables a cell to move from one place to another. They are also involved in cytoplasmic streaming which results in the continuous motion of organelles characteristic of plant cells.

The Microtrabecular Lattice: Intracellular Framework

The fluid portion of the cytoplasm (the cytosol), long thought to be homogeneous and structureless, actually contains an irregular three-dimensional network of extremely slender protein threads. Known as the **microtrabecular lattice,** this network extends throughout the cytoplasm and is attached to the cell membrane. Microtubules, microfilaments, endoplasmic reticulum, mitochondria, and other organelles are inte-

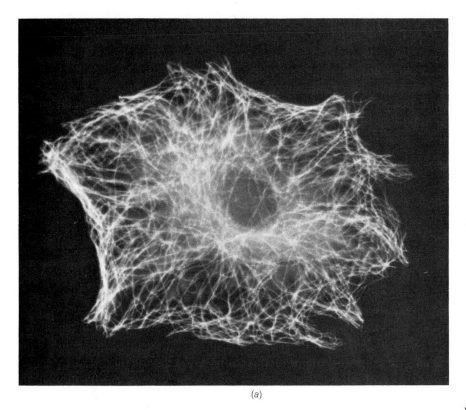

(a)

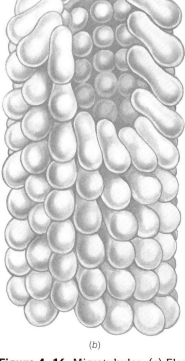

(b)

Figure 4–16 Microtubules. (a) Electron micrograph of a human cell showing the extensive distribution of microtubules throughout the cell. This cell was stained with fluorescent antitubulin, permitting the tubulin that makes up the microtubules to be viewed. (b) Structure of a microtubule. Part of the microtubule has been split in half lengthwise. Note the spiral arrangement of the component units and their two-lobed "barbell" shape. (a, courtesy of Drs. Keigi Fujiwara, Hugh Randolph Byers, and Elena McBeath, Department of Anatomy, Harvard Medical School.)

grated with the lattice and suspended in it (Fig. 4–17). In fact, the lattice links the microtubules, microfilaments, and organelles into a structural and functional unit.

The microtrabecular lattice changes in shape and contracts in various regions, continuously redistributing the organelles, and is perhaps responsible for many cellular responses. For example, the movement of pigment granules within the pigment cells in the skin of fishes, amphibians, and reptiles permits the animal to lighten or darken its skin color rapidly to match the color of its surroundings (Fig. 4–17). Such movement, which is under hormonal and nervous control, is accomplished rapidly by the microtrabecular lattice. Shortening and thickening of microtrabecular threads results in movement of the pigment granules to the center of the cell. Elongation of the fibers disperses the pigment granules throughout the cell.

Centrioles

Many types of cells possess a pair of tiny **centrioles,** organelles that function in nuclear division (Chapter 9). A centriole is a hollow cylinder composed of nine triple microtubules (Fig. 4–18). Centrioles are located within a dense area of cytoplasm, the centrosome, which is generally located near the nucleus.

Cilia and Flagella

Many cells have movable whiplike structures projecting from the free surface. If a cell has one, or only a few, of these appendages and they are relatively long in proportion to the size of the cell, they are termed **flagella.** If a cell has many short appendages, they are called **cilia.** Flagella and cilia are very similar in structure, and both function either in moving the cell along through a watery environment or in moving liquids and

Figure 4–17 (*a*) Model of the microtrabecular lattice drawn about 200,000 times its actual size. The lattice is traversed by microtubules which appear to be attached to the trabeculae of the lattice. (*b, c*) The green anole lizard, *Anolis carolinensis*, changes color not, as one might guess for the purpose of camouflage, but according to its activity. Basking in the sun, the lizard is usually brown. When threatened or aggressive, the animal turns green. The color change can also be the result of other environmental cues, including changes in temperature and moisture. Any of these factors act to release hormones that in turn influence the movement of pigment in the microtrabecular lattice. (*a*, drawing based on work of Keith Porter. *b, c*, Luci Giglio.)

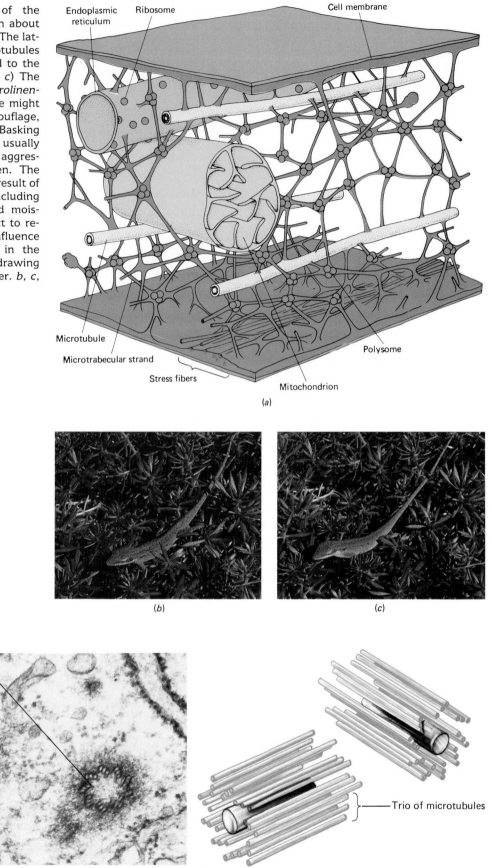

(*a*)

(*b*)

(*c*)

(*a*)

(*b*)

Trio of microtubules

Figure 4–18 Centrioles. (*a*) Electron micrograph of a pair of centrioles from monkey endothelial cells. Note that one centriole has been cut longitudinally and one transversely. (*b*) A line drawing of the centrioles. (*a*, B.F. King, School of Medicine, University of California, Davis/BPS.)

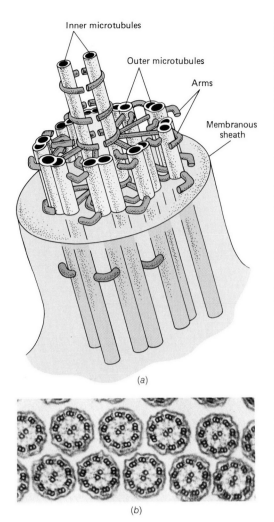

(a)

(b)

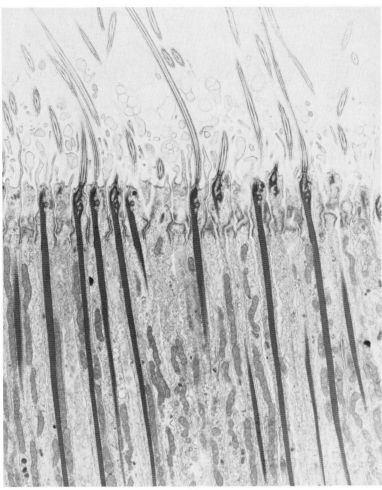

(c)

Figure 4–19 Cilia. (*a*) Structure of a cilium. (*b*) Electron micrograph of cross sections through cilia showing the 9 + 2 arrangement of microtubules. (*c*) Electron micrograph of the bases of the cilia that cover the gills of the primitive chordate *Amphioxus*. Notice the striated rootlets that penetrate deep into the cell, and the long wormlike mitochondria arranged along them. Presumably the mitochondria provide the energy for the ciliary contraction (approximately ×12,000). (*b*, Omikron, Photo Researchers, Inc; *c*, Dr. M.C. Holley)

particles across the surface of the cell. Flagella or cilia are commonly found on one-celled and small multicellular organisms and on the sperm cells of animals. They are the principal means of locomotion of such cells (see Fig. 4–2a). Cilia commonly occur on the cells lining the internal ducts of animals; their beating assists in moving materials through these passageways (e.g., through the respiratory passageways).

Each cilium or flagellum consists of a slender, cylindrical stalk covered by an extension of the cell membrane and containing a cytoplasmic matrix with groups of microtubules embedded in it (Fig. 4–19). Nine pairs of microtubules are arranged around the circumference, and two single microtubules are located in the center. This 9 + 2 arrangement is characteristic of all cilia and flagella except those found in organisms that belong to kingdom Monera. Microtubules are thought to be important in the movement of the cilium or flagellum, perhaps by complex sliding movements along one another that bend the entire organelle.

At the base of the stalk of a cilium or flagellum there is a **basal body,** a structure similar to a centriole. The basal body is the structure from which the stalk arises. Cilia develop as the cell matures (Fig. 4–20).

Vesicles and Vacuoles

We have already described several types of vesicles, for example, those that are released from the Golgi complex filled with a secretory product. The term vesicle is used to describe many kinds of small membranous bags found within the cell.

(a)

(b)

(c)

(d)

Figure 4–20 How cilia grow. (a) shows a carpet of cilia lining the trachea of a rat and (b) shows one of the millions of cells bearing those cilia at an early stage in its life. The new cilia project like spines from the cell surface. In (c) they are much longer, and in (d) they form a pattern like a crown or flower on the top of the cell. In the center of the radiating cilia a group of finger-like microvilli (Chapter 5) are apparent (a, b, and c, courtesy of Dr. Ulf Nordin, *Acta Otolaryngol.* 94, 1982; d, courtesy of Drs. James A. Papp and Joseph T. Martin, *Amer. J. Anat.* 169, 1984.)

Although the terms vesicle and vacuole are sometimes used interchangeably, vacuoles are actually larger structures, sometimes produced by the merging of many vesicles. A **vacuole** may be defined as a bubble-like space usually filled with watery fluid and bordered by a membrane. Present in many types of cells, vacuoles are most common in plant cells and the cells of protists. Most protozoa have food vacuoles containing food undergoing digestion (Fig. 4–21), and many have contractile vacuoles, which remove excess water from the cell.

More than half the volume of a plant cell may be occupied by a large central vacuole containing stored food, salts, pigments, and wastes. Plants lack waste disposal systems and often utilize vacuoles as a stor-

Figure 4–21 The protozoan *Chilodonella.* Inside its body are vacuoles containing ingested diatoms (small, photosynthetic, plantlike protists). From the number of diatoms scattered about its insides, one might judge that *Chilodonella* has a rather voracious appetite (approximately ×150). (Walker England, Photo Researchers, Inc.)

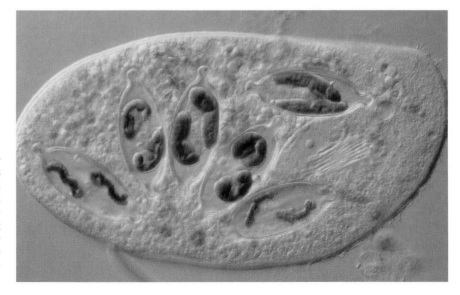

age place for toxic materials. Such waste products often aggregate and form small crystals.

The Cell Nucleus

Of all the structures within the cell, the nucleus is the most prominent (Fig. 4–22). Perhaps for this reason investigators guessed that the nucleus served as the control center for the cell even before techniques to

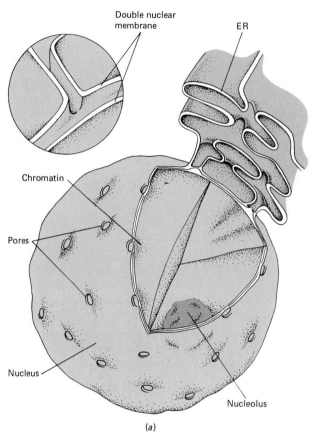

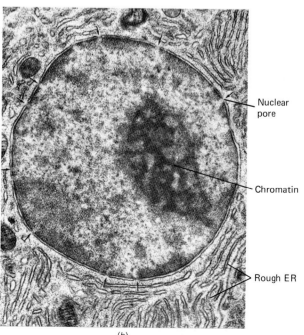

Figure 4–22 The cell nucleus. (*a*) A portion of the nucleus has been cut away to show the interior. (*b*) Electron micrograph showing the nucleus of a pancreatic acinar cell. Note the two membranes that form the nuclear envelope. Arrows indicate nuclear pores (approximately ×40,000). (*c*) Scanning electron micrograph of the surface of the nuclear membrane (approximately ×60,000). The outermost of the two nuclear membranes is shown above, but it has been broken partly to expose part of the inner membrane, shown below. (*b*, Don Fawcett; *c*, courtesy of Dr. Daniel Branton, University of California, Berkeley.)

prove that hypothesis were available. During recent years all kinds of experiments have been performed that have confirmed the vital role of the nucleus. In one such experiment the researcher surgically removed the nucleus from a living ameba. The enucleated ameba was unable to eat or grow, and it died after a few weeks. However, if after a day or two an enucleated ameba was given a new nucleus, it made a complete recovery. These and other experiments show that the nucleus is essential to the well-being of the cell.

THE NUCLEAR ENVELOPE

The nucleus is separated from the surrounding cytoplasm by a **nuclear envelope,** a double membrane that regulates the flow of materials into and out of the nucleus. In some cells the outer nuclear membrane appears to be continuous with the endoplasmic reticulum and the Golgi complex. The two membranes of the nuclear envelope are fused at intervals forming **nuclear pores**—channels through which the interior of the nucleus communicates with the cytoplasm. However, passage of materials through the pores is thought to be selective.

THE NUCLEOLUS

The nucleus may contain one or more large, prominent **nucleoli** (singular, **nucleolus**). The nucleolus is not walled off from the rest of the nucleus by any kind of membrane. Rather, it is a dense region rich in RNA and protein—materials used to manufacture ribosomes. A nucleolus develops around a certain region of one chromosome, the nucleolar organizer, which contains the genes for synthesizing ribosomal RNA (the type of RNA found in ribosomes). During nuclear division, the nucleolus becomes disorganized and its components disperse. When division is completed, the nucleolus reorganizes.

CHROMATIN AND CHROMOSOMES

In a cell that is not dividing, an irregular network of granules and strands termed **chromatin** is evident (Fig. 4–22). The chromatin consists of protein and uncoiled DNA that may be actively synthesizing substances needed by the cell. When a cell begins the process of nuclear division (mitosis), the chromatin coils and condenses into discrete rod-shaped bodies, the **chromosomes** (Fig. 4–23). (You might think of the condensing of chromatin to form a discrete chromosome as being somewhat like taking two intertwined Slinky toys that have been stretched out as far as they can extend, then releasing them slowly to recoil into a small, tightly wound double helix.)

Each chromosome contains several hundred **genes** arranged in a specific linear order; the genes, in turn, are composed of the nucleic acid DNA. Chemically coded within the DNA of the genes are instructions for producing all the proteins needed by the cell. These proteins determine what the cell will look like and what functions it will perform. The chromosomes serve as a chemical cookbook for the cell, whereas the genes might be compared to the individual recipes. When condensed, the chromosomes may be compared to a closed cookbook: The recipes are all inside but the pages cannot be read. When the chromosomes elongate forming chromatin, the book is open and the instructions can be followed. Chromosomes will be discussed in more detail in Chapter 9.

Figure 4–23 Scanning electron micrograph of a chromosome from a hamster cell. Just prior to division, the loose threads of DNA that make up the chromosomes shrivel into the knotted coils you see here (magnification ×15,000). (Courtesy of Drs. Susanne M. Gollin and Wayne Wray, Kleberg Cytogenetics Laboratory, Department of Medicine, Baylor College, and Biology Department, The Johns Hopkins School of Medicine.)

CHAPTER SUMMARY

I. The cell is considered the basic unit of life because it is the smallest self-sufficient unit of living material, and because, as stated in the cell theory, organisms are composed of cells and their products.

II. Most cells are microscopic, but their size and shape vary according to their function.

III. Biologists have learned much about cellular structure by studying cells with light and electron microscopes and by using chemical techniques.

IV. Prokaryotic cells lack a nucleus with a distinct nuclear membrane and other membranous organelles. Eukaryotic cells have distinct membrane-bound nuclei and a variety of other organelles. Plant cells differ from animal cells in that they possess rigid cell walls, plastids, and large vacuoles; the cells of complex plants lack centrioles.

V. The cell is bounded by a cell membrane, and most eukaryotic cells have elaborate organelles specialized to perform specific intracellular functions.

    A. The endoplasmic reticulum (ER) is a system of internal membranes that transport and store materials within the cell and divide the cytoplasm into compartments.

        1. The smooth ER produces steroids in certain cells and performs a variety of metabolic functions in other cells.

        2. The rough ER is studded along its outer walls with ribosomes that manufacture proteins.

    B. The Golgi complex concentrates secretions that are produced in the ER and packages them for export from the cell. It adds carbohydrate components to some compounds and also produces lysosomes.

    C. Lysosomes function in intracellular digestion; peroxisomes contain enzymes that break down hydrogen peroxide.

    D. Mitochondria are the power plants of the cell; the cristae of the inner membrane contain enzymes needed for cellular respiration.

    E. Cells of algae and plants contain plastids: chloroplasts that contain chlorophyll, pigment-filled chromoplasts, and colorless leukoplasts.

    F. Microtubules and microfilaments help maintain the shape of the cell and play a role in cellular movement. These structures are integrated with the microtrabecular lattice.

    G. Centrioles, cilia, and flagella are composed of microtubules. Cilia and flagella move cells through the surrounding medium or move the surrounding medium past the cells.

    H. In eukaryotic cells the nucleus, control center of the cell, is bounded by a double-layered nuclear envelope.

        1. The nucleolus functions in the assembly of ribosomes.

        2. When a cell begins to divide, chromatin condenses, forming long, threadlike chromosomes. Chromatin and chromosomes are composed of DNA and proteins and contain the hereditary units, the genes.

Post-Test

1. The ability of a microscope to reveal fine detail is known as _____ _____ .

2. The sites of protein synthesis in the cell are the _____ .

3. In certain cells steroids are produced by the _____ _____ _____ .

4. The _____ _____ concentrates and packages cellular secretions.

5. Many of the reactions of cellular respiration take place within the _____ .
6. _____ are plastids that contain chlorophyll.
7. In eukaryotic cells cilia and flagella are composed of _____ in a 9 + 2 arrangement.
8. _____ are important in cell movement, such as cytoplasmic streaming.
9. The control center of the cell is the _____ ; it is bounded by a _____ _____ _____ .
10. In a cell that is not dividing the uncoiled DNA and protein are evident as _____ ; when the cell begins to divide, this material condenses to form discrete _____ .
11. Each chromosome contains several hundred _____ _____ arranged in a specific linear order.
12. In addition to a cell membrane, plant cells are bounded by a _____ _____ .
13. Many plant cells have a large central _____ .
14. Prokaryotic cells lack a _____ _____ .

_____ as well as other _____ organelles.
15. The size and shape of a cell is related to the type of _____ _____ .
16. Label the diagram of the cell. See Figure 4–7 in the text for the correct labels.

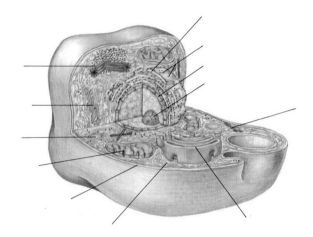

Review Questions

1. Imagine that a mutant cell was produced that lacked mitochondria. What would be its fate? Why?
2. Explain the following statement: All living cells can trace their ancestry back to ancient times. Can you trace a muscle cell in your own body back to a cell of a grandparent?
3. Compare mitochondria with chloroplasts.
4. A cell from a gastric (stomach) gland secretes the enzyme pepsin. Trace in sequence its production, transport, packaging, and release from the cell, naming each of the organelles involved.
5. Why are lysosomes sometimes referred to as the self-destruct system of the cell? Do you think this name is justified? Why?

6. What is the relationship between chromatin and chromosomes? Between chromosomes and genes?
7. Draw a typical animal cell and a typical plant cell and label as many structures as you can in each. What are the fundamental differences between the two cell types?
8. How does a bacterial cell differ from an animal cell?
9. Define (a) magnification and (b) resolving power. What are the advantages of the electron microscope over the ordinary light microscope?
10. What do centrioles, cilia, and flagella have in common?

Readings

Alberts, B., et al.: *Molecular Biology of the Cell.* New York, Garland Publishing, 1983. A thorough presentation of cell structure and function. A well-illustrated and easy-to-read reference textbook.

Avers, J. *Cell Biology,* 2nd ed. New York, Van Nostrand, 1981. A fine presentation of the details of cell structure and function.

de Duve, C. *A Guided Tour of The Living Cell.* San Francisco, 1985. W.H. Freeman Company. (2 vols). The discoverer of the lysosome discusses every organelle.

Fawcett, D.W. *The Cell,* 2nd ed. Philadelphia, W.B. Saunders, 1981. A study of cell structure through an exciting collection of electron micrographs.

Grivell, L.A. "Mitochondrial DNA," *Scientific American,* March 1983, 78–88. A discussion of the genetic system of mitochondria.

Hoagland, M. *The Roots of Life: A Layman's Guide to Genes, Evolution and the Ways of Cells.* New York, Avon, 1979. A paperback written for the general public by a working scientist.

Holtzman, E., and A.B. Novikoff. *Cells and Organelles,* 3rd ed. Philadelphia, Saunders College Publishing, 1984. An integrated approach to the structural, biochemical, and physiological aspects of the cell.

Lazarides, E., and J.P. Revel. "The Molecular Basis of Cell Movements," *Scientific American,* May 1978, 100–112. The role of microfilaments in cell movement.

Margulis, L. *Symbiosis in Cell Evolution.* San Francisco, W.H. Freeman, 1981. A fascinating summary of the hypothesis that the cell organelles of eukaryotic organisms evolved from prokaryotes, written from a position of strong personal advocacy.

Chapter 5

THE CELL MEMBRANE: INTERACTING WITH THE ENVIRONMENT

Learning Objectives

After you have read this chapter you should be able to:

1. Discuss the importance of the cell membrane to the cell, describing the various functions it performs.
2. Describe the structure of the cell membrane.
3. Describe the functions of membrane proteins.
4. Identify microvilli and give their function.
5. Describe the plant cell wall and its function.
6. Contrast desmosomes, tight junctions, and gap junctions.
7. Contrast the physical with the physiological processes by which materials are transported across cell membranes.
8. Solve simple problems involving osmosis. For example, predict whether cells will swell or shrink under various osmotic conditions.
9. Summarize the currently accepted hypothesis of how small hydrophilic molecules are transported through the cell membrane.
10. Compare exocytosis and endocytosis.

To carry out life processes, the cell must maintain an appropriate internal environment. It must regulate its own composition, providing constant conditions despite changes in the outside world. Such regulation is possible because all cells, even the simplest ones, are physically separated from the external environment by a limiting **cell membrane,** or **plasma membrane.**

WHY DO CELLS NEED CELL MEMBRANES?

The cell membrane is not an inanimate wall. Quite the contrary, it is a complex mechanism that permits many selective interactions between the cell and its environment. Among its many functions we can list the following:

1. *The cell membrane regulates the passage of materials into and out of the cell.* Each cell differs from its surroundings in physical properties and chemical composition. If the cell membrane were completely **permeable,** substances would pass freely into and out of the cell. The cell would then reflect the chemical composition of the surrounding medium, with disastrous results. This does not happen because the cell membrane is **selectively permeable,** which means that it can prevent the entrance of certain substances while permitting, even facilitating, the entrance of others, such as needed nutrients. The cell membrane also prevents the loss of certain substances from the cell but facilitates the exit of specific products and wastes. This ability to regulate passage of materials enables the cell to maintain a constant internal environment despite changes in the external environment.
2. *The cell membrane receives information that permits the cell to sense changes in its environment and to respond to them.* Receptor proteins in the cell membrane receive chemical messages from other cells. Hormones, growth factors, and neurotransmitters (chemicals released by nerve cells) are among the substances that combine with such receptors. When a compound combines with a receptor, the membrane is stimulated to send a signal into the cell, resulting in some type of response.
3. *The cell membrane maintains structural and chemical relationships with neighboring cells.* Certain proteins in the cell membrane permit cells to recognize one another, to adhere to each other when appropriate, and to exchange materials.
4. The cell membrane *protects the cell, may function in secretion of substances,* and *may be involved in cell movement,* and *in some cells it is important in transmitting impulses.*

THE STRUCTURE OF THE CELL MEMBRANE

The cell membrane is so thin, about 6 to 10 nanometers, that it can be seen only with the electron microscope (Fig. 5–1). According to the currently accepted hypothesis of cell membrane structure, the **fluid mosaic model,** the membrane consists of a rather fluid **lipid bilayer** (a double layer of lipid) in which a variety of globular proteins are embedded.

The Lipid Bilayer

The lipid components of the cell membrane include phospholipids, glycolipids, and cholesterol. All of these have an important structural feature in common. They are asymmetrical, elongated molecules that have

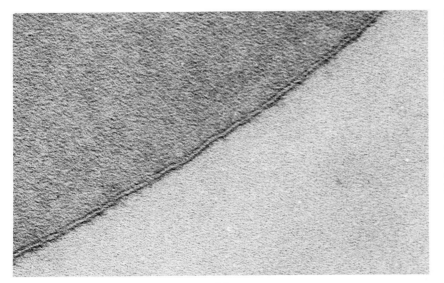

(a)

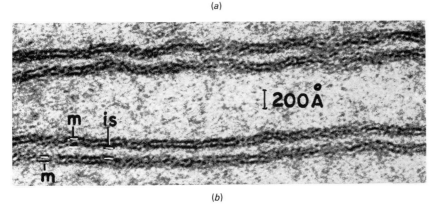

(b)

Figure 5–1 The cell membrane. (a) The cell membrane (also called the plasma membrane) is the interface between the cell and its environment. (b) Electron micrograph of a portion of the cell membranes of adjacent intestinal cells (approximately ×240,000). In each membrane, the dark lines represent the hydrophilic heads of its lipid molecules, whereas the light zone represents the hydrophobic tails. m, membrane; is, intercellular space. (a, Omikron, Photo Researchers, Inc.)

one highly polar, **hydrophilic** (water-loving) portion and one nonpolar, **hydrophobic** (water-hating) portion. Each lipid molecule has a hydrophilic head group, such as choline, and two hydrophobic tails, which are usually fatty acid chains.

In the cell membrane the nonpolar, hydrophobic fatty acid chains of the phospholipids meet and overlap in the middle, while the polar, hydrophilic heads are directed toward the outside of the membrane. When viewed with the electron microscope, the cell membrane is seen as two dark lines separated by an intermediate light zone. The dark lines apparently represent the hydrophilic heads of the lipids, and the light zone is produced by the hydrophobic tails (Fig. 5–1). In diagrams the hydrophilic head groups are represented by circles and the hydrophobic tails by two wavy lines (Fig. 5–2).

The formation of a lipid bilayer from phospholipids is a rapid, spontaneous process. The driving force for this process is the hydrophobic interactions of the hydrocarbon chains. When these lipids are placed in a watery medium,[1] the polar heads show an affinity for water, but the hydrocarbon tails avoid it. Thus the molecules become oriented to form a bilayer. No covalent bonds link adjacent lipid molecules to one another. Only the hydrophobic forces hold the molecules in place. Still, the bilayer is strong enough to be a membrane. Lipid bilayers are self-sealing as well as self-assembling. If a hole is made in the membrane, the bilayer will seal itself, thus repairing the tear.

---

[1] Even cells of terrestrial organisms dwell in a watery medium—the various body fluids, such as tissue fluid.

Figure 5–2 Schematic representation of the architecture of the cell membrane according to the current fluid mosaic model. The hydrophilic heads of the lipid molecules are shown in yellow, their hydrophobic tails in black.

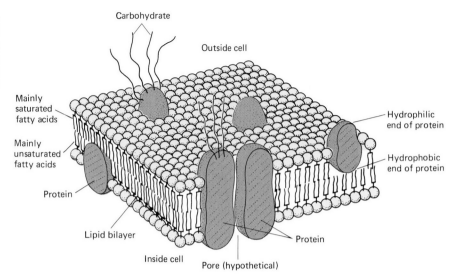

The lipid bilayer is very impermeable to ions and polar molecules, with the exception of water. Water molecules pass in and out of the cell through the lipid bilayer with ease.

Membrane Proteins

The lipid bilayer is a fluid matrix in which the proteins may move about. The lipids serve as a barrier to polar molecules and ions, whereas the membrane proteins carry out specific functions of the membrane, such as chemical transport, energy transfer, and transmission of messages. Membranes may differ in the numbers and kinds of proteins present; membranes with different functions contain different proteins. Furthermore, cell membranes are dynamic structures, and their chemical composition and molecular arrangement may change with varying conditions.

Most of the membrane proteins are associated with the inner, cytoplasmic surface of the membrane (see Focus on Splitting the Lipid Bilayer). The proteins that protrude from the outer surface (away from the cytoplasm) are mainly glycoproteins, that is, proteins to which sugar residues are attached. Glycoproteins on the cell membrane appear to serve as the cell's communication system with its outer environment, both with other cells and with messenger molecules, such as hormones.

Microvilli

On the free surface of many types of cells (e.g., the cells lining the intestine) the cell membrane is marked by numerous tiny evaginations (outpocketings) known as **microvilli** (Fig. 5–3). Microvilli enormously increase the surface area of the cell available for absorption of materials from the cell's environment. The number of microvilli can rapidly increase or decrease in response to environmental conditions or to changes in the metabolic needs of the cell.

CELL WALLS

Cells of plants, algae, bacteria, and fungi have thick cell walls that lie just outside their cell membranes. The cell wall, consisting of carbohydrates secreted by the cell, protects the cell and gives it a characteristic

Focus on . . .

SPLITTING THE LIPID BILAYER

High-resolution electron micrographs of cell membranes typically show a three-layered structure of lines in a dense-light-dense pattern (Fig. 5–1). The two dense lines correspond to the polar heads of the lipid bilayer, and the light area between them corresponds to the hydrophobic region of the fatty acid chains. The cell membranes of animal, plant, and microbial cells and the membranes of a great many subcellular organelles all appear to have this three-layered structure.

Cells can be rapidly frozen in liquid nitrogen and then fractured with a microtome knife. A small amount of ice is evaporated from the fracture surface ("etched") and then a small amount of platinum or gold is deposited on that surface, forming a replica of the fractured surface. When the metallic coating is examined in the transmission electron microscope, one

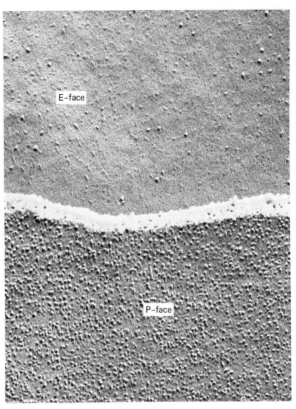

A freeze fracture made of the membrane from a cell of the eye of a monkey. Notice the greater number of proteins present in the P-face of the membrane. (Don Fawcett, Photo Researchers, Inc.)

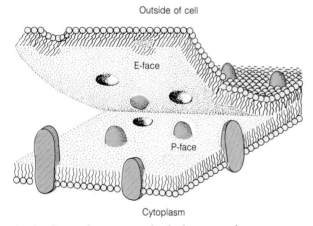

In the freeze-fracture method, tissue specimens are often broken along the hydrophobic interior of the lipid bilayer in such a way that the two monolayers of the cell membrane are separated and can be photographed separately. (1) The inner monolayer is an outwardly directed half-membrane (the so-called P-face) that presents the majority of globular proteins, seen here to project from it. (2) The outer monolayer is an inwardly directed half-membrane (the so-called E-face). Although relatively smooth, it nevertheless shows some protein particles.

can see that the fracture typically splits each cell membrane into two half-membranes in the middle of the lipid bilayer (see figure). By this means, termed **freeze fracture** or freeze etch, one can examine the interior of the split membrane. The two faces are not identical. The face nearer the cytoplasm contains many particles, and the face near the outside of the cell contains many pits (see figure). The particles are integral proteins (proteins that extend into the lipid bilayer), and the pits are spaces where the proteins had been.

shape (Fig. 5–4). The rigid plant cell wall is pierced in many places by tiny pores through which water and dissolved materials can pass.

Cellulose, the main component of the plant cell wall, is usually present in the form of long, threadlike fibers. The cell wall has been compared to reinforced concrete in which the cellulose fibers are like steel rods and the matrix material around them is the concrete. Compounds present in the plant cell wall include pectins (gelling agents used in the preparation of jams and jellies) and lignin, a compound that provides rigidity. Cells whose main function is support contain a great deal of lignin in their walls. In fact, wood consists mainly of cell walls in which cellulose is reinforced by lignin.

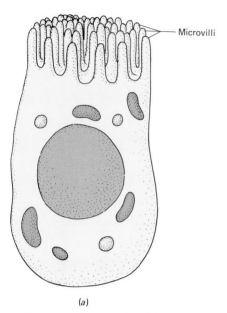

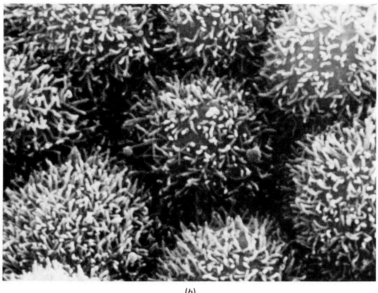

(a)

(b)

Figure 5–3 Microvilli. (*a*) Microvilli, present on the free surfaces of many kinds of cells, greatly increase the surface area for absorption of materials. (*b*) Scanning electron micrograph of ovarian epithelial cells from a mouse, showing microvilli (approximately ×65,000). (*b*, courtesy of Dr. E. Anderson, Harvard Medical School.)

The cell walls of prokaryotes are not composed of cellulose. They contain a unique complex carbohydrate known as **peptidoglycan,** which will be discussed in Chapter 14. Many bacteria have a slimy polysaccharide capsule that lies just outside the cell wall and serves as an additional protective layer.

CELL JUNCTIONS

Some types of cells communicate directly with one another. Adjacent plant cells may actually be joined by extensions of cytoplasm that pass through openings in the cell walls and cell membranes. These cytoplasmic extensions, called **plasmodesmata,** provide a pathway for the passage of water, ions, nutrients, and other materials from one cell to another.

Figure 5–4 Plant cells are usually rectangular. However, this amaryllis protoplast was made circular when its walls were dissolved by an enzyme. Enzymes enable biologists to take a plant cell apart without disrupting the individual components or "organelles." This, in turn, allows the functions of an individual organelle to be studied (approximately ×600). (William J. Marin, Jr., Brookhaven National Laboratory.)

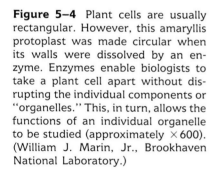

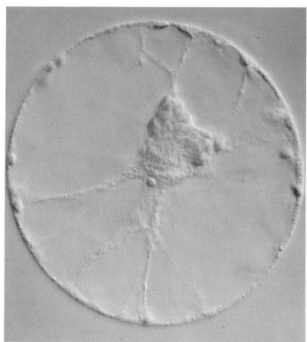

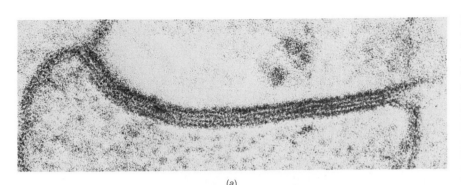

(a)

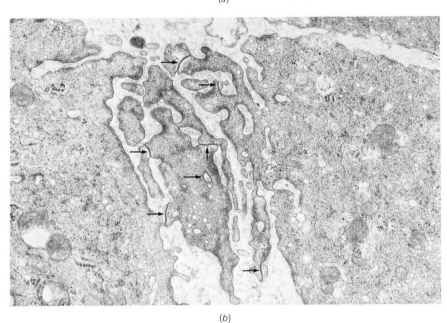

(b)

Figure 5–5 Gap junctions. (a) Electron micrograph of a single gap junction between smooth muscle cells in the wall of the uterus. The increase in the number of these gap junctions that occurs with advancing pregnancy may be the trigger that finally begins labor by permitting uterine smooth muscle cells to pass impulses readily from one to the other. (b) A lower magnification of the wall of the uterus showing an unusually large number of gap junctions (arrows) in one view (approximately ×36,000). (Courtesy of Puri and Garfield, from *Biology of Reproduction* 27:967–975, 1982.)

Serving a somewhat similar function as plant cell plasmodesmata, **gap junctions** connect the cytoplasmic compartments of animal cells (Fig. 5–5). When a marker substance is injected into one of the cells connected by gap junctions, the marker passes rapidly into the adjacent cell. Gap junctions are thought also to provide for electrical communication between certain types of cells. In some species the gap junction (Chapter 22) permits the transmission of an electrical impulse from one cell to the next, perhaps in the form of a surge of potassium ions. Such an arrangement is found in the electrical organs of the electric eel and the electric catfish. Cardiac muscle cells are also connected by gap junctions that permit the rapid transmission of neural impulses from one cell to the next so that all the muscle fibers in the ventricles of the heart can contract one after the other in a rapid regular fashion.

In tissues where cells are in close contact with one another, specialization of the cell membrane may provide for the adhesion of neighboring cells. Adjacent epithelial cells, especially those of the upper layer of the skin, are anchored together by button-like plaques, called **desmosomes,** on the two opposing cell surfaces (Fig. 5–6). Even with the desmosomes, a tiny intercellular space about 24 nanometers wide separates the two opposing cell surfaces.

In certain tissues, cells are held together by **tight junctions,** connections so tight that materials cannot pass between the cells. In a tight junction the two cell membranes are actually fused, and there is no intercellular space (Fig. 5–7). Cells connected by tight junctions constitute a

Figure 5–6 Electron micrograph showing a desmosome (*D*) between two cells of the ovarian epithelium of a rabbit (approximately ×70,000). (Courtesy of Dr. Everett Anderson, *Journal of Morphology* 150:135–166, 1976.)

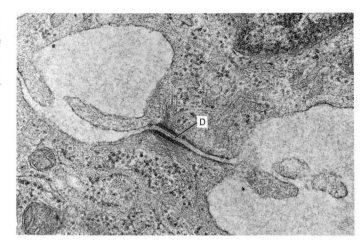

continuous barrier and are found where a sharp physical separation between two body compartments is essential. A classic example is the "blood-brain barrier," which involves the lining of the cerebral blood vessels. The cells lining these blood vessels are joined by tight junctions whose role is essentially protective, for they block the passage of certain molecules from the blood into the brain tissue.

HOW DO MATERIALS PASS THROUGH MEMBRANES?

Whether a membrane will permit the molecules of any given substance to pass through it depends on the structure of the membrane and the size and charge of the molecules. A membrane is said to be permeable to a given substance if it permits the substance to pass through and impermeable if it prevents passage of the substance. A **selectively permeable** membrane allows some, but not other, substances to pass through it. Responding to varying environmental conditions or cellular needs, the cell membrane may present a barrier to a particular substance at one time and then actively promote its passage at another.

Materials move passively through cell membranes—as they do through nonliving materials—by physical processes, such as diffusion and osmosis. However, in living cells materials can also be moved actively by physiological processes, such as active transport, phagocytosis, or pinocytosis (Table 5–1). Such physiological processes require the expenditure of energy by the cell.

Figure 5–7 Freeze-fracture replica illustrating a tight junction. P-face (P) and complementary E-face (E). Rabbit ovarian epithelium (approximately ×50,000). (Courtesy of Dr. E. Anderson, Harvard University School of Medicine.)

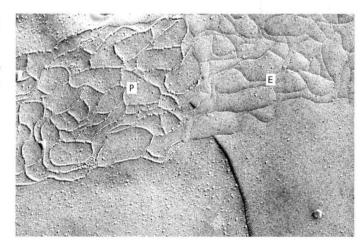

Table 5–1
MECHANISMS FOR MOVING MATERIALS THROUGH CELL MEMBRANES

Process	How it works	Energy source	Example
Physical Process			
Diffusion	Net movement of molecules (or ions) from region of greater concentration to region of lower concentration	Random molecular motion	Movement of oxygen in tissue fluid
Facilitated diffusion	Carrier protein in cell membrane accelerates movement of relatively large molecules from region of higher to region of lower concentration	Random molecular motion	Movement of glucose into cells
Osmosis	Water molecules diffuse from region of higher to region of lower concentration through differentially permeable membrane	Random molecular motion	Water enters red blood cell placed in distilled water
Physiological Process			
Active transport	Protein molecules in cell membrane transport ions or molecules through membrane; movement may be against concentration gradient (i.e., from region of lower to region of higher concentration)	Cellular energy	Pumping of sodium out of cell against concentration gradient
Endocytosis			
Phagocytosis	Cell membrane encircles particle and brings it into cell by forming vacuole around it	Cellular energy	White blood cells ingest bacteria
Pinocytosis	Cell membrane takes in fluid droplets by forming vesicles around them	Cellular energy	Cell takes in needed solute dissolved in tissue fluid
Exocytosis	Cell membrane ejects materials; vesicle filled with material fuses with cell membrane	Cellular energy	Secretion of mucus

Diffusion

Some substances pass into or out of cells by simple physical diffusion. All molecules in liquids and gases tend to move (diffuse) in all directions until they are spread evenly throughout the available space (Fig. 5–8). **Diffusion** may be defined as the movement of molecules from a region of higher concentration to one of lower concentration brought about by the kinetic energy of the molecules. Molecules tend to move down a **concentration gradient,** that is, from where they are more concentrated to where they are less concentrated. Diffusion depends upon the

Figure 5–8 The process of diffusion. When a small lump of sugar is dropped into a beaker of water, its molecules dissolve (*a*) and begin to diffuse (*b* and *c*). Over a long period of time, diffusion results in an even distribution of sugar molecules throughout the water (*d*).

(*a*)

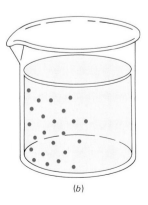
(*b*)

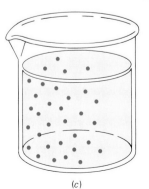
(*c*)

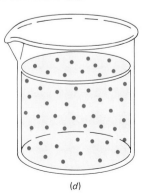
(*d*)

random movement of individual molecules, propelled by collision with other vibrating molecules or with the side of the container. As it diffuses, each individual molecule moves in a straight line until it bumps into something—another molecule or the side of the container. Then it rebounds in another direction. Molecules continue to move even when they have become uniformly distributed throughout a given space. However, as fast as some molecules travel in one direction, others travel in the opposite direction, so that on the whole all the molecules remain uniformly distributed; an equilibrium exists.

The rate of diffusion is a function of the size and shape of the molecules, the temperature, and whether the medium is a solid, liquid, or gas. Small particles move faster than larger ones at the same temperature. Heat energy causes molecules to move more rapidly, so as temperature rises, the rate of diffusion increases. The molecules of any number of substances will diffuse independently of each other within the same solution; eventually, all become uniformly distributed.

You might demonstrate the diffusion of gases by opening a bottle of ammonia on a front-row desk of your classroom. Students in the second row would begin to smell ammonia within a few moments because some molecules of ammonia would have left the bottle and begun to diffuse through the air. Some time later the odor would be evident throughout the room. If the room was closed and there were no air currents, the molecules would eventually distribute themselves evenly throughout the room.

Diffusion is important in living systems. A large variety of substances are distributed throughout the cytoplasm by diffusion, and this process is also responsible for moving a great many substances in and out of the cell across the cell membrane. Oxygen, carbon dioxide, water, and numerous other small ions and molecules can diffuse into or out of the cell readily. However, diffusion through the cell membrane is limited by both molecular size and the compatibility of substances with the lipid membrane.

In **facilitated diffusion,** carrier proteins in the cell membrane assist the passage of small solute molecules through the membrane. The proteins serve as a passive conveyor belt that permits the substance to pass in either direction down a concentration gradient. The carrier proteins are thought to combine temporarily with the solute molecule and to accelerate its movement through the membrane.

Osmosis

Osmosis is a special kind of diffusion—the diffusion of water molecules across a differentially permeable membrane from a region where water molecules are more concentrated to one where they are less concentrated. Cell membranes selectively regulate the passage of most solutes, but water is able to move rather freely in and out of the cell. When living cells are placed in a solution that has a solute concentration equal to that in the cells, the water molecule concentration is also equal, and therefore water molecules move in and out of cells at the same rate. The net movement of the water molecules is zero (Fig. 5–9). Such a solution is described as **isotonic** to the cells, that is, of equal solute concentration.

Cells—especially of single-celled organisms—may find themselves in solutions that are of greater or lesser solute concentration relative to the solute concentration within the cytoplasm. If the solution has a greater solute concentration, it is said to be **hypertonic** (above strength) to that of the cell, whereas if it has a lesser concentration, it is **hypotonic** (under strength) compared with the cell. Note that the terms hypertonic

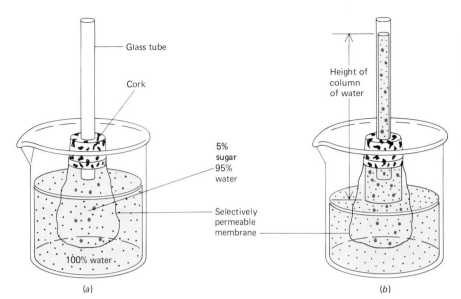

Figure 5–9 Osmosis. (*a*) A 5% sugar solution is placed in a sac made of a selectively permeable membrane and suspended in water. The sac is attached to a glass tube. The membrane permits passage of the water molecules but not of the larger sugar molecules. Therefore, the water molecules pass across the membrane *into* the sac, since the water can be thought of as less concentrated inside the sac than outside it. This causes the column of liquid in the glass tube to rise. (*b*) When equilibrium is reached, the pressure of the column of water just equals, and is a measure of, the osmotic pressure of the sugar solution.

and hypotonic are relative to each other (Table 5–2). A 5% solution is hypertonic to a 2% solution but hypotonic to a 10% solution.

Suppose that we place some living cells in distilled water that contains 100% water molecules. If the total number of solute molecules in the cells amounted to 1% of the total molecules present, then water molecules would account for only 99% of the total. Since water molecules, like solute molecules, tend to move from a region where they are more concentrated to a region where they are less concentrated, they diffuse inward across the cell membrane (Fig. 5–10). Although the solute molecules have a tendency to diffuse in the opposite direction, the cell membrane prevents them from "leaking out" to any great extent. Instead, they may be thought of as being trapped within the cell and exerting an **osmotic pressure** upon the less-concentrated solution on the other side of the membrane. So much water may enter the cell that it swells and bursts.

On the other hand, when cells are placed in a solution that is hypertonic to them, water tends to flow out of them. The cells may become dehydrated, shrink, and die. Can you explain this in terms of the relative concentrations of water molecules in the two solutions? Remember, when a solution contains more solute molecules, it has proportionately fewer water molecules, so the solvent and solute concentrations are reciprocally related.

Although the excretory systems of multicellular animals like ourselves usually are able to maintain their body fluids in an isotonic state, freshwater protists have a continuous osmotic problem. Their watery

Table 5–2
OSMOTIC TERMINOLOGY

Solute concentration in solution A	Solute concentration in solution B	Tonicity	Solute diffusion	Solvent diffusion
Greater	Less	A hypertonic to B B hypotonic to A	A to B	B to A
Less	Greater	B hypertonic to A A hypotonic to B	B to A	A to B
Equal	Equal	Isotonic	Equal	Equal

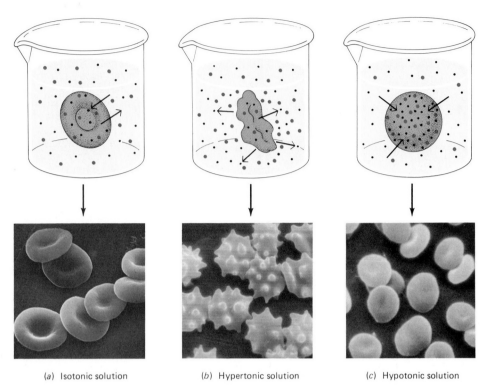

(a) Isotonic solution   (b) Hypertonic solution   (c) Hypotonic solution

Figure 5–10 Osmosis and the living cell. (a) A cell is placed in an isotonic solution. Because the concentration of solutes (and thus water molecules) is the same in the solution as in the cell, the net movement of water molecules is zero. (b) A cell is placed in a hypertonic solution. This solution has a greater solute concentration (thus a lower water concentration) than the cell and therefore exerts an osmotic pressure on the cell. This results in a net movement of water molecules out of the cell, causing the cell to dehydrate, shrink, and perhaps die. (c) A cell is placed in a hypotonic solution. The solution has a lower solute (and thus a greater water) concentration than the cell. The cell contents thus exert an osmotic pressure on the solution, drawing water molecules inward. There is a net diffusion of water molecules into the cell, causing the cell to swell and perhaps even to burst. (Micrographs of human red blood cells courtesy of Dr. R.F. Baker, University of Southern California Medical School.)

surroundings are hypotonic to them, and water tends to pass into their cells. Some of these organisms possess an adaptation that solves the problem—a contractile vacuole that takes up the excess water and pumps it out of the cell.

Plant cells are also adapted to the hypotonic water that often bathes them. Their rigid cell walls enable them to withstand, without bursting, pressure exerted by the water that seeps in, filling their central vacuoles. As this internal pressure, called **turgor pressure,** increases, it forces water molecules back out of the cell (Fig. 5–11). The turgor pressure levels off when the outward passage of water molecules equals the rate of inward movement of water molecules. When conditions become dry and the plant cell does not have enough water, the central vacuole decreases in size and the cell shrinks. Thus the plant wilts.

Active Transport

A cell requires certain substances—potassium ions, for example—in greater concentration than they are present in the cell's surroundings. The potassium concentration is about 35 times greater inside the cell

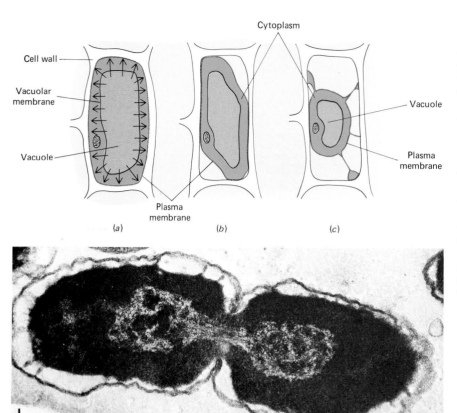

(a)   (b)   (c)

Cytoplasm

Cell wall

Vacuolar
membrane

Vacuole

Plasma
membrane

Vacuole

Plasma
membrane

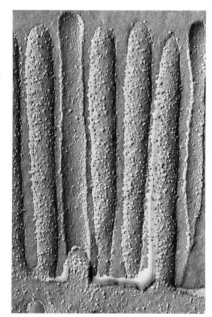

Figure 5–11 (a) A turgid plant cell. In a hypotonic solution the central vacuole fills with water because the vacuole contents are hypertonic to the surrounding fluid. The cell is kept from bursting by the tough cell wall, but the high internal pressure causes the cell to become turgid, something like an automobile tire full of compressed air. (b) When conditions become dry or if the surrounding solution becomes hypertonic, water leaves the central vacuole. First this causes loss of turgidity (wilting), followed in extreme cases by an actual decrease in size of the central vacuole so that (c) the cell shrinks. (d) Section through a dividing *Escherichia coli*, a bacterium that has shrunk (plasmolyzed) due to a hypertonic environment. (Courtesy of M.E. Bayer, Institute for Cancer Research/ BPS.)

(d)

than outside. Other substances—for example, sodium ions—are more concentrated in the environment than could be tolerated inside the cell. Given the opportunity to do so, diffusion would quickly eliminate such differences in solute concentration and the cell would die. To maintain a steady state, the cell must prevent diffusion from occurring in some cases, or even reverse its direction. In **active transport** the cell moves materials from a region of lower concentration to a region of higher concentration. Working uphill this way against a concentration gradient requires energy. Expenditure of energy in active transport is an example of how even cells that appear to be resting are actually performing work just to remain alive.

The question of how hydrophilic compounds, such as glucose and amino acids, can pass through a hydrophobic lipid bilayer has puzzled biologists (Fig. 5–12). The currently accepted hypothesis suggests that transport proteins extend entirely through the membrane. Four or more transport protein molecules grouped in contact with each other provide a water-filled pore about 1 nanometer in diameter through which hydrophilic molecules can pass (Fig. 5–13). An active site on the protein specifically binds to one kind of molecule—glucose, for example. When the molecule being transported binds with the transport protein, it is thought that the transport protein changes its shape. As a result, the molecule being transported is forced through the channel and ejected on the other side.

Endocytosis and Exocytosis

In diffusion and in active transport individual molecules and ions pass through the cell membrane. Larger quantities of material—particles of food, or even whole cells—must sometimes be moved in or out of the cell.

Figure 5–12 Tiny granules are visible embedded in the membranes of the microvilli of one of the cells from the lining of the small intestine of a mouse. Some of the granules are thought to be proteins involved in the transport of digested food into the cell. (Courtesy of Drs. Tohru Arima and Torno Yamamoto, *Cell Tiss. Res.* 233, 1983.)

Figure 5–13 Active transport. (a) According to current theory, a group of transport proteins forms a fluid-filled pore through which hydrophilic molecules can pass. When the molecule being transported binds to an active site on one of these proteins, it is thought that the transport protein changes shape. (b) The molecule is then forced through the channel and ejected on the other side.

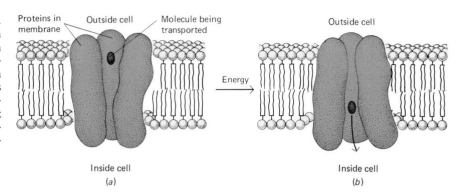

Such cellular work requires the cell to expend energy. In **exocytosis** (Fig. 5–14), a cell ejects waste products or specific secretion products, such as mucus or hormones. Generally, the material to be ejected is enclosed within a membrane, and the vesicle thus formed fuses with the cell membrane. The vesicle then opens at the point of fusion and the enclosed material is released to the exterior without the loss of other cell contents.

In **endocytosis** materials are taken into the cell. In one type of endocytosis, termed **phagocytosis** ("cell eating"), the cell ingests large solid particles, such as food (Fig. 5–15). Folds of the cell membrane move outward and enclose the particle to be ingested, forming a vacuole around it. The vacuole, still attached to the cell membrane, bulges into the cell interior. The membrane then tightens like a drawstring purse and fuses together, leaving the vacuole floating freely in the cytoplasm.

In **pinocytosis** ("cell drinking") the cell takes in dissolved materials. Tiny droplets of fluid are trapped by folds of the cell membrane. These folds pinch off into the cytoplasm as tiny vesicles of fluid (Fig. 5–16). The contents of these vesicles are slowly transferred into the cytoplasm, and the vesicles themselves may become smaller and smaller until they appear to vanish.

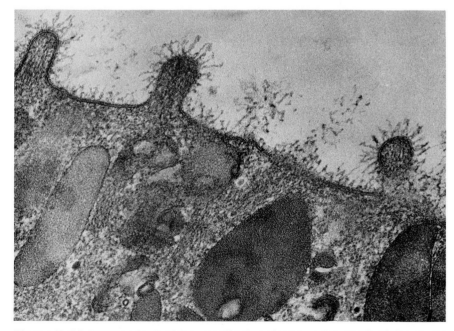

Figure 5–14 Exocytosis. A high-magnification electron micrograph of the upper surface of a secreting cell (approximately ×125,000). Secretion granules can be seen in the cytoplasm approaching the cell membrane. The filaments projecting diffusely from the cell surface are of unknown significance but may be proteins. (J.F. Gennaro, Photo Researchers, Inc.)

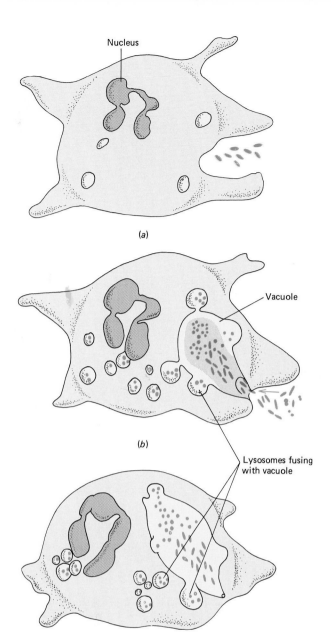

(a)

(b)

Vacuole

Lysosomes fusing
with vacuole

(c)

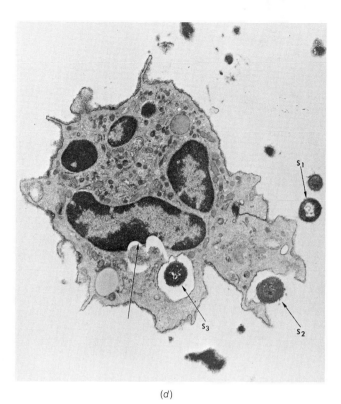

(d)

Figure 5–15 Phagocytosis. (*a*) In phagocytosis the cell ingests large solid particles, such as bacteria. Folds of the cell membrane surround the particle to be ingested, forming a small vacuole around it. (*b*) This vacuole then pinches off inside the cell. (*c*) Lysosomes may fuse with the vacuole and pour their potent digestive enzymes onto the digested material. (*d*) A white blood cell in the presence of *S. pyogenes* (approximately $\times 23,000$). One bacterium (S_1) is free, one bacterium (S_2) is being phagocytized, and a third bacterium (S_3) has been phagocytized and is seen within a vacuole (phagosome). Note that near the vacuole (see arrow) the white blood cell's own nucleus has been partly digested. (J.G. Hadley, Battelle/BPS.)

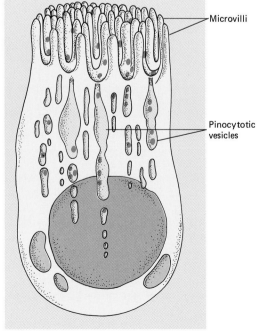

Microvilli

Pinocytotic
vesicles

Figure 5–16 In pinocytosis tiny particles of fluid are trapped by folds of the cell membrane, which then pinch off into the cytoplasm as little vesicles of fluid. The content of these vesicles is slowly transferred to the cytoplasm across their membrane linings.

CHAPTER SUMMARY

I. The cell membrane functions to (1) regulate the passage of materials into and out of the cell, (2) receive information that permits the cell to sense changes in its environment and respond to them, (3) communicate with other cells, and (4) protect the cell.

II. The plasma membrane consists of a fluid lipid bilayer in which a variety of proteins are embedded.
   A. The nonpolar, hydrophobic fatty acid chains of the membrane phospholipids meet and overlap in the middle, while the polar, hydrophilic heads are directed toward the outside of the membrane.
   B. Glycoproteins on the cell membrane communicate with other cells and with messenger molecules known as hormones. Other proteins may function in transport of materials through the membrane, in energy transfer, or in transmission of messages.
   C. Microvilli increase the surface area of the cell for absorption of materials from the environment.

III. Cell walls are found outside the cell membranes of plant cells, algae, bacteria, and fungi. They consist of carbohydrates secreted by the cell. Cell walls protect the cell and give it a characteristic shape.

IV. Adjacent cells are connected and held together by specializations of the cell membrane, such as desmosomes, tight junctions, and gap junctions.

V. The differentially permeable cell membrane allows the passage of some, but not other, substances.
   A. Some ions and molecules pass through the cell membrane by simple diffusion; others pass by facilitated diffusion, in which a carrier protein helps move a molecule through the membrane.
   B. Osmosis is a kind of diffusion in which molecules of water pass through a differentially permeable membrane from a region where water is more concentrated to a region where water is less concentrated.
   C. In active transport the cell expends energy to move ions or molecules against a concentration gradient.
   D. In endocytosis (phagocytosis and pinocytosis) materials, such as food, may be moved into the cell; a portion of the cell membrane envelops the material, enclosing it in a vacuole that is then released inside the cell. In exocytosis the cell ejects waste products or secretes substances, such as mucus.

Post-Test

1. A membrane that permits passage of some materials but not of others is described as _____ _____.

2. The lipid components of the cell membrane have a polar, _____ portion and a nonpolar, _____ portion.

3. In an electron micrograph the light zone of the cell membrane represents the _____ _____ of the lipid molecules.

4. _____ increase the surface area of the cell for absorption of materials.

5. The main component of the plant cell wall is _____ _____ .

6. _____ are button-like plaques that hold epithelial cells tightly together.

7. Cardiac muscle cells are connected by _____ _____ _____ , which permit rapid transmission of impulses from one cell to the next.

8. Red dye poured into a beaker of water spreads throughout the water. This is an example of _____ _____ .

9. The diffusion of water through a differentially permeable membrane from a region of greater concentration to a region of lesser concentration of water molecules is termed _____ .

10. A solution with a greater solute concentration than a tissue is said to be _____ to the tissue.

11. Cells will neither swell nor shrink if they are placed in _____ solutions.

12. Waste products may be ejected from a cell through the process of _____ .
13. A white blood cell engulfs a bacterium through _____ .
14. In pinocytosis a cell takes in _____ _____ _____ .

15. In _____ _____ the cell moves ions from a region of lower concentration to a region of higher concentration.

Review Questions

1. Imagine a cell with a cell wall but no cell membrane. In what ways would it be handicapped?
2. The cell membrane has been described as a "fluid mosaic." Is this a good description? Why?
3. The cell membrane has also been described as having "protein icebergs in a lipid sea." Explain why this is a good description or why it is not.
4. Why is it advantageous for cells lining the digestive tract to be equipped with microvilli?
5. The blood-brain barrier prevents the passage of certain materials from the blood into the brain tissue. What type of junctions would be found holding the cells of the barrier together? Why?
6. Describe two ways in which cells communicate with one another.
7. What problems would a cell face if its membrane were permeable rather than differentially permeable?

8. A 0.9% sodium chloride solution is isotonic to red blood cells. A laboratory technician accidentally places a sample of red blood cells in a 1.8% sodium chloride solution. What happens? Explain.
9. Why do carrot and celery sticks become limp after a time? How could they be made crisp once more? Explain in terms of turgor pressure.
10. A saltwater ameba transferred to fresh water forms a contractile vacuole. In what way is this adaptive? Explain. Would a freshwater ameba form a contractile vacuole if placed in salt water?
11. Carrier proteins are utilized in both facilitated diffusion and active transport, yet these processes are basically very different. Contrast them.

Readings

Holtzman, E., and A.B. Novikoff. *Cells and Organelles,* 3rd ed. Philadelphia, Saunders College Publishing, 1984. An integrated approach to the structural, biochemical, and physiological aspects of the cell.

Unwin, N., and R. Henderson. "The Structure of Proteins in Biological Membranes," *Scientific American* February 1984, 78–94. A discussion of the configurations of membrane proteins that permit them to be embedded in lipids yet function in the watery medium that surrounds the membrane.

PART II
LIFE AND THE FLOW OF ENERGY

Producers capture energy from sunlight and incorporate it into their organic compounds. Some of that energy is transferred to the consumers that feed on the producers. (Robert V. Fuschetto, Photo Researchers, Inc.)

Chapter 6
THE ENERGY OF LIFE

Learning Objectives

After you have finished studying this chapter you should be able to:

1. Define the term *energy* and contrast potential and kinetic energy.
2. State the first and second laws of thermodynamics and discuss their applications to living organisms and to the ecosphere.
3. Distinguish between endergonic and exergonic reactions and explain how they may be coupled so that the second law of thermodynamics is not violated.
4. Describe the energy dynamics of a reaction that is in equilibrium.
5. Explain the function of enzymes and describe how they work.
6. Describe factors, such as pH and temperature, that influence enzymatic activity.
7. Compare the action and effects of the various types of enzyme inhibitors (e.g., competitive and noncompetitive inhibitors).
8. Describe the chemical structure of ATP and its role in cellular metabolism.
9. Describe the role in metabolism of hydrogen and electron acceptors, such as NAD.

Every activity of a living cell or organism requires energy. Movement, synthesis of needed compounds, manufacture of new cells, and transport of materials are just a few of the life processes that require energy (Fig. 6–1). Because an organism has no way of creating new energy or of recycling the energy that it has used, life depends upon a continuous input of energy. This means that there is a one-way flow of energy through any individual organism and through the ecosphere. Producers trap light energy from the sun during photosynthesis and incorporate some of that energy in the chemical bonds of carbohydrates and other organic compounds. Then, a portion of that chemical energy can be transferred to the consumers that eat the producers and to the decomposers that feed on all of them, sooner or later (Fig. 6–2).

An organism must be considered an open system with respect to energy because of the one-way energy flow through it. Energy is captured, temporarily stored, and then used to perform biological work. During these processes it is converted to heat and dispersed into the environment. Because this energy cannot be reused, organisms must continuously obtain fresh supplies of it. But just how do they do so? In this chapter we will focus on some principles of energy capture, storage, transfer, and use. In the following two chapters we will explore some of the main metabolic pathways used by cells in their continuous quest for energy.

WHAT IS ENERGY?

Energy may be defined as the ability to cause various types of changes, or more simply, as the capacity to do work. Here we are concerned with the ability of living systems to do biological work. It takes the same amount of energy to light a 75-watt bulb as to keep your brain in operation while you read these words. And at this very moment you are expending considerable amounts of energy to maintain your breathing, concentrate urine in your kidneys, digest food, circulate your blood, and maintain countless other metabolic activities. These are all forms of biological work.

Energy can exist in several forms. These include heat, electrical, mechanical, chemical, sound, and radiant energy (the energy of electromagnetic waves, such as radio waves, visible light, x-rays, and gamma rays).

Potential and Kinetic Energy

Energy can be described as potential or kinetic. **Potential energy** is stored energy; it has the capacity to do work owing to its position or state.

Figure 6–1 This bobcat is expending energy in an effort to capture the snowshoe hare. If caught and eaten, the hare will provide nutrients containing energy for future activity. For its part, the hare is expending a great deal of energy in its effort to escape becoming an energy source for the bobcat. (Animals, Animals, Stouffer Prod.)

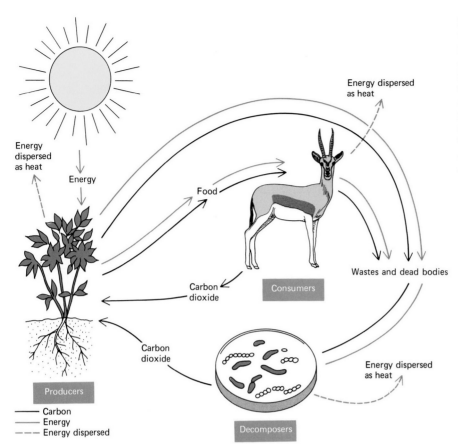

Figure 6-2 Flow of energy through the ecosphere. Carbon and many other chemical elements are continuously recycled. Energy, on the other hand, cannot be recycled. During every energy transaction some energy is dispersed as heat. For this reason a constant energy input from the sun is required to keep the ecosphere in operation. The producers act as antennae by which solar energy is absorbed to power our Spaceship Earth.

In contrast, **kinetic energy** is the energy of motion. An example of the conversion of potential energy to kinetic energy is the release of a drawn bow (Fig. 6-3): The tension in the bow and string represents stored energy; when the string is released, this potential energy is released so that the motion of the string propels the arrow. It would require the input of additional energy to draw the bow once again and restore the potential energy. Most of the actions of a complex organism involve a complex series of energy transformations. For example, to prepare for a running event, athletes eat foods that build up their reserves of glycogen. During the event an athlete's body continuously converts the energy stored in the glycogen into the kinetic energy used to run the race.

Measuring Energy

To study energy transformations, scientists must be able to measure energy. How is this done? Heat is a convenient form in which energy can be measured because all other forms of energy can be converted into heat. In fact, the study of energy has been named **thermodynamics,** that is, heat dynamics. Although several units may be used in measuring energy, the most widely used unit in biological systems is the **kilocalorie (kcal).** A kilocalorie is equal to 1000 calories. A **calorie** is the heat required to raise the temperature of 1 gram of water from 14.5° to 15.5° C. Nutritionists use the kcal in measuring the potential energy of foods and usually refer to this unit as a Calorie (with a capital C).

THE LAWS OF THERMODYNAMICS

All activities of our universe—from the life and death of cells to the life and death of stars—are governed by two laws of energy.

POTENTIAL

KINETIC

1. The polarized charges inside and outside the membrane of the resting muscle cell represent potential electrical energy. When a neuron releases a chemical that stimulates depolarization of the charges, this in turn stimulates the muscle to *initiate* contraction.

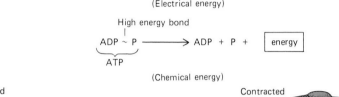

Motor neuron

Motor neuron

(Electrical energy)

2. Chemical energy is stored in the high energy phosphate bonds of a molecule called ATP. Breaking these bonds releases energy to carry out muscle contraction.

High energy bond

$$ADP \sim P \longrightarrow ADP + P + \boxed{energy}$$

ATP

(Chemical energy)

3. This energy released inside each muscle cell causes submicroscopic thick filaments to pull on thin filaments by means of cross bridges. The thin filaments move past the thick ones, causing the cell to shorten.

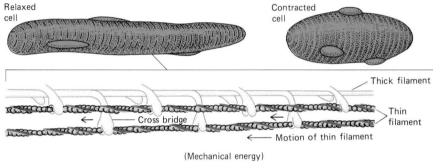

Relaxed cell

Contracted cell

Thick filament

Cross bridge

Thin filament

Motion of thin filament

(Mechanical energy)

4. The archer's muscles contract, pulling on the bones, which then move and transmit the muscular force to the bow.

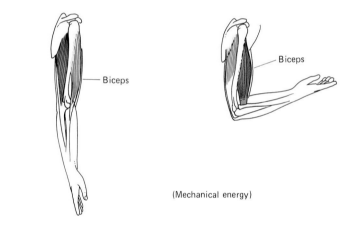

Biceps

Biceps

(Mechanical energy)

5. The archer has drawn the bow. The energy stored in the drawn bow is potential mechanical energy. When the archer releases her fingers the string moves forward and the arrow is propelled toward a target.

108

Figure 6–3 *See opposite page for legend.*

The First Law

The **first law of thermodynamics,** known also as the law of conservation of energy, states that during ordinary chemical or physical processes energy can be transferred and changed in form, but can be neither created nor destroyed. The universe is a closed system when it comes to energy. As far as we know, the energy present when the universe formed an estimated 20 billion years ago is all that can ever exist.

Although an organism can neither make nor destroy energy, it can capture some from the environment and use it for its own needs. Organisms can also transform energy from one form to another. Photosynthetic organisms transform light energy to electrical energy and then to chemical energy stored in chemical bonds. Some of that chemical energy may later be transformed by an animal that eats the plant or alga to the mechanical energy of muscle contraction or to another needed form. As these many transformations occur, some of the energy is converted to heat energy and dissipated to the environment. Although this energy can never again be used by the organism, it is not really "lost" because it can be accounted for in the surrounding physical environment (Fig. 6–4).

The Second Law

Given all possible forms of arrangement, order is an extremely unlikely state. From your own experiences you might have observed that creating order requires an input of energy—or work. For example, if you spill the many pieces of a jigsaw puzzle onto the floor, it is highly unlikely that the pieces will spontaneously reassemble themselves into the original picture. Similarly, if you drop a crystal vase, the fragments of glass will not suddenly jump back into place to reconstruct the vase. In fact, it would be very difficult for you to gather every small shard of glass and fit these fragments together to mend the vase. Out of the multitude of possible ways in which the pieces can be assembled, only one represents the highly ordered form that is the vase or the finished puzzle.

The **second law of thermodynamics** states that disorder in the universe is continuously increasing. This law explains that in any process requiring or releasing energy, some energy is dissipated as heat and is no longer available to do work. Although the total amount of energy in the universe is not decreasing with time, the energy available to do work does decrease because useful forms of energy are continuously degraded to heat, a less useful form of energy. In fact, heat is actually the energy of random molecular motion. Heat can be made to do work when there is a temperature gradient, that is, a difference in potential.

The term **entropy** refers to the energy that has become so randomized and uniform throughout a system that it is no longer available to do work. It is a measure of the disorder of a system. According to the second law of thermodynamics, entropy in the universe is continuously increasing. Eventually all energy will be random and uniform in distribution. With only this useless form of energy, no work could be performed. The universe will have run down. We are in no immediate danger, however, because this depressing state is not scheduled to occur for several billion years.

Figure 6–3 Potential energy is stored energy. Kinetic energy is the energy of motion. Shooting a bow and arrow illustrates some of the energy transformations that take place in living systems generally. The potential chemical energy released by cellular respiration and stored temporarily in the substance ATP is converted to mechanical energy, which draws the bow, and to waste heat (not shown). The energy, once again potential, that is stored in the drawn bow is released as the arrow speeds toward its target. In all of these transformations, energy is neither created nor destroyed, and in each the remaining total useful energy is less than in any of the steps that preceded it.

1. An automobile travels down the highway. Its energy is highly directional.

2. As the car travels, the surface of the road heats up, along with the surrounding air and the car's tires.

3. The engine of the car is turned off and the car begins to slow.

4. All the energy of the car's forward motion has been converted into heat and the car is now at a halt.

Figure 6–4 The concentrated energy of motion is converted into the dispersed energy of heat—also basically motion but that of millions of randomly moving molecules. Though all the car's kinetic energy is lost to use, none of it is lost in the sense of being destroyed—it is merely dispersed and diluted past the point of recovery.

Because of the second law of thermodynamics, no process requiring energy is ever 100% efficient. Much of the energy is dispersed as heat and is transferred to the environment, so that there is an increase in entropy. Cellular energy utilization is about 55% efficient, with the other 45% of the energy being lost as heat. Such biological processes are actually quite efficient compared with most machines made by human beings. For example, a gasoline engine is only about 17% efficient.

Because they are highly organized, living organisms are very unstable. In fact, life is a constant struggle against the second law of thermodynamics. Survival of individual organisms, as well as ecosystems, depends upon continuous energy input. Thus, producers must carry on photosynthesis, and consumers and decomposers must eat.

FREE ENERGY AND COUPLED REACTIONS

Free energy is energy that is available to do work under conditions of constant temperature and pressure. **Spontaneous reactions,** reactions that can occur without outside input, release free energy and therefore can perform work. These reactions are also called **exergonic reactions** (Fig. 6–5). Because energy is released, the products contain less energy

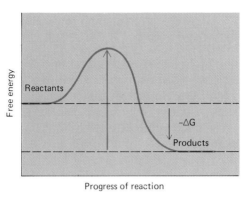

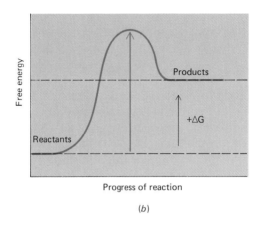

Figure 6–5 Free-energy diagram. (*a*) In exergonic reactions free energy is released, and the product has less energy than the reactants. (*b*) In endergonic reactions there is a net input of free energy, so the products contain more energy than the reactants. Note that even the exergonic reaction requires some input of energy to get started. This initial investment energy is termed activation energy.

than the reactants. Reactions that are not spontaneous require a net input of free energy and are said to be **endergonic.** In these reactions the products contain more energy than the reactants.

Many of the metabolic reactions—protein synthesis, for example—in a living organism are endergonic. These reactions are driven by the energy released from exergonic reactions. Indeed, in the living cell endergonic and exergonic reactions are **coupled;** the thermodynamically favorable exergonic reaction provides the energy needed for the thermodynamically unfavorable endergonic reaction. As we will see in a later section, exergonic and endergonic reactions are often linked to ATP formation and breakdown.

EQUILIBRIUM

Over a ten-year period the population of a city may remain the same. Although some people have moved into town, others have moved out, or perhaps died. However, the net change in the population is zero. We might say that the population in this city is in a state of dynamic equilibrium. In an **equilibrium** the rate of change in one direction is exactly equal to the rate of change in the opposite direction.

Consider a reaction in terms of the numbers of each type of molecule involved. At the beginning of a reaction only the reactant molecules may be present. These molecules move about and collide with one another with sufficient energy to react. As more and more product molecules are produced, fewer and fewer reactant molecules are left. As the number of product molecules increases, they collide more frequently, and some have sufficient energy to initiate the reverse reaction. The reaction proceeds in both directions simultaneously and reaches an equilibrium in which the rate of the reverse reaction is about the same as the rate of the forward reaction. Reversibility of chemical reactions permits cells to control their release of free energy according to their needs. It also allows cells to resynthesize many large compounds for use in metabolism.

When the free energy difference between the products and reactants of a chemical reaction is zero, the reaction is at equilibrium. Any change, such as a change in temperature or pressure, that affects the reacting system may cause the equilibrium to shift. The reaction may then proceed in a specific direction until once again the free-energy difference is zero and a new equilibrium has been established.

When there is little free-energy difference on the two sides of a chemical equation, the direction of a reaction will be governed mainly by the concentrations of the reactants and products. The reaction tends to proceed in the direction that will minimize the difference in concentration between the substances on the two sides of the equation. If one substance, say, the product, is continuously removed as it is formed, the reaction will indeed proceed to completion.

ENZYMES: CHEMICAL REGULATORS

The thousands of chemical reactions that take place within the living cell must be precisely controlled and coordinated. These reactions interlock somewhat as a multitude of assembly processes interlock on a modern-day production line. In the manufacturing of an automobile, for example, it might not be possible to complete a car without some seemingly minor part, such as a ball bearing or special bolt. Very elaborate planning and scheduling are necessary to ensure that all parts are delivered to the plant by the time they are needed and that each worker performs a small part of the overall job at precisely the right time.

It is this concept of control that we wish to emphasize. Although energy can be released from glucose and other fuel molecules by burning, cells could not survive such high temperatures nor effectively utilize energy that is released in short sudden bursts. Cells require a slow, steady release of energy that they must be able to regulate to meet metabolic energy requirements. Accordingly, fuel molecules are slowly oxidized and energy is extracted in small amounts during cellular respiration, which includes sequences of 30 or more reactions. In fact, most cellular metabolism proceeds by a series of steps, so a given molecule may go through as many as 20 or 30 chemical transformations before it reaches some final state. And then the apparently completed molecule may be pre-empted by yet another chemical pathway so as to be totally transformed or consumed in the course of metabolism. The changing needs of the cell require a system of flexible chemical control. The key elements of this control system are the remarkable enzymes.

What Enzymes Do

An **enzyme** is an organic catalyst, a compound that greatly increases the speed of a chemical reaction without being consumed itself. All enzymes are proteins (though, of course, not all proteins are enzymes). Like all catalysts, an enzyme affects the rate of a reaction by lowering the energy needed to activate the reaction. Even a strongly exergonic reaction that releases more than enough energy as it proceeds is prevented from beginning by an energy barrier. This is because before new chemical bonds can be formed, existing ones must be broken. The energy required to overcome this barrier and get the reaction going is called **activation energy.** An enzyme greatly reduces the activation energy necessary to initiate a chemical reaction.

An enzyme can only promote a chemical reaction that could be made to proceed without it. There is nothing in the action of a catalyst of any kind that could change the operation of the second law of thermodynamics, so enzymes do not influence the direction of a chemical reaction or the final concentrations of the molecules involved. They simply *speed up* reactions.

Consider the enzyme carbonic anhydrase as an example of what enzymes do. Like most enzymes it is identifiable as an enzyme by its distinctive **-ase** ending. Carbonic anhydrase promotes the interconversion of carbonic acid with carbon dioxide and water:

$$CO_2 + H_2O \underset{\longleftarrow}{\overset{\text{Carbonic anhydrase}}{\longrightarrow}} H_2CO_3$$

Carbon dioxide   Water   Carbonic acid

This reaction takes place in many cells and tissues, especially in the kidneys and red blood cells. Carbonic acid ionizes, forming bicarbonate ions, the form in which most of the carbon dioxide in the blood is transported. The reaction shown above could proceed without the enzyme but would be uselessly slow. When carbonic anhydrase is present, the reaction proceeds about 10 million times faster. Amazingly, a single molecule of this enzyme can promote the conversion of an estimated 600,000 molecules of carbon dioxide into carbonic acid each second.

How Enzymes Work

The chemicals upon which an enzyme operates are referred to as its **substrates.** Enzymes are very **specific.** An enzyme will catalyze only a few closely related chemical reactions, or in many cases only one particular reaction. Enzymes form temporary chemical compounds with their substrates. These complexes then break up, releasing the product and regenerating the original enzyme molecule for reuse.

Enzyme + Substrate 1 + Substrate 2
$$\longrightarrow \text{Enzyme-substrate complex}$$

Enzyme-substrate complex $\longrightarrow$ Enzyme + Product(s)

Note that the enzyme itself is not permanently altered or consumed by the reaction.

Why does the enzyme-substrate complex break up into chemical products different from those that participated in its formation? As shown in Figure 6–6, each enzyme has one or more regions called **active sites,** which in the case of a few enzymes have been shown to be actual indentations in the enzyme molecule. These active sites are located close to one another on the enzyme's surface, so during the course of a reaction, substrate molecules occupying these sites are temporarily brought together and react with one another. It is thought that when the enzyme and substrate bind together, the shape of the enzyme molecule changes slightly. This produces strain in critical bonds in the substrate molecules so that these bonds break. The new chemical compound thus formed has little affinity for the enzyme and moves away from it. An enzyme can be thought of as a molecular lock into which only specifically shaped molecular keys—the substrates—can fit (Fig. 6–6).

Unlike a lock and key, however, the enzyme and its substrate seem not to be exactly complementary shapes. A recent model of enzyme action, known as the **induced-fit model,** is based on data indicating that the active sites of an enzyme are not rigid. When the substrate binds to

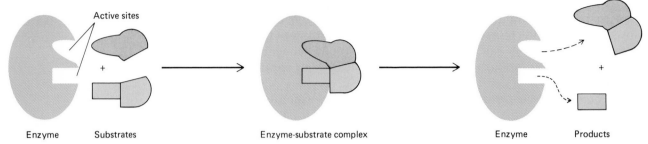

Active sites

Enzyme   Substrates

Enzyme-substrate complex

Enzyme   Products

Figure 6–6 Lock-and-key mechanism for enzyme action. The substrates fit the active sites of the enzyme molecule much as keys fit locks. However, in this model the lock acts on the key rather than the other way around. Note that when the products separate from the enzyme, the enzyme is free to catalyze production of additional products. The enzyme is not permanently changed by the reaction. In recent years the lock-and-key concept of enzyme action has undergone some modification, as you can see from Fig. 6–7.

the enzyme, it may induce a change in shape in the enzyme molecule, resulting in an optimum fit for the substrate-enzyme interaction. The change in shape of the enzyme molecule can put strain on the substrate. This stress may help bonds to break, thus promoting the reaction (Fig. 6–7).

Some enzymes—for example, pepsin, secreted by the stomach—consist only of protein. Other enzymes have two components, a protein referred to as the apoenzyme and an additional chemical component called a **cofactor.** The cofactor of some enzymes is a metal ion that is often a functional part of the active site. Many of the trace elements required by organisms function as cofactors. Together, the cofactor and apoenzyme work very efficiently. Hydrogen peroxide can be split by inorganic catalysts, such as free iron atoms. However, it would take 300 years for an iron atom to split the same number of molecules of hydrogen peroxide split in one second by a molecule of the enzyme catalase, which contains a single iron atom.

An organic, nonpolypeptide compound that serves as a cofactor is called a **coenzyme.** Many coenzymes are synthesized from vitamins, particularly from the B vitamins. The coenzyme serves as an adapter, permitting the enzyme to accept a substrate for which it would, by itself, have little affinity (Fig. 6–8).

Regulating Enzymatic Action

Enzymes regulate the chemistry of the cell, but what controls the enzymes? One mechanism of enzyme control simply depends upon the amount of enzyme produced. The synthesis of each type of enzyme is directed by a specific gene. The gene, in turn, may be switched on by a signal from a hormone or by some other type of cellular product. When the gene is switched on, the enzyme is synthesized. Then the amount of enzyme present influences the rate of the reaction. Up to a maximum value, the rate of an enzyme-dependent reaction increases as the concentration of the enzyme increases.

Figure 6–7 Comparison of models of enzyme action. (a) The lock-and-key model. (b) The induced-fit model. Chemical reactions are favored when substrate molecules get close enough to one another to react, when they are presented to each other in the right orientation, and when their existing chemical bonds are strained. Enzymes often do all three of these things. The straining of the substrate's bonds is accomplished by the apparent fact that most active sites are a bit bigger than the size of their substrate molecules. Accordingly, when a fit is forced upon them, the active site exerts a kind of pull on the substrate, helping to pull it apart. To be sure, the fit of the enzyme and substrate must not be too poor, or they will have no affinity for one another.

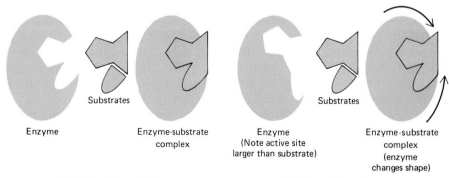

Enzyme   Substrates   Enzyme-substrate complex

Enzyme (Note active site larger than substrate)   Substrates   Enzyme-substrate complex (enzyme changes shape)

(a) Lock-and-key model

(b) Induced fit model

Another important method of enzymatic control depends upon the activation of enzyme molecules that are present in an inactive form in the cytoplasm. In the inactive form the active sites of the enzyme are inappropriately shaped, so the substrates do not fit. Among the factors that influence the shape (conformation) of the enzyme are acidity and alkalinity and the concentration of certain salts.

Some enzymes possess an activator site (also called an allosteric site). When a molecular activator, such as the substance cyclic AMP, occupies the activator site, the shape of the enzyme molecule changes, making the active sites better suited for binding with the substrate.

Enzyme Optima

Enzymes generally work best under certain narrowly defined conditions referred to as optima. These include appropriate temperature, pH, and salt concentration. For example, pepsin, the protein-digesting enzyme of the stomach, works best at the strongly acid pH of 2. In contrast, the starch-digesting enzyme amylase in saliva and pancreatic juice has a pH optimum of 8.5 (slightly alkaline). Strong acids or bases irreversibly inactivate most enzymes by permanently changing their molecular conformation.

Enzymatic reactions occur very slowly or not at all at low temperatures, but activity resumes when the temperature is raised to normal. The rates of most enzyme-regulated reactions increase with increasing temperature, within limits. Temperatures greater than 50° or 60° C rapidly inactivate most enzymes by altering their secondary and tertiary level structures. The enzyme is said to be denatured. (An everyday example of this is cooking an egg white; it changes in consistency as the protein is denatured.)

Enzyme Inhibition

Enzymes may be rendered less effective or even nonfunctional by certain chemical agents known as **inhibitors.** Enzyme inhibition may be reversible or irreversible. Reversible inhibitors can be competitive or noncompetitive. In **competitive inhibition** the inhibitor competes with the normal substrate for the active site of the enzyme (Fig. 6–9). A competitive inhibitor usually is chemically similar to the normal substrate and so fits the active site and binds with the enzyme. However, it is not similar enough to the normal substrate to take its place effectively. The enzyme cannot act upon it to form reaction products. A competitive inhibitor occupies the active site only temporarily and does not damage the enzyme irreversibly.

In **noncompetitive inhibition** the inhibitor binds with the enzyme at a site other than the active site. Such an inhibitor renders the enzyme inactive by altering its shape. Many important noncompetitive inhibitors are metabolic substances that help regulate enzyme activity by combining reversibly with the enzyme.

Many poisons are **irreversible inhibitors** that permanently inactivate or even destroy the enzyme. Nerve gases are irreversible inhibitors that poison the enzyme cholinesterase, essential in the normal function of nerves and muscles. A number of insecticides and drugs are irreversible inhibitors (Fig. 6–10). The antibiotic penicillin and its chemical relatives inhibit a bacterial enzyme necessary for bacterial cell wall construction. Unable to produce new cell walls, susceptible bacteria cannot multiply effectively (Fig. 6–10). Since human body cells do not have cell walls (and so do not employ the susceptible enzyme), penicillin is harmless to humans, except for the occasional allergic patient.

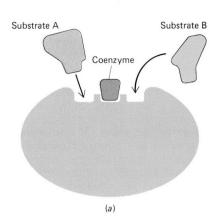

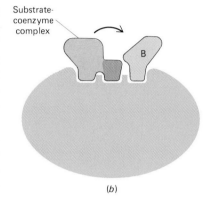

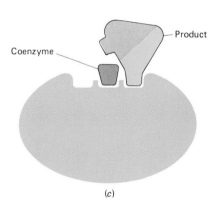

Figure 6–8 Coenzyme action. Some enzymes are not able to attach directly to the substrates whose chemical reactions they catalyze. Such enzymes employ accessory coenzymes to serve as adaptors, aiding the attachment of one or more substrates to the enzyme's active sites. One substrate combines first with the coenzyme to form a coenzyme-substrate complex. Then the coenzyme-substrate complex combines with the second substrate. This complex yields the products and releases the coenzyme. Some familiar vitamins serve as coenzymes for vital enzymes of cellular metabolism.

Figure 6–9 Competitive and noncompetitive inhibition. (*a*) In competitive inhibition the inhibitor competes with the normal substrate for the active site of the enzyme. A competitive inhibitor occupies the active site only temporarily. (*b*) In noncompetitive inhibition, the inhibitor binds with the enzyme at a site other than the active site, altering the shape of the enzyme and so inactivating it. Noncompetitive inhibition may be reversible. A somewhat similar action is sometimes used by cells to control enzyme action.

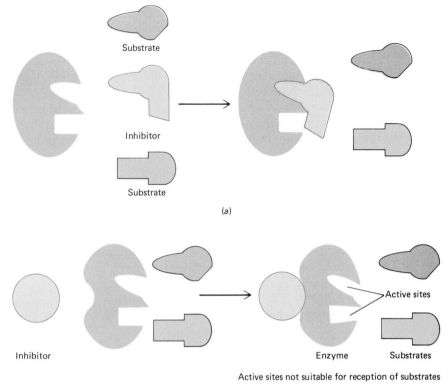

(*a*)

(*b*)

PACKAGING ENERGY—ATP

In all living cells energy is temporarily packaged within a remarkable chemical compound called **adenosine triphosphate,** or more simply, **ATP.** This compound stores large amounts of energy for very short periods of time. We may think of ATP as the energy currency of the cell. When you work, you earn money, and so you might say that your energy is symbolically stored in the money you earn. In the same way, the energy of the cell may be stored in ATP. When you have earned extra money, you might deposit some in the bank. Similarly, the cell deposits energy as lipid in fat cells or as glycogen in the liver and muscle. Moreover, just as you dare not make less money than you spend, so, too, the cell must avoid energy bankruptcy, which would mean its death. Finally, just as you (alas) do not keep what you make very long, so too the cell is forever spending its ATP.

ATP, a nucleotide, consists of three main parts (Fig. 6–11): (1) a nitrogen-containing base, adenine, which also occurs in DNA, (2) ribose, a five-carbon sugar, and (3) three inorganic phosphate groups that are identifiable as phosphorus atoms surrounded by oxygen atoms. Inorganic phosphate is usually designated P_i. Notice that the phosphate groups are attached to the end of the molecule in a series, rather like three passenger cars behind a locomotive. The couplings, that is, the chemical bonds attaching the last two phosphates, also resemble those of a train in that they are readily attached and detached.

When the third phosphate is removed, the remaining molecule is called **adenosine diphosphate,** or simply **ADP.** When two phosphate groups are removed, the molecule that remains is **adenosine monophosphate,** AMP. Conversely, when a phosphate is attached to AMP, it becomes ADP, and when a phosphate is added to ADP, ATP is produced. Addition of a phosphate group to a molecule is referred to as **phosphorylation.** Note that these reactions are readily reversible.

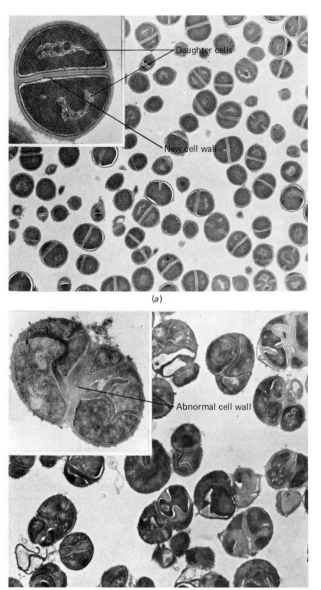

(a)

(b)

Figure 6–10 Antibiotic damage to bacterial cell walls. (a) Normal bacteria. Inset shows the new cell wall laid down between daughter cells of a dividing bacterium. (b) Damaged bacteria. The insets are magnified approximately ×54,000. (Courtesy of Victor Lorian and Barbara Atkinson, with permission of *The American Journal of Clinical Pathology.*)

Figure 6–11 (a) Chemical structure of adenosine triphosphate (ATP). (b) A computer-generated model of ATP. The red balls are oxygen atoms; blue = nitrogen, green = carbon, yellow = phosphate, white = hydrogen. Note the hydrogen atom attached to the last oxygen in the triphosphate group. At different pHs, this and other oxygen atoms might be bonded with hydrogen or be present in ionized form. (b, Courtesy of Computer Graphics Laboratory, University of California, San Francisco.)

$$AMP + P_i + Energy \rightleftharpoons ADP$$

$$ADP + P_i + Energy \rightleftharpoons ATP$$

As these equations indicate, energy is required to add a phosphate to either the AMP or the ADP molecule. Conversely, since energy can

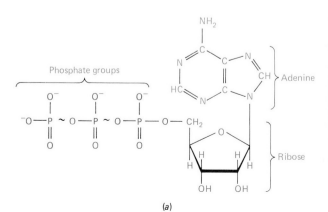

(a)

(b)

Figure 6–12 ATP is an important link between endergonic and exergonic reactions in living cells.

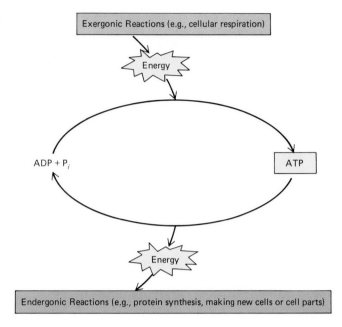

neither be created nor destroyed, the energy is released or transferred to another molecule when the phosphate is detached. Thus ATP is an important link between exergonic (energy-releasing) and endergonic (energy-requiring) reactions.

The high energy bonds in ATP are often indicated by wavy lines attaching the last two phosphates to the rest of the molecule (Fig. 6–12). These bonds contain more than twice the energy of an average chemical bond because an unusually large amount of energy is required to join the second and third phosphates.

Energy released from exergonic reactions is packaged in ATP molecules for use in endergonic reactions (Fig. 6–12). This energy may be used to produce fats or glycogen, molecules stockpiled for long-term energy storage. The cell contains a pool of ADP, ATP, and phosphate in a state of equilibrium. Large quantities of ATP cannot be stockpiled in the cell. In fact, recent studies suggest that a bacterial cell has no more than a one-second supply of ATP. Thus ATP molecules are used almost as quickly as they are produced. Every second, thousands of ATPs are made from ADP and phosphate and an equal number are broken down, yielding their energy to whichever life processes may require them (Fig. 6–13). In this way ADP and phosphate are ceaselessly recycled, shuttling back and forth in a blur of energy between cellular respiration and consumption.

Most ATP is manufactured during the electron transport phase of cellular respiration in the mitochondria (Chapter 8). However, ATP can also be produced at what is called the substrate level. Imagine that a particular substrate is chemically combined with phosphate to form substrate-phosphate, which with the help of an enzyme, substrate-phosphatase, can donate its phosphate to ADP. In accepting the phosphate, ADP becomes ATP.

Figure 6–13 The chemical energy of ATP is converted to light energy in the light organs of this deep-sea angler fish, *Caulophryne jordani*. The energy transformation actually occurs within luminous bacteria that inhabit the light organs of the fish. (Peter David, Seaphot.)

$$\text{Substrate-phosphate} + \text{ADP} \xrightarrow{\substack{\text{Substrate-}\\ \text{phosphatase}}} \text{Substrate} + \text{ATP}$$

Why does the phosphate leave the original substrate compound and join ADP? Chemical reactions, according to the second law of thermodynamics, tend to proceed in the direction of higher to lower potential energy. Thus, because there is more potential energy in the substrate-phosphate than in ATP, the reaction proceeds toward the production of

ATP. Only a few compounds in the cell meet these requirements, however, so not much ATP is synthesized in this way. Still, muscle cells make considerable use of this method of ATP production.

HYDROGEN AND ELECTRON TRANSFER

Energy processing by cells involves the transfer of chemical energy through the flow of electrons and protons. Generally, there is a sequence of redox (oxidation-reduction) reactions that take place as hydrogen or its electrons are transferred from one compound to another. You should recall from Chapter 2 that oxidation is a chemical reaction in which a substance loses electrons. In living cells oxidation almost always involves the removal of a hydrogen atom (which contains an electron) from a compound. Reduction is a chemical reaction in which a substance gains electrons (or hydrogens).

Electrons released during an oxidation reaction cannot exist in the free state in living cells. That is why every oxidation must be accompanied by a reduction in which the electrons are accepted by another atom or molecule. The oxidized molecule gives up energy, while the reduced molecule receives energy.

When hydrogen atoms are removed from an organic compound, they take with them some of the energy that had been stored in their chemical bonds. The hydrogen along with its energy is transferred to a hydrogen acceptor molecule, which is generally a coenzyme. One of the most frequently encountered hydrogen acceptor coenzymes is **nicotinamide adenine dinucleotide,** more conveniently referred to as **NAD.** NAD is a coenzyme that can temporarily package large amounts of free energy. Here is a generalized equation showing the transfer of hydrogen from a compound we will call X to NAD:

$$X H_2 + NAD^+ \longrightarrow \underset{\text{Oxidized}}{X} + \underset{\text{Reduced}}{NAD-H + H^+}$$

Note that the NAD is reduced when it combines with hydrogen. Some of the energy stored in the bonds holding the hydrogens to molecule X has been transferred by this reaction to the NADH. This energy can now be used for some metabolic process or it can be transferred through a complex series of reactions to ATP. As will be discussed in Chapters 7 and 8, transfer of energy in both photosynthesis and cellular respiration involves sequences of redox reactions. The acceptor compounds that make up these sequences keep energy flowing.

Other important hydrogen or electron acceptor compounds are **flavin adenine dinucleotide (FAD)** and the cytochromes. FAD is a nucleotide that accepts hydrogens and their electrons. The cytochromes are proteins that contain iron. The iron component accepts electrons from hydrogen and then transfers them to some other compound. Cytochromes, like NAD and FAD, are important in cellular respiration and will be discussed further in Chapter 8. Such electron acceptors provide a mechanism by which the cell can slowly and efficiently capture energy from fuel molecules (Fig. 6–14).

CHAPTER SUMMARY

   I. Life depends upon a continuous input of energy. Producers capture energy during photosynthesis and incorporate some of that energy in the chemical bonds of organic compounds. Some of this energy can then be transferred to consumers and decomposers.

  II. Energy may be defined as the capacity to do work.

Figure 6–14 The total energy released by a falling object is the same whether it is released all at once or in a series of steps. Similarly, in cellular metabolism the energy of an electron liberated from a foodstuff is the same whether it is released all at once, or, as is actually the case, gradually as it passes to successive electron acceptors. Such acceptors function in both photosynthesis and cellular respiration. As part of a complicated scheme involving the diffusion of protons across membranes these acceptors permit the *controlled* extraction of some of the electrons produced by these processes.

$$E = e_1 + e_2 + e_3 + e_4 + e_5$$

   A. Potential energy is stored energy; kinetic energy is energy of motion.

   B. A common unit used to measure energy is the kilocalorie.

III. The first law of thermodynamics states that energy can be neither created nor destroyed but can be transferred and changed in form. The second law of thermodynamics states that disorder in the universe is continuously increasing.

   A. The first law explains why organisms cannot produce energy but must borrow it continuously from somewhere else.

   B. The second law explains why no process requiring energy is ever 100% efficient; in every energy transaction some energy is dissipated as heat. The term entropy refers to the energy that is no longer available to do work.

IV. Spontaneous reactions release free energy and can therefore perform work.

   A. Reactions that release free energy are exergonic reactions; endergonic reactions require a net input of free energy.

   B. In the living cell endergonic and exergonic reactions are coupled.

  V. In an equilibrium the rate of change in one direction is exactly the same as the rate of change in the opposite direction; the free energy difference between the reactants and products is zero.

VI. An enzyme is an organic catalyst; it greatly increases the speed of a chemical reaction without being consumed itself.

   A. An enzyme lowers the activation energy necessary to get a reaction going.

   B. Enzymes bring substrates into close contact so that they can more easily react with one another.

   C. Some enzymes consist of an apoenzyme and a cofactor. An organic cofactor is called a coenzyme.

   D. A cell can regulate enzymatic activity by controlling the amount of enzyme produced and by regulating conditions that influence the shape of the enzyme.

   E. Enzymes work best at specific temperatures, pH, and salt concentrations.

   F. Most enzymes can be inhibited by certain chemical substances. Reversible inhibition may be competitive or noncompetitive.

VII. ATP is the energy currency of the cell; energy is temporarily stored within its chemical bonds.

A. ATP is formed by the phosphorylation of ADP, a process that requires a great deal of energy.
B. ATP is a link between exergonic and endergonic reactions.
VIII. NAD, FAD, and the cytochromes are hydrogen or electron acceptor compounds that transfer hydrogen (or electrons) along with its energy.

Post-Test

1. Stored energy is referred to as _____ _____ .
2. The study of energy is called _____ .
3. The law of conservation of energy is also known as the _____ _____ _____ _____ .
4. Energy that can be used to do work is referred to as _____ energy.
5. _____ reactions require a net input of free energy.
6. In _____ the rate of change in one direction is equal to the rate of change in the opposite direction.
7. Organic catalysts that regulate chemical reactions are _____ .
8. The energy required to break existing bonds and start a reaction is known as _____ _____ _____ .
9. The chemical substances upon which an enzyme acts are referred to as the _____ .
10. Organic compounds that serve as cofactors are called _____ .
11. In _____ inhibition the inhibitor renders the enzyme inactive by altering its shape.
12. When ADP is phosphorylated, the compound formed is _____ .
13. A chemical reaction in which a substance gains _____ (or hydrogens) is referred to as reduction.
14. The number of phosphate groups in ATP is _____ .
15. The _____ are electron acceptor compounds that consist of protein and iron.

Review Questions

1. Trace the various forms that energy takes from sunlight to the heat released during muscle contraction.
2. Give three examples of (a) potential energy, (b) kinetic energy.
3. When a consumer eats a producer, it cannot obtain for itself all the energy captured by the producer during photosynthesis. Why not? Explain in terms of the second law of thermodynamics.
4. Imagine that you could redesign your body so that you could create all the energy you require. What advantages would this ability confer on you?
5. Why is it significant that endergonic and exergonic reactions can be coupled?
6. What is activation energy? How does an enzyme affect activation energy?
7. Explain the lock and key analogy of enzymatic action.
8. How are enzymes affected by (a) pH, (b) temperature, (c) inhibitors?
9. In what way does ATP serve as a link between exergonic and endergonic reactions? From what is ATP synthesized?
10. What role does NAD^+ play in cellular metabolism?

Readings

Atkins, P.W. *The Second Law.* W.H. Freeman and Company, San Francisco, 1984. A basic, understandable introduction to thermodynamics with an extensive section denoted to its biological implications.

Clayton, R.K. *Photosynthesis: Physical Mechanisms and Chemical Patterns.* Cambridge, England, Cambridge University Press, 1980. An important reference resource for information about current concepts and experiments dealing with the photosynthetic process.

Clayton, R.K., and W.R. Sistron (eds). *The Photosynthetic Bacteria.* New York, Plenum Press, 1978. A multiauthor book summarizing research on many aspects of the photosynthetic process in bacteria.

Cloud, P. "The Biosphere," *Scientific American,* September 1983, 176–189. A fascinating discussion of the relationship between microbial, animal, and plant life on earth and the physical environment—and the energy transfer pathways that knit it all together.

Dickerson, Richard E. "Cytochrome c and the Evolution of Energy Metabolism," *Scientific American,* March 1980, 136–153.

Lehninger, A.L. *Bioenergetics: The Molecular Basis of Biological Energy Transformations,* 2nd ed. Menlo Park, Calif., Benjamin/Cummings, 1971. A classic presentation of energy transformations in living systems. An unusually clear exposition of a somewhat difficult subject.

Chapter 7
PHOTOSYNTHESIS: CAPTURING ENERGY

Outline

I. Light and life
II. Absorbing light: chlorophyll
III. The reactions of photosynthesis
  A. The light-dependent reactions
    1. The photosystems
    2. Cyclic photophosphorylation
    3. How proton gradients form ATP: chemiosmosis
  B. The light-independent reactions: carbon fixation
    1. Carbon dioxide fixation: some details
    2. The four-carbon pathway
Focus on chlorophyll structure

Learning Objectives

After you have read this chapter you should be able to:

1. Write a summary reaction for photosynthesis, explaining the origin and fate of each substance involved.
2. Describe the internal structure of a chloroplast.
3. Summarize the events of the light-dependent reactions of photosynthesis, explaining the role of light in the activation of chlorophyll.
4. Contrast cyclic and noncyclic photophosphorylation.
5. Describe how proton gradients form ATP according to the chemiosmotic theory.
6. Summarize the events of the light-independent reactions of photosynthesis.
7. Discuss the advantages of the C_4 pathway.

Our technological society will someday run out of fossil fuels. What energy sources will be used then to power our industries, homes, and vehicles? One seriously discussed alternative source of energy is hydrogen. However, there are practical difficulties in storing and transporting this highly reactive gas, so hydrogen may have to be combined with carbon to form artificial hydrocarbon fuels similar to those already in use, such as gasoline or propane. This would involve some sacrifice in energy content, but since these fuels have a high hydrogen content, they would still serve well, as indeed they do now. Even if we had no carbon in elemental form with which to combine the hydrogen, hydrogen-carrying fuels could still be produced, at some further energy sacrifice, from such common carbon sources as carbon dioxide. Modification of the process would permit the production of still other energy-rich substances, such as alcohol or glycerine, which are important in industry.

Although making practical use of hydrogen fuel has not yet become a reality in human society, the biological world has long made use of this energy source. The vast chemical industry of the biosphere has for millennia captured solar energy by photolysis (breaking down with light energy) of water, released the resulting oxygen, and incorporated the hydrogen into energy-rich carrier compounds. These compounds form the matter of biological commerce because they also can be incorporated into the body structure of organisms as structural components. The energy of these hydrogen-containing compounds can be released once again by allowing them to react once more with the formerly released oxygen. The details of these metabolic processes form the subject of this chapter.

Plants and algae are producers that obtain their energy from the sun. They are uniquely capable of converting solar energy into stored chemical energy—the process of **photosynthesis.** Each year these remarkable organisms produce more than 200 billion tons of food. The chemical energy stored in this food fuels the chemical reactions that sustain life. All consumers obtain their energy by consuming plants and algae or by eating other consumers that have fed on plants and algae. Almost all life, therefore, is ultimately dependent upon photosynthesis.

Chemical energy is a stable energy form; it may be stored for long periods of time. A striking example is the energy in coal and oil, which was stored from photosynthesis millions of years ago.

LIGHT AND LIFE

Since life on our planet depends upon light, it seems appropriate to discuss the nature of light and how it permits photosynthesis to occur. Light is a very small portion of a vast, continuous spectrum of radiation, the electromagnetic spectrum (Fig. 7–1). All radiations in this spectrum behave as though they travel in waves. At the end of the spectrum are gamma rays with very short wavelengths (measured in nanometers). A **wavelength** is the distance from one wave peak to the next. At the other end of the spectrum are low-frequency radio waves, with wavelengths so long that they are measured in kilometers.

Within the spectrum of visible light, violet has the shortest wavelength and red the longest. Ultraviolet light has a still shorter range of wavelengths and infrared a still longer one. Light behaves not only as a wave but also as a particle. It is composed of particles of energy called **photons.** The energy of a photon is different for light of different wavelengths. The shorter the wavelength, the more energy light has, and the longer the wavelength, the lower the energy. In other words, the energy of the photon is inversely proportional to the wavelength.

Why does photosynthesis depend upon visible light rather than upon some other wavelength of radiation? We can only speculate, but

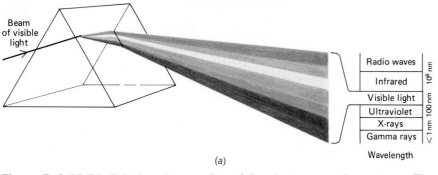

Figure 7–1 Visible light is only a portion of the electromagnetic spectrum. The green pigment chlorophyll absorbs light mainly in the blue, violet, and red regions of the spectrum. During photosynthesis light energy is used to synthesize organic compounds. (J.N.A. Lott, McMaster University/BPS.)

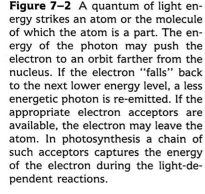

Figure 7–2 A quantum of light energy strikes an atom or the molecule of which the atom is a part. The energy of the photon may push the electron to an orbit farther from the nucleus. If the electron "falls" back to the next lower energy level, a less energetic photon is re-emitted. If the appropriate electron acceptors are available, the electron may leave the atom. In photosynthesis a chain of such acceptors captures the energy of the electron during the light-dependent reactions.

one reason may be that most of the radiation reaching our planet from the sun is within this portion of the electromagnetic spectrum. Another consideration is that only radiation within the visible light portion of the spectrum excites certain types of biological molecules, moving electrons into higher energy levels.

Photons interact with atoms in a variety of ways, but all the ways depend on the electron structure of the atom. As you should recall, an atom consists of an atomic nucleus surrounded by one or more energy levels containing electrons. In the hydrogen atom there is only a single electron that occupies the first energy level. The lowest energy state an atom possesses is called the **ground state,** but energy can be added to an electron so that it will attain a higher energy level. When an electron is raised to a higher energy level than its ground level, the atom is said to be excited.

When an electron is raised to a higher energy state it may soon return to its ground level (Fig. 7–2). In this case energy is usually dissipated as heat or as light of a longer wavelength. This is referred to as fluorescence. Or the excited electron may be lost, leaving the atom with a net positive charge. In this instance the electron may be accepted by a reducing agent. This is what occurs in photosynthesis.

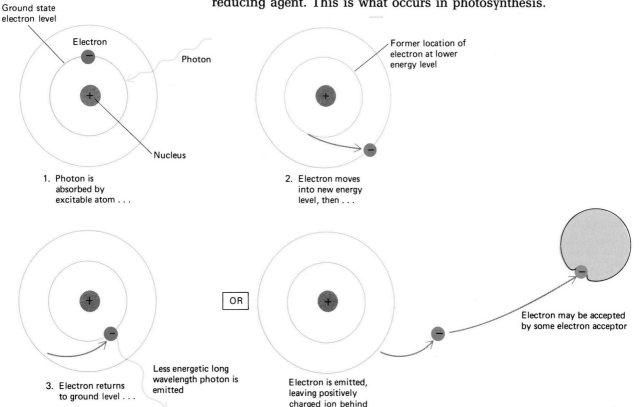

ABSORBING LIGHT: CHLOROPHYLL

Have you ever wondered why most plants are green? The reason is that their leaves reflect most of the green light that strikes them. If green light is reflected, most of it is not being absorbed or used. **Chlorophyll,** the main pigment plants use in photosynthesis, absorbs light primarily in the blue, violet, and red regions of the spectrum. (See Focus on Chlorophyll Structure.) Actually, there are several kinds of chlorophyll. The most important are chlorophyll *a* and chlorophyll *b*. Plants (and algae) also have accessory photosynthetic pigments, such as carotenoids, that absorb different wavelengths of light.

Chlorophyll and other photosynthetic pigments are located within the membranes of **thylakoids,** tiny flattened sacs within cells capable of

Focus on . . .

CHLOROPHYLL STRUCTURE

Chlorophyll is the principal pigment used in photosynthesis. A pigment may be defined as any substance that absorbs light. In the accompanying diagram of chlorophyll *a*, note the porphyrin ring structure (also found in hemoglobin). Also note the magnesium atom in the center of the ring. The magnesium is the part of the molecule that is actually excited by light. The long phytol tail of the molecule is a long chain of carbon and hydrogen atoms that give the otherwise flat chlorophyll molecule a shape somewhat like a long-handled skillet.

The phytol chain is hydrophobic; it repels water and is attracted to lipid. The lamellae of the chloroplast, like other membranes, are composed of lipids, and apparently the phytol chains of chlorophyll molecules are embedded in these membranes. Because of their shape, many chlorophyll molecules can be fitted together like a stack of saucers. Each thylakoid is filled with a myriad of precisely oriented chlorophyll molecules, something like the plates in a storage battery. This arrangement probably permits the generation and utilization of minute electric currents that power photosynthesis.

The CH_3 group shown in blue distinguishes chlorophyll *a* from chlorophyll *b*, which has a slightly different structure in this region.

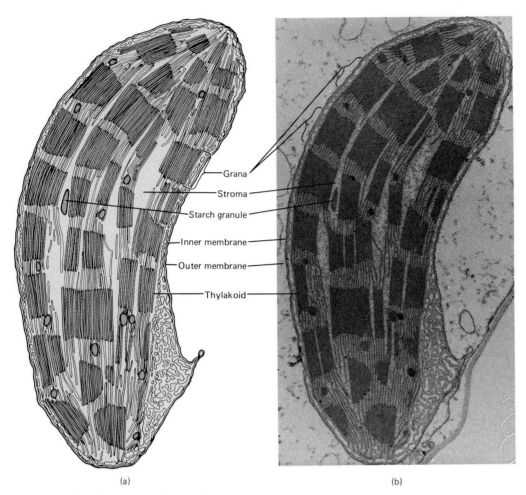

(a) (b)

Figure 7–3 Electron micrograph and diagram of a chloroplast from a leaf cell of a timothy plant, *Phleum pratense*. Note grana, which consist of stacks of thylakoids. (L.K. Shumway, Photo Researchers, Inc.)

carrying on photosynthesis. In prokaryotes that photosynthesize, thylakoids often occur as extensions of the cell membrane and may be arranged around the periphery of the cell. In photosynthetic eukaryotes, thylakoids are found within **chloroplasts** (Chapter 4). All the enzymes required for the light-dependent reactions of photosynthesis are associated with the thylakoid membranes.

Let us examine a chloroplast and its component thylakoids. Surrounded by two membranes, the chloroplast has an interior packed with stacks of thylakoids (Fig. 7–3). These stacks are referred to as **grana.** Each granum looks something like a stack of coins, and each "coin" is a thylakoid (Fig. 7–4). Some thylakoid membranes extend from one granum to another. The fluid-filled region of the chloroplast outside the thylakoids is called the **stroma.** Most of the enzymes required for the light-independent reactions of photosynthesis are found in the stroma.

THE REACTIONS OF PHOTOSYNTHESIS

The principal raw materials for photosynthesis are water and carbon dioxide. Using the energy that chlorophyll molecules trap from sunlight, water is split, its oxygen is liberated (Fig. 7–5), and its hydrogen is combined with carbon dioxide to produce carbohydrate molecules. The reactions of photosynthesis may be summarized as follows:

$$6CO_2 + 12H_2O \xrightarrow[\text{chlorophyll}]{\text{Light energy}} C_6H_{12}O_6 + 6O_2 + 6H_2O$$

Carbon   Water   Glucose   Oxygen   Water
dioxide

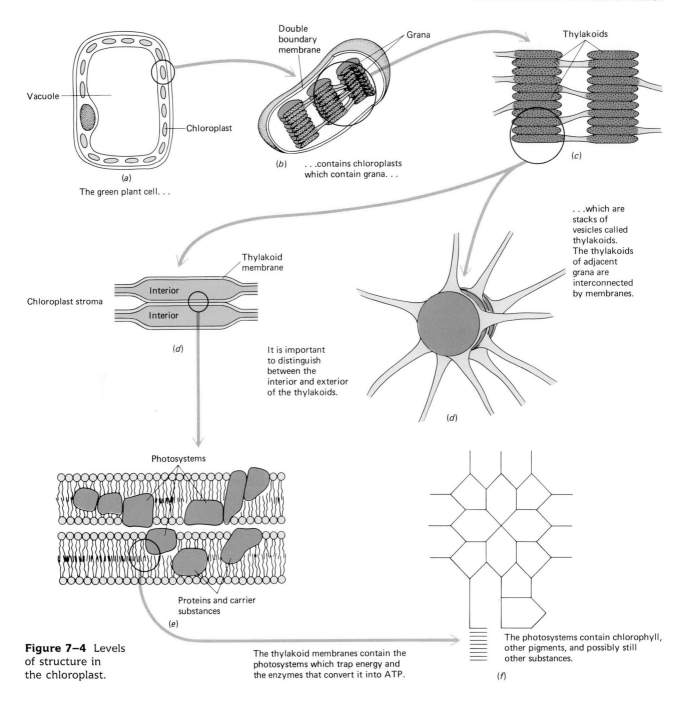

Vacuole

Chloroplast

(a)
The green plant cell. . .

Double boundary membrane

Grana

(b) . . .contains chloroplasts which contain grana. . .

Thylakoids

(c)

. . .which are stacks of vesicles called thylakoids. The thylakoids of adjacent grana are interconnected by membranes.

Thylakoid membrane

Chloroplast stroma

Interior

Interior

(d)

It is important to distinguish between the interior and exterior of the thylakoids.

(d)

Photosystems

Proteins and carrier substances

(e)

The photosystems contain chlorophyll, other pigments, and possibly still other substances.

(f)

Figure 7–4 Levels of structure in the chloroplast.

The thylakoid membranes contain the photosystems which trap energy and the enzymes that convert it into ATP.

Figure 7–5 On sunny days the oxygen released by aquatic plants may sometimes be visible as bubbles in the water. This plant (*Elodea*) is actively carrying on photosynthesis, as evidenced by the oxygen bubbles. (E.R. Degginger.)

Table 7–1
SUMMARY OF PRINCIPAL REACTIONS OF PHOTOSYNTHESIS

Reaction series	Summary of process	Needed materials	End products
Light-dependent reactions (take place in thylakoid membranes)	Energy from sunlight used to split water, manufacture ATP, and reduce NADP		
(1) photochemical reactions	Chlorophyll energized; reaction center gives up energized electron to electron acceptor	Light energy; pigments, such as chlorophyll	Electrons
(2) electron transport	Electrons are transported along chain of electron acceptors in thylakoid membranes; electrons reduce NADP; splitting of water provides some of H^+ that accumulates inside thylakoids	Electrons, NADP, H_2O, electron acceptors	$NADPH + H^+$, O_2, H^+
(3) chemiosmosis	H^+ are permitted to move across the thylakoid membrane down proton gradient; they cross membrane through special channels formed by protein complex CF_0-CF_1; energy released is used to produce ATP	Proton gradient, $ADP + P_i$	ATP
Light-independent reactions (take place in stroma)	Carbon fixation; carbon dioxide is combined with organic compound	Ribulose bisphosphate, CO_2, ATP, $NADPH + H^+$	Carbohydrates, $ADP + P_i$, $NADP^+$

This equation describes what happens during photosynthesis but not how it happens. The "how" is much more complex and involves many chemical reactions. Although a detailed examination of photosynthesis is beyond the scope of a beginning biology course, we will describe the principal reactions (Table 7–1). For convenience we can divide the reactions of photosynthesis into the light-dependent and the light-independent reactions.

The Light-Dependent Reactions

The light-dependent reactions can take place only in the presence of light. During this phase of photosynthesis several important events occur.

1. Chlorophyll absorbs light energy, which is immediately converted to electrical energy. The electrical energy is represented in the flow of electrons from the chlorophyll molecule. We say that the chlorophyll becomes temporarily energized.
2. Some of the energy of the energized chlorophyll is used to make ATP. During this process electrical energy is transformed to chemical energy.
3. Some of the photon energy trapped by the chlorophyll is used to split water, a process known as **photolysis.** Oxygen from the water is released. Some of the oxygen is used by the plant for cellular respiration, but most of it is released into the atmosphere.
4. Hydrogen from the water combines with the hydrogen carrier molecule $NADP^+$, forming NADPH (reduced NAD). Here again, electrical energy is converted to chemical energy.

We may summarize the light-dependent reactions as follows:

$$12H_2O + 12NADP + 18ADP + 18P_i \xrightarrow[\text{chlorophyll}]{\text{Light energy}}$$
$$6O_2 + 12NADPH + 12H^+ + 18ATP$$

Thus, in the light-dependent reactions the energy from sunlight is used to make ATP and to reduce NADP. Some of the captured energy is temporarily stored within these two compounds.

THE PHOTOSYSTEMS

According to the currently accepted model, chlorophyll molecules and associated electron acceptors are organized into units called **photosystems** (Fig. 7–6). There are two photosystems, each containing up to 400 molecules of chlorophyll. Photosystem I contains a reactive pigment (perhaps a special form of chlorophyll *a*) known as P700 because one of the peaks of its light absorption spectrum is at 700 nanometers. Photosystem II utilizes a pigment, P680, whose absorption maximum is at a wavelength of 680 nanometers.

All chlorophyll molecules of a photosystem apparently serve as antennae to gather solar energy. When they absorb light energy, it is passed from one chlorophyll molecule to another until it reaches the special P700 or P680 pigment molecule, referred to as the **reaction center.** Only this molecule is able to give up its energized electron to an electron acceptor compound.

In photosystem I the energized electron is first transferred to a compound known as bound ferredoxin, then to a second electron acceptor, and finally to the electron acceptor NADP$^+$ (Fig. 7–7). When NADP$^+$ is in an oxidized acceptance condition, it is positively charged. When it accepts electrons, the electrons unite with protons to form hydrogen, so the reduced form of NADP$^+$ is NADPH. The reduction of NADP$^+$ requires two electrons. Thus two photons of light must be absorbed by photosystem I to form one NADPH. (Ferredoxin, however, can only transfer one electron at a time.) When an electron is transferred to ferredoxin, photosystem I becomes positively charged. It could never emit another electron, whether or not it became excited, unless one somehow were first restored to it. That needed electron is donated by photosystem II.

Like photosystem I, photosystem II is activated by a photon and gives up an electron to a chain of electron acceptors (Fig. 7–7). The missing electrons are replaced by electrons from water. When P680 absorbs light energy, it becomes positively charged and exerts a strong pull on the electrons in water molecules. Water is split (photolysis) into its components—protons (H$^+$), electrons, and oxygen. The protons from water are transferred to NADP, which explains how the NADP is converted to NADPH + H$^+$. The oxygen split from the water is released into the atmosphere.

The electron emitted from photosystem II passes from one acceptor to another through a chain of easily oxidized and reduced compounds. (Among these are a coenzyme related to vitamin K; iron-containing pigments known as cytochromes; and a copper-containing protein.) As electrons are transferred along this chain of electron acceptors, they become less and less energized. Some of the energy released is used to establish a proton gradient, which leads to the synthesis of ATP (see next section). An electron emitted from photosystem II is eventually donated to photosystem I.

The process just described is known as **noncyclic photophosphorylation.** It is not a cycle because there is a one-way flow of electrons to

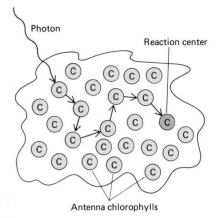

Figure 7–6 A photosynthetic unit. The many chlorophyll molecules (*c*) in the unit are excited by photons and transfer their excitation energy to the specially positioned chlorophyll molecule (*dark green*), the reaction center.

Figure 7–7 Light-dependent reactions (noncyclic photophosphorylation). Light energy absorbed by photosystem I results in the transfer of electrons from photosystem I to NADP. When photosystem II is activated by absorbing a photon, electrons are passed along an electron acceptor chain and are eventually donated to photosystem I. Photosystem II is responsible for the photolytic dissociation of water and the production of atmospheric oxygen.

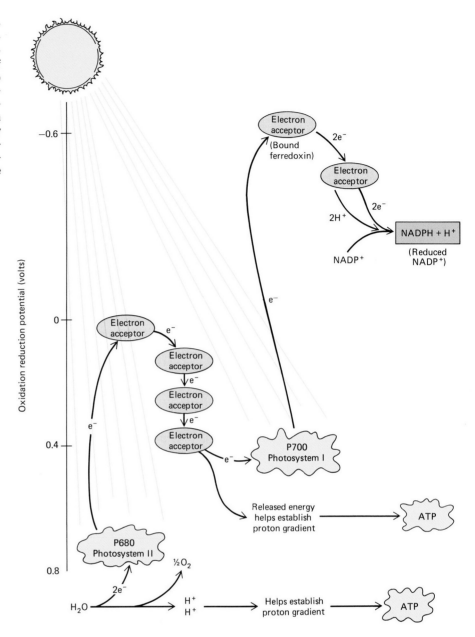

NADP, and it is called photophosphorylation because electrons obtain energy from photons and then contribute that energy to the phosphorylation of ADP (producing ATP). For every two electrons that enter the pathway of noncyclic photophosphorylation, there is an energy yield of two ATP molecules and one NADPH.

CYCLIC PHOTOPHOSPHORYLATION

Only photosystem I is involved in cyclic photophosphorylation. In this pathway the electrons originate from photosystem I and eventually are returned to the same photosystem (Fig. 7–8). For every two electrons that enter the pathway, one ATP molecule is produced. NADPH is not produced, which means that there is a far lower energy yield and that carbon dioxide is not reduced to produce carbohydrates. Photolysis does not occur in cyclic photophosphorylation, so oxygen is not produced in this process.

Cyclic photophosphorylation occurs in plant cells when there is too little NADP to accept electrons from ferredoxin. Biologists think that this

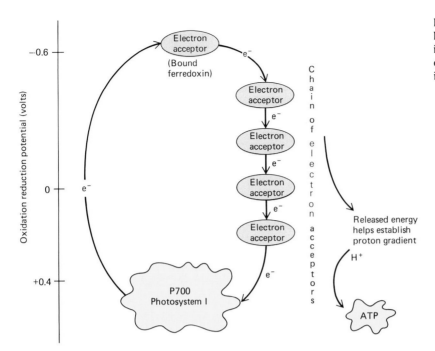

Figure 7–8 Cyclic photophosphorylation. Note that only photosystem I is involved. Photolysis does not occur, so neither oxygen nor NADPH is produced.

process was used by ancient bacteria to produce ATP from light energy. A reaction pathway analogous to cyclic photophosphorylation in plants is present in modern photosynthetic bacteria.

HOW PROTON GRADIENTS FORM ATP: CHEMIOSMOSIS

The photosystems and electron acceptors are embedded in the thylakoid membrane. According to the **chemiosmotic theory,** energy released from electrons traveling through the chain of acceptors is used to pump protons across the thylakoid membrane into the thylakoid. These protons accumulate within the thylakoids. Because protons and hydrogen ions (H^+) are the same, this accumulation of protons causes the pH of the thylakoid interior to fall. In fact, the pH approaches 4 in bright light. This produces a pH difference of about 3.5 units across the thylakoid membrane—more than a thousand-fold difference in hydrogen ion concentration.

In accordance with the general principles of diffusion, the highly concentrated hydrogen ions inside the thylakoid tend to diffuse out. However, they are prevented from doing so because the thylakoid membrane is impermeable to them except at certain points bridged by a remarkable protein, the **CF_0-CF_1 complex,** also called **ATP synthetase.** This complex extends across the thylakoid membrane, projecting from the membrane surface both inside and outside. These proteins actually form channels through which protons *can* leak out of the thylakoids (Fig. 7–9).

Note that the concentrated proton solution inside the thylakoids represents a low entropy state. If the protons can move out of the thylakoids into the surrounding space, the resulting random and more or less even distribution of protons within the chloroplast would be a high entropy state. The second law of thermodynamics identifies this situation as one in which there is a potential for useful work resulting from the change in entropy. What happens is that as the protons pass through the CF_0-CF_1 complex, energy is released, and in some way the complex causes ATP to be synthesized. Just how this is done is now the subject of active investigation.

Figure 7–9 The proton gradient across the thylakoid membrane.

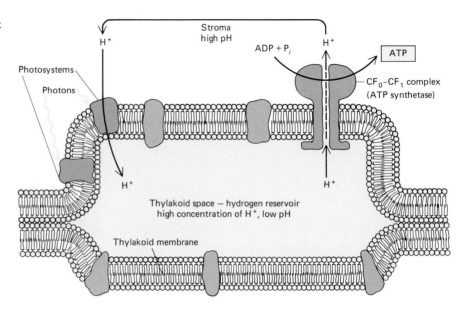

The Light-Independent Reactions: Carbon Fixation

Although the light-independent reactions do not *require* light, they do depend upon the products of the light-dependent reactions. In the light-independent reactions, the energy of the reduced NADPH and the ATP produced during the light-dependent phase of photosynthesis is used to manufacture carbohydrate from the carbon dioxide of the air and hydrogen split off from water. The process may be summarized as follows:

$$12NADPH + 12H^+ + 18ATP + 6CO_2 \longrightarrow$$
$$C_6H_{12}O_6 + 12NADP^+ + 18ADP + 18P_i + 6H_2O$$

The reactions of the light-independent phase actually proceed by way of a cycle known as the **Calvin cycle.** With each complete cycle a portion of a carbohydrate molecule is produced. Six turns of the cycle are required to make just one glucose molecule.

During the light-independent phase, chemical energy from the ATP and NADPH produced during the light-dependent phase is transferred to the chemical bonds of carbohydrate molecules. This form of energy packaging is more suitable for long-term storage. Some of the carbohydrate molecules produced during the light-independent phase are used as fuel molecules. Others are used to manufacture various types of organic compounds needed by the plant cells. For example, by adding such minerals as nitrates and sulfur from the soil, plant cells can produce proteins.

CARBON DIOXIDE FIXATION: SOME DETAILS

As you read the following description of the Calvin cycle, follow the reactions illustrated in Figure 7–10. We will consider that the cycle begins with a pentose sugar phosphate called **ribulose phosphate.** In the first chemical transformation, ATP from the light-dependent reactions is expended to add a second phosphate to the five-carbon skeletons of three ribulose phosphates. This chemical reaction converts them to **ribulose bisphosphate (RuBP).** A key enzyme then combines three molecules of carbon dioxide with three molecules of RuBP. This is called **CO_2 fixation.**

Instantly each molecule splits into two three-carbon molecules called phosphoglycerate (PGA). There are now six three-carbon molecules. With the energy from more ATP from the light-dependent phase the PGA molecules are converted to molecules of diphosphoglycerate (DPGA). Next, DPGA is converted to **glyceraldehyde-3-phosphate,**

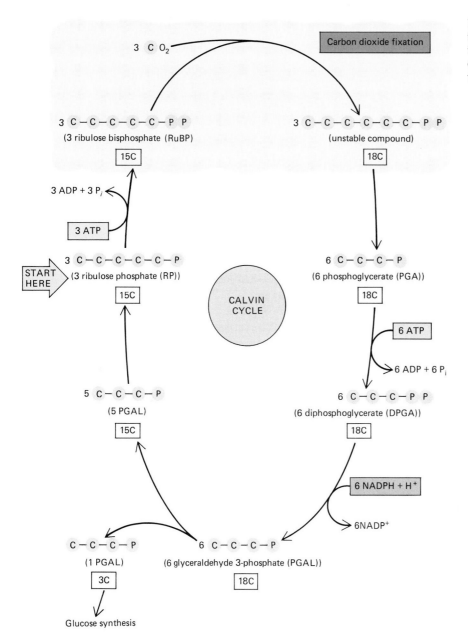

Figure 7–10 The Calvin, or light-independent, reactions. Carbon dioxide is fixed; that is, brought into chemical combination by these reactions.

known simply as **PGAL.** In this conversion hydrogen is donated by reduced NADP formed during the light-dependent reactions.

Now the plant cell is in a position to harvest the three CO_2 molecules it has put into the process, not in their original form but as carbohydrate. One of the six PGAL molecules leaves the system to be used in carbohydrate synthesis and perhaps ultimately in the synthesis of other organic molecules, such as proteins or fats. Each of these three-carbon molecules of PGAL is essentially half a hexose (six-carbon sugar) molecule. They are joined in pairs to produce, usually, glucose or fructose. In some plants glucose and fructose are joined to produce sucrose, table sugar. This we harvest from sugarcane, sugar beets, or maple sap. The plant cell might use glucose to produce cellulose or package it as starch.

Notice that although one of the six PGAL molecules was removed from the cycle, five of them remained. This represents 15 carbon atoms in all. Through an ingenious series of reactions, these 15 carbons and their associated atoms are rearranged into three molecules of the very five-carbon compound, ribulose phosphate, with which we started. Now this same ribulose phosphate is in a position to begin the process of CO_2 fixation and eventual PGAL production once again.

Figure 7–11 (*a*) Overview of the C_4 series of reactions in photosynthesis. If, as usual, CO_2 is present in low concentration, the C_4 system readily absorbs it and in effect concentrates it for use by the C_3 system to which it is pumped. Since the C_4 system consumes some energy, ultimately made available only by photosynthesis, this system is important to the plant only at high light intensities when the stomata can be kept open despite those high light intensities. Thus, to be effective, the C_4 system requires an abundance of both water and light. Yet under these conditions, it can fix more carbon than the C_3 system can fix by itself. (*b*) Comparison of the leaf structures of a C_3 and a C_4 plant.

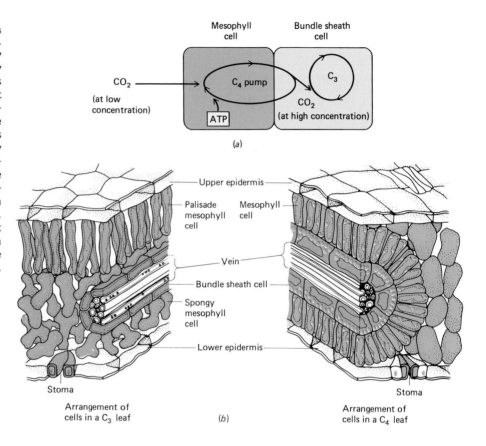

In summary, the inputs required for the light-independent reactions are three molecules of CO_2, hydrogen from the photolysis of water, and ATP. In the end the three carbons from the CO_2 can be accounted for by the harvest of a three-carbon (half-hexose) molecule. The remaining organic compounds are used to synthesize the pre-existing substances with which yet another three CO_2 molecules may combine once again.

THE FOUR-CARBON PATHWAY

Most plants are referred to as C_3 plants because the first product of CO_2 fixation is a three-carbon compound, PGA. Some plants, referred to as C_4 plants, are able to fix carbon dioxide in the four-carbon compound known as **oxaloacetate** (Fig. 7–11). After the oxaloacetate is produced, it moves to adjacent cells, where it is oxidized with the release of carbon dioxide. This carbon dioxide then reacts with ribulose bisphosphate and finally enters the Calvin cycle.

What is the advantage of the C_4 pathway? Its essential feature is that it concentrates carbon dioxide in the cells that carry on photosynthesis by the C_3 pathway. Carbon dioxide enters the leaf through tiny pores called stomata that open and close in response to such factors as water content and light intensity. When the stomata are closed, the supply of carbon dioxide is greatly diminished, and in C_3 plants photosynthesis is significantly slowed. However, the enzyme **phosphoenolpyruvate (PEP) carboxylase,** which catalyzes the formation of oxaloacetate, has a high affinity for carbon dioxide and binds it effectively even when carbon dioxide is at low concentration. Thus in C_4 plants photosynthesis, and ultimately growth, can continue efficiently even under certain adverse conditions. Many quick-growing and aggressive plants possess the C_4 pathway, among them sugarcane, corn, and crabgrass. In crop plants the C_4 pathway makes high yields possible. If this pathway could be incorporated into more of our crop plants by genetic manipulation, we might well be able to greatly increase food production in some parts of the world.

CHAPTER SUMMARY

I. During photosynthesis light energy is captured by chlorophyll and used to chemically combine the hydrogen from water with carbon dioxide to produce carbohydrates; oxygen is released as a by-product.

II. Light behaves as both a wave and a particle. Its particles of energy, called photons, can excite pigment molecules, such as chlorophyll.

III. Chlorophyll and other photosynthetic pigments are found within the membranes of thylakoids.

   A. In prokaryotes that photosynthesize, thylakoids are found as extensions of the cell membrane.

   B. In eukaryotes thylakoids are found within chloroplasts. The thylakoids are arranged in stacks called grana.

IV. During the light-dependent reactions of photosynthesis chlorophyll absorbs light and becomes energized. Some of the energy of the energized chlorophyll is used to make ATP, some is used to split water. Hydrogen from the water is transferred to NADP.

   A. Chlorophyll molecules and associated electron acceptors are organized into photosystems. Only the reaction center, a special pigment molecule, actually gives up its energized electron to an electron acceptor compound.

   B. When photosystem I emits an electron, it is transferred to an electron acceptor called ferredoxin and then to $NADP^+$.

   C. In noncyclic photophosphorylation the electrons emitted by photosystem II are passed through a chain of electron acceptors. A proton flow is set up that provides the energy for ATP synthesis. Electrons from photosystem II are donated to photosystem I.

      1. As electrons pass through the chain of electron acceptors in photosystem II, protons follow them and accumulate in the thylakoids.

      2. The protons can leak out of the thylakoids only through channels in the CF_0-CF_1 protein complex.

      3. As the protons leak through the protein complex, energy is released and is used to synthesize ATP.

   D. In cyclic photophosphorylation electrons from photosystem I are eventually returned to the same photosystem; for every two electrons that enter the pathway, one ATP is produced. No NADPH is produced.

V. During the light-independent reactions, energy stored within ATP and reduced NADP during the light-dependent reactions is used to chemically combine carbon dioxide with hydrogen.

   A. The light-independent reactions proceed via the Calvin cycle.

   B. In the Calvin cycle, carbon dioxide is combined with ribulose phosphate, a five-carbon sugar. With each turn of the cycle, one carbon atom enters the cycle. At each turn of the cycle ribulose phosphate is regenerated.

   C. Three turns of the cycle result in the synthesis of a three-carbon compound known as PGAL. Two molecules of PGAL (produced with six turns of the cycle) can be used to produce a molecule of glucose.

VI. In the C_4 pathway the enzyme PEP carboxylase binds CO_2 effectively even when CO_2 is at low concentration. The carbon dioxide is fixed in oxaloacetate. When the oxaloacetate is oxidized, the CO_2 released enters the Calvin cycle.

Post-Test

1. Light is composed of particles of energy called _____ .

2. Chlorophyll is located within the membranes of _____ .

3. In photolysis some of the energy captured by chlorophyll is used to _____ _____ .

4. Only the reaction center of a photosystem actually gives up its _____ to the _____ _____ compound.

5. In photophosphorylation the _____ from photons is used to add phosphate to _____ , producing _____ .

6. In _____ photophosphorylation there is a one-way flow of electrons to NADP.

7. For every two electrons that enter the noncyclic photophosphorylation pathway, two _____ molecules and one _____ are produced.

8. In cyclic photophosphorylation the electrons that originate from photosystem I are eventually transferred _____ .

9. According to the chemiosmotic theory, energy released from electrons is used to pump _____ across the _____ membrane.

10. As protons pass through the CF_0-CF_1 complex, _____ is released and _____ is synthesized.

11. The inputs for the light-independent reactions are _____ , _____ , and _____ .

12. The process of _____ _____ involves the chemical combination of carbon dioxide with ribulose bisphosphate.

13. _____ turns of the Calvin cycle are required to produce one glucose molecule.

14. When carbon dioxide is a limiting factor, plants that have the _____ pathway have an advantage.

Review Questions

1. What is the role of light in photosynthesis? Explain.
2. Explain the specific role of chlorophyll in photosynthesis.
3. Photosynthetic prokaryotes do not have chloroplasts. How do they manage?
4. List the principal light-dependent reactions of photosynthesis.
5. What is the function of antenna chlorophylls? What is the function of the reaction center of a photosystem?
6. How is oxygen produced during photosynthesis?
7. According to the chemiosmotic theory, how is ATP produced?
8. Summarize the events of the light-independent reactions.
9. What advantage do C_4 plants have over C_3 plants?

Readings

Bjorkman, O., and J. Berry. "High-Efficiency Photosynthesis," *Scientific American,* October 1973, 80–93. Description of the C_4 pathway, the structure of the leaf cells and the biochemistry of the pathway of carbon dioxide.

Dickerson, Richard E. "Cytochrome c and the Evolution of Energy Metabolism," *Scientific American,* March 1980, 136–153.

Hatch, M.D., and N.K. Boardman. *Photosynthesis.* New York, Academic Press, 1981. A detailed description of the chemistry and physics underlying the photosynthetic process. A valuable source book.

Hinckle, P.C., and R.E. McCarty, "How Cells Make ATP," *Scientific American,* March 1978, 104–123. An interesting presentation of the chemiosmotic theory and how it may explain both photosynthesis and oxidative phosphorylation.

Holum, J.R. *Fundamentals of General, Organic, and Biological Chemistry.* New York, John Wiley and Sons, 1982. An excellent reference book for those who would like to learn more about basic chemistry and its application to biology.

Lehninger, A.L. *Principles of Biochemistry.* New York, Worth, 1982. A standard biochemistry text with a detailed but very clear account of bioenergetics, photosynthesis, and cellular respiration.

Miller, K.R. "The Photosynthetic Membrane," *Scientific American,* October 1979, 102–113. Describes the structure of the thylakoid membrane and how it is adapted to convert light energy to chemical energy.

Stoeckenius, W. "The Purple Membrane of Salt-Loving Bacteria," *Scientific American,* June 1976, 38–50. A curious photosynthetic mechanism that uses a pigment rather like the visual purple of animals instead of chlorophyll in capturing the energy of light.

Stryer, L. *Biochemistry,* 2nd ed. San Francisco, W.H. Freeman, 1981. A well-illustrated, readable text that covers the molecules of life and the concepts of cellular energetics from the ground up.

Chapter 8

ENERGY-RELEASING PATHWAYS: CELLULAR RESPIRATION AND FERMENTATION

Learning Objectives

After you have read this chapter you should be able to:

1. Write a summary reaction for cellular respiration, giving the origin and fate of each substance involved.
2. Summarize the events of glycolysis, giving the key organic compounds formed and the number of carbon atoms in each; indicate the number of ATP molecules used and produced and the transactions in which hydrogen loss occurs.
3. Summarize the events of the citric acid cycle, beginning with the conversion of pyruvate to acetyl-CoA; indicate the fate of carbon-oxygen segments and of hydrogens removed from the fuel molecule.
4. Summarize the operation of the electron transport system, including the reactions by which a gradient of protons is established across the inner mitochondrial membrane; explain how the proton gradient drives ATP synthesis.
5. Compare aerobic respiration with anaerobic pathways in terms of ATP formation, final hydrogen acceptor, and end products; give two specific examples of anaerobic pathways.

During photosynthesis light energy from the sun is converted into chemical energy stored within the bonds of organic molecules, such as carbohydrates. Most cells depend upon these fuel molecules for the continuous input of energy required to carry on life processes (Fig. 8–1). Although some ATP is manufactured during the light-dependent reactions of photosynthesis, that ATP is used up during the light-independent reactions. Thus even plant cells must depend upon energy-releasing pathways that enzymatically break down organic molecules. Those pathways are collectively referred to as **cellular respiration,** the process by which a living cell breaks down a fuel molecule to carbon dioxide and water, releasing the energy trapped within its chemical bonds.

THE REACTIONS OF CELLULAR RESPIRATION

One of the principal pathways of cellular respiration is that of the common nutrient glucose. The overall reaction for the metabolism of glucose follows:

$$C_6H_{12}O_6 + 6O_2 \longrightarrow 6CO_2 + 6H_2O + Energy$$

Glucose   Oxygen   Carbon dioxide   Water

The cellular oxidation of glucose to carbon dioxide and water cannot occur in a single reaction, for there are no enzymes that can catalyze the direct attack of oxygen on glucose. Instead, the oxidation of glucose occurs in a sequence of reactions that can be grouped into four phases (Fig. 8–2; Table 8–1):

1. Glycolysis, the conversion of the 6-carbon glucose molecule to two three-carbon molecules of pyruvate[1] with the formation of two ATPs.
2. Formation of acetyl coenzyme A. Pyruvate is degraded to a two-carbon fuel molecule and combines with coenzyme A, forming acetyl coenzyme A; carbon dioxide is released.
3. The citric acid cycle, which converts acetyl coenzyme A to carbon dioxide and removes electrons and hydrogens.
4. The electron transport system and chemiosmotic phosphorylation. The hydrogens and electrons removed from the fuel molecule during the preceding phases are transferred along a chain of electron acceptor compounds. As the electrons are passed from one electron acceptor to another, a proton gradient is established, which sets the stage for ATP production by chemiosmosis.

Figure 8–1 Some of the organic compounds obtained in food are oxidized during cellular respiration. The energy released is used to carry on life processes. (Animals, Animals, David C. Fritts.)

---

[1] Pyruvate and many other compounds in glycolysis and the citric acid cycle exist as ions at the pH found in the cell. They sometimes associate with H^+ to form acids; for example, pyruvate forms pyruvic acid. In some textbooks these compounds are presented in the acid form.

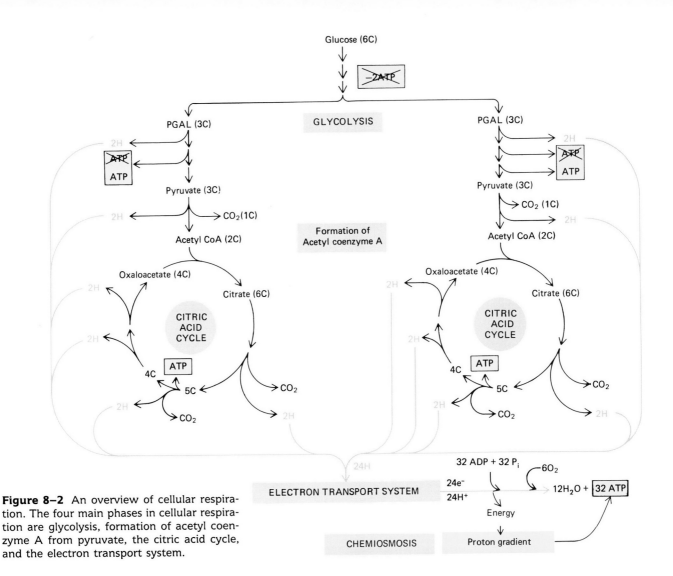

Figure 8-2 An overview of cellular respiration. The four main phases in cellular respiration are glycolysis, formation of acetyl coenzyme A from pyruvate, the citric acid cycle, and the electron transport system.

Table 8-1
SUMMARY OF CELLULAR RESPIRATION

Phase	Summary	Needed materials	End products
Glycolysis (takes place in cytoplasm)	Series of about 10 reactions during which glucose is degraded to pyruvate; net profit of 2 ATPs; hydrogens released; can proceed anaerobically	Glucose, 2 ATPs	Pyruvate, ATP, hydrogen
Formation of acetyl CoA (takes place in mitochondria)	Pyruvate is degraded, and combined with CoA to form acetyl CoA; CO_2 is released	Pyruvate, CoA, NAD	Acetyl CoA, NAD—H, CO_2
Citric acid cycle (takes place in mitochondria)	Series of reactions in which fuel molecule (part of acetyl coenzyme A) is degraded to hydrogen and carbon dioxide; aerobic	Acetyl CoA, H_2O, hydrogen acceptors (e.g., NAD)	CO_2, H, reduced hydrogen acceptors, CoA
Electron transport and chemiosmosis (take place in mitochondria)	Chain of several electron transport molecules; hydrogens (or their electrons) are passed along chain; energy released is used to generate proton gradient across inner mitochondrial membrane; as protons move through ATP synthetase in membrane, ATP is synthesized; for each pair of hydrogens that enter chain, maximum of 3 ATPs can be synthesized; aerobic	Hydrogen, ADP, P_i, oxygen	ATP, water

Glycolysis

The sequence of reactions that convert glucose to pyruvate with the production of ATP is termed **glycolysis.** Each reaction is regulated by a specific enzyme, and there is a net gain of two ATP molecules. The reactions of this pathway take place in the cytoplasm, and these reactions can proceed anaerobically, that is, in the absence of oxygen.

In a highly simplified explanation of glycolysis we can divide the process into two major phases. In the first phase the glucose molecule is split to form two molecules of the three-carbon compound glyceraldehyde-3-phosphate, or **PGAL.** (Note that this is the same compound formed during the light-independent reactions of photosynthesis.) This transformation requires energy, so the cell must invest two molecules of ATP in order to initiate the oxidation of glucose. We may summarize this step of glycolysis as follows:

$$\underset{\substack{\text{6-carbon} \\ \text{compound}}}{\text{Glucose}} + 2\text{ATP} \rightarrow\rightarrow\rightarrow \underset{\substack{\text{3-carbon} \\ \text{compound}}}{2\text{PGAL}} + 2\text{ADP} + 2\text{P}$$

Several arrows are used to indicate that the equation summarizes a sequence of several reactions.

In the second phase of glycolysis two hydrogen atoms are removed from each PGAL (oxidizing the PGAL), and certain other atoms are rearranged so that each molecule of PGAL is transformed into a molecule of pyruvate. During these reactions enough chemical energy is released from the fuel molecule to produce four ATP molecules.

$$\underset{\substack{\text{3-carbon} \\ \text{compound}}}{2\text{PGAL}} + 4\text{ADP} + 4\text{P}_i \rightarrow\rightarrow\rightarrow\rightarrow \underset{\substack{\text{3-carbon} \\ \text{compound}}}{2\text{Pyruvate}} + 4\text{H} + 4\text{ATP}$$

Note that in the first phase of glycolysis two molecules of ATP are consumed, but in the second phase four molecules of ATP are produced. Thus glycolysis yields a net energy profit of two ATP molecules.

The two hydrogen atoms removed from each PGAL immediately combine with the hydrogen carrier molecule, NAD:

$$\underset{\text{Oxidized NAD}}{\text{NAD}^+} + 2\text{H} \longrightarrow \underset{\text{Reduced NAD}}{\text{NAD—H} + \text{H}^+}$$

The fate of these hydrogen atoms will be discussed with the electron transport system and with anaerobic pathways.

The reactions of glycolysis take place in the cytoplasm. Such necessary ingredients as ADP, NAD, and phosphates float freely in the cytoplasm and are used as needed. For a more detailed description of glycolysis, see Focus on Glycolysis.

Formation of Acetyl CoA

Pyruvate molecules produced during glycolysis move into the mitochondria, and essentially all subsequent reactions of cellular respiration take place within these tiny power plants. Each three-carbon molecule of pyruvate is degraded by the removal of a carbon-oxygen segment and two hydrogen atoms. The carbon-oxygen segment is removed as CO_2, which eventually is expired (by way of the lungs in complex animals). The pyruvate molecule is converted to a two-carbon acetyl group, which reacts with a large, complex coenzyme called **coenzyme A.** This results in the formation of a compound called **acetyl coenzyme A (CoA).**

$$2\text{pyruvate} + 2\text{CoA} + 2\text{NAD}^+ \longrightarrow$$
$$2\text{acetyl—CoA} + 2\text{NAD—H} + 2\text{H}^+ + 2\text{CO}_2$$

Focus on . . .

GLYCOLYSIS

1. Glucose, a stable six-carbon sugar, receives a phosphate group from an ATP molecule. The ATP serves as a source of both phosphate and the energy needed to attach the phosphate to the glucose molecule. (Once the ATP is spent, it becomes ADP and joins the ADP pool of the cell until turned into ATP again.) Now the glucose has been phosphorylated, thereby becoming glucose-6-phosphate. (Note the phosphate attached to its carbon atom 6.)

2. An enzyme now rearranges the hydrogen and oxygen atoms of the molecule to form fructose-6-phosphate.

3. Next, another ATP donates a phosphate to the molecule, forming fructose-1,6-bisphosphate. So far, two ATP molecules have been invested in the process without any being produced.

glucose (6C)

Hexokinase

ATP
ADP

glucose-6-P (6C)

fructose-6-P (6C)

ATP
ADP

fructose-1 6-bisP (6C)

(Continued)

Focus on . . .

GLYCOLYSIS (Continued)

4. Fructose-1,6-bisphosphate is then split into two three-carbon sugars, glyceraldehyde-3-phosphate (PGAL) and dihydroxyacetone phosphate. Dihydroxyacetone phosphate is enzymatically converted to glyceraldehyde-3-phosphate for further metabolism in glycolysis.

5. Glyceraldehyde-3-phosphate reacts with an SH (sulfhydryl) group in the enzyme, glyceraldehyde-3-phosphate dehydrogenase, forming an H—C—OH group that can undergo dehydrogenation with NAD^+ as hydrogen acceptor. The product of the reaction is phosphoglycerate, still bound to the enzyme. Phosphoglycerate reacts with inorganic phosphate to yield 1,3-diphosphoglycerate and free enzyme. In this oxidation-reduction reaction, glyceraldehyde-3-phosphate has been converted to a diphosphoglycerate, and the newly added phosphate group is attached with an energy-rich bond.

6. The energy-rich phosphate reacts with ADP to form ATP. This transfer of energy from a compound with an energy-rich phosphate is referred to as a substrate-level phosphorylation.

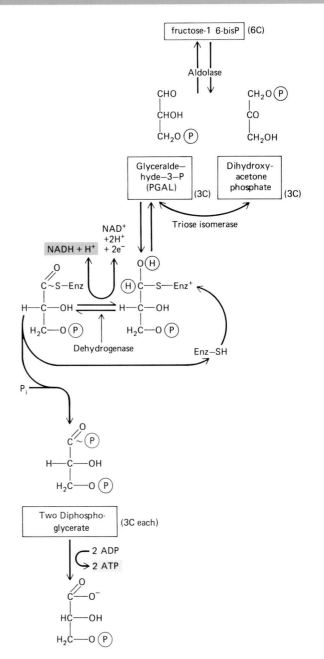

Coenzyme A is manufactured in the cell from one of the B vitamins, pantothenic acid. Fats and amino acids can also be degraded to acetyl CoA and can enter the respiratory pathway at this point. The six carbon atoms from the original glucose molecule may now be accounted for as follows: four of the carbons are located in the two acetyl CoA molecules, and the other two were released as carbon dioxide.

7. The 3-phosphoglycerate is rearranged to 2-phospho-glycerate by the enzymatic shift of the position of the phosphate group.

8. Next, an energy-rich phosphate is generated by the removal of water rather than by the removal of hydrogen atoms. The product is phosphoenol pyruvate (sometimes abbreviated PEP).

9. Phosphoenol pyruvate can transfer its phosphate group to ADP to yield ATP and pyruvate. This is the second energy-rich phosphate group generated at the substrate level in the metabolism of glyceraldehyde-3-phosphate to pyruvate.

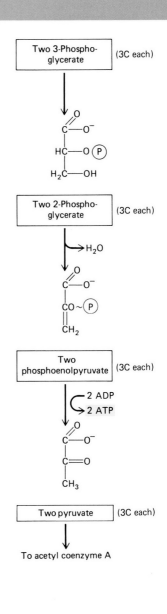

Each glucose metabolized yields two molecules of glyceraldehyde-3-phosphate. A total of four ATPs is produced as one molecule of glucose is metabolized to pyruvate. However, two ATP molecules were utilized at the beginning of the process, one to convert glucose to glucose-6-phosphate and the second to convert fructose-6-phosphate to fructose-1,6-bisphosphate. The net yield in the process is two ATPs (four ATPs produced minus two ATPs used up in the reaction).

The Citric Acid Cycle

The **citric acid cycle** (also known as the **Krebs cycle** after Sir Hans Krebs, who worked it out) is the principal series of reactions in which hydrogen atoms are removed from fuel molecules and transferred to carrier molecules, such as NAD, for further processing in the electron transport sys-

Figure 8–3 The citric acid cycle. During this series of reactions, acetyl coenzyme A, produced from glucose and other organic compounds, is metabolized to yield carbon dioxide and hydrogen. The hydrogen is immediately combined with NAD or FAD and is fed into the electron transport system.

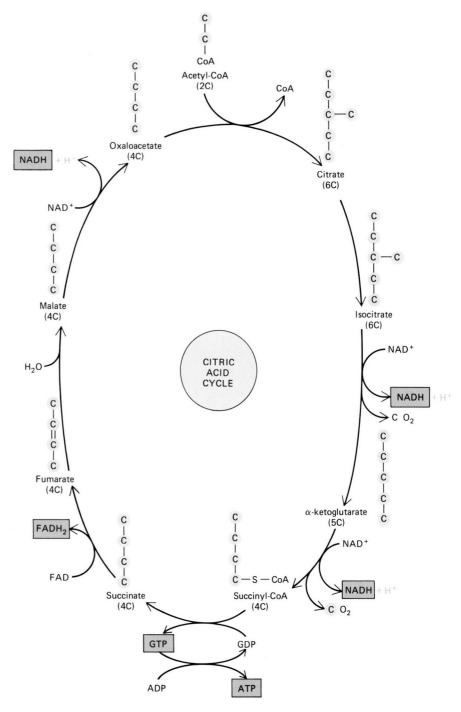

tem. The reactions of the citric acid cycle are summarized here and in Figure 8–3. For a more detailed description of the individual steps, see Focus on the Citric Acid Cycle.

Acetyl CoA enters the citric acid cycle by combining with a four-carbon compound, oxaloacetate (already present in the mitochondrion), to form **citrate,** a six-carbon compound.

$$\text{Oxaloacetate} + \text{Acetyl CoA} \longrightarrow \text{Citrate}$$

4-carbon compound   2-carbon compound   6-carbon compound

The citrate now goes through a series of chemical transformations so that it loses first one, then another carbon-oxygen segment in the form of CO_2. At the end of a complete cycle a four-carbon oxaloacetate is all that

THE CITRIC ACID CYCLE

1. Pyruvate is degraded by the removal of a carbon-oxygen segment as carbon dioxide and by the removal of two hydrogen atoms. The two-carbon fragment that results, called acetyl, is linked to a molecule called coenzyme A. (The hydrogens are picked up by NAD.)
2. Coenzyme A is then released for reuse, and the acetyl fragment enters the citric acid cycle as it combines with a four-carbon oxaloacetate molecule to form citrate, a six-carbon compound with three acid (carboxyl) groups.
3.,4. By two intermediate reactions the atoms of citrate are rearranged to form first *cis*-aconitate and then isocitrate.
5. Now a carbon-oxygen group is split off from the six-carbon compound and is released as carbon dioxide. Two hydrogens are removed at the same time, and the resulting fuel molecule is called alpha-ketoglutarate.
6. Another carbon dioxide molecule and two more

hydrogen atoms are split off from the fuel molecule, leaving a four-carbon compound called succinate.
7. During the complex conversion of succinate to fumarate, enough energy is generated to produce one ATP at the substrate level. During this reaction hydrogens are also released, but they are picked up by the hydrogen carrier molecule FAD rather than by NAD.
8. With the addition of water, fumarate is converted to malate.
9. As malate is converted to oxaloacetate two more hydrogen atoms are removed and are picked up by NAD.

Note that with each turn of the cycle, three carbons are split off from the fuel molecule and released as carbon dioxide. These represent the three carbons that enter the cycle as pyruvate. With two turns of the cycle, six carbons are split off, representing the six carbons of the original glucose.

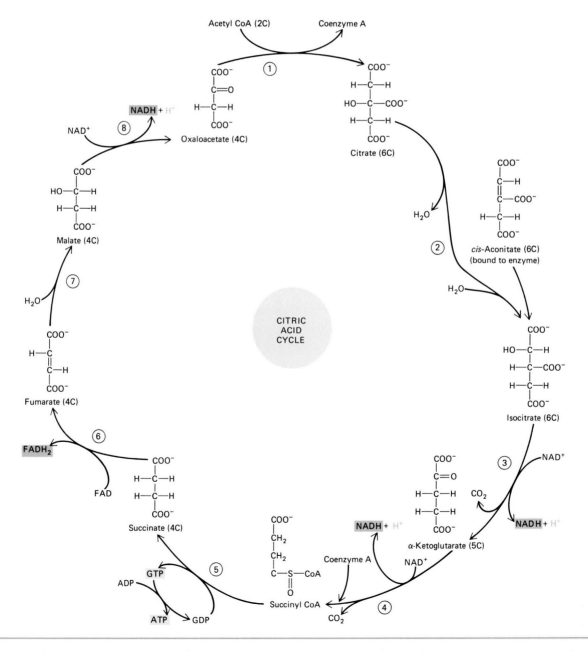

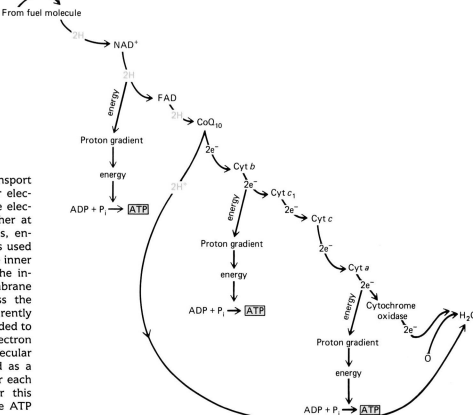

Figure 8-4 The electron transport system. As hydrogens or their electrons are transferred from one electron carrier molecule to another at lower and lower energy levels, energy is released. This energy is used to transport protons across the inner mitochondrial membrane to the intermembranous space. A membrane potential is established across the inner membrane, and this apparently is the source of the energy needed to synthesize ATP. The final electron (and hydrogen) acceptor is molecular oxygen, so water is produced as a product of these reactions. For each pair of hydrogens that enter this pathway, a maximum of three ATP molecules are produced.

is left, and the cycle is ready for another turn. By this time the original pyruvate has lost all of its three carbons, or at least the equivalent, and may be regarded as having been completely consumed. Two turns of the cycle are necessary to process the two pyruvate molecules derived from the original glucose. Only one molecule of ATP is produced directly (by a substrate-level phosphorylation) with each turn of the citric acid cycle. How, then, is most of the ATP produced?

The Electron Transport System and Chemiosmotic Phosphorylation

Now let us consider the fate of all the hydrogens removed from the fuel molecule during glycolysis, acetyl CoA formation, and the reactions of the citric acid cycle. These hydrogens are first transferred to the hydrogen acceptors, NAD or FAD. NAD is the first hydrogen acceptor in the series of compounds that make up the electron transport system (Fig. 8-4). As the hydrogens (or their electrons) are transferred from one to another of a series of some 15 different electron acceptor molecules, the hydrogen protons become separated from their electrons. The protons (H^+) are released into the surrounding medium and set up an electrochemical gradient across the inner mitochondrial membrane. This gradient provides the energy for ATP synthesis (Fig. 8-5). This chemiosmotic mechanism is similar to the mechanism employed by the chloroplast to make ATP.

The electron acceptors of the electron transport chain are embedded within the inner membrane of the mitochondrion. Among the electron acceptors in the chain are **cytochromes,** a group of closely related proteins characterized by a central atom of iron. It is the iron that combines

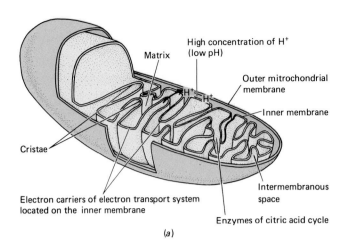

Matrix

High concentration of H$^+$
(low pH)

Outer mitochondrial
membrane

Inner membrane

Cristae

Intermembranous
space

Electron carriers of electron transport system
located on the inner membrane

Enzymes of citric acid cycle

(a)

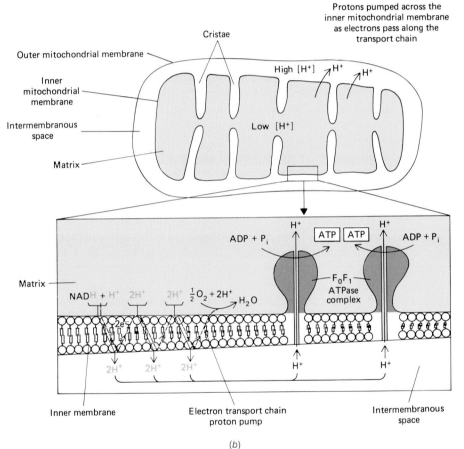

Cristae

Protons pumped across the
inner mitochondrial membrane
as electrons pass along the
transport chain

Outer mitochondrial membrane

Inner
mitochondrial
membrane

High [H$^+$] H$^+$ H$^+$

Intermembranous
space

Low [H$^+$]

Matrix

H$^+$ ATP ATP H$^+$

ADP + P$_i$ ADP + P$_i$

F$_0$F$_1$
ATPase
complex

Matrix

NADH + H$^+$ 2H$^+$ 2H$^+$ $\frac{1}{2}$O$_2$ + 2H$^+$ H$_2$O

2e$^-$

2H$^+$ 2H$^+$ 2H$^+$ H$^+$ H$^+$

Inner membrane

Electron transport chain
proton pump

Intermembranous
space

(b)

Figure 8–5 Chemiosmosis. (*a*) According to the chemiosmotic theory, the electron transport chain in the inner mitochondrial membrane is a proton pump. (*b*) The energy released during electron transport from substrate to oxygen is used to transport protons, H$^+$, from the mitochondrial matrix to the intermembranous space, where a high concentration of protons accumulates. There are three sites in the electron transport chain in which protons are transported. The protons are prevented from diffusing back into the matrix through the inner membrane everywhere except through special pores in ATP-synthetase (ATPase) in the membrane. The flow of the protons through the ATPase generates ATP at the expense of the free energy released as the protons pass from a region of high concentration (outside) to a region of lower concentration (inside).

with the electrons from the hydrogen atoms. Cytochrome molecules accept only the electron from the hydrogen, rather than the entire atom. The last cytochrome in the chain passes the two electrons on to molecular oxygen. Simultaneously, the electrons reunite with protons. The chemical union of the hydrogen and oxygen produces water.

Oxygen is the final hydrogen acceptor in the electron transport system. This vital role of oxygen explains why we must breathe. What happens when the cells are deprived of oxygen? When no oxygen is available to accept the hydrogen, the last cytochrome in the chain is stuck with it. When that occurs, each acceptor molecule in the chain may remain stuck with electrons, and the entire system may be blocked all the way back to NAD. No further ATPs can be produced by way of the electron transport

Table 8-2
THE ENERGY YIELD FROM THE COMPLETE OXIDATION OF GLUCOSE

Net ATP profit from glycolysis		2ATP*
Also from glycolysis,	$2NADH_2 \longrightarrow$	4–6ATP
2 pyruvate to 2 acetyl CoA	$2NADH_2 \longrightarrow$	6ATP
2 acetyl CoA through citric acid cycle		2ATP (substrate level)
	$6NADH_2 \longrightarrow$	18ATP
	$2FADH_2 \longrightarrow$	4ATP
Total ATP profit		36–38ATP

*These are the only two ATPs that can be generated anaerobically. Production of all other ATPs depends upon the presence of oxygen.

system. Most cells of complex organisms cannot live long without oxygen because the amount of energy they can produce in its absence is insufficient to sustain life processes.

Lack of oxygen is not the only factor that may interfere with the electron transport system. Some poisons, including cyanide, inhibit the normal activity of the cytochrome system. Cyanide combines electrons tightly with one of the cytochrome molecules so that they cannot be passed along to the next acceptor in the chain.

As electrons are transferred along the acceptors in the electron transport chain, they are passed down an energy hill. Sufficient energy is released at three points to transport protons across the inner mitochondrial membrane, and ultimately to synthesize ATP. The passage of each pair of electrons from NADH to oxygen yields three ATPs.

Protons pumped outside the inner mitochondrial membrane can flow back to the matrix of the mitochondrion only through special channels in the inner membrane. These channels occur in the enzyme ATP synthetase, also known as the F_0-F_1 complex. As the protons move down an energy gradient, the energy released is used to synthesize ATP (Table 8-2). The ATP synthetase acts as a turbine converting one form of energy into another. The flow of electrons is tightly coupled to the phosphorylation process and will not proceed unless phosphorylation can also occur. Because the phosphorylation of ADP to form ATP is coupled with the oxidation of electron transport compounds, this entire process is referred to as **oxidative phosphorylation.**

REGULATION OF CELLULAR RESPIRATION

Cellular respiration requires a steady input of nutrient fuel molecules and oxygen. Under normal conditions these materials are adequately provided and do not affect the rate of respiration. Instead, the rate of cellular respiration is regulated by the amount of ADP and phosphate available. In a resting muscle cell, for example, ATP synthesis continues until all the ADP has been converted to ATP. Then, when there are no more acceptors of phosphate, phosphorylation must stop. Since electron flow is tightly coupled to phosphorylation, the flow of electrons also stops. When an energy-requiring process like muscle contraction occurs, ATP is split to yield ADP and inorganic phosphate plus energy. The ADP formed can then serve as an acceptor of phosphate and energy to become ATP once again. Oxidative phosphorylation continues until all the ADP has again been converted to ATP. Because phosphorylation is tightly coupled to electron flow, the cell possesses a system of control that can regulate the rate of ATP production and adjust it to the momentary rate of energy utilization.

Although the reactions of glycolysis take place in the cytoplasm, those of the citric acid cycle and the electron transport system occur within the mitochondria. The outer membrane of the mitochondrion engages in active transport and is thought to regulate the intake of materials. Most respiratory enzymes are associated with the inner mitochondrial membrane. It is thought that these enzymes must be exactly located with respect to one another so that hydrogens and electrons may be passed from one to another in the multitudinous and still somewhat bewildering metamorphoses of cellular respiration. No assembly line yet built can match the mitochondrial disassembly line for speed and efficiency. What a pity we cannot see it as it must be, vibrating with thousands of elaborate transformations every second, pouring forth streams of ATP, water, and carbon dioxide as it labors along with thousands like it in every one of the billions of cells of the body.

ANAEROBIC PATHWAYS

When sufficient oxygen is not available, the electron transport system and the citric acid cycle cannot operate, and metabolism may proceed **anaerobically,** that is without oxygen. Recall that the net profit of two ATPs produced during glycolysis does not require the presence of oxygen. When oxygen is present, the hydrogens removed from the fuel molecule during glycolysis pass through the electron transport system. In the absence of oxygen or of required enzymes this pathway is blocked, and the hydrogens must be disposed of in some other way.

Certain types of bacteria engage solely in anaerobic metabolism, a clear advantage to those that inhabit the soil or polluted waters where oxygen is in short supply. Other bacteria are able to shift to anaerobic pathways when deprived of oxygen. In **anaerobic respiration,** an inorganic compound, such as nitrate, serves as the final acceptor of electrons. Some of the bacteria important in recycling nitrogen carry on anaerobic respiration. In **fermentation** an intermediate compound in carbohydrate metabolism is the final acceptor of electrons from NADH. Both anaerobic respiration and fermentation depend upon the reactions of glycolysis.

When deprived of oxygen, yeast cells split carbon dioxide off from pyruvate. Hydrogen from $NADH + H^+$ is transferred to the resulting compound, forming **ethyl alcohol,** or drinking alcohol (Fig. 8–6). Such anaerobic reactions are the basis for the production of beer, wine, and other alcoholic beverages. Yeast cells are also used in the baking industry to produce the carbon dioxide that causes dough to rise.

An alternative pathway for hydrogens removed from the fuel molecule during glycolysis is to transfer them to the pyruvate molecule. When hydrogen atoms are added to pyruvate, **lactate,** the ionic form of lactic acid, is formed.

$$\text{Pyruvate} + 2H \longrightarrow \text{Lactate}$$

Lactate is produced when bacteria sour milk or ferment cabbage to form sauerkraut. It is also produced during muscle activity in the muscle cells of humans and other complex animals. During strenuous physical activity, such as running, the amount of oxygen delivered to the muscle cells may be insufficient to keep pace with the rapid rate of fuel oxidation. Not all of the hydrogen atoms accepted by NAD can be processed in the usual manner because there is a shortage of oxygen. Instead, these hydrogens are transferred to pyruvate to form lactate. As lactate accumulates in muscle cells, it contributes to **muscle fatigue.**

Lactate acidifies the blood, which stimulates respiration, indirectly resulting in rapid breathing. The increased oxygen intake is necessary so

Figure 8–6 Comparison of anaerobic with aerobic pathways. When oxygen is not available, fermentation occurs in certain bacteria, yeast, and muscle cells.

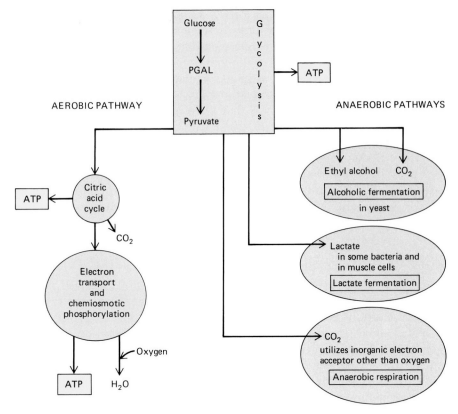

that a portion of the lactate can be oxidized in order to reconvert the rest to glucose. By running faster than the circulatory system can supply oxygen to the muscles, the athlete incurs an oxygen debt that must be repaid when the exertion is over.

Anaerobic metabolism is very inefficient because the fuel is only partially oxidized. Alcohol, the end product of fermentation by yeast cells, can be burned and can even be used as automobile fuel. Obviously, it contains a great deal of energy that the yeast cells were unable to extract using anaerobic methods. Lactate, a three-carbon compound, contains even more energy than the two-carbon alcohol. In contrast, during aerobic respiration all available energy is removed because fuel molecules are completely oxidized. A net profit of only two ATPs can be produced anaerobically from one molecule of glucose, compared with up to 38 when oxygen is available (Table 8–3). The two ATPs produced during glycolysis represent only about 5% of the total energy in a molecule of glucose. About 55% of the energy in a glucose molecule can be captured by using aerobic pathways. The rest is lost as heat. This compares favorably with the finest machines we can make. In humans and other homeothermic (warm-blooded) animals some of the heat produced

Table 8–3 COMPARISON OF AEROBIC AND ANAEROBIC PATHWAYS		
	Anaerobic	**Aerobic**
Maximum ATP profit	2 ATPs	36–38 ATPs
Final hydrogen acceptor	Pyruvate	Molecular oxygen
End products	Ethyl alcohol in yeast, certain bacteria, lactate in muscle cells	CO_2, H_2O

during respiration and other metabolic activities is utilized to maintain a constant body temperature.

The inefficiency of anaerobic metabolism necessitates a large supply of fuel. By rapidly degrading many fuel molecules a cell can compensate somewhat for the small amount of energy that can be gained from each. To perform the same amount of work, an anaerobic cell must consume up to 20 times as much glucose or other carbohydrate as a cell metabolizing aerobically. Skeletal muscle cells, which often metabolize anaerobically for short periods, must therefore store large quantities of glucose in the osmotically inactive form of glycogen.

CHAPTER SUMMARY

I. During cellular respiration a fuel molecule, such as glucose, is oxidized, forming carbon dioxide and water with the release of energy.

II. During glycolysis a molecule of glucose is degraded, forming two molecules of pyruvate.
   A. A net profit of two ATP molecules is gained during glycolysis.
   B. Four hydrogen atoms are removed from the fuel molecule.

III. Pyruvate molecules are combined with coenzyme A, producing acetyl coenzyme A. Carbon dioxide is released during this process.

IV. Two carbons from the original glucose molecule are released as carbon dioxide. The four remaining carbons are present in the two acetyl coenzyme A molecules. Acetyl CoA enters the citric acid cycle by combining with a four-carbon compound (oxaloacetate), forming citrate, a six-carbon compound.
   A. With two turns of the citric acid cycle the two acetyl CoAs are completely degraded.
   B. Carbon dioxide is released and hydrogens are transferred to NAD or FAD. Only one ATP is produced directly with each turn of the cycle.

V. Hydrogen atoms (or their electrons) removed from fuel molecules are transferred from one electron acceptor to another down a chain of acceptor molecules that make up the electron transport system. The final acceptor in the chain is molecular oxygen, which combines with the hydrogen to form water.
   A. According to the chemiosmotic theory, energy liberated in the electron chain is used to establish a proton gradient across the inner mitochondrial membrane.
   B. The flow of protons back through the membrane (via the enzyme ATP synthetase) releases energy, which is used to synthesize ATP.

VI. In the electron transport system phosphorylation is tightly coupled to electron flow. The rate of ATP synthesis depends upon the available ADP and phosphate.

VII. In anaerobic pathways the hydrogens are used to convert pyruvate to ethyl alcohol or lactate. There is a net gain of only two ATPs per glucose molecule, compared with about 38 ATPs per glucose in aerobic respiration.

Post-Test

1. The pathway through which glucose is converted to pyruvate is referred to as _____ .

2. The reactions of glycolysis take place within the _____ , whereas the citric acid cycle takes place within the _____ .

3. During the citric acid cycle the fuel molecule is oxidized to _____ and _____ .

4. The first electron acceptor in the electron transport system is _____ ; the last is _____ .

5. When protons move down an energy gradient in cellular respiration energy is released and used to synthesize _____ .

6. During strenuous muscle activity, the pyruvate in muscle cells may accept hydrogen to become _____ .

7. When deprived of oxygen yeast cells transfer hydrogens to pyruvate forming _____ _____ _____ and releasing _____ _____ _____ .

8. A net profit of only _____ ATPs can be produced anaerobically from one molecule of glucose, compared with the _____ produced in aerobic respiration.

9. The anaerobic process by which alcohol or lactate is produced as a product of glycolysis is referred to as _____ .

10. Anaerobic metabolism is inefficient because the fuel molecule is only partially _____ .

Review Questions

1. What is the role of oxygen in cellular respiration?
2. How does the chemiosmotic theory relate to cellular respiration? How does a proton gradient contribute to ATP synthesis?
3. Why are mitochondria referred to as the power plants of the cell? Be specific.
4. Why are ATPs consumed during the first steps of glycolysis?
5. What are the products of glycolysis?
6. In what form is the fuel molecule when it feeds into the citric acid cycle? What are the products of the citric acid cycle?
7. Trace the fate of hydrogens removed from the fuel molecule during glycolysis when oxygen is present.
8. Trace the fate of hydrogens removed from the fuel molecule when sufficient oxygen is not available in muscle cells.
9. Draw a mitochondrion and indicate the locations of the various events of electron transport and chemiosmosis.

Readings

Dickerson, Richard E. "Cytochrome c and the Evolution of Energy Metabolism," *Scientific American,* March 1980, 136–153.

Hinckle, P.C., and R.E. McCarty. "How Cells Make ATP," *Scientific American,* March 1978, 104–123. An interesting presentation of the chemiosmotic theory and how it may explain both photosynthesis and oxidative phosphorylation.

Holtzman, E., and A.B. Novikoff. *Cells and Organelles,* 3rd ed. Philadelphia, Saunders College Publishing, 1984. An integrated approach to the structural, biochemical, and physiological aspects of the cell.

Lehninger, A.L., *Principles of Biochemistry.* New York, Worth, 1982. A standard biochemistry text with a detailed but clear account of bioenergetics, photosynthesis, and cellular respiration. A fine source book for students.

Shulman, R.G. "NMR Spectroscopy of Living Cells," *Scientific American,* January 1983, 86–93. Spectroscopy makes it possible to study metabolic processes in intact living cells.

Stryer, L. *Biochemistry,* 2nd ed. San Francisco, W.H. Freeman, 1981. A well-illustrated, readable text that covers the molecules of life and the concepts of cellular energetics from the ground up.

PART III
THE CONTINUITY OF LIFE: CELL DIVISION AND GENETICS

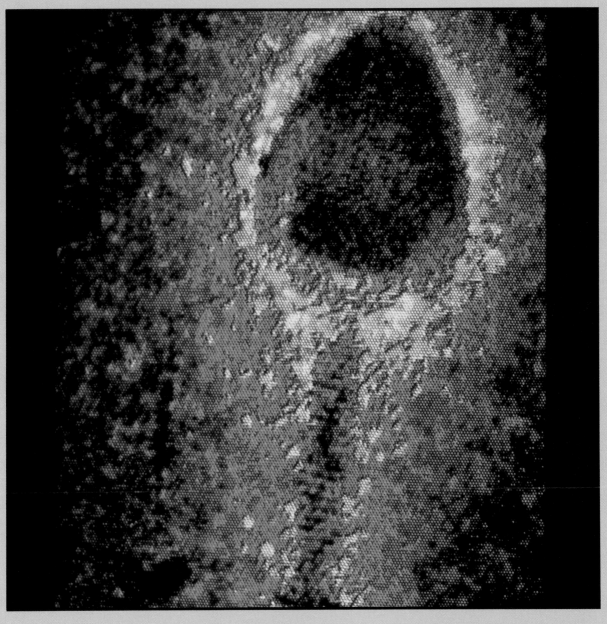

Bacteriophage T4, a virus that infects bacteria, is a focus of genetic research. (Thomas Broker, Cold Spring Harbor Laboratory.)

Chapter 9

CHROMOSOMES, MITOSIS, AND MEIOSIS

Outline

Learning Objectives

After reading this chapter you should be able to:

1. Describe the structure of a chromosome and give the function of a gene.
2. Identify the stages in the cell cycle and describe the main events of each.
3. Distinguish between haploid and diploid and define homologous chromosomes.
4. Describe the events occurring in each stage of mitosis with emphasis on the behavior of chromosomes.
5. Contrast the events of mitosis and meiosis.
6. Cite the similarities and differences between spermatogenesis and oogenesis.
7. Describe the structure of a mature sperm.
8. Describe cell division and the three mechanisms of genetic recombination (transformation, conjugation, and transduction) that take place in bacteria.

Human beings have long been aware that one of the unique characteristics of living things is their ability to reproduce their kind . . . "like begets like." Oak trees always give rise to new oak trees, never to rose bushes. This process, called **heredity,** involves the transfer of biological information from parent to offspring.

Resemblances between parents and offspring are close but are usually not exact. The offspring of a particular set of parents differ from each other and from their parents in many respects. Some variations are inherited, whereas others are due to the effects of such environmental factors as temperature, moisture, food, or light on the development of the organism. The expression of inherited characteristics may be strongly influenced by the environment in which the individual develops, but heredity determines the limits of those variations.

The branch of biology concerned with heredity is called **genetics.** Since its inception at the beginning of the present century, the science of genetics has advanced with great speed and is now developing at an even more accelerated pace, largely as a result of the science of molecular genetics, established in the 1950s.

CHROMOSOMES

The length of the DNA in the cell nucleus is thought to approach a meter and the total length of all the DNA in a human body might stretch several times to the sun and back. Printing the information contained in the microscopic nucleus of a single human cell in book form would require an estimated 600 volumes. The information contained within a bacterium would require much less space; a single edition of the *New York Times* might easily encompass it.

Chromosome Structure

Chromosomes are packages in which genetic information is stored. As discussed in Chapter 4, eukaryotic chromosomes are suspended in the semifluid ground substance of the nucleus, the nucleoplasm. Except during nuclear division the chromosomes appear as an irregular network of dark-staining granules and strands referred to as **chromatin.** The double helix of DNA with its associated proteins is coiled and looped numerous times. Chromatin contains about 60% protein, 35% DNA, and 5% RNA.

Just before and during nuclear division the chromatin condenses dramatically and becomes visible under the microscope as elongated, discrete chromosomes (Fig. 9–1). Each chromosome consists of a highly coiled and compacted DNA molecule wound around groups of small,

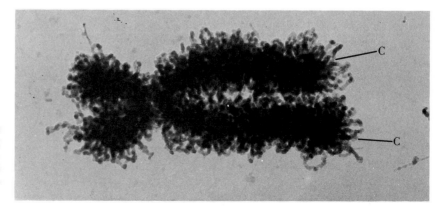

Figure 9–1 A human chromosome isolated and photographed during cell division (approximately × 50,000). Note the two chromatids (*C*) attached at their centromeres. (Courtesy of E.J. DuPraw.)

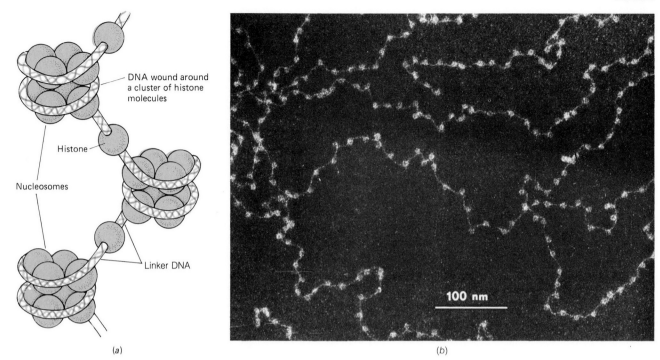

(a)

(b)

basic proteins called **histones** (Fig. 9–2). The repeating units of histone surrounded by DNA are called **nucleosomes.**

Some chromosomes are short, others long. All have an elaborate banding pattern and usually are provided with an assortment of knobs and constrictions. At a fixed point along its length, each chromosome has a small, light-staining, circular zone, called a **centromere,** that controls the movement of the chromosome during cell division. As the chromosome becomes shorter and thicker just before cell division occurs, the centromere region becomes accentuated and appears as a constriction.

The single bacterial chromosome is a simple loop of DNA suspended within a nuclear region (nucleoid). Histones are not associated with bacterial DNA, but other basic proteins are thought to be present.

Genes

Each chromosome may contain hundreds, even thousands, of genes. Although the number varies, a typical mammalian cell is estimated to have about 50,000 genes. A **gene** is a sequence of DNA that contains the information to produce a specific RNA molecule. Many of the RNA molecules then code for a specific polypeptide chain. A gene, then, may be thought of as a unit of hereditary information. Each gene encompasses a region of the chromosome extending over about six nucleosomes.

How Many Chromosomes?

Every organism of a given species contains a characteristic number of chromosomes in each of its somatic (nonsex) cells. We have 46, for example, and the crayfish more than 100. A certain species of roundworm (*Ascaris megalocephala*) has only two chromosomes and a marine protist has been found that has about 1600.

In complex organisms each chromosome occurs as a member of a pair, so the total number of chromosomes is normally even. The 46 chromosomes of the human cell, for example, comprise 23 pairs. Each pair is

Figure 9–2 The units of histone surrounded by DNA in the chromosome are called nucleosomes. (*a*) Hypothetical structure of a nucleosome. Each nucleosome bead is thought to contain a set of eight histone molecules that form a protein core around which the double-stranded DNA is wound. The DNA wound around the histone consists of 146 nucleotide pairs; another segment of DNA, 60 nucleotide pairs long, links the nucleosome beads. This linker DNA and the nucleosome bead make up the nucleosome. It is thought that one type of histone extends to cover the linker DNA and to contact beads of adjacent nucleosomes. This histone appears to be responsible for packing nucleosomes and may help link them to one another. (*b*) Electron micrograph of nucleosomes prepared from the nucleus of a chicken red blood cell (×170,000). As in the diagram, notice the nucleosome beads (normally packed much more closely together) connected by fine filaments. (Courtesy of D.E. Olins and A.L. Olins.)

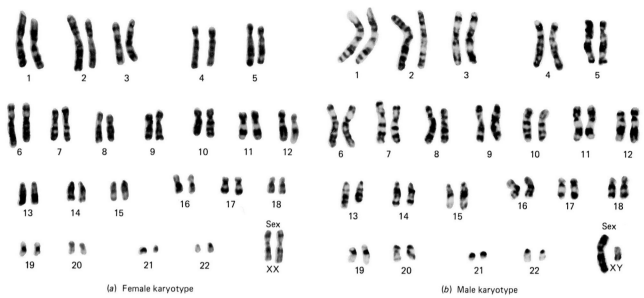

(a) Female karyotype

(b) Male karyotype

Figure 9–3 Human chromosomes. Notice the bands of each chromosome. By careful study of such chromosome photographs, known as karyotypes, it is possible to diagnose some hereditary diseases. Such diagnoses can often be made before birth from cells sloughed off into the amniotic fluid surrounding the fetus. The fluid can be sampled by the process known as amniocentesis.

sufficiently distinctive that biologists can count them, identify them and identify certain abnormalities (Fig. 9–3). Chromosome shapes and patterns can sometimes be used to infer relationships among species.

The members of a given pair of chromosomes are referred to as **homologous chromosomes.** Both members carry information governing the same traits, though this is not necessarily the same information. For example, members of a pair of homologous chromosomes might carry genes that specify hemoglobin, but one might have the information for normal hemoglobin structure, and the other might specify the abnormal hemoglobin structure that results in sickle-cell anemia.

A cell with two members of all pairs of chromosomes is called **diploid,** or **2n.** In human beings, the diploid number is 46 chromosomes (23 pairs). The gametes (eggs and sperm) cannot be diploid, or the zygote (fertilized egg) resulting from their fusion would have at least twice as many chromosomes as it should. A special type of cell division called **meiosis** reduces the number of chromosomes to one of each pair, resulting in a **haploid,** or **n,** sperm or egg. When two such haploid gametes join in fertilization, the normal diploid number of chromosomes is restored.

Each cell of a multicellular organism ordinarily has exactly the same number and kinds of chromosomes. If a cell should receive more or less than the proper number of chromosomes by some malfunctioning of the cell division process, the resulting cell may show marked abnormalities (such as a tendency to become a tumor) or may even be unable to survive. However, there are exceptions. For example, liver cells are polyploid, containing four complete haploid sets of chromosomes. About one-third of all flowering plants and almost three-fourths of all grasses seem to have originated from their ancestral species by polyploidy.

THE CELL CYCLE

Since chromosomes are the packages in which genetic information is stored, the behavior of chromosomes during the cell cycle is important. In cells that are capable of dividing, the **cell cycle** is the period from the beginning of one division to the beginning of the next division. The cell cycle may be represented in diagrams as a circle (Fig. 9–4). The length of time between two successive divisions, represented by a complete revolution of the circle, is the generation time, **T.** The generation time can

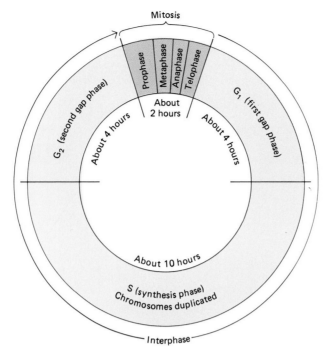

vary over wide ranges, but is usually several hours long in plant and animal cells.

Cell division involves two main processes, mitosis and cytokinesis. The actual division of the cytoplasm to form two cells is **cytokinesis.** However, before a eukaryotic cell can divide, its nucleus must undergo **mitosis,** a complex division that precisely distributes a complete set of chromosomes to each daughter nucleus. It is mitosis that ensures that each new cell contains the identical number and types of chromosomes present in the original mother cell.

By convention the cell life cycle is divided into five major stages, or phases. These phases are not distinct from one another but blend almost imperceptibly, one into the next.

Interphase

Most of the life of the cell is spent in **interphase,** actively synthesizing materials needed for its growth and maintenance. Because this stage of the cell's life cycle occurs between the phases of successive mitoses, it is called interphase (between phases). In about 1950 it was realized that chromosomes undergo replication during the interphase, then separate and are distributed to the daughter nuclei during mitosis. Interphase may be divided into three subphases. The time between mitosis and the beginning of DNA replication is termed the G_1 phase, or **first gap phase.** During the G_1 phase the cell grows and carries out certain processes, such as increased activity of enzymes involved in DNA synthesis, that make it possible for the cell to enter the **synthesis phase (S phase)** and become committed to a future cell division. Numerous experiments have shown that DNA is synthesized during the S phase. At this time the DNA content about doubles, and half the resulting DNA can be shown by isotopic tagging to be new (Fig. 9–5).

Following the completion of the S phase the cell enters a **second gap phase, G_2.** The cell employs a number of organelles specialized for mitosis that it needs at no other time. In G_2 the materials needed to manufacture these organelles are manufactured and stockpiled. At this

Figure 9–5 Photomicrograph of
mouse cells in which thymidine (a
DNA precursor) has been labeled
with tritium (a radioisotope). The
bright grains indicate replicated DNA
in interphase cells. Only the nuclei
are clearly visible. (Jonathan G.
Izant.)

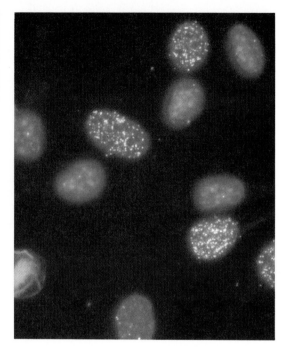

time there is an increase in protein synthesis as the final steps in the
cell's preparation for division take place.

Mitosis

The completion of G_2 is marked by the beginning of mitosis, the **M
phase.** Mitosis itself may be divided into four stages: **prophase, metaphase, anaphase,** and **telophase** (Figs. 9–6 through 9–8).

PROPHASE

The first stage of mitosis, **prophase,** begins when the chromatin threads
begin to condense and appear as chromosomes. In forming the mitotic
chromosomes, the long threads of DNA are condensed and coiled into
much shorter bundles, permitting the chromosomes to separate and pass
into the daughter cells without tangling. The DNA thread of each human
chromosome is several centimeters long, but at mitosis it is condensed
into a chromosome that is only 5 to 10 micrometers in length. The DNA
threads undergo several orders of coiling and folding in forming the mitotic chromosomes.

As prophase proceeds, the chromosomes become shorter and
thicker and are individually visible under the light microscope. The chromosomes have not yet separated completely from their duplicates and
are referred to at this point as **chromatids.** Twin chromatids remain attached to one another at their centromeres.

As described in Chapter 4, **centrioles** are hollow cylinders composed of nine triple microtubules. These organelles are characteristic of
protist and animal cells but are not found in the cells of plants or most
fungi. Their exact function in cell division is still not altogether clear
after more than a century of careful study. Early in prophase, the members of the two pairs of centrioles separate and migrate toward opposite
poles of the cell. Clusters of microtubules, composed mainly of protein,
extend outward in all directions from the centrioles. These clusters of
microtubules are called **asters.** In animal cells microtubules extend from
the region surrounding the centrioles and form a **mitotic spindle** (Fig.

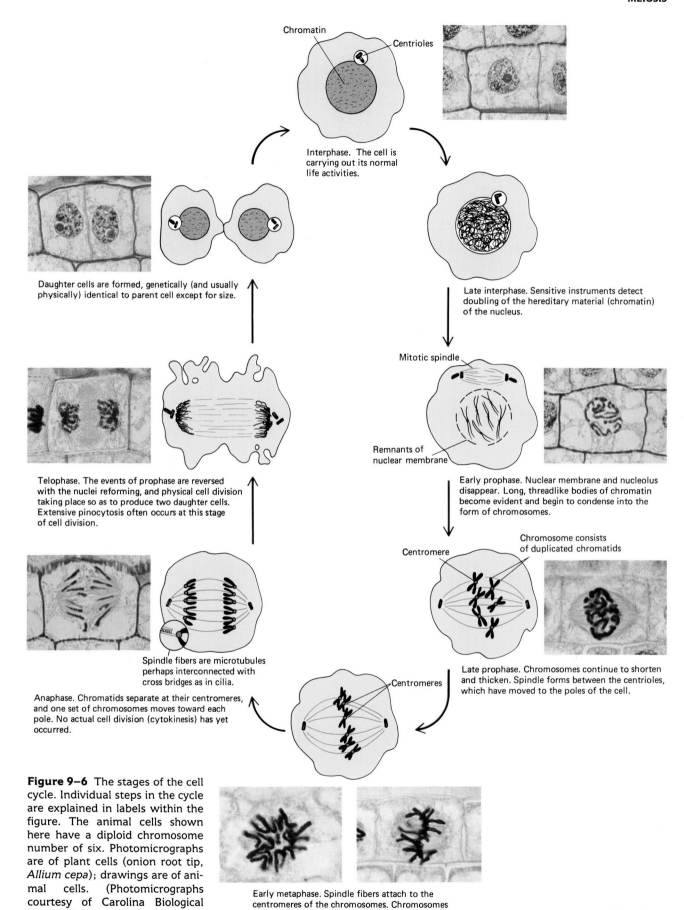

Figure 9–6 The stages of the cell cycle. Individual steps in the cycle are explained in labels within the figure. The animal cells shown here have a diploid chromosome number of six. Photomicrographs are of plant cells (onion root tip, *Allium cepa*); drawings are of animal cells. (Photomicrographs courtesy of Carolina Biological Supply Company.)

Chromatin

Centrioles

Interphase. The cell is carrying out its normal life activities.

Late interphase. Sensitive instruments detect doubling of the hereditary material (chromatin) of the nucleus.

Mitotic spindle

Remnants of nuclear membrane

Early prophase. Nuclear membrane and nucleolus disappear. Long, threadlike bodies of chromatin become evident and begin to condense into the form of chromosomes.

Daughter cells are formed, genetically (and usually physically) identical to parent cell except for size.

Telophase. The events of prophase are reversed with the nuclei reforming, and physical cell division taking place so as to produce two daughter cells. Extensive pinocytosis often occurs at this stage of cell division.

Chromosome consists of duplicated chromatids

Centromere

Late prophase. Chromosomes continue to shorten and thicken. Spindle forms between the centrioles, which have moved to the poles of the cell.

Spindle fibers are microtubules perhaps interconnected with cross bridges as in cilia.

Anaphase. Chromatids separate at their centromeres, and one set of chromosomes moves toward each pole. No actual cell division (cytokinesis) has yet occurred.

Centromeres

Early metaphase. Spindle fibers attach to the centromeres of the chromosomes. Chromosomes line up along the equatorial plane of the cell.

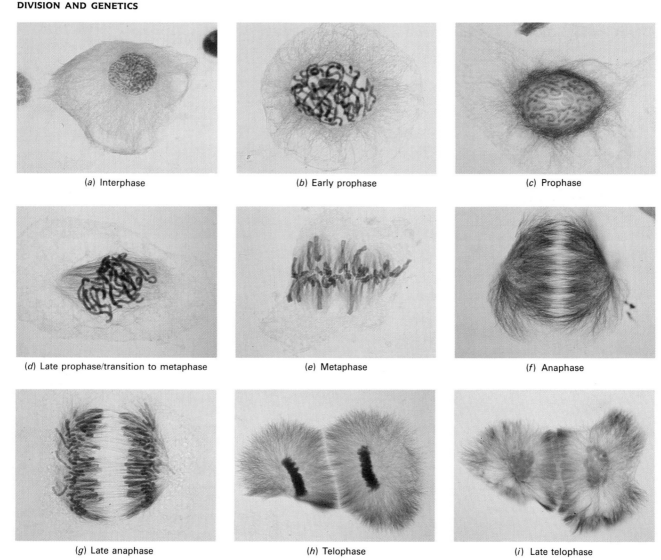

(a) Interphase (b) Early prophase (c) Prophase

(d) Late prophase/transition to metaphase (e) Metaphase (f) Anaphase

(g) Late anaphase (h) Telophase (i) Late telophase

Figure 9-7 Interphase and the stages of mitosis in plant cells (*Haemanthus*) prepared with stains. (Andrew S. Bajer, University of Oregon.)

9–8). Although plant cells lack centrioles and asters, they do form mitotic spindles.

During prophase the nuclear envelope breaks down, permitting the nuclear contents to mingle with the cytoplasm. Toward the end of prophase the condensed chromatids attach to the spindle fibers at their centromeres and are pulled to the equator of the cell, midway between the two poles and perpendicular to the axis of the spindle. The nucleolus also becomes disorganized in prophase, and there is some reason to believe that the ribosomes that have been stored and manufactured there are released into the cytoplasm by the breakdown of the nuclear membrane that occurs at this point.

METAPHASE

The short period during which the chromatids are lined up along the equatorial plane of the cell constitutes **metaphase.** The mitotic spindle is complete and can be seen to be composed of numerous fibers that extend from pole to pole. They end near the centrioles but do not actually touch them. The spindle is gel-like in consistency and is more viscous than the surrounding cytoplasm.

During metaphase each chromatid is completely condensed and appears thick and discrete. Because metaphase chromosomes can be

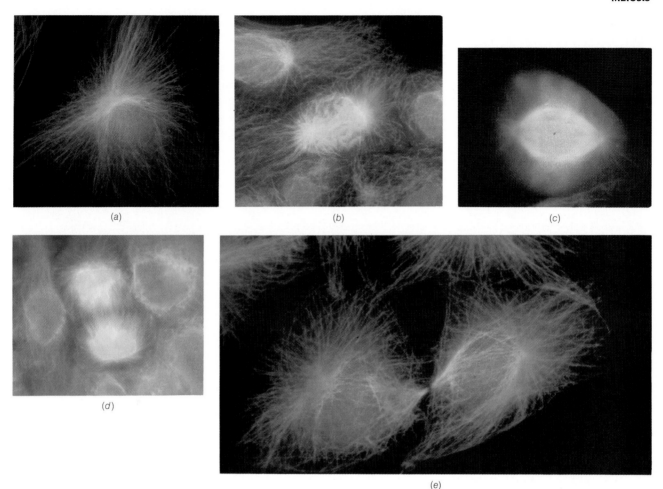

(a) (b) (c)

(d)

(e)

seen more clearly than those at any other stage, they are sometimes photographed and studied clinically to determine possible chromosome abnormalities (see Fig. 10–15).

ANAPHASE

Anaphase begins with the sudden splitting apart of the centromeres of sister chromatids. Each chromatid is now an independent chromosome. The separated chromosomes now begin to move toward opposite poles. The chromosomes move toward the poles with the centromeres (attached to the spindle fibers) leading the way and the arms of the chromosomes trailing behind. Anaphase ends when a complete set of chromosomes has arrived at opposite ends of the cell.

TELOPHASE

The final stage of mitosis, **telophase,** is characterized by a return to interphase conditions. The chromosomes begin to elongate by uncoiling. A new nuclear membrane forms around each set of chromosomes, produced at least in part from lipid components of the old nuclear membrane. Nucleoli reappear, while spindle fibers disappear.

Cytokinesis

Cytokinesis, actual division of the cytoplasm to yield two daughter cells, usually accompanies mitosis. Cytokinesis begins during later anaphase and extends through telophase. The division of an animal cell is accom-

Figure 9–8 Interphase and mitosis in human cells grown in culture. These cells are stained with fluorescent dyes. Chromosomes are stained orange, and the microtubules, yellow-green. In interphase (a) the microtubules are not yet organized into a mitotic spindle. During prophase (b) the asters move toward opposite poles of the cell. In metaphase (c) the mitotic spindle is well defined, and the chromosomes are lined up along the equatorial plane. In late anaphase (d) the complete set of chromosomes is moving toward each end of the cell, and in late telophase (e) there is a complete set of chromosomes at each end of the cell. Cytokinesis has begun, evidenced by the split between the groups of microtubules. (Jonathan G. Izant.)

plished by a cleavage furrow that encircles the surface of the cell in the plane of the equator. At the furrow the cell membrane is pulled inward by a ring of microfilaments composed mainly of the contractile protein actin. The furrow gradually deepens and separates the cytoplasm into two daughter cells, each with a complete nucleus.

In plant cells cytoplasmic division occurs by the formation of a **cell plate** between the daughter cells. It assembles in association with microtubules from the spindle and vesicles that break off from the endoplasmic reticulum and Golgi complex.

Focus on . . .

FACTORS THAT AFFECT THE CELL CYCLE

The frequency of mitosis varies greatly not only among cells from different species but among cells from different tissues within a single organism. In any multicellular organism the frequency of mitosis must be closely controlled, or growth abnormalities or even tumors may result. The evidence indicates that the frequency of mitosis is basically governed by some kind of intracellular biological clock. It may be influenced by such external factors as the temperature of the surroundings.

When conditions are optimal, a bacterium can divide every 20 minutes. Among the more rapidly dividing cells of complex organisms are the cells that line the digestive tract, cells in the skin, and the stem cells that give rise to blood cells. These cell types divide rapidly throughout life. In contrast, cell divisions in the central nervous system usually cease in the first few months of life (although divisions of mature nerve cells have been experimentally obtained under laboratory conditions).

In nearly all animal cells the production of substances that control the entrance of the cell into the S phase or the M phase depends on stimulation by growth-promoting substances present in the blood. These **growth factors** are small proteins and appear to act specifically on some kinds of cells but not others. For example, a special nerve growth factor is essential for the mitosis of sympathetic nerve cells.

Substances that inhibit mitosis, called **chalones,** counter the action of the growth factors. The various chalones are also specific and affect only the type of tissue in which they are produced. For example, the chalone produced by skin cells inhibits mitosis by neighboring skin cells. Damaged skin cells are thought to synthesize less chalone, and therefore, cells in the vicinity of a wound are released from this inhibition. They begin dividing, producing new tissue to provide for the healing of the wound. When enough healthy cells have been produced, they synthesize enough to inhibit further mitotic division and turn off the wound-healing process.

The cell cycle can also be affected by certain drugs. Colchicine is a drug used to block cell division in eukaryotic cells. This substance binds with a microtubule protein and interferes with the normal function of the mitotic spindle. The chromosomes cannot separate appropriately and move to the opposite ends of the cell. As a result, the cell may end up with an extra set of

chromosomes. This can be done not only with ordinary body cells but also with those destined to become gametes, and in due course, zygotes that can give rise to another generation. This fact is of great commercial importance, especially in connection with the development of new varieties of ornamental and agricultural plants. Plants consisting of cells with extra sets of chromosomes tend to be larger and more vigorous than normal plants and may have such qualities as increased fruit sugar content in tissues or doubled flower petals.

Some drugs used in cancer therapy block cell division or specifically injure dividing cells. Because cancer cells divide much more rapidly than most normal body cells, they are most affected by these drugs. However, these drugs do affect some types of normal body cells, especially those that multiply rapidly, such as those lining the digestive tract, and those that produce the continuously abraded layers of the skin and its derivatives. This is why such drugs may produce side effects, like causing hair loss.

Some common antibiotics interfere with bacterial cell division by inhibiting synthesis of needed proteins, or in the case of penicillin, by preventing the production of the new cell wall (Chapter 6). Since human cells lack such a cell wall, they are unaffected by the penicillin.

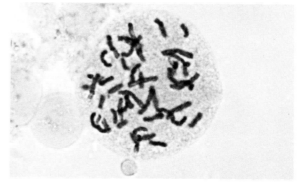

When a cell is treated with colchicine, mitosis is arrested in metaphase during which the chromosomes are thickest. The colchicine spreads the chromosomes apart, making them easier to count. (Cytogenetics Laboratory, University of California, San Francisco.)

MEIOSIS

Mitosis ensures that each daughter cell will receive exactly the same number and kind of chromosomes that the mother cell had. In contrast, meiosis, a special type of cell division reserved for the production of gametes and spores, reduces the number of chromosomes so that each new cell is haploid. The process of meiosis assures that the chromosome number of a species remains the same from generation to generation, but to do this it must separate the members of each homologous pair of chromosomes. In so doing it separates any contrasting genetic traits each might bear and distributes them to separate gametes or spores. The result is that no two offspring even of the same parents are likely to be exactly alike.

How Does Meiosis Differ from Mitosis?

The events of meiosis are somewhat similar to the events of mitosis, but there are several important differences (Fig. 9–9).

1. In meiosis there are two successive nuclear and cell divisions, producing a total of four cells. In mitosis there is only one nuclear division, and cytokinesis typically occurs once yielding two daughter cells.
2. Each of the four cells produced in meiosis contains the haploid number of chromosomes, that is, only one of each homologous pair. In mitosis each daughter cell contains the diploid number of chromosomes.
3. During meiosis the homologous chromosomes containing ge-

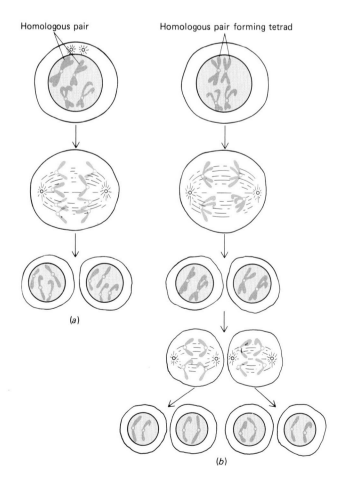

Figure 9–9 Meiosis compared with mitosis. The diploid number for the cell shown here is four. (*a*) Mitosis. Note that each daughter cell has an identical set of four chromosomes (two pairs), which is the diploid number. (*b*) Meiosis. Two divisions are taking place, giving rise to four daughter cells. Each daughter cell has only two chromosomes, one of each pair. The chromosomes shown in blue originally came from one parent; those shown in red came from the other parent. Note that in the prophase of the first meiotic division, homologous chromosomes come together, forming tetrads.

netic information from each parent are thoroughly shuffled, and
one of each pair is randomly distributed to each new cell; the
cells produced possess new combinations of chromosomes
(Figs. 9–9 and 9–11). After mitosis the daughter cells contain an
identical complement of chromosomes, and this chromosome
complement is identical to that of the mother cell.

The Stages of Meiosis

Meiosis consists of two cell divisions, logically named the first and second meiotic divisions, or simply meiosis I and meiosis II. Each division
includes stages comparable to the four stages of mitosis, namely prophase, metaphase, anaphase and telophase.

The **first meiotic division** separates the members of each homologous pair of chromosomes and distributes them into daughter cells.
These chromosomes are in duplicated pairs, that is, they are chromatids.
The **second meiotic division** separates the chromatids into individual
chromosomes, which then enter separate, haploid daughter cells. Since
it is easier to follow these events in an organism that possesses only a
few chromosomes than in, say, a crayfish with its 100-plus, we will discuss a hypothetical organism with a diploid number of only four chromosomes (Fig. 9–10).

THE FIRST MEIOTIC DIVISION

As in mitosis, the chromosomes duplicate themselves during the S phase
before meiosis actually begins. Recall that when a chromosome is duplicated, it consists for a time of two chromatids joined by their centromeres. To be considered a complete chromosome, the structure must
possess an unshared centromere.

During prophase of the first meiotic division, while the chromatids
are still elongated and thin, the homologous chromosomes come to lie
close together side by side along their entire length, a process called
synapsis (Fig. 9–10). Since the diploid number in our example is four,
there are two homologous pairs. One of each pair, the **maternal chromosomes,** were originally inherited from the organism's mother, whereas
the other member of each pair, the **paternal chromosomes,** were contributed by the father. Since each chromosome is doubled at this time and
actually consists of two chromatids, synapsis results in the coming together of four chromatids, forming a complex known as a **tetrad.** The
number of tetrads equals the haploid number of chromosomes. In human
cells there are 23 tetrads (and a total of 92 chromatids) at this stage.

SYNAPSIS, LINKAGE, AND CROSSING-OVER. All the genes located on a particular chromosome are **linked** together and therefore will tend to be inherited together. However, this tendency for linked genes to stay together is

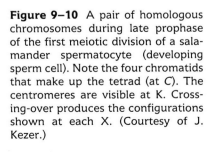

Figure 9–10 A pair of homologous
chromosomes during late prophase
of the first meiotic division of a salamander spermatocyte (developing
sperm cell). Note the four chromatids
that make up the tetrad (at C). The
centromeres are visible at K. Crossing-over produces the configurations
shown at each X. (Courtesy of J.
Kezer.)

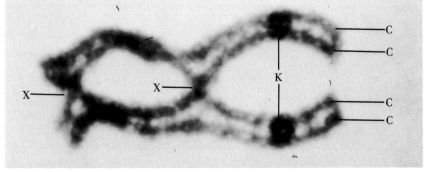

not absolute. During synapsis homologous chromosomes become wrapped about one another, and genetic material may be exchanged between these homologous chromosomes. This process, known as **crossing-over,** involves breaking the maternal and paternal chromatids and rejoining the broken segments reciprocally (Fig. 9–10). The exchange of genetic material between homologous chromosomes is called **genetic recombination.** The new combinations of genes greatly enhance the prospects for variety among offspring of sexual partners.

In many species the prophase of the first meiotic division is an extremely extended phase during which the cell grows and synthesizes nutrients for the future embryo. In many types of gametes, the chromosomes assume unusual configurations during this phase. Hundreds of pairs of loops may project from the chromatid axis, giving an appearance similar to that of an old-fashioned brush used to clean soot from oil-lamp chimneys. From this comes the name **lampbrush chromosomes.** The loops are sites of intense RNA synthesis (Fig. 9–11).

While the events characteristic of prophase I of meiosis are occurring, other events that are also similar to those of mitotic prophase take place. The centrioles move to opposite poles, a spindle forms between the centrioles, and the nuclear membrane dissolves.

SEPARATING HOMOLOGOUS CHROMOSOMES. The tetrads line up along the equator of the spindle and the cell is said to be in metaphase. Both chromatids of one chromosome are oriented toward the same pole. Their joint centromere is attached to the spindle fibers of only one of the two poles. Each pair of centromeres of homologous chromosomes is separated so that only the ends of the homologous chromatids contact one another (Fig. 9–12).

During anaphase of the first meiotic division, the homologous chromosomes of each pair, but not the daughter chromatids, separate and randomly move toward opposite poles. The chromatids are still united by their centromeres. This differs from mitotic anaphase, in which the centromeres separate and the daughter chromosomes pass to opposite poles. In the telophase of the first meiotic division in our example, there would be two double chromosomes at each pole, that is, four chromatids. In humans there would be 23 doubled chromosomes (46 chromatids) at each pole. During telophase the nuclei reorganize, the chromatids begin to elongate, and cytokinesis (cell division) generally takes place.

THE SECOND MEIOTIC DIVISION

During the interphase that follows there is no S phase, for no further chromosome replication takes place. In most organisms meiotic interphase is very brief; in some organisms it is absent. Since the chromatids do not completely elongate between divisions, the prophase of the second meiotic division is also brief. Prophase II is similar to a mitotic prophase; there is no pairing of homologous chromosomes (indeed, only one of each pair remains in the cell), and there is no genetic recombination.

During metaphase II the chromatids again line up on the equator. The metaphases of the first and second division can be distinguished because in the first the chromatids are arranged in bundles of four (tetrads), and in the second the chromatids are arranged in groups of two (dyads). During anaphase II, the centromeres split and the daughter chromatids, now complete chromosomes, separate and move to opposite poles. Thus in the telophase of the second meiotic division, there is one of each kind of chromosome, the haploid number at each end of the cell. Nuclear membranes then form, the chromosomes gradually elongate, forming chromatin threads, and cytokinesis occurs.

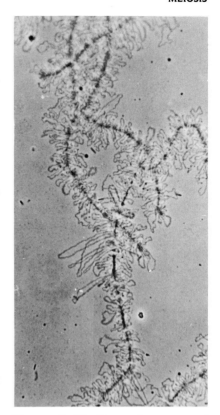

Figure 9–11 Photomicrograph of chromosomes from the oocyte of the newt *Triturus viridescens* (approximately ×1100) showing the loops radiating from the central thread. The presence of these loops gave the structures their name, lampbrush chromosomes. (Courtesy of Dr. Dennis Gould.)

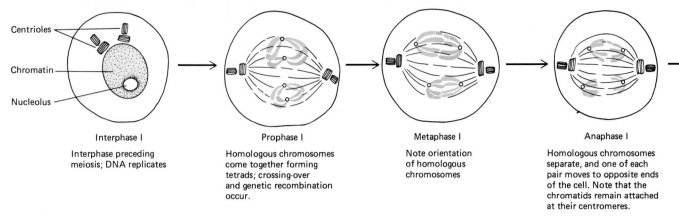

Interphase I	**Prophase I**	**Metaphase I**	**Anaphase I**
Interphase preceding meiosis; DNA replicates	Homologous chromosomes come together forming tetrads; crossing-over and genetic recombination occur.	Note orientation of homologous chromosomes	Homologous chromosomes separate, and one of each pair moves to opposite ends of the cell. Note that the chromatids remain attached at their centromeres.

Figure 9–12 The stages of meiosis. The diploid number for the cell shown here is four.

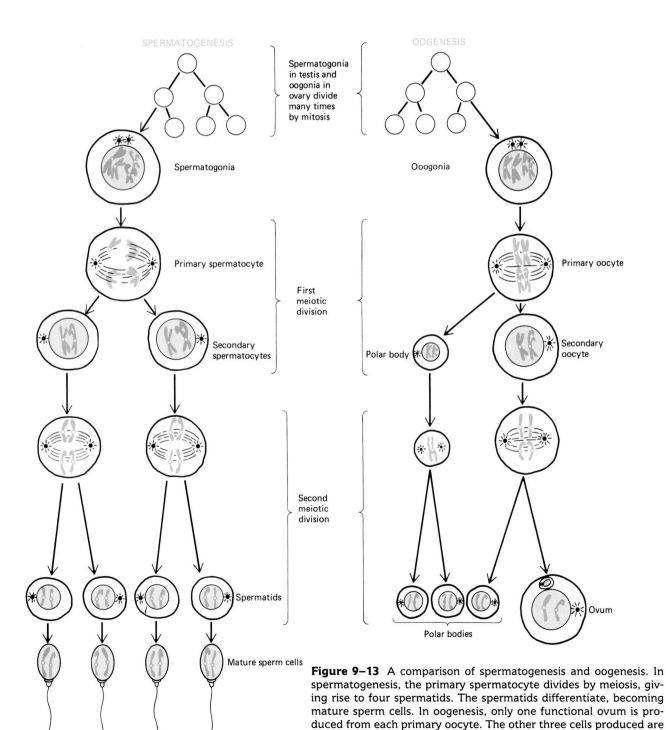

Figure 9–13 A comparison of spermatogenesis and oogenesis. In spermatogenesis, the primary spermatocyte divides by meiosis, giving rise to four spermatids. The spermatids differentiate, becoming mature sperm cells. In oogenesis, only one functional ovum is produced from each primary oocyte. The other three cells produced are polar bodies that degenerate.

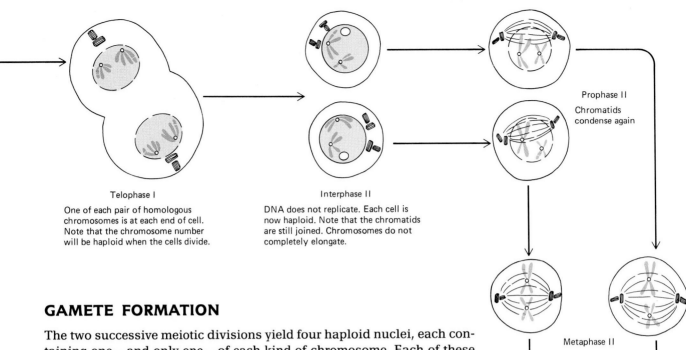

Telophase I

One of each pair of homologous chromosomes is at each end of cell. Note that the chromosome number will be haploid when the cells divide.

Interphase II

DNA does not replicate. Each cell is now haploid. Note that the chromatids are still joined. Chromosomes do not completely elongate.

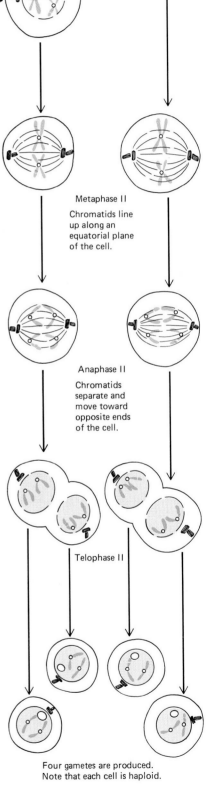

Prophase II

Chromatids condense again

Metaphase II

Chromatids line up along an equatorial plane of the cell.

Anaphase II

Chromatids separate and move toward opposite ends of the cell.

Telophase II

Four gametes are produced. Note that each cell is haploid.

GAMETE FORMATION

The two successive meiotic divisions yield four haploid nuclei, each containing one—and only one—of each kind of chromosome. Each of these cells has a different combination of genes. In plants meiosis generally occurs in structures called **sporangia** and gives rise to reproductive cells called **spores.** Plants that develop from the spores are themselves haploid and eventually give rise to eggs and sperm by mitosis. (Plant reproduction will be discussed in Chapter 18.) In animals meiosis takes place in specialized reproductive structures, the **gonads,** and results in formation of sperm and eggs by a maturational process that follows or accompanies the events of meiosis. Gametogenesis, or gamete formation, is basically similar in all organisms, but there are some differences in detail.

Spermatogenesis

In most animals the sperm and egg are distinctively different, with the sperm being much smaller than the egg but highly mobile. In humans the sperm has, in addition to the nucleus-containing **head,** a **middle piece** containing a spiral mitochondrion, and a **tail** that is little more than a flagellum. This little missile of hereditary information is guided to the egg by chance. Many sperm take a wrong turn, and out of millions only one meets the mark in time. The random lack of guidance of sperm cells allows us to assume random distribution of traits in genetic problems.

In humans and other vertebrates **spermatogenesis,** or sperm production, occurs in the male gonad, or **testis.** More specifically, spermatogenesis takes place within the walls of a vast tangle of hollow tubules, the **seminiferous tubules,** that make up the testis (Fig. 31–4). These tubules are lined partially with primitive, undifferentiated cells called **spermatogonia.** The spermatogonia divide during embryonic development and childhood just like all other body cells, by simple mitosis, producing yet more spermatogonia but not as yet any mature sperm. At adolescence this begins to change and some spermatogonia undergo spermatogenesis, while the remainder continue to divide mitotically. In humans there is no definite breeding season in which this takes place, but in many other animals spring or fall is the only time not only for male spermatogenesis but also, naturally, for egg cell production by females.

The spermatogonia grow into larger cells called the **primary spermatocytes** (Fig. 9–13). These cells undergo a first meiotic division to form **secondary spermatocytes.** The second meiotic division follows, with each secondary spermatocyte giving rise to two **spermatids;** in all,

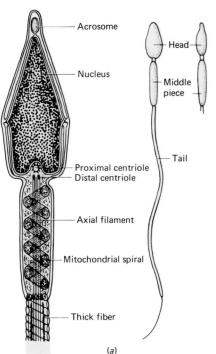

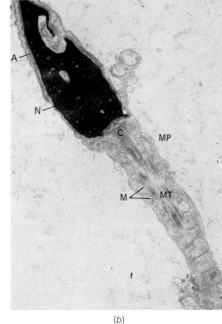

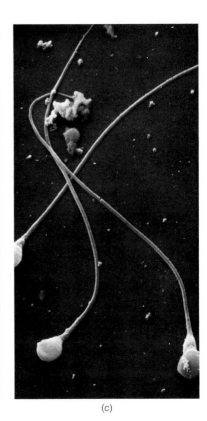

(a)

(b)

(c)

Figure 9–14 Sperm cell structure. (a) A mammalian sperm. *Left*, a head and middle piece. Structures shown are those that would be visible through the electron microscope. Middle and right, top and side views of a sperm. (b) Electron micrograph of a human sperm cell (approximately ×37,500). A, acrosome; N, nucleus; MP, midpiece; M, mitochondria; MT, microtubules; C, centrioles. (c) Scanning electron micrograph of human sperm (approximately ×2400). (b, courtesy of Dr. Lyle C. Dearden; c, David M. Phillips, the Population Council.)

four are derived from the original primary spermatocyte. The spermatid is a fairly large cell with much more cytoplasm than a mature sperm possesses.

The haploid spermatid matures into an actual sperm cell by a rather complicated process. The nucleus shrinks, and part of the Golgi complex becomes the **acrosome,** which is thought to produce enzymes that aid the sperm in penetrating the egg (Fig. 9–14). The two centrioles nestle in a small depression on the nuclear surface. One of these, the **distal centriole,** gives rise to the flagellum of the sperm. Although somewhat longer than most, the sperm flagellum is typical of eukaryotes, with the usual 9 + 2 arrangement of microtubules. Finally, most of the cytoplasm is discarded to be phagocytized and destroyed by the nutritive **Sertoli cells** that are also present in the tubule wall.

There is some variation among the sperm of various mammals. The rat sperm, for example, has a hooklike head. Invertebrate sperm can vary radically from mammalian sperm. Crustacean sperm, for instance, have no tails at all and indeed no effective means of locomotion. Should they come into contact with the egg, however, they can grasp its surface by means of three pointed projections, like grappling hooks. Their middle piece is built like a spring, uncoiling to push the nucleus through the outer membrane of the egg. The intestinal parasitic nematode *Ascaris* has an ameboid sperm (Fig. 9–15), and some plant sperm are ciliated. In many vascular plants, the sperm is virtually reduced to a nucleus contained within the pollen grain.

Oogenesis

Oogenesis, the production of ova, or eggs, begins in the female gonad, the **ovary.** In early fetal life the immature, diploid sex cells known as **oogonia** divide extensively by mitosis. In humans and other vertebrates

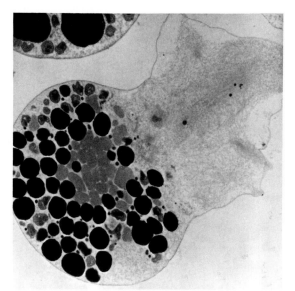

Figure 9–15 The intestinal parasite *Ascaris* has an unusual ameboid sperm cell. It lacks a flagellum and moves by means of pseudopods. Visible here are the cytoskeleton, mitochondria, and dark granules of unknown function that may store food. (M.K. Abbas and G.D. Cain, *Histochemistry* 81:59–65, 1984.)

each oogonium comes to be surrounded by a group of **follicle cells.** The entire structure is known as a **follicle** (Fig. 9–16). By the third month of human fetal development the oogonia begin to develop into **primary oocytes,** which are comparable to the male primary spermatocytes. The ovaries of an infant contain about 400,000 of these, which by then are in the prophase of their first meiotic division. These cells remain in prophase until sexual maturity is reached.

Approximately 13 years later, one of the primary oocytes completes its first meiotic division, and upon maturity is released from the ovary by **ovulation.** About a month later, the same thing happens to another, and a month after that, to another. So it continues until the end of reproductive life in the later 40s or 50s. An egg ovulated at age 47 has been in prophase for 47 years.

As far as the nucleus is concerned, the meiotic events of oogenesis (see Fig. 9–13) are exactly the same as those of spermatogenesis, but instead of discarding as much mass as possible (as does the sperm), the egg retains as much as possible for use in early embryonic development. This is evident in the first meiotic division, in which the distribution of cytoplasm between the two daughter cells is unequal. The **secondary oocyte** that results is very large, and its sister cell, the **first polar body,** contains virtually nothing but nucleus and soon dies.

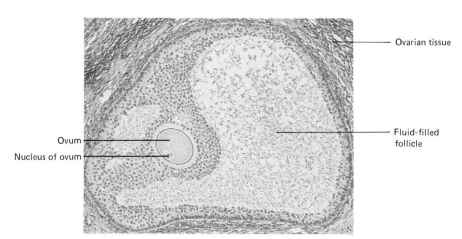

Ovarian tissue

Fluid-filled follicle

Ovum

Nucleus of ovum

Figure 9–16 Photomicrograph of ovum residing in a follicle of the human ovary (approximately × 40). (Biophoto Associates, Photo Researchers, Inc.)

The secondary oocyte immediately proceeds to the second meiotic division but halts in metaphase until it is fertilized. This, of course, normally takes place after ovulation in most animals. When the second meiotic division finally is completed, again it is unequal, and a **second polar body** is formed to dispose of superfluous chromosomes with a minimum amount of cytoplasm. Many eggs, particularly those of such animals as birds and reptiles, which incubate their eggs externally, have a great deal of **yolk** in addition to cytoplasm. The yolk serves as food storage for the developing embryo and, for a short time, for the newly hatched young. Each primary oocyte gives rise, in the end, to just one egg, in contrast to spermatogenesis, in which one primary spermatocyte gives rise to four sperm.

CELL DIVISION AND REPRODUCTION IN PROKARYOTES

The much simpler structure of prokaryotes extends to their hereditary material, so they do not require the elaborate processes of precise distribution and shipment of packages of genetic material needed by eukaryotes. In bacteria the DNA is found mainly in a single long, circular molecule. This bacterial chromosome is not associated with histones and other proteins that are bound to eukaryotic chromosomes, and its structure is less elaborate.

When stretched out to its full length, the bacterial chromosome is about 100 times longer than the cell itself. A small amount of genetic information may be present as smaller DNA molecules, called **plasmids,** which replicate at the same time but independently of the chromosome. Bacterial plasmids often bear genes involved in resistance to antibiotics.

Bacteria generally reproduce asexually by **transverse binary fission,** in which the cell develops a transverse cell wall and then divides into two daughter cells (Fig. 9–17). The transverse wall is formed by an ingrowth of both the cell membrane and the cell wall. When the newly formed cell wall does not separate completely into two walls, a chain of bacteria may be formed. (Some prokaryotes, such as the chlamydiae, do not possess a cell wall at all.)

The duplication of the chromosome and the division of the cell often get out of phase, so a bacterial cell may have from one to four (or even more) identical chromosomes. Although the bacterial cell does not have a mitotic spindle, a copy of the single chromosome must be distrib-

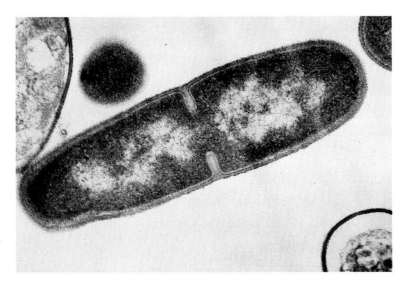

Figure 9–17 Dividing *Bacillus licheniformis* bacteria (approximately × 136,000). Notice the septum forming between the two daughter cells. (Lee D. Simon, Photo Researchers, Inc.)

uted to each new daughter cell. This appears to be facilitated by a connection between each chromosome and the cell membrane.

It may not seem astonishing that under ideal conditions a bacterium can divide every 20 minutes, but if that rate of reproduction could continue unabated, one bacterium could give rise to 250,000 within six hours. Although bacteria do not usually find conditions quite so ideal when they invade our bodies, this fact helps explain why some bacterial diseases can progress as fast as they do, especially in people whose immune mechanisms or other defenses have been compromised.

Although complex sexual reproduction involving fusion of gametes does not occur in prokaryotes, genetic material is sometimes exchanged between individuals. Such genetic recombination can take place by three mechanisms: transformation, conjugation, and transduction. In **transformation** fragments of DNA released by a broken cell are taken in by another bacterial cell. This mechanism has been used experimentally to show that genes can be transferred from one bacterium to another and that DNA is the chemical basis of heredity.

In **conjugation** two cells of different mating types (the equivalent of sexes?) come together, and genetic material is transferred from one to another. Conjugation has been most extensively studied in the bacterium *Escherichia coli,* where there are F^+ strains and F^- strains. F^+ individuals contain a plasmid known as the **F factor,** which sometimes is capable of organizing special hollow **pili** (Fig. 9–18). These pili can serve as conjugation bridges that pass from the F^+ to the F^- cell. The F pili are long and narrow and have a channel through which fragments of DNA may pass from one bacterium to the other. Pili apparently occur only in bacteria that contain plasmids, which suggests that pili may really be not bacterial organelles but special structures induced by the plasmids so that they can be passed from one host cell to another. If this interpretation is correct, then plasmids are actually genetic parasites comparable to viruses. However that may be, sometimes chromosomal DNA gets through the pilus as well. Genes for antibiotic resistance can be passed from one bacterial cell to another, even of a different species, by this route. Both transformation and conjugation occur only rarely among bacteria, but because there are a great many bacteria, these processes must nevertheless have a significant effect on the genetic makeup and evolutionary processes that occur in bacterial populations.

In the third process of gene transfer, **transduction,** bacterial genes are carried from one bacterial cell into another within a **bacteriophage,** a bacterial virus. Whether this can be considered a reproductive process is a matter of debate. What happens is that a virus's DNA becomes intermixed with a bit of the host's DNA, which the virus disrupted when it infected the host bacterium. If that chimerical virus happens to infect another bacterium without killing it (Chapter 11), then the fragment of the previous host's DNA that it carries may become integrated into that of the new host. This new bit of genetic information is then duly replicated along with that of the new host every time that cell divides.

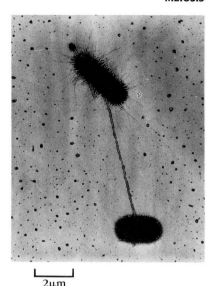

$2\mu m$

Figure 9–18 F pilus connecting *Escherichia coli* bacteria during conjugation. DNA is transferred during this process. Bacterial viruses have attached to the pilus and are visible as tiny bumps. (Courtesy of C. Brinton, Jr., and J. Carnahan.)

CHAPTER SUMMARY

I. In the production of a new generation, genetic information is transferred from parent to offspring; this is termed heredity. Genetics is the study of heredity.

II. Each eukaryotic chromosome consists of nucleosomes joined by connecting DNA.

  A. A gene is a unit of hereditary information that codes for a specific RNA molecule and ultimately for a specific polypeptide.

B. Each somatic cell of every organism of a given species has a characteristic number of pairs of chromosomes, the diploid (2n) number. The chromosomes differ in length and shape and other morphological features.

III. The cell cycle is the period from the beginning of one division to the beginning of the next division.

A. Interphase can be divided into the first gap phase, the synthesis phase, and the second gap phase.

1. During the G_1 phase the cell grows and prepares for the S phase.
2. DNA is replicated during the S phase.
3. During the G_2 phase there is an increase in protein synthesis.

B. During mitosis a complete set of chromosomes is distributed to each end of the cell, and a daughter nucleus is formed about each set.

1. During prophase chromatids condense, the nucleolus disappears, the nuclear membrane breaks down, and the mitotic spindle begins to form.
2. During metaphase the chromatids line up along the equator of the cell; the mitotic spindle is complete.
3. During anaphase the chromosomes separate and move toward opposite poles of the cell.
4. During telophase a nuclear membrane forms around each set of chromosomes, nucleoli reappear, the chromosomes elongate, the spindle disappears, and cytokinesis generally takes place.

IV. In the formation of gametes—eggs and sperm—a special kind of cell division termed meiosis occurs. This consists of a sequence of two cell divisions in which the chromosome number is reduced to one-half, the haploid (n) number. Each gamete contains one of each kind of chromosome, that is, one complete set of chromosomes. In fertilization a haploid set of chromosomes from the sperm and a haploid set of chromosomes from the egg unite, forming the diploid number in the zygote.

A. During the prophase of the first meiotic division, the members of a pair of chromosomes undergo synapsis and crossing-over, and genetic recombination occurs.

B. The members of each pair of homologous chromosomes separate during the anaphase of the first meiotic division and are distributed to different daughter cells.

C. During the second meiotic division the two chromatids of each homologous chromosome separate, and each is distributed to a daughter cell.

V. In spermatogenesis the meiotic divisions convert each spermatocyte to four spermatids, which then undergo a process of growth and change and become the functional sperm. A human sperm consists of a head capped with an acrosome, a middle piece containing mitochondria, and a long flagellum.

VI. In oogenesis meiosis divides the cytoplasm of the oocyte unequally: The mature ovum receives nearly all of the cytoplasm and yolk, and three small polar bodies receive mainly nuclear material.

VII. The bacterial chromosome is a simple loop of DNA, sometimes accompanied by much smaller loops of DNA, the plasmids.

A. Bacteria reproduce asexually by transverse binary fission without mitosis.

B. Genetic recombination in bacteria occurs by transduction, transformation, and conjugation.

Post-Test

1. _____ is the branch of biology that is concerned with heredity.
2. Except during nuclear division eukaryotic chromosomes appear as long, thin, dark-staining threads called _____ .
3. A human somatic cell with 46 chromosomes would be described as _____ , or _____ .
4. A sperm cell with only one of each pair of chromosomes is described as _____ , or _____ .
5. The members of a given pair of chromosomes are referred to as _____ chromosomes.
6. The period from the beginning of one cell division to the beginning of the next division is known as the _____ _____ .
7. In _____ a complete set of chromosomes is distributed to each new daughter nucleus.
8. The actual division of the cytoplasm to form two cells is called _____ .
9. During the _____ stage of its life cycle a cell grows and carries out its normal functions.
10. DNA is replicated during the _____ phase of interphase.
11. Twin chromatids are attached at their _____ _____ .
12. During _____ the chromatids are lined up along the equatorial plane of the cell.
13. During _____ the chromatids separate and begin to move toward opposite poles.
14. _____ are substances that inhibit mitosis by countering the action of growth factors.
15. In meiosis there are _____ nuclear and cell divisions, producing a total of _____ cells.
16. Synapsis results in the formation of complexes of chromosomes known as _____ .
17. During _____ _____ segments of DNA of homologous chromatids are broken and exchanged.
18. The exchange of DNA segments between homologous chromatids may result in _____ _____ _____ .
19. In vertebrates spermatogenesis takes place in the seminiferous _____ of the _____ .
20. In the developing sperm cell the acrosome develops from the _____ _____ .
21. In oogenesis the first meiotic division gives rise to a secondary _____ and a _____ .
22. In bacterial _____ two cells of different mating types exchange genetic material.

Review Questions

1. How does meiosis differ from mitosis?
2. Define the following terms: (a) diploid, (b) haploid, and (c) homologous chromosomes.
3. Describe the stages in meiosis and indicate when and how synapsis and separation of homologous chromosomes occur.
4. How do you suppose certain liver cells become tetraploid?
5. Describe the structure of a chromosome. What are nucleosomes?
6. What is the relationship between genes and chromosomes? What are the functions of genes?
7. Two very different species may have the same diploid chromosome number. How can this be explained?
8. Compare the essential features of spermatogenesis and oogenesis.
9. Describe the process by which a spermatid differentiates into a mature sperm.
10. How is the diploid number reestablished in fertilization?
11. Why do you think cytoplasmic division is unequal in oogenesis?
12. An organism has a diploid chromosome number of 10. (a) How many pairs of chromosomes would it have in a typical body cell, such as a muscle cell? (b) How many tetrads would form in the prophase of the first meiotic division? (c) How many chromatids would be present in the prophase of the first meiotic division? (d) How many chromosomes would be present in a mature sperm cell?
13. Compare eukaryote chromosomes with bacterial "chromosomes."
14. Describe bacterial conjugation and transduction.

Readings

Holtzman, E., and A.B. Novikoff. *Cells and Organelles,* 3rd ed. Philadelphia, Saunders College Publishing, 1984. The fourth part—"Duplication and Divergence: Constancy and Change"—includes a readable, detailed discussion of chromosome structure as well as of mitosis and meiosis.

Kornberg, R.D., and A. Klug. "The nucleosome," *Scientific American,* 244 (2):52–64, 1981.
Lewin, R. "Do Chromosomes Cross-Talk?" *Science,* 214: 1334–1335, 1981. Discussion of chromosome makeup.
Mazia, D. "The Cell Cycle," *Scientific American,* January 1974. This article focuses on the stages of the cell cycle.

Chapter 10

PATTERNS OF INHERITANCE

Outline

Learning Objectives

After you have read this chapter you should be able to:

1. Define the basic terms relating to genetic inheritance, for example, *gene, dominance, recessiveness, codominance, chromosome, homozygous, heterozygous, alleles, homologous, genotype,* and *phenotype.*
2. Relate the inheritance of genetic traits to the behavior of chromosomes in meiosis.
3. Summarize the concepts of homologous chromosomes and allelic genes.
4. Distinguish between homozygous and heterozygous genotypes. Relate genotype to phenotype in terms of dominance.
5. Describe the inheritance of sex-linked genes.
6. Solve simple problems in genetics involving monohybrid and dihybrid crosses by applying Mendel's laws.
7. Solve simple problems in genetics involving incomplete dominance, polygenes, and multiple alleles.
8. Recognize a state of genetic linkage and, given an example, be able to solve simple genetic problems involving sex linkage.
9. Describe the inheritance of the Rh and ABO factors and outline the mechanisms of Rh disease.
10. Summarize the characteristics of selected genetic diseases.
11. Summarize the characteristics of selected chromosomal disorders.
12. Summarize the Lyon hypothesis of the inactivation of X chromosomes.
13. Define *consanguinity* and give its principal genetic implications.

You are not a duplicate of your mother or a carbon copy of your father, or even an exact mixture of the characteristics of them both. Just how were the distinctive plans for the immensely complicated machinery of your body drawn up and expressed? What mechanisms of inheritance determine whether you are male or female, tall or short, light- or dark-skinned, whether you have type A or type O blood, whether you have normal vision or are colorblind, and countless other of your characteristics?

THE FATHER OF GENETICS

The roots of much modern biology extend into the 19th century, and genetics is no exception. The foundation of our modern knowledge of genetics was laid at that time by an obscure Austrian monk, Gregor Mendel, who lived in the town of Brünn, Austria, which is now Brno, Czechoslovakia (Fig. 10–1). "Natural philosophy," as biology was called in the 19th century, was usually not a profession but a hobby; only clergymen, physicians, wealthy people, or others with some leisure time were able to pursue it. But Mendel was no dilettante. The monastery also functioned somewhat as a modern agricultural research station does, and the breeding of improved varieties of cattle, crop plants, and even honeybees was part of its mission.

Figure 10–1 Gregor Mendel, the father of genetics. (V. Orel, Mendelianum of the Moravian Museum.)

Mendel was active in the natural history society of Brünn and presented his findings in a series of research reports, which were published in the society's journal. His findings were revolutionary; despite this, they were almost universally ignored. Whatever the reasons for the neglect, Mendel did not gain the recognition rightly due him for more than 30 years, when biologists first rediscovered these principles and then rediscovered his papers during a prepublication literature search.

Mendel worked with garden peas, which exist in a number of distinct varieties differing in such characteristics as height, flower color, seed coat color, and seed coat texture. Pea plants also are normally self-fertilized, so simple surgery of the male flower parts makes the plant incapable of being fertilized at all except by artificial means. In this way, the crossing of varieties can be closely controlled: The emasculated flower can no longer fertilize itself and cannot be fertilized naturally by an unknown plant, and thus "illegitimate" offspring cannot occur.

MENDEL'S LAWS

Mendel's breeding experiments led him to certain conclusions about the mechanisms of heredity, which later scholars restated as **Mendel's laws of inheritance.**

1. Heredity is transmitted by unit factors (now called genes), which exist in pairs.
2. When gametes are formed, the two genes of each pair separate from one another, and each gamete receives only one gene of each pair. This is known as the **law of segregation.**
3. When two alternative forms of the same gene are present in an individual, only one of the alternatives is usually expressed. This concept is known as the **law of dominance.**
4. If one considers two or more independent characteristics in a cross, such as flower color and seed coat texture, each characteristic is inherited without relation to other traits. All possible combinations of independent characteristics thus will occur in the gametes. This is the **law of independent assortment.**

INTRODUCING GENES

Mendel did not think in terms of genes, although his observations were to lead to the formulation of this concept. As we have discussed, a gene may be defined as a region of DNA containing the information necessary to manufacture a specific polypeptide (or in some cases just a particular type of RNA). This concept can be illustrated with a common genetic abnormality, albinism. One kind of **albino** is a person or animal who lacks the genetic information to produce the body pigment melanin. This pigment is the protein responsible for most of the color of hair, skin, and eyes. Without it an organism would appear completely unpigmented, with white hair and perhaps pink eyes, through which hemoglobin of the blood would show unmasked.

In this kind of albinism it has been shown that all that is lacking is the ability to make **tyrosinase,** an enzyme needed to make melanin from the amino acid tyrosine. Without that enzyme no pigment can be formed. But what is meant by a lack of tyrosinase? It is possible, by immunological techniques, to demonstrate that many tyrosinase-negative persons have at least a version of the tyrosinase enzyme. However, it is not a correct version and is not functional. These albinos lack the genetic information needed to produce functional tyrosinase.

HOW GENES BEHAVE

Recall that two members of a pair of chromosomes are said to be homologous. They have genes for similar traits arranged in similar order. The gene for each trait occurs at a particular point in the chromosome called a **locus** (plural, loci). In guinea pigs, for example, if one chromosome of a pair contains a gene for coat color, so will the other chromosome of the pair. Genes governing variations of the same trait that occupy corresponding loci on homologous chromosomes are known as **alleles** (see Fig. 10–2).

The term allele emphasizes that there are two or more alternative forms of the gene at a specific locus in homologous chromosomes. Each of these forms can be assigned a letter as its symbol. It is customary to designate the dominant gene, the one that expresses itself, with a capital letter, and the recessive allele (the gene that does not express itself in the presence of a dominant allele) with a lowercase letter. Thus, the letter *B* could be used for black coat color and the letter *b* for brown when specifying the alleles that determine coat color in guinea pigs.

A Monohybrid Cross

The usage of genetic terms and some of the basic principles of genetics can be illustrated by considering a simple **monohybrid cross,** that is, a cross between two individuals that differ with respect to a single characteristic. The mating of a genetically pure brown male guinea pig with a genetically pure black female guinea pig is illustrated in Figure 10–3. During meiosis in the spermatocytes in the male, the two *bb* alleles separate, so each sperm has only one *b* allele. In the formation of ova in the female, the *BB* alleles separate, so each ovum has only one *B* allele. The fertilization of this egg by a *b*-bearing sperm results in an animal with the alleles *Bb*, that is with one allele for brown coat and one allele for black coat. What color would you expect this animal to be?

Suppose that two black guinea pigs each having alleles for both brown and black coat color were crossed. Half of the gametes of each would have alleles for black coat color, and the other half would have alleles for brown coat color. The chance that two alleles for brown coat

Figure 10–2 Homologous chromosomes and alleles.

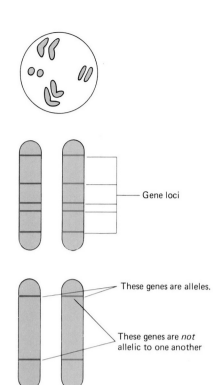

Each diploid cell has two sets of chromosomes, the 2n number. Members of one set can be paired with members of the other set. The members of a given pair correspond in shape, size, and type of genetic information, and are referred to as homologous chromosomes.

— Gene loci

Each chromosome is made up of perhaps thousands of genes. The genes occupy definite physical locations on the chromosomes known as *gene loci*

These genes are alleles.

These genes are *not* allelic to one another

Since diploid organisms possess pairs of homologous chromosomes, the genes borne in corresponding loci of the pair also occur in pairs. If genes occupy the same locus on each of a pair of chromosomes, they are said to be alleles. Allelic genes code for the same polypeptide, and so govern the *same trait*

Gene for seed color (purple)

— Gene for seed color (yellow)

However, even though allelic genes govern the same traits, they need not necessarily contain the same information

— Gene for taste of kernel (starchiness)

Gene for taste of kernel (sweet)

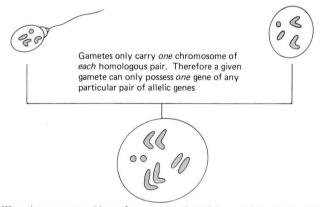

Gametes only carry *one* chromosome of *each* homologous pair. Therefore a given gamete can only possess *one* gene of any particular pair of allelic genes

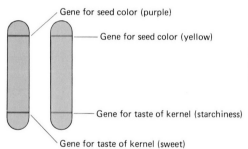

When the gametes combine to form a zygote, it, and the resulting embryo, will have homologous pairs of chromosomes, but of each pair, one member will be *maternal* in origin, and one *paternal* in origin. Each pair will bear allelic genes.

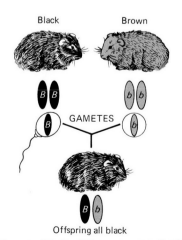

Black   Brown

GAMETES

Offspring all black

Figure 10–3 When a genetically pure black guinea pig is mated with a brown guinea pig, all the offspring are black.

color would meet would be the product of their frequencies of occurrence, that is, $0.50 \times 0.50 = 0.25$.

Similarly the chance that two gametes with alleles for black coat color might meet is 0.25. However, since black and brown gametes can meet each other in two ways (brown from the mother and black from the father *or* black from the mother and brown from the father), their likeli-

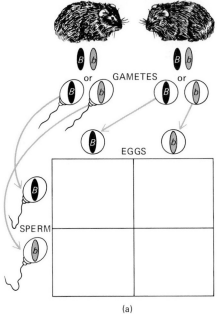

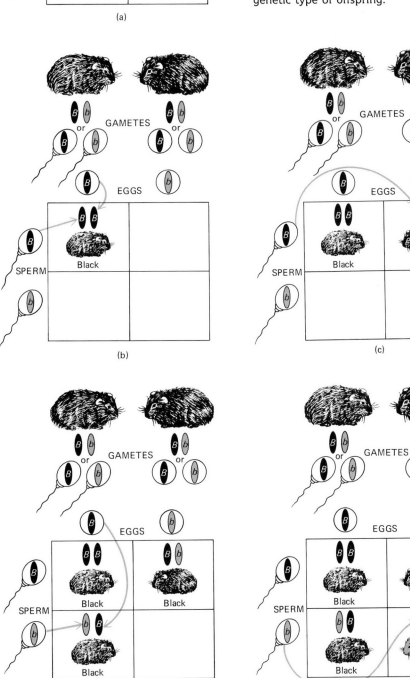

Figure 10-4 The Punnett square method of predicting probable genotypes. Note that the letters in each square indicate the genotype of one genetic type of offspring.

(a)

(b)

(c)

(d)

(e)

hood of meeting is 0.25 + 0.25 = 0.50. This means that if the number of offspring is great enough, three-quarters of the offspring (0.50 + 0.25) will have a black coat phenotype and about one-quarter (0.25) will have brown coats.

The probable combinations of eggs and sperm may be represented in a checkerboard or **Punnett square,** as illustrated in Figure 10–4. The types of eggs can be represented across the top and the types of sperm indicated along the left side. The squares are filled in with the resulting zygote combinations, so that the letters in each square indicate the genotype of one genetic type of offspring.

The generation with which a particular genetic experiment is begun is called the **parental generation,** or P_1. Offspring of this generation are referred to as the first filial generation, or F_1. Those resulting when two F_1 individuals are bred constitute the second filial generation, or F_2.

Homozygous and Heterozygous Organisms

If both genes specify brown coat, then the alleles are identical and the organism is said to be **homozygous** for coat color. In another guinea pig both alleles may specify black coat. This second animal is also homozygous for coat color. But it often happens that one of the alleles carries instructions for brown coat and the other carries instructions for black coat. In that case, the individual is said to be **heterozygous** for coat color. Despite this, only one of the two contrasting genes will express itself.

Dominant and Recessive Genes

When one member of an allelic pair tends to dominate the other completely, so that it alone is expressed in the heterozygous condition, it is said to be **dominant.** If, however, a gene is expressed *only* when homozygous, it is termed **recessive.** In guinea pigs the allele for black coat color is dominant, and the allele for brown coat color is recessive. A guinea pig must have two alleles for brown coat in order to be brown.

Many genes are not clearly dominant or recessive, and when found together, tend to reach a compromise. Such traits are **incompletely dominant.** If *both* are *fully* expressed, they are said to be **codominant.** Human blood types, which we will soon discuss, provide an example of codominance.

The expression and interaction of genes can be much more complex than we are able to indicate in this brief treatment. For instance, a single gene will probably have multiple effects, a quality referred to as **pleiotropy,** rather than a single effect. Some genes are able to suppress the expressions of other, nonallelic genes. This is referred to as **epistasis.** The genes thus suppressed are said to be **hypostatic.** Finally, even a dominant gene may be expressed only in certain instances, a condition called **incomplete penetrance.**

Genotype and Phenotype

An individual's genetic makeup is its **genotype.** However, the genotype is not always detectable in one who is heterozygous for the trait. So we have the term **phenotype,** which refers to the appearance of the individual with respect to a certain inherited trait. Two persons with normal skin pigmentation might very well be genetically different. One might have two alleles for normal pigmentation, while the other might have an allele for normal pigmentation and an allele for albinism without any physical evidence of the latter.

Figure 10–5 Incomplete dominance in Japanese four-o'clocks. Red is incompletely dominant to white in some types of flowers. A plant with the genotype *Rr* has pink flowers.

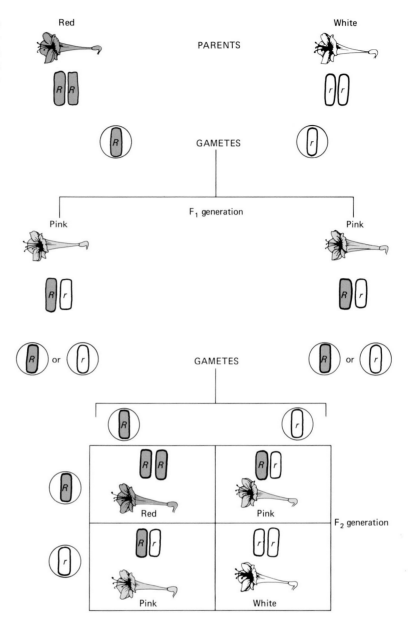

INCOMPLETE DOMINANCE

Not every pair of genes consists of a dominant and a recessive allele. For example, when red- and white-flowered Japanese four-o'clocks are crossed, the offspring do not have red or white flowers. Instead, all the F_1 offspring have pink flowers. How can we explain that? Does this result in any way prove that Mendel's assumptions about inheritance are wrong? Quite the contrary, for when two of these pink-flowered plants are crossed, offspring appear in the ratio of one red-flowered to two pink-flowered to one white-flowered plant (Fig. 10–5).

In this instance, as in other aspects of science, finding results that differ from those predicted simply prompts scientists to reexamine and modify their assumptions to account for the new exceptional results. The pink-flowered plants are clearly the heterozygous individuals, and neither the red allele nor the white allele is completely dominant. When the heterozygote has a phenotype that is intermediate between those of its two parents, the genes are said to show incomplete dominance. In these crosses the genotypic and phenotypic ratios are identical.

Incomplete dominance is not unique to Japanese four-o'clocks. Red- and white-flowered sweet pea plants also produce pink-flowered plants when crossed. In both cattle and horses, reddish coat color is incompletely dominant to white coat color. The heterozygous individuals have roan-colored coats (reddish with white spots). If you saw a white mare nursing a roan colt, what would you guess was the coat color of the colt's father? Is there more than one possible answer?

INDEPENDENT ASSORTMENT

We now know that Mendel's law of independent assortment applies only to traits carried on nonhomologous chromosomes. If different genes are carried on the same chromosome, they will indeed tend to be inherited together. However, if they occur on *different* chromosome pairs, they will be inherited independently.

To illustrate this principle, let us consider two pairs of human traits. The ability to curl the tongue into a tube is a dominant trait; the absence of this talent is recessive. Attached or absent earlobes are recessive; free earlobes are dominant.

A mating that involves individuals differing in two traits is referred to as a **dihybrid cross.** If a man homozygous for tongue curling (*CC*) marries a woman without this trait (*cc*), all their children (*Cc*) may be expected to have this ability, since it is dominant and they are heterozygous for it. Similarly (but quite unrelated), if he has free earlobes (*EE*) and she is homozygous for attached earlobes (*ee*), the children will have free earlobes (*Ee*). One could write the children's genotype for both sets of traits as *CcEe*. Since the two sets of traits are due to genes found on separate chromosome pairs, they will be inherited (or *assorted*) independently of each other.

Let us suppose that—improbable as it is—some of these children were to marry others who have genotypes identical to their own. For reasons of convenience, let us suppose that the total number of grandchildren produced was 16.

As you see by reference to Figure 10–6, four kinds of gametes are possible with respect to these genes—that is, since each gamete will have one of each pair of alleles, a total of four gametic genotypes is possible, given the genetic makeup of the F_1 generation. If, by chance, the chromosome bearing gene *E* is sorted during meiosis into the same gamete as the chromosome bearing gene *c*, the resulting sperm or egg will have the genes *Ec*. Similarly, if the chromosomes bearing *e* and *C* are assorted together, the gamete will have the genes *eC*. Other gametes would be *EC*, and still others would be *ec*. *There are no other possible combinations than these four.* Since allelic genes are always borne on chromosomes that are *separated* from one another in meiosis, a gamete can ordinarily have no more than *one* copy of each allele. Since there are four possible kinds of gametes, working out a cross of that kind will require a box of 16 squares. Count the phenotypes. If you do so properly, the proportions, or ratios, of phenotypes to one another that you will predict among the offspring will fall into a 9:3:3:1 ratio.

POLYGENIC INHERITANCE

Some traits are governed in their expression by more than one pair of allelic genes known as **polygenes.** Sometimes such polygenes are located on different chromosomes and therefore assort independently. Not many such traits are known, for they are difficult to investigate. It is likely, however, that many human characteristics are inherited in this fashion

Figure 10–6 Independent assortment is illustrated in this dihybrid cross. When two heterozygous individuals are crossed, the ratio of phenotypes is 9:3:3:1.

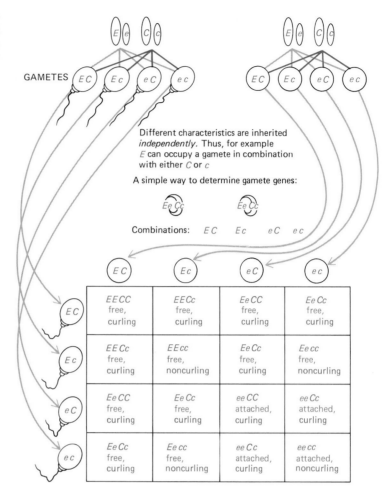

and will eventually be proved to be so. One of the better-attested instances of polygenic inheritance is human skin color.

It is well known that if a person of extreme white complexion marries a person of equally extreme dark complexion, their offspring will be intermediate in color. If that were all there was to it, it would be an obvious case of codominance. To test that assumption, let us consider the offspring of two such intermediate persons. The assumption of simple codominance predicts a 1:2:1 ratio among their offspring. However, what we will observe instead (if their family is sufficiently large, or if we observe offspring in several such families) is a complex distribution not easily expressed by any ratio, ranging from white to very dark with the majority of intermediate color (Fig. 10–7).

The explanation for this seems to be that human skin color is governed by at least three independent allelic pairs of genes, each pair located on a different chromosome. Thus, the cross dark × white is *AABBCC* × *aabbcc*. The F₁ offspring can be only *AaBbCc*. What about the F₂? We have worked out some of the genotypes of the F₂ for you in Figure 10–7. It is up to you to determine their phenotypes, which is not difficult. The individual with the genotype *aabbcc* would be so light that his blood would show through the skin. *AABBCC* would be very dark indeed. Intermediate degrees of darkness can be predicted by counting the number of capital letters in each genotype. Most of us, including most who are called "blacks," would fall into one of the intermediate genotypes.

Many human traits that appear to be inheritable but have a confusing pattern of inheritance eventually may be shown to be produced by polygenes. Still others may be strongly influenced by polygenes but not

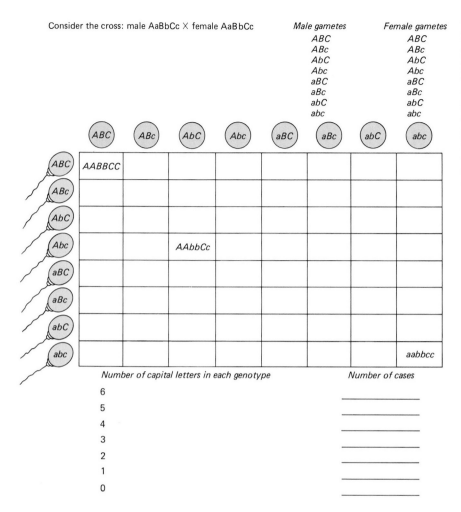

Consider the cross: male AaBbCc × female AaBbCc

Number of capital letters in each genotype   Number of cases

6   _____
5   _____
4   _____
3   _____
2   _____
1   _____
0   _____

Figure 10–7 Polygenic inheritance. Human skin color is an example of polygenic inheritance, which involves the additive interaction of at least three separate, incompletely dominant sets of genes on three separate chromosomes. We will let *A*, *B*, and *C* represent the genes for dark skin color and *a*, *b*, and *c* the light alleles. Since dominance is incomplete, skin color can be estimated by counting the number of capital letters in each genotype.

entirely attributable to them. Height, body build, susceptibility to certain mental and physical diseases, and what we call "intelligence" *may* be such characteristics.

SEX-LINKED GENES

Most chromosomes occur in pairs whose members are like each other. These are called **autosomes.** However, in humans and many other animals there is a pair of chromosomes, the **sex chromosomes,** that are alike in females but different in males. Sex is chromosomally determined, and in a sense it is inherited. Two X chromosomes, as they are called, produce a female. An X and a Y chromosome produce a male (Fig. 10–8).

The Y chromosome is the smallest of all chromosomes and, as far as anyone knows, in human beings contains only genes conferring male sex. Although the Y chromosome pairs with the X chromosome in meiosis, the Y and X chromosomes are not truly homologous, except for a short segment that may exist solely for purposes of meiosis. Thus genes found on the X chromosomes have no known alleles on the Y chromosome. Therefore, in the male any allele that lies on the X-linked chromosome will be expressed whether or not such a gene is dominant or recessive in the XX female. One cannot properly refer to such an X-linked gene in the male as either homozygous or heterozygous. It must be called by a special name: **hemizygous.**

Since no important Y-linked genes are known in men, it is the X chromosomes, present in both men and women, that are meant when the term **sex linkage** is used. The vast majority of sex-linked genes are reces-

Figure 10–8 The inheritance of sex. Sex is determined by the sperm at the moment of fertilization. When the egg is fertilized by an X-bearing sperm, the offspring will be female. When the egg is fertilized by a Y-bearing sperm, the zygote will contain an X and a Y chromosome and the offspring will be male.

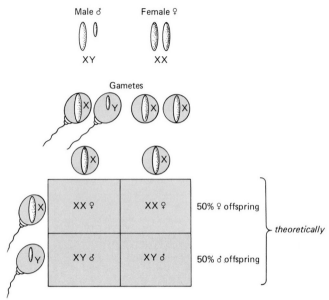

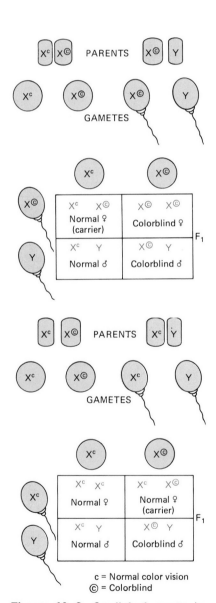

Figure 10–9 Sex-linked genes in humans are usually borne on the X chromosome. Two crosses involving color blindness are shown here. Each circled Ⓒ represents a gene for color blindness. c represents the gene for normal vision.

sive, so that in a female, they must be homozygous to be expressed. One practical consequence is that while females may carry recessive sex-linked traits, these traits usually find expression only in their sons. (A son receives his X chromosome from his mother, *never* from his father.)

In order to be expressed in a female, a recessive sex-linked trait must be present on *both* chromosomes, and so must be inherited from *both* parents. A colorblind girl, to take an example, must have a colorblind father and a mother who is at least heterozygous for color blindness (Fig. 10–9). Such a combination is unusual. Yet a colorblind boy need only have a mother who is heterozygous for the trait. His father can be normal.

Careful analysis of the foregoing will indicate that the X chromosome does not necessarily contain sex-determining genes for femaleness. They could occur on any chromosome. Sex-determining genes must, however, occur on the Y chromosome, for it is solely responsible for the difference between girls and boys. It follows from this that the X chromosome is free to bear many genes that have nothing to do with sex, particularly since the X chromosome occurs in both sexes. In contrast, the Y chromosome is far less likely to carry any genes not directly related to sex.

Suppose for a moment that a woman who is heterozygous for both brown tooth enamel and color blindness marries a man who also bears these traits on his single X chromosome. Admittedly, this combination is hardly likely, but comparable crosses have been made hundreds of times with fruit flies, and the results are predictable.

For our example, it is assumed that in the woman both abnormal traits occur on the same X chromosome. This will produce, as you can see from Figure 10–10, a female chromosomal makeup, or **karyotype,** in which only one gene for color blindness occurs, along with only one gene for brown tooth enamel (number 1). Since color blindness is recessive, the phenotype does not reflect it. The *dominant* brown enamel is, however, expressed. Number 4, also female, will have brown teeth and will be colorblind. Number 2, a male, will share these shortcomings. Only number 3, a male karyotype, will be completely normal.

It should be obvious that the genes for color blindness and brown tooth enamel in our example stay together; they are *not* independently assorted. In fact, the inheritance of multiple genes borne on a single chromosome is very much like that of a *single* gene, at least, as we shall

Focus on . . .

ROYAL BLOOD

"Our poor family seems persecuted by this awful disease," wrote Queen Victoria in her journal, "the worst I know." The disease to which Her Majesty referred was hemophilia. It evidently originated with a mutation in one of her X chromosomes, and has been passed on to almost every family of European royalty. It was even a contributory factor in the Russian Revolution. Czar Nicholas had married Princess Alexandra of Germany, a granddaughter of Queen Victoria. The couple had a number of daughters but just one son, Prince Alexander. This heir to the throne had hemophilia. The monk Rasputin, a quack, claimed to be able to preserve his life and thereby gained great influence over the czarina and, through her, over the czar. The resulting public policies are said by some to have contributed to the Russian Revolution of 1917.

Hemophilia is a disorder of the clotting mechanism of the blood produced by the absence of necessary globulin. If the globulin is artificially provided by injection, the sufferer may live indefinitely. In the absence of this treatment, hemophilia is almost invariably fatal in early youth, though in exceptional cases the patient may reach reproductive age. Typically, death is caused by a fatal stroke or bleeding to death, either internally or externally. Should the hemorrhage take place in the brain, death is quick. Long-term sufferers are susceptible to arthritis arising from bleeding in joint cavities and also, more recently, to AIDS (Chapter 25) contracted from the globulin-containing blood products upon which they are dependent.

An X-linked recessive gene produces the disorder and is inherited in a fashion similar to that of red-green color blindness. As with color blindness, hemophilic disease is much more likely to occur among males than females,[1] but a few cases of homozygous female hemophiliacs are now known. No respecter of rank, hemophilia also occurs in those who are not of royal descent. It should not be confused with the numerous disorders of the clotting mechanism which superficially resemble it but which are not genetic, or not X-linked.

[1]Can you guess why? Hint: Formerly, hemophiliacs seldom or never reached reproductive age.

see, if they are located close together. The condition where genes are borne on the same chromosome is termed one of **linkage.** Linked genes tend to be inherited together[1]. After all, genes are just particular configurations of DNA, and we know that DNA is passed from ancestral cells to their progeny in the packages called chromosomes. It follows that if two or more genes happen to occur in the same chromosomal package, they will tend to be handled together during meiosis and thus will tend to be inherited together.

AUTOSOMAL LINKAGE

Genes also tend to be inherited together if they occur together on a nonsex chromosome, or autosome. Let us take another example from human genetics for which there is ample evidence. Some persons secrete proteins associated with their blood type in the saliva. The trait can be detected even before birth by an examination of the cells sloughed off the embryo in the amniotic fluid. The gene that produces this trait is dominant and is called the secretor gene; we shall call it *S*. The recessive allele is *s*.

We next introduce a serious autosomal hereditary disease involving progressive paralysis and the wasting of muscles, known as myotonic dystrophy, *m.* The dominant normal condition we shall call *M.* Since the locations, or loci, of the *S,s* and *M,m* genes are found on the same chromosome, they are said to be linked.

Suppose that a couple, suspecting that their unborn child may be dystrophic, consults a physician for advice. By careful investigation of the couple and their relatives the physician can deduce that their geno-

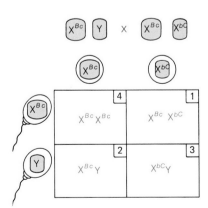

Figure 10–10 A number of characteristics have been shown to be X-linked in humans. They are inherited along with the chromosome that bears them. Therefore, they tend to be inherited together.

---

[1] Genetic linkage and sex linkage are not the same thing. A sex-linked gene is one that is carried on (that is, linked to) the X chromosome (at least in human genetics); linked genes occur on the same chromosome, thus being linked together. Of course, linked genes can be found on the X chromosome, in which case they are not only linked but sex-linked as well.

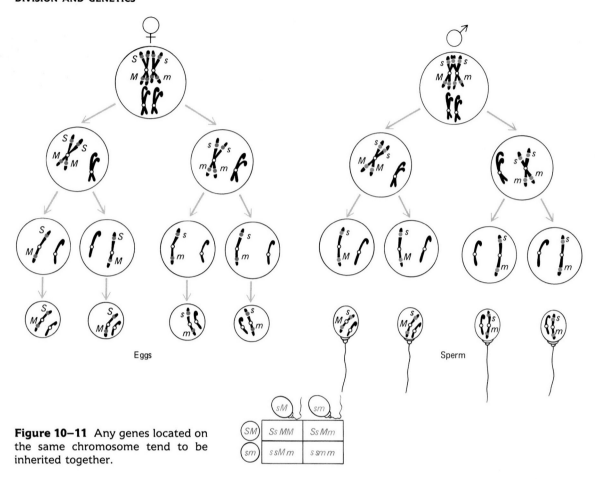

Figure 10–11 Any genes located on the same chromosome tend to be inherited together.

types and the probable genotypes of their children are as shown in Figure 10–11.

If these genes were independently assorted, the male, who has the genotype *SsMm,* would give rise to *four* genetically distinct kinds of gametes. As it is, he produced only two. The possible combinations of his gametes and those of his wife are such that only four genotypes are possible in the offspring, instead of the eight that independent assortment would produce. Notice that in this couple's particular case, if a dystrophic child is born, he *must* be a nonsecretor. The physician has an easy task if the fetus is a secretor—the parents can rest reasonably assured that the child will be normal. But risk increases if the child is a nonsecretor.

Thus far you may have gained the impression that linked genes *always* stay together. As a matter of fact, though they usually do keep company, there are exceptions, and a few individuals (probably about 10% in the foregoing example) will not conform to what we would predict on the basis of genetic linkage. For example, some from the very cross we just studied might turn out to be dystrophic secretors, an "impossible" combination of traits. How can they be accounted for? The solution to this puzzle was determined with the microscope by watching the behavior of chromosomes in meiosis. Recall that when gametes are formed, homologous chromosomes come together during synapsis and crossing over occurs. As that happens, the chromosomes may exchange parts, as shown in Figure 10–12. Suppose that had occurred in the body of the female member of the couple whose case we have been considering. In that particular case, it would turn out *opposite* to our original prediction and *only* a secretor baby would become dystrophic. By studying the results of crossing over in a great many crosses, detailed chromosome maps of the location of specific genes in a number of species have been made (see Focus on Gene Maps and Genetic Linkage).

BLOOD TYPES—A SPECIAL CASE

Although, as the saying has it, all blood is red, there is a surprisingly great diversity of other characteristics, one of which, **blood type,** is of enormous medical and legal importance. Through knowledge of blood types, one may transfuse blood safely, solve tangled inheritance disputes, and identify drops of blood left at the scene of a crime.

Blood type is determined by the kind or kinds of antigens (proteins) occurring on the surface of red blood cells. These have the potential for stimulating the production of antibodies (proteins that combine with specific antigens). The antibodies are capable of reacting with the complementary antigens. Such a reaction can produce an abnormal clumping of blood cells called **agglutination,** or their complete breakdown by **hemolysis.**

The Rh Factors

The Rh series of blood types has been known only since the 1940s. There are several somewhat codominant varieties, but all may be thought of as producing either a *positive* or a *negative* phenotype. For our purposes we will lump the several alleles into two groups—*R*, which produces positive phenotypes, and *r*, which produces negative. *R* is dominant. Thus a person with Rh-positive blood has the genotype *RR* or *Rr*. The negative phenotype can only be *rr*. About 85% of the population of the United States has Rh-positive blood.

A person with Rh-positive blood has Rh antigens associated with the blood-cell membranes, and *no* anti-Rh antibodies in the plasma. An Rh-negative person possesses no Rh antigens and no natural anti-Rh antibodies. But if he were to receive a transfusion of Rh-positive blood, he would quickly develop anti-Rh antibodies, which would then hemolyze the cells of the *next* transfusion of Rh-positive blood. The first such transfusion would, however, be relatively harmless in itself.

Although several kinds of maternal–fetal blood type incompatibilities are known, **Rh incompatibility** is probably the most important (Fig. 10–13). If a woman is Rh-negative and her husband is Rh-positive, the fetus may be Rh-positive. A small quantity of blood from the fetus may pass through some defect in the placenta (especially during the birth process) and into the mother's blood, sensitizing her white blood cells, which then produce antibodies to the Rh antigens. When this woman becomes pregnant again, sensitized white blood cells produce antibodies that may pass through the placenta into the fetal blood and cause clumping of the red blood cells. Breakdown products of the hemoglobin released into the circulation damage many organs, including the brain. This disease is known as **erythroblastosis fetalis.** In extreme cases so many fetal red blood cells are destroyed that the fetus dies before birth.

When Rh-incompatibility problems are suspected, blood can be exchanged while the baby is still within the mother's uterus, but this is a risky procedure. Rh-negative women are now treated just after childbirth (or at termination of pregnancy by miscarriage or abortion) with an anti-Rh preparation. This drug apparently clears the Rh-positive cells from the mother's blood very quickly, thus minimizing the chance of sensitizing her own white blood cells. As a result, her cells do not produce the antibodies that could harm her next baby.

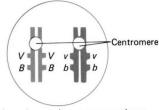

Two homologous chromosomes undergo synapsis in meiosis.

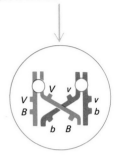

Crossing over between a pair of chromatids.

First meiotic division

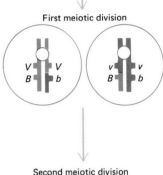

Second meiotic division

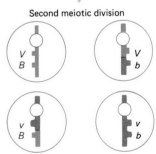

Four haploid gametes produced; here two crossover and two noncrossover gametes occur.

Figure 10–12 Crossing over and genetic recombination. Crossing over, the exchange of segments between chromatids of homologous chromosomes, permits recombination of genes— vB and Vb, for example. The farther apart the genes are located on a chromosome, the greater the probability that an exchange of segments between them will occur.

Focus on . . .

GENE MAPS AND GENETIC LINKAGE

The farther away from one another the loci of different linked genes are on the chromosomes, the more frequently they tend to cross over. By counting enough instances of crossing over, one can estimate the relative distances between linked genes. Using this principle, genetic maps have been constructed that give the order and relative distances of all known genes on them.

More than 1000 gene loci have been identified on the four chromosomes of the fruit fly (*Drosophila melanogaster*). Some of these gene loci are shown in the genetic map illustrated below. Note that both the nor-

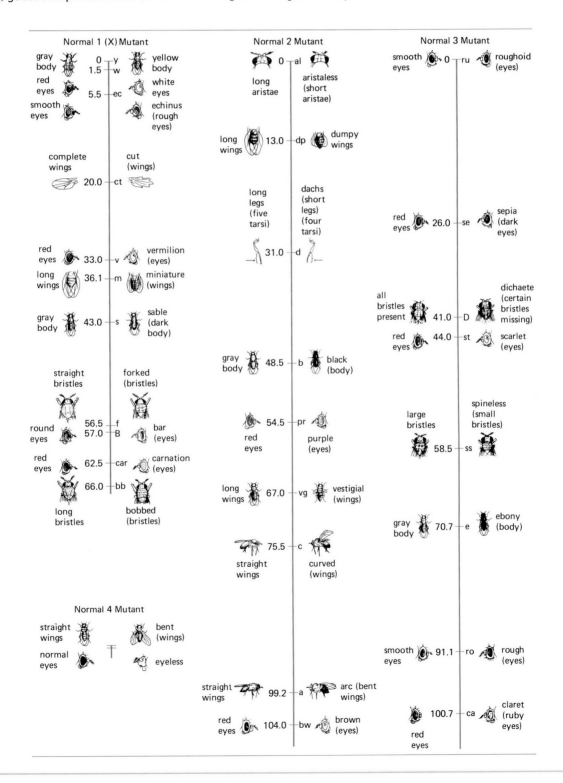

mal and the mutant phenotypes are indicated. Similar maps have been constructed for some bacteria and fungi, and for corn, mice, and humans.

An additional mapping technique developed in recent years involves the experimental fusion of human cells with those of mice in tissue culture. For some reason, the resulting "hybrid" cells tend to retain the mouse chromosomes but progressively eject the human chromosomes over the course of several cellular generations. But if just one human chromosome were to remain in such a cell, any human gene products that could be detected would have to result from genes present on that particular chromosome.

Why is all this attention paid to the location of genes? For one thing, it can help prove the genetic nature of some disease states. If a disease can be shown to be associated with a nondisease trait known to be genetic, then it would seem that the disease itself must be inherited. Second, precise genetic maps will facilitate the development of advanced genetic engineering techniques for use in eukaryotes. Third, genetic maps are needed for the development of new varieties of agricultural plants and, increasingly, of animals as well. Last, genetic maps are of importance in keeping track of the evolutionary history of closely related species and populations of animals and plants, including human beings. By careful comparison of genetic maps, it should be possible to learn a great deal about the movements and migrations of our preliterate ancestors who lived before such a thing as history was dreamed of.

The ABO System and Multiple Alleles

Multiple alleles exist when a particular gene, occupying one given locus, exists in more than two forms. Thus an organism can have no more than two of a given set of multiple alleles at a time, but more than two such alleles are known to exist. An important example is afforded by the ABO series of blood types. This series of blood types was the first to be discovered—as a result of early experiments in blood transfusion, which were unpredictably but all too often fatal.

LANDSTEINER'S LAW

As we have seen, in the case of blood cells when complementary antibodies and antigens come into contact, the blood cells clump together **(agglutination)** or break open **(hemolysis).** Both these reactions can take

Figure 10–13 Rh incompatibility can cause serious problems when an Rh-negative woman and an Rh-positive man produce Rh-positive offspring. (a) Some Rh$^+$ red blood cells (RBCs) leak across the placenta from the fetus into the mother's blood. (b) The woman produces special antibodies (D antibodies) in response to antigens on the fetal red blood cells. (c) Some of these antibodies cross the placenta and enter the blood of the fetus, causing red blood cells to rupture and release hemoglobin into the circulation. The fetus may develop erythroblastosis fetalis.

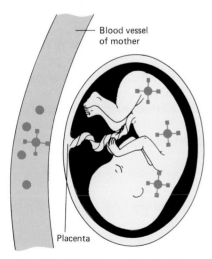

A few Rh+ RBCs leak across the placenta from the fetus into the mother's blood

(a)

The mother produces anti-Rh antibodies in response to Rh antigen on Rh+ RBCs

(b)

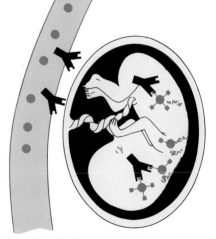

Anti-Rh antibodies cross the placenta and enter the blood of the fetus. Hemolysis of Rh+ blood occurs. The fetus may develop erythroblastosis fetalis.

(c)

●   Rh— RBC of mother

Rh+ RBC of fetus with Rh antigen on surface

Anti-Rh antibody made against Rh+ RBC

Hemolysis of Rh+ RBC

Table 10–1
ABO BLOOD TYPES

Phenotype	Genotype	Antigens	Antibodies	How inherited
A	AA, AO	A	Anti-B	Dominant
B	BB, BO	B	Anti-A	Dominant
AB	AB	A and B	None	Codominant
O	OO	None	Anti-A and B	Recessive

place in laboratory glassware and presumably also in the body of a living person. For reasons not fully understood, the usual kinds of ABO antibodies develop normally even without known exposure to antigens capable of provoking them. However, agglutination is the reaction most likely to occur in laboratory glassware, and hemolysis is the characteristic response *in vivo.* Hemolysis results in the release of hemoglobin in the plasma. This in turn can produce kidney damage that may lead to death. Such reactions do not ordinarily take place in one's own blood because, as the hematologist Karl Landsteiner discovered, the *antibodies of the plasma are never complementary to the antigens of the cells* **(Landsteiner's law).** There is no such assurance when the bloods of two individuals are mixed, as in a transfusion.

If blood is transfused into a patient whose plasma contains complementary antibodies hostile to the cells of the donor, damage to these cells is bound to result. The reverse can also happen. If blood is transfused into a recipient and the donated *plasma* is incompatible with the recipient's cells, trouble will eventually result. Initially, though, such transfusions may be harmless, since the half pint or so of potentially dangerous plasma is rapidly diluted in the recipient's blood volume to harmless levels.

THE GENETICS OF THE ABO SERIES

Table 10–1 gives the basic facts of inheritance for both the antibodies and the antigens of the ABO series. Three alleles are known for the ABO series, which by their interaction can produce blood types A, B, AB, and O—depending upon the exact combination of genes in the particular allelic pair possessed by an individual (Fig. 10–14). A normal person can possess no more than two of the ABO alleles. Type O gene is recessive to all other genes, and A and B exhibit codominance with respect to each other. Thus two genotypes can give rise to type A, and two to type B. AB and O each have only one possible genotype.

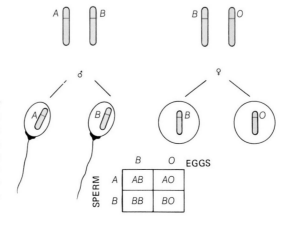

Figure 10–14 Multiple alleles for blood types. Three different alleles exist for ABO blood type. In this cross each parent produces only two kinds of gamete, but between them, there are three types of gametes with respect to ABO blood type. Four possible genotypes occur among their offspring.

GENES AND DISEASE

More than 150 human disorders involving enzyme defects have been linked with genetic mutations (see Table 10–2). These disorders are sometimes referred to as inborn errors in metabolism. Most are inherited as autosomal recessive traits. Sickle-cell anemia, phenylketonuria (PKU), cystic fibrosis, and Tay-Sachs disease are well-known diseases that have been linked to gene defects.

Many recessive genetic diseases can now be detected in the heterozygous state by modern enzyme analysis and other advanced chemical techniques. It is also possible to detect abnormalities in fibroblasts (cells of connective tissue) found in the amniotic fluid, for these fibroblasts originate in the fetus. Thus through genetic screening it can often be determined even before birth whether a fetus will suffer from a genetic disease.

CHROMOSOMES AND DISEASE

Chromosome abnormalities are called **aneuploidies.** An aneuploidy may involve an extra chromosome so that one of the person's homologous sets consists of three chromosomes instead of the normal two. (Having three of a kind is called **trisomy;** Fig. 10–15; Table 10–3.)

These aneuploidies appear to arise through defects of meiosis, in which chromosomes fail to separate. This is called **nondisjunction.** In still other cases, a part of a chromosome may break off and attach to another chromosome. The gamete then may have chromosomes that are abnormally short or long. Such **translocations** may be functionally much the same as a trisomy if they result in the presence of too much genetic material in a zygote.

A trisomy might well produce a 50% increase in the activities of the products of most of the involved genes, and recent studies have confirmed this prediction in the case of some genes. So it is little wonder that the body chemistry and development of its possessor will be unbalanced. People who have a trisomy of the No. 21 or 22 chromosomes or of the sex chromosomes may survive into adulthood. Trisomies of the other chromosomes are much rarer because they are much more likely to be fatal. Quite possibly, trisomy of the No. 21 or 22 is less likely to be fatal because these chromosomes are the smallest and possess the fewest genes.

It is striking how little disorder is produced by abnormalities of the sex chromosomes, particularly by X-chromosome aneuploidy despite the

Figure 10–15 Down's syndrome, a disease usually associated with trisomy of the 21st chromosome. (*a*) A photomicrograph of the chromosomes (called a karyotype) of a child with Down's syndrome. Note the extra chromosome 21. (*b*) A two-year-old boy with Down's syndrome. (Courtesy of Mr. and Mrs. Beny Peretz.)

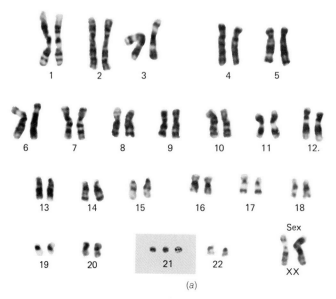

(a)

(b)

Table 10-2
SOME IMPORTANT GENETIC DISORDERS*

Name of disorder	Mode of inheritance	Description	Treatment, if any	Comments
Cystic fibrosis (CF)	Autosomal recessive	High level of sweat electrolytes, pulmonary disease, cirrhosis of liver, pancreatic malfunction, and especially, nonsecretion of digestive enzymes; females sometimes reproduce; life expectancy 12–16 years, with some living into 30s and 40s; commonest in persons of northern European extraction	Symptomatic, with emphasis on digestive enzyme replacement and control of respiratory infections	CF kills more children than diabetes, rheumatic fever, and poliomyelitis combined; exists in different degrees of severity; thick mucus interferes with lung clearance; recent research indicates that CF may become almost routinely detectable prenatally by examination of amniotic fluid
Hemoglobinopathic disease, e.g., sickle-cell anemia (SCA)	Group of autosomal recessive or incompletely dominant traits	Abnormalities of red blood cells caused by presence of certain inappropriate amino acids at crucial locations in hemoglobin molecule	Varies with type of disease. Some, e.g., hereditary methemoglobinemia, may require no treatment. Some can hardly be treated at all; SCA can be treated to some degree	These traits are similar and related but not allelic; microcytic anemia is commonest in Mediterranean populations, sickle-cell anemia in some black populations; when heterozygous, SCA offers some protection against malaria
Tay-Sachs disease	Autosomal recessive	Abnormal accumulation of lipids due to deficiency of the enzyme hexoseaminidase A; destroys central nervous system in young children		Especially prevalent among Jews of Eastern European ancestry

relatively large size of the X chromosome. The probable explanation is that even in normal females one of the X chromosomes in each cell is not functional, or at least not fully functional. This is known as the **Lyon hypothesis.** In early development one of the two chromosomes is inactivated. Which one is inactivated appears to be a random choice. Thus for the most part a female possesses only the enzymes or other proteins produced by half of her X chromosomes, which in many cases will be the same as the amount produced by a male's single X chromosome. Interestingly, in the majority of investigated cases, when a female is heterozygous for a sex-linked trait, half her cells display it and half her cells do not, showing that half her cells have inactivated the paternal X chromosome, and half have inactivated the maternal X chromosome (Fig. 10–16).

The inactivated X chromosome even seems to be visible in certain female body cells. Thus female granulocytes (a type of white blood cell) often possess a nuclear "drumstick" projection never found in normal male granulocytes (Fig. 10–17), and female epithelial cells (cells that line body cavities and cover body surfaces) display nuclear **Barr bodies**

Name of disorder	Mode of inheritance	Description	Treatment, if any	Comments
Phenylketonuria (PKU)	Autosomal recessive	Deficiency of liver phenylalanine hydroxylase leads to a depression of the levels of other amino acids, leading in turn to mental deficiency	A low-phenylalanine diet minimizes symptoms; most states have extensive PKU screening programs, in which newborns are tested for excessive blood phenylalanine, or for presence of metabolic products in urine	Since melanin is synthesized from tyrosine, tyrosine deficiency caused by phenylalanine-hydroxylase deficiency results in light coloring
Tyrosinase-negative oculocutaneous albinism (T⁻ albinism)	Autosomal recessive	Absence of pigmentation due to functional absence of tyrosinase; visual acuity 20/200 or less; marked susceptibility to skin cancer	Avoidance of sunlight	Somewhat more common among blacks than whites
Tyrosinase-positive oculocutaneous albinism (T⁺ albinism)	Autosomal recessive	Reduction of pigmentation due to malabsorption of tyrosine by body cells; if racial background is heavily pigmented, some pigmentation will survive, though in some cases phenotype is virtually identical with T⁻; pigmentation and visual acuity improve with age		Highest incidence in American Indians, less in blacks, least in whites; Hybrid T⁺–T⁻ individuals appear normal

*This table is not complete. For detailed information a book of human genetics should be consulted, such as *The Metabolic Basis of Inherited Disease,* by John B. Stanbury, et al. (McGraw-Hill, New York, 1983). It should also be noted that a number of common diseases, such as diabetes mellitus, though not genetic in the usual sense, appear to have a genetic component influencing susceptibility.

Figure 10–16 Calico cat. In some cells, according to the Lyon hypothesis, one X chromosome is inactivated, and in others, the other X chromosome is inactivated. In calico cats, some genes influencing coat color are X linked. Only the cells with active X chromosomes bearing a "black" gene produce black hairs. Only those whose active X chromosome has an "orange" gene can produce orange hairs. Calico cats must be either female or abnormal, XXY males, since to display this trait they must have two X chromosomes.

Table 10–3
SOME CHROMOSOME ABNORMALITIES

Aneuploidy	Common name	Description
Trisomy 13		Multiple defects, with death by 1–3 months
Trisomy 15		Multiple defects, with death by 1–3 months
Trisomy 18		Ear deformities, heart defects, spasticity, and other damage. Death by 1 year
Trisomy 21	Down's syndrome (mongolism)	Overall frequency is about 1 in 700 live births. True trisomy usually found among children of older (40+) mothers, but translocation resulting in equivalent of trisomy may occur in children of younger women. A 35-year-old mother has 1:200 risk of having Down's syndrome child. A 40-year-old mother has 1:50 chance of doing so, and the risk at 44 is 1:20. Epicanthic skin fold, though not same as that in Mongolian race, produces superficial resemblance; hence older name "mongolian idiocy." Varying degrees of mental retardation, usually IQ below 70, though more intelligent exceptions are known. Short stature, protruding furrowed tongue, transverse palmar crease, cardiac deformities common. Patients usually die by age 30–35; 50% die by age 3 or 4. Unusually susceptible to leukemia and respiratory infections. Females are fertile if they live to sexual maturity, producing 50% Down's syndrome offspring
Trisomy 22		Similar to Down's syndrome but with more skeletal deformities
XO	Turner's syndrome	Female with short stature, webbed neck, sometimes slight mental retardation. Ovaries degenerate in late embryonic life, leading to rudimentary sexual characteristics. Similar disorders occur sometimes in XX individuals, perhaps resulting from abnormalities of X-chromosome inactivation and, very rarely, in XY individuals
XXY	Klinefelter's syndrome	Male with slowly degenerating testes, enlarged breasts, and developing eunuchoidism
XYY		Unusually tall male, heavy acne, some tendency to mild mental retardation
XXX		Despite triploid X chromosomes, usually fertile, normal females
Short 5	Cri-du-chat	Microcephaly, severe mental retardation. In infancy, cry resembles that of cat. Defective chromosome is heterozygous.

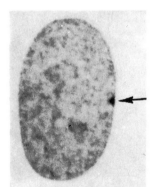
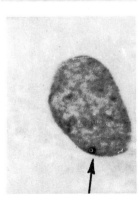

Figure 10–17 Barr body in human fibroblasts cultured from the skin of a female (approximately ×2200). (Courtesy of Dr. Ursula Mittwoch, Galton Laboratory, University College, London.)

not found in normal male cells. These appear to be the condensed and inactivated second X chromosome.

CONSANGUINITY: MARRIAGE BETWEEN RELATIVES

Marriage between relatives is referred to as **consanguinity,** which is derived from a Latin word meaning "related," or "brother" or "sister." The risk of bearing a child with a major congenital abnormality of genetic origin is 70% greater if the parents are, for example, first cousins than if they are unrelated. The death rate for children of such consanguineous matings is between three and four times higher than among a comparison population. Almost half the offspring of brother-sister unions are more or less grossly handicapped by genetic disease.

No mystery enshrouds the increased risk of malformed offspring from consanguineous matings. The problem can be traced to a greater chance of sharing harmful genes. All of us no doubt shared the same ancestors, hence the same genes, in the distant past. But with time the pool of human genes became diverse as a result of mutations and evolutionary processes, so that today each of us is genetically unique. Only identical twins have identical genes. Each of us carries an estimated eight harmful recessive genes. If you marry a nonrelative, the chance is remote that you will select someone who shares any of the same harmful genes. There is little chance that any of your offspring will be unfortunate homozygotes, receiving a double dose of harmful recessive genes. However, if you selected your first cousin as your mate, you would have a common set of grandparents, and so in all probability one of every eight of your genes would be identical. Your offspring would be homozygous for one-sixteenth of all their gene traits. Second cousins share one thirty-second of their genes.

CHAPTER SUMMARY

I. Much of the science of genetics was founded by the 19th-century research of Gregor Mendel. As currently understood, the basic principles describing the behavior of genes and their interactions may be summarized as follows.
A. The basic unit of heredity is the gene.
B. Fundamentally, a gene acts by directing the production of a polypeptide and ultimately a protein.
C. Characteristics of these proteins determine the traits of the organism.
D. Genes are associated with chromosomes and occur on them.
E. The behavior of genes is similar to the behavior of chromosomes.
F. Body cells possess both chromosomes of a given pair, that is, they are diploid. Gametes, however, are haploid, that is, each gamete possesses but one member of each chromosomal pair.
G. The production of any protein or peptide is governed by two genes present in each and all body cells. These genes are alleles.
H. Allelic genes are carried on homologous chromosomes, which, like them, are paired.
I. One member of each homologous pair of chromosomes is maternal in origin. The other member is paternal in origin.
J. Members of a gene pair may be alike (homozygous) or unlike (heterozygous).
K. When members of a gene pair are unlike, the traits they govern may be determined by one of them if it is clearly dominant, or by both of them. A gene that is expressed only when homozygous is said to be recessive.
L. Normally, no offspring can receive any more or less than one representative of a homologous pair of chromosomes from each parent.
M. The member of a pair of chromosomes (and its contained genes) that a gamete receives is governed entirely by chance.
N. The particular combination of maternal and paternal chromosomes received by an offspring is therefore also governed by chance.

O. The inheritance of a particular chromosome has no influence on the inheritance of any other chromosome not homologous to it.

P. The inheritance of any gene has no influence on the inheritance of any other gene borne on a nonhomologous chromosome.

Q. The X chromosome is exceptional in that it has no homologous mate in the male. A gene located on the X chromosome of a male will be expressed even if it is recessive.

II. Allelic genes sometimes exist in multiple form, as in the ABO series of blood types. Genetically, O is recessive to both A and B, but A and B are codominant with respect to each other. Blood-group inheritance may be used to rule out possible candidates in disputed paternity and similar suits.

III. The Lyon hypothesis proposes that female X chromosomes are inactivated at random in early embryonic life, so each cell possesses but one functional X chromosome, of maternal or paternal origin.

IV. Genes borne on the same chromosome tend to be inherited together, a condition known as linkage.

A. However, members of homologous chromosome pairs may exchange parts (along with associated genes) when crossing-over occurs in meiosis.

B. The farther apart linked genes are, the more frequently they cross over. This, plus newer techniques based on cell fusion, permits the development of chromosomal gene linkage maps.

Post-Test

1. The modern science of genetics was founded in the _____ century by _____ _____, whose work was nevertheless almost entirely forgotten until its rediscovery after his death.

2. A unit of hereditary information is called a _____, which is somewhat more exactly defined as the hereditary information necessary to produce a particular _____.

3. The expression of a gene in the physical or other characteristics of an organism is its _____.

4. Genes are inherited independently if they occur on non-_____ chromosomes.

5. Genes are sex-linked (in human beings) if they occur on the _____ chromosome. A father cannot pass such a sex-linked trait to his _____.

6. The ABO series of blood types is a system of _____ alleles in which the genes for types A and B are mutually _____, but both of them are _____ in their expression to O.

7. It has sometimes been suggested that consanguineous marriages are genetically acceptable if no hereditary disease can be demonstrated in the family. However, it is believed that almost everyone carries _____ genes for genetic disease, so that two persons with a near mutual ancestor are more likely to have some of the _____ genes for such disease.

8. If two normally pigmented persons produce an albino baby, the parents must have been _____ for this _____ trait.

9. A type AB baby is born into a family whose wife is type B and whose husband is type O. The baby <u>can/cannot</u> be shown to have a different father.

10. According to the Lyon hypothesis, females and XXY males _____ their second X-chromosome.

11. Down's syndrome is produced by _____ of the No. 21 chromosome and is especially common among the children of older _____.

Review Questions

1. What are alleles?
2. What is meant by the terms *homozygous, heterozygous,* and *hemizygous?*
3. Why can one receive only one member of a homologous pair of chromosomes from each parent? What would happen if you tried to solve genetic problems without realizing this?
4. Define *dominant, recessive, codominant, incompletely dominant.*
5. What are polygenes? Give an example.
6. What usually makes a gene dominant? Why do you suppose that most known genetic defects are recessive?
7. What chromosome or chromosomes bear sex-linked traits?
8. What is the Lyon hypothesis?
9. Give an example of genetic screening.
10. What is an aneuploidy? Give an example.

11. Why do you think sex-chromosome aneuploidies are common, whereas most possible aneuploidies are rare or unknown?

12. Why are female hemophiliacs virtually unknown? What about female victims of red-green color blindness?

13. Physicians are concerned about possible hemolytic disease of the newborn (erythroblastosis fetalis) when the father is Rh-positive and the mother is Rh-negative. They are not concerned, however, when the father is Rh-negative and the mother is Rh-positive. Why not?

14. What is meant by linkage? By crossing-over? How is it possible to detect linkage among genes?

15. Mendel originated the principle of independent assortment on the basis of several crosses involving genes that today are known, in fact, to be linked in some cases. How could linked genes appear to be inherited independently?

16. A certain kind of white coat is dominant in cats. Such cats are also deaf. (a) What is the proper name for such multiple genetic effects? (b) If two white cats have normally colored offspring, what is their genotype?

17. A color-blind girl has a normally sighted mother. What was the phenotype of her father? the genotype of her mother?

18. The wife of a type AB man has a type O baby. Should he entertain any suspicions?

19. Hemophilia is X-linked and recessive. A hemophilic man has a son. He plans never to have any other children, but he worries that he may pass this genetic disease on to his grandchildren through his son. Are these fears justified? Why or why not?

Readings

Corcos, A., and F. Monaghan. Some Myths About Mendel's Experiments. *The American Biology Teacher,* 47: 233–236 (April 1985). Mendel's laws were only implied in his papers. He himself may not have fully appreciated their importance.

Edward, J.H. *Human Genetics.* New York, Methuen, 1978. A good text of general genetics with many examples and illustrations drawn from the genetics of human beings.

Feldman, M.W., and R.C. Lewontin. "The Heritability Hang-up," *Science,* 190: 1163–1168. A discussion of how the inheritance of complex and subtle traits may be established.

Jansen, Jann D. *Child in a White Fog.* New York: Vantage Press, 1975. What it is like to have a child who is severely afflicted with cystic fibrosis, a hereditary disease.

Kolata, G.B. "Thalassemias: Models of Genetic Diseases," *Science,* 210: 300–302 (17 October 1980). Sickle-cell anemia is not the only one.

Langer, A. "Practical Genetics in Office Practice," *Hospital Medicine,* 18: 109–117 (August 1982). Excellent introduction to practical medical genetics. Especially useful summary of symptoms of Down's syndrome.

Lewontin, R. *Human Diversity.* San Francisco, W.H. Freeman, 1982. The principles of genetics as applied to human variation.

Mange, A.P., and E.J. Mange. *Genetics: Human Aspects.* Philadelphia, Saunders College Publishing, 1980.

Menozzi, P., Piazza, A., and L. Cavalli-Sforza. "Synthetic Maps of Human Gene Frequencies in Europeans." *Science,* 201: 786–792 (1 September 1978). Inferring prehistoric population dynamics from present frequencies of human genes.

Suzuki, D.T., Anthony, J.F.G., and R.C. Lewontin. *An Introduction to Genetic Analysis,* 2nd ed. San Francisco, W.H. Freeman, 1981. A stiff but excellent introduction to general genetic principles.

Woolf, C.M., and F.C. Dukepoo. "Hopi Indians, Inbreeding and Albinism," *Science,* 164: 30–37 (4 April 1969). A typical example of the increased prevalence of a genetic trait in a small, genetically isolated population.

Chapter 11
DNA: THE MOLECULAR BASIS OF INHERITANCE

Outline

I. The Secret Formula of Life
   A. Miescher and the nucleus
   B. The X-ray eye
   C. Building models
II. The structure of DNA
III. DNA replication
Focus on The Mermaid's Wineglass and the Secret of Life
Focus on DNA and the Transfer of Genetic Information

Learning Objectives

After you have read this chapter you should be able to:

1. Outline the history of the scientific investigation of the nucleic acids, giving the major contributions of Miescher, Watson, and Crick.
2. Describe or diagram the basic chemical structure of a nucleic acid strand and distinguish chemically between DNA and the varieties of RNA.
3. Given the base sequence of one strand of DNA, predict that of a complementary strand of DNA.
4. Summarize the process of DNA replication.

Biologically speaking, no diamond is as remarkable as a fingernail. A product of life, that fingernail is composed, like the diamond, of huge aggregations of carbon atoms. Unlike the diamond, however, it contains other atoms as well—hydrogen, oxygen, nitrogen, phosphorus, and sulfur—in an arrangement no less precise than that of the atoms of the diamond. A diamond grows in accordance with mathematics and the geometric principles consequent upon the nature and angles of its chemical bonds. The fingernail's growth is consistent with such principles but is not simply a consequence of them. It is assembled within a living organism and is the product of a sophisticated information storage and retrieval system that programs the exact sequence of its components.

THE SECRET FORMULA OF LIFE

When in 1828 the chemist Friedrich Wöhler heated the inorganic compound ammonium carbonate, he synthesized urea, an organic compound that had never been observed except as the product of a living thing. With that accomplishment, much of the notion of a special force "vital to life" seemed to disappear. Yet for some time life continued to hold many secrets about itself from the investigations of its researchers. Until recently the deepest of these secrets was the mechanism necessary to put the chemicals of life together in their exceedingly improbable, relatively entropic patterns.

Miescher and the Nucleus

Among the buildings of the 19th-century University of Tübingen stood a castle-like structure containing a chemical laboratory advanced for its day, but resembling to the modern eye a whitewashed laundry room in a sunnier-than-average cellar. Here, a young Swiss student, Friedrich Miescher, pursued doctoral research in chemistry.

For his thesis problem Miescher had chosen the chemical makeup of the cell nucleus. In two centuries of microscopic investigation of tissues, it had become evident that all were composed of cells or cell products. By the mid-19th century cells were considered, as they still are, the ultimate units of life.

Early microscopists speculated that the nucleus might be in some way the cell's center of control, perhaps the place where the mysterious vital force of life might be concentrated. Later experiments seemed to confirm these intuitions, for it was shown that when the nucleus was removed from the cell, invariably the cell died. Experiments also showed that the nucleus controlled the appearance and function of the cell (see Focus on the Mermaid's Wineglass and the Secret of Life).

Miescher understood that if he were to investigate these tiny structures, it would be necessary to prepare them almost like a reagent, for the crude methods of analysis then available could never be applied to a single cell nucleus. He needed purified and isolated cell nuclei by the millions.

Daily Miescher collected pus-soaked dressings from the hospitals of Tübingen, which were filled with soldiers wounded in the Franco-Prussian War. Back in his laboratory he arranged to have the cells digested by gastric juice, which left most of the nuclei untouched. After washing and purification, these could be chemically analyzed. Miescher's analyses disclosed that one of the most important constituents of the nuclei was a very weak acid that contained phosphorus, carbon, oxygen, and nitrogen. Miescher named it *nuclein* almost casually, not realizing its importance. We know that Miescher's nuclein is one of the most important substances in the universe, and indeed it contains the very secret of life. Today we call it DNA.

Focus on . . .

THE MERMAID'S WINEGLASS AND THE SECRET OF LIFE

Introducing *Acetabularia*

In the imaginations of the more romantically inclined biologists, the little seaweed *Acetabularia* resembles a mermaid's wineglass. Less imaginatively, it has been described as a little green toadstool measuring, at most, 2 or 3 inches in length. Although it is a typical alga of tropical seas, it also occurs in some subtropical waters that are both shallow and somewhat rocky.

Nineteenth-century biologists discovered that this insignificant underwater plant consists of a single giant cell. Small for a seaweed, *Acetabularia* is gigantic for a cell. It consists of (1) a rootlike holdfast, (2) a long cylindrical stalk, and (3) at sexual maturity, a cuplike cap. The nucleus is found in the holdfast, about as far away from the cap as it can be. In due course, the nucleus divides by meiosis, and its progeny of pronuclei swim up the stalk into the cap, where they become the pronuclei of sex cells. These will be released upon maturity to swim away in search of partners. Although there are several species of *Acetabularia,* with caps of different shapes, all species function similarly.

Acetabularia. (Charles Seaborn.)

Hammerling's and Brachet's Experiments

If the cap of *Acetabularia* is removed experimentally just before reproduction, another one will grow after a few weeks. Such behavior, common among lower organisms, is called **regeneration.** This fact attracted the attention of investigators, especially A. Hämmerling and Jean Brachet, who in the early decades of this century became interested in the relationship that might exist between the nucleus and the physical characteristics of the plant. Because of its great size, *Acetabularia* could be subjected to surgery impossible with smaller cells. These investigators and their colleagues performed a brilliant series of experiments that in many ways laid the foundation for much of our modern knowledge of the nucleus. In most of these experiments they employed two species of *Acetabularia,* *A. mediterranea,* which has a smooth cap, and *A. crenulata,* with a cap broken up into a series of finger-like projections.

The kind of cap that is regenerated depends upon the species of *Acetabularia* used in the experiment. As you might expect, *A. crenulata* will ordinarily regenerate a *crenulata* cap, and *A. mediterranea* will regenerate a *mediterranea* cap. But it is possible to graft two capless plants of different species together by telescoping their stalks into one another—not as easy as it may sound, incidentally. After this union, they will regenerate a common cap that has characteristics intermediate between those of the two species involved. Thus, there is evidently something about the lower part of the cell that controls cap shape.

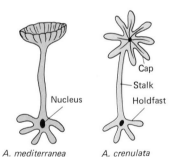

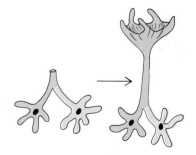

Stalk Exchange

It is also possible to attach a section of *Acetabularia* to a holdfast that is not its own by telescoping the cell walls of the two into one another. In this way the stalks and holdfasts of different species may be intermixed.

First, we take *A. mediterranea* and *A. crenulata* and

remove their caps. Now we sever the stalks from the holdfasts. Finally, we exchange the parts.

What happens? Not, perhaps, what you would expect. The caps that regenerate are characteristic not of the species that donated the holdfasts but of those that donated the stalks.

However, if the same caps are removed still again, this time the caps that regenerate will be characteristic of the species that donated the holdfasts. That will continue to be the case no matter how many times the regenerated caps are removed.

From all this we may deduce that the ultimate control of the cell is vested in the holdfast, for from now on, no matter how often the caps of these grafted plants are removed, they are always regenerated according to the species of the holdfast. However, initially there is a time lag before the holdfast gains the upper hand. The simplest explanation for this delay is that the holdfast produces some cytoplasmic temporary messenger substance whereby it exerts its control, and that initially the grafted stems still contain enough of that substance from their former holdfasts to regenerate a cap of the former shape. But this still leaves us with the question of just what it is about the holdfast that accounts for its dictatorship. An obvious suspect is the nucleus.

Nuclear Exchange

If the nucleus is removed and the cap cut off, a new cap will regenerate. *Acetabularia,* however, is usually able to regenerate only once without a nucleus. If the nucleus of an alien species is now inserted and the cap is cut off once again, the new cap will be characteristic of the species of the nucleus. If more than one kind of nucleus is inserted, the regenerated cap will be intermediate in shape between the species that donated the nuclei.

There is only one reasonable explanation for these observations: The control of the cell exerted by the holdfast is attributable to the nucleus that is located there. It is hard to imagine any way that could have been demonstrated with a "higher" organism than this simple protist. The information it provided helped to pave the way for the epochal discoveries of those who have worked out the role of the nucleic acids in the control of all cells and all cellular life.

Summing it all up:

1. Ultimate control of the cell is exercised by the nucleus. In the end, the form of the cap is determined by the presence of the nucleus in the holdfast. In the long run, the only thing that can successfully compete for control with a nucleus is another nucleus.
2. Some control of the form of the cap is exercised by the nonnuclear parts of the cell, presumably the cytoplasm or something in it. But since that substance can exercise control for just one regeneration, it must be limited in quantity and unable to reproduce itself without the nucleus. Perhaps it is perishable as well.
3. The source of the messenger substance must be the nucleus.

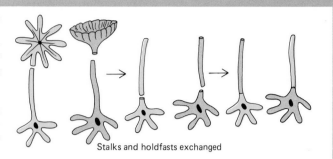

Stalks and holdfasts exchanged

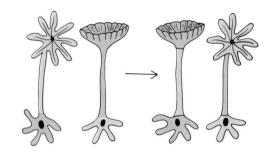

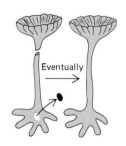

Eventually

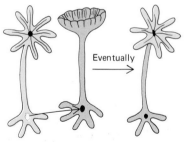

Eventually

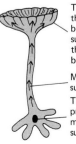

The characteristics of the cell are governed by the messenger substance, and therefore ultimately by the nucleus.

Messenger substance

The nucleus produces the messenger substance

From Miescher's time the story of DNA broadens like tributaries of a river (see Focus on DNA and the Transfer of Genetic Information). Eventually, many fragments of knowledge were integrated by a pair of scholars, James Watson and Francis Crick, who in 1953 described the exact molecular structure of Miescher's nuclein. In doing so they took the crucial step that led to the foundation of the modern science of molecular biology.

Focus on . . .

DNA AND THE TRANSFER OF GENETIC INFORMATION

In 1928 the British investigator Frederick Griffith reported the results of what then impressed most biologists as a very eccentric experiment. Bacteriologists had long known of the existence of weak strains of vir- ulent pathogenic (disease-causing) bacteria that can be raised on laboratory media. These strains are so quickly destroyed by a host's immunological defenses that they are incapable of causing disease. Griffith isolated two

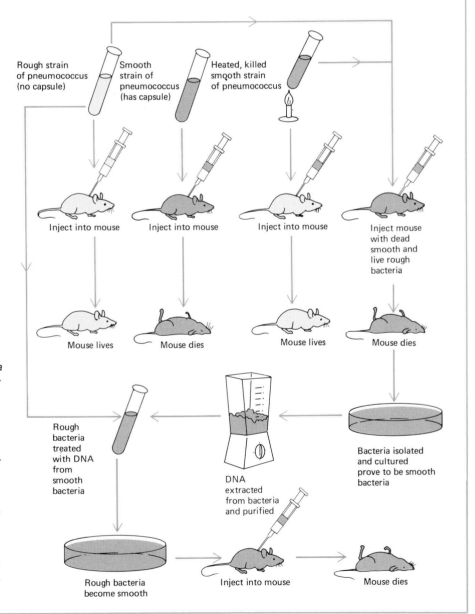

The experiments of Frederick Griffith and Oswald Avery. Griffith demonstrated the transfer of genetic information from dead, heat-killed bacteria to living bacteria of a different strain. Although neither the rough strain of Pneumococcus nor the heat-killed smooth strain could kill a mouse, a combination of the two did. Autopsy of the dead mouse showed the presence of living, smooth-strain pneumococci. Later, Avery demonstrated the restoration of virulence to rough bacteria by treating them with DNA from smooth bacteria— proving that it was the DNA that carried the genetic information necessary for the bacterial transformation.

Rough strain of pneumococcus (no capsule)

Smooth strain of pneumococcus (has capsule)

Heated, killed smooth strain of pneumococcus

Inject into mouse

Inject into mouse

Inject into mouse

Inject mouse with dead smooth and live rough bacteria

Mouse lives

Mouse dies

Mouse lives

Mouse dies

Rough bacteria treated with DNA from smooth bacteria

DNA extracted from bacteria and purified

Bacteria isolated and cultured prove to be smooth bacteria

Rough bacteria become smooth

Inject into mouse

Mouse dies

distinct strains of pneumococcus, a type of bacteria that causes pneumonia. One of these, a strain of pneumococcus lacking its usual protective polysaccharide capsule, forms *rough* colonies on agar in the laboratory. The normal, virulent strain produces glistening, *smooth* colonies as a result of its mucoid capsule.

Pursuing an immunological investigation with no obvious relationship to inheritance, Griffith injected dead smooth pneumococci into mice *along with* live rough microbes. The mixture should have been harmless. Instead, many of the mice died. Griffith then cultured the blood of the dying mice and recovered live smooth pneumococci from it; these were fully virulent when injected into other mice. Even the descendants of such pneumococci could kill still *other* mice!

Griffith's experiments could be explained, after a fashion. Some form of inheritable information had passed from the dead smooth pneumococci to the live rough ones. The information needed to make a capsule, lacking from those laboratory mutants, had somehow passed to them from the cadavers of the smooth microbes.

In the early 1940s, Griffith's remarkable experiments were repeated with modifications by Oswald Avery, a researcher at the Rockefeller Institute of New York City. Avery reasoned that some transforming principle, some chemical substance, had conveyed genetic information from the dead bacteria to the living. He and his research team first suspected that the substance was part of the polysaccharide coat of the pneumococcus, a reasonable belief, but wrong. Further experiments by these Rockefeller Institute scientists discovered that it was something from the *interior* of the dead cells that produced capsular transformation. Through a lengthy series of experiments they eliminated everything else and were left with DNA. They found that they could produce fibers of smooth pneumococcal DNA capable of transforming nonvirulent strains to virulent ones. In some manner, DNA carried genetic information.

The X-ray Eye

Much of the progress in analyzing the structure of macromolecules (very large molecules) has come from the application of a sophisticated form of photography called x-ray diffraction analysis. However large a molecule such as DNA may be, one cannot reasonably expect to see it in detail. Even the most powerful and sophisticated electron microscopes have unsurpassable limits set by the lengths of the waves of radiation by which they operate.

When x-rays of extremely short wavelength are passed through a crystal, they are scattered by the atoms of the crystal somewhat as light rays are scattered by the particles of dust or water in a fog. Because of the regularity of the atomic arrangement in a crystal, the scattering of the x-rays is itself regular rather than diffuse, and the scattered radiation interferes with itself to produce a characteristic pattern that can be photographed—a kind of molecular fingerprint distinctive for any crystalline substance. The photograph, however, is no more an image of its molecules than a fingerprint is an image of a person. Still, the mathematical analysis of such patterns obtained at different angles of exposure enables the researcher to obtain information from which the shape of the molecule may be inferred.

Maurice Wilkins and Rosalind Franklin had already employed x-ray diffraction to investigate the structure of DNA when James Watson and Francis Crick became friends at Cambridge University. Watson, who had recently obtained his PhD from Indiana University, wished to gain experience in the rapidly developing new field of molecular biology. Crick's professional education had been delayed by World War II, so that at an age greater than the usual, he was a doctoral candidate in the field of protein structure. Viewing Franklin's observations from their own perspective, Watson and Crick concluded that DNA must be a helical molecule, somewhat like the keratin protein of hair and nails, whose structure had recently been discovered by Linus Pauling. Although inferring the helical structure of DNA was a difficult piece of work, this was not the most difficult part of what they set out to accomplish. The hard part was to show that the helical shape was consistent with DNA's chemical makeup.

Building Models

The chemical makeup of DNA had become vastly better understood since Miescher's day. It seemed clear that the backbone of the DNA molecule consisted of alternating sugar and phosphate units. Attached somehow to this backbone were four kinds of bases—adenine, thymine, guanine, and cytosine. It had been shown that the amount of thymine in a sample

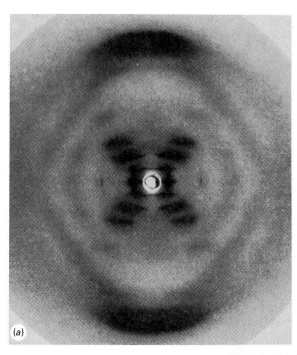

(a)

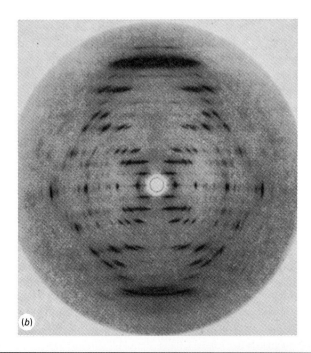

(b)

Figure 11–1 X-ray diffraction photographs of suitably hydrated fibers of DNA, showing the so-called B configuration. (a) Pattern obtained using the sodium salt of DNA. (b) Pattern obtained using the lithium salt of DNA. This pattern permits a most thorough analysis of DNA. The diagonal pattern of spots (reflections) stretching from 11 o'clock to 5 o'clock and from 1 o'clock to 7 o'clock provides evidence for the helical structure of DNA. (c) A modern computerized reconstruction of the appearance of a DNA molecule if it could be viewed from the top. This detailed picture has been built up from a myriad of clues afforded by indirect evidence like those early x-ray diffraction photographs on which Watson and Crick based so much of their initial thinking. (a, b Courtesy of Biophysics Research Unit, Medical Research Council, King's College, London; c, Courtesy of Computer Graphics Laboratory, University of California, San Francisco.)

(c)

of DNA was equal to the amount of adenine, and that the amount of cytosine similarly was equal to the amount of guanine. After several false starts, Watson and Crick found that these facts suggested that DNA was not a single, but a **double** helix, held together by bases that articulated with one another, every adenine with a thymine, every guanine with a cytosine.

Watson and Crick next built a model of the DNA molecule as they understood it. This model was built of accurately scaled models of atoms and groups of atoms, with both sizes and bond angles properly proportioned according to their prototypes. If everything could be made to fit, their understanding of the DNA molecule would have been shown quite conclusively to have been the right one. When finally a great double helix stood in their laboratory, they knew they had been right. It was a great moment of discovery, perhaps one of the greatest in the history of science. They had discovered nothing less than the secret of life—a molecule whose structure preprogrammed the composition of every component in, and all the functions of, every cell on earth (Fig. 11–1).

The genetic code, submicroscopic as it is, is a vehicle ample for storing the library of information necessary to construct a human or a blind worm living in a human intestine. Obediently following the directions in that library, cells produced the construction of muscles, bone, and nerves that bears your social security number. Another library, much smaller, produced the worm, or one smaller yet, one of the bacteria in your mouth.

THE STRUCTURE OF DNA

DNA is composed of molecular subunits called nucleotides. Each nucleotide consists of (1) a five-carbon sugar, deoxyribose, (2) a phosphate group, and (3) a nitrogen-containing compound called a base. Cytosine (C) and thymine (T) are single-ring pyrimidine bases. Adenine (A) and guanine (G) are double-ring purine bases. These bases project like the rungs of a ladder more or less at right angles from the sugar-phosphate backbone of the DNA molecule. It is the sequence in which these bases appear that determines the genetic code. Thus the sequence AGGTCCATCCG bears a different set of information than the sequence TGGAACTAGTCC.

In nature DNA exists in the form of *two* strands, attached to each other by means of their bases (Fig. 11–2) as follows:[1]

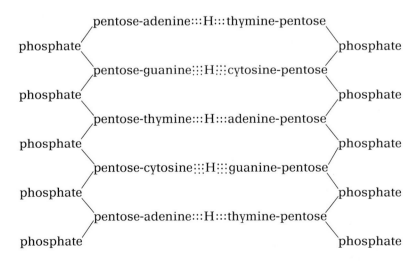

---

[1] H represents a hydrogen bond.

Figure 11–2 Structure of DNA. Note that the two sugar-phosphate chains run in opposite directions. This orientation permits the complementary bases to pair.

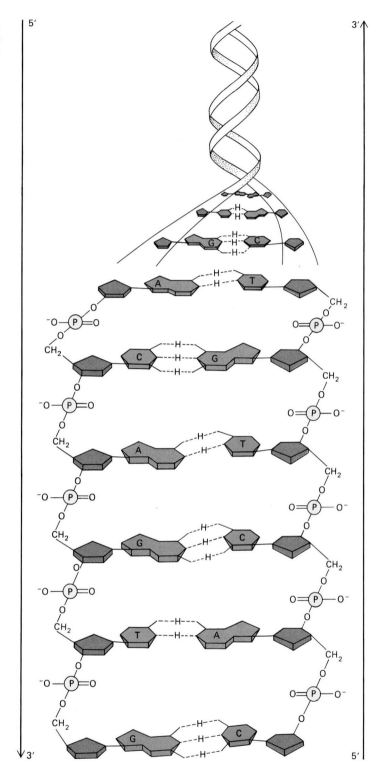

The ladder-like double strand of DNA is twisted into a **double helix,** somewhat like a spring or a spiral staircase (Fig. 11–3). In eukaryotes a strand of **histone protein** is present within the helix.

An important point to remember is that if you know the sequence of one of the two strands, you can predict the base sequence of the other. This is because adenine on one strand normally pairs only with thymine on the other strand, and cytosine ordinarily pairs only with guanine. No other pairing relationship is normally possible due to the nature of the hydrogen bonds that form between the bases. This concept of **base**

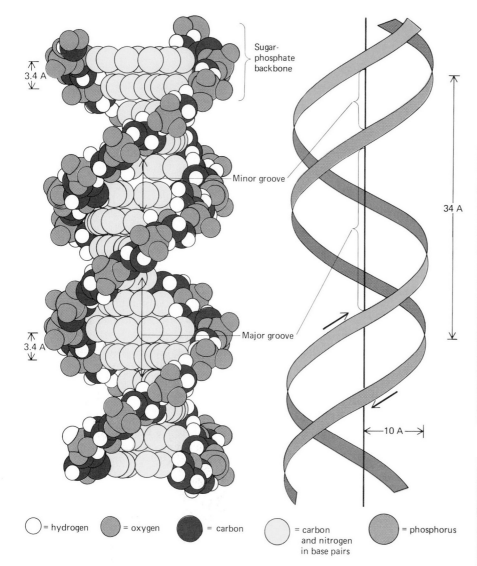

= hydrogen = oxygen = carbon = carbon and nitrogen in base pairs = phosphorus

Figure 11–3 According to our modern understanding, most DNA exists in this double-helical form. Notice that the molecule has two "backbones" united by their bases to form a *double* helix.

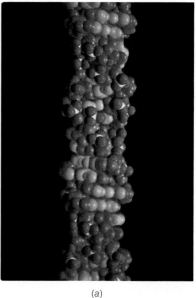

(a)

(b)

Figure 11–4 B-DNA and Z-DNA compared. As can easily be seen, the shape of Z-DNA (a) is much less regular than that of B-DNA (b). Z-DNA appears to be genetically inactive in that it is not transcribed. However, its presence in areas of the genome that are known to be controllable implies that it plays a role in the control of genes, especially eukaryote genes. (Computer reconstructions by Dr. Richard J. Feldmann.)

pairing explains why the two chains of DNA are complementary to each other; that is, the sequence of nucleotides in one chain dictates a complementary sequence of nucleotides in the other. Moreover, the two strands extend in opposite directions and have their terminal phosphate groups at opposite ends of the double helix; because of this they are said to be **antiparallel.**

The techniques employed by Watson, Crick, and their colleagues produced an "average" picture of the DNA strand. Their model showed regularly spaced base pairs, and their double helix curved smoothly as it twisted to the right. This type of DNA is known as **B-DNA.** Today it is understood that within most DNA helices are regions whose coiling varies from the standard. The number of nucleotides per turn can vary, for example. A strikingly different form is **Z-DNA,** in which the sugar-phosphate backbone zigzags to the left (Fig. 11–4). Since Z-DNA occurs most frequently in chromosomal regions known to be regulatory in function, it is inferred that the information in Z-DNA is not translated into protein but rather serves to regulate the expression of nearby genes.

DNA REPLICATION

The most distinctive properties of the genetic material are that it carries information and undergoes **replication;** that is, makes an exact duplicate of itself. When a cell prepares to divide, its DNA chains separate, and

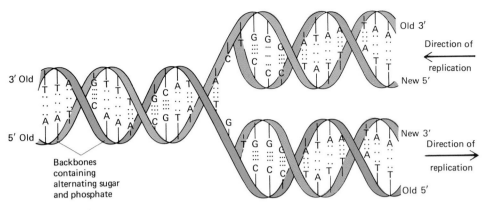

Figure 11–5 Mechanism of the replication of DNA. The two strands of the DNA double helix are shown separating. The left one, which runs from the 5′ phosphate to 3′ OH, is being copied, starting from the bottom. The newly synthesized chain begins with a 5′ phosphate. In the new chain, as in the old, adenine forms base pairs with thymine, and cytosine forms base pairs with guanine.

each one is used as a template for a new chain of complementary nucleotides. Thus two new chains are synthesized. The result is two double-chain molecules, each identical to the original double-helix and each containing one of the original chains (Fig. 11–5). Because each of the two original chains is not directly copied but is conserved and used as part of a new chain, this method of information copying is known as **semiconservative replication.**

In prokaryotes double-stranded DNA is generally found as closed circles. Replication begins when an enzyme opens the circular DNA molecule at a specific initiation site. Then other enzymes separate the two DNA chains. Replication is **bidirectional**—proceeding in both directions from the initiation site. Replication is also bidirectional in the linear DNA of eukaryotic cells. As the two strands of replicating DNA separate, they form a kind of Y-shaped region known as the **replication fork.** Such replication forks may occur in several places simultaneously in a eukaryotic DNA chain (Fig. 11–6).

After separating, the DNA strands act as templates for the assembly of new complements to themselves. This is the task of **DNA polymerase** enzymes, which would not be able to accomplish this without the origi-

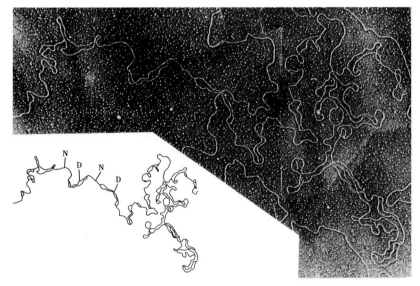

Figure 11–6 Electron micrograph of DNA replication in action. Stretches of replicated DNA (*D*) alternate with stretches not yet replicated (*N*). The sketch is a tracing of the photograph. Note the many replication forks. (Courtesy H.J. Kriegstein and D.S. Hogness.)

nal strands themselves being available for direction. As new complementary strands are produced, the original strand separates further until finally there are two identical double strands of DNA. At that point physical cell division occurs, separating the two daughter cells. Each one is left with an identical chromosome, and each is therefore genetically identical to the other, barring mutation.

All this can be compared to making a copy of a computer disc, a disc containing not only directions for what the computer is expected to do but also a program for the computer to build another computer. The specificity of the base pairing of DNA is necessary for a regular helix to be produced, but it is by no means assured a priori by the properties of the bases. Improper pairing can and does occur, producing a local distortion in the structure of the helix when this happens. The usual fidelity and regularity of the base pairing relationships are owed to DNA polymerase, which must be one of the most important enzymes in the universe, for proper DNA synthesis could not take place without it.

DNA polymerase appears actually to be a complex of enzymes, or one of a closely associated "team." It has a twofold function. Not only must it catalyze the polymerization of nucleotides, but it must also do this in a particular way. An ordinary enzyme causes a thermodynamically possible reaction to proceed more rapidly than it otherwise would but does not otherwise influence it, so the nature of its products will almost always be the same. Not so with DNA polymerase. At one time the product will have a freshly added, terminal guanine-nucleotide, another time it will have a terminal cytosine-nucleotide, and still another time it will have a terminal thymine- or adenine-nucleotide. That means the DNA polymerase enzyme actually has *four* specific and mutually exclusive substrates, and that it switches its preferences among them in accordance with the bases pre-existing on the single strand of DNA that is serving as a template for the production of the new strand.

Thus, if the DNA polymerase is resting on a pre-existing strand of DNA that has, at that point, a projecting guanine base, the DNA polymerase will behave as a cytosine-nucleotide polymerase and will add only a cytosine-nucleotide to the forming strand at that point, and none other. It will then advance to the next base of the pre-existing strand, which could be, say, thymine. At that point the DNA polymerase enzyme will behave as an adenine-nucleotide polymerase, adding an adenine-nucleotide to the previously added cytosine-nucleotide. In this way each base on the lengthening new strand will assuredly be complementary to its mate on the pre-existing strand.

Consider all the things this remarkable enzyme must do. It must recognize a specific base on a pre-existing strand of DNA and somehow modify its active site to accept the nucleotide whose base is complementary to that pre-existing one. It must then bring that nucleotide to the proper end of the new, elongating strand of DNA, it must bring about the necessary dehydration reaction to add it to the new strand, and next it must release the reaction products. The enzyme then moves one base further on the templating pre-existing strand so that it can repeat the process all over again. And it continues to do this tirelessly at a rate of about 200 bases per second for hundreds or thousands of bases every time the cell divides, and occasionally at other times as well.

The synthesis of DNA occurs in the nuclei of cells of higher organisms only during the S phase of the cell cycle (Chapter 9), when chromosomes are in their extended form and are not readily visible. Some sort of biological signal must initiate DNA synthesis at this time and turn it off at other times. The enzyme (the DNA polymerase) and the substrates (the four nucleotides) are present in the cell all the time. The explanation currently considered most likely is that some sort of change in the DNA template initiates the synthesis of DNA at the appropriate time in the cell cycle and then turns it off.

CHAPTER SUMMARY

I. The basic chemical makeup of DNA was discovered in the 19th century by Friedrich Miescher.

II. James Watson and Francis Crick worked out the basic structure of DNA in the 1950s.
  A. Watson and Crick suggested that the backbone of DNA consisted of alternating sugar and phosphate units to which were attached four bases.
  B. These investigators then built a model of DNA showing that the molecule is a double helix.

III. DNA is composed of molecular subunits called nucleotides; each nucleotide consists of deoxyribose (a sugar), a phosphate group, and a purine or pyrimidine base.

IV. DNA duplicates itself by the process of semiconservative replication, in which each of the two original chains is used as half of a new chain.
  A. Replication is bidirectional, proceeding in both directions.
  B. DNA synthesis takes place at replication forks, Y-shaped regions where the two strands of DNA separate and where DNA synthesis occurs.
  C. DNA polymerase catalyzes the polymerization of nucleotides, adding the appropriate nucleotide to the chain.

Post-Test

1. Miescher discovered nuclein, which was really _____ .

2. Watson and Crick showed that the DNA molecule is for the most part shaped like a _____ _____ .

3. Each nucleotide chemically consists of a _____ , a _____ _____ , and a purine or pyrimidine _____ .

4. A sequence of bases AACGGTCA on one strand of DNA would necessitate the sequence _____ in the complementary strand.

5. In the DNA of all known organisms, the amount of adenine equals the amount of _____ , and the amount of _____ equals the amount of _____ .

6. As the two strands of replicating DNA separate, they form a Y-shaped region known as the _____ .

7. Avery showed that the genetic message was contained within _____ .

8. The process of DNA replication is described as _____ because the two original strands of DNA are retained in the products.

9. DNA synthesis is catalyzed by the enzyme _____ _____ .

10. Experiments with *Acetabularia* indicate that the ultimate control of the cell is exercised by the _____ _____ .

Review Questions

1. Who was Miescher?
2. How did Watson and Crick "prove" their concept of doubly helical DNA to be correct?
3. What is *Acetabularia*? What was discovered about the function of the cell nucleus using this organism, and how was it done?
4. A strand of DNA has the base sequence AATTGACTGCAGT. What is the base sequence of the complementary strand?
5. How did Griffith's experiment help show that the transformation of harmless bacteria into virulent ones involves a change in genetic information?
6. How is the molecular structure of DNA uniquely adapted to its function? Explain.
7. What is meant by semiconservative replication?

Readings

Bendiner, E. "Avery: Making Sense of a 'Stupid' Nucleotide," *Hospital Practice,* October 1982, 195–219. A fascinating character study of an important participant in a little-appreciated chapter in the history of the double helix.

Grivell, L.A. "Mitochondrial DNA." *Scientific American,* March 1983, 78–88. A discussion of the genetic system of mitochondria.

Judson, H.F. *The Eighth Day of Creation.* New York, Simon and Schuster, 1979. Aside from the famous (and controversial) *Double Helix,* this is perhaps the most readable and authoritative history of nucleic acid research to date.

Watson, J.D. *The Double Helix.* New York, Atheneum, 1968. Not everyone thinks this personal account to be historically accurate, and without doubt it is a highly colored and biased personal history. But there is nothing like it. One feels that, in essence if not in detail, this is how it really was to make a Nobel Prize–winning discovery.

Chapter 12
GENE FUNCTION AND REGULATION

Outline

Learning Objectives

After you have read this chapter you should be able to:

1. Describe the process of transcription.
2. Summarize the sequence of events that occur in translation.
3. Describe the function of transfer RNA.
4. List the functions of the ribosome.
5. Summarize the processes of initiation, elongation, and termination in protein synthesis.
6. Draw a diagram illustrating an operon and describe how it functions.
7. Contrast enzyme induction and repression in prokaryotes.

The genes contain the code for making all the proteins needed by the cell. Cellular proteins are always needed, even in cells that are not growing or producing any protein product for export. For one thing, the changing needs of a cell require differing emphases to be placed from time to time on particular biochemical pathways, and this requires more or new enzyme molecules to promote the reactions of those pathways. For another, proteins that have been damaged by other substances or which contain errors must be replaced, and the disposal of these requires an ongoing process of protein degradation and remanufacturing.

Hundreds of kinds of protein are constantly in production, each molecule requiring on the average 20 minutes to make, with large numbers of molecules being produced simultaneously. All this requires not only a multitude of individual protein factories, the ribosomes (Fig. 12–1) but also a whole associated economy of raw material suppliers, subcontractors, and management systems, which might well be thought of as a kind of corporate manufacturing conglomerate—the Ribosome Protein Manufacturing Company. How genes make proteins and how genes are regulated are the subjects of this chapter.

THE GENETIC CODE

DNA determines the sequence of amino acids in a protein and, therefore, the identity of each protein. You should recall that each protein consists of a specific sequence of amino acids. The amino acid sequence of hemoglobin is very different from the amino acid sequence of insulin, for example. But how does a sequence of nucleotides in DNA specify a sequence of amino acids in a protein?

There are only four types of nucleotides in DNA, but there are 20 kinds of amino acids that commonly occur in proteins. How can a code specifying 20 amino acids be based on just four nucleotides? The single nucleotide cannot be the fundamental unit of the genetic code nor, by similar logic, could a unit composed of two nucleotides. However, units composed of three nucleotides can form 64 possible combinations, more

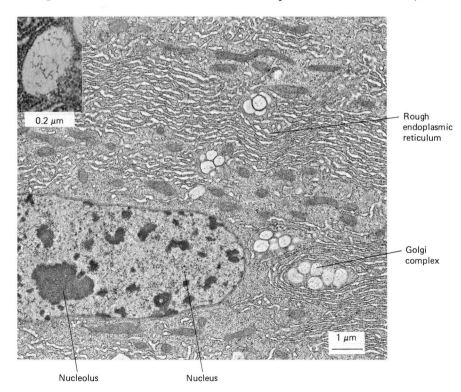

0.2 μm

Rough endoplasmic reticulum

Golgi complex

1 μm

Nucleolus   Nucleus

Figure 12–1 Very concentrated rough endoplasmic reticulum from the silk gland of a silkworm moth larva. Because silk is a kind of protein, the silk gland possesses an extensive rough endoplasmic reticulum equipped with myriads of ribosomes whose principal task is to translate genetic information carried by RNA from the DNA of the nucleus (prominent in this electron micrograph) into the fibroin silk protein. The fibroin accumulates in the vesicles within the Golgi complex. The higher-magnification inset gives some idea of how concentrated and closely spaced the ribosomes of this rough endoplasmic reticulum are. (Dr. H. Akai, Ministry of Agriculture, Forestry and Fisheries, Yatabe, Japan, *Experientia #39,* 1983)

than enough codes to specify 20 amino acids. Indeed, several triple combinations can be used and are used to specify one amino acid, and some of them have no amino acid equivalent at all. Each sequence of three nucleotides—or more simply, three bases—is known as a **triplet.** These triplets are the basic units of genetic information.

The triplets must be handled accurately by the cell's genetic readout machinery. It takes no more than a single base change to completely alter the meaning of a triplet in a strand of DNA, with the result that an inappropriate amino acid would be incorporated into a protein at the point specified by the changed triplet.

THE ROLE OF RNA

Proteins are not synthesized directly on the DNA. In fact, they are manufactured in the cytoplasm along the ribosomes. The way DNA actually produces protein is indirect (Fig. 12–2) and involves another nucleic acid, **ribonucleic acid,** or **RNA.** RNA differs from DNA chemically in that it contains the sugar **ribose** rather than deoxyribose. And though it, too, possesses four bases, it has the base **uracil (U)** in place of thymine. Uracil behaves chemically like thymine, forming hydrogen bonds with adenine.

There are three main varieties of RNA—messenger RNA (mRNA), ribosomal RNA (rRNA), and transfer RNA (tRNA). As far as we know, all three are made in the same way by DNA and all are involved in protein production, but in different ways. **Messenger RNA** copies genetic instructions and carries them out into the cytoplasm to the ribosomes. The synthesis of RNA along DNA is known as **transcription,** for in this process information is transcribed from one kind of storage code (DNA nucleotides) to another (RNA nucleotides). The nucleotide triplet language in which the information is conveyed is not changed in this process. Guided by the information within the mRNA, the ribosomes, composed of **ribosomal RNA** and proteins, assemble the amino acids delivered to them by **transfer RNA** and synthesize proteins. The process of protein synthesis on the ribosomes is termed **translation** because the language of nucleic acids, composed of the four types of bases, is translated into the different language of proteins, based on the 20 amino acids.

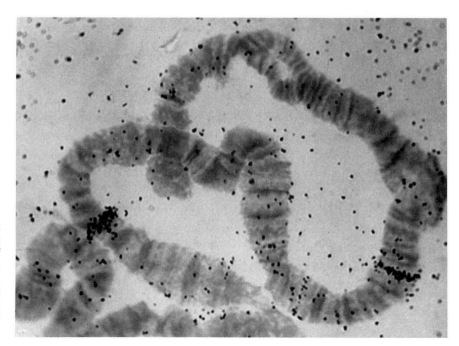

Figure 12–2 Production of RNA (transcription) along specific portions (genes) of DNA in a chromosome of the fruit fly *Drosophila melanogaster*. The dark particles are radioactively labeled tRNA. Note the two regions along the DNA where the particles are concentrated. (Courtesy of T.A. Grigliatti.)

TRANSCRIPTION

During transcription genetic information is transferred from a DNA template to an RNA molecule. The base sequence in the RNA is complementary, not identical, to the base sequence in the DNA. Transcription is controlled by the enzyme **RNA polymerase,** actually a large enzyme complex. RNA polymerase appears to move along a strand of DNA like a railroad locomotive on a track, constructing a single strand of RNA as it goes. This remarkable enzyme, responding to cues built into the base sequence of DNA, recognizes which of the two complementary strands it should transcribe, what kinds of genes it should transcribe (and which should be left untranscribed), where it should begin transcription, and where it should end transcription. Without this last ability, RNA polymerase would travel around the circular bacterial chromosome indefinitely, like a toy train that had been left on and forgotten, spinning off endless amounts of RNA as it did so.

Transcription begins when a subunit of RNA polymerase binds to a **promoter,** a specific site on the DNA where transcription can be initiated. The promoter consists of a special sequence of several base pairs. Like DNA, RNA is composed of nucleotides with distinguishable molecular ends. An mRNA chain is always synthesized by adding nucleotides to the 3′ end, so the process begins at the 5′ end and proceeds in the 5′→3′ direction (Fig. 12–3). One RNA polymerase complex can synthesize an entire RNA molecule. Ribonucleotides are matched to complementary bases along the DNA template being transcribed. This process continues until the RNA polymerase comes to a specific nucleotide sequence in the DNA that it recognizes as a stop signal. The RNA molecule separates from the DNA strand, which is now free to transcribe another copy of the RNA strand.

Only one of the DNA strands—the **coding strand,** or **plus strand**—is transcribed into RNA. The other strand, the **minus strand,** functions in DNA replication and in repair. Which of the two strands serves as the coding strand varies from region to region. One strand may be transcribed in one region, and the other strand may code farther along.

TRANSLATION

In **translation** the information transcribed within an mRNA molecule is used to synthesize a specific polypeptide. The sequence of amino acids in a polypeptide chain is determined by the sequence of bases in mRNA. A sequence of three bases in mRNA corresponding to a triplet in DNA is known as a **codon.**

The Role of Transfer RNA

Transfer RNA serves as a kind of adaptor molecule that carries amino acids to their place of assembly. Each tRNA molecule is relatively small, elaborately folded and convoluted, and capable of forming a temporary chemical compound with a particular and specific amino acid. The tRNA thus can function as a kind of tag. By calling for the tRNA tag one would obtain the amino acid to which it was attached. The feature that permits tRNA to function in this manner is the trio of bases that project from one of the hairpin curves of its molecule (Fig. 12–4). This sequence of three bases, called an **anticodon,** associates with a specific codon along the mRNA. For instance, the anticodon guanine-uracil-adenine on a tRNA molecule can become associated with the mRNA strand only where a complementary codon occurs; that is, where the sequence cytosine-adenine-uracil occurs.

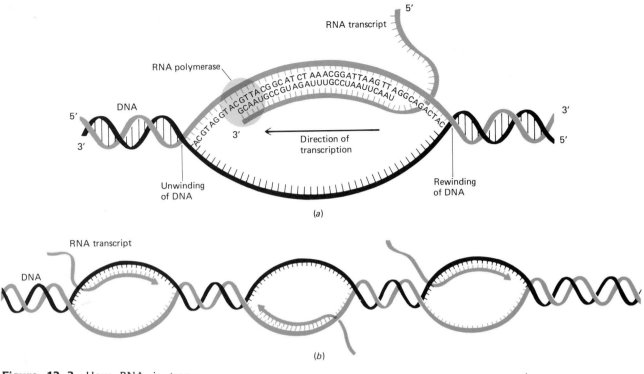

(a)

(b)

Figure 12–3 How RNA is transcribed from a DNA template. (a) A DNA double helix is unwound by RNA polymerase, giving the enzyme access to the nucleotide sequence. The DNA is then rewound behind the moving transcription complex. Initiation and termination sites, encoded in specific DNA sequences, determine where transcription starts and stops. RNA synthesis depends on base-pairing rules similar to those for DNA synthesis: Adenine pairs with uracil, cytosine pairs with guanine. (b) Only one of the two strands is transcribed for a given gene, but the opposite strand may be transcribed for a neighboring gene. (c) A diagram of the reaction catalyzed by RNA polymerase. The exposed DNA strand on the right is being copied. Each incoming nucleotide is selected for its ability to base-pair with the DNA template. The nucleotide is then added to the 3' end of the growing RNA chain.

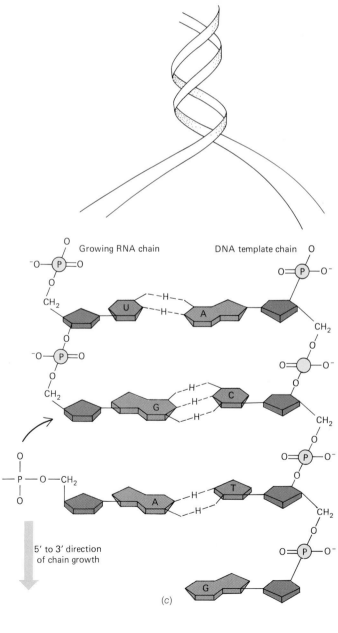

(c)

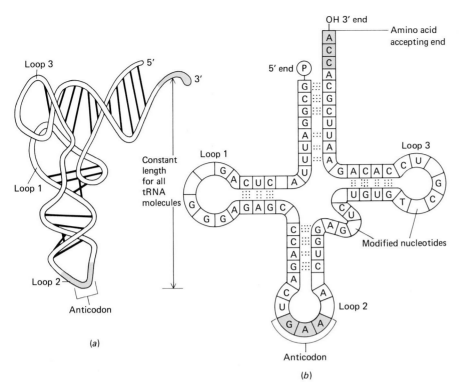

Figure 12–4 Two representations of the structure of a typical tRNA molecule; tRNA molecules are the compounds that "read" the genetic code. A diagram of the actual shape of a tRNA molecule is shown in (a). Its three-dimensional shape is determined by intramolecular hydrogen bonds between base-paired regions, which are most clearly observed in the two dimensional clover leaf form depicted in (b). One loop contains the triplet anticodon that forms specific base pairs with the mRNA codon. The amino acid is attached to the terminal ribose at the 3′ OH end, which has the nucleotide sequence CCA. Each tRNA also has guanylic acid (G) at the 5′ end (at P) and contains several modified nucleotides. The pattern of folding permits a constant distance between anticodon and amino acid in all tRNAs examined.

Each kind of tRNA must be attached to the proper amino acid, the one specified by its anticodon. What assures a proper match? A series of about 20 enzymes known as **amino acyl–tRNA synthetases** (one for each kind of amino acid) assures that the proper amino acid is united with the appropriate kind of tRNA. Like any industrial transaction, this one has a cost. One ATP must be expended to accomplish it. For this reason, an amino acyl–tRNA synthetase has three active sites—one for ATP, one for the amino acid, and one that fits any of several correct kinds of tRNA but no incorrect ones.

The Role of Ribosomes

Messenger RNA cannot produce protein by itself; specialized organelles, the **ribosomes,** convert the information that has been copied into the mRNA into the structure of an actual protein product. Each ribosome is a particle about 25 nanometers in its widest dimension—about a millionth of an inch. It is composed of two subunits different in size and weight. The subunits are thought to occur together only when the ribosome is actually functioning and are probably held together by hydrogen bonds. The small subunit consists of a molecule of rRNA and one molecule each of 21 kinds of protein. Messenger RNA binds to the small subunit. The large subunit comprises two molecules of rRNA and 34 different protein molecules. Enzymes needed to catalyze the formation of peptide bonds between the amino acids are associated with the large subunit.

A ribosome has three functions. First, it must bind mRNA. Then, it must ensure that the genetic information of that mRNA is properly "read out." Finally, the ribosome must act as a catalyst, establishing the peptide bonds among the amino acids of the polypeptide that it ultimately produces. Messenger RNA usually binds a number of ribosomes—as many as 30 or 40—into a temporary structure called a **polyribosome** (Fig. 12–5). Each ribosome in the cluster independently produces a polypeptide from the same mRNA.

The ribosome serves as an enzymatic matchmaker that ensures not only that the peptide bond is properly formed between adjacent amino

Figure 12–5 Transcription and translation in a bacterial cell. (*a*) Electron micrograph of two strands of DNA, one inactive and the other actively producing mRNA, which is attached to polyribosomes. (*b*) Diagrammatic representation of the processes of transcription and translation. (*a*, Courtesy of O.L. Miller, Jr.; *b*, after Miller, Hamkalo, and Thomas, Jr.)

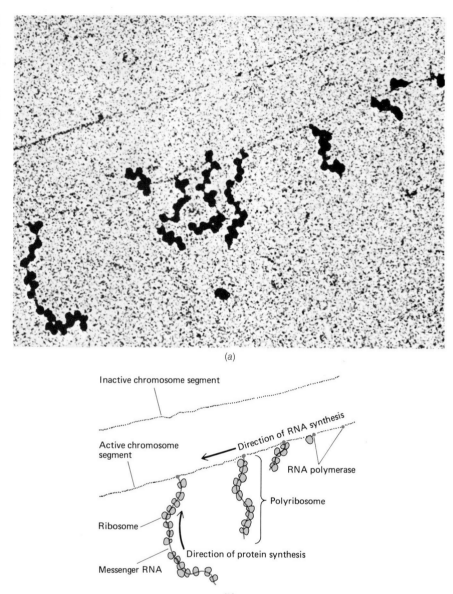

(a)

Inactive chromosome segment

Active chromosome segment

Direction of RNA synthesis

RNA polymerase

Polyribosome

Ribosome

Direction of protein synthesis

Messenger RNA

(b)

acids, but also that the proper tRNA species do indeed attach to the mRNA strand where they are "supposed" to. The importance of the ribosome in all this can be appreciated when we consider that many antibiotic drugs have the bacterial ribosome as their target. Our ribosomes are not as sensitive to the action of these antibiotics, which are therefore much less toxic to us than to our prokaryotic parasites.

Suppose we consider a strand of mRNA that begins with the triplets AUG/GCA/AAC/UCG. If, for some reason, the ribosome were to "ignore" the initial base, we would have the sequence UGG/CAA/ACU . . . The amino acids specified by the first sequence would be methionine-alanine-asparagine-serine. Those specified by the second sequence would be tryptophan-glycine-threonine, etc. (These can be determined by looking at a standard table; see Table 12–1.)

An error (which one might think would often occur) would cause the amino acid sequence of a strand of protein to be completely and uselessly altered. The cell has a number of mechanisms, however, that tend to protect it against such errors, including the requirement that in order to be translated, mRNA strands must begin with a particular sequence of bases.

Table 12–1
THE GENETIC CODE: THE SEQUENCE OF NUCLEOTIDES IN THE TRIPLET CODONS OF mRNA THAT SPECIFY A GIVEN AMINO ACID

First position (5' end)	Second position	Third position (3' end)			
		U	C	A	G
U	U	Phe	Phe	Leu	Leu
	C	Ser	Ser	Ser	Ser
	A	Tyr	Tyr	Terminator	Terminator
	G	Cys	Cys	Terminator	Trp
C	U	Leu	Leu	Leu	Leu
	C	Pro	Pro	Pro	Pro
	A	His	His	$Glu–NH_2$	$Glu–NH_2$
	G	Arg	Arg	Arg	Arg
A	U	Ileu	Ileu	Ileu	Met
	C	Thr	Thr	Thr	Thr
	A	$Asp–NH_2$	$Asp–NH_2$	Lys	Lys
	G	Ser	Ser	Arg	Arg
G	U	Val	Val	Val	Val
	C	Ala	Ala	Ala	Ala
	A	Asp	Asp	Glu	Glu
	G	Gly	Gly	Gly	Gly

Overview of Translation

The ribosome functions as a large enzyme complex, virtually an enzymatic machine shop. It is equipped with a variety of active sites and control sites that permit it to function appropriately. The most important are the A- and P-sites, which have the responsibility of bringing two amino acids together: the new one that is about to be added to the end of the polypeptide chain, and the terminal amino acid of that chain, that is, the one most recently added. The new amino acid really is an amino acid, but the previously added one is now part of a peptide, and is more properly referred to as a **residue.**

The tRNA attached to the new amino acid is accepted by the ribosomal **A-site,** and the tRNA of the peptide amino acid is grasped by the **P-site.** A peptide bond is then formed between the two amino acids, and the tRNA of the peptide amino acid is released. The newly incorporated amino acid is the new terminal amino acid of the peptide, and the P-site of the ribosome now reaches for its tRNA, which is still attached to it. Now the vacated A-site of the ribosome is ready to accept another amino acid–tRNA unit whose anticodon is complementary to the next codon of the mRNA and to repeat the whole process.

To sum it up (Fig. 12–6), translation involves:

1. Acceptance of an amino acid–tRNA complex by the A-site.
2. Establishment of a peptide bond between the A-site amino acid and the P-site amino acid residue.
3. Release of the tRNA formerly occupying the P-site.
4. **Translocation** (movement) of the ribosome down the mRNA strand by one codon.
5. Acceptance of another amino acid–tRNA complex by the A-site, and so on until the mRNA strand is completely processed or the ribosome encounters a codon with no amino acid equivalent, which would cause it to stop translating.

Figure 12–6 Translation. A ribosome progresses along an mRNA strand, adding amino acids to a growing polypeptide chain as called for by the sequence of codons in the mRNA. (a) An amino acid–tRNA complex has bound to an empty A-site next to an occupied P-site. Base pairs have formed between the anticodon and the codon. (b) The growing polypeptide chain is detached from the tRNA molecule in the P-site and joined by a peptide bond to the amino acid linked to the tRNA at the A-site. (c) The released tRNA joins the cytoplasmic pool of tRNA, free to bind with another amino acid. As the ribosome moves three nucleotides along the mRNA molecule, the growing peptide chain (still attached to a tRNA) is transferred to the P-site.

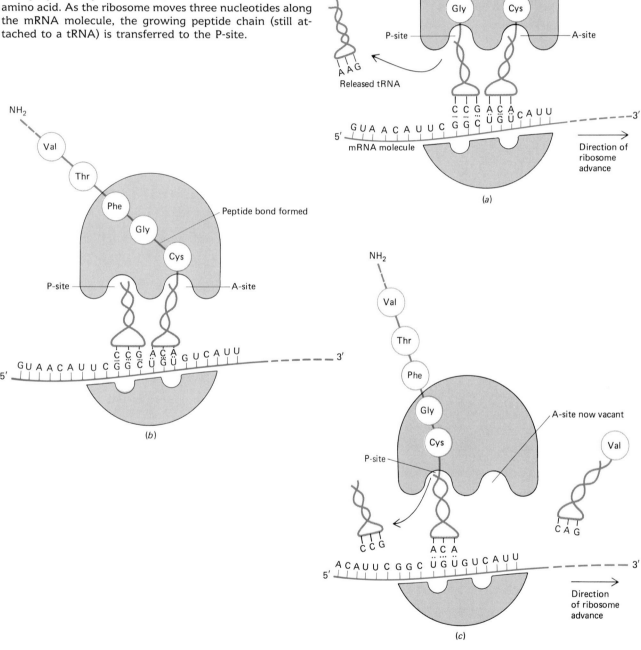

Starting Production: Initiation

Initiation[1] is the process that takes place before the peptide bond forms between the first two amino acids of the polypeptide. It involves the binding of the mRNA to the ribosome, forming a complex with the first amino

---

[1] Initiation is a bit simpler and better understood in bacteria than in eukaryotes, so we will describe bacterial initiation. However, eukaryote initiation is similar in principle.

acid–tRNA. When ribosomal proteins and RNA are first produced, they spontaneously associate to form both the large subunit and the small; evidently, under normal conditions the components fit together in just one way so that the ribosome is self-assembling. The large and small subunits must dissociate, however, for initiation to take place. The **initiation factor 3** protein **(IF-3)** is able to separate the ribosome into its subunits. This protein attaches to a special control site on the small subunit of the ribosome and releases the small subunit from the large subunit.

Before an mRNA molecule can be translated, it must form an **initiation complex** with the small ribosomal subunit. This complex contains, in addition to the mRNA and small subunit, a molecule of a protein, **initiation factor 2 (IF-2)** and a special amino acid–tRNA complex, formylated methionine–tRNA (Fig. 12–7). An amino acid must be placed in the P-site before anything else can be done. AUG is the first codon that can be recognized by the ribosome even if some other codons precede it. By consulting Table 12–1, you can see that AUG stands for the amino acid methionine.

Some of the tRNA molecules that carry methionine are able to react with a special enzyme that adds what is called a formyl group to the methionine amino acid. The formyl group is added to the amino group so as to block it. This allows such a formylated methionine–tRNA to occupy the P-site instead of the A-site as a new amino acid–tRNA normally would do. However, *formylated* methionine is used only if the AUG codon is the *initial* one. AUG codons in other parts of the mRNA strand do not add formyl methionine–tRNA; instead they add plain methionine-tRNA to the growing peptide chain. Recall (Chapter 3) that when proteins are formed, the carboxyl group of an amino acid that is already part of the peptide is joined to the amino group of the next amino acid so as to form a peptide bond. That means that a peptide (or protein) will always have an amino end and a carboxyl end, but it is the carboxyl end that is active in elongation, not the amino end. Since the amino group of formylated methionine is blocked, it cannot form a peptide bond with a preceding amino acid. Formylated methionine does have a perfectly normal carboxyl end, so it can well serve as the *first* amino acid in a chain.

The binding of the initiation complex to the P-site of the small subunit evidently nullifies the effect of the IF-3 protein, so the large and small subunits now reunite. This releases both IF-2 and IF-3 from the complex, and now the complete ribosome contains an amino acid in its P-site and is ready to start the process of peptide elongation.

Elongation of the Polypeptide Chain

Any amino acid–tRNA can enter the A-site except the initiator, formyl methionine-tRNA. Which one actually does so depends, however, on the codon being read at the time by the A-site. For an amino acid–tRNA to enter the A-site, it must be part of a complex, the **elongation complex.** The elongation complex consists of a protein, **Ef-Tu,** an energy-charged molecule similar to ATP that is called **GTP,** and the amino acid–tRNA itself. The whole complex then is *amino acid–tRNA|Ef-Tu|GTP.*

Only the amino acid–tRNA part of the elongation complex actually enters the A-site, and as soon as it has done so, the GTP part is broken down to GDP much as ATP breaks down to yield ADP. The energy released is used to form the peptide bond between the amino acid in the A-site and the formyl methionine that is still in the P-site. Then the remaining GDP and Ef-Tu are released, the ribosome translocates the mRNA strand (which requires the energy of another molecule of GTP), and the new amino acid shifts to the P-site. The formyl methionine that

Figure 12–7 Initiation of translation. Each black rectangle represents an amino acid. **1.** The IF-3 protein dissociates the ribosomal subunits.

2. An initiation complex forms when the IF-2 protein combines with formyl methionine-tRNA, and then with the mRNA to be translated and the small ribosomal subunit.

3. After the formyl methionine (the first amino acid of the peptide) binds to the P-site, IF-2 and IF-3 are released. A second amino acid-tRNA complex can now bind to the empty A-site.

4. The process of peptide elongation continues with the addition of new amino acids.

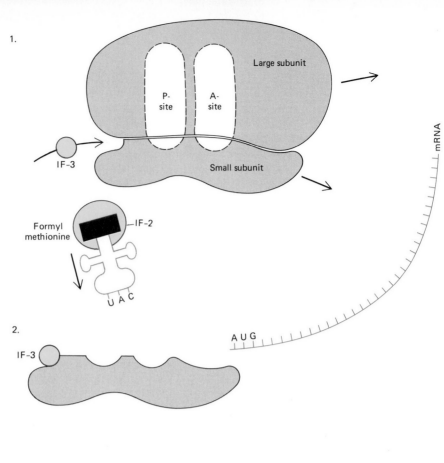

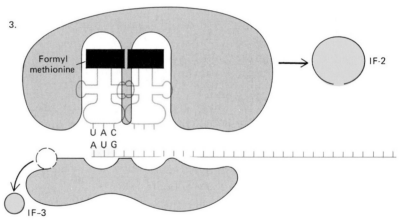

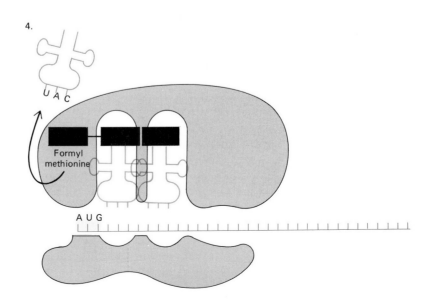

formerly occupied the P-site is now thrust outside the ribosome, though it is still attached to the new amino acid by a peptide bond. A special enzyme may now cleave that bond, splitting off the formyl methionine for reuse in the initiation of another protein.

Finishing Up: The Termination Stage

Termination is accomplished by special STOP signals, UGA, UAG, and UAA, at least in bacteria.These codons have no amino acid meaning, but they are recognized by several **release factors.** The release factors evidently break the bond between the polypeptide and the tRNA occupying the A site. The polypeptide is released from the ribosome. An instant later the ribosome releases the mRNA strand and is ready to start translating another one. If, however, the STOP signal occurs in the middle of an mRNA strand, the ribosome may not actually release the strand but probably initiates the translation of the next peptide that is encoded on it.

The Power Bill

Notice that in addition to the energy cost of synthesizing any of the amino acids themselves, the cell must expend three molecules of ATP and GTP to add an amino acid to a peptide—one to form the amino acid–tRNA, one to form the peptide bond, and one to translocate the mRNA strand to prepare for the next cycle. All this makes protein synthesis one of the most energy-expensive processes in the entire cellular economy.

Shipping the Product

Since proteins are often sharply localized in their internal cellular distribution, there must be a way to ensure that they reach their proper destination. This involves a kind of coding. On the amino end of the strand, for example, some proteins have a distinctive sequence of amino acids that ensures the attachment of the strand to any adjacent membranes of the endoplasmic reticulum (in eukaryotes) even as the protein is being made. The ER then evidently transports the protein to its ultimate destination, presumably using the amino acid sequence of its amino end as a guide.

REGULATING GENES

One does not usually think of a gene as something to be controlled. One has type O blood, or AB or B, or A—not type O today and type AB tomorrow. Yet it takes no more than a few hours in the sun to demonstrate that for many of us, skin color is not exactly permanent. The ultraviolet radiation of sunlight calls into action a quantitative genetic potential that the skin cells had but did not express until it was called out by the appropriate stimulus.

The ready availability of *Escherichia coli* and its ability to grow well under artificial conditions have led to its extensive study, so today this bacterium is probably the best understood organism on the face of the earth. Our knowledge of its gene-regulating mechanisms reflects that fact, thanks largely to the studies and the literature synthesis performed by the French investigators François Jacob and Jacques Monod.

The bacteria *Escherichia coli* inhabit the mammalian gut in countless billions. For a short time after each meal the bacteria wallow in nutrients, predigested by enzymes. The bacteria at that time need little in the way of digestive enzymes and produce less of them, channeling their protein production in other directions, especially the synthesis of macro-molecules. Under these conditions *E. coli* cells can divide as frequently as several times an hour. During fasting, however, they may divide only once a day, for then they cannot rely on predigested food and must either digest or synthesize all the food materials they need for themselves. It would pay our little passengers to be able to turn on and off the genes involved in the production of their many kinds of digestive and synthetic enzymes, as each situation would require.

Feedback Inhibition

When given a carbon and a nitrogen source, the *E. coli* bacterium is able to manufacture the amino acid tryptophan. If tryptophan is already avail-able in the surrounding medium, however, the bacterium will use that ready-made form rather than manufacture it from simple inorganic raw materials. There are four steps in the biosynthetic pathway that produce tryptophan from its precursors. Each of these steps is catalyzed by a specific enzyme, and the end product will form only if *all* enzymes are functional. Should even the first of them be inoperable, tryptophan can-not be manufactured.

If a chemical engineer were to design a control system for the regu-lation of tryptophan production, she would probably choose to place the cutoff valve at the beginning rather than the end of the process, both to avoid tying up resources unnecessarily and to avoid the accumulation of useless intermediate products in the machinery. Similarly, the first en-zyme in the *E. coli* biosynthetic pathway for tryptophan is the one that becomes nonfunctional when it is necessary or desirable to turn off tryp-tophan production. The way this is accomplished depends upon a proper fit between the active site of an enzyme and its substrate, and that de-pends upon the tertiary structure of the protein that constitutes that en-zyme. Tertiary and other higher-order protein configurations depend upon such factors as bond angles and electrostatic intramolecular forces, which are often in a delicate thermodynamic balance. A small alteration in the properties of a protein can change its physical shape, sometimes drastically. When this happens to an enzyme molecule, the active site may be affected, which would inhibit its ability to catalyze.

If tryptophan were present in large quantities, it could affect the configuration of one or more of the enzymes in its biosynthetic pathway, a process known as **feedback inhibition.** It has been shown that the *first* enzyme in this pathway is indeed affected by tryptophan in just this way. The enzyme apparently has a second site, the **allosteric site,** in addition to its active site. When tryptophan is present in large quantities, trypto-phan molecules bind to this allosteric site, changing the configuration of the enzyme so that it can no longer bind properly to its substrates. This cuts off the biosynthetic pathway at the source, preventing the bacterium from synthesizing tryptophan when this substance is abundant.

The Operon: Genes Associated for Transcription

Even more sophisticated control mechanisms operate directly on the level of gene transcription. According to the operon concept, protein syn-thesis is regulated by a coordinated group of genes called an operon. An **operon** is a unit of gene expression including regulatory genes, control elements (the sites where the regulator proteins act), and a set of struc-tural genes. **Structural genes** are those that code for the proteins needed by the cell, either enzymes or proteins that have structural functions. The **regulator gene** codes for a repressor protein that regulates the structural

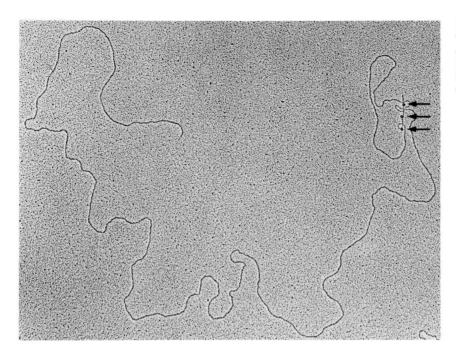

Figure 12–8 RNA polymerase binds to promoters along the chromosome of the *E. coli* phage T7 (arrows). Elongating strands of RNA can just be made out at these points. (Courtesy of Dr. Th. Koller.)

genes. Such a repressor protein functions by binding with a control element known as the **operator site** thought to lie adjacent to the set of structural genes. The operator site overlaps the promoter site for binding RNA polymerase, so when the repressor binds to the operator, the promoter site is blocked (Fig. 12–8). As a result, RNA polymerase cannot initiate transcription of the associated structural genes.

End-Product Repression

If it is true that the best control is exercised early in the biosynthetic pathway, then it is even better to prevent the biosynthetic pathway from coming into existence—not by inhibiting enzyme action but by inhibiting enzyme *formation* in the first place. This saves all the materials and energy that would otherwise go into the production of these enzymes at times when they are not useful.

The *E. coli* chromosome has five structural genes that code for the five enzymes of the tryptophan biosynthetic pathway, or put differently, five genes for tryptophan production. (These genes are physically adjacent to one another—a condition called linkage in a eukaryote, but perhaps better called close linkage in a prokaryote, since given its single chromosome, almost all of the genes of a prokaryote must be considered linked to some extent.)

The RNA polymerase enzyme necessary for mRNA production can only begin transcription at a promoter. The promoter region for the tryptophane biosynthesis genes occurs just ahead of those genes on the *E. coli* chromosome. Closely adjacent to all the other genes in this associated group is a **regulatory gene.** It produces the mRNA that in turn produces a repressor protein. The repressor protein (Fig. 12–9) is potentially able to bind to the operator site, blocking the promoter region. By itself, however, the repressor is unable to do this. To do so it must be slightly modified in its configuration. Perhaps you have guessed that in this case the agent that modifies the configuration of the repressor protein is none other than tryptophan. The end-product, tryptophan, is a **corepressor.** It helps inhibit its own synthesis when present in sufficient quantity. This process is called **end-product repression.**

Figure 12–9 Stylized drawing showing the relationship between a typical repressor protein (shown in green) and the DNA molecule. Notice that the alpha-helices of the protein interpenetrate the major groove of the DNA double helix. Since RNA polymerase must attach to the promoter site in this major groove, the presence of the protein blocks transcription. (From Y. Takeda, D.H. Ohlendorf, W.F. Anderson, and B.W. Matthews, "DNA-Binding Proteins," *Science* 221:1020–1026 © 1983 by the American Association for the Advancement of Science. Used by permission.)

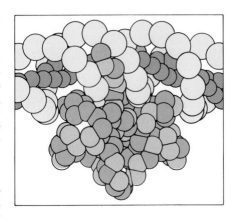

In summary, when tryptophan is present, not only does it inhibit the biosynthetic pathway of its synthesis by interacting with a crucial enzyme in that pathway, but it also eventually prevents the *formation* of those enzymes by corepression. This prevents RNA polymerase from producing the kinds of mRNA that make the tryptophan biosynthetic enzymes.

Enzyme Induction

An old farmer (not known for his sanitary practices) once left a pail of cream overlong in the barn and found that it had collected a number of drowned flies. No matter. He simply strained out the flies, and considered the cream as good as new. Consider an *E. coli* bacterium that no doubt was one of many on the feet and bodies of those flies. If it originally came from the intestine of a mature cow, it had never been exposed to the milk sugar lactose in its life. Neither had, perhaps, a thousand generations of its ancestors—a lapse of time that, if converted to a human scale, would stretch back to early prehistoric times. Despite this, if that cream were to be microscopically examined, it would be found filled with exuberantly growing and dividing bacteria, all abundantly provided with the two enzymes necessary for the consumption of lactose— a permease needed to transport it into the cell, located in the cell membrane, and beta-galactosidase in the cytoplasm, which breaks the disaccharide lactose into its component hexoses, glucose and galactose. Yet just a few generations back—minutes in real time—not one of the bacteria transported into that bucket of cream would have had those enzymes, nor would their ancestors for a thousand generations before them.

What explanation are we to imagine? Because of its complexity, the information needed to produce the enzymes could not have originated spontaneously but must have been present, though unexpressed, even before the bacteria were exposed to the lactose-rich medium. Presumably it had been present ever since the cow from which the bacteria came had been a calf, living on its mother's milk. The hidden genes were replicated a thousand times without once revealing their presence. Along with them, whatever factor was keeping them hidden had also been maintained.

The explanation is just the opposite of end-product repression. In end-product repression the repressor protein was effective only in the presence of a particular substance, but in **enzyme induction** the repressor is only effective in the *absence* of a substance, in this case lactose. Thus, the regulator gene that produces this repressor protein is replicated whenever the rest of the bacterial genome is, so in all generations

1. Normal situation, operon "off." Transcription and translation of the regulator gene produces a repressor protein that binds to the operator and blocks RNA polymerase from transcribing the operon. The RNA polymerase promoter site (not shown) overlaps the operator site.

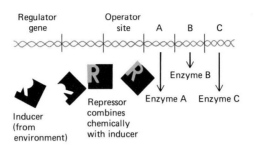

2. Inducer (from environment), which is usually a substrate of enzyme A, B, or C, inactivates repressor. Operator-promoter site now is free and transcription by RNA polymerase turns the operon "on."

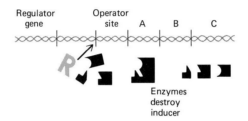

3. When enzymes have completed their task and the substrate (inducer) is consumed, the repressor is free to combine again with the operator site and turn the operon "off" again.

Figure 12–10 A typical mechanism for regulating prokaryotic gene expression. The operon is the basic unit of gene control. It consists of a group of structural genes, associated regulatory genes, and operator sites.

of *E. coli,* that repressor protein will be manufactured and will prevent the expression of the structural enzyme-producing genes.

When lactose is present, the lactose combines with the repressor substance, modifying it so that it cannot bind to the promoter region of the bacterial chromosome associated with these structural genes. Now the RNA polymerase of the cell is unimpeded in its action and immediately begins the production of the mRNA that will result in the production of the appropriate enzymes (Fig. 12–10).

Should the lactose once again come to be in short supply, these events are reversed. The repressor protein is no longer able to combine with the absent lactose, so it is now available for binding to the promoter region of the chromosome.

Why is it necessary to repress the formation of these enzymes when the substrate is not present? Clearly, it takes energy and raw materials to manufacture those enzymes, energy and raw materials that might better be used for other things when the proper substrate for the enzyme is not present. The slight disadvantage that this unprofitable diversion of resources might represent could be the margin between life and death, or at least between effective and ineffective competition for the bacterium that did not switch them off when they were not needed.

GENE CONTROL IN MULTICELLULAR ORGANISMS

In nine months the tiny human zygote constructs a marvel of complexity alongside which the most complex computer is simple. Moreover, that complex machine not only builds itself but also operates itself, repairs itself, and even reproduces itself. Only in the last few years has this very matter of self-design, self-construction, and self-control of living things begun to yield its secrets to biologists.

Figure 12–11 A possible mechanism for the control of the secretion of hormones in a multicellular organism. The steroid hormone ecdysone unlocks certain genes in the cells of insect tissues. In preparation for molting (ecdysis), the insect's prothoracic glands secrete ecdysone, which is carried by the blood to various body cells. The hormone molecules pass through the cell membrane and (1) are accepted by receptor proteins in the cytoplasm. (2) The receptor-hormone complex enters the nucleus and (3 and 4) binds to a pair of primary response genes, which then begin the process of transcription. The mRNA they produce is translated to produce primary response proteins. (5) Some of the primary response proteins are inhibitory and prevent transcription of the complex. These are presumably part of negative feedback regulatory systems. Some, however, turn on the secondary response genes, which are transcribed and (6) in due course translated. (7) The secondary response proteins may then leave the cell and function in ecdysis. For example, epidermal cells may produce enzymes capable of digesting the lower layers of the old cuticle.

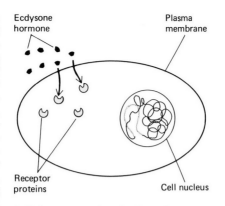

1. Ecdysone passes through cell membrane. It is accepted by receptor proteins in the cytoplasm.

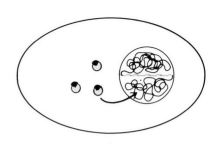

2. Receptor-ecdysone complex attaches to a segment of nuclear DNA

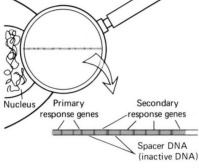

3. The nuclear DNA contains a portion that embodies specialized genes

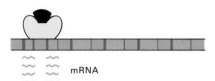

4. When the receptor-ecdysone complex attaches to the primary response genes transcription of these genes is activated. In due course the resulting mRNA is translated. It forms primary response proteins:

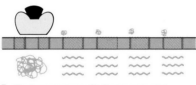

5. If a large amount of primary response protein is made, one of these newly synthesized proteins binds to the primary response genes and turns them off by displacing the receptor-ecdysone complex. (This is evidently a regulatory provision). Other primary response proteins cause the secondary response genes to be transcribed.

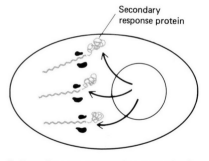

6. Secondary response proteins are translated from secondary response mRNA.

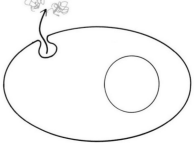

7. In some cases the secondary response proteins are exported from the cell.

The ancestry of any of your cells—say a muscle cell—could be traced through many cell generations back to a single, fertilized egg, the zygote. But a very different cell—perhaps an epithelial cell—in your body could also trace its ancestry back to the very same zygote. Except for sex cells, all the descendants of that primitive zygote were produced by mitosis, which, as you know, ensures that all daughter cells have exactly the same DNA. That would seem to indicate that all these very different cells share the same genetic information and that in this regard they are identical.

Yet obviously they are not identical at all. How can this be? The generally accepted view is that although the various differentiated cells of an organism probably have the same genetic information, different parts of that information are expressed in different tissues. Whatever is inappropriate—and that must be a vast fund—is repressed, not only in each cell, but also in the descendants of that cell. In the case of a differentiated cell that actively divides, such as an epithelial cell, the very suppression of much of its heredity has become hereditary.

There is yet another aspect of this matter. Some cells express some of the information they contain only at certain times. An example of this is the thyroid gland. It secretes its hormone only when stimulated to do so by another hormone released into the blood from the pituitary gland (Fig. 12–11).

It would seem that some of the genes function only when the cell is stimulated in just the proper way. How is this suppression accomplished, and what makes it temporary or permanent? We know very few of the answers in the case of eukaryotes. The mechanisms they employ are much too poorly understood to discuss here in any detail, but they might involve the alkaline nuclear proteins called **histones,** or alternative forms of DNA, such as left-coiling varieties (Z-DNA), or both of these plus other mechanisms not yet imagined. With one or two exceptions, the kind of enzyme induction that occurs in prokaryotes is not known in eukaryotes, and no operon or comparable mechanisms have been described in them.

However, the work of Barbara McClintock of the Cold Spring Harbor Laboratories has cast some light on the matter of eukaryote gene regulation. She has discovered that in maize (corn) the expression of certain genes governing kernel color depends upon their location in the chromosome. The most reasonable interpretation of this appears to be that when these genes are located near a repressor, they are controlled or suppressed entirely, but when they become separated from their control regions, they produce their effects unrestrainedly. Other researchers have suggested that cellular genes known to predispose toward the development of cancer may actually have originated as quite normal genes regulating cellular growth. These **oncogenes** may become separated from their control regions, perhaps by an irregularity of meiosis or mitosis, and then produce the uninhibited growth characteristic of tumors.

INTRONS AND EXONS

Most eukaryotic genes contain very long base sequences that do not end up in mature RNA. Some of these **interrupted genes** have several such sets of sequences. Newly transcribed mRNA contains these "garbage" sequences but they are enzymatically snipped out of the mRNA strand, and the portions containing useful information are joined together to make mature mRNA. The sequences in the DNA that are retained in the mature RNA are called **exons,** and the discarded ones are called **introns** (Fig. 12–12).

Perhaps intron-exon sequences may play a role in controlling the flow of information from nucleus to cytoplasm and in controlling cellular differentiation. The fact that many introns are near-copies of the adjacent

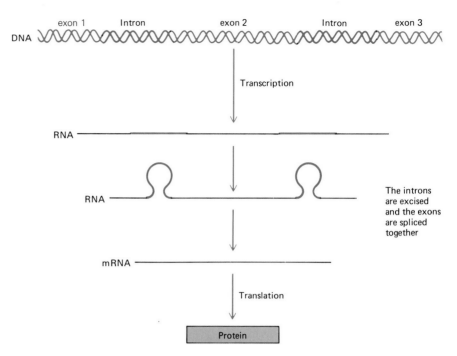

Figure 12–12 Introns are sequences that must be cut out of RNA before RNA can be translated into a protein. Are the sequences in the DNA that code for them "genetic junk," or do they have a function?

exons has led to certain evolutionary proposals. For instance, it has been suggested that introns may serve as the raw material from which many new functional genes are constructed, or from which those lost by mutation (see Focus on Genetic Mistakes) may perhaps be reconstructed. One school of thought even denies introns any function whatever, arguing that they are genetic parasites in effect, replicated along with meaningful genetic information whenever the cell divides. In the present state of our knowledge we are unable to decide which, if any, of these proposals is correct.

Focus on . . .

GENETIC MISTAKES

What would happen if through some accident one of the bases of a DNA strand were changed to the wrong one? Or if, during replication, a base that was not truly complementary to its template equivalent were erroneously placed in a daughter strand?

This error would be transmitted to all strands of DNA replicated from the one in which it originally occurred. Thus all the descendants of that cell would share its genetic error. Such an error would also lead to a corresponding defect in the complementary strand of mRNA and ultimately to the *wrong* amino acid in a protein. In fact, if the error were to result in what is known as a "non-sense" sequence of bases—that is, a sequence that has no amino acid equivalent—the protein would suffer a break at that point. Since particular sequences of bases are needed to initiate translation, the chances are that the remainder of the strand would

not be transcribed at all. That would probably result in a protein only a fraction of its proper length. However, if the sequence did have an amino acid equivalent, the changed protein might function much as it did before, or it might be considerably disturbed.

Such an abrupt change in the genetic information of a cell is called a **mutation.** The hereditary disease sickle-cell anemia is caused by a mutation that results in a defect in the hemoglobin molecule. Just one inappropriate amino acid residue is substituted in two of the polypeptide chains that constitute this vital molecule. The inappropriate amino acids cause adjacent hemoglobin molecules to form abnormal chemical linkages with one another under conditions of low oxygen. This in turn causes the formation of an abnormal crystal-like structure in the hemoglobin that distorts the red blood cell, clogging capillaries and making the cell a ready prey to the macrophages whose task is to remove damaged or worn out blood cells. So many cells are destroyed that anemia results.

Just one amino acid can make a difference of life and death in this way. All human genetic diseases must have originated as mutations. To be sure, beneficial traits also originate by mutation. Yet they are rare, and the reason is not hard to understand. It is after all unlikely that the random removal or replacement of a transistor will improve a high-fidelity amplifier. So, too, a random change in the genetic material is unlikely to improve the vastly more complicated function of a cell, or of the organism of which it is a part.

Every population of organisms ever investigated has been shown to be subject to mutation. The wonder is that we do not observe more mutations than we do. Fortunately, cells possess enzymes whose function is to repair or remove damaged sections of DNA so that most mutations that do occur are not propagated. Just the same, some survive. Mutations are caused by a variety of known agents—certain chemicals, for example, including a number of common food additives and some natural agents that occur in our environment. Ionizing radiation, in the form of x-rays or radioactive fallout, also produces mutations. Some of these mutations result in cancer, some in abnormal transmissable genes.

It appears that under present conditions the effects of these agents are small (although they would not be small in the event of a nuclear war or large-scale nuclear accident). Yet the population exposed to them is large. The total number of mutations they produce must therefore be large and is probably increasing. We risk much when we add to the already substantial genetic burden carried by the human species.

It is sometimes argued that complete removal of mutagenic substances from the environment would be prohibitively expensive. There is probably a threshold, according to this school of thought, beneath which a mutagenic agent causes no harm. All we need to do is reduce the concentration of a mutagen in the environment below this threshold and all will be well.

The biggest objection to this view is that the most diligent search has failed to disclose the existence of such a threshold for any known mutagen. What is observed instead is what is called a linear dose-response relationship. The smaller the dose, the smaller the effect; yet any dose at all does produce some effect. Moreover, even if thresholds do exist, we are totally ignorant of what they might be. It seems only prudent to attempt to reduce the exposure of the human population to known mutagens to the utmost extent possible. At the same time it is imperative to discover any unknown mutagens that may exist among the thousands of supposedly innocuous substances with which we so carelessly saturate our environment and our bodies.

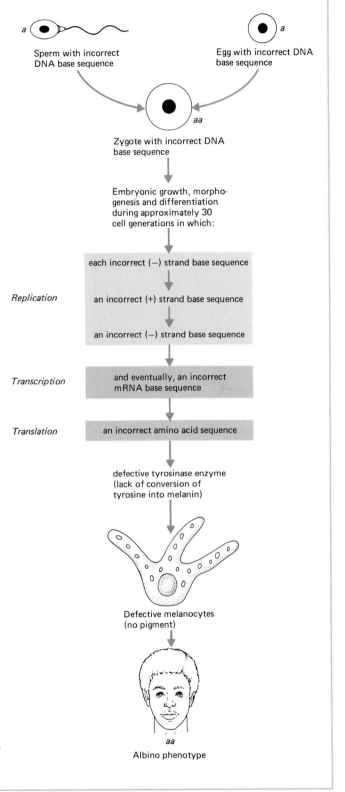

Sperm with incorrect DNA base sequence

Egg with incorrect DNA base sequence

Zygote with incorrect DNA base sequence

Embryonic growth, morphogenesis and differentiation during approximately 30 cell generations in which:

each incorrect (−) strand base sequence

Replication

an incorrect (+) strand base sequence

an incorrect (−) strand base sequence

Transcription

and eventually, an incorrect mRNA base sequence

Translation

an incorrect amino acid sequence

defective tyrosinase enzyme (lack of conversion of tyrosine into melanin)

Defective melanocytes (no pigment)

Albino phenotype

CHAPTER SUMMARY

I. The genetic information of DNA is stored in the form of groups of three bases.

A. The base sequence in each such triplet specifies a particular amino acid that will eventually be incorporated into a polypeptide.

B. In gene expression the DNA strands unwind. The plus strand serves as a template for the formation of a mRNA molecule (transcription).

C. The mRNA strand bears the information content of the DNA into the cytoplasm.

D. The triplets of the mRNA are called codons.

II. The mRNA becomes associated with ribosomes in the cytoplasm.

A. Transfer RNA is attracted to the mRNA strand. Each species of tRNA has a particular anticodon and binds with a particular amino acid.

B. Each kind of tRNA becomes attached by its anticodon to the complementary codon of the mRNA strand.

C. The amino acids associated with the tRNA molecules are joined by action of the ribosome.

D. The mRNA, tRNA, and polypeptide now separate. The sequence of tRNAs determines the sequence of amino acids in the finished peptide.

III. Recent research advances have made some headway against the mystery of how genes are regulated in their expression.

A. Mechanisms postulated by Jacob and Monod in their operon theory of gene regulation can account for the responses of some prokaryotic cells to changes in their environment.

B. According to this proposal, regulator genes produce a repressor substance that binds to an operator site. This represses the structural genes by blocking the promoter.

C. In the presence of an appropriate substrate, the repressor substance is neutralized and the structural genes produce an appropriate series of enzymes to deal with the substrate.

D. The mechanisms of end-product repression appear to operate by similar mechanisms, but allosteric inhibition involves the direct repressive action of some substance on the enzyme itself. Several levels of control mechanism redundancy assure appropriate regulation of bacterial metabolism under almost any circumstances.

IV. Some important gene regulation definitions:

A. *Operon.* In prokaryotes, a group of functionally related genes, including a group of structural genes, associated regulator genes, and operator sites.

B. *Regulator gene.* The segment of DNA in an operon that produces a repressor substance that controls the rate of synthesis by associated structural genes. The repressor substance binds to the operator site.

C. *Repressor substance.* A small protein that locks the operator gene in the "off" position; it is capable of combining with inducer substance from environment, losing its effectiveness when it does so.

D. *Operator site.* In prokaryotes, a section of DNA that acts as a switch, turning off structural genes in the presence of repressor substance.

E. *Inducer substance.* In prokaryotes, an environmental trigger (such as a food substance that is broken down by the enzymes produced by the structural genes) that combines with the repressor substance, thus unlocking the structural gene or genes of the operon.

F. *Structural gene.* In prokaryotes, a section of DNA in an operon that codes for any polypeptide other than a repressor.

Post-Test

1. The three main functional types of RNA are _____ _____, _____, and _____ .
2. Information is transferred from DNA in the nucleus to the ribosomes by _____ .
3. Each ribosomal subunit consists of _____ and _____ .
4. A sequence of three nucleotides in DNA, a _____ _____, corresponds to the _____ in mRNA.
5. The synthesis of RNA on a DNA template is referred to as _____ .
6. Transcription begins when the enzyme _____ _____ binds to a _____ .
7. A polyribosome consists of 30 or so _____ .

8. The tRNA attached to a new amino acid is accepted by the ribosomal _____ .
9. Three steps in the formation of a polypeptide along a ribosome are _____ , _____ , and _____ .
10. _____ are units of gene expression, including structural genes, control elements, and operator sites.
11. The regulator gene codes for a _____ protein that regulates _____ _____ .
12. In enzyme _____ the repressor is only effective in the absence of a particular substance.

Review Questions

1. Summarize the complete sequence of information transfer from DNA to finished protein, including the role of ribosomes.
2. If DNA directs the synthesis of protein, how can it be said that DNA determines the nature of the cell's chemical constituents other than protein?
3. What is an operon? Describe its operation.
4. Compare feedback inhibition, end-product repression, and enzyme induction as they occur in bacterial cells. Why does a bacterium need all these mechanisms? Why wouldn't just one or two suffice?

5. In almost all known cases, mammals cease to produce the intestinal enzyme lactase in adulthood and cannot therefore digest milk when mature. Some human beings constitute the exception. In Europeans and some other cattle-raising peoples the lactase enzyme persists through life. What kind of mutation could account for this odd racial characteristic?
6. Contrast DNA replication with transcription of messenger RNA.
7. What is the relationship among triplets, codons, and anticodons?

Readings

Brown, D. "Gene Expression in Eukaryotes," *Science,* 211: 667–674 (13 February 1981). A summary of fairly recent thinking on the regulation of eukaryote gene expression—still much a mystery.

Chambon, P. "Split Genes," *Scientific American,* May 1981, 60–71. The role of introns and exons in the genetic message-coding for a protein.

Cohen, S.N., and J.A. Shapiro. "Transposable Genetic Elements," *Scientific American,* February 1980, 40–49. Genes do not always stay where they belong—they can even pass the prokaryote-eukaryote barrier. This poses evolutionary questions and opens up new frontiers of genetic engineering.

Dickerson, R.E., et al. "The Anatomy of A-, B-, and Z-DNA," *Science* 216: 475–484 (30 April 1982). Newly characterized and unfamiliar forms of DNA that may function in gene control.

Kolata, G. "Cell Biology Yields Clues to Lung Cancer." *Science,* 218: (1 October 1982). By studying lung cancer cells in culture, researchers are gaining insights into how the cells multiply and how they can be prevented from multiplying.

Kolata, G. "Fetal Hemoglobin Genes Turned On in Adults," *Science,* 218: 1295–1296 (24 December 1982). Accidental discovery of a drug that affects the suppression of fetal hemoglobin production, which forms the basis of a potential method of treatment of sickle-cell anemia.

Marx, J.L. "Cancer Cell Genes Linked to Viral *onc* Genes," *Science,* 216: (14 May 1982). A discussion of the hypothesis that cancers of both viral and nonviral origin may be caused by the inappropriate activation of similar cellular genes.

McClintock, B. "The Significance of Responses of the Genome to Challenge," *Science,* 226: 792–801 (16 November 1984). McClintock's Nobel Prize lecture.

Oppenheimer, J.H. "Thyroid Hormone Action at the Cellular Level," *Science,* 203: 971–979 (9 March 1979). A classic problem—control of eukaryote gene function by hormones—seems near solution for the thyroid gland.

Watson, J.D. *Molecular Biology of the Gene,* 4th ed. Menlo Park, Calif., Benjamin Cummings, 1986.

(See also readings for Chapter 10.)

Chapter 13
GENETIC FRONTIERS

Outline

I. Recombinant DNA techniques
   A. Cleaving DNA
   B. Constructing recombinant DNA molecules
   C. Introducing recombinant DNA molecules into a host cell
   D. Isolating a single gene
   E. Expression of the recombinant DNA
II. Future genetic engineering
III. Repairing genetic defects
IV. Gene insertion in eukaryotes
   A. Giant mice
   B. Engineering plants
V. Recombinant RNA
VI. Some worries
Focus on Probing for Genetic Disease
Focus on Molecular Genetics and Cancer
Focus on Localizing a Gene by Cell Fusion

Learning Objectives

After you have read this chapter you should be able to:

1. Describe the primary techniques utilized in recombinant DNA experiments.
2. Summarize the problems involved in isolating, identifying, and cloning a single human gene.
3. Describe the action of restriction endonucleases and their importance in recombinant DNA experiments.
4. Identify the role of plasmids in recombinant DNA experiments.
5. Describe the special measures that have been employed to introduce genes experimentally into plant cells and animal cells.
6. Discuss the role of oncogenes in the development of cancer.
7. Summarize the ethical and other objections that have been raised against recombinant DNA studies and give potential practical and research applications.

Techniques of selective breeding have long been used to build a better cow, or honeybee, or corn plant. However, these results come slowly and fall short of producing some of what one might like to accomplish: cow's milk low in saturated fats, for example, or a cow that could live in the tsetse fly–infested regions of Africa. And selective breeding really leaves us helpless in the face of genetic disease. One cannot help a sufferer of sickle-cell anemia or of cystic fibrosis by simply suggesting that he or she should not have been born.

Recent advances in biochemical and genetic techniques promise an unprecedented control of heredity. One of the most exciting areas of biology today is **genetic engineering,** the manipulation of DNA outside an organism so that new genetic strains of organisms with new characteristics can be constructed.

RECOMBINANT DNA TECHNIQUES

Recombinant DNA techniques permit the formation of new combinations of genes by isolating genes from one organism and introducing them into either a similar or an unrelated organism. By these methods foreign DNA can be inserted not only into the simple single cells of bacteria but also into cells derived from the bodies of complex organisms. These techniques permit an investigator to modify the genetic complement of a bacterium so that it can produce substances, such as insulin, that normally it cannot produce; consume substrates, such as spilled oil, that normally it cannot consume; and do other things that probably have not yet even been imagined.

Recombinant DNA techniques allow the isolation of genes from the tremendously complex metabolic machinery of eukaryotic cells. The genes can then be inserted into bacteria, where they can be studied far more conveniently. Finally, they can be reinserted into the eukaryotic organism after extensive deliberate modification, possibly after being combined with other genes or parts of genes from other donor organisms.

In a way, genetic engineering is not new; organisms have mechanisms for exchanging genes, and "experiments" in recombinant DNA and gene transfer have apparently always occurred (Fig. 13–1). However, these natural experiments have occurred randomly, and certainly not with human goals and desires in view. For deliberate genetic engineering to be effective, we must be able not only to *change* genes but also to change them in a controlled fashion. How genes operate has been understood in principle since the 1950s, but applying that knowledge to genetic engineering has been slow until recently.

The greatest problems that must be solved to make genetic engineering practical have only been clearly identified in the last ten years or so. Among these are finding genes with desired characteristics; transferring genes from one organism to another, a process called **vectoring;** and ultimately, making genes to order. As yet, the first two of these problems are well on their way to solution, but work on the third is not so far advanced. Thus far it has usually been necessary to find a pre-existing living source for the desired gene.

Cleaving DNA

To begin the process of genetic engineering, the investigator must first break up the cell's DNA into more manageable fragments by the use of enzymes known as **restriction endonucleases** (Fig. 13–2). Normally, the bacterium uses these enzymes as defense weapons against viral infection. A virus may penetrate a potential host cell but there find itself under

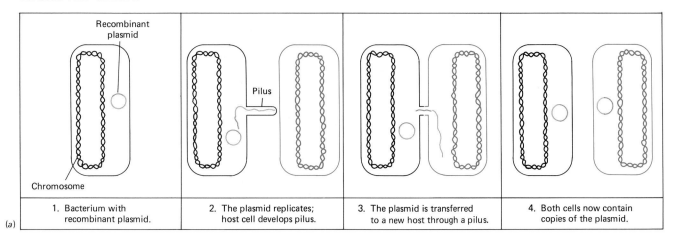

(a)

| 1. Bacterium with recombinant plasmid. | 2. The plasmid replicates; host cell develops pilus. | 3. The plasmid is transferred to a new host through a pilus. | 4. Both cells now contain copies of the plasmid. |

(b)

| 1. This bacterial cell contains DNA of a temperate retrovirus. | 2. If the cell is stressed, the viral DNA is transcribed. | 3. Note that one of the viral RNA strands has inadvertently incorporated a gene of host origin. | 4. When mature viruses form, some contain a copy of the host gene. | 5. When the host cell bursts, the viruses are released. The viral RNA with the information from its former host's gene enters a new host. | 6. Reverse transcriptase incorporates viral DNA and former host's gene into new host's DNA. |

Figure 13–1 Mechanisms of natural DNA recombination. (a) During conjugation plasmids, together with genes for antibiotic resistance, can be transferred from one bacterium to another. A special organelle, the pilus, forms and serves as a conjugation bridge through which DNA passes (see Fig. 9–18). (b) In transduction genes are transferred from one bacterial chromosome to another within a bacteriophage. There is evidence that genes may sometimes be transferred among different bacterial *species* by these mechanisms.

attack by restriction endonucleases, which seek out certain **recognition sites** in the viral DNA (if it is a DNA virus, that is) and then chop the alien strand into little, harmless bits. Bacteria produce methyl groups that cover their own recognition sites. In this way the bacterial DNA is protected from the action of these enzymes—otherwise the bacterium would swiftly destroy its own genome. (Recall that the term genome refers to all of the genetic material in the cell.) Eukaryotic DNA possesses recognition sites in random locations throughout its DNA. For that reason restriction endonuclease obtained from bacteria is able to fragment eukaryotic DNA (though under natural conditions it probably never has the opportunity to do so).

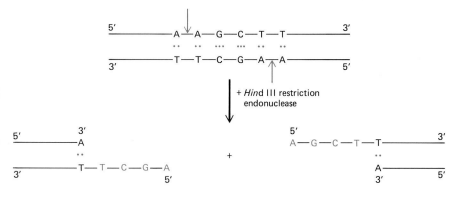

Figure 13–2 The cleavage of DNA by the restriction endonuclease *Hind* III. The enzyme, an endonuclease, cleaves bonds within the DNA double helix. The products have short single-strand stubs (indicated in color) that are antiparallel complements of each other.

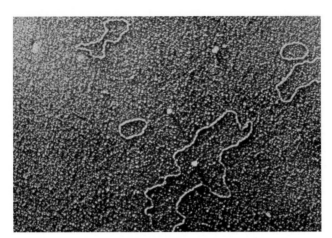

Figure 13–3 Electron micrograph of plasmids, circular bacterial chromosomes often used in genetic engineering to transfer desired genes into bacteria. (Thomas Broker/Cold Spring Harbor Laboratory.)

Certain kinds of restriction endonucleases leave some fragments with **sticky ends,** potentially able to recombine with other, similar fragments of DNA from perhaps some other organism altogether. The reason for the stickiness involves the way in which the endonuclease cleaves the DNA. It leaves one strand longer than the other, and since the recognition sites are always the same for each specific endonuclease no matter what kind or area of DNA is cleaved, they all have the same base sequence. Some of these sequences are **palindromic;** that is, they read the same in both directions, like the word "radar." For this reason, even though they are all the same, they are all complementary and, if treated with the correct ligase enzyme, will attach to one another.

Constructing Recombinant DNA Molecules

Restriction endonucleases split the potential donor DNA into many fragments. The next step is to construct recombinant DNA molecules. Recall that some bacteria, including *Escherichia coli,* have a small accessory chromosome known as a **plasmid** (Fig. 13–3). When present, plasmids are replicated along with the main chromosome and are distributed to daughter cells, as the chromosome is.

To construct recombinant DNA molecules, the investigator treats plasmids from *E. coli* cells with restriction endonuclease (Fig. 13–4). The originally circular plasmids now become double strands of DNA, with palindromic sticky ends. These are then mixed (under the proper conditions) with sticky DNA fragments from the donor cell. By the use of ligase and other enzymes, the ends of the donor and plasmid DNA are joined. In principle this produces a linear strand of DNA, but that strand itself has sticky ends, so the two ends can be joined with one another. The result? Reconstituted circular plasmids, this time containing some fragment of donor DNA. In the vast majority of cases this fragment will *not* be the desired one. Yet by chance, a very few can be expected to possess exactly what is desired.

Introducing Recombinant DNA Molecules into a Host Cell

The recombinant DNA molecules are next implanted into bacterial cells, thus producing numerous strains of genetically engineered bacteria. The large variety of recombinant molecules produced in this way are collectively referred to as a **library,** or gene bank. One of the DNA combinations in this library might be the target gene capable of producing the desired protein. If that one can be picked out from among all the others,

Figure 13–4 The process of enzymatically inserting a mammalian gene, the gene for rat insulin, into a plasmid from the bacterium *E. coli*. The plasmid is placed back in the bacterial cell, and that cell is cloned to obtain many copies of the insulin gene. The cloned cells produce large amounts of insulin when they are incubated in the growth medium.

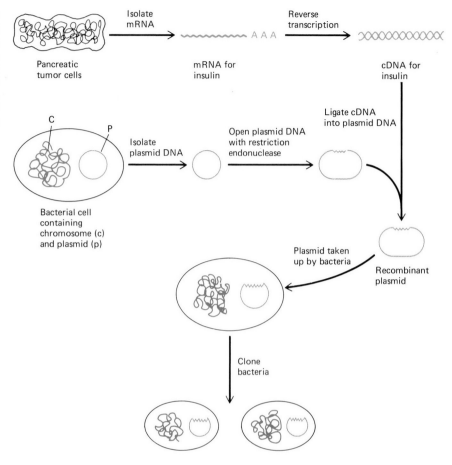

it can be cultured, thus automatically cloning the desired genes which have artificially been implanted in it.

Since the early days of electron microscopy it has been known that some strains of bacteria are able to produce a hairlike **pilus** with which they temporarily attach themselves to other bacteria of the same species. It has been shown that pili are developed only by bacteria that have plasmids (see Fig. 13–1). During conjugation, two bacteria come together and a plasmid passes through the hollow pilus from one bacterium to another.

Plasmids can be of help to molecular biologists, for they can serve as genetic vectors carrying desired genes into cells and propagating them. When desired genes are inserted into plasmids, and these altered plasmids implanted in bacterial hosts, the plasmid and its genes are replicated along with the entire bacterial population descended from that ancestral host cell. Should the gene code for such a protein as human thyroid-stimulating hormone (TSH), that protein accumulates inside the bacterium from which it can later be harvested. Since *E. coli* can host as many as 20 such plasmids, at least 20 TSH genes could be accommodated at one time, collectively producing 20-fold the amount of the hormone that a single gene might.

To get the patchwork plasmids into host bacteria does not necessarily require a pilus. There are several less natural but more convenient ways, involving, for example, the removal of bacterial cell walls so that the plasmids can penetrate their new hosts easily. To be sure, not all the bacteria will be infected, but if the plasmid contains genes conferring antibiotic resistance, then one need only treat the bacterial population with that antibiotic. Only bacteria that have been infected will resist the antibiotic; the rest will conveniently die.

Another way to implant genes in a population of host cells involves bacteriophages (bacterial viruses), which actually can become part of the main bacterial chromosome and can thus carry new genes into it. In a newer technique, the desired DNA is mixed with a calcium salt, which is then converted to calcium phosphate. The DNA is then coprecipitated with the insoluble calcium phosphate, which protects it from damage. The DNA itself enters bacteria without a phage or plasmid vector. Several newer, less involved techniques also are on the verge of application.

Isolating a Single Gene

A needle could be found in a haystack of any size if one had a magnet or, better, a metal detector. Fortunately, the genetic equivalent of the metal detector is available to us and can be used to find just the DNA base sequence desired among the vast tangle of genes in the cell nucleus. This device is called a genetic probe.

To select the cells with the desired genes from the genetic library, the researcher may spread a sample of the bacterial culture on plates. If it is dilute enough, on the average, only one cell falls onto a particular location. When it reproduces it gives rise to a clone—a colony of genetically identical cells—which can be propagated further in a liquid medium and then assayed for the presence of the desired gene. One way—not usually a very feasible one—might be to analyze each lot for protein production. Even if this were practical, the process could take months.

However, a more workable solution is available. Host cells containing the target gene can be identified by using a **genetic probe,** a radioactively labeled segment of mRNA or artificially synthesized single-stranded DNA complementary to the target gene.

Suppose we wanted to identify the genes that code for insulin. Because we know the amino acid sequence of insulin, we could synthesize the mRNA that produces it. This mRNA will attach itself to, or *hybridize* with, complementary DNA that contains the base sequence needed for production of the protein. If the synthesized mRNA is radioactively labeled, it will identify the sequence of the DNA that is responsible for the production of insulin.

Thus, if after such treatment any radioactive DNA can be detected in that of a particular clone of bacterial cells, it indicates that this DNA had hybridized with the radioactive probe. The culture from which that DNA came is the one that will be able to produce the desired protein. This probe hybridizes only with DNA from a bacterial cell that contains the base sequence needed for production of the desired protein. (See Focus on Probing for Genetic Disease.)

Expression of the Recombinant DNA

Even though a gene has now been isolated and identified in our example, large-scale propagation of the bacterial strain would not necessarily produce the desired protein. The necessary gene will not be transcribed unless it can be associated with a set of regulatory and promoter genes.

Suppose, for example, that we wanted to engineer bacteria that could produce thyroid-stimulating hormone (TSH). It would be necessary to arrange recombination of plasmids containing TSH genes with bacterial DNA followed by the reinsertion of those plasmids into new hosts. Then the resulting strains would have to be tested for actual TSH production. If, for instance, one of them that by chance had the TSH gene were to land in a plasmid adjacent to the *lac*-operon, the presence of lactose in the culture medium would turn on TSH production. We could then turn off or regulate TSH manufacture at will by controlling the concentration of lactose.

Focus on . . .

PROBING FOR GENETIC DISEASE

Genetic engineering techniques can be used in the characterization and detection of genetic disease. The earliest studies of this sort were carried out on sickle-cell anemia, since that disease produces a thoroughly characterized protein abnormality. It will soon be evident why that is important.

Sickle-cell hemoglobin differs slightly from normal hemoglobin in its amino acid sequence. That means that the DNA responsible for the disease must also differ from the normal in its base sequence. Knowing the genetic code, we can easily determine what the base sequence of the abnormal DNA must be.

With a DNA synthesizer it is possible to make DNA that contains this sequence of bases, plus those on either side. It is likely that such a combination is unique in the genome. The artificial DNA is also radioactively labeled, which means, you may recall, that radioactive isotopes are incorporated into some of it. This is now the **probe DNA.**

If the probe DNA is allowed to hybridize with that from a person with sickle-cell anemia, it will unite with his or her DNA at the point where the abnormality lies. Furthermore, even if a person is heterozygous for the disease, the abnormal DNA for which he or she is haploid will hybridize with the probe DNA. This procedure allows us to detect normal carriers of a wide variety of genetic diseases.

Actually, the DNA probe method of detection is not of much use in sickle-cell anemia because the abnormal hemoglobin such people possess can be more simply detected by other methods in either a homozygous or a heterozygous state. However, such abnormal hemoglobin is present only in blood. If we wish to determine whether a fetus will suffer from the disease, we must examine not the blood but the fibroblasts in the amniotic fluid. These tell us nothing by means of the older techniques but can be shown to be heterozygous or homozygous for the disease by the DNA probe method.

An important side benefit of the use of probe DNA is that it allows us to determine on which chromosome the abnormal gene occurs. This is done by hybridizing probe DNA with a chromosome preparation from a person with sickle-cell anemia and observing which of the chromosomes becomes radioactively labeled as a result. Few genetic diseases are as well understood as sickle-cell anemia, but as more and more abnormal proteins are identified and sequenced, an increasing number of hereditary disorders will be detectable in healthy carriers and fetuses.

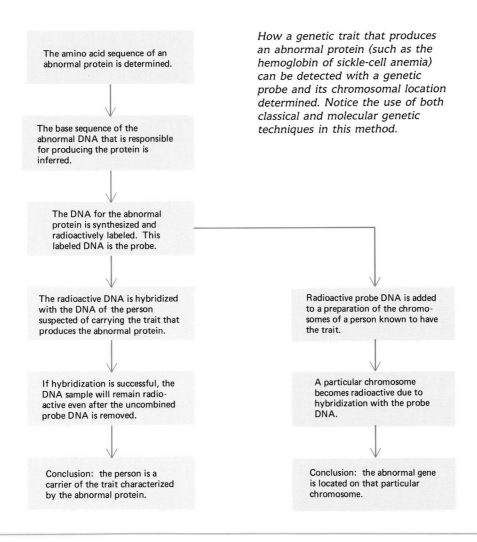

The amino acid sequence of an abnormal protein is determined.

The base sequence of the abnormal DNA that is responsible for producing the protein is inferred.

The DNA for the abnormal protein is synthesized and radioactively labeled. This labeled DNA is the probe.

The radioactive DNA is hybridized with the DNA of the person suspected of carrying the trait that produces the abnormal protein.

If hybridization is successful, the DNA sample will remain radioactive even after the uncombined probe DNA is removed.

Conclusion: the person is a carrier of the trait characterized by the abnormal protein.

Radioactive probe DNA is added to a preparation of the chromosomes of a person known to have the trait.

A particular chromosome becomes radioactive due to hybridization with the probe DNA.

Conclusion: the abnormal gene is located on that particular chromosome.

How a genetic trait that produces an abnormal protein (such as the hemoglobin of sickle-cell anemia) can be detected with a genetic probe and its chromosomal location determined. Notice the use of both classical and molecular genetic techniques in this method.

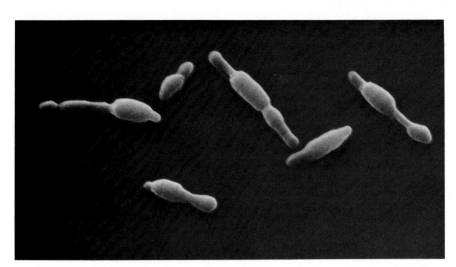

Figure 13–5 Scanning electron micrograph of *E. coli* cells "bulging" with human insulin; see also Figure 1–19. (Courtesy of Dr. Daniel C. Williams and the Lilly Microscope Laboratory.)

Although human insulin is now produced commercially by engineered *E. coli* bacteria, the use of *E. coli* for commercial protein production has the drawback that *E. coli* will not normally secrete these products. The insulin accumulates in the cell and must be removed by harsh mechanical disruption (Fig. 13–5). In the future, *Bacillus subtilis* may be used instead, since this bacterium naturally secretes protein into the culture fluid.

FUTURE GENETIC ENGINEERING

The next advance in genetic engineering will probably involve the actual planned synthesis of totally new genes, designed in advance to possess exactly the right combination of desired properties. Although this has not yet been accomplished, it is sometimes possible to combine existing genes in new associations that have properties much different from any that the donor organisms ever had.

Recent studies have demonstrated the feasibility of introducing sequences into *E. coli* from the bacterium *Pseudomonas,* resulting in the production of the plant dye indigo. This was done accidentally, and resulted from the presence of certain genes in the *Pseudomonas* plasmid that produced enzymes for the later parts of the metabolic pathway leading to the dye. Other genes in the *E. coli* recipient were able to provide an earlier part of the chain of intermediates. The really significant aspect of this study may be that the indigo-producing gene was not derived from a plant source but was put together by combining certain *E. coli* genes already present with some that occurred naturally in a *Pseudomonas* plasmid. The equivalent of a eukaryote gene was thus constructed from prokaryote components! One is tempted to say that the experimenters patched the bacterial blue genes.

REPAIRING GENETIC DEFECTS

One of the goals of genetic engineering is the implantation of genes into eukaryotic cells, with the hope of correcting human genetic disease. In some cases, however, the problem may be due to the action of regulatory genes rather than the production of enzymes that have direct physiological effects. (See Focus on Molecular Genetics and Cancer.)

One of the best examples of this involves the treatment of the hemoglobin diseases sickle-cell anemia and thalassemia. Such diseases result from an abnormal structure of the hemoglobin molecule. In sickle-

Focus on . . .

MOLECULAR GENETICS AND CANCER

A cancer cell is an abnormal cell that lacks biological inhibitions. It appears that cancer is almost a genetic disease, transmitted not so much between generations of hosts as between generations of cells. In addition to lacking the normal contact inhibition that stops cell division when its reasonable limits have been reached, cancer cells are immortal. Not that a cancer cell cannot die, but it cannot die of old age. Normal human fibroblasts, for example, can be maintained in tissue culture for perhaps fifty generations before they become enfeebled, yet the HeLa* cancer cells isolated from a patient in the 1950s are today alive and well in tissue culture vessels the world over, still going strong after thousands of cell generations. Given the proper conditions and care, there seems to be no reason why descendants of today's HeLa cells might not still be multiplying in the year 2100 or beyond.

Recombinant DNA techniques are currently being used in analyses of the process by which normal cells become transformed into cancer cells. These and other studies indicate that carcinogenesis in humans and other mammals is a multistage process, one that involves several independent steps. Yet carcinogenesis seems also to be a basic genetic change that produces a cascade of processes, which in turn bring about the multitude of specific properties that a cancer cell must have to become malignant. It begins to appear that cancer cells owe such traits to possession of at least one and probably several genes that are known as **oncogenes.**

In a typical study, a human line of cancer cells from a bladder tumor was shown to have undergone a mutation in a specific codon (a change from G to T), so the amino acid glycine in position 12 of this gene's protein product was replaced by valine. This one change was shown to be critical to the conversion of the normal cell's normal gene into the cancer cell's oncogene. It didn't take much, perhaps because the malignant gene

somehow duplicated itself and multiple copies came into existence, thus amplifying the effects of the original oncogene.

Oncogenes and oncogene-like DNA sequences are widespread among living things. It is thought that **proto-oncogenes** function in normal cellular growth and embryonic development. The fact that cancers of many kinds occur more frequently in people with chromosomal abnormalities or are associated with specific known chromosomal defects suggests that proto-oncogene *position* may be important. Perhaps when dissociated from their normal regulatory regions, such genes turn into actual oncogenes.

Oncogenes that are of demonstrable animal origin are found in some viruses, especially retroviruses, which are able to transmit them to new cellular victims. Still other viruses carry oncogene sequences that do not seem to have originated in this way but that also produce cancer. It seems likely, however, that several oncogenes, each requiring a separate induction, may usually be necessary to produce cancer. This could explain the observed multistep process of cancer induction required for the production of most experimentally developed cancers, and it could also explain the well-known fact that a tendency to cancer is often hereditary.

Imagine, for instance, that each one of a patient's body cells contained an oncogene inherited from his parents. A viral infection might implant another, and late in life, mitotic malfunction might cause the original gene to be duplicated in one of the patient's continually dividing cell lines. Carcinogenic food additives produced yet another oncogene in one cell in that line, and that completed its transformation into a cancer cell. Over the course of several years that ancestral cancer cell refused to obey normal contact inhibition and divided without hindrance until an obvious tumor was formed and the cancer diagnosed.

The accumulation of the results of these and other random events would certainly take time. Perhaps that is why most cancers occur with greatest frequency in elderly people.

*HeLa cells are easily cultured and are used the world over for numerous kinds of research in cell biology.

cell anemia, for example, hemoglobin differs from the normal by a single amino acid in a critical location in two of its chains. However, persons with these diseases are known to have once produced normal fetal hemoglobin (hemoglobin produced before birth). Normally, however, the genes for making fetal hemoglobin are permanently turned off at about the time of birth. Some hemoglobin disorders have been successfully, albeit somewhat accidentally, treated with a drug that partially restores the production of fetal hemoglobin. As far as anyone knows, despite its difference from regular hemoglobin in a few amino acid sequences, fetal hemoglobin works quite well in adults.

The discovery of a drug that will restore fetal hemoglobin production in adults would involve the manipulation of a pre-existing regulatory gene. This is a simpler matter by far than the replacement of a defective gene. Still, it is gene replacement that is often needed for the conquest of hereditary disease.

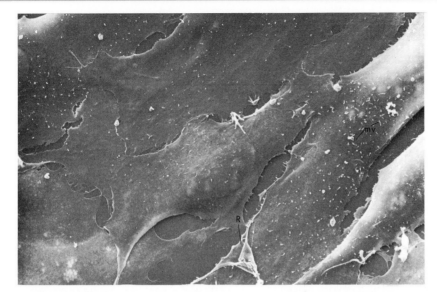

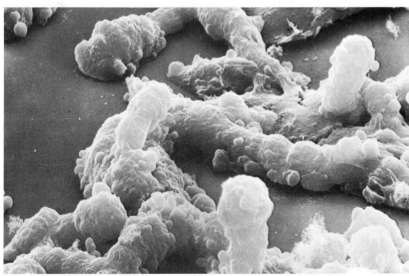

Scanning electron micrographs of cultured cells. (top) Normal cells from a hamster embryo. A few microvilli (mv) and ruffles (R) are visible. (bottom) The same cell type after transformation by a cancer-causing virus (approximately ×3145). Note that the cells have blebs, or bubbles, on their surfaces and aggregate to form several layers. (Courtesy of R.D. Goldman.)

GENE INSERTION IN EUKARYOTES

There is a class of RNA viruses, the **retroviruses,** that make DNA intermediates of themselves by reverse transcription. Sometimes these DNA intermediates become integrated into the host cell genome, where they are replicated along with host DNA until (usually at some time of threat to the host cell) they may burst from the unfortunate host cell. The fact of importance of genetic engineering is that when they do so, such retroviruses sometimes incorporate fragments of host DNA and are for that reason potential vectors for deliberate gene transfer.

The monkey tumor virus (SV40) has been used for this purpose. In an early experiment researchers managed to incorporate a rabbit β-globin gene into the virus. When the virus was introduced to cultured monkey kidney cells, some of them acquired the ability to produce the distinctive rabbit form of the β-globin protein. The potential danger in-

Focus on . . .

LOCALIZING A GENE BY CELL FUSION

Although computerization of medical records makes the detection of linkage among human genes possible, the fact remains that humans cannot be experimentally bred and tend to have small families. Cell fusion techniques can help in localizing genes. These techniques depend upon the fact that some forms of virus cause cells to fuse with one another as a way of infecting new host-cells. The **Sendai virus** is particularly noted for this property and can be inactivated so as to be harmless to the cells that it causes to fuse. The second fact on which cell fusion depends is that when human and mouse cells are fused, with every cell generation there

is a tendency for the compound cells that result to lose human chromosomes. If a certain biochemical marker trait occurs in a cell line that has just one human chromosome, that trait must be determined by a gene on that chromosome.

Notice that both classical and biochemical genetics were teamed to make this discovery. Methods similar to this were employed to identify the chromosome that contains the gene for Huntington's chorea. This is a serious, dominant hereditary disease that affects the nervous system and usually kills its victims in middle age.

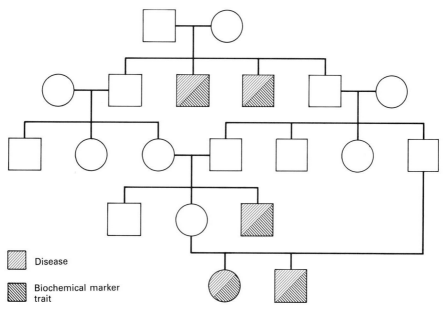

◪ Disease

◨ Biochemical marker trait

A small, inbred community (such as an isolated village or distinctive religious sect) is intensively studied. In one lineage it is observed that a genetic disease only occurs in people who also—incidentally—have another biochemical peculiarity or marker. Though the disease and the marker trait are unrelated in function, the fact that they are inherited together probably means that they occur on the same chromosome.

volved in the use of virus vectors in the treatment of human genetic disease is of course the production of viral disease. However, it may be possible to develop harmless viruses that will penetrate but not damage the cells that are exposed to them. (See Focus on Localizing a Gene by Cell Fusion.)

Giant Mice

Another, somewhat more recent, method of gene insertion is to use DNA directly to carry genetic information from one cell to another. There are several ways to do this, the most recent being microinjection of DNA into the recipient cell. In a typical study of this sort[1] the human gene for

---

[1] "Metallothionein—Human GH Fusion Genes Stimulate Growth of Mice," by Richard D. Palmiter et al., *Science*, 18 November 1983, 809–814.

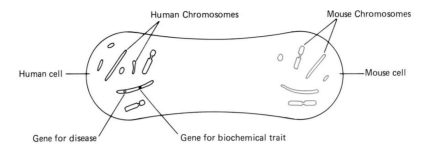

Human Chromosomes

Mouse Chromosomes

Human cell

Mouse cell

Gene for disease

Gene for biochemical trait

Cells from one of the disease victims from this lineage are fused with mouse cells.

Hybrid cell contains both human and mouse chromosomes. However, as the hybrid cell reproduces, different human chromosomes are transmitted to different daughter cells so that sometimes certain human chromosomes fail to be transmitted. Mouse chromosomes, however, are always transmitted to daughter cells. After many cell generations, clones of mouse cells can be isolated, each of which has a single, different human chromosome remaining.

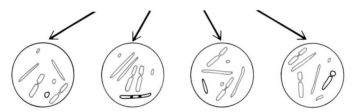

The clone that tests positively for the biochemical peculiarity must be the one which contains the chromosome bearing the gene for that peculiarity. That chromosome must also be the one that bears the gene for the disease.

growth hormone production was isolated from a library of human DNA, with the intention of transferring it to a fertilized mouse egg (Fig. 13–6). To ensure high activity of this gene, its DNA was combined with a portion of the mouse DNA ordinarily responsible for producing the metallothionein protein, a substance whose production is stimulated by the presence of heavy metals, such as zinc. The metallothionein protein itself was not desired, but its regulator gene, now associated with the gene for human growth hormone, would turn on the production of this hormone when exposed to small amounts of heavy metals. The metallothionein regulator could therefore be used as a kind of switch whereby experimenters could turn human growth hormone production on and off at will.

The metallothionein and human growth hormone complex was then injected into mouse eggs. When these became embryos, the experimenters exposed them to small amounts of zinc. In those in which the gene transplant had been successful, growth was enhanced. One mouse, from

Figure 13–6 How to make a giant mouse. (Photograph courtesy of R.L. Brinster.)

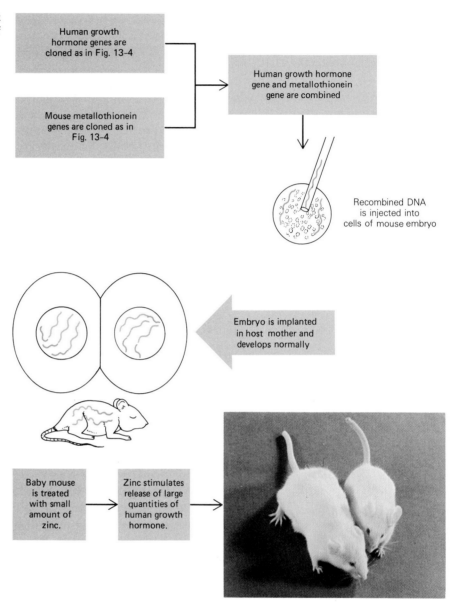

an egg that had received two copies of the human growth hormone gene, grew to more than double the normal size. As might be expected, such mice are also able to transmit their enhanced growth capability genetically to their offspring. What use might be made of giant mice is a puzzle, but one can imagine all sorts of potentially practical applications of similar experiments in such areas as the breeding of domestic meat animals.

Engineering Plants

Plants have been selectively bred for centuries, if not millennia. The success of such efforts depends upon the presence of pre-existing genes, either in the variety of plant being selected or in closely related wild or domesticated plants. The desirable traits must also be transferable by crossbreeding. Even primitive varieties of cultivated plants may have certain traits, such as disease resistance, that could be advantageously introduced into varieties more suited to modern needs. However, the rarer varieties of agricultural plants, especially primitive ones, are swiftly becoming extinct. This greatly reduces the size of the potential gene pool

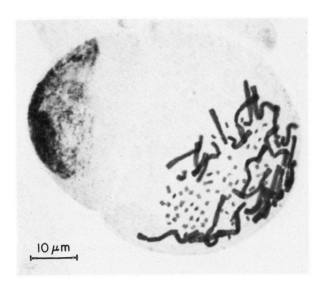

Figure 13–7 The hybrid cell undergoing mitosis was derived by protoplast fusion between soybean (*Glycine max*) and vetch (*Vicia hajastana*). The large chromosomes are derived from the vetch and the small ones from the soybean. (Courtesy of F. Constabel et al.: *C.R. Acad. Sci.* 285: 319–322, 1977.)

from which agricultural researchers may draw. Indeed, just when their genetic resources are most needed, wild plants of *all* kinds are threatened with extinction as the last available agricultural land is brought into cultivation to feed the exploding human population.

If genes could be introduced into plants from strains or species with which they do not ordinarily interbreed, or if totally synthetic genes could be introduced into them, the agricultural researcher's task would be greatly eased. Much research funding has been made available to plant geneticists because of the economic and humanistic potential of increased plant yields. Geneticists working with plants are perhaps also at greater liberty to experiment with new techniques than those working with animals, for manipulation of plant genes does not demand the same type of ethical considerations.

Unfortunately, a suitable vector for the introduction of genes into plant cells has proved very difficult to find. Until recently, most genetic introductions have been performed by removing the plant cell wall to produce naked protoplasts (see Fig. 5–4). If two such protoplasts are chemically induced to fuse, there is a chance that the resulting cell may give rise to mature, differentiated plants with some of the desirable genetic traits of the "parent" plants. Although natural reproductive barriers can sometimes be overcome by such methods (potatoes and tomatoes have been "crossed" in this way), protoplast fusion is essentially artificially facilitated sexual reproduction (Fig. 13–7). Like the more traditional techniques of sexual propagation, it is essentially a hit-or-miss procedure with results that cannot be predicted, let alone guaranteed.

A more recent technique employs the crown gall bacterium, *Agrobacterium tumefaciens,* which produces plant tumors (Fig. 13–8). It does this by introducing a special plasmid into the cells of its host. The plasmid induces the abnormal growth (by forcing the plant cells to produce abnormal quantities of growth hormone). The plasmid also diverts the metabolism of the host cells to produce substances known as opines, simple derivatives of amino and keto acids. These opines are specifically preferred food substances for the bacterium.

The ability of the bacterium to introduce plasmids into a eukaryotic host suggests that desirable genes can be incorporated into that plasmid. It is possible to "disarm" the plasmid so that it does not induce tumor formation. The cells into which such a denatured virus is introduced are essentially normal except for the genes that the experimenter has inserted. It has been shown that genes placed in the plant genome in

Figure 13–8 A crown gall tumor growing on a tree. The growth of this tumor is induced by a plasmid carried by *Agrobacterium tumefaciens.* (E.J. Cable, Tom Stack and Associates.)

this fashion are transmitted sexually via seeds to the next generation, but they can also be propagated asexually if desired.

An additional, very interesting complication of plant genetic engineering is the substantial genome of the chloroplasts. Chloroplasts are pivotal in photosynthesis, and photosynthesis is the basis of plant productivity. Obviously, it would be useful to develop techniques for changing the portion of the chloroplast genetic information that resides within the chloroplast itself. Methods of chloroplast genetic engineering are currently the focus of intense research interest.

RECOMBINANT RNA

It has proved possible to insert genes into RNA somewhat as is done with DNA and to have the resultant recombinant RNA replicate itself directly without a DNA intermediate. Why might this be desirable? DNA is usually thought necessary to produce RNA, but bear in mind that it is RNA, not DNA, that actually directs protein synthesis. RNA does not necessarily *need* DNA; indeed, many RNA viruses are capable of self-replication without a DNA intermediate. Imagine that a strand of desired mRNA can be made by DNA in a minute. In ten minutes, ten strands will be manufactured. On the other hand, imagine a strand of mRNA that can replicate itself every minute. In ten minutes, there will be not ten but 2^{10} strands of mRNA. The more strands of messenger RNA that are available, the more protein can be made (Fig. 13–9). The very simplicity of a self-replicating RNA system is attractive. Without the need for DNA it might prove more practical one day to carry out protein synthesis totally in vitro, under much better control than is possible in a complicated cellular medium. (As yet, unfortunately, due presumably to the presence of unknown promoting factors, the in vitro production of mRNA or protein is *much* slower than what can be accomplished inside living cells.)

Recombinant RNA technology is thus far based upon a small bacterial virus known impersonally as Q-beta. When infecting a host cell, Q-beta produces an enzyme, logically named Q-beta replicase, that immediately replicates the viral RNA. The two resulting particles then replicate themselves, and so on in an intracellular viral population explosion that proceeds exponentially to produce a cellful of virus.

In theory, a gene inserted in a strand of RNA should be rapidly replicated by the Q-beta replicase enzyme to produce a large number of copies, all of which could then be used to produce protein simultaneously. The problem has been, however, that Q-beta replicase is very specific; to avoid replicating the RNA of its bacterial host, the enzyme is activated only by certain recognition codes built into the Q-beta viral RNA. But this fact also suggested a potential solution: link the desired sequence to be replicated with a fragment of viral RNA containing the required recognition code. This was done by preparing a DNA plasmid that incorporated the recognition sequences plus a desired gene. When this was transcribed, the RNA equivalent was automatically created. In the presence of Q-beta replicase the RNA immediately made multiple copies of itself.

SOME WORRIES

While acknowledging the potential uses of recombinant DNA techniques as important and beneficial, many scientists regard the potential misuses as being at least equally potent. There is the possibility that an organism might be produced that would have undesirable ecological or other effects, not by design but by accident. Totally new strains of bacte-

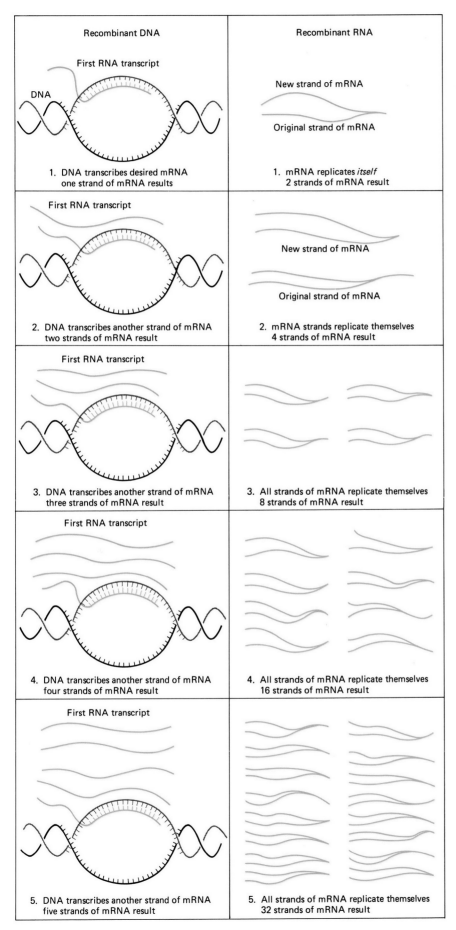

Recombinant DNA

First RNA transcript

DNA

1. DNA transcribes desired mRNA
   one strand of mRNA results

First RNA transcript

2. DNA transcribes another strand of mRNA
   two strands of mRNA result

First RNA transcript

3. DNA transcribes another strand of mRNA
   three strands of mRNA result

First RNA transcript

4. DNA transcribes another strand of mRNA
   four strands of mRNA result

First RNA transcript

5. DNA transcribes another strand of mRNA
   five strands of mRNA result

Recombinant RNA

New strand of mRNA

Original strand of mRNA

1. mRNA replicates *itself*
   2 strands of mRNA result

New strand of mRNA

Original strand of mRNA

2. mRNA strands replicate themselves
   4 strands of mRNA result

3. All strands of mRNA replicate themselves
   8 strands of mRNA result

4. All strands of mRNA replicate themselves
   16 strands of mRNA result

5. All strands of mRNA replicate themselves
   32 strands of mRNA result

Figure 13-9 The advantage in recombinant RNA technology. Recombinant RNA replicates itself and so multiplies at an exponential rate. Unfortunately, in vitro methods of encouraging translation have yet to be perfected.

ria or other organisms, with which the world of life has no previous experience, might be similarly difficult to control.

The real problem, however, might lie in the permanence of the results of genetic manipulation. Most of what we humans do is quite temporary, however severe the consequences of our actions may be in the short term. Real permanence involves the genetic information of organisms, for *that* is self-propagating in an active fashion that our environmental manipulations cannot match. Only the extinction of plants and animals has had permanent consequences for us to date, and of course that also involves genetic information, specifically its loss. Now, however, it is possible to *create* genetic information and incorporate it into self-reproducing life forms that, once created, might well persist for geological epochs.

Recent history has failed to bear out these genetic worries. Experiments over the past decade have demonstrated that, at least until now, the experiments can be carried out in complete safety and that some of the apprehensions voiced when recombinant DNA techniques were first introduced can perhaps be laid to rest for the present. It must be acknowledged, however, that these very apprehensions led to seemingly safe experimental design. Moreover, the dialog that has led us to this point is one that could have been held only in a free and open society. The standards that resulted from it will only be observed by conscientious researchers with high standards of professional ethics. It is by the preservation and propagation of such freedom and such ethics that we have the surest safeguard against the misuse of genetic engineering, or for that matter, any product of modern technology.

CHAPTER SUMMARY

I. Recombinant DNA techniques enable investigators to isolate, identify, and manipulate—cleave, splice, and recombine—genes from the cells of organisms ranging from viruses and bacteria to plants, animals, and humans.
  A. If the gene is inserted into a suitable host DNA, it may be transcribed and translated, leading to the production of a protein not previously produced by the host organism.
  B. Restriction endonucleases cleave double-stranded DNA at specific sites often characterized by a specific, palindromic sequence of deoxynucleotides.
  C. Segments of DNA can be rejoined by the enzyme DNA ligase.
  D. Many bacteria contain small, accessory, circular, double-stranded helical chromosomes called plasmids.
    1. Plasmids may be transferred from one bacterium to another during conjugation.
    2. Plamids undergo replication independently of the host chromosome.
  E. DNA synthesized in the laboratory can be inserted into a plasmid and there undergoes transcription and translation under appropriate conditions.
II. Eukaryotic genes can be introduced into eukaryotic cells as well as into bacterial cells. It may be possible, with these techniques, to replace defective genes or to develop specialized varieties of organisms that can produce enzymes for cleaning oil spills, enhance crop yields, or grow in new areas of the world.
  A. Vectoring of new genes into eukaryotic cells has been accomplished by cell fusion, by using retroviruses or by using special plasmids that are transferable to plant cells.

B. When such new genes are successfully introduced into nuclear DNA, they are passed on to offspring by sexual reproduction in classical Mendelian (Chapter 10) fashion.

III. Cancer appears to be caused often by mutations or transpositions of certain growth-regulating genes called oncogenes.

Post-Test

1. Techniques for manipulating specific genes in the laboratory are termed _____ _____ _____ methods.
2. The enzymes found in bacteria that cleave DNA at specific places are called _____ _____. They function in the intact cell as a defense against _____ invasion.
3. The places (that is, the specific sequences of nucleotides) that are attacked by those enzymes are called _____ _____.
4. Small accessory double-helical chromosomes called _____ occur in many bacteria.
5. These typically contain genes for resistance to _____ _____.
6. Single eukaryotic genes can be isolated by _____ _____ them with labeled _____ that is complementary to the desired gene.

7. The gene can then be incorporated into a plasmid that has first been treated with _____ _____ _____ to break it and to produce "sticky" ends.
8. The plasmid is then rejoined by means of _____ _____ enzymes and incorporated into a bacterium. The bacterium is then _____.
9. The _____ _____ disease has provided biologists with a vector capable of introducing recombinant genes into plants. This disease is produced by a _____ that injects a special _____ into its host cells.
10. The principal advantage of recombinant RNA as opposed to DNA is that the viral RNA system employed is _____ - _____, so that much more of the desired _____ is produced in a short period of time.

Review Questions

1. What is genetic engineering? Give some specific examples.
2. Describe how genetic information from a eukaryote could be implanted in a bacterium. What is to be gained by this?
3. What is restriction endonuclease? Where is it likely to attack a strand of DNA? What practical application of this is employed in recombinant DNA research?
4. What has the main difficulty been thus far in introducing new genes into plant cells? How has this difficulty been overcome?
5. Cigarette smoke has apparently been shown to activate oncogenes. How are oncogenes affected by such mutagenic chemicals, by radiation, or by viral transduction to produce cancer?
6. Why was it necessary to include metallothionein genes in the sequence that implanted functional human growth hormone genes in giant mice?
7. What are the potential advantages of using recombinant RNA to produce protein artificially? Why are these advantages as yet unrealized?
8. What are the principal potential ethical problems of genetic engineering? What safeguards could be employed to guard against potential harm that would not excessively hamper practical and theoretical research?

Readings

Demain, A.L. "Industrial Microbiology," *Science,* 214: (27 November 1981). A look at the applications of new genetics techniques to industrial microbiology.

Gilbert, W., and L. Villa-Komaroff. "Useful Proteins from Recombinant Bacteria," *Scientific American,* April 1980, 74–96. Description of the techniques of cutting and splicing recombinant DNA by which bacteria are programmed to manufacture insulin, interferon and other proteins useful to us.

Novick, R.P. "Plasmids," *Scientific American,* December 1980. Are plasmids subcellular organisms poised on the threshold of life? The author of this interesting article presents evidence to support this hypothesis.

Watson, J.D., and J. Tooze. *The DNA Story.* San Francisco, W.H. Freeman, 1981. A collection of documents, such as newspaper and scientific articles, debating the moral issues of recombinant DNA and the desirability of regulating research in this area.

Watson, J.D., J. Tooze, and D.T. Kurtz. *Recombinant DNA: A Short Course.* San Francisco, W.H. Freeman, 1983. The most succinct and probably the best available summary of current recombinant DNA research.

The October 1985 issue of *Scientific American* is devoted to molecular genetics.

PART IV
THE DIVERSITY
OF LIFE

A cone-headed grasshopper, *Copiphora cornuta*, from South America. (Chip Clark.)

Chapter 14

THE DIVERSITY OF LIFE; MICROBIAL LIFE AND FUNGI

Learning Objectives

After you have studied this chapter you should be able to:

1. Offer two justifications for the use of scientific names and classifications of organisms.
2. Arrange the Linnaean categories in hierarchical fashion, and completely classify an organism such as the human being or the domestic cat.
3. Determine in which kingdom an organism belongs and summarize the basic characteristics of the members of each kingdom.
4. Describe the structure of a virus and trace the steps that take place in the process of viral infection.
5. Describe the structure of bacteria, identify the principal types, and review their ecological role.
6. Compare cyanobacteria with bacteria.
7. Briefly compare the structure and habits of members of the four phyla of protozoa (Sarcodina, Flagellata, Ciliata, Sporozoa) and summarize the life-style of *Amoeba proteus*.
8. Describe the algal protists.
9. Describe the structure and life-style of a fungus, summarize the importance of the fungi, and compare the classes discussed in the chapter.
10. Trace the life cycle of (a) a club fungus, (b) a cellular slime mold, and (c) a plasmodial slime mold.

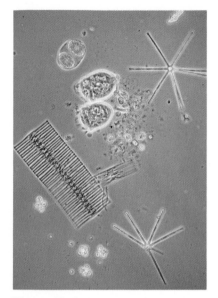

Figure 14–1 A sample of freshwater microorganisms (plankton). Such variety once seemed beyond hope of systematic classification. However, a modern planktonologist could identify and classify each one of the microorganisms shown here. (J. Robert Waaland, University of Washington/BPS.)

Millions of distinguishable kinds of organisms inhabit our planet. To make some order out of this diversity, and to enable us to communicate knowledge about these life forms, a system of classifying them is necessary (Fig. 14–1). Precise descriptions and designations of organisms are essential to biologists in their work. How can ecologists trace the flow of nutrients through a food chain, for example, unless they can identify and record the relative numbers of each species of organism present and their changes over a period of time? The science of naming and classifying organisms is known as **taxonomy.**

CLASSIFYING ORGANISMS: THE BINOMIAL SYSTEM OF NOMENCLATURE

In the 18th century Carolus Linnaeus, a Swedish botanist and natural historian, developed our modern system of taxonomy. The Linnaean system is known more formally as the **binomial system of nomenclature,** because each organism is assigned a two-part name. The first part of the name designates the **genus** (plural, genera) and the second part, the **species.** For example, the common house cat belongs to the genus *Felis* and the species *catus*. Its proper scientific name is *Felis catus*. By convention, the genus name is written first and is always capitalized; the species name is written second and is not capitalized. Both names are underlined or italicized.

Scientific names generally are composed from Greek or Latin roots, or from Latinized versions of the names of persons, places, or characteristics. This practice permits taxonomy to be a truly universal scientific study, for many scientifically important organisms do not have common names, and the names of those that do often vary in different locations and languages. A researcher in Russia can know exactly which organisms were used in a published study by an Australian, and if it is possible to obtain them, can repeat or extend the Australian's experiments.

The narrowest category in the Linnaean system is the species, and the broadest is the kingdom. In between there exists a range of categories that form the hierarchy given in Table 14–1. A species is a group of organisms that can interbreed in their natural environment and are reproductively isolated from other organisms. Closely related species are assigned to the same genus, and closely related genera may be grouped together in a single **family.** For example, the family Felidae includes all catlike animals—genus *Felis* (the house cat genus), genus *Panthera* (the tiger genus), and three or four others (Fig. 14–2). Families are grouped into **orders;** for example, order Carnivora is composed of animals that

Table 14–1
CLASSIFICATION OF THE DOMESTIC CAT, THE HUMAN BEING, AND CORN

Category	Classification of cat	Classification of human being	Classification of corn
Kingdom	Animalia	Animalia	Plantae
Phylum (Division)	Chordata	Chordata	Tracheophyta
Subphylum (Subdivision)	Vertebrata	Vertebrata	Spermatophytina
Class	Mammalia	Mammalia	Angiospermae
Order	Carnivora	Primates	Commelinales
Family	Felidae	Hominidae	Poaceae
Genus	*Felis*	*Homo*	*Zea*
Species	*catus*	*sapiens*	*mays*

Kingdom: Plantae, ANIMALIA, etc.

Phylum: Echinodermata, Arthropoda, Mollusca, Cnidaria, CHORDATA, etc.

Class: Aves, Amphibia, Reptilia, MAMMALIA, etc.

Order: Primates, Perissodactyla, Artiodactyla, Insectivora, CARNIVORA, etc.

Family: Canidae, Ursidae, Mustelidae, Viverridae, FELIDAE, etc.

Genus: *Panthera, FELIS,* etc.

Species: *concolor, CATUS,* etc.

Figure 14–2 The principal categories used in classifying an organism.

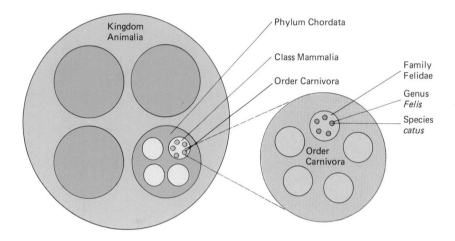

habitually eat meat—including family Felidae, family Ursidae (bears), and several others. Orders are grouped into **classes,** classes into **phyla,** and phyla into **kingdoms.** In classifying plants, the term **division** is generally used rather than phylum.

HOW MANY KINGDOMS?

For hundreds of years biologists regarded living things as falling into two broad categories, plants and animals. With the development of microscopes, it became increasingly obvious that many organisms did not fit very well into either the plant or the animal kingdom. For instance, the bacteria and cyanobacteria do not have a nuclear membrane. This difference from all other organisms is far more fundamental than the differences between plants and animals. In the light of our present knowledge it is difficult to consider bacteria and cyanobacteria as plants, as was done formerly. Certain organisms—for example, the protist *Euglena*—seem to possess characteristics of both plants and animals. In fact, most single-celled organisms seem to have more in common with one another than with either multicellular plants or multicellular animals.

These and other considerations have led to the five-kingdom system of classification used by many biologists today. As outlined in Chapter 1 and in Table 14–2, the five kingdoms currently recognized are kingdoms Monera, Protista, Fungi, Plantae, and Animalia (Fig. 14–3).

DIVERSITY AND TIME

According to geologists, the earth formed about 4.6 billion years ago, and the first cells appeared about a billion years later. The diverse forms of life that grace our planet have evolved from those cells over the past 3.5 billion years. In their attempt to trace the evolution of organisms, geolo-

Table 14–2
FIVE KINGDOMS: MONERA, PROTISTA, FUNGI, PLANTAE, AND ANIMALIA

Kingdom	Characteristics	Ecological role and comments
Monera	Prokaryotes (lack distinct nuclei and other membranous organelles); single-celled; microscopic	
Bacteria	Cell walls composed of peptidoglycan; cells are spherical (cocci), rod-shaped (bacilli), or coiled (spirilla)	Decomposers; some chemosynthetic autotrophs; important in recycling nitrogen and other elements. A few are photosynthetic, usually employing hydrogen sulfide as hydrogen source; some pathogenic; some utilized in industrial processes
Cyanobacteria (blue-green algae)	Specifically adapted for photosynthesis and use water as hydrogen source; chlorophyll and associated enzymes organized into layers in cytoplasm; some can fix nitrogen	Producers; blooms (population explosions) associated with water pollution
Protista	Eukaryotes; mainly unicellular or colonial	
Protozoa	Microscopic; heterotrophic; depend upon diffusion to support their metabolic activities	Important part of zooplankton; near base of many food chains; some are pathogenic
Eukaryotic algae	Photosynthetic; sometimes hard to differentiate from protozoa; some have brown pigment in addition to chlorophyll	Very important producers, especially in marine and freshwater ecosystems; part of phytoplankton
Fungi	Plantlike but cannot carry on photosynthesis; heterotrophic; absorb nutrients	Decomposers; some pathogenic
Slime molds	Animal or protozoan characteristics during part of life cycle; fungal traits during remainder	
True fungi (molds, yeast)	Body composed of threadlike hyphae; rarely discrete cells; hyphae may form tangled masses called mycelia, which infiltrate fungus's food or habitat	Some used as food; yeast used in making bread and alcoholic beverages; some used to make industrial chemicals or antibiotics; responsible for much spoilage and crop loss
Plantae	Multicellular; complex; adapted for photosynthesis; photosynthetic cells have chloroplasts; plants have reproductive tissues or organs; pass through distinct developmental stages and alternations of generations; cell walls of cellulose; cells have indeterminate growth; often no fixed body size or exact shape	Almost entire ecosphere depends upon plants in their role as primary producers; one of most important sources of oxygen in earth's atmosphere
Animalia	Multicellular heterotrophs, many of which exhibit advanced tissue differentiation and complex organ systems; most able to move about by muscular contraction; extremely and quickly responsive to stimuli, with specialized nervous tissue to coordinate responses; determinate growth	Almost sole consuming organisms in biosphere; some specialized as herbivores, carnivores, or detritus feeders

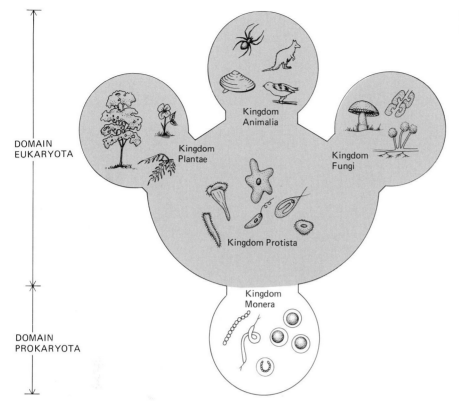

Figure 14–3 The five-kingdom system of classification.

gists have divided time into **eras, periods,** and **epochs** (Table 14–3). These divisions are based on the strata of rock that make up the earth's crust. The strata were formed over millions of years from mud in the bottoms of oceans, lakes, and other bodies of water. Each layer contains fossils that represent the types of organisms that inhabited the earth at the time the stratum was formed. As a rule, the upper strata are the more recently deposited, which affords us a means of estimating at least their relative age. Radioactive dating, a method used to assign an age to each stratum more objectively, will be discussed in Chapter 34.

VIRUSES

Even the five-kingdom theory cannot comfortably accommodate the **viruses** (Fig. 14–4). All other life is at least cellular, but viruses are not. Indeed, it may be seriously questioned whether they are alive at all in the conventional sense of the term. Most viruses consist of a nucleic acid core, RNA or DNA, surrounded by a protein coat. Ordinarily a virus drifts passively until, by chance, it attaches itself to a host cell by means of the protein coat, and its nucleic acid is injected into the host, perhaps by a contraction of the virus coat. Once inside, the viral nucleic acid subverts the metabolic machinery of the host cell and typically replicates itself many times. When the cell finally bursts open, swarms of viruses are released (Fig. 14–5).

Some biologists have suggested that viruses represent a primitive form of life because of their simplicity. However, since all viruses are totally dependent upon living cells in order to reproduce, other biologists think that viruses are descendants of cellular organisms that have become highly specialized as parasites. Still others hypothesize that viruses were originally fragments of DNA or RNA broken off from the nucleic acids of cellular organisms.

Viruses are no longer considered the simplest form of life. **Viroids** are even smaller and simpler than viruses. Each viroid consists of a very

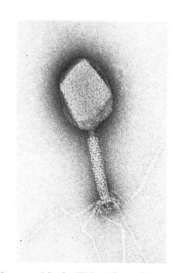

Figure 14–4 This virus (approximately ×275,000), a bacteriophage, is a complex combination of helical and polyhedral shapes. (Courtesy of Dr. Lyle C. Dearden.)

Table 14–3
LIFE AND THE GEOLOGICAL TIME TABLE*

Era	Period	Epoch	Time from beginning of period to present (millions of years)	Duration (millions of years)	Geological conditions	Plants and microorganisms	Animals
Cenozoic (Age of Mammals)	Quaternary	Recent	(Last 10,000 years)		End of last Ice Age; warmer climate	Decline of woody plants; rise of herbaceous plants	Age of *Homo sapiens*
		Pleistocene	1.9	1.9	Four Ice Ages; glaciers in Northern Hemisphere; uplift of Sierras	Extinction of many species	Extinction of many large mammals
	Tertiary	Pliocene	6	4	Uplift and mountain-building; volcanoes; climate much cooler	Grasslands develop; forests decline; flowering plants	Large carnivores; many grazing mammals; first known human-like primates
		Miocene	25	19	Climate drier, cooler; mountain formation		Many forms of mammals evolve
		Oligocene	38	13	Rise of Alps and Himalayas; most land low; volcanic activity in Rockies	Spread of forests; flowering plants, rise of monocotyledons	Apes evolve; all present mammal families are represented
		Eocene	54	16	Climate warmer	Gymnosperms and angiosperms dominant	Beginning of Age of Mammals; modern birds
		Paleocene	65	11	Climate mild to cool; continental seas disappear		Evolution of primate mammals
Mesozoic (Age of Reptiles)	Cretaceous		135	70	Two major land masses begin to separate; Rockies form; other continents low; large inland seas and swamps	Rise of angiosperms; gymnosperms decline	Dinosaurs reach peak, then become extinct; toothed birds become extinct; first modern birds; primitive mammals
	Jurassic		181	46	Climate mild; continents low; inland seas; mountains form	Ferns and gymnosperms common	Large, specialized dinosaurs; first toothed birds; insectivorous marsupials
	Triassic		230	49	Many mountains form; widespread deserts	Gymnosperms and ferns dominate	First dinosaurs; egg-laying mammals

Paleozoic (Age of Old Animals)				
Permian	280	Glaciers; Appalachians form; continents rise	Conifers evolve	Modern insects appear; mammal-like reptiles; extinction of many Paleozoic invertebrates
Pennsylvanian	320	Lands low; great coal swamps	Forests of ferns and gymnosperms	First reptiles; spread of ancient amphibians; many insect forms
Mississippian	345	Climate warm and humid; later cooler	Club mosses and horsetails dominant; gymnosperms	Ancient sharks abundant; many echinoderms
Devonian	405	Glaciers; inland seas	Terrestrial plants established first forests; gymnosperms appear	Age of Fishes; amphibians appear; wingless insects and millipedes appear
Silurian	425	Continents mainly flat; flooding	Vascular plants appear; algae dominant	Fish evolve; marine arachnids dominant; first insects
Ordovician	500	Sea covers continents; climate warm	Marine algae dominant; terrestrial plants first appear?	Invertebrates dominant; first fish appear
Cambrian	600	Climate mild; lands low; oldest rocks with abundant fossils	Algae dominant	Age of marine invertebrates; most modern phyla represented
Precambrian	3800	Planet cooled; glaciers; earth's crust forms; mountains form	Primitive algae and fungi, marine protozoans	Toward end, marine invertebrates

Evidence of first bacterial cells   3.5 billion years ago

Origin of the earth   4.6 billion years ago

Origin of the universe   15–20 billion years ago

*You may want to study this table starting from the bottom and working your way up through time.

Figure 14–5 (*a*) Sequence of viral infection of a cell. (*b*) Bacteriophages infecting a bacterial cell, *Escherichia coli*. The heads, tails, and base plate of most of the viruses are clearly visible. (Courtesy of Dr. Lee D. Simon, Institute for Cancer Research, Philadelphia.)

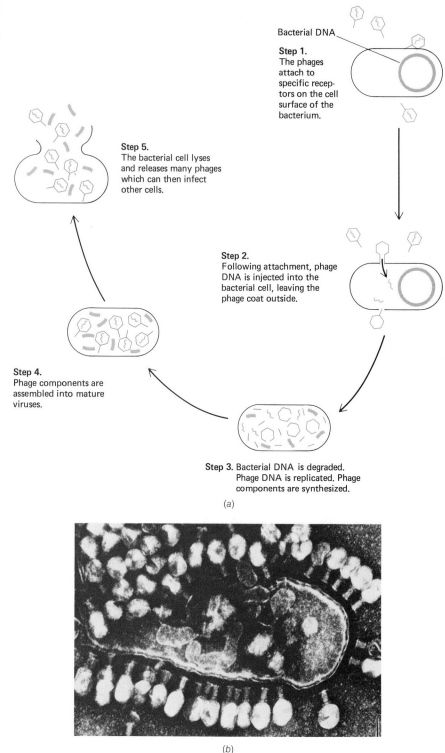

Bacterial DNA

Step 1.
The phages attach to specific receptors on the cell surface of the bacterium.

Step 5.
The bacterial cell lyses and releases many phages which can then infect other cells.

Step 2.
Following attachment, phage DNA is injected into the bacterial cell, leaving the phage coat outside.

Step 4.
Phage components are assembled into mature viruses.

Step 3. Bacterial DNA is degraded. Phage DNA is replicated. Phage components are synthesized.

(*a*)

(*b*)

short strand of RNA without any sort of protective coat (Fig. 14–6). Viroids have been linked to several plant diseases, including potato spindle-tuber disease and a disease that causes stunting of chrysanthemums. Recently, a disease-causing agent even smaller than a viroid has been discovered. Tentatively named a **prion**, this "organism" appears to consist only of a protein. The prion is thought to cause scrapie, a neurological disease of sheep and, perhaps, the human disorder Alzheimer's disease. (Very recent evidence indicates that prions may carry genetic information, but may trigger abnormal cellular responses.)

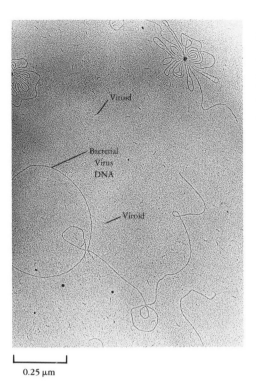

Figure 14–6 A viroid is a rodlike structure consisting of a single-stranded circular molecule of RNA. This electron micrograph compares the size of a viroid with molecules of bacteriophage DNA. (Courtesy of T. Koller and J.M. Sogo, Swiss Federal Institute of Technology, Zurich.)

0.25 μm

KINGDOM MONERA

As you have learned already, the bacteria and cyanobacteria are assigned to kingdom Monera. These prokaryotes lack nuclear membranes, mitochondria, and other membrane-bound organelles. Most are unicellular organisms, but many species form colonies or ribbons of independent cells.

Monerans may have been the only living organisms on our planet for more than 2 billion years. Fossils thought to be about 3.5 billion years old have been found that resemble modern photosynthetic bacteria. In contrast, the oldest eukaryotic fossils are thought to be only about a billion years old.

Bacteria

Bacteria are the oldest, the most abundant, and the most adaptable organisms known. They make their home in fresh and salt water, as deep as 5 meters in soil (and much deeper in oil wells), and even in hot springs and in the ice of glaciers. They are found in the air, in food, and in and on the bodies of plants and animals, both living and dead. Despite their small size (from 1 to 10 micrometers in length and from 0.2 to 1 micrometer in width) the total weight of all the bacteria in the world is greater than that of all other organisms combined!

Their wide distribution and sheer numbers are testimony to the biological success of the bacteria. Among the secrets of this success are their small size, astounding reproductive capability, rapid rate of mutation, and ability to live almost anywhere. When conditions are especially hostile, bacteria of many species can form protected spores and remain in a state of suspended animation for long periods until environmental conditions become favorable again.

A bacterial cell contains about 5000 kinds of chemical compounds. How these are synthesized, what each does, and how they interact are complex biochemical problems that have been the focus of much research. Because there is so much uniformity in basic biochemical proc-

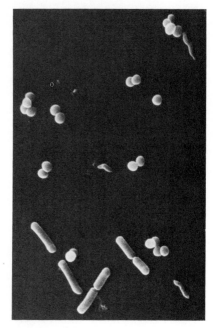

Figure 14–7 Three types of bacteria, illustrating the coccus, bacillus, and spirillum shapes (approximately ×8300). The coccus bacteria are *Staphylococcus aureus*, a common cause of boils and wound infections. (Courtesy of Daniel C. Williams and Lilly Research Laboratory.)

esses, the knowledge that has been gained from studying bacteria has been applied to the more complex cells of humans and other organisms.

IDENTIFYING BACTERIA

Bacteria can be identified on the basis of their shape, staining properties, and type of metabolism. Some bacteria assume varied shapes, but others can be classified as spherical, rod-shaped, or spiral (Fig. 14–7). Spherical bacteria, or **cocci** (singular, coccus), occur singly in some species, in groups of two in others, or in long chains or irregular clumps that look like bunches of grapes. Rod-shaped bacteria, called **bacilli** (singular, bacillus), may occur as single rods or as long chains of rods. Spiral bacteria are known as **spirilla** (singular, spirillum).

On the basis of the Gram staining procedure, bacteria are divided into two major groups. Bacteria that retain crystal violet stain during laboratory staining procedures are referred to as **gram-positive,** and those that do not retain the stain are **gram-negative.** The difference in staining properties is accounted for by differences in the cell wall structure. It is a significant distinction, reflecting differences in the structure and chemistry of the two groups of bacteria.

STRUCTURE OF BACTERIA

Most bacterial species exist as single-celled forms, but some are found as colonies or ribbons of loosely joined cells. The cell wall surrounding the cell membrane provides a strong, rigid framework that supports the cell, protects it, and maintains its shape. The great strength of the bacterial cell wall may be attributed to the properties of **peptidoglycan,** a unique macromolecule found only in prokaryotes (see Fig. 4–5). The antibiotic penicillin interferes with peptidoglycan synthesis, especially in gram-positive bacteria. The result is fragile cell walls that cannot protect the cell (see Fig. 6–10).

Enzymes needed for the operation of the electron transport system (which in eukaryotic cells are found in mitochondria) are located along the cell membrane. The dense cytoplasm of the bacterial cell contains ribosomes and storage granules. Although the membranous organelles of eukaryotic cells are absent, in some bacterial cells the cell membrane appears to be elaborately folded inwardly. Until recently these complex extensions of the cell membrane, known as **mesosomes,** were thought to be the sites of cellular respiration. However, some microbiologists now think that mesosomes do not really exist in the living cell, but are artefacts formed during fixation.

The DNA is found mainly in a single, circular molecule known as a chromosome (Chapter 9). No histones or other proteins are associated with it. Plasmids, smaller DNA molecules that replicate independently of the chromosome, may be present.

Many types of bacteria have flagella, but these are single-stranded, unlike the flagella of eukaryotes, which have two microtubules in the center surrounded by nine pairs of microtubules. Many gram-negative bacteria have hundreds of hairlike appendages called **pili** (Fig. 14–8). These structures help the bacteria adhere to the cells they infect, or to other bacteria during mating.

TYPES OF BACTERIA

Among the many groups of bacteria are the eubacteria, spirochetes, mycoplasmas, and rickettsias. Most **eubacteria,** or true bacteria, are harm-

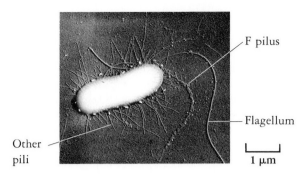

F pilus

Flagellum

Other
pili

1 μm

Figure 14–8 Pili of *Escherichia coli* (approximately ×11,980). Note that there are many kinds. The long *F* pilus is thought to be used in the transfer of DNA. (Courtesy of C. Brinton, Jr.)

less decomposers of great ecological importance. However, some are notorious for the diseases they cause.

Spirochetes are slender, spiral-shaped bacteria with flexible cell walls. The group has been given a bad name by one of its members, *Treponema pallidum,* which causes syphilis.

Mycoplasmas are tiny bacteria bounded by a pliable cell membrane but lacking a typical bacterial cell wall. These bacteria are smaller than some viruses. Mycoplasmas may be the simplest form of life capable of independent growth and metabolism (Fig. 14–9).

Rickettsias are small bacteria that lack the ability to carry on metabolism independently and so are obligated to live as intracellular parasites. Such organisms are referred to as obligate parasites. Most rickettsias parasitize such arthropods as fleas, lice, ticks, and mites. The few species pathogenic to humans (and other animals), such as the one that causes Rocky Mountain spotted fever, are transmitted by the bites of mosquitoes and other arthropods.

BACTERIAL FRIENDS AND FOES

Most bacteria are free-living decomposers that get their nourishment from dead organic matter. Bacteria possess enzymes that enable them to break down all types of organic material. With the help of the fungi they are responsible for decomposing wastes and dead organisms, permitting nutrients to be returned to the soil and recycled.

Some species of bacteria carry on photosynthesis. Bacterial photosynthesis differs in two important ways from photosynthesis carried on by algae, cyanobacteria, or plants. First, their chlorophylls absorb light most strongly in the near infrared portion of the light spectrum rather than in the visible light range. This enables them to carry on photosynthesis in red light that would appear very dim to human eyes. Second, water is not used as a hydrogen donor, so oxygen is not produced. Instead fatty acids, alcohols, or other compounds are used as hydrogen donors. The sulfur bacteria use sulfur compounds, such as hydrogen sulfide (originating from mineral sources or the decomposition of organic matter), as hydrogen donors.

$$6CO_2 + 12H_2S \longrightarrow C_6H_{12}O_6 + 6S_2 + 6H_2O$$

Carbon   Hydrogen   Glucose   Sulfur   Water
dioxide   sulfide

Some bacteria are **chemosynthetic autotrophs.** They produce their own food from simple inorganic ingredients using energy obtained from oxidizing inorganic compounds. Some of these bacteria play important roles in the nitrogen cycle (Chapter 19).

Humans have taken advantage of the abilities of certain bacteria to produce such chemicals as ethanol (drinking alcohol) and acetic acid

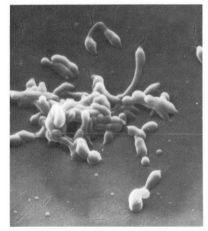

Figure 14–9 Scanning electron micrograph of mycoplasmas. The irregular shape of a mycoplasma is due to its lack of a cell wall. (Courtesy of S. Razin.)

(vinegar). Bacteria are employed in the production of many foods, including buttermilk, yogurt, butter, cheese, pickles, and sauerkraut. They are used in curing cocoa, tea, and tobacco and in producing certain antibiotics, such as streptomycin and bacitracin.

Some species form symbiotic relationships with other organisms. Only a relatively few species of bacteria are **pathogens,** disease-causing organisms, but these can be life-threatening foes. Among the harmful types of bacteria are some species of streptococci that cause "strep throat," scarlet fever, wound infections, and skin and ear infections; species of staphylococci that cause boils and skin infections and infect wounds; species of *Clostridia* that cause tetanus, gas gangrene, and food poisoning; and species of *Hemophilus* that can cause meningitis and infections of the respiratory tract and the ear.

Cyanobacteria

The **cyanobacteria** (formerly known as the blue-green algae) (Fig. 14–10) are found in ponds, lakes, swimming pools, moist soil, and on the bark of trees. Some are found in the oceans, and a few species even inhabit hot springs. All are microscopic, but most form long ribbons of cells or globular colonies. About half of the cyanobacteria are blue-green. Others may be brown, black, purple, yellow, or even red. The Red Sea gets its name from red cyanobacteria, which sometimes occur there in such great numbers that they color the water red.

Most cyanobacteria are adapted to carry on photosynthesis. Like algae and plants, they generate oxygen during photosynthesis, and they

Focus on . . .

THE ARCHAEBACTERIA

All bacteria appear fundamentally similar when studied with the microscope. However, the archaebacteria are biochemically very different from the other bacteria. One of their most striking distinguishing features is the absence of peptidoglycan in their cell wall. This macromolecule is present in the cell wall of all other bacteria. Many archaebacteria inhabit extreme environments, such as salt ponds, acid waters, or hot sulfur springs. Some archaebacteria live in sewage, swamps, and the digestive tracts of humans and other animals—habitats where organic material is decomposed under anaerobic conditions. These anaerobic bacteria produce methane (natural gas) from carbon dioxide and hydrogen (see figure).

Since life on earth possesses a certain degree of biochemical unity, the biochemical and metabolic differences between the archaebacteria and other bacteria suggest that these groups must have diverged from each other long ago, relatively early in the history of life on earth. This hypothesis is supported by the fact that many of the extreme conditions to which the archaebacteria are adapted resemble conditions thought to have existed on the primitive earth. Because the archaebacteria are as biochemically unlike other prokaryotes as the prokaryotes in general are unlike the eukaryotes, some taxonomists have proposed that the archaebacteria be classified in a new kingdom.

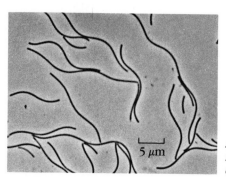

A methane-producing bacterium, Methanospirillum hungatei. (Courtesy of J.G. Ferry.)

5 μm

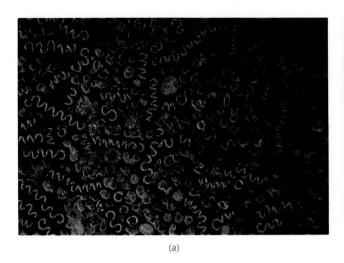

(a)

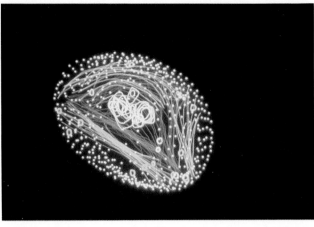

(b)

utilize chlorophyll *a*. Many species can fix nitrogen; that is, they can convert atmospheric nitrogen to nitrites, a form that can be utilized by plants. The cyanobacteria in the rice paddies of Southeast Asia enable farmers to grow rice on the same land for many years without adding nitrogen fertilizer. In some regions cyanobacteria are deliberately added to the soil to increase crop yields.

Because they can tolerate extremes of salinity, temperature, and pH, some cyanobacteria thrive in polluted lakes and bays. They are largely inedible to fish and may reproduce so prolifically that they form "blooms" in the water. The bloom may cause the water to become very turbid, limiting the penetration of sunlight. As a result of the crowding and shading, many of the cyanobacteria as well as other organisms die. Their decomposition by bacteria consumes large amounts of oxygen from the water, which may result in spectacular fish kills. Some cyanobacteria also produce toxic metabolic products that kill fish or any animals that take in the water.

KINGDOM PROTISTA

Kingdom Protista consists of unicellular eukaryotic organisms. Some protists are animal-like and are classified as protozoa. Others carry on photosynthesis and are referred to as algal protists.

Protozoa

Taken from two Greek words, the term **protozoa** means "first animals" and reflects their presumed position as ancestors of all multicellular animals. About 30,000 species of protozoa have been identified. They make their homes in water or in damp soil. All are one-celled heterotrophs, and most phagocytize food particles in the surrounding water by encasing them in food vacuoles. Although microscopic in size and completely unknown to the human species for most of our existence, these organisms affect us all. They form an important component of the **zooplankton,** the small, often microscopic animal-like organisms that drift passively about the ocean and fresh waters in response to wind and water currents. As part of the zooplankton, they serve as a vital link in many food chains. Much of the economically important deposits of chalk and limestone were laid down by shelled protozoans. Some, such as the amebas of amebic dysentery, are parasites of human beings. Protozoa are generally divided into four phyla on the basis of their mode of locomotion.

Figure 14–10 Cyanobacteria. (*a*) Living cyanobacteria (approximately × 60). *Anabaena spiroides*, a spiral-shaped cyanobacterium, and *Microcystis aeruginosa*, an irregularly shaped cyanobacterium. These species are often toxic. (*b*) Computer-generated reconstruction of a unicellular cyanobacterium. The white dots represent the outer surface of the cell, while the white and yellow circles represent the locations of various types of specialized inclusion bodies, which store nutrients. The inclusion bodies are built up with nutrients when the environment is favorable and are broken down when it is unfavorable. The green and blue contour lines represent the intracellular positions of two photosynthetic thylakoid membranes that entirely surround the central portion of the cytoplasm. (*a*, Tom Adams; *b*, D.L. Balkwill, S.A. Nierzwicki-Bauer, and S.E. Stevens, Jr. Reproduced from *The Journal of Cell Biology*, 1983, vol. 97, pp. 713–722 by copyright permission of the Rockefeller University Press.)

Figure 14–11 A giant ameba, *Chaos carolinense*, ingesting *Pandorina*, a multicellular alga. Though unicellularity does impose a limit on the size an organism can attain, a few unicellular organisms, such as *Chaos*, are actually larger than some of the simple multicellular organisms. Note the pseudopods extending to surround the prey. (Michael Abbey, Photo Researchers, Inc.)

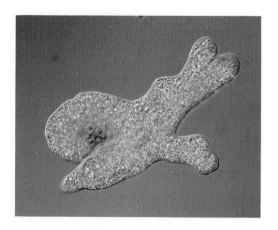

PHYLUM SARCODINA (AMEBAS)

Members of the phylum Sarcodina are amebas that move by means of **pseudopods,** or "false feet." These protists flow along, extending their cytoplasm in the direction in which the cell is moving (Fig. 14–11). The transitory stream of cytoplasm is a pseudopod.

Some sarcodines (such as foraminiferans, Fig. 14–12) secrete or build shells about themselves. Such shells have accumulated on the ocean bottoms, forming vast deposits that have provided excellent fossil records. Some of these deposits ended up as part of the dry land when, through geological changes, the oceans shifted in position. The white cliffs of Dover are a striking example of a deposit of billions of shells from these tiny organisms. Several members of phylum Sarcodina are parasites, such as the ameba responsible for amebic dysentery (not to be confused with the harmless freshwater ameba commonly studied in biology classes).

PHYLUM FLAGELLATA

Members of **phylum Flagellata** propel themselves by means of their whiplike flagella. Most flagellates are free-living organisms that either capture prey or absorb nutrients through their cell membrane. A few flagellates, such as the trypanosomes responsible for sleeping sickness, are parasitic in human beings and other animals (Fig. 14–13).

Euglena is a flagellate that has chloroplasts and carries on photosynthesis. However, it lacks a cell wall and moves freely by means of a long flagellum (Fig. 14–14). Although it is photosynthetic, *Euglena* re-

Figure 14–12 The shell of a foraminiferan *Poneroplia perfusus*. (Eric V. Gravé, Photo Researchers, Inc.)

Figure 14–13 Scanning electron micrograph of *Trypanosoma cruzi* being engulfed by a large ruffled macrophage, which is part of the host organism's defense system. (From S.G. Reed, T.G. Douglass, and C.A. Speer: "Surface interactions between macrophages and *Trypanosoma cruzi*," *American Journal of Tropical Medicine and Hygiene*, 31: 723–729, 1982.)

quires certain nutrients, such as amino acids. If deprived of light, the chloroplasts degenerate and *Euglena* must then obtain its food from the environment.

PHYLUM CILIATA

Cilia are the organelles of locomotion used by members of **phylum Ciliata.** Some species of ciliates claim the most complex cells known. The *Paramecium* is a familiar member of phylum Ciliata (Fig. 14–15). Within its single cell are highly specialized organelles for carrying out metabolic functions. Like some other ciliates, *Paramecium* has small organelles called trichocysts that discharge a long, threadlike structure toward its prey to aid in capture. Trichocysts are also used defensively against threatening enemies.

PHYLUM SPOROZOA

Phylum Sporozoa consists of parasitic protozoa that have no special means of locomotion. They move by gliding along or by changing body shape. These protists get their name from their sporelike infective stages. The most important members of class Sporozoa are the plasmodia, which cause malaria. *Plasmodium* spends part of its highly complex life cycle within the salivary glands of the female *Anopheles* mosquito. When an infected mosquito bites a human being, it injects some of the parasites into the blood, thereby causing malaria (Fig. 14–16).

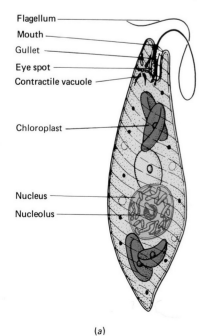

Flagellum
Mouth
Gullet
Eye spot
Contractile vacuole

Chloroplast

Nucleus
Nucleolus

(a)

(b)

Figure 14–14 *Euglena.* (*a*) This protist has both plantlike and animal-like traits. It has at various times been classified in the plant kingdom (with the algae) and in the animal kingdom (when protozoans were considered animals). (*b*) Living euglenids; note eyespots. (*b*, Tom Adams.)

Figure 14–15 Paramecia, members of the ciliated protozoa. (*a*) A diagram of *Paramecium caudatum*. (*b*) A scanning electron micrograph of *Paramecium multimicronucleatum*. (*c*) *Paramecium* discharging its trichocysts, specialized organelles that produce a substance that hardens into entangling threads when the organism is disturbed. (*b*, courtesy of Dr. Eugene Small. *c*, courtesy of Dr. H. Plattner, *Experimental Cell Research*, 1984.)

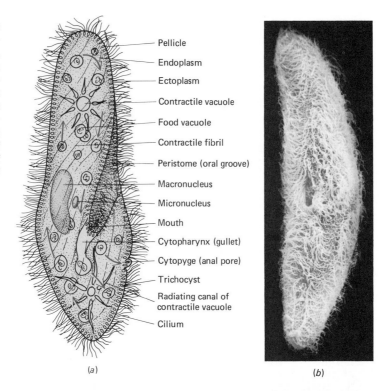

Pellicle
Endoplasm
Ectoplasm
Contractile vacuole
Food vacuole
Contractile fibril
Peristome (oral groove)
Macronucleus
Micronucleus
Mouth
Cytopharynx (gullet)
Cytopyge (anal pore)
Trichocyst
Radiating canal of contractile vacuole
Cilium

(*a*)

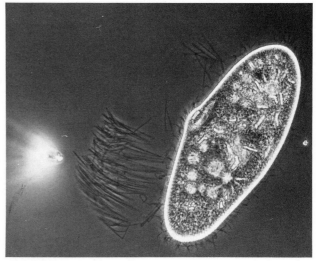

(*b*)

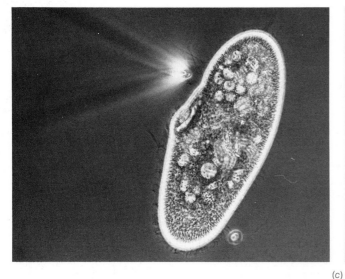

(*c*)

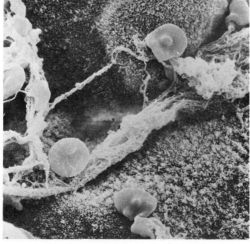

Figure 14–16 Scanning electron micrograph of red blood cells infected by malarial parasites. The blebs are caused by the presence of the sporozoan. (Courtesy of P.R. Walter, Y. Garin, and P. Blot. *The American Journal of Pathology*, 109: 330, 1982.)

LIFE-STYLE OF A PROTOZOAN—*AMOEBA PROTEUS*

How does a one-celled animal like the ameba carry on all the metabolic processes essential to life? Within the confines of its single cell the ameba possesses all the intricate equipment needed for survival and propagation. In a way, its small size is an advantage because gases and other materials can be exchanged readily with its watery environment.

To obtain food, the ameba engulfs tiny organisms in the water. As food is ingested, it is encased in a bit of cell membrane, which pinches off to form a food vacuole. Lysosomes attach to the food vacuole and pour digestive enzymes onto the food. The digested food then is absorbed through the vacuole membrane into the surrounding cytoplasm. The indigestible portion that remains in the food vacuole is simply pushed out of the cell and left behind as the ameba moves on.

Food vacuoles move about in the cytoplasm so that nutrients are circulated to various parts of the cell. Metabolic wastes are excreted by diffusion or discharged through special contractile vacuoles, which pump excess water out of the cell. Freshwater amebas are hypertonic to the surrounding medium and have a constant problem ridding themselves of excess water, which continuously enters by osmosis. Marine amebas, which normally have no contractile vacuoles, will form them when placed in fresh water.

Responsiveness in the ameba depends upon the innate irritability of its cytoplasm. The ameba moves toward or away from a variety of specific stimuli. When hungry, it responds positively to food by engulfing it.

Under favorable conditions an ameba grows to adult size in about three days. Reproduction then takes place by the asexual process of binary fission, which is essentially typical cell division (including mitosis). The two new amebas immediately become independent.

Algal Protists

Algal protists are unicellular algae. Multicellular algae are classified as plants. The algal protists carry on photosynthesis and are part of the **phytoplankton,** the free-floating, microscopic, photosynthetic organisms of aquatic environments.

The **dinoflagellates** are mobile, photosynthetic organisms. Certain members of this group—for example, *Gymnodinium brevis*—are responsible for red tide. Under certain conditions these dinoflagellates multiply explosively, forming great blooms that turn the water red. Their enormous numbers deplete the water of oxygen, especially at night, when they do not carry on photosynthesis. This oxygen depletion may cause the death of fish and other organisms. Some species contain a powerful nerve poison that kills fish, and during episodes of red tide huge fish kills may occur, resulting in thousands of dead fish washing up on the beaches.

Diatoms are the most numerous photosynthetic marine organisms. They are also found in fresh water. These tiny organisms produce cell walls that become impregnated with silicon. The cell wall usually consists of two pieces that form the top and bottom of a glass box. The cell walls of marine diatoms have intricate patterns produced in part by the arrangement of little pores through which water and gases pass (Fig. 14–17). Many diatoms contain carotenoids, pigments that impart a yellow or golden brown color. The cell walls of dead diatoms form a major part of the sediment of the ocean floor. Diatomaceous earth, obtained from sedimentary rocks composed largely of these cell walls, is used industrially in insulating material, swimming pool filters, and even in toothpaste.

(a)

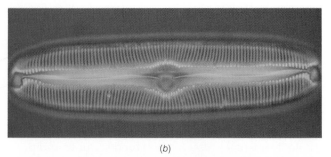

(b)

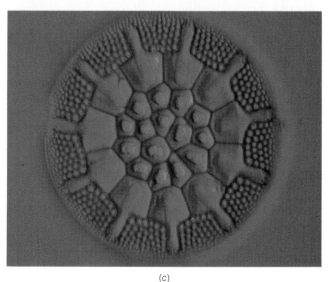

(c)

Figure 14–17 The structure of diatoms viewed through different methods of microscopy. (a) Scanning electron micrograph of a diatom. The frustule (shell) is built like a pillbox, with the lid overlapping the base. When the diatom divides, the new lid of one cell forms inside its base; and the new base of the other cell forms inside its lid. For a time, the two new cells remain interlocked. (b) A diatom viewed through interference optics microscopy. (c) Light micrograph of *Astrolampra affinis*. (a, courtesy of Dr. F.E. Round. b, Jonathan G. Izant. c, M. Murayama, Murayama Research Laboratory/BPS.)

KINGDOM FUNGI

What does the delicious mushroom have in common with the black mold that forms on stale bread and the mildew that collects on damp shower curtains? All of these life forms belong to the **kingdom Fungi,** a diverse group of more than 200,000 species. Although they vary greatly in size and shape, all of the fungi are eukaryotes. Fungi were formerly classified as members of the plant kingdom partly because their cells are encased in cell walls. However, they differ from plants (and algae) in that (1) the cell wall is composed of chitin rather than cellulose, and (2) fungi lack chloroplasts and cannot carry on photosynthesis.

Structure of a Fungus

The body structure of a fungus varies from the unicellular **yeasts** (Fig. 14–18) to the multicellular **molds,** a term used loosely to include the mildews, rusts and smuts, mushrooms and puffballs, and slime molds. Yeasts reproduce both asexually, mainly by budding but also by fission, and sexually, by forming spores that develop into new individuals. Each bud that separates from the mother yeast cell can grow into a new yeast. Some species group together to form colonies.

A mold consists of long, branched, threadlike strings of cells called **hyphae** (singular, hypha) that form a tangled mass or tissue-like aggregation known as a **mycelium** (Fig. 14–19). The cobweb-like mold sometimes seen on bread consists of the mycelia of mold colonies. What is not seen is the extensive mycelia that grow down into the substance of the bread. The color of the mold results from the reproductive spores that are produced in large numbers on the mycelia. Some hyphae are divided by **septa** (walls) into individual cells containing one or more nuclei. Others

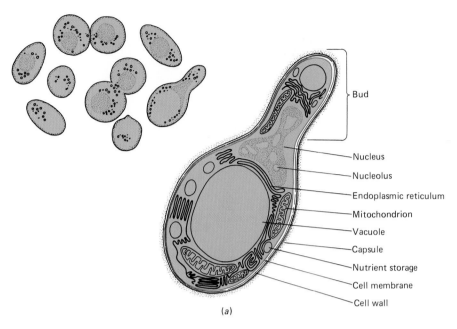

(a)

(b)

Figure 14–18 Yeasts are unicellular fungi that can reproduce asexually, mainly by budding. (a) Budding cells of the common bread yeast. (b) Scanning electron micrograph of yeast cells. (b, courtesy of D. Von Wettstein, *Experientia* 39(7) 1983.)

are **coenocytic,** undivided by septa. These are something like an elongated, multinucleated giant cell. Many fungi, particularly those that cause disease in humans, are dimorphic (have two forms). They can change from the yeast form to the mold form in response to changes in temperature, nutrients, or other environmental factors.

Life-Style of a Fungus

Fungi are heterotrophs that absorb their food through the cell wall and cell membrane. Some are saprobes, absorbing food from organic wastes or dead organisms; others are parasites, absorbing food from the living bodies of their hosts (Fig. 14–20). Most fungi stay in one place, although their reproductive cells may be motile.

Fungi grow best in dark, moist habitats, but they are found wherever organic material is available. Moisture is necessary for their growth, and they can obtain water from the atmosphere as well as from the me-

Figure 14–19 Germination and growth of a typical mold.

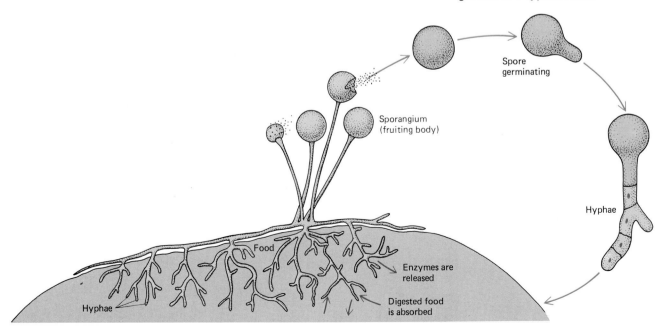

Figure 14–20 A parasitic fungus. After the ant ingests a spore of the fungus, the fungus begins to grow inside its body, absorbing nutrients from it. Somehow, the fungus affects the nervous system of the ant: The ant climbs high in the tree, where it dies. New fungal spores are then widely dispersed in the environment after the development of specialized structures called fruiting bodies. (L.E. Gilbert, University of Texas, Austin/BPS.)

dium on which they live. When the environment becomes very dry, fungi survive by going into a resting stage or by producing spores that are resistant to desiccation. Although the optimum pH for most species is about 5.6, fungi can tolerate and grow in a pH range of about 2 to 9. Fungi are not as sensitive to high osmotic pressures as bacteria. They can grow in concentrated salt solutions or sugar solutions, such as fruit jelly, that discourage or prevent bacterial growth. Fungi also thrive over a wide temperature range. Even refrigerated food is not immune to fungal invasion.

When a fungal spore comes into contact with an appropriate substrate, perhaps an overripe peach that has fallen to the ground, it germinates and begins to grow. A threadlike hypha emerges from the tiny spore. Soon a tangled mat of hyphae infiltrate the peach, while other hyphae extend upward into the air. Cells of the hyphae secrete potent digestive enzymes into the peach, degrading its organic compounds to small molecules that the fungus can absorb. Fungi are very efficient at converting nutrients into new cell material. Nutrients not immediately needed can be stored in the mycelium. Some excess nutrients may be excreted into the surrounding medium.

Reproduction

Fungal reproduction occurs in a variety of ways, asexually by fission, budding, or spore formation, and sexually by means that are characteristic for each group. Spores are usually produced on hyphae that project into the air (aerial hyphae) above the food source. This arrangement permits the spores to be blown by the wind and distributed to new areas. The spores of terrestrial fungi are generally nonmotile cells dispersed by wind or by animals. Spores of aquatic fungi typically have flagella.

Some spores form directly from hyphal cells, whereas others are produced within specialized branches of hyphae. In some fungi, such as mushrooms, the aerial hyphae form large, complex fruiting bodies, specialized reproductive structures where spores are produced. The familiar part of a mushroom or toadstool is actually a large fruiting body. The bulk of the organism is a nearly invisible network of hyphae buried out of sight in the rotting material it invades.

The nuclei of most fungal cells are haploid; only the zygote is diploid. A sexual spore is produced by sexual reproduction in which two compatible nuclei come together and fuse, creating a diploid zygote. Then meiosis occurs, restoring the haploid state to the spore nucleus. In some species, sexual reproduction takes place between mycelia of different mating types referred to as + and −. These may be compared to the two sexes of other organisms.

Importance of Fungi

Like the bacteria, the fungi serve as decomposers in the ecosphere. Without them life on earth would eventually become impossible, for nutrients would remain locked up in wastes and the bodies of dead organisms and would not be available for living organisms.

Some fungi, known as **mycorrhizae,** enter into symbiotic relationships with the roots of complex plants. Such relationships occur in more than 90% of all families of complex plants. The fungus benefits the plant by decomposing organic material in the soil, thus making the minerals available to the plant. The roots supply sugars, amino acids, and some other organic substances needed by the fungus. Many forest trees die from malnutrition when transplanted to grassland soils even though essential nutrients are abundant. When forest soil containing fungi is added to the soil around these trees, they quickly assume a normal growth pattern.

Some fungi form symbiotic relationships with algae or cyanobacteria, forming compound organisms known as **lichens** (Fig. 14–21). The photosynthetic organism produces food for both members of the lichen. Very often lichens are the first organisms to inhabit bare rocky areas, and they play an important part in the formation of soil. Lichens gradually etch the rocks to which they cling, facilitating disintegration by wind and rain.

The virtues of the fungi have not escaped the ants. Several species of ants grow domesticated strains of fungi on mixtures of vegetable matter and their own droppings. There is even one species of ant whose young queen is provided with a special pocket under her head in which she carries a culture of the fungus to be eaten by her colony throughout its future.

There are many industrial uses for domesticated strains of fungi. Yeast is indispensable in the processes of alcoholic fermentation, cheese ripening, and bread making. Other fungi are used to produce citric acid of high quality and other industrial chemicals. The antibiotic penicillin is a product of the blue mold, *Penicillium.* Study of the *Neurospora* fungus has produced almost as much genetic knowledge as have studies of fruit flies. And many a gourmet could wax eloquent on the mushroom, the morel, and the truffle.

On the negative side, fungi are responsible for great economic losses each year. Equipped with some of the most powerful digestive

Figure 14–21 A multicolored lichen from Bylot Island, in northeastern Canada. (E.R. Degginger.)

(a)

(b)

Figure 14–22 (a) The ascomycete *Claviceps purpurea* infects the flowers of cereals. The fungus produces a structure called an ergot where a seed would normally form in the grain head. When livestock eat this grain or when humans eat bread made from infected rye, they may be poisoned by the toxic substances in the ergot. Lysergic acid, one of the compounds in ergot, is an intermediate in the synthesis of lysergic acid diethylamide (LSD). (b) Apple cedar rust on apple leaves, a serious plant disease. (Carolina Biological Supply Company.)

enzymes known, fungi are capable of reducing wood, fiber, and food to their constituents with staggering efficiency. Various molds and rots produce untold damage to stored goods and building materials every year. A major effort of the chemical industry supplies fungicides to combat them (only more or less effectively).

Fungus diseases of crop plants are mentioned in many ancient writings, and it is safe to assume that they have been with us since the beginnings of agriculture (Fig. 14–22). Great crop losses have been sustained from fungus infections, some of which have changed the course of history. Potato blight, for example, was responsible for the great Irish famine of the 1840s. About a million people starved to death and 2 million emigrated from Ireland as a consequence.

Most fungi that invade humans and other animals are opportunists that cause infections only when the body's immunity is lowered. Ringworm, athlete's foot, and thrush are examples of human fungus infections. Fungi do not cause as much animal disease as do bacteria, but fungal diseases can be hard to eradicate.

Types of True Fungi

The true fungi are divided into classes based on their characteristic reproductive structures. The **water molds** (class Oomycetes) (Fig. 14–23) are distinguished by their flagellated asexual spores, which can swim to new locations. Although some members of this group feed on plant and animal debris, a few are parasites of plants and animals. Members of this class include the fungi that cause potato blight and downy mildew on grapes. And you may have seen the cottony mycelia of a water mold covering the bodies of aquarium fish.

The **zygomycetes** (class Zygomycetes) are terrestrial fungi that produce sexual resting spores called **zygospores.** Their hyphae are coenocytic, and their cell walls consist mainly of chitin. Most are saprobes that live in the soil on decaying plant or animal matter. Bread molds are members of this group (Fig. 14–24).

The **sac fungi** (class Ascomycetes) produce their spores in little sacs called **asci** (singular, ascus). Among the diverse members of this class are yeasts, powdery mildews, most of the blue-green, brown, and red molds that cause food to spoil, and the edible morels and truffles (Fig. 14–25). Some members of this group cause serious plant diseases, including Dutch elm disease, chestnut blight, powdery mildews that ruin fruits and ornamental plants, and ergot disease of rye plants.

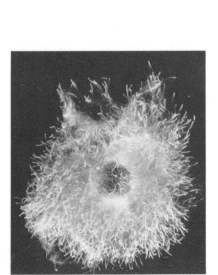

Figure 14–23 *Saprolegnia*, a member of class Oomycetes, growing on hemp seed. (J. Robert Waaland, University of Washington/BPS.)

The **club fungi** (class Basidiomycetes) include the most familiar of the fungi—mushrooms (Fig. 14–26), toadstools, bracket fungi, and puffballs, and also some important plant parasites, the rusts and smuts. All of these fungi develop **basidia,** enlarged, club-shaped hyphal cells. At the tip of each basidium four **basidiospores** develop. Each fungus produces millions of basidiospores, and each basidiospore has the potential, should it happen upon an appropriate environment, to give rise to a new mycelium (Fig. 14–27). Hyphae of this mycelium consist of cells that contain one nucleus; that is, they are **monokaryotic.** If in the course of its growth a hypha encounters another hypha of a different mating type, the two hyphae join, forming a mycelium in which each cell contains two nuclei, one of each mating type. These cells are described as **dikaryotic.**

The hyphae of the mycelium grow extensively and eventually form **fruiting bodies,** more formally known as **basidiocarps.** The fruiting body is the structure that we know as a mushroom (or toadstool) (Fig. 14–28). Each fruiting body consists of intertwined hyphae matted together. The lower surface of the fruiting body consists of many thin, perpendicular plates, called gills, that extend radially from the stalk to the edge of the cap. The basidia develop on the surface of these gills. Within each basidium is a cell containing two nuclei that originated from the two different hyphae. These nuclei now fuse, making the cells diploid. This process may be thought of as fertilization. Each cell then undergoes meiosis, forming four haploid nuclei that develop into basidiospores.

Figure 14–24 The bread mold *Rhizopus nigricans.* (Carolina Biological Supply Company.)

Slime Molds

Slime molds are interesting specimens frequently used by researchers studying developmental processes. They are one of the simpler organisms that exhibit cellular differentiation and morphogenesis (development of body form). Unlocking their secrets may provide important clues to the mechanisms of development in complex animals. Some biologists classify slime molds as protists because for part of their life cycle they lack cell walls and resemble amebas, ingesting food rather than absorbing it.

Cellular slime molds begin their life history as independent, multiplying "amebas." During this stage the cells feed upon soil bacteria, grow, and multiply by binary fission. After the food supply in their immediate vicinity is depleted, they swarm together, forming tiny heaps of cells (Fig. 14–29). The signal for this aggregation process is the release of cyclic AMP, a compound that also serves as an intermediate in the action of several vertebrate hormones. The cells of each heap arrange

Figure 14–25 The edible morel, *Morchella deliciosa,* a common ascomycete. (E.R. Degginger.)

Figure 14–26 *Gyrodon meruloides,* a member of class Basidiomycetes. (E.R. Degginger.)

Figure 14–27 Life cycle of a basidiomycete. A mushroom develops from the mycelium, a mass of white, branching threads found underground. A compact "button" appears and grows into a fruiting body, or mushroom. On the undersurface of the fruiting body are gills, thin perpendicular plates that extend radially from the stem. Basidia develop on the surface of these gills and produce basidiospores, which are shed. If these spores reach a suitable environment, they give rise to new mycelia.

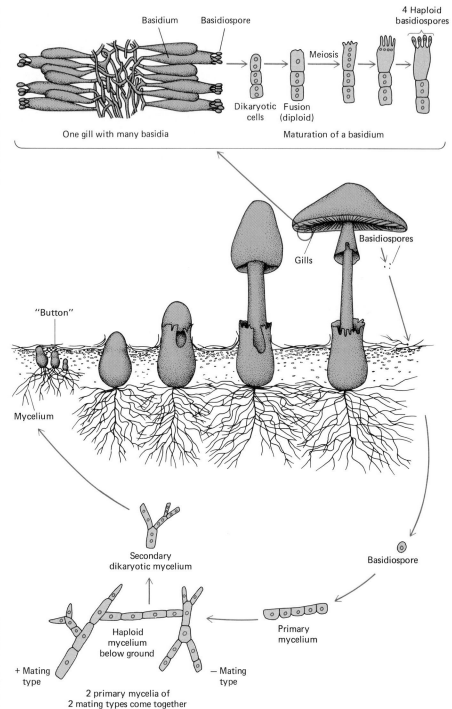

Figure 14–28 *Amanita phalloides*, the death angel. About 2 ounces (50 grams) of this mushroom could kill a 150-pound (68 kilograms) man. (Courtesy of Leo Frandzel.)

themselves to form a slug-shaped organism composed of a few hundred to more than 100,000 cells. This slug behaves like a multicellular animal, creeping about and responding in a coordinated manner to such stimuli as light and heat. Within the slug cellular differentiation takes place, and two distinct cell types can be identified.

After a period of migration, which may involve search for a favorable habitat in which to continue its life cycle, the slug rounds up and forms a fruiting body. Cells that had formed the anterior one-third of the migrating slug now become stalk cells, while those from the posterior two-thirds of the slug differentiate into spore cells. In this stage the slime mold consists of a thin stalk supporting a mass of hundreds of spores. These fruiting bodies look very much like tiny lollipops standing on end.

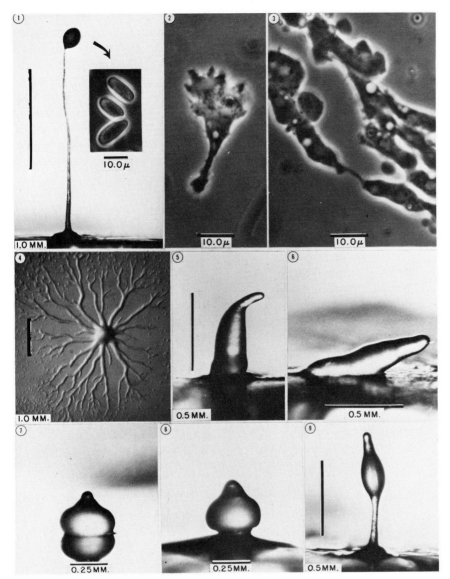

Figure 14–29 The life cycle of the cellular slime mold *Dictyostelium discoideum*. (*1*) Mature fruiting body releases spores. (*2*) Each spore cell opens liberating an amebalike individual which eats, grows, and reproduces by binary fission. (*3*) After the food supply (usually soil bacteria) in their immediate vicinity has been depleted, amebas stream together. (*4*) Amebas aggregating resulting in the formation of a heap of cells (*5*). Cells of each heap organize themselves, forming a slug-shaped, multicellular organism (*6*). After a period of migration, which may involve search for a favorable habitat in which to continue its life cycle, the slug begins to form a fruiting body (*7*). Cells that form the anterior third of the slug differentiate to become stalk cells, (*8*), while those from the posterior two-thirds form spore cells. (*9*) Supported by a thin stalk, a mass of spores is raised above the surface. (Courtesy of Dr. J. Gregg, *The Fungi.* New York, Academic Press.)

Stalk cells die with each generation, but the spores are dispersed. When conditions are favorable, each spore cell liberates a new "ameba," and the life cycle begins anew.

The **plasmodial slime molds** spend part of their life cycle as thin, streaming masses of protoplasm that creep along decaying leaves or wood (Fig. 14–30). They feed upon bacteria and small organic particles.

Figure 14–30 A plasmodial slime mold, *Tubifera ferruginosa*. (E.R. Degginger.)

At this stage the organism, referred to as a plasmodium, may cover an area a meter or more in diameter. Its shape varies as it grows and moves.

When the food supply dwindles or conditions become dry, little mounds form on the plasmodium, each of which sends up a stalk with a spore sac at its tip. The spores are extremely resistant to unfavorable environmental conditions, and years may pass before they germinate. Each spore produces one to four flagellated cells, which eventually become amebalike and function as undifferentiated gametes. After fertilization the diploid zygote divides many times by mitosis to form a new plasmodium. Most plasmodial slime molds are decomposers, but a few are plant parasites.

CHAPTER SUMMARY

I. In the binomial system of nomenclature every organism is given a two-part name designating its genus and species. The complete hierarchy includes kingdom, phylum (division in plants), class, order, family, genus, species.

II. According to the five-kingdom system of classification, all organisms except the viruses can be assigned to one of the following kingdoms: Monera, Protista, Fungi, Plantae, or Animalia.

III. Viruses do not fall into any of the five kingdoms; a typical virus consists of a nucleic acid core and a protein coat.

IV. Kingdom Monera consists of the bacteria and cyanobacteria; these are prokaryotes that differ from all other organisms in that they possess no nuclear membrane, mitochondria, or other membrane-bound organelles.

   A. Bacteria are classified according to shape: cocci, bacilli, and spirilla. Bacteria may also be identified as gram-positive or gram-negative.

   B. Among the many groups of bacteria are the eubacteria, spirochetes, mycoplasmas, rickettsias, and archaebacteria.

   C. Most bacteria are free-living decomposers; some are photosynthetic; others are chemosynthetic; a few are pathogens.

   D. Cyanobacteria are photosynthetic; some can fix nitrogen.

V. Kingdom Protista consists of unicellular eukaryotic organisms: protozoa and algal protists.

   A. Protozoa are single-celled heterotrophs that ingest or absorb food. They are divided into four phyla on the basis of their mode of locomotion.

      1. Phylum Sarcodina consists of the amebas; these protozoans move by forming temporary pseudopods.

      2. Phylum Flagellata is composed of protozoans that move by means of their whiplike flagella.

      3. Members of phylum Ciliata move by means of cilia; the paramecium is a member of this group.

      4. Phylum Sporozoa consists of parasitic protozoa that move by gliding or by changing body shape.

   B. Algal protists are unicellular algae. This group includes euglenoids, dinoflagellates, and diatoms.

VI. The fungi are eukaryotic decomposers or parasites with chitinous cell walls. Some are unicellular yeasts; others are multicellular molds.

   A. A mold consists of hyphae that form mycelia.

   B. Fungi are ecologically important as decomposers. Some fungi are mycorrhizae that live in symbiotic relationships with the roots of complex plants. Certain fungi are responsible for important plant diseases, such as potato blight, downy mildew of grapes, and Dutch elm disease.

C. Some of the major groups of true fungi are the water molds, the zygomycetes, the sac fungi, and the club fungi.

D. The cellular and the plasmodial slime molds are fungi that spend part of their life cycle as mobile ameba-like organisms.

Post-Test

1. Closely related genera may be grouped together in a single _____ .
2. Orders are grouped into _____ ; phyla are grouped into _____ .
3. The mold that produces penicillin is *Penicillium notatum. Penicillium* is this organism's _____ name; *notatum* is its _____ name.
4. A _____ is an organism even smaller than a virus; it consists only of a very short strand of RNA.

Select the kingdom in column B to which each organism in column A belongs:

Column A	*Column B*
_____ 5. eubacteria	a. Monera
_____ 6. mushroom	b. Protista
_____ 7. oak tree	c. Fungi
_____ 8. virus	d. Plantae
_____ 9. sporozoan	e. Animalia
_____ 10. euglena	f. none of the above
_____ 11. *Escherichia coli*	

Using the same choices above in column B, match the following characteristics:

_____ 12. one-celled eukaryotic organisms
_____ 13. cells without membrane-bound organelles

_____ 14. decomposers with chitinous cell walls
15. A rod-shaped bacterium is referred to as a _____ _____ , a spherical bacterium as a _____ .
16. Peptidoglycan is found in the _____ _____ of _____ .
17. *Treponema pallidum* is the spirochete that causes _____ .
18. Members of phylum Sarcodina move by means of _____ .
19. The most numerous photosynthetic marine organisms are the _____ .
20. A lichen consists of an alga (or cyanobacterium) in symbiotic association with a _____ .
21. The sac fungi produce _____ within asci; the club fungi produce the same structures within _____ .
22. Bread molds are fungi of the class _____ .
23. Hyphae that lack septa are referred to as _____ _____ .
24. Red tide is caused by a type of protist called a _____ .
25. Malaria is caused by a member of phylum _____ _____ .

Review Questions

1. Since it involves collecting and classifying organisms, taxonomy has been compared to stamp collecting—a scholarly hobby of little practical value. What do you think?
2. Write from memory the complete hierarchy of Linnaean classification, from kingdom to species. Classify a human being.
3. Local populations of a species that differ consistently from other populations of the same species in some way are often called subspecies. Which is more objective—the concept of species or of subspecies?
4. Draw a series of diagrams to illustrate a virus infecting a cell. Which, if any, properties of life does a virus possess?
5. Contrast the cyanobacteria with the bacteria. Why are these organisms placed in the same kingdom?
6. Why do some biologists think that the archaebacteria should be classified in a new kingdom?
7. In what important way do the rickettsias differ from most eubacteria?
8. How does bacterial photosynthesis differ from algal or plant photosynthesis?
9. Compare the four phyla of protozoa.
10. How do fungi differ from algae? How are they alike?
11. Draw and label a diagram illustrating the structure of a mold.
12. Describe the life cycle of a typical club fungus, such as a mushroom.
13. Some biologists classify slime molds as protists rather than as fungi. How do they justify this?

Readings

Ahmadjian, V. "The Nature of Lichens," *Natural History,* March 1982. A beautifully illustrated summary of the algae-fungal partnership.

Anagnostakis, S. "Biological Control of Chestnut Blight." *Science,* 215: 29 January (1982). An interesting account of the history of chestnut blight fungus in the United

States and of virus-like agents that may be useful in controlling it.

Blakemore, R.P., and R.B. Frankel. "Magnetic Navigation in Bacteria," *Scientific American,* December 1981. Certain aquatic bacteria have internal compasses that orient them in the earth's magnetic field.

Butler, P.J.G., and A. Klug. "The Assembly of a Virus," *Scientific American,* November 1978. An account of the assembly of the tobacco mosaic virus.

Corliss, J.O. "Consequences of Creating New Kingdoms of Organisms." *Bioscience,* May 1983. The objections of a holdout against the five-kingdom scheme of taxonomy.

Costerton, J.W., G.G. Geesey, and K.J. Cheng. "How Bacteria Stick," *Scientific American,* January 1978. The first step in infection.

Diener, T.O. "The Viroid—A Subviral Pathogen," *American Scientist,* September–October 1983. A discussion of the origin, structure, and pathogenesis of viroids with speculation about prions.

Dobzhansky, T., et al. *Evolution.* San Francisco, W.H. Freeman, 1977. An introduction to evolution that includes chapters on taxonomy.

Eldredge, N., and J. Cracraft. *Phylogenetic Patterns and the Evolution Process.* New York, Columbia University Press, 1980. A discussion of the interaction between taxonomy and evolutionary theory.

Gardner, E.J. *History of Biology.* Minneapolis: Burgess Publishing Company, 1965. Chapter 8 summarizes the work of Linnaeus and that of his lesser-known contemporaries and predecessors in the establishment of our system of classification.

Krogmann, D.W. "Cyanobacteria (Blue-Green Algae)—Their Evolution and Relation to Other Photosynthetic Organisms," *Bioscience,* February 1981. A good example of the application of modern taxonomic techniques.

Leedale, G.F. "How Many Are the Kingdoms of Organisms?" *Taxon,* Vol. 23. Discusses problems with assigning organisms to discrete kingdoms.

Moore, R.T. "Proposal for the Recognition of Super Ranks," *Taxon,* 23: 650–652, 1974. A suggestion for using the term *dominium* to comprise the viruses, prokaryota, and eukaryota.

Nester, E.W., N.N. Pearsall, J.B. Roberts, and C.E. Roberts. *The Microbial Perspective.* Philadelphia: Saunders College Publishing, 1982. A very readable and interesting presentation of microbiology, including bacteria, viruses, fungi, and protozoa. The emphasis is on the role of microorganisms in human health and disease.

Palleroni, N.J. "The Taxonomy of Bacteria," *Bioscience,* June 1983. Modern approaches to microbial taxonomy are discussed following a discussion of the history of bacterial classification systems.

Vidal, G. "The Oldest Eukaryotic Cells," *Scientific American,* February 1984. A review of the fairly recent discoveries of eukaryotic fossils in rock formations dated as old as 1.4 billion years.

Wernick, R. "From Ewe's Milk and a Bit of Mold: A Fromage Fit for a Charlemagne," *Smithsonian,* February 1983. The story of the production of Roquefort cheese; an exercise in practical microbiology.

Whittaker, R.H. "New Concepts of Kingdoms of Organisms." *Science,* 163: 150, 1969. A proposal for classifying living things according to a five-kingdom system.

Whittaker, R.H. "On the Broad Classification of Organisms," *Quarterly Review of Biology,* 34: 210, 1959. The earliest rumble of what was to become the five-kingdom revolution.

Whittaker, R.H., and L. Margulis. "Protist Classification and the Kingdoms of Organisms," *Biosystems,* Vol. 10, 1978.

Woese, C. R., and G.E. Fox. "Phylogenetic Structure of the Prokaryote Domain: The Primary Kingdoms," *Proceedings of the National Academy of Sciences,* 74 (11): 5088–5090, 1977. Significant for its proposal of both the domain concept and the suggestion that the domain Prokaryota should contain two kingdoms, the eubacteria and the methanogens.

Woese, C.R. "Archaebacteria," *Scientific American,* June 1981, 98–122. An approach to the evolutionary significance of the archaebacteria in which they are considered a third major group along with the prokaryotes and eukaryotes.

Chapter 15
PLANT LIFE

Learning Objectives

After you have studied this chapter you should be able to:

1. Compare the plantlike green algae with the protist-like green algae and summarize the characteristics of the three divisions of plantlike algae.
2. List the characteristics of the bryophytes and describe their general adaptations to the terrestrial environment.
3. Compare bryophytes and ferns from the standpoint of the dominance of the sporophyte or gametophyte generation and vascularity. Describe the complete life cycle of each.
4. Distinguish between gymnosperms and angiosperms, between monocots and dicots. Identify each from a specimen or diagram, or from the characteristics of its seed.
5. Discuss the possible or proposed evolutionary origin of and relationships among the plants.

What we commonly call **plants,** as opposed to the plantlike protists, are multicellular photosynthetic organisms. A colonial protist may exhibit some degree of cellular specialization, but a real plant has identifiable, specialized body parts, and these parts are composed of differentiated tissues. However, with this in mind we should point out that each higher algal division contains some representatives that are protist-like. For instance, some of the green algae are single-celled and might be taken for protists except that the green algae have cellulose walls and a single chloroplast.

THE HIGHER ALGAE

Most of the **higher algae** (or lower plants) are multicellular and exhibit some degree of body specialization. They also have plantlike patterns of growth. These plants are referred to here as "lower" not because they should be considered inferior but because they are not as highly differentiated as the "higher" plants.

The plantlike algae consist of three divisions: the Chlorophyta, the Phaeophyta, and the Rhodophyta. Because of their many protist-like representatives, the Chlorophyta (green algae) may almost be thought of as straddling the protist and plant kingdoms. They have often been suggested as possible ancestors of the vascular plants[1] because their cells and life cycles greatly resemble those of the latter. By contrast, the remaining classes of algae are so different from both the green algae and the higher plants that most biologists feel there is no close connection.

Multicellular Green Algae (Division Chlorophyta)

We know of about 7000 species of **chlorophytes, multicellular green algae,** living in a variety of environments, including large and small bodies of water, damp soil, the undersides of thin desert rocks, and even the surfaces of old snowdrifts. The chloroplasts of these algae possess both chlorophyll *a* and *b* and store food in the form of starch or, sometimes, oil. A very pretty example, the nonmotile freshwater forms called **desmids,** look like little green snowflakes, just large enough to be visible to the naked eye (Fig. 15–1). Few are as spectacular as *Volvox,* a hollow colonial form that rolls through the water like a tiny green planet, driven by unresting flagella (Fig. 15–2). Each cell of this spherical colony has flagella, a chloroplast, and a little red eyespot that gives it a rudimentary ability to sense light. Within the volvox colony may be several asexually produced smaller daughter colonies, which will one day be released by the death and disintegration of the mother colony.

The filamentous green algae are beautiful on a microscopic scale—strings of crystalline cylinders, each with its own chloroplasts, often star-shaped or formed like tiny intertwined helices. Present by the millions, though, they make up the slimy green scum that covers the surface of stagnant ponds and polluted streams (see Focus on the Reproductive Life of *Spirogyra*).

Some green algae, the **siphonous algae,** have multinucleate cells. One siphonous alga, *Acetabularia,* has an important place in the history of molecular and cellular biology, as was discussed in Chapter 11.

One of the clearest indications of pollution (often from excess nutrients in sewage) in marine environments is a heavy growth of a lettuce-like plant in the shallows, sometimes in quantities sufficient to impede

---

[1] "The origin of the life cycle of land plants" by Linda Graham. *American Scientist* 73:178–186, March/April 1985.

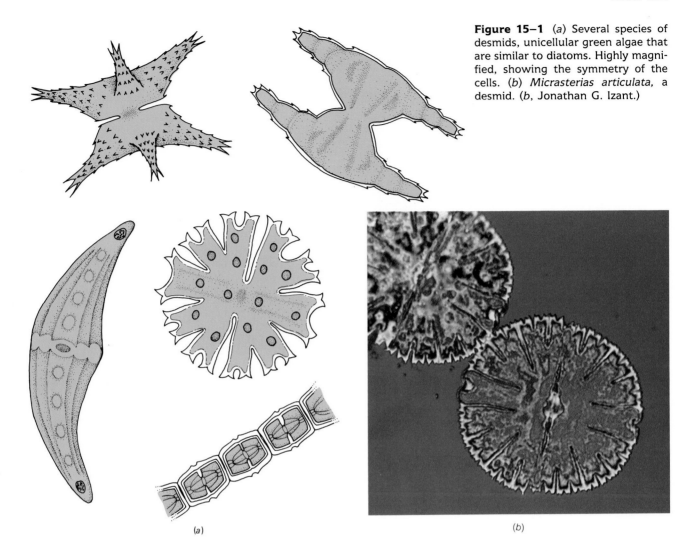

Figure 15–1 (a) Several species of desmids, unicellular green algae that are similar to diatoms. Highly magnified, showing the symmetry of the cells. (b) *Micrasterias articulata*, a desmid. (b, Jonathan G. Izant.)

(a)

(b)

Figure 15–2 *Volvox* reproducing. A daughter colony can be seen leaving a mother colony through a rent in its wall. (James Bell, Photo Researchers, Inc.)

Focus on . . .

THE REPRODUCTIVE LIFE OF *SPIROGYRA*

Spirogyra, one of the filamentous green algae, is distinguished by the possession of one to several (depending on species) spiral chloroplasts, for which it is named. *Spirogyra* can reproduce both sexually and asexually. The asexual is the usual mode, whose advantage is that it is very rapid, enabling this plant to become an aquatic pest, able to overrun a polluted backyard pond in a single season. Each cylindrical cell divides transversely, so the filaments elongate continuously, breaking apart from time to time.

It is striking that the cells of the filament are haploid. In the autumn two filaments of this alga may chance to lie side by side. This, however, is not enough by itself to lead to sex, for they must be of different **mating types** to conjugate. Since the sexes of *Spirogyra* are not distinguishable except by other *Spirogyra,* we say that it is **isogamous,** meaning that its sex cells (which in this case are the same as ordinary body cells) are alike.

If the necessary conditions are fulfilled, little dome-shaped protuberances develop on the sides of two adjacent cells. These elongate and fuse with one another to form the **conjugation tube.** The contents of one cell withdraws from the cell walls, becomes round, and creeps through the conjugation tube to fuse with the other cell. What determines the creeper and what the creepee, no one knows. In any case, the resulting zygote becomes a diploid **zygospore,** which never exists as a real generation in its own right but divides in the spring by meiosis to form four haploid nuclei, three of which promptly degenerate. The resulting haploid vegetative cell then divides repeatedly to form a new filament.

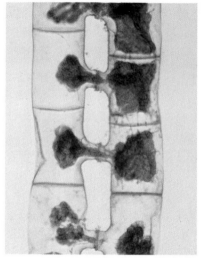

Sexual reproduction in Spirogyra occurs by fusion of nonflagellated gametes. Conjugation tubes form between cells of adjacent filaments of different mating types. Gametes of the donor filament migrate through the conjugation tubes and fuse with other gametes. The resulting zygotes form thick cell walls and become dormant. (Carolina Biological Supply Company.)

boats and swimmers. Sooner or later, the abnormally stimulated growth dies off, decays, stinks, and releases further pollution into the water (Fig. 15–3). An individual specimen of this seaweed looks somewhat like a

Figure 15–3 A shallow portion of Tampa Bay, seen here at low tide. Excessive enrichment of the habitat by sewage and other wastes has encouraged the excessive growth of *Ulva,* forming this continuous "lawn" of plants.

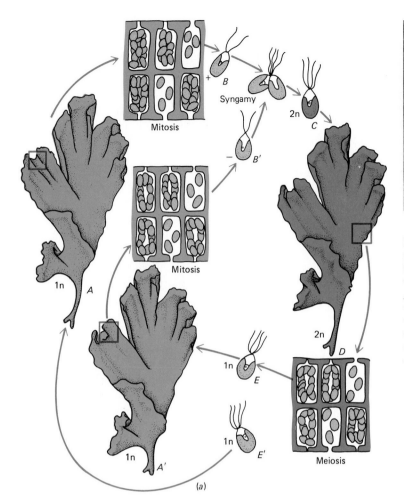

(b)

Figure 15–4 Diagrammatic life history of *Ulva*, showing alternation of generations that are physically indistinguishable, but one of which has twice as many chromosomes per cell as the other. *A* and *A'*, plus and minus gametophytes, respectively. *B* and *B'*, plus and minus gametes. One plus and one minus gamete must meet in order to produce a zygote. *C*, zygote. *D*, sporophyte plant. *E* and *E'*, spores produced by meiosis of the two mating types. (*a*, after Norstog, K., and R.W. Long. *Plant Biology*. Philadelphia, W.B. Saunders, 1976. *b*, E.R. Degginger.)

lettuce leaf and is therefore known as **sea lettuce,** or scientifically as *Ulva lactuca*. The "leaf" or *blade* of sea lettuce is just two cells thick, and the entire plant is anchored to the bottom or perhaps to a rock by a rootlike *holdfast*.

What cannot be determined by inspection is whether the plant is haploid or diploid. In *Ulva* there is an alternation of generations, a bit like that typical of higher plants. The technical term for the *Ulva* pattern is **isomorphic,** whereas the higher plants have a **heteromorphic** habit, in which the generations are physically as well as chromosomally different from each other. *Ulva* produces flagellated isogametes that cannot really be identified as either sperm or egg. These fuse to form the diploid generation. The diploid *Ulva* matures, then produces haploid spores by meiosis.These are called **zoospores** because they can swim by means of flagella, an animal-like characteristic. The zoospores, if fortunate, grow into haploid plants. These plants produce isogametes, which fuse to form diploid plants (Fig. 15–4).

The Brown Algae (Division Phaeophyta)

The multicellular marine organisms known as **brown algae** have organs somewhat comparable to the stems, leaves, and roots of higher plants. However, these structures must be considered analogous rather than homologous to those higher plant parts. By that we mean that the organs of a brown alga cannot be said to have the same embryonic or evolutionary origin as those of terrestrial plants, but they may *function* in somewhat comparable ways. Some of the brown algae also have vascular tis-

Figure 15–5 *Laminaria*, a typical brown alga. These complex algae somewhat resemble the vascular plants in the differentiation of their body parts. (J. Robert Waaland, University of Washington/BPS.)

sues of sorts, needed because of their very large size, which requires some special provision for transport of materials in their bodies.

What the brown algae have that the green algae do not is **fucoxanthin,** the pigment that produces the characteristic brown color. Of course, they also have chlorophyll. Many of the brown algae are large, some of the largest (giant kelps) being almost the size of sapling trees. The intertidal species have slimy polysaccharides that help keep them moist between tidal immersions, and these are often of commercial value. **Algin,** for example, is often added to ice cream of low cream content to make it seem smooth to the palate.

The leaflike **blades** of brown algae usually are equipped with gas-filled bladders or other flotational devices (Fig. 15–5). The stemlike **stipes** bend with waves and currents rather than resisting, as do the stems of land plants that are impregnated with woody materials. The rootlike **holdfasts** need not absorb water and nutrients as do the roots of land plants—after all, a marine alga is surrounded by a nutrient solution—but they are necessary to keep the plant from floating off.

Most phaeophytes have a well-defined alternation of generations. Many are more or less isomorphic but some are highly heteromorphic.

The Red Algae (Division Rhodophyta)

Like the brown algae, the red algae of the division Rhodophyta are almost entirely marine organisms. Most are small and filamentous or flattened. Their life cycle involves the alternation of heteromorphic generations. The unique feature of the rhodophytes is their pigmentation. No other group of eukaryotes has the red pigment **phycoerythrin** or the blue pigment **phycocyanin.** (Cyanobacteria have them both, although not in the same proportions as the rhodophytes.) These accessory pigments enable the red algae to trap and utilize the blue and violet wavelengths of light that predominate in the marine environment. As a result, they can grow in deeper waters than the brown and green algae.

Perhaps the ecologically most important representatives of the rhodophytes are the **coralline algae** (Fig. 15–6), which have skeletons of calcium carbonate. These skeletons enable the coralline algae to build reefs, much as corals themselves do. In fact, many tropical reefs are composed of more coralline algae than of coral. Like the brown algae, the rhodophytes yield commercially valuable carbohydrates, such as **agar,** used by microbiologists as a medium for growing bacteria, and **carrageenin,** a food additive.

THE NONVASCULAR LAND PLANTS (DIVISION BRYOPHYTA)

The bryophytes include about 16,000 species of mosses and liverworts. They inhabit moist areas, for water is necessary for their sperm to reach the egg. These plants lack true stems, roots, leaves, and for the most part vascular tissue. Still, bryophytes are more highly differentiated than algae, and their bodies consist of several kinds of tissue. For example, they have rootlike structures called **rhizoids** that attach the bryophyte to the ground. Because they must depend upon diffusion to distribute all substances within the body, their size is limited. The biggest mosses, for example, are only about a foot high.

The complex alternation of generations in bryophytes may be understood by studying Figures 15–7 and 15–8. Briefly, a diploid spore-producing **sporophyte** generation alternates with a separate and distinct haploid **gametophyte** generation.

Figure 15–6 Coralline algae. These are algae that secrete a calcium carbonate shell and play a major role in the formation of coral reefs. (Peter Scoones, Seaphot.)

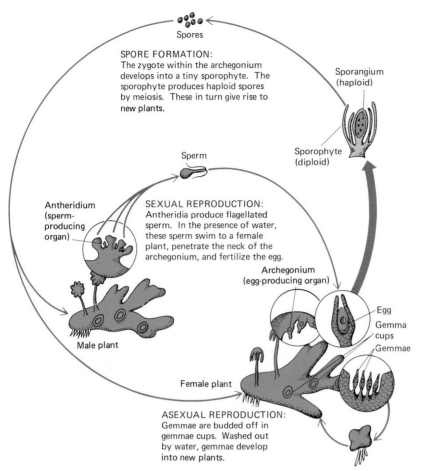

SPORE FORMATION:
The zygote within the archegonium develops into a tiny sporophyte. The sporophyte produces haploid spores by meiosis. These in turn give rise to new plants.

Spores

Sperm

Sporangium (haploid)

Sporophyte (diploid)

Antheridium (sperm-producing organ)

SEXUAL REPRODUCTION:
Antheridia produce flagellated sperm. In the presence of water, these sperm swim to a female plant, penetrate the neck of the archegonium, and fertilize the egg.

Archegonium (egg-producing organ)

Egg

Gemma cups

Gemmae

Male plant

Female plant

ASEXUAL REPRODUCTION:
Gemmae are budded off in gemmae cups. Washed out by water, gemmae develop into new plants.

Figure 15–7 Life cycle of a liverwort. The sporophyte generation produces spores, whereas the gametophyte generation produces eggs and sperm. Gemmae are specialized structures from which new individuals arise by asexual reproduction.

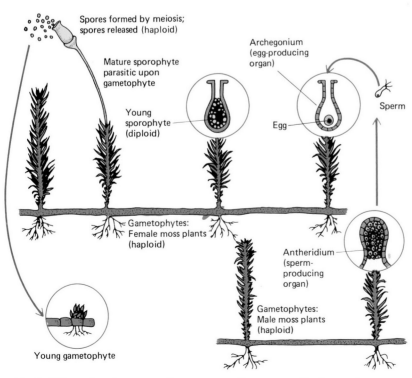

Spores formed by meiosis; spores released (haploid)

Mature sporophyte parasitic upon gametophyte

Young sporophyte (diploid)

Archegonium (egg-producing organ)

Egg

Sperm

Gametophytes: Female moss plants (haploid)

Antheridium (sperm-producing organ)

Young gametophyte

Gametophytes: Male moss plants (haploid)

Figure 15–8 Life cycle of a moss. Masses reproduce asexually by underground runners.

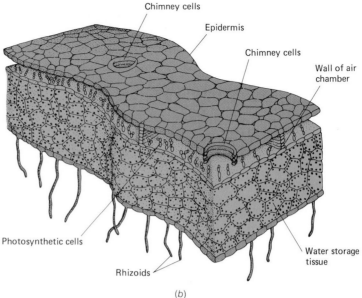

Figure 15–9 (a) *Marchantia*, a liverwort. (b) Anatomy of a liverwort. The upper surface (epidermis) is one cell thick and punctured by many pores through which gas exchange occurs. In some species these pores are surrounded by little "chimneys" whose function may be to limit water loss through the pores. The rhizoids absorb some water, there is a waxy protective outer layer called the cuticle (not shown), and starch is stored in some of the deeper tissues. All these features are adaptations for terrestrial life. (a, L. Egede–Nissen/BPS.)

Liverworts are most important in moist tropical habitats, especially rain forests, but do occur in temperate climates in wet habitats and can sometimes be seen in greenhouses. Some of them are prostrate in habit, looking like little leaves stuck to the ground without anything resembling a stem (Fig. 15–9). Some, however, especially in the tropics, grow on trees without harming them, with groups of more or less erect leaflike parts projecting into the air.

Mosses (Fig. 15–10) are ecologically somewhat more significant than liverworts. They are important pioneering organisms in many ecosystems and often are found growing on trees. Mosses resist dry conditions very effectively—not so much by any water conservation mechanisms as by simply tolerating desiccation.

THE VASCULAR PLANTS (DIVISION TRACHEOPHYTA)

A plant living in a terrestrial environment must face problems beyond the experience of aquatic plants. Most of these problems are connected with the shortage of water. The plant must have some means to retard the evaporation of water, such as a waxy cuticle or cork-filled bark. Yet there must also be some way to absorb water and needed minerals from the environment, usually the soil. The plant must be able to support the organs of photosynthesis perhaps many meters from the source of nutrients in the earth. Finally, it must find some way to get the gametes together. In the absence of all-surrounding water to buoy it up, to carry materials to all its parts, and to carry its gametes and zoospores about, a terrestrial plant must have special organs and adaptations, such as stems, roots, leaves, flowers, and vascular tissues, that are unknown or very little developed among algae or less complex plants.

The tracheophytes are the other of the two great divisions of the "higher," or terrestrial, plants. They differ strikingly from the bryophytes not only in vascularity but usually in their mode of reproduction as well. Among the tracheophytes with mutually independent alternating generations, the sporophyte is far the more conspicuous. In fact, among the gymnosperms and flowering plants, the sporophyte is the only stage that can be seen with the naked eye. Indeed, until the 19th century, no one thought that seed plants *had* a gametophyte generation.

Figure 15–10 Carpet of moss plants. It is possible that all these gametophytes are clones derived from a single progenitor. (J. Robert Waaland, University of Washington/BPS.)

(a) (b) (c)

The vascular plants have a circulatory system composed of the tissues phloem and xylem, which will be discussed in Chapter 17. These vascular tissues permit the tracheophytes to attain great size and also permit them to survive under conditions that many bryophytes could not tolerate. Their reproductive adaptations are at least equally important, freeing many of the tracheophytes from dependence on large amounts of water for the dispersal of their gametes.

Figure 15–11 Representative ferns. (a) Young plant showing characteristic "fiddlehead" growth pattern of new leaves. (b) Cinnamon fern, showing brown reproductive spikes and green sterile leaves (the spikes bear the sporangia). (c) Tree fern, *Cyathea.* These Southern Hemisphere plants are especially common in New Zealand (where this photograph was taken) but have been introduced widely into subtropical parts of the world, where they are often grown as ornamentals. (a, Carolina Biological Supply Company. b, J.N.A. Lott, McMaster University/BPS. c, B.J. Miller, Fairfax, Virginia/BPS.)

The Ferns and Horsetails

Among the simplest of the tracheophytes, the ferns (Class Filicopsida) (Fig. 15–11) reproduce both asexually and sexually by means of an alternation of generations comparable to that of the bryophytes. Although the gametophyte, the gamete-producing generation, is the more conspicuous in the bryophytes, the sporophyte, the spore-producing generation, is the more conspicuous among the ferns. Because the male gamete must swim to reach the female, the fern requires water. Not much water is required, however; even dew will suffice. Ferns possess vascular tissues comparable to those of the angiosperms, and they have true roots, stems, and leaves. Ferns can attain considerable size—some tropical representatives grow as large as small trees—but the familiar small ferns and bracken are a more important part of many plant communities.

Many ferns are capable of asexual reproduction, usually by means of underground runners or stems. Bracken, for instance, spreads so rapidly in old fields that it can become a serious pest.

The horsetails and lycopods are lesser known allies of the ferns (Fig. 15–12) that are of little present ecological importance. Yet in the past giant horsetails were extremely important members of plant communities, and their remains form much of the coal we mine today.

The Seed Plants

Ferns are dispersed by means of their extremely light spores, which can be carried for hundreds of miles by the wind. The seed plants employ a different strategy, in which a miniature plant is encased in a package

(b)

(a)

Figure 15–12 *Equisetum*, horsetail plants. Shown here are two species with different growth habits. (a) *Equisetum telematia*, the common horsetail, with unbranched, non-green, nonphotosynthetic "fertile" shoots bearing conelike reproductive organs (strobili) from which spores are released. There are also separate, highly branched, green, photosynthetic "sterile" shoots. Both types arise from an underground rootlike rhizome. (b) A dense growth of *Equisetum* on a bank. (a, J. Robert Waaland, University of Washington/BPS. b, Charles Seaborn.)

that also contains the food it will need for its early nourishment. It is the seed that is dispersed by a variety of adaptations; there are no spores as such. In some ways this is clumsier than spores and would seem to offer no advantage over them, except perhaps that of avoiding temporarily unfavorable conditions through delayed germination. Perhaps the real secret of the seed plant's success is not so much seeds as pollen. Each pollen grain produces and in some cases contains the male gamete, so that in seed plants the lightly packaged gametes are distributed by the air—or even more efficiently by animal vectors, such as bees. This permits the genetic interaction of parents of diverse heredity over vast distances. Ferns, with their water-dependent gametes, could not possibly manage this.

The seed plants are somewhat informally divided into two groups: the gymnosperms and the angiosperms. Though convenient, these categories may not represent "natural" taxonomic groups in that the gymnosperms in particular may be **polyphyletic,** that is, derived from several origins.

GYMNOSPERMS

Most **gymnosperms** are medium to large evergreen shrubs and trees. This group includes the **conifers,** the familiar cone-bearing evergreens such as pines and spruces. Ginkgo trees (Fig. 15–13) and the palmlike cycads are also gymnosperms. Most gymnosperms have needle-like or scalelike leaves designed to conserve moisture in cold, dry, or windy habitats (Figs. 15–14 and 15–15) and bear their usually naked seeds in cones. Gymnosperms possess well-developed vascular tissues, they have true roots, stems, and leaves, and they reproduce by means of seeds.

Gymnosperms are of tremendous ecological and economic importance. Most construction lumber is produced from gymnosperms, and

Figure 15–13 *Ginkgo biloba* tree, showing "fruit." Ginkgo trees once were a dominant terrestrial tree; however, they are missing from all known geological deposits later than the Mesozoic and have come very close to extinction. Native only to China today, they are now propagated in urban areas around the world because of their resistance to air pollution. The "fruit," however, produces an obnoxious and odorous mess as it rots on the ground. (Courtesy of Lloyd Black.)

Figure 15–14 *(left)* *Welwitschia mirabilis* in the Namib desert in southwestern Africa. Note the female cones. This unusual gymnosperm absorbs condensation from the surfaces of its leaves and stores water in a large, subterranean, turnip-like root. (F.J. Odendaal, Stanford University/BPS.)

they dominate the entire northern forested area—much of Canada and Russia and certain other areas, such as the northwestern United States.

The familiar evergreen trees and shrubs are considered to be the equivalent of the sporophyte generation of ferns, although they do not produce actual spores. The gametophyte generation is initially microscopic—reduced to a few cells parasitic upon the sporophyte, which are found in the sexual organs of the tree. The nutritive storage tissue of the seed is, however, a haploid gametophyte derivative. One might almost say that the gametophyte generation of gymnosperms is so highly reduced as to be almost theoretical. The reproduction of gymnosperms will be considered in more detail in Chapter 18.

ANGIOSPERMS

Most economically important food plants belong to the **angiosperms** (Class Angiospermopsida). These vascular plants reproduce by means of seeds and are distinguished from the gymnosperms principally by their possession of the sexual organs called **flowers** (Fig. 15–16). As with gymnosperms, the conspicuous visible plant is considered a sporophyte. Some angiosperms are **evergreen** (always covered with leaves), but most of them are either annual (that is, they live for only one season) or **deciduous** (that is, they lose their leaves seasonally). Their seeds are often surrounded by a hard or fleshy envelope called a **fruit.** A fruit consists mainly of the ovary wall of the flower, though other flower parts may also participate in its formation. Fruits are frequently modified to provide for seed dispersal to new locations—for instance, by the digestive system of animals.

Angiosperms are further divided into the subclasses **Monocotyledonae** and **Dicotyledonae** (or monocots and dicots). In monocots the seed has but a single small embryonic leaf, known as a **cotyledon,** and a large **endosperm,** a food storage tissue. In dicots the seed has two large cotyledons and hardly any endosperm. Dicots typically have large

Figure 15–15 Coniferous gymnosperms are the typical trees of northern latitudes. It is likely that their ability to retain their leaves and carry out photosynthesis when the growing season is over for angiosperms gives them a critical advantage in some seasonably arid habitats (or cold habitats that are physiologically arid).

Figure 15–16 *(right)* Indian pipe, a flowering plant that is not a producer. It does not carry on photosynthesis but derives its energy from other organic matter, as do fungi. (Leo Frandzel.)

Figure 15–17 (*a*) Pasture rose, a dicot. (*b*) Maple flower, a wind-pollinated dicot. (*c*) Palm flower, a monocot. (*d*) Castor oil plant leaf. Note the typical dicot-branching veins. (*e*) Palm leaf. Note parallel monocot venation. (*a*, Leo Frandzel.)

(*a*) (*b*) (*c*)

(*d*) (*e*)

four- or five-part flowers and a layer of embryonic tissue, the cambium, that permits secondary growth of the stem in diameter. Monocots tend to have parallel leaf veins. Dicots usually have divergently branching or netlike leaf veins (Fig. 15–17). Their flower parts are usually present in multiples of three. Monocots seldom exhibit secondary growth of the stem, and some (the grasses) grow from the base rather than from the top of the plant. (Angiosperm reproduction will be more fully discussed in Chapter 18.)

Vascular tissues are typically organized in bundles, which in ferns and monocots are arranged more or less haphazardly throughout the substance of the stem. In dicots the vascular bundles are arranged in a ring within the stem. In woody dicots the vascular bundles form a more or less continuous layer under the bark, usually extending into the heart of the stem. In these, certain of the vascular tissues also have other functions, such as skeletal support and the storage of food.

THE FAMILY TREE OF PLANTS

How did the vascular and nonvascular plants originate? Although a matter of controversy, it is now widely believed that the vascular plants were directly derived from algal ancestors rather than from bryophyte ances-

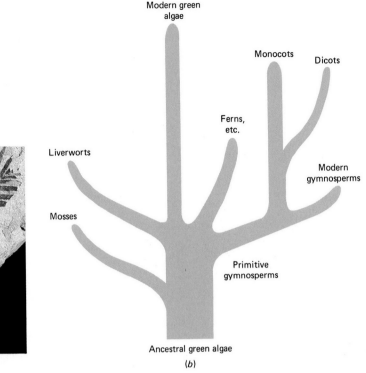

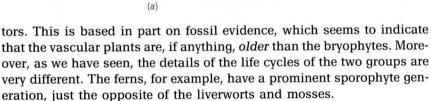

(a)

(b)

Figure 15–18 (a) A fossil cycad, *Sphenozamites*, from the Jurassic period. (b) Diagram of suggested course of plant evolution. (a, E.R. Degginger.)

tors. This is based in part on fossil evidence, which seems to indicate that the vascular plants are, if anything, *older* than the bryophytes. Moreover, as we have seen, the details of the life cycles of the two groups are very different. The ferns, for example, have a prominent sporophyte generation, just the opposite of the liverworts and mosses.

Concerning vascular plants, it was formerly thought that the seed plants were simply derived from the fossil seed ferns. Although this cannot be ruled out, it is odd that the oldest gymnosperms are even older than the seed ferns, which some botanists are now inclined to view as gymnosperms themselves. The angiosperms are often thought to have been derived from the gymnosperms, but not from the conifers. Some of the gymnosperms, such as the ginkgo tree, are broad-leaved and deciduous. The conifers are in most ways more specialized than such nonconiferous gymnosperms. Yet there exist today no broad-leaved gymnosperms that can easily be viewed as angiosperm ancestors, so most botanists think that angiosperms must have been derived from broadleaved gymnosperms that are not only extinct but very poorly represented in the fossil record. The details of reproduction in angiosperms and gymnosperms are also quite different. However, the current consensus among botanists might resemble the proposed evolutionary relationships among the plants that are summarized in Figure 15–18.

The major divisions of the plant kingdom and their characteristics are outlined in Table 15–1.

CHAPTER SUMMARY

I. The plant kingdom includes the algal divisions Chlorophyta, Phaeophyta, and Rhodophyta, and the terrestrial divisions Bryophyta and Tracheophyta.

II. The chlorophytes are in large part protist-like, but many of them possess differentiated bodies with isomorphic alternation of generations, in which haploid and diploid forms succeed one another.

III. The phaeophytes, the brown algae, are noted for the differentiation of their tissues and body parts.

Table 15–1
THE PLANT KINGDOM

Kingdom and division	Characteristics	Comments
Plants	Multicellular; complex; adapted for photosynthesis; have reproductive structures; pass through distinct developmental stages; alternation of generations—sporophyte generation produces spores, which develop into gametophyte plants, which produce gametes	Producers
Chlorophytes	Unicellular, colonial, and multicellular, many with distinct alternation of generations (isomorphic)	Some similar to protists
Phaeophytes	Almost all multicellular, with pronounced alternation of generations, differentiated bodies and, in few cases, vascular tissues	Important intertidal and shallow marine forms
Rhodophytes	Almost all multicellular with alternation of generations (heteromorphic); deeper-living, with unique accessory photosynthetic pigments	Sometimes thought to have independent origin from other plants
Bryophytes (mosses, liverworts)	Small because they lack true stems, roots, leaves, and any sort of vascular tissue (and must depend upon diffusion for distribution of materials); gametophyte is dominant form; smaller sporophyte is attached to gametophyte in mosses and is partially parasitic upon it	Often pioneer organisms in colonization of new habitats
Tracheophytes (vascular plants)	Have vascular tissue that enables them to attain large size; have true roots, stems, and leaves; multicellular embryos; sporophyte dominant form	Adapted for life on land
(1) Ferns	Gametophyte develops as tiny independent plant; spores are found in little clusters on underside of some sporophyte leaves	
(2) Seed plants	Embryo is encased within seed equipped with nourishment; embryo may lie dormant for long periods until conditions are favorable; gametophyte reduced to small structure on sporophyte plant	
(a) Gymnosperms	Most have needle-like or scalelike leaves; bear their usually naked seeds in cones	Dominate entire northern forested area; produce much lumber
(b) Angiosperms (flowering plants)	Possess sexual organs called flowers; seeds often surrounded by fleshy or hard envelope called fruit (consisting mainly of ovary wall); most either annual (live only one season) or deciduous (lose their leaves seasonally); two types: monocots (seed has single small embryonic leaf and large endosperm, leaves have parallel veins), and dicots (seed has two large embryonic leaves and little endosperm, leaves have netlike veins)	Provide food for humans and other consumers

IV. The rhodophytes, the red algae, also have heteromorphic alternation of generations and possess unique accessory photosynthetic pigments that permit them to grow at greater depths than other algae.

V. Bryophytes are small plants lacking vascular tissues, true leaves, true stems, and true roots.

    A. Alternation of generations is marked in the bryophytes, and the gametophyte generation is the more conspicuous.

    B. Bryophytes include mosses, liverworts, and a few other groups.

VI. Tracheophytes are fully vascular and possess true roots, stems, and leaves. They include the ferns and the seed plants.

    A. In the ferns two independent generations alternate. The sporophyte generation is the more conspicuous.

B. The gross gymnosperm or angiosperm body is also considered to be a sporophyte phase of the life cycle. Among these seed plants the gametophyte generation is generally held to be represented by portions of the pollen grain and the female reproductive structures.

C. A typical angiosperm body consists of roots, stems, and leaves. Angiosperms reproduce by means of sexual organs known as flowers.

D. Angiosperms are divided into two subclasses, monocots and dicots.
   1. In the monocots the endosperm is large, the embryo has a single leaf (cotyledon), the vascular bundles are scattered throughout the stem, the leaf veins are parallel, the flower parts are in multiples of three, and secondary stem growth is absent.
   2. In the dicots the endosperm is consumed in early embryonic life, and food is usually stored in the embryonic leaves. Dicots have two cotyledons, the vascular bundles are arranged in concentric circular layers, the main leaf veins are divergent, the flowers have four or five parts, and secondary stem growth is pronounced.

Post-Test

1. The division Chlorophyta is made up of the _____ _____ .

2. In conjugation, *Spirogyra* produces a kind of sex organelle, the _____ _____ . Because the conjugants are indistinguishable they are called _____ -gametes.

3. The red algae are distinguished by their _____ photosynthetic pigments, which otherwise occur only in the _____ .

4. Some of the green algae share the characteristic of alternation of _____ with the higher plants, but in the green algae this arrangement is _____ -morphic, in contrast to the _____ -morphic pattern characteristic of land plants.

5. The mosses and liverworts constitute the division _____ . It is believed that these are small mainly because they lack _____ tissues.

6. Among the bryophytes, an inconspicuous, diploid _____ generation alternates with a larger, haploid _____ generation.

7. Ferns possess the vascular tissues _____ and _____ , plus undeniable roots, stems and leaves. Despite this, they reproduce by means of _____ and swimming _____ .

8. The _____ , which have no true fruit, and the _____ , which do, constitute the _____ plants.

9. The male gametophyte of gymnosperms is considered to be the _____ _____ .

10. The gymnosperm seed possesses tissue derived from the haploid gametophyte that serves the same nutritive function as does the _____ of angiosperms.

Review Questions

1. Compare the major features of the three classes of green algae.
2. What reasons might there be to suppose that the red algae have originated independently of the green algae?
3. What features of the red algae set them apart from all other plants?
4. Compare the adaptations of the mosses and ferns.
5. In your opinion, is *Volvox* best considered a protist or a plant? Explain.
6. Prepare a table comparing the major features of all the classes of tracheophytes.
7. What differences in means of fertilization exist between the bryophytes and the tracheophytes?
8. Give at least five ways in which the dicotyledons differ from the monocotyledons, comparing endosperm, leaf and flower structure, and vascular tissue.
9. In what ways are ferns like the seed plants? In what ways do they differ?
10. What similarities are there between the higher green algae and the bryophytes? the tracheophytes?
11. What evidence can be used to argue that the bryophytes and the tracheophytes originated independently?

Readings

Christopher, T.A. "The Seeds of Botany," *Natural History,* March 1981, 50–56. The establishment of the first Western botanical garden in Padua, Italy.

Good, R. *Features of Evolution in the Flowering Plants.* New York, Dover, 1974 (first published 1956). A technical but thought-provoking and iconoclastic discussion of theories of spermatophyte evolution.

Gray, A. *Gray's Manual of Botany,* 8th ed. New York, Van Nostrand Reinhold, 1950. The classic summary of the vascular plants.

Klein, R.M. *The Green World.* New York, Harper and Row, 1979. An eminently readable and brief introduction to botany.

Perry, F. *Flowers of the World.* New York, Crown, 1972.

Pickett-Heaps, J. *New Light on the Green Algae,* Burlington, N.C., Carolina Biological Supply Company, 1982. It is not easy to obtain detailed information on the microanatomy and life-style of algae. Here is a convenient source.

Raven, P.H., R.F. Evert, and H. Curtis. *Biology of Plants,* 3rd ed. New York, Worth Publishers, 1981. An excellent general botany textbook.

Ray, P., T.A. Steeves, and S.A. Fultz. *Botany.* Philadelphia, Saunders College Publishing, 1983. This comprehensive botany textbook includes a fine presentation of the evolution of plants. The very best available.

Swaminathan, M.S. "Rice," *Scientific American,* January 1984. Rice, along with wheat and maize, is a member of the grass family upon which the human species largely depends. This article describes the advances in plant genetics that have greatly increased its yield per acre.

Thomas, B. *The Evolution of Plants and Flowers.* New York, St. Martin's Press, 1981. A beautifully illustrated and imaginative summary of current thinking about plant origins and evolutionary development.

Tippo, O., and W.L. Stern. *Humanistic Botany.* New York, Norton, 1977. A fine and fascinating discussion of what plants mean to the human species.

Watson, E.V. *Mosses.* London, Oxford University Press, 1972. A brief Oxford Biology Reader that debunks much of what "everyone knows" about mosses.

Chapter 16
ANIMAL LIFE

Outline

Learning Objectives

After you have studied this chapter you should be able to:

1. List the characteristics common to most animals and, using these characteristics, develop a brief definition of an animal.
2. Discuss the classification and proposed relationships of the animal phyla on the basis of (a) symmetry, (b) type of body cavity, (c) pattern of embryonic development (that is, compare protostomes with deuterostomes).
3. Classify a given animal in the appropriate phylum and class.
4. Identify the distinguishing characteristics of each phylum.
5. Describe the body plan and life-style of one member of each phylum.
6. Summarize the life cycle of each parasitic worm described in the chapter, including tapeworm and *Ascaris.* Identify adaptations that these animals possess for their parasitic life-style.
7. Identify factors contributing to the great biological success of the insects.
8. List the three main characteristics of a chordate.
9. List and give examples of each of the seven classes of vertebrates, and describe the course of their evolution according to contemporary evolutionary theory.
10. Distinguish among the three main groups of living mammals and identify specific mammals that belong to each group.

K P C O F G S

In addition to the more than a million known species of animals on our planet, perhaps several million more remain to be identified. Most of the members of the animal kingdom are classified in about 29 phyla. The animals most familiar to us—dogs, birds, fish, frogs, snakes—are **vertebrates,** animals with backbones. However, vertebrates account for only about 5% of the species of the animal kingdom. The majority of animals are the less familiar **invertebrates**—animals without backbones. The invertebrates include such diverse forms as coral, worms, and butterflies.

WHAT IS AN ANIMAL?

Oddly, it is difficult to define what we mean by animal. It is easy enough to distinguish between a cow and a rose bush, but there are so many diverse animal forms that exceptions can be found to almost any definition (Fig. 16–1). Still, there are some characteristics that describe at least most animals:

1. All animals are multicellular eukaryotes.
2. The cells of an animal exhibit a division of labor. In all but the simplest animals, cells are organized to form tissues, and tissues are organized to form organs. In most animal phyla specialized organ systems carry on specific functions.
3. Animals are heterotrophs; they ingest their food and then digest it inside the body, usually within a digestive system.
4. Most animals are capable of locomotion at some time during their life cycle. However, some animals, for example the sponges, are **sessile** (firmly attached to a substrate) as adults.
5. Most animals reproduce sexually, with large, nonmotile eggs and small, flagellated sperm. Sperm and egg unite to form a fertilized egg, or **zygote,** which goes through a series of embryonic stages before developing into a larva or immature form. Some animals can also reproduce asexually.

Figure 16–1 Despite their diversity, most members of the animal kingdom share several distinctive traits. (a) The nearly transparent body of this freshwater crustacean, *Simocephalus vetulus,* or water flea, shows a complex set of organ systems. (b) As heterotrophs, all animals must feed either on producers or on other animals that eat producers. The cheetah is an example of an animal that lives virtually entirely off other consumers. The young Thompson's gazelle is an example of an animal that relies directly on producers as its source of food. (a, Herman Eisenbeiss, Photo Researchers, Inc. b, E.R. Degginger.)

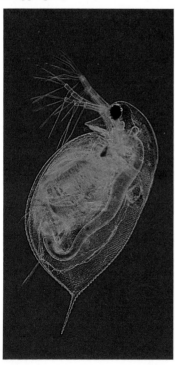

(a)

(b)

THEIR PLACE IN THE ENVIRONMENT

Animals are consumers. They feed upon a wide variety of organisms, but ultimately they depend upon producers for their raw materials, energy, and oxygen. Animals are also dependent upon decomposers for recycling nutrients.

Animals are distributed in virtually every corner of the earth, wherever plants are near to serve as their trophic base. Of the three environments—salt water, fresh water, and land—the sea is the most hospitable and has long been thought to be where life began. Seawater is isotonic to the tissue fluids of most marine animals, so there is little problem in maintaining fluid and salt balance. The buoyancy of seawater supports its inhabitants, and the temperature is relatively constant. **Plankton** (the organisms that are suspended in the water and float with its movement) consists of tiny animals, plants, and protists that provide a ready source of food.

ANIMAL RELATIONSHIPS

Most biologists agree that the evolutionary origin of animals is obscure. Animals are thought to have arisen from the Protista, but from which group no one knows. Similarly, the relationships among the various phyla are a matter of conjecture. However, there is no scarcity of hypotheses, and a few of the more widely held will be presented in this section.

The animal kingdom may be divided into two branches, or subkingdoms: the **Parazoa,** which are the sponges, and the **Eumetazoa,** which include all the other animals. This distinction is made because the sponges are so different that most biologists think they are not directly ancestral to any other animal phylum.

The Eumetazoa are often divided on the basis of symmetry. Two phyla, the cnidarians (jellyfish and relatives) and the ctenophores (comb jellies), are radially symmetrical and are included in the **Radiata.** In **radial symmetry** similar structures are regularly arranged as spokes from a central axis. Multiple planes can be drawn through the central axis, each dividing the organism into two mirror images. All other animals are bilaterally symmetrical (at least in their larval stages) and belong to the **Bilateria.** A bilaterally symmetrical animal is divided into roughly identical right and left halves when sliced down the midline of the body. See Figure 16–2 and Focus on Symmetry and Body Plan.

Acoelomates and Coelomates

A widely held system for relating the animal phyla to one another is based upon the type of body cavity, or **coelom.** To understand what a true coelom is, it is necessary to digress briefly into the animal's embryonic origins. The structures of most animals develop from three embryonic tissue layers, called **germ layers.** The outer layer, called the **ectoderm,** gives rise to the outer covering of the body and to the nervous system. The inner layer, or **endoderm,** lines the digestive tract. **Mesoderm,** the middle layer, extends between the ectoderm and endoderm and gives rise to most of the other body structures, including the muscles, bones, and circulatory system.

In the simplest Eumetazoa (the Cnidaria and Platyhelminthes) the body is essentially a double-walled sac surrounding a single digestive cavity, the **gastrovascular cavity,** with a single opening to the outside—the mouth. There is no body cavity, so these animals are referred to as **acoelomates** (Fig. 16–3). The more complex animals usually have a **tube-**

Figure 16–2 Types of body symmetry in animals. (*a*) In radial symmetry multiple planes can be drawn through the central axis, each dividing the organism into two mirror images. (*b*) Most animals exhibit bilateral symmetry. A midsagittal cut (a lengthwise vertical cut through the midline) divides the animal into right and left halves. The diagram also illustrates various ways in which the body can be sectioned so that its internal structure can be studied. Many cross sections and other types of sections are used in illustrations throughout this book to show relationships among tissues and organs.

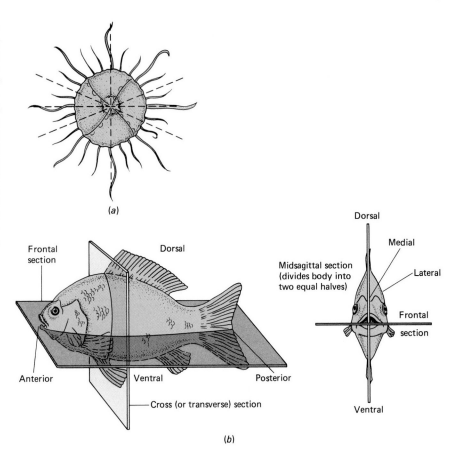

within-a-tube body plan. The inner tube, the digestive tract, is lined with tissue derived from endoderm and is open at both ends—the mouth and the anus. The outer tube or body wall is covered with tissue derived from ectoderm. Between the two tubes is a second cavity, the body cavity. If the

Focus on . . .

SYMMETRY AND BODY PLAN

Most animals exhibit **bilateral symmetry,** a type of symmetry in which the body can be divided through only one plane (which goes through the midline of the body) to produce roughly equivalent right and left halves that are mirror images. Cnidarians (jellyfish, sea anemones, and their relatives) and adult echinoderms (sea stars, sea urchins, and their relatives) have **radial symmetry.** In them, similar body parts are arranged around a central body axis. Radial symmetry is considered an adaptation for a sessile life-style, for it enables the organism to receive stimuli equally from all directions in the environment.

Bilateral symmetry is considered an adaptation to motility. The front, or **anterior,** end of the animal generally has a head, where sense organs are concentrated, and it is this end that receives most environmental stimuli. The **posterior,** or rear, end of the animal may be equipped with a tail for swimming, or it may just follow along.

To locate body structures, it is helpful to define some basic terms and directions (see Fig. 16–2). The back surface of an animal is its **dorsal** surface; the belly

side is its **ventral** surface. (In animals that stand on two limbs, as humans do, the term posterior refers to the dorsal surface and anterior to the ventral surface.) A structure is said to be **medial** if it refers to the midline of the body and **lateral** if it is toward one side of the body. For example, in a human the ear is lateral to the nose. The terms **cephalic** and rostral (and superior, in humans) refer to the head end of the body; the term **caudal** refers to structures closer to the tail. (In human anatomy, the term inferior is used to refer to structures located relatively lower in the body.)

In a bilaterally symmetrical animal we can distinguish three planes (flat surfaces that divide the body into specific parts). The **midsagittal** plane (or section) divides the body into equal right and left halves. Any section or plane cut parallel to the midsagittal plane is described simply as a **sagittal** section; it divides the body into unequal right and left parts. A **frontal** (or coronal) section, or plane, divides the body into anterior and posterior parts. A **transverse,** or **cross,** section cuts at right angles to the body axis.

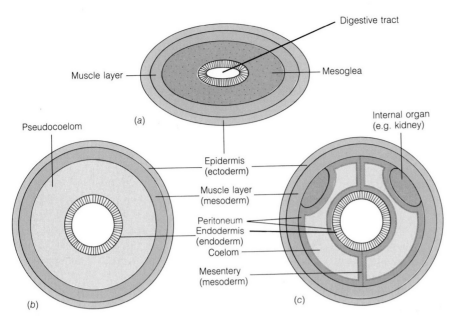

body cavity lies between the mesodermal surface and the surface of the endoderm, it is called a **pseudocoelom.** If the body cavity lies *within* the mesoderm, so that it is completely lined by mesodermal surfaces, the body cavity is a true **coelom.** Only animals with true coeloms are referred to as **coelomates.** The tree shown in Figure 16–4 indicates the relationships of the major phyla of animals based on the type of coelom they possess.

Protostomes and Deuterostomes

A different family tree is shown in Figure 16–5. This important scheme is based partly on the pattern of embryonic development. According to this view, animals are more properly divided into two groups: the proto-

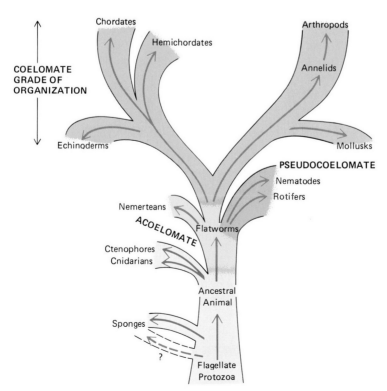

Figure 16–4 Proposed evolutionary relationships are illustrated by this phylogenetic tree indicating acoelomate, pseudocoelomate, and coelomate phyla. (After Barnes.)

Figure 16–5 A phylogenetic tree based on protostome-deuterostome characteristics. (After Barnes.)

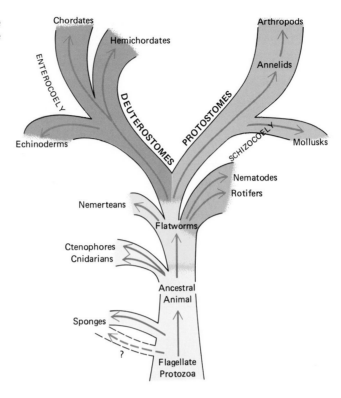

stomes and the deuterostomes. These groups reflect two main lines of evolution. Early during embryonic development a group of cells move inward to form an opening called the **blastopore.** In most of the mollusks, annelids, and arthropods, this opening develops into the mouth. This group makes up the **protostomes** (meaning "first, the mouth"). In echinoderms (for example, the sea star) and chordates (the phylum that includes the vertebrates), the blastopore develops into the anus. The opening that develops into the mouth forms later in development. These animals are the **deuterostomes** ("second, the mouth"). Moreover, among protostomes the coelom forms entirely from a cavity that develops inside the mesodermal tissue (schizocoely), whereas at least some deuterostomes form the coelom from pockets that develop in the wall of the gut (enterocoely). In this scheme the deuterostomes are thought to have diverged from the main protostome line at a point in evolution considerably after the flatworms.

PHYLUM PORIFERA (THE SPONGES)

Porifera, meaning "to have pores," aptly describes the sponges, which make up phylum Porifera (Fig. 16–6). A sponge resembles a sac perforated by tiny holes. Although the sponge is a multicellular organism, its cells are so loosely associated that they do not form definite tissues. Sponges occupy aquatic, mainly marine habitats. Living sponges may be bright red, orange, green, purple, or quite drab. They are generally asymmetrical and may be flat or shaped like fans, balls, or vases.

Sponges secrete skeletal structures called **spicules** and are divided into classes on the basis of the type of skeleton they secrete. Members of class Calcispongiae have a chalky skeleton of calcium carbonate spicules. Members of class Hexactinellida are glass sponges with skeletons of six-rayed silica spicules. Sponges that belong to class Demospongiae have a skeleton of spongin (a protein material) fibers. What we recognize as a bath sponge is actually a dried spongin skeleton from which all living tissue has been removed.

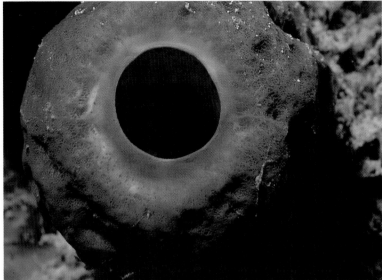

(b)

(a)

Figure 16–6 Sponges vary widely in size, shape, and color. (a) Tube sponge. (b) View of the open end (osculum) of a large sponge collected in the Virgin Islands. (a, Charles Seaborn. b, Robin Lewis, Coastal Creations.)

Body Plan of a Sponge

In a simple sponge, water enters through hundreds of tiny pores, passes into the central cavity, or **spongocoel,** and flows out through the sponge's open end (osculum) (Fig. 16–7). Water is kept moving by the action of flagellated cells that line the spongocoel. Each of these cells is equipped with a tiny collar (an extension of the cell membrane) that surrounds the base of the flagellum. Appropriately, these cells are known as **collar cells.** In some types of sponges the body wall is extensively folded and there are complicated system of canals.

Several types of cells make up the body of a sponge, and there is a division of labor, with certain cells specialized to function in nutrition, support, or reproduction. The collar cells, which make up the inner layer of the sponge, create the water current that brings food and oxygen to the cells and carries away carbon dioxide and other wastes. They also trap food particles. The outer layer of the sponge is composed of epidermal cells that are capable of contraction. Between the outer and inner cell layers is a gelatin-like layer (the mesoglea) supported by skeletal spicules. The mesoglea is not a layer of cells, but ameba-like cells wander about here, and some of these cells secrete the spicules.

Life-Style of a Sponge

Although larval sponges are ciliated and able to swim about, the adult sponge remains attached to some solid object on the sea bottom and is incapable of locomotion. Since the sponge cannot swim about in search of food, it is adapted for trapping and eating whatever food the sea brings to it. The sponge is a filter feeder. As water circulates through the body, food is trapped along the sticky collars of the collar cells. Digestion is intracellular, the job of individual cells. The ameba-like cells help dis-

Figure 16–7 (*a*) Part of a sponge colony. (*b*) A simple sponge cut open to expose its cellular organization. (*c*) A photomicrograph of a portion of a sponge (*Grantia*) showing spicules (the light structures). (Carolina Biological Supply Company.)

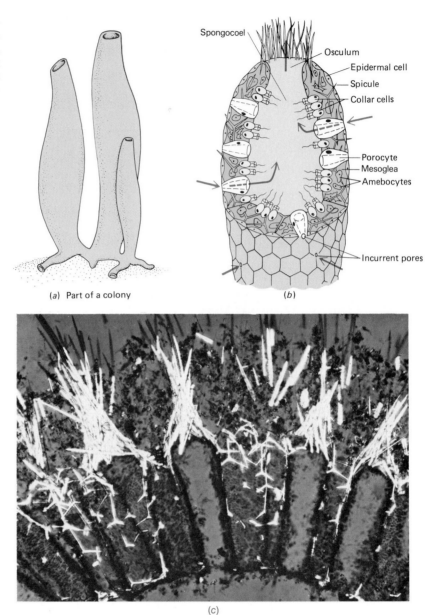

(*a*) Part of a colony

(*b*)

Spongocoel
Osculum
Epidermal cell
Spicule
Collar cells
Porocyte
Mesoglea
Amebocytes
Incurrent pores

(c)

tribute food to other cells of the sponge. Undigested food is simply eliminated into the water.

Oxygen from the water diffuses throughout the sponge. Respiration and excretion are carried on by individual cells. Each cell of the sponge body is irritable and can react to stimuli. However, there are no sensory or nerve cells that would enable the animal to react as a whole. Behavior appears limited to the basic metabolic necessities, such as procuring food and regulating the flow of water through the body. Pores and the large opening at the sponge end may be closed by contraction of surrounding cells.

Sponges can reproduce asexually. A small fragment or bud may break free from the parent sponge and give rise to a new sponge. Such fragments may remain to form a colony with the parent sponge. Sponges also reproduce sexually. Most sponges are **hermaphroditic,** meaning that the same individual can produce both egg and sperm. Some of the ameba-like cells develop into sperm cells, others into egg cells. Even hermaphroditic sponges can cross-fertilize with other sponges, however. Fertilization and early development take place within the jelly-like middle layer. Embryos eventually move into the spongocoel and leave the parent along with the stream of excurrent water. After swimming about for a while, the larva finds a solid object, attaches to it, and settles down to a sessile life.

Sponges have a remarkable ability to regenerate. When the cells of a sponge are separated from one another, they reaggregate to form a complete sponge again. When clusters of cells are isolated from one another in separate containers of seawater, each cluster will reorganize and regenerate to form a new sponge. If the disaggregated cells of two sponge species are mixed together, the cells sort themselves out and reorganize separate sponges of the original species.

PHYLUM CNIDARIA (COELENTERATES)

Most of the 10,000 or so species of phylum Cnidaria are marine. They are grouped in three classes: (1) **Class Hydrozoa,** including the hydras and the Portuguese man-of-war; (2) **Class Scyphozoa,** the jellyfish; and (3) **Class Anthozoa,** the sea anemones and true corals (Fig. 16–8). All the cnidarians have stinging cells called **cnidocytes,** from which they get

Figure 16–8 Representative cnidarians. (*a*) A member of class Hydrozoa, *Obelia* forms a colony of polyps. (*b*) *Cyanea capillata*, the lion's mane jellyfish, a member of class Scyphozoa. The bell of this jellyfish can measure 2 to 3 meters; its tentacles trail more than 10 meters. (*c*) This tree coral, *Dendronephthya*, photographed in the South Pacific, is an anthozoan. (*d*) Polyps from the coral *Montastrea cavernosa* extended for feeding. (*e*) Opal bubble coral, *Plerogyra sinuosa*, from the South Pacific. (*a*, Carolina Biological Supply Company. *b* and *d*, Charles Seaborn. *c* and *e*, Robin Lewis, Coastal Creations.)

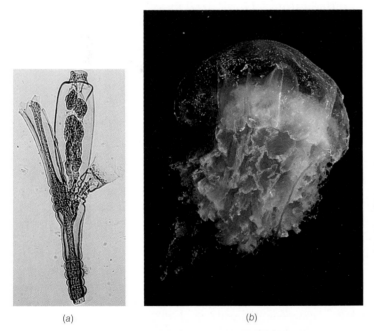

(a)　(b)

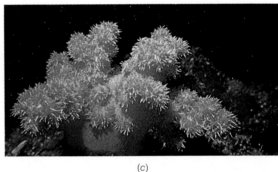

(c)

(d)

(e)

Figure 16-9 The polyp and medusa body forms characteristic of phylum Cnidaria are basically similar. (*a*) The polyp form as seen in *Hydra*. (*b*) The medusa form is basically an upside-down polyp, as shown by inverting a jellyfish. (*c*) In the anthozoan polyp the gastrovascular cavity is characteristically divided into chambers by vertical partitions.

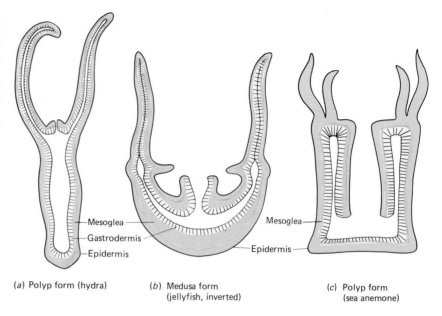

Mesoglea
Gastrodermis
Epidermis

Mesoglea
Epidermis

(*a*) Polyp form (hydra)　　(*b*) Medusa form (jellyfish, inverted)　　(*c*) Polyp form (sea anemone)

their name (cnidaria is from a Greek word meaning "sea nettles"). The cnidarian body is radially symmetrical and is organized as a hollow sac with the mouth and surrounding tentacles located at one end. The mouth leads into the digestive cavity, called the **gastrovascular cavity.** The mouth is the only opening into this cavity and so must serve for both ingestion of food and egestion of wastes. (The former name of this phylum, Coelenterata, was derived from the fact that the only cavity in the body serves as the digestive cavity; Greek *coel,* hollow, and *enteron,* gut).

Body Plan of a Cnidarian

Much more highly organized than the sponge, a cnidarian has two definite tissue layers. The outer **epidermis** and the inner **gastrodermis** are composed of several types of epidermal cells. These layers are separated by a gelatin-like **mesoglea,** which is not cellular, but sometimes contains a few cells.

Cnidarians have two body shapes, the polyp and the medusa (jellyfish) (Fig. 16-9). The **polyp** form, represented by *Hydra,* resembles an upside-down, slightly elongated jellyfish. Some cnidarians have the polyp shape during their larval stage and later develop into the **medusa** form. Though many cnidarians live a solitary existence, others group into colonies. Some colonies—for example, the Portuguese man-of-war—consist of both polyp and medusa forms.

Life-Style of a Hydra

The freshwater hydra is a solitary polyp seldom more than 1 centimeter long (Fig. 16-10). Although capable of locomotion, an adult hydra generally finds a rock, a twig, or even a leaf to which to attach and waits for dinner to come along. When a likely prospect happens to brush by one of its tentacles, the hydra's stinging cells respond. Coiled within each stinging cell is a "thread capsule," or **nematocyst.** Each stinging cell has a small projecting trigger on its outer surface that responds to touch and to chemicals dissolved in the water ("taste") and causes the nematocyst to fire its thread. Some types of nematocyst threads are sticky; others are long and coil around the prey; a third type is tipped with a barb or spine and can inject a protein toxin that paralyzes the prey. The tentacles encircle the prey and stuff it through the mouth into the gastrovascular cavity, where digestion begins. The partially digested fragments are taken up by

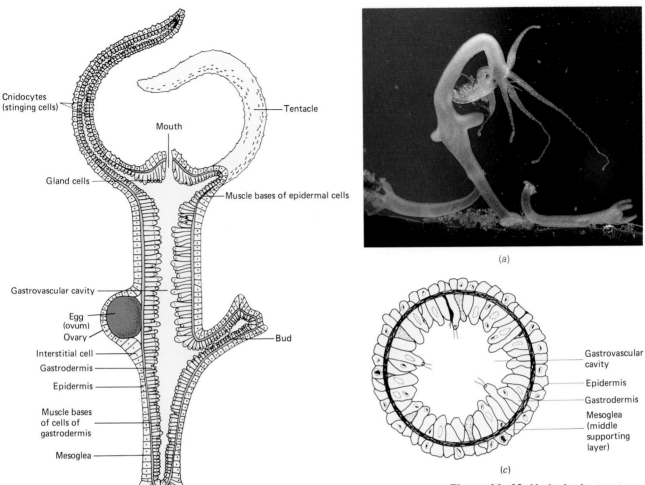

Cnidocytes (stinging cells)

Mouth

Tentacle

Gland cells

Muscle bases of epidermal cells

Gastrovascular cavity

Egg (ovum)

Ovary

Bud

Interstitial cell

Gastrodermis

Epidermis

Muscle bases of cells of gastrodermis

Mesoglea

Epidermal gland cells

(b)

(a)

Gastrovascular cavity

Epidermis

Gastrodermis

Mesoglea (middle supporting layer)

(c)

Figure 16–10 *Hydra* body structure. (a) A brown hydra, *Hydra oligactis*, capturing a small crustacean. Note the buds on the hydra's body. One bud has already detached as a separate animal. (b) This hydra is cut longitudinally to reveal its internal structure. Asexual reproduction by budding is represented on the animal's left. Sexual reproduction is represented by the ovary on the animal's right. Male hydras develop testes, which produce sperm. (c) Cross section of a hydra. (a, Tom Branch, Photo Researchers, Inc.)

pseudopods of the gastrodermis cells, and digestion is completed within food vacuoles in those cells.

Respiration and excretion occur by diffusion, for the body of a hydra is small enough that no cell is far from the surface. The cnidaria are the simplest animals to possess true nerve cells. These cells are simply arranged, forming irregular **nerve nets** connecting sensory cells in the body wall with muscle and gland cells. The coordination achieved is of the crudest sort; there is no aggregation of nerve cells to form a brain or nerve cord, and an impulse set up in one part of the body passes in all directions more or less equally.

Hydras reproduce asexually by budding during periods when environmental conditions are optimal. However, they differentiate as males and females in the fall or when pond water becomes stagnant. Females develop an ovary that produces a single egg, and males form a testis that produces sperm. After fertilization the egg becomes covered with a shell, leaves the parent, and remains within the protective shell throughout the winter.

PHYLUM PLATYHELMINTHES (FLATWORMS)

As their name implies, flatworms are flat, elongated, legless animals. The Platyhelminthes are divided into three classes: (1) class Turbellaria (Fig. 16–11), the free-living flatworms, including the planarian and its

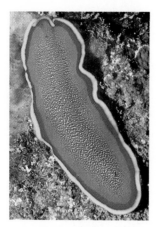

Figure 16–11 A marine turbellarian, *Pseudoceros fwerksii*, from Hawaii. Turbellarians are voracious predators, feeding primarily on other worms. (Charles Seaborn.)

relatives; (2) class Trematoda, the flukes, which are either internal or external parasites; and (3) class Cestoda, the tapeworms, the adults of which are intestinal parasites of vertebrates.

Getting a Head and Other Advances

Along with their bilateral symmetry, flatworms have a definite anterior and posterior end. This is a great advantage to any mobile organism. With a concentration of sense organs in the part of the body that first meets the environment, the animal is able to detect an enemy quickly enough to escape. The animal is also more likely to see or smell prey quickly enough to capture it. The beginning of **cephalization,** the development of a head, is an important advance in flatworms.

Flatworms have three definite tissue layers. In addition to an outer epidermis, derived from ectoderm, and an inner endodermis, derived from endoderm, the flatworm has a middle tissue layer, derived from mesoderm, which makes up most of the body. The flatworms are also the simplest animals that have well-developed **organs,** functional structures of two or more kinds of tissue. The hydra, recall, has loosely organized tissues, such as nervous tissue, but they are not organized into organs.

The simple brain consists of two masses of nervous tissue, called **ganglia,** in the head region. The ganglia are connected to two nerve cords that extend the length of the body. In this ladder-type nervous system, a series of nerves connect the cords like the rungs of a ladder.

Another important characteristic of the flatworms is their excretory system. It consists of two excretory tubes extending the length of the body and giving off branches called **protonephridia.** Each protonephridium ends in a **flame cell,** a collecting cell equipped with cilia that channels fluid containing wastes into the system of excretory tubules.

The gastrovascular cavity, when present, is often extensively branched. It has only one opening, the mouth, usually located on the middle of the ventral surface.

Life-Style of a Planarian

Planarians are free-living, mainly marine, flatworms, though some are freshwater and a few are terrestrial. The common planarians usually studied in biology laboratories are found burrowing in the mud at the edges of ponds or on the underside of rocks or leaves. The common American planarian, *Dugesia,* is about 15 millimeters long, with what appear to be crossed eyes and distinct "ears" called **auricles** (Fig. 16–12). The auricles actually serve as organs of smell, or more precisely, chemoreception.

Planarians are carnivorous and trap small animals in a mucous secretion. The digestive system consists of a single opening (the mouth), a pharynx, and a branched intestine. A planarian can project its pharynx (the first portion of the digestive tube) outward through its mouth, using it like a vacuum cleaner to suck up small pieces of the prey. Extracellular digestion takes place in the intestine by enzymes secreted by gland cells. Digestion is completed after the nutrients have been absorbed into individual cells. Undigested food is eliminated through the mouth, there being no separate anus. The highly branched gastrovascular cavity does help distribute food to all parts of the body, so that each cell is within range of diffusion.

The planarian's flattened body ensures that gases can also reach or leave all the cells by diffusion. There are no specialized respiratory or circulatory structures. Although some excretion takes place by diffusion, a very simple excretory system with flame cells is present.

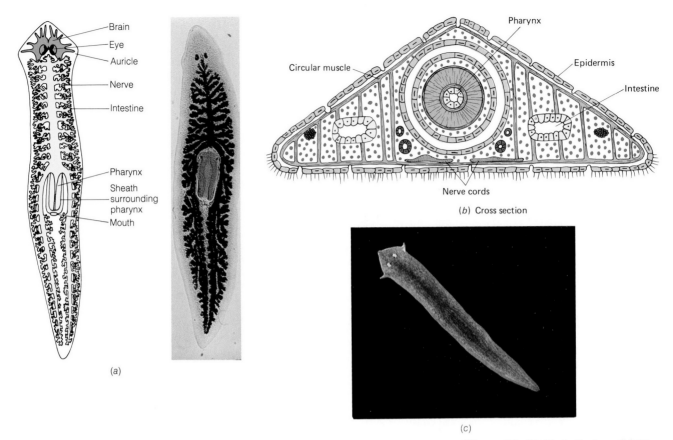

(b) Cross section

(a)

(c)

Figure 16–12 Turbellarians. (a) The common American planarian *Dugesia*. Note the branched intestine. (b) Cross section of a planarian. (c) A living dugesia. (a, Carolina Biological Supply Company. c, Tom Adams.)

Planaria can reproduce either asexually or sexually. In asexual reproduction an individual constricts in the middle and divides into two individuals. Each regenerates its missing parts. Sexually these animals are hermaphroditic. During the warm months of the year, each is equipped with a complete set of male and female organs. Two planaria come together in copulation and exchange sperm cells so that their eggs are cross-fertilized.

Adaptations for a Parasitic Life-Style—the Flukes and Tapeworms

The flukes parasitic in human beings are the blood flukes, widespread in China, Japan, and Egypt, and the liver flukes, common in China, Japan, and Korea. Other parasitic flukes infest hosts throughout the animal kingdom. Blood flukes of genus *Schistosoma* infect about 200 million people who live in tropical areas. Both blood flukes and liver flukes have complicated life cycles, involving a number of different forms, alternation of sexual and asexual generations, and parasitism on one or more intermediate hosts, such as snails and fishes (Fig. 16–13). When dams are built, the resulting marshy areas often provide habitats for the aquatic snails that serve as intermediate hosts in the fluke life cycle. The snails also thrive in rice paddies and natural ponds.

The body plan of the flukes is basically like that of the free-living flatworms, but they possess certain adaptations that enable them to succeed as parasites. They are equipped with one or more. suckers with which to cling to the host, and the mouth is anterior rather than ventral. The reproductive organs are extremely complex.

Tapeworms are long, flat, ribbon-like animals. As adults, members of the more than 1000 species live as parasites in the intestines of probably every kind of vertebrate, including humans. Sometimes thought of as

Figure 16–13 (a) The common liver fluke of sheep, *Clonorchis sinensis*. (b) Internal structure of a fluke. (c) Life cycle of a blood fluke, a schistosome. (a, Carolina Biological Supply Company.)

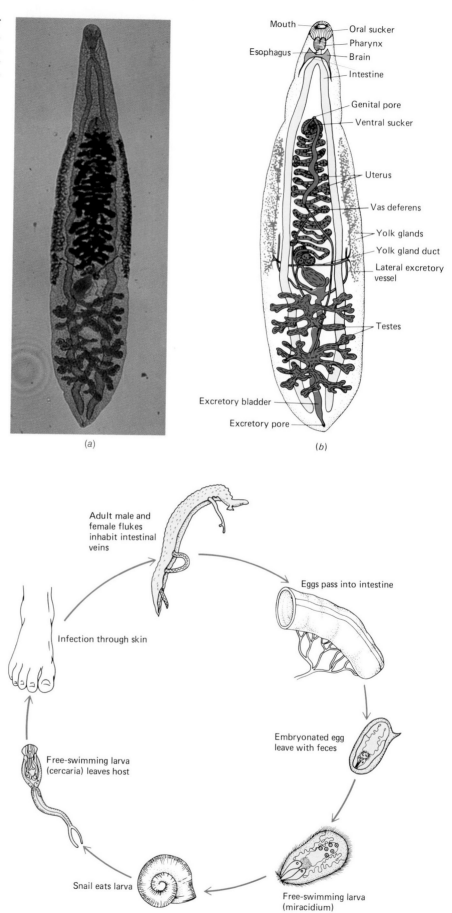

(a)

Mouth
Oral sucker
Pharynx
Esophagus
Brain
Intestine

Genital pore
Ventral sucker

Uterus

Vas deferens

Yolk glands
Yolk gland duct
Lateral excretory vessel

Testes

Excretory bladder
Excretory pore

(b)

Adult male and female flukes inhabit intestinal veins

Eggs pass into intestine

Infection through skin

Embryonated egg leave with feces

Free-swimming larva (cercaria) leaves host

Snail eats larva

Free-swimming larva (miracidium)

(c)

the most degenerate members of phylum Platyhelminthes, the tapeworms are actually strikingly specialized for their parasitic mode of life.

Among their many adaptations are suckers and sometimes hooks on the head (scolex), which enable the parasite to maintain its attachment to the host's intestine (Fig. 16–14). Their reproductive adaptations and abilities are extraordinary. The body of the tapeworm consists of a long chain of segments called **proglottids.** Each segment is an entire reproductive machine equipped with both male and female organs and containing as many as 100,000 eggs. Since an adult tapeworm may have 2000 segments, its reproductive potential is staggering. A single tapeworm may produce 600 million eggs in one year. Segments farthest from the tapeworm's head contain the ripest eggs and are shed daily, leaving the host's body with the feces.

Figure 16–14 Some species of tapeworms are armed with powerful hooks that enable these parasites to maintain their attachment to the host. (Courtesy of Fred H. Whittaker and Bioscience.)

Tapeworms do lack certain organs. They absorb food directly through their body walls from the host's intestine, and they have no mouths and no digestive systems of their own. Neither do they possess any sense organs or brains. Some tapeworms have rather complex life cycles, spending the larval stage within the body of an intermediate host and their adult lives within the body of a different, final host.

As an example, let us consider the life cycle of the beef tapeworm, so named because human beings become infected when they eat poorly cooked beef containing the larva (Fig. 16–15). The microscopic tapeworm larva spends part of its life cycle encysted within the muscle tissue of beef. When a human being ingests infected meat, the digestive juices break down the cyst, releasing the larva. The larva attaches itself to the intestinal lining and within a few weeks matures into an adult tapeworm, growing to a length of perhaps 50 feet. The parasite reproduces sexually within the human intestine and sheds proglottids filled with ripe eggs. Once established within a human host, the tapeworm makes itself very much at home and may remain there for the rest of its life, as long as ten years. A person infected with a tapeworm may suffer pain or discomfort,

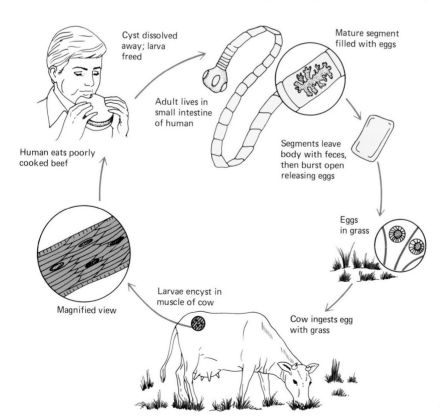

Cyst dissolved away; larva freed

Mature segment filled with eggs

Adult lives in small intestine of human

Segments leave body with feces, then burst open releasing eggs

Human eats poorly cooked beef

Eggs in grass

Magnified view

Larvae encyst in muscle of cow

Cow ingests egg with grass

Figure 16–15 Life cycle of the beef tapeworm, a parasitic flatworm.

increased appetite, weight loss, and other symptoms, or may be totally unaware of its presence.

For the life cycle of the beef tapeworm to continue, its eggs must be ingested by an **intermediate host,** a cow. (This requisite explains why we are not completely overrun by tapeworms, and why the tapeworm must produce millions of eggs to ensure that least a few will survive.) When a cow eats grass or other food contaminated with infected human feces, the eggs hatch in the cow's intestine. The larvae make their way into muscle, where they encyst, then await release by a **final host,** perhaps a human eating rare steak.

Two other tapeworms that infect human beings are the pork tapeworm and the fish tapeworm. The pork tapeworm infects us when we eat poorly cooked infected pork, and the fish tapeworm is contracted by ingesting raw or poorly cooked infected fish. Like most parasites, tapeworms tend to be species-specific; that is, each type can infect only certain species.

PHYLUM NEMERTINEA (PROBOSCIS WORMS)

Phylum Nemertinea is a relatively small group of animals (about 550 species) that are important to biologists mainly because they are the simplest animals to possess definite organ systems (Fig. 16–16). Almost all are marine, although a few inhabit fresh water or damp soil. They have long, narrow bodies, from 5 centimeters to 20 meters long. Some are vivid orange, red, or green, with black or colored stripes. Their most remarkable organ—the **proboscis,** from which they get their common name—is a long, hollow, muscular tube which they eject from the anterior end of the body and use for seizing food or for defense. The proboscis secretes mucus and may be equipped with a barb and with poison-secreting glands.

An important advance of the nemerteans is a tube-within-a-tube body plan. The digestive tract is a complete tube with a mouth at one end for taking in food and an anus at the other for eliminating wastes. This is, of course, in contrast to the cnidarians and planarians, whose food enters and wastes leave by the same opening.

A second advance of the nemerteans is the separation of digestive and circulatory functions. These animals are the most primitive organisms to have a separate circulatory system. It is rudimentary, consisting simply of muscular tubes—the blood vessels—extending the length of the body and connected by transverse vessels. Some of these primitive forms have red blood cells filled with hemoglobin, the same red pigment that transports oxygen in human blood. Nemerteans have no heart, and the blood is circulated through the vessels by the movements of the body and the contractions of the muscular blood vessels.

Despite the complexity of nemerteans, compared with the flatworms, they are not considered to be ancestral to other, more complex phyla. However they provide some idea of what the immediate ancestors of such organisms as the annelids and mollusks may have been like.

PHYLUM NEMATODA (ROUNDWORMS)

The **nematodes** are of great ecological importance because they are widely distributed in the soil, the sea, and fresh water, and because they are so numerous. A spadeful of soil may contain more than a million of these mainly microscopic white worms, which thrash around coiling and uncoiling. Though many are free-living (Fig. 16–17), others are important parasites in plants and animals. Among the human parasites be-

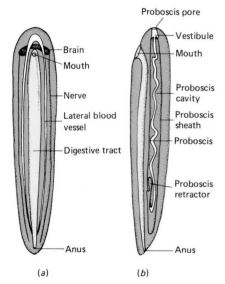

Figure 16–16 Structure of a typical nemertean. (a) Dorsal view of the digestive, circulatory, and nervous systems. (b) Lateral view of the digestive tract and proboscis. Note the complete digestive tract that extends from mouth to anus, giving this animal a tube-within-a-tube body plan.

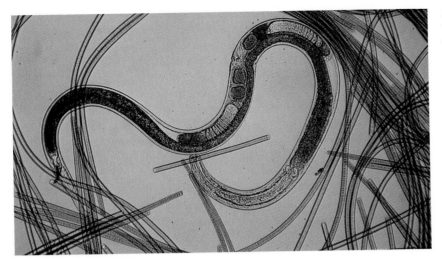

Figure 16–17 A free-living nematode among the cyanobacteria *Oscillatoria*, its typical food. (Tom Adams.)

longing to phylum Nematoda are the hookworms, the intestinal roundworm *Ascaris*, pinworms, trichina worms, and filaria worms.

The elongate, cylindrical, threadlike nematode body is pointed at both ends and covered with a tough cuticle (Fig. 16–18). Nematodes are the most primitive animals to have a body cavity, although it not a true coelom. Because the body cavity is not completely lined with mesoderm, it is referred to as a pseudocoelom. Like the proboscis worms, the nematodes exhibit bilateral symmetry, a complete digestive tract, three defi-

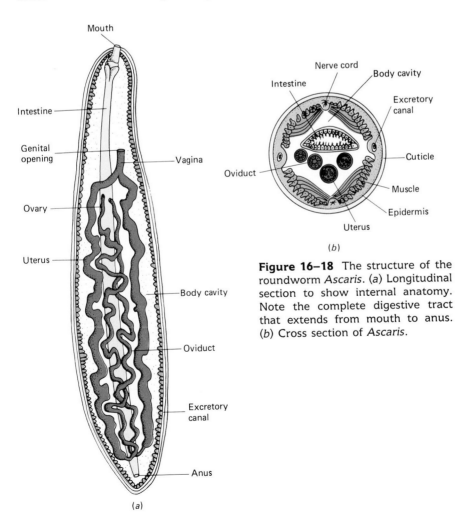

Figure 16–18 The structure of the roundworm *Ascaris*. (*a*) Longitudinal section to show internal anatomy. Note the complete digestive tract that extends from mouth to anus. (*b*) Cross section of *Ascaris*.

nite tissue layers, and definite organ systems. However, they lack circulatory structures. The sexes are usually separate, and the male is smaller than the female.

Ascaris lumbricoides is an example of a parasitic roundworm. A whitish worm about 25 centimeters long, *Ascaris* is a common intestinal parasite of human beings. The nematode spends its adult life in the human intestine, where it makes its living by sucking in partly digested food. Like the tapeworm, it must devote a great deal of effort to reproduction in order to ensure survival of its species. The sexes are separate, and copulation takes place within the host. A mature female may lay as many as 200,000 eggs a day.

Ascaris eggs leave the human body with the feces and, where sanitation is poor (that is, in most of the world), find their way into the soil. In many parts of the world human wastes are used as fertilizer, a practice that encourages the survival of *Ascaris* and many other human parasites. People are infected when they ingest ascaris eggs. The eggs hatch in the intestine, and the larvae then journey through the body before settling in the small intestine. During this migration the larvae can cause a great deal of damage to the lungs and other tissues. Very heavy infestations can also produce malnutrition or even intestinal blockage.

PHYLUM MOLLUSCA

Mollusks are among the most familiar invertebrates for almost everyone has walked along the shore collecting their shells. **Phylum Mollusca,** with its more than 100,000 living species, is the second largest of all the animal phyla. It includes the clams, oysters, octopods, snails, slugs, and the largest of all the invertebrates—the giant squid, which may achieve a weight of several tons (Fig. 16–19). Although most mollusks are marine, there are snails and clams that live in fresh water and many species of snails and slugs that inhabit the land.

Mollusks are soft-bodied animals usually covered by a dorsal shell. A broad, flat muscular **foot,** located ventrally, can be used for locomotion. Most of the organs make up a **visceral mass,** located above the foot. The **mantle,** a heavy fold of tissue, covers the visceral mass and, in most species, contains glands that secrete the shell. The mantle generally overhangs the visceral mass, forming a mantle cavity that often contains gills.

There are four principal classes of mollusks:

1. **Class Polyplacophora** consists of the chitons, primitive marine animals with segmented shells. Their heads are small, and they lack eyes and tentacles.
2. **Class Gastropoda** includes the snails, slugs, and whelks, animals whose bodies and shells are coiled. A few species, such as the nudibranchs and garden slugs, lack shells. Gastropods have well-developed heads with eyes and one or two pairs of tentacles. Most are marine, but some live in fresh water or on land. The land snails are among the few types of terrestrial invertebrates.
3. **Class Bivalvia** includes the clams and oysters and their relatives. (The clam is discussed in the next section.) Members of this class have two-part shells hinged dorsally and opening ventrally. The head is not distinct, but the foot is usually large and used for burrowing.
4. **Class Cephalopoda** includes the squids and octopods, which are active predators. These animals have large heads with long tentacles and conspicuous eyes. (See Focus on Adaptations of Squids and Octopods.)

(a)

(b)

(c)

Figure 16–19 There are many beautiful forms of mollusks. (*a*) A flamingo tongue, *Cyphoma gibbosum*, photographed in the Virgin Islands. (*b*) A bay scallop, *Argopecten irradians*, photographed in a sea grass bed in Tampa Bay. (*c*) Lightning whelk, *Busycon contrarium*. (*a* and *b*, Robin Lewis, Coastal Creations. *c*, James H. Carmichael, Jr., Coastal Creations.)

Body Plan of a Clam

The soft body of the bivalve is laterally compressed and completely enclosed by two shells hinged dorsally and opening ventrally (Fig. 16–20). This arrangement allows the hatchet-shaped foot to protrude ventrally for locomotion. Apertures are also present for flow of water into and out of the mantle cavity. Extensions of the mantle, called siphons, permit the animal to obtain water relatively free of sediment. There is an **incurrent siphon** for water intake and an **excurrent siphon** for water output. Large, strong muscles attached to the shell enable the animal to open or close its shell.

The inner pearly layer of the bivalve shell is secreted in thin sheets by the epithelial cells of the mantle. Composed of calcium carbonate and known as mother-of-pearl, it is valued for making jewelry and buttons. Should a bit of foreign matter lodge between the shell and the epithelium, the epithelial cells are stimulated to secrete concentric layers of calcium carbonate around the intruding particle. This is how a pearl is formed.

All of the organ systems typical of complex animals are present in the clam. The digestive system is a coiled tube extending from mouth to anus. The **open circulatory system** (typical of most mollusks) does not

Focus on . . .

ADAPTATIONS OF SQUIDS AND OCTOPODS

In contrast to most other mollusks, the cephalopods are active, predatory animals. They are fast-swimming organisms, adapted for an entirely different life-style than their filter-feeding relatives. Some biologists consider members of this group the most advanced of the invertebrates.

The octopus has no shell, and the shell of the squid is reduced to a small "pen" in the mantle. The cephalopod foot is divided into tentacles, ten in squids, eight in octopods. The tentacles, or arms, surround the central mouth of the large head. Cephalopods have large, well-developed eyes that form images. Although they develop differently, the eyes appear strikingly like vertebrate eyes and function in much the same way.

The tentacles of squids and octopods are covered with suckers for seizing and holding prey. The mouth is equipped with a rasplike structure (radula), something like a belt of teeth, and with two strong, horny beaks used to kill prey and tear it to bits. The mantle is thick and muscular and fitted with a funnel. By filling the mantle cavity with water and ejecting it through the funnel, the animal can attain rapid jet propulsion in the opposite direction.

Besides their speed, many cephalopods have two other important adaptations that enable them to escape from their predators, which include the whales and moray eels. One is the ability to confuse the enemy by rapidly changing colors. By expanding and contracting pigment cells in its skin, the cephalopod can display an impressive variety of mottled colors. Another defense mechanism is the ink sac, which produces a thick black liquid. This is released in a dark cloud when the animal is alarmed. While its enemy pauses, temporarily blinded and confused, the cephalopod easily escapes. The ink has been shown to paralyze the chemical receptors of some predators.

The octopus feeds on crabs and other arthropods, catching and killing them with a poisonous secretion of

Eastern Pacific octopus searching for mollusks on a coral head. (J.W. Porter, University of Georgia/BPS.)

its salivary glands. During the day, the octopus usually hides among the rocks, and in the evening it emerges to hunt for food. Its motion is incredibly fluid, giving little hint of the considerable strength in its eight arms.

Small octopods survive well in aquaria and have been studied extensively. They have a high degree of intelligence and can make associations among stimuli. Their very adaptable behavior resembles that of the vertebrates more closely than the stereotyped patterns of behavior seen in other invertebrates.

have a complete circuit of blood vessels. The heart pumps blood into a single blood vessel, which may branch into other vessels. Eventually, blood flows into a network of large spaces that make up the **hemocoel** (blood cavity). In this type of system the blood bathes the tissues directly. The blood finds its way into vessels that conduct it to the gills, where gas exchange takes place, then back to the heart.

Three pairs of ganglia serve as neural control centers. Neurons connect the ganglia with one another and with the various organs. Sense receptors include organs of equilibrium and cells sensitive to light and touch.

Life-Style of a Clam

The clam is a filter feeder. It obtains food by straining the seawater brought in over the gills by the siphon. The water is kept in motion by the beating of cilia on the surface of the gills. This stream of water carries food particles, trapped in the mucus that has been secreted by the gills, to the mouth.

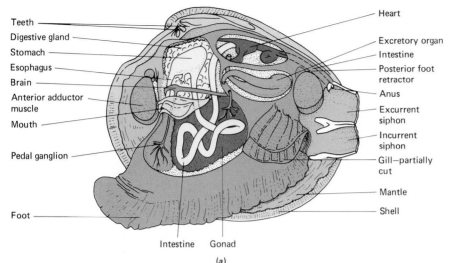

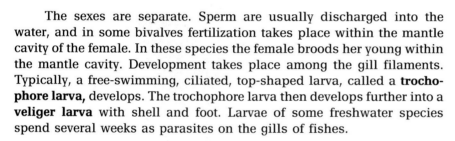

Teeth

Digestive gland

Stomach

Esophagus

Brain

Anterior adductor muscle

Mouth

Pedal ganglion

Foot

Intestine   Gonad

(a)

Heart

Excretory organ

Intestine

Posterior foot retractor

Anus

Excurrent siphon

Incurrent siphon

Gill—partially cut

Mantle

Shell

Stomach

Mouth

Nephridium

Mesodermal cells

Intestine

Anus

(b)

Figure 16–20 Body plan and life stages of a clam. (*a*) Internal anatomy of a clam. (*b*) Trochophore larva, the first larva stage of a marine mollusk. This type of larva is also characteristic of annelids.

The sexes are separate. Sperm are usually discharged into the water, and in some bivalves fertilization takes place within the mantle cavity of the female. In these species the female broods her young within the mantle cavity. Development takes place among the gill filaments. Typically, a free-swimming, ciliated, top-shaped larva, called a **trochophore larva**, develops. The trochophore larva then develops further into a **veliger larva** with shell and foot. Larvae of some freshwater species spend several weeks as parasites on the gills of fishes.

PHYLUM ANNELIDA (SEGMENTED WORMS)

Phylum Annelida are worms whose bodies are partitioned internally and externally into ringlike segments. This phylum includes the earthworms, leeches, and many marine and freshwater worms. Both the body wall and the internal organs are segmented. The segments are separated from one another by transverse partitions, called **septa.** The bilaterally symmetrical, tubular body may consist of more than 100 segments. Some structures, such as the digestive tract and certain nerves, run the length of the body, passing through successive segments. Other structures are repeated in each segment (Fig. 16–21). Segmentation is an advantage because not only is the coelom divided into segments, but each segment has its own muscles, enabling the animal to elongate one part of its body while shortening another part. (The annelid's hydrostatic skeleton, which enables it to do this, is discussed in Chapter 21.) In the annelid the individual segments are almost all alike, but in many segmented

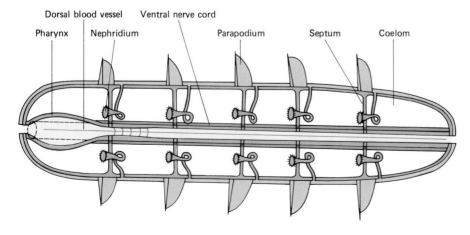

Dorsal blood vessel   Ventral nerve cord

Pharynx   Nephridium   Parapodium   Septum   Coelom

Figure 16–21 A generalized annelid. Note that the body is segmented, and there is a serial repetition of body parts.

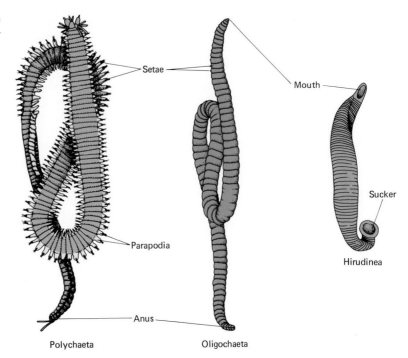

animals—arthropods and chordates—different segments and groups of segments are specialized to perform different functions. In some groups the specialization may be so pronounced that the basic segmentation of the body plan may be obscured.

Bristle-like structures called **setae** aid in locomotion. Annelids have well-developed coeloms, closed circulatory systems, and complete digestive tracts extending from mouth to anus. Respiration takes place through the skin or by gills. Typically, a pair of excretory structures called **nephridia** are found in each segment. The nervous system generally consists of a simple brain composed of a pair of ganglia and a double ventral nerve cord. A pair of ganglia and lateral nerves are repeated in each segment.

The 10,000 or so species are assigned to three main classes (Fig. 16–22).

1. **Class Polychaeta** ("many hairs") includes marine worms, such as sandworms and tubeworms (Fig. 16–23). These animals swim freely in the sea, burrow in the mud near the shore, or live in tubes formed by cementing bits of shell and sand together with mucus and other secretions from the body wall. Each body segment has a pair of paddle-shaped appendages called **parapodia** that extend laterally and function in locomotion. These fleshy structures bear many stiff setae. Most polychaetes have well-developed heads (or **prostomia**) bearing eyes and antennae. The prostomium may also be equipped with tentacles, bristles, and palps (feelers).

2. **Class Oligochaeta** includes the earthworm and some freshwater worms. These worms lack well-developed heads and have no parapodia. Fewer setae are present than in the polychaetes. The earthworm is discussed in the next section.

3. **Class Hirudinea** (Fig. 16–24) comprises the leeches. Most are blood-sucking parasites that inhabit freshwater, although some tropical species are terrestrial and drop on their hosts from foliage. Prominent muscular suckers, present at each end of the body, are used for clinging to the prey. Leeches lack both setae and appendages but have a very complex coelomic cavity.

(a)

(b)

Figure 16–23 Polychaete annelids. (*a*) *Hermodice carunculata*, a West Indian fireworm. (*b*) The Christmas tree worm, *Spirobranchus giganteus*, photographed in a Florida coral reef. (*a*, Charles Seaborn. *b*, James H. Carmichael, Coastal Creations, Inc.)

Body Plan of an Earthworm

Lumbricus terrestris, the common earthworm, is about 8 inches long. Its body is divided into more than 100 segments separated externally by grooves and internally by septa. The mouth is located in the first segment, the anus in the last.

The body wall contains an outer layer of circular muscles and an inner layer of longitudinal muscles (Fig. 16–25). The earthworm moves forward by contracting its circular muscles to elongate the body, grasping the ground or walls of the burrow with its setae (it has no parapodia), and then contracting its longitudinal muscles to draw the posterior end

Figure 16–24 *Helobdella stagnalis*, a blood-sucking leech of mammals. The dark area within its swollen body is recently ingested blood. Another species, the medicinal leech, *Hirudo medicinalis*, was widely used by physicians to absorb ''bad blood'' from their patients. Today, leeches are being used in some modern surgical procedures because they release a substance that prevents clotting in the immediate area where they are attached. (Tom Adams.)

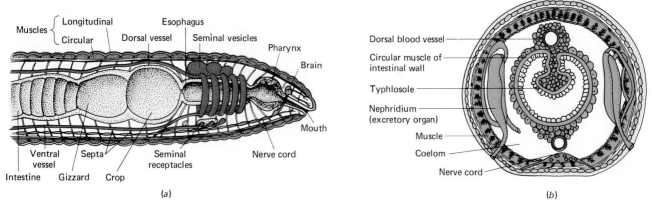

Figure 16–25 (a) Diagrammatic sagittal section of the anterior portion of an earthworm. (b) Cross section of the earthworm.

forward. Locomotion proceeds in waves. The muscles work against the hydrostatic skeleton provided by the coelomic fluid within the coelom of each segment. Each segment except the first bears four pairs of setae supplied with tiny muscles that can move the seta in and out and change its angle. Thus in earthworm locomotion the body is extended, anchored by the setae, then contracted.

The earthworm's soil meal is processed in the complex digestive system. Food is swallowed through the muscular pharynx and passes through the esophagus to the stomach. The stomach consists of two parts—a thin-walled crop where food is stored and a thick-walled muscular gizzard where food is ground to bits. The rest of the digestive system is a long, straight intestine where food is digested and absorbed. The surface area of the intestine is increased by a dorsal, longitudinal fold called the **typhlosole.** Wastes pass out of the intestine to the exterior through the anus.

The efficient, closed circulatory system consists of two main blood vessels that extend longitudinally. In the region of the esophagus, five pairs of blood vessels propel blood from the dorsal to the ventral blood vessel.

Gas exchange takes place through the earthworm's moist skin, and oxygen is usually transported by the respiratory pigment hemoglobin present in the blood plasma. The excretory system consists of paired organs repeated in almost every segment of the body. Each organ, called a nephridium, consists of a ciliated funnel (nephrostome) opening into the next anterior coelomic cavity and connected by a tube to the outside of the body (see Fig. 16–21). Wastes are removed from the coelomic cavity partly by the beating of the cilia and partly by currents set up by the contraction of muscles in the body wall.

The nervous system consists of a pair of **cerebral ganglia** that serve as a brain and a **subpharyngeal ganglion** just below the pharynx. A ring of nerve fibers connects the ganglia. From the lower ganglion a double ventral nerve cord extends beneath the digestive tract to the posterior end of the body.

Life-Style of an Earthworm

An earthworm literally eats its way through the soil, ingesting its own weight in soil and decaying vegetation every 24 hours. During this process the soil is turned and aerated, and nitrogen wastes from the earthworm enrich it. This is why earthworms are vital to the formation and maintenance of fertile soil.

Like other oligochaetes, earthworms are hermaphroditic. During copulation two worms, heading in opposite directions, press their ventral surfaces together and sperm from each worm passes into the female

Figure 16–26 Two earthworms, genus *Lumbricus*, copulating. (R.K. Burnard, Ohio State University/BPS.)

reproductive system of the other worm (Fig. 16–26). Development takes place within a cocoon produced by the parent.

PHYLUM ARTHROPODA (ANIMALS WITH JOINTED FEET)

Arthropods are the most diverse and biologically successful of all animals. There are more of them (about 800,000 described species), they live in a greater variety of habitats, and they can eat a greater variety of food than the members of any other phylum (Fig. 16–27).

Arthropods have paired, jointed appendages from which they get their name (arthropod means "jointed foot"). These function as swimming paddles, walking legs, mouth parts, or accessory reproductive organs for transferring sperm. Another important arthropod characteristic is the hard, armor-like **exoskeleton** composed of chitin that covers the entire body and appendages. The exoskeleton provides protection against predators and against excessive loss of moisture. It also gives support to the underlying soft tissues. Distinct muscle bundles somewhat comparable to individual vertebrate muscles attach to the inner surface of the exoskeleton. These act upon a system of levers that permit the extension and flexion of parts at the joints. The exoskeleton has certain disadvantages, however. Body movement is somewhat restricted, and in order to grow, the arthropod must shed this outer shell periodically and grow another larger one. This process, known as **molting,** leaves the arthropod temporarily vulnerable to predators.

The arthropod body, like that of the annelid, is segmented. In some arthropod classes, however, segments become fused together or lost during development. The bodies of most arthropods are divided into three regions: the head, thorax, and abdomen. There is an incredible range of variations in body plan and in the shape of the jointed appendages in the numerous species.

Most of the aquatic arthropods have a system of gills for gas exchange. The land forms, in contrast, usually have a system of fine, branching air tubes, called **tracheae,** which conduct air to the internal organs. A few (some spiders and land crabs) have structures comparable to lungs. Arthropods have a variety of well-developed sense organs: complicated eyes, such as the compound eyes of insects; organs of hearing; organs in the antennae sensitive to touch and chemicals; and cells on the surface of the body that are sensitive to touch.

The open circulatory system includes a dorsal, tubular heart that pumps blood into a dorsal artery, and sometimes other arteries. From the arteries blood flows into large spaces, which collectively make up the hemocoel. Blood in the hemocoel bathes the tissues directly. Eventually

Figure 16–27 The arthropods are considered the most successful animals. (*a*) The Merostomata include only a few closely related species that have survived to the present day. Seasonally, horseshoe crabs, *Limulus polyphemus,* return to beaches for mating. (*b*) The vast majority of the arachnids prey on other small invertebrates. In this photograph a garden spider is shown wrapping its prey, a webworm. The worm has already been killed or paralyzed by the spider's venomous bite. (*c*) Most members of class Crustacea are aquatic or semiaquatic. Here a banded coral shrimp, *Stenopus hispidus,* is eating another shrimp. (*d*) A May beetle, *Melolontha melolontha,* preparing for a landing. A major reason for the success of the insects is their ability to fly. This has allowed them to become widely dispersed through almost all terrestrial environments. (*e*) A millipede, a member of class Diplopoda. (*a,* Peter J. Bryant, University of California, Irvine/BPS. *b,* E.R. Degginger, *c,* James Carmichael, Coastal Creations. *d,* Stephen Dalton, Photo Researchers, Inc. *e,* Carolina Biological Supply Company.)

blood finds its way back into the heart through openings, referred to as **ostia,** in its walls.

There is much disagreement concerning arthropod classification. Here we will use a scheme that divides the phylum into three living subphyla.

1. In **subphylum Chelicerata** the arthropod body is divided into a cephalothorax (fused head and thorax) and an abdomen. There

are no antennae and no chewing mandibles, mouthparts that are characteristic of other arthropod subphyla. Instead, the first pair of appendages, immediately anterior to the mouth, are the **chelicerae,** used to manipulate food and pass it to the mouth. The second pair of appendages, called **pedipalps,** are modified to form different functions in various groups. Posterior to the pedipalps there are usually four pairs of legs.

Two classes in this subphylum are **class Merostomata** and **class Arachnida.** The only living merostomes are the horseshoe crabs. The 60,000 or so species of class Arachnida include the spiders, scorpions, mites, ticks, and harvestmen (daddy long-legs). Arachnids have six pairs of jointed appendages. The first pair are the chelicerae, fanglike structures used to penetrate prey and suck out its body fluids. In some arachnids the chelicerae are used to inject poison into the prey. The second pair of appendages are the pedipalps, used by spiders to hold and chew food, and modified as sense organs for tasting the food in some species. The other four pairs of legs are used for walking. Most arachnids are carnivorous and prey upon insects and other small arthropods (Fig. 16–27). Many types of arachnids have glands that secrete silk, used for making webs.

2. **Subphylum Crustacea** includes the lobsters, crabs, and shrimp. These arthropods have **biramous appendages**—appendages with two jointed branches at their ends. They also are the only group to have two pairs of **antennae,** sensory organs for touch and taste. Crustaceans are also characterized by **mandibles,** located on each side of the ventral mouth and used for biting and grinding food. There are usually five pairs of walking legs. Other appendages may be specialized for swimming, sperm transmission, carrying eggs and young, or sensation.

3. The insects, centipedes, and millipedes are grouped together in **subphylum Uniramia** because they all possess unbranched (uniramous) appendages. They also bear only a single pair of antennae rather than two pairs, as in crustaceans. The insects belong to **class Insecta. Class Chilopoda** includes the centipedes (hundred-legged worms), and **class Diplopoda** includes the millipedes (thousand-legged worms). Both centipedes and millipedes have heads and elongated bodies with many segments, each bearing legs. Each centipede has one pair of legs on each segment, but a millipede has two pairs of legs per segment.

The Insects—Secrets of Success

With more than 750,000 described species, the insects are the most successful group of animals on our planet in terms of number of species, as well as number of individuals and diversity. Insects are primarily terrestrial animals, but some species live in fresh water, a few are truly marine, and others inhabit the shore between the tides.

An insect may be described as an **articulated** (jointed), **tracheated** (having tracheal tubes for gas exchange), **hexapod** (having six feet). The insect body consists of three distinct parts—**head, thorax,** and **abdomen.** Three pairs of legs and usually two pairs of wings emerge from the thorax. One pair of antennae protrude from the head. A complex set of mouth parts are present and may be adapted for piercing, chewing, sucking, or lapping. Excretion is accomplished by two or more slender **Malpighian tubules,** which receive metabolic wastes from the blood and, after concentrating them, discharge them into the posterior part of the intestine.

Figure 16–28 Life cycle of the monarch butterfly. (*a*, E.R. Degginger. *b–e*, courtesy of Leo Frandzel.)

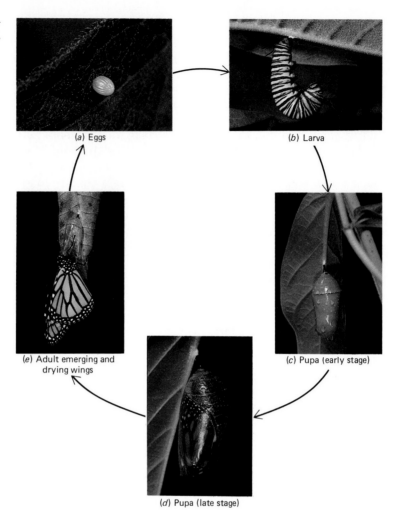

(*a*) Eggs

(*b*) Larva

(*e*) Adult emerging and drying wings

(*c*) Pupa (early stage)

(*d*) Pupa (late stage)

The sexes are separate and fertilization takes place internally. During development there are several molts. In some orders there are several developmental stages, called **nymphal stages,** and gradual **metamorphosis** (change in body form) to the adult form. In others there is a **complete metamorphosis** with four distinct stages in the life cycle: egg, larva, pupa, and adult (Fig. 16–28).

There are more species of insects than of all other classes of animals combined. What they lack in size, insects make up in sheer numbers. It has been calculated that all the insects in the world would weigh more than all of the remaining terrestrial animals on earth. What are the secrets of insect success? One important factor is their body plan, which can be modified and specialized in so many ways that insects have been able to adapt to an incredible number of life-styles. They have filled almost every variety of ecological niche. One of their most important adaptations is their ability to fly. Unlike other invertebrates, which creep slowly along (or under) the ground, the insects fly rapidly through the air. Their wings and their small size facilitate their wide distribution.

The insect body is well protected by the tough exoskeleton, which also helps prevent water loss by evaporation. Other protective mechanisms include mimicry, protective coloration, and aggressive behavior. Metamorphosis divides the insect life cycle into different stages, a strategy that has the advantage of placing larval forms into their own niches so that in most cases they do not have to compete with adults for food or habitats.

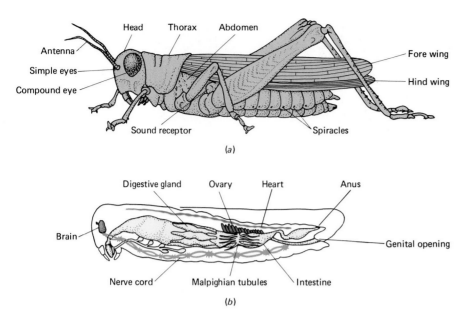

Figure 16–29 (a) External anatomy of the grasshopper. Note the three pairs of articulated legs. (b) Internal anatomy of the grasshopper.

Life-Style of a Grasshopper

We will consider the structure and activities of the grasshopper as representative of the insect world (Fig. 16–29). The grasshopper body consists of head, thorax, and abdomen. Two pairs of wings and three pairs of legs are present. Its jointed legs provide for much greater ability to maneuver than is possible for an annelid. A complex set of mouthparts is adapted in the grasshopper for chewing. In other insects mouthparts may be specialized for sucking or lapping.

The leaves ingested by the grasshopper are processed in its complete digestive tract. Absorbed nutrients are distributed by an open circulatory system. A tubular heart pumps blood into the head region through a large artery. The blood then flows freely in the hemocoel, bathing the tissues directly. Eventually the blood finds its way back into the heart chamber and is pumped out again.

Air enters the tracheae through tiny openings, called **spiracles,** that pierce the body wall. The Malpighian tubules remove wastes from the blood in the hemocoel and empty them into the intestine. This excretory system is well adapted for water conservation, and wastes are discharged as rather dry crystals.

The simple brain is continuous with a paired **ventral nerve cord.** Like other insects, the grasshopper has efficient sense organs. Prominent among these are two types of eyes, compound eyes for image formation and simple eyes for light perception. A well-developed endocrine system regulates growth and development. Very complex behavior patterns are known in insects, but these operate on an instinctive or programmed level (see Chapter 33).

The male grasshopper is equipped with a copulatory organ for transferring sperm into the female reproductive tract. The adults usually die soon after the eggs are laid. Grasshoppers do not undergo complete metamorphosis. The young grasshoppers resemble their parents but lack wings. They gradually grow in size and develop into adults. As in other arthropods, the grasshopper periodically outgrows its nonliving exoskeleton. At such times the skeleton is shed and replaced by a new, larger one.

Figure 16–30 Some representatives of the five classes of phylum Echinodermata. (*a*) The golden crinoid, a feather star, from the Pacific Ocean. (*b*) Blue sea star, *Linckia laevigate* (class Asteroidea), of the South Pacific. (*c*) Brittle stars (Ophiuroidea) living on the surface of a sponge. (*d*) A slate pencil urchin, *Heterocentrotus mammilatus* (Echinoidea), photographed in a Hawaiian coral reef. (*e*) A sea cucumber, *Parastichopus californica*, from the waters of the Pacific Northwest. (*a*, Robin Lewis, Coastal Creations. *b–e*, Charles Seaborn.)

PHYLUM ECHINODERMATA (SPINY-SKINNED ANIMALS OF THE SEA)

All of the members of phylum Echinodermata inhabit the sea. About 6000 living and 20,000 extinct species have been identified. The living species are divided into five principal classes (Fig. 16–30). **Class Crinoidea** includes the sea lilies and feather stars; **class Asteroidea,** the sea stars; **class Ophiuroidea,** the brittle stars; **class Echinoidea,** the sea urchins and sand dollars; and **class Holothuroidea,** the sea cucumbers.

The echinoderms are in many ways unique in the animal kingdom. Although their larvae have bilateral symmetry, the adults have **pentaradial symmetry.** This means that the body is arranged in five parts around a central disc, where the mouth is located. A thin, ciliated **epider-**

mis covers the endoskeleton, which consists of small calcareous plates (composed of $CaCO_3$), typically bearing spines that project outward. The phylum name Echinodermata, meaning "spiny-skinned," reflects this trait.

A characteristic found only in echinoderms is the **water vascular system,** a network of canals through which seawater circulates. Branches lead to numerous tiny **tube feet,** which extend when filled with fluid and serve in locomotion, obtaining food, and in some forms, gas exchange. The water vascular system is a hydraulic system. To extend a foot, a rounded muscular sac **(ampulla)** at the upper end contracts forcing water into the tube of the foot. At the bottom of the foot is a suction type structure that adheres to the substratum. The foot can be withdrawn by contraction of muscles in its walls, which forces water back into the ampulla.

Echinoderms have well-developed coeloms, complete digestive systems, but only rudimentary circulatory systems and no specialized excretory structures. There are a variety of respiratory structures in the various classes, including the dermal gills in the sea stars and respiratory trees in sea cucumbers. The nervous system is simple, usually consisting of nerve rings about the mouth with radiating nerves.

Life-Style of a Sea Star

A **sea star** (or starfish) consists of a central disc from which radiate 5 to 20 or more **arms,** or **rays** (Fig. 16–31). In the center of the underside of the disc is the mouth. The endoskeleton consists of a series of calcareous plates that permit some movement in the arms. Around the base of the delicate skin gills, used in gas exchange, are tiny pincer-like spines **(pedicellaria).** Operated by muscles, these keep the surface of the animal free of debris and the larvae of other marine organisms.

Most sea stars are carnivorous, feeding upon crustaceans, mollusks, annelids, and even other echinoderms. To attack a clam or oyster, the sea star mounts it, assuming a humped position as it straddles the edge opposite the hinge. Then, with its tube feet attached to the two shells, it begins to pull. The sea star uses many of its tube feet at a time but can change and use new groups when some tube feet get tired. By applying a steady pull on both shells over a long period, the sea star tires the powerful muscles of the clam so that they are forced to relax, opening the shell.

Figure 16–31 (*a*) The sea star *Asterias* viewed from above with the arms in various stages of dissection. (1) Upper surface with a magnified detail showing the features of the surface. The end is turned up to show the tube feet on the lower surface. (2) Arm is shown in cross section. (3) Upper body wall of arm has been removed. (4) The upper body wall and digestive glands have been removed, and the ampullas and plates are shown in magnified view. (5) All the internal organs have been removed except the retractor muscles, showing the inner surface of the lower body wall. (*b*) Photomicrograph showing spines and pedicellaria on surface of an echinoderm.

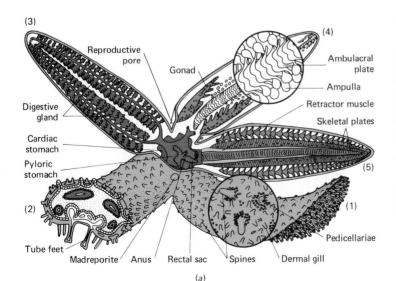

(a)

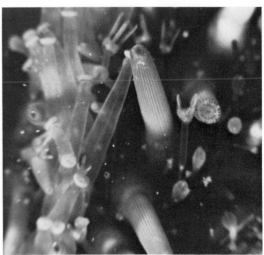

(b)

To begin its meal, the sea star projects its stomach through its mouth and into the soft body of its prey. Digestive enzymes are secreted into the shellfish so that it is partly digested while still in its own shell. The soft parts of the clam are digested to the consistency of a thick soup and pass into the sea star's body for further digestion by enzymes secreted from glands located in each arm. Its water vascular system does not enable the sea star to move rapidly, but since it usually preys upon slow-moving or stationary clams and oysters, speed of attack is not necessary, as it is for most predators.

The blood-circulatory system in sea stars is poorly developed and probably of little help in circulating materials. Instead, this function is assumed by the coelomic fluid, which fills the large coelom and bathes the internal tissues. Metabolic wastes pass to the outside by diffusion. The nervous system consists of a ring of nervous tissue encircling the mouth and a nerve cord extending from this into each arm. There is no aggregation of nerve cells that could be called a brain.

PHYLUM CHORDATA

Phylum Chordata, the phylum to which we humans belong, contains three subphyla. **Subphylum Urochordata,** the tunicates, includes the sea squirts, filter-feeding marine animals. **Subphylum Cephalochordata** consists of the lancelets, small, translucent, fishlike animals (Fig. 16–32). **Subphylum Vertebrata** consists of the animals with backbones.

Chordates share three characteristics that distinguish them from all other groups (Fig. 16–33):

1. *All chordates have a notochord during some time in their life cycle.* The notochord is a dorsal longitudinal rod that is firm yet flexible and supports the body.
2. *All chordates have a dorsal tubular nerve cord.* The nerve cord differs from that of other animals not only in its position but in being single rather than double, and hollow rather than solid.
3. *All chordates have pharyngeal gill grooves during some time in their life cycle.* In some chordates the grooves perforate and become functional gill slits, but in terrestrial animals they become modified to form entirely different structures more suitable for life on land.

(a)

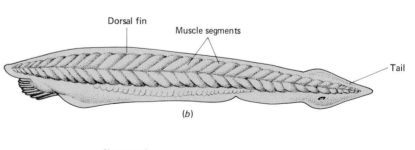

(b)

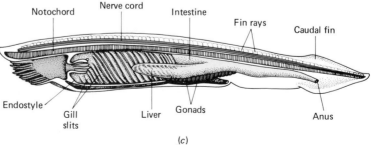

(c)

Figure 16–32 Nonvertebrate chordates. (*a*) Clear tunicates, *Clavelina picta*, photographed in the Virgin Islands. (*b*) External view and (*c*) longitudinal section of amphioxus, a member of the subphylum Cephalochordata. (*a*, courtesy of Robin Lewis, Coastal Creations.)

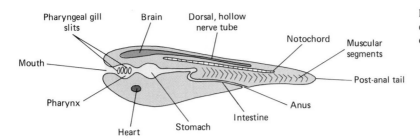

Figure 16–33 A generalized chordate illustrating the main chordate characteristics.

What Is a Vertebrate?

Vertebrates are chordates but are distinguished from all other chordates in having a backbone, or **vertebral column,** which forms the chief skeletal axis of the body. This flexible support develops around the notochord and largely replaces the notochord in most species. The vertebral column consists of cartilaginous or bony segments called **vertebrae.** Dorsal projections of the vertebrae enclose the nerve cord along its length. Anterior to the vertebral column, a **cranium,** or braincase, encloses and protects the brain, the enlarged anterior end of the nerve cord.

The cranium and vertebral column are part of the endoskeleton, which, in contrast to the nonliving exoskeleton of invertebrates, is a living tissue that can grow with the animal. Another vertebrate characteristic is *pronounced* cephalization. Although not all of these characteristics are exclusive to vertebrates, these animals also have closed circulatory systems with two-, three-, or four-chambered ventral hearts; paired kidneys; complete digestive tracts and large digestive glands (liver and pancreas); muscles attached to the skeleton for movement; brains that are regionally differentiated for specialized functions; 10 or 12 pairs of cranial nerves that emerge from the brain; an autonomic division of the nervous system that regulates involuntary functions of internal organs; well-developed organs of special sense (eyes; ears that serve as organs of equilibrium and in some, for hearing as well; organs of smell and taste); two pairs of appendages; and separate sexes.

The vertebrates are less diverse and much less numerous and abundant than the insects but rival them in their adaptation to an enormous variety of life styles. Most exceed the insects in the ability to receive stimuli and react appropriately to them. The 43,000 or so living species are assigned to seven classes: class Agnatha, the jawless fish, such as lamprey eels; class Chondrichthyes, the sharks and rays with cartilaginous skeletons; class Osteichthyes, the bony fish; class Amphibia, frogs, toads, and salamanders; class Reptilia, lizards, snakes, turtles, and alligators; class Aves, birds; and class Mammalia, the mammals.

Fish

The earliest known fish were small, armored, jawless freshwater fish that lived on the bottom and filtered their food from the water. Today, the only living members of **class Agnatha** are the lamprey eels and hagfishes (Fig. 16–34). Many species of adult lampreys are parasites on other fish; they are the only parasitic vertebrates. The **placoderms,** a group of primitive jawed fishes that lived during the Paleozoic era and are now extinct, are thought to be the ancestors of both the cartilaginous and the bony fish.

Cartilaginous fish **(class Chondrichthyes)** include sharks, rays, and skates (Fig. 16–35). As their name implies, their skeletons are composed of cartilage. The dogfish shark is commonly used in biology classes to demonstrate basic vertebrate characteristics in a simple, uncomplicated form.

(a)

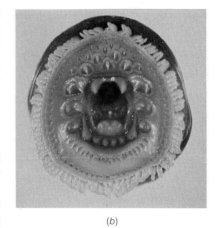

(b)

Figure 16–34 Class Agnatha. (*a*) A West Coast sea lamprey holds onto a rock with its mouth. (*b*) Suction-cup mouth of adult lamprey. Note the rasplike teeth. (*a*, Carolina Biological Supply Company. *b*, courtesy of Dr. Kiyoko Uehara, *Cell and Tissue Research.*)

(a)

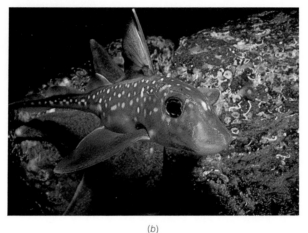

(b)

Figure 16–35 Chondrichthyes. (a) Dorsal view of a skate, *Raja binoculara.* (b) Ratfish, *Hydrolagus colliei.* (Charles Seaborn.)

The fish most familiar to us are the bony fish (class Osteichthyes) (Fig. 16–36). Numbering about 20,000 species, the bony fish are the most numerous vertebrates. They vary greatly in color and shape and range in size from the Philippine goby, which is only about 10 millimeters (0.4 inch) long, to the ocean sunfish, which may reach 907 kilograms (2000 pounds). The bodies of most bony fish are covered with overlapping bony scales (which develop from the dermis of the skin). Most bony fish have both median and paired fins, with fin rays of cartilage. A lateral protective flap of the body wall, the **operculum,** extends posteriorly from the head and covers the gills.

Bony fish generally are oviparous (that is, they lay eggs) and fertilize their eggs externally. Perhaps because the eggs and young offspring so often become food for other animals, fish lay impressive numbers of eggs. The ocean sunfish is said to lay over 300 million! Many species of fish build nests for their eggs and even watch over them.

Biologists think that the fishes diverged into two major groups during the Devonian period. The **ray-finned fishes** gave rise to most modern bony fishes, and the **lobe-finned fishes,** which include the lungfishes, are generally credited with giving rise to the land vertebrates (Fig. 16–37). It is thought that there were frequent seasonal droughts during the Devonian period. Ponds dried up, and fishes with some adaptation for breath-

Figure 16–36 The lionfish, *Pterois volitans*, is a native of Pacific coral reefs. The first several rays of its dorsal fins are armed with a toxin (similar in action to cobra venom) poisonous to humans as well as other animals. The lionfish is generally sluggish. The long filaments of its other fins, along with the flesh surrounding its mouth, serve as both camouflage and a means of attracting its prey. (Charles Seaborn.)

Figure 16–37 Ancestors of this coelacanth are thought to have given rise to the amphibians. The paired fins show the basic plan of a jointed series of bones that could evolve into the limbs of a terrestrial vertebrate. Living coelacanths are difficult to observe. Its members (genus *Latimeria*) inhabit deep ocean waters; when brought to the surface, the fish cannot survive the change in atmospheric pressure. (Peter Scoones, Seaphot)

ing air, such as lungs, had a tremendous advantage for survival. Their fleshy lobe fins could support their weight and enabled them to emerge onto dry land and make their way to another pond or stream. This ability to move about, however awkwardly, on dry land also gave these animals the advantage of a new food source—terrestrial insects or vegetation. Natural selection could have favored those individuals best adapted for making their way on land, resulting ultimately in the evolution of the amphibians.

Even the ancestors of the ray-finned fishes are believed to have had lungs, but in them the lungs became modified as swim bladders, hydrostatic organs that may also store oxygen in some species. By secreting gases into the swim bladder or absorbing gases from it, the fish can change the density of its body and so hover at a given depth of water.

Amphibians

Modern **amphibians** include the frogs, toads, salamanders, and the legless, wormlike caecilians (Fig. 16–38). Although some amphibians are quite successful as land animals and can live in dry environments, most spend at least their early lives in an aquatic environment and return to the water to reproduce. Eggs and sperms are generally laid in the water, and the embryos develop into larvae, which in the case of frogs and toads are called **tadpoles.** These have tails and gills and feed on aquatic plants. After a time the larva undergoes metamorphosis. The gills and gill slits

Figure 16–38 Modern amphibians. (*a*) Eastern spadefoot toad, *Scaphiopus holbrooki*. (*b*) The red salamander, *Pseudotriton ruber*, belongs to the family Plethodontidae. The plethodonts spend their entire lives in fairly moist or humid environments. They have lost their lungs and now rely almost entirely on their moist skin and the membranes that line their mouth and pharynx as organs for gas exchange. The red salamander is common in the eastern United States. (*a*, James H. Carmichael, Coastal Creations. *b*, Carolina Biological Supply Company.)

(*a*)

(*b*)

disappear, the tail is resorbed, and the limbs emerge. When these structural modifications are complete, the amphibian can move onto the land.

Adult amphibians do not depend solely on their primitive lungs for the exchange of respiratory gases. Their moist skin, which lacks scales and is richly supplied with blood vessels, also serves as a respiratory surface. Glands within the skin secrete mucus, which helps keep the body moist and also makes the animal slippery, enabling it to escape from predators. Some amphibians also have skin glands that secrete poisonous substances to discourage predators.

The amphibian heart has three chambers, two **atria** (receiving chambers) and a single **ventricle** (a chamber that pumps blood into the arteries). Although there is some mixing of oxygen-rich and oxygen-poor blood, there is a double circuit of blood vessels. One circuit directs blood to the various tissues and organs of the body, while the other circuit conducts blood to the lungs and skin to be recharged with oxygen.

Reptiles

Reptiles are terrestrial animals that do not have to return to water to reproduce. Many adaptations make this life-style possible. The female's oviduct secretes a protective leathery shell around the egg, which helps prevent the developing embryo from drying out (see Fig. 1–1a). Since sperm cannot penetrate this shell, fertilization must occur within the body of the female before the shell is added. This internal fertilization requires an additional adaptation—a copulatory organ for transferring sperm from the body of the male into the female reproductive tract.

As the embryo develops within the protective shell, a membrane called the **amnion** forms and surrounds the embryo. The amnion secretes amniotic fluid, providing the embryo with its own private pond. This fluid keeps the embryo moist and also serves as a shock absorber should the egg get bounced about. (In a few reptiles, the egg develops within the body of the mother; these animals are ovoviviparous.)

The reptile body is covered with hard, dry, horny scales that protect the animal from drying out and from predators. This dry skin cannot serve as an organ for gas exchange. In compensation, the lungs are better developed than the saclike lungs of amphibians. Most reptiles have a three-chambered heart, but the ventricle is partly divided by an incomplete partition.

Like fish and amphibians, reptiles lack metabolic mechanisms for regulating body temperature. They are **poikilothermic,** depending upon heat from the environment, rather than on what little they may be able to generate for themselves. However, they do have behavioral adaptations that enable them to maintain a body temperature higher than that of the environment. You may have observed lizards basking in the sun. What you may not have known is that the lizard was waiting for its body temperature to rise so that its metabolic rate would increase and it could actively hunt for food. When the body temperature of a reptile is cold, the metabolic rate is low and it is very sluggish (see Fig. 1–5b). This probably explains why reptiles are more abundant in warm than in cold climates.

Modern reptiles include the turtles, lizards, snakes, alligators, and crocodiles (Fig. 16–39). Lizards and snakes are the most common. These animals have rows of scales that overlap like shingles on a roof, forming a protective armor that must be shed periodically. Snakes have a flexible, loosely jointed jaw structure that permits them to swallow animals many times larger than the diameter of their own jaws.

There is some debate regarding the origin of the reptiles. Some biologists hypothesize that they evolved from the amphibians, whereas oth-

(a)

(b)

Figure 16–39 Modern reptiles. (a) The desert collared lizard, *Crotaphytus collaris*. (b) An eastern diamondback rattlesnake, *Crotalus adamanteus*. (a, Charles Seaborn. b, James H. Carmichael, Coastal Creations.)

ers think they evolved from a group of lungfish different from those that gave rise to the amphibians. There are more extinct than living species of reptiles. The Mesozoic era (thought to have ended about 60 million years ago) is known as the Age of Reptiles. During that time reptiles were the dominant terrestrial animals, occupying an impressive variety of ecologic niches comparable to those of the mammals today. Although some were small, others were the largest, most monstrous animals that ever stalked the earth.

The reptiles were the dominant land animals for almost 200 million years. In addition to the dinosaurs, swimming and flying reptiles were abundant. Then, quite suddenly during the Cretaceous period, many of them, including all the dinosaurs, disappeared from the fossil record. Many theories have been proposed to explain the sudden extinction of the large reptiles. Some blame the mammals, which may have been competing with them and feasting on their eggs. Others blame changes in climate, for their environment was getting colder at the time they disappeared. Another theory suggests that about 65 million years ago a large extraterrestrial body, such as a large comet or meteorite, hit or passed near the earth. Perhaps the impact raised a massive dust cloud that blocked out sunlight for several years, resulting in mass extinctions. Gradually, as the dust settled, an iridium-rich layer of clay was deposited, marking the boundary between the Cretaceous and Tertiary periods. (The element iridium in the earth's crust is thought to come from extraterrestrial sources, so the presence of this layer is evidence for this theory.) Whatever the cause, by the end of the Mesozoic, these large reptiles had disappeared, leaving only three orders of reptiles. Long before their decline, however, the reptiles gave rise to both the birds and the mammals.

Birds

About 9000 species of **birds** have been described and classified in 27 orders (Fig. 16–40). Birds live in a wide variety of habitats and can be found on all the continents, most islands, and even the open sea. The largest living birds are the ostriches of Africa, which may be 2 meters (7 feet) tall and weigh 136 kilograms (300 pounds), and the great condors of the Americas, with wingspreads up to 3 meters (10 feet). The smallest known bird is Helena's hummingbird of Cuba, which is less than 6 centi-

(a)

(b)

Figure 16–40 Modern birds. (a) The bald eagle, *Haliaeetus leucocephalus.* The pesticide DDT, once widely used, resulted in thin eggshells from which few eaglets hatched, and the bald eagle became a greatly endangered species. Now these birds are becoming more numerous in many sections of the United States. (b) The roseate spoonbill, *Ajaia ajaja*, is a wading bird that uses its spoonlike beak to gather shellfish and aquatic insects from tidal areas. (James H. Carmichael, Coastal Creations.)

meters (2.3 inch) long and weighs less than 4 grams (0.1 ounce). Beautiful and striking colors are found among the birds. Many birds, especially females, are protectively colored by their plumage. Brighter colors are usually assumed by the male during the breeding season to help him attract a mate.

Birds are beautifully adapted for flight. They are the only animals that have feathers. Thought to have evolved from reptilian scales, feathers are flexible and very strong for their light weight. They protect the body, decrease water loss through the body surface, decrease the loss of body heat, and aid in flying by presenting a plane surface to the air.

The anterior limbs of birds are usually modified for flight; the posterior pair for walking, swimming, or perching. Not all birds fly. Some, such as penguins, have small, flipper-like wings used in swimming.

In addition to feathers and wings, birds have many other adaptations for flight. They have compact, streamlined bodies, and the fusion of many bones gives the rigidity needed for flying. Their bones are strong but very light: Many bones are hollow, containing large air spaces. The jaw is light, and instead of teeth, there is a lightweight, horny beak. The very efficient lungs have thin-walled extensions, called air sacs, that occupy spaces between the internal organs and within certain bones. Birds, like mammals, have four-chambered hearts and a double circuit of blood flow so that oxygen-poor blood is recharged with oxygen in the lungs before being pumped out to nourish the tissues.

The very efficient respiratory and circulatory systems provide sufficient oxygen to the cells to permit a high metabolic rate. This is necessary for the tremendous muscular activity required for flying. Some of the heat generated by metabolic activities is used to maintain a constant body temperature. This ability permits metabolic processes to proceed at constant rates and enables birds to remain active in cold climates. Birds and mammals are the only modern animals that can maintain a constant body temperature internally by means of metabolic mechanisms. They are sometimes called warm-blooded but **homeothermic** is the preferred term.

Birds have become adapted to a variety of environments, and various species have very different types of beaks, feet, wings, tails, and behavioral patterns. All birds must eat frequently because they do not store much fat. Their choice of food varies widely from species to species and may include seeds or fruits, worms, mollusks, insects, arthropods, fish, or even small mammals and dead animals.

Birds have a well-developed nervous system with a brain that is proportionately larger than the brain of reptiles. Birds rely mainly on sight, and their eyes are proportionately larger than those of other vertebrates. They have excellent binocular and color vision: A bird can spot a worm many yards away. Hearing is also well developed.

In striking contrast to the silent reptiles, birds have developed the voice. Most birds have short simple calls that signal danger or influence social feeding, flocking, or interaction between parent and young. Songs are usually more complex than calls and are performed mainly by males. They are related to reproduction, attracting and keeping a mate, and claiming and defending territory. One of the most fascinating aspects of bird behavior is the annual migration that many birds make.

Mammals

Mammals are the animals most familiar to us. Their distinguishing features are the presence of hair, **mammary glands,** which produce milk for the young, and the differentiation of teeth into incisors, canines, premolars, and molars. A muscular diaphragm helps move air in and out of the lungs. Like the birds, but unlike other vertebrate groups, mammals are homeotherms—they maintain a constant body temperature. This process is supported by the covering of hair, which serves as insulation, by the four-chambered heart and double circulation, and by the presence of sweat glands. The nervous system is more highly developed than in any other group. The cerebrum is especially large and complex with an outer gray region called the cerebral cortex.

EARLY MAMMALS

Mammals are thought to have evolved from a group of reptiles called **therapsids** probably during the Early Triassic period some 200 million years ago. The therapsids were carnivores with differentiated teeth (a mammalian trait) and legs adapted for running. Some of them may have been homeothermic, and some may even have had fur. The fossil record indicates that the earliest mammals were small, about the size of a mouse or shrew.

How did the mammals manage to coexist with the reptiles during the 160 million or so years that the reptiles ruled the earth? Many adaptations permitted the mammals to compete for a place on the earth, but above all it is thought that the early mammals specialized in being inconspicuous. They lived in trees and were nocturnal, searching for food (mainly insects and plant material and perhaps reptile eggs) while the reptiles slept. Such a life-style is suggested by the large eye sockets in fossil species, which indicate that these animals had the large eyes characteristic of present-day nocturnal primates, and also by the fact that the eyes of almost all modern mammals have only rod cells—photosensitive cells adapted for night vision (primates, such as human beings, are an exception). By bearing their young alive, mammals avoided the hazard of having their eggs consumed by predators. And by nourishing the young and caring for them, the parents could offer both protection and an education, which probably focused upon how not to get eaten. Clearly, though, these feeble little creatures were not really a match for the reptiles of their day and could probably not have rendered the dinosaurs extinct by competition.

As reptiles died out, for whatever reason, the mammals began to move into their abandoned territories and ecological niches. Larger forms and numerous varieties evolved, and mammals became widely distributed and adapted to an impressive variety of ecological niches.

MODERN MAMMALS

Today mammals inhabit virtually every corner of the earth—on the land, in the water, and even in the air. Their sizes range from the tiny pigmy shrew, weighing less than an ounce, to the blue whale, which may weigh more than 90,000 kilograms (100 tons) and is the largest animal ever known. Three main subclasses of mammals are the **egg-laying mammals,** called **monotremes;** the pouched mammals, or **marsupials;** and the **placental mammals.**

The duck-billed platypus (*Ornithorhynchus*) and the spiny anteater (*Tachyglossus*) are monotremes native to Australia (Fig. 16–41). The females lay eggs, which may be carried in a pouch on the abdomen or kept warm in a nest. When the young hatch, they are nourished with milk from the mammary glands.

Marsupials include kangaroos and opossums (Fig. 16–42). Embryos begin their development in the mother's uterus (womb), where they are nourished by yolk and from fluid in the uterus. After a few weeks, still in a very undeveloped stage, the young are born and crawl to the **marsupium** (pouch), where they complete their development. Each of the young

attaches itself by its mouth to a mammary gland nipple and is nourished by its mother's milk. Like the monotremes, the marsupials are found mainly in Australia. Only the opossum is common in North America, although a few species inhabit South America. At one time marsupials may have inhabited much of the world but were replaced by the placental mammals. Australia became geographically isolated from the rest of the world before placental mammals reached it, and there the marsupials remained the dominant type of mammal. Their evolution proceeded in many directions, fitting them for many different life-styles, paralleling the evolution of placental mammals elsewhere. Thus in Australia and adjacent islands we find marsupials that correspond to our placental wolves, bears, rats, moles, flying squirrels, and even cats (Fig. 16–43). There are also a number of forms without placental counterparts, such as the kangaroo and wallaby.

Most familiar to us are the placental mammals, in which an organ of exchange called the **placenta** develops, enabling the young to remain within the body of the mother until embryonic development is complete. The young are born at a more mature stage than in the marsupials. Indeed, among some species the young can walk around and begin to interact with other members of the group within a few minutes of birth.

There are about 17 living orders of placental mammals (see Focus on Placental Mammals). The probable family tree of the vertebrates is

Figure 16–42 The kangaroo, a marsupial native to Australia. (*a*) An adult. (*b*) Young marsupials are born in a very immature state. (*c*) The young continue to develop in the safety of the marsupium. (*a* and *c*, courtesy of Robin Lewis, Coastal Creations. *b*, photograph by Robert Anderson, reprinted with permission of Hubbard Scientific Company.)

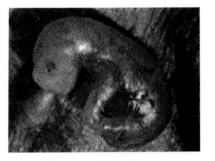

(b)

(a)

(c)

Figure 16–43 Koalas are marsupials specialized for their way of life. They are tree-living animals and are restricted in their diet to leaves of a very few species of eucalyptus. This young koala has outgrown the pouch and will ride on its mother's back until it is old enough to go out on its own. (E.R. Degginger.)

illustrated in Figure 16–44. Human beings, along with the lemurs, monkeys, and apes, belong to the order Primates. Although primates are not highly specialized animals, they have one outstanding feature—the unparalleled development of the brain. This specialization reaches its peak in the human being. Among all organisms, only we have the ability, and with it the responsibility, to control (at least in some measure) our own destiny and the fate of the other organisms that make up the ecosphere.

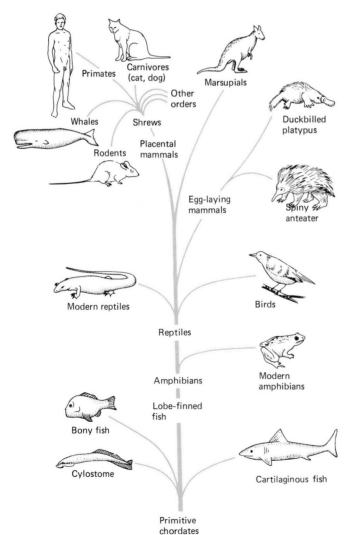

Figure 16–44 A vertebrate family tree.

Table 16–1
COMPARISON OF THE PRINCIPAL ANIMAL PHYLA

Phylum	Level of organization	Symmetry	Digestion	Circulation	Gas exchange
Porifera (pore bearers) Sponges	Multicellular but tissues loosely arranged	Radial or none	Intracellular	Diffusion	Diffusion
Cnidaria Hydra, jellyfish, coral	Tissues	Radial	Gastrovascular cavity with only one opening; intra- and extracellular digestion	Diffusion	Diffusion
Platyhelminthes (flatworms) Planarians, flukes, tapeworms	Organs	Bilateral; rudimentary head	Digestive tract with only one opening	Diffusion	Diffusion
Nemertinea Proboscis worms	Organ systems	Bilateral	Complete digestive tract with mouth and anus	At least two pulsating blood vessels; no heart; blood cells with hemoglobin	Diffusion
Nematoda (roundworms) *Ascaris,* hookworms, nematodes	Organ systems	Bilateral	Complete digestive tract with mouth and anus	Diffusion	Diffusion
Mollusca Clams, snails, squids	Organ systems	Bilateral	Complete digestive tract	Open system	Gills and mantle
Annelida (segmented worms) Earthworms, leeches, marine worms	Organ systems	Bilateral	Complete digestive tract	Closed system	Diffusion through moist skin; oxygen circulated by blood
Arthropoda (jointed-footed animals) Crustaceans, insects, spiders	Organ systems	Bilateral	Complete digestive tract	Open system	Trachea in insects; gills in crustacea; book lungs or tracheae in spider group
Echinodermata (spiny-skinned animals) Sea stars, sea urchins, sand dollars	Organ systems	Embryo bilateral; adult radial	Complete digestive tract	Open system; reduced	Skin gills
Chordata Tunicates, Lancelets, Vertebrates	Organ systems	Bilateral	Complete digestive tract	Closed system; ventral heart	Gills or lungs

Waste disposal	Nervous system	Reproduction	Other characteristics
Diffusion	Irritability of cytoplasm	Asexual by budding; sexual, may be hermaphroditic	Filter feeders; skeleton of chalk, glass, or spongin; larvae swim by cilia; adults incapable of locomotion
Diffusion	Nerve net; no centralization of nerve tissue	Asexual by budding; sexual, separate sexes	Have cnidocytes along tentacles
Protonephridia; flame cells and ducts	Simple brain; two nerve cords; ladder-type system; simple sense organs	Asexual, by fission; sexual, hermaphroditic, but some cross-fertilize	Three definite tissue layers; no body cavity; many parasitic
Two lateral excretory canals with flame cells	Simple brain; two nerve cords; with cross nerves; simple sense organs	Asexual, by fragmentation; sexual, sexes separate	No body cavity; proboscis for defense and capturing prey
Excretory canals	Simple brain; dorsal and ventral nerve cords; simple sense organs	Sexual, sexes separate	Have pseudocoelom; many parasitic
Kidneys	Three pairs of ganglia; simple sense organs	Sexual, sexes separate; fertilization in water	Soft-bodied; usually have shell and ventral foot for locomotion
Pair of metanephridia in each segment	Simple brain; ventral nerve cord; simple sense organs	Sexual; hermaphroditic but cross-fertilize	Earthworms till soil
Malpighian tubules in insects; antennal (green) glands in crustaceans	Simple brain; ventral nerve cord; well-developed sense organs	Sexual, sexes almost always separate	Hard exoskeleton; most successful group of animals
Diffusion	Nerve rings; no brain	Sexual, sexes separate	Water vascular system
Kidneys and other organs	Dorsal nerve cord with brain at anterior end	Sexual, sexes separate	(1) Notochord, (2) Hollow tubular, dorsal nerve cord, (3) Pharyngeal grooves

Focus on . . .

PLACENTAL MAMMALS

Different placental mammals have adaptations that allow them to fly, swim, and occupy a multitude of terrestrial environments. Their range of adaptations, and in fact sometimes their exterior appearance, are comparable to those of the reptiles during the Mesozoic era. The phenomenon of an ancestral organism giving rise to a group of organisms occupying a wide variety of habitats is called adaptive radiation. It is a subject we will study in greater detail in the chapters on evolution.

Erethizon dorsatum, a porcupine. Porcupines are members of order Rodentia, the largest order of mammals. Porcupine females bear only one offspring in a season. For mammals, young porcupines are unusually well able to care for themselves; at the age of two days they can climb trees and find food. (Charles Seaborn.)

Ursus maritimus, a polar bear, photographed in the Kane Basin in the Arctic. Polar bears are members of order Carnivora, together with dogs, cats, raccoons, and seals. (E.R. Degginger.)

Leontopatheus rosalia, the golden lion tamarin monkey. Only 150 of these primates survive in their native coastal rain forest in Brazil, which has been reduced to 2% of its original acreage. (Charles Seaborn.)

A common dolphin, Delphinus delphis. The streamlined body and overall fishlike form are adaptations strikingly similar to those possessed by some of the oceanic reptiles of the Mesozoic. Dolphins belong to order Cetacea. (E.R. Degginger.)

CHAPTER SUMMARY

I. Animals are eukaryotic, multicellular heterotrophs whose cells exhibit a division of labor; they are generally capable of locomotion at some time during their life cycle; and they generally reproduce sexually.

II. Most animal phyla exhibit bilateral symmetry; only two phyla have radial symmetry. Acoelomates are animals that lack a body cavity; pseudocoelomates have a body cavity derived from the cavity of the embryo; coelomates have a true coelom lined with mesoderm. Animal phyla can also be grouped as protostomes or deuterostomes.

III. Phylum Porifera consists of the sponges, the simplest animals.
   A. Sponges are divided into classes on the basis of the type of skeleton they secrete.
   B. The sponge body is a sac perforated by tiny holes through which water enters; as water circulates through the sponge, materials are exchanged by diffusion.

IV. Phylum Cnidaria includes the hydras, jellyfish, and corals.
   A. Cnidarians are characterized by radial symmetry, stinging cells, definite tissue layers, and a nerve net.
   B. Cnidarians have a polyp or medusa body form.

V. Phylum Platyhelminthes includes the planarians, flukes, and tapeworms.
   A. Flatworms have bilateral symmetry, three definite tissue layers, simple brains and nervous systems, and other well-developed organs, including protonephridia with flame cells. They exhibit cephalization.
   B. The flukes and tapeworms have suckers and other adaptations for their parasitic life-styles.

VI. Members of phylum Nemertinea (proboscis worms) have tube-within-a-tube body plans, complete digestive tracts with mouth and anus, and separate circulatory systems.

VII. Phylum Nematoda, the roundworms, includes species of great ecological importance and species parasitic in plants and animals.
   A. Nematodes have pseudocoeloms and complete digestive tracts.
   B. Some nematodes parasitic in humans include Ascaris, hookworms, trichina worms, and pinworms.

VIII. Phylum Mollusca includes clams, oysters, snails, squids, and ocotopods.
   A. Mollusks are soft-bodied animals usually covered by shells; a mollusk has a ventral foot for locomotion, and a mantle that covers the visceral mass.
   B. Squids and octopods are active predatory animals with large heads, long tentacles, and prominent eyes.

IX. Phylum Annelida, the segmented worms, includes many marine worms, earthworms, and leeches.
   A. Annelids have long bodies segmented internally as well as externally; setae aid in locomotion.
   B. Polychaetes have bristled parapodia that function in locomotion; leeches lack setae and appendages, but are equipped with suckers.

X. Phylum Arthropoda, the largest phylum, includes horseshoe crabs, spiders, scorpions, mites, crustaceans, insects, millipedes, and centipedes.
   A. Arthropods are animals with jointed appendages and armorlike exoskeletons.
   B. Insects are the most successful group of animals, with more species and greater numbers than any other group.

XI. Phylum Echinodermata is made up of the sea stars, sea urchins, sand dollars, and sea cucumbers.
  A. Echinoderms have spiny skins and exhibit pentaradial symmetry.
  B. Other unique features of the echinoderms are the water vascular system and tube feet.
XII. Phylum Chordata includes subphylum Vertebrata, the animals with backbones.
  A. At some time in its life cycle a chordate has a notochord, a dorsal tubular nerve cord, and pharyngeal gill grooves.
  B. The vertebrate classes include jawless fish, cartilaginous fish, bony fish, amphibians, reptiles, birds, and mammals.
    1. Mammals have hair, mammary glands, differentiated teeth, and maintain a constant body temperature.
    2. Three main subclasses of mammals are the egg-laying monotremes, the marsupials, and the placental mammals.

Post-Test

1. Animals without backbones are _____ .
2. An animal that is divided into roughly right and left halves when sliced down the midline may be described as _____ _____ .
3. An acoelomate animal lacks a _____ _____ .
4. The two phyla that are deuterostomes are the _____ and the _____ .
5. Sponges secrete skeletal structures called _____ _____ .
6. A hermaphroditic animal can produce _____ _____ and _____ .
7. A distinctive feature of the cnidarians is the presence of cnidocytes that are _____ _____ _____ .

21. Humans become infected with trichina worms by _____ ; and with _____ by walking barefoot in contaminated soil.
22. Reptiles are thought to have given rise to the _____ and the _____ .
23. The vertebrates that have a three-chambered heart and moist skin are the _____ .
24. The amnion is an adaptation to _____ life; it secretes a fluid that _____ .
25. Monotremes are mammals that _____ _____ ; marsupials possess _____ _____ .

Select the most appropriate match in column B for each entry in column A.

Column A

_____ 8. nerve net
_____ 9. tube feet
_____ 10. notochord
_____ 11. protonephridia with flame cells
_____ 12. jointed appendages; exoskeleton
_____ 13. soft-bodied with ventral foot and mantle
_____ 14. sponge
_____ 15. snail
_____ 16. sea star
_____ 17. bird
_____ 18. earthworm
_____ 19. grasshopper
_____ 20. crayfish

Column B

a. phylum Cnidaria
b. phylum Platyhelminthes
c. phylum Mollusca
d. phylum Arthropoda
e. phylum Echinodermata
f. phylum Chordata
g. none of the above

Review Questions

1. Which animals are considered the most successful? Describe some of their characteristics that contribute to this success.
2. What advances do the members of phylum Platyhelminthes exhibit over those animals that belong to phylum Cnidaria? In what ways are they alike?
3. Which group of animals (a) exhibit radial symmetry, (b) have flame cells, (c) have the most primitive brains, (d) are the simplest to have complete digestive tubes?
4. Distinguish between insects and spiders.
5. List the three distinguishing characteristics of the chordates.
6. Briefly describe the seven classes of vertebrates.
7. Describe the three main groups of mammals.
8. Which groups of animals are able to depend upon diffusion for gas exchange? Why?
9. Describe the life cycle (draw a diagram if you can) of the tapeworm.
10. Give two distinguishing characteristics of (a) phylum Annelida, (b) phylum Arthropoda, (c) phylum Mollusca, (d) phylum Echinodermata.
11. Following generally accepted evolutionary principles, draw a hypothetical ancestral tree of the animal kingdom.
12. What do echinoderms and chordates have in common?
13. Which phyla are acoelomate? coelomate?
14. Locate each of the following in a specific type of animal and give its function: (a) amnion, (b) placenta, (c) swim bladder, (d) spicules, (e) mantle, (f) setae, (g) tracheae.

Readings

Alldredge, A.L., and L.P. Madin. "Pelagic Tunicates: Unique Herbivores in the Marine Plankton," *Bioscience,* September 1982, 655–663. An account of the ecology and unique adaptations of pelagic tunicates.

Barnes, R.D. *Invertebrate Zoology,* 4th ed. Philadelphia, Saunders College Publishing, 1980. This comprehensive textbook discusses the life processes of each invertebrate phylum.

Barth, R.H., and R.E. Broshears. *The Invertebrate World.* Philadelphia, Saunders College Publishing, 1982. A brief but interesting parade through the phyla, emphasizing the highlights of invertebrate zoology. The protozoans are included.

Daly, H.V., J.T. Doyen, and P.R. Ehrlich. *Introduction to Insect Biology and Diversity.* New York, McGraw-Hill, 1978. An introduction to the study of insects that emphasizes the major features of insects as living systems.

Fingerman, M. *Animal Diversity,* 3rd ed. Philadelphia, Saunders College Publishing, 1981. A very brief summary of both the invertebrates and the vertebrates with emphasis on evolutionary relationships.

Gardiner, M.S. *The Biology of Invertebrates.* New York, McGraw-Hill, 1972. A functional rather than a systematic approach to invertebrate zoology. The book is organized around life processes rather than by phyla.

Orr, R.T. *Vertebrate Biology,* 5th ed. Philadelphia, Saunders College Publishing, 1982. This introduction to vertebrate biology focuses on biological processes.

Romer, A.S., and T.S. Parsons. *The Vertebrate Body,* 6th ed. Philadelphia, Saunders College Publishing, 1986. A well-respected, classic textbook that takes a comparative approach to life processes in vertebrates.

Vaughan, T.A. *Mammalogy,* 3rd ed. Philadelphia, Saunders College Publishing, 1986. An introduction to the mammals; a systematic approach.

Villee, C.A., W.F. Walker, and R.D. Barnes. *General Zoology,* 6th ed. Philadelphia, Saunders College Publishing, 1984. An introductory zoology textbook with evolutionary emphasis.

Welty, J.C. *The Life of Birds,* 3rd ed. Philadelphia, Saunders College Publishing, 1982. An introduction to the biology of birds.

PART V
PLANT STRUCTURE AND FUNCTION

Columbine, Cascade Mountains. (Art Wolfe.)

Chapter 17
THE PLANT BODY

Learning Objectives

After you have studied this chapter you should be able to:

1. Distinguish between meristematic and permanent tissues in plants.
2. Describe and give the functions of the following types of permanent tissues: surface tissues, fundamental tissues, and vascular tissues.
3. Describe and give the function of the component parts of the roots and shoots of plants.
4. Describe the structure of a leaf and give the functions of the epidermis, mesophyll, and veins.
5. Summarize the role of guard cells in the water economy of the plant, and describe how they function and are controlled.
6. Describe the arrangement of vascular tissues in the stems of monocots and dicots.
7. Summarize the functions of phloem and xylem.
8. Describe the structure of a typical root.
9. Describe the typical respiratory adaptations of roots.

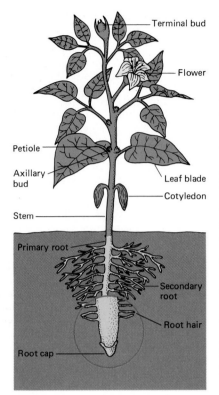

Figure 17–1 A young bean plant showing some of the plant organs. The root usually comprises all the underground portions of the plant, although in some plants, parts of the stem dwell beneath the soil surface. The shoot consists of the stem, leaves, and associated above-ground portions.

We have surveyed the main kinds of plants, and you have no doubt noted many references to such structures as leaves and roots, to vascular tissues, and to reproductive modes. In this chapter and the next two we will briefly examine how complex plants live, develop, and reproduce, with emphasis on the functioning of their tissues, how growth occurs and how it is regulated, and how fertilization produces seeds and fruit.

PLANT ORGANS AND PLANT SYSTEMS

Plants have fewer types of organs and organ systems than animals. The bodies of complex plants are composed of two major parts, the root and the shoot (Fig. 17–1). These are generally considered the organ systems of the plant body. They can be distinguished by the arrangement of the vascular tissues, by the way lateral shoots and branch stems are formed, and by the presence of leaves on the shoot. The two, of course, are intimately connected, and many of the tissues, such as xylem and phloem, are essentially continuous from root tip to shoot tip.

The **roots** perform two functions: They hold the plant firmly in the soil, and even more important, they are the sole means most land plants have to absorb water and to mine nutrients from the soil. The roots of all the plants in the world extract several thousand metric tons of minerals from the soil every minute.

The shoot consists of several organs—the stem, the foliage (leaves), and the reproductive organs (flowers and fruits). The stem functions in support and transport of nutrients, the leaves in photosynthesis, and the flowers, fruits and seeds in reproduction.

Leaves are generally the main and often the only significant photosynthetic organs of the plant. Most leaves are thin and flat for maximum absorption of light energy and efficient internal diffusion of gases. Arranged on the plant so that they interfere minimally with one another's light supply, the leaves of plants form an intricate green mosaic, bathed in sunlight and atmospheric gases (Fig. 17–2).

The **stem** is the plant's midsection and middleman, conducting commerce between the leaves and the roots. Water and minerals travel upward; carbohydrates travel downward (and sometimes upward, as we will see). Just as important, the stem supports both its own weight and that of the leaves, and by its growth thrusts these above the leaves and stems of its competitors.

Though these are the basic functions of the parts of the plant body, they are by no means the only ones. In some species (such as the African baobab tree, Fig. 17–3) the stems are modified for food storage, and in

Figure 17–2 Leaf mosaic. The leaves of this ivy are growing in such a way that each shades the other minimally, promoting the maximal absorption of solar energy for the entire plant.

Figure 17–3 The African baobab tree stores large volumes of water and starch in its stem tissues, which are often eaten by hungry elephants. (Courtesy of Dr. Pat Gill.)

others (such as the potato) the stems lie underground, swollen with food. Leaves can be modified for protection (in the case of thorns), for support (in the tendrils of some vines), and for water storage (in the case of many succulent and salt marsh plants). In some curious species, like the Venus' flytrap, leaves catch and digest animal prey. Leaves may even be missing altogether, leaving photosynthesis to the flattened stems (as in cacti), or even to the seedpods (as in one eccentric variety of garden peas).

PLANT TISSUES

Plants also have fewer basic types of tissues than animals possess, and biologists are not in complete agreement regarding their classification. Some cells may appear as intermediates between cell types, and a given cell may change from one type to another during its life. However, plant tissues can be assigned to one of two major categories, meristematic tissues and permanent tissues. Meristematic tissues, composed of immature cells, undergo cell division; permanent tissues are composed of mature, differentiated cells.

Meristematic Tissues

The **meristematic tissues** of plants have no direct equivalent in the animal body because, in general, animals possess **determinate growth,** and plants, **indeterminate growth.** What we mean by this is that although there is a basic similarity in the overall shape and maximum size of many specimens of the same kind of plant, there is no real limit to how many branches, let us say, that a plant might have, the places in which they might branch, or how long they might individually be. Moreover, a plant grows continually and usually throughout life. Thus, plants particularly need a permanently embryonic, essentially totipotent tissue that can give rise to adult differentiated tissues and structures throughout the life of the plant. At least in dicots, meristematic tissues are found in every part of the adult plant body.

Meristematic tissues are usually made up of very small cells with thin walls and few vacuoles (Fig. 17–4). Under the appropriate hormonal or other stimuli, these cells are able to differentiate into all other adult tissues. Meristems are of two types: primary and secondary. The **primary meristem** is present in the initial bud, and the **secondary meristem** is incorporated into the mature body parts derived from that bud.

Figure 17–4 Meristematic tissue is composed of immature cells that grow, divide, and differentiate, giving rise to other types of plant tissues. This tissue, from the root tip of an onion (*Allium cepa*), shows cells in various stages of mitosis. (Courtesy of Triarch.)

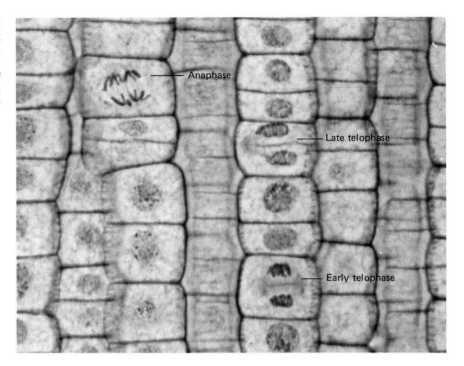

Anaphase

Late telophase

Early telophase

Permanent Tissues

Permanent (or mature) **tissues** are composed of mature, specialized cells. They include surface tissues, fundamental tissues, and vascular tissues (Table 17–1).

SURFACE TISSUES

Surface tissues make up the protective outer covering of the body of the plant. The surface tissue of roots and stems in young plants and herbaceous plants is the **epidermis.** This is also the surface tissue of leaves. Surface tissues protect the underlying cells from drying, from mechanical abrasion, from invasion by parasitic fungi, bacteria, and protists, and often from attack by insects or larger herbivores (for example, think of the stinging hairs of nettle leaves—Fig. 17–5).

The epidermis of leaves and the cork layers of stems and roots are examples of surface tissues specialized to protect underlying tissues. The epidermis is often only one cell thick, and epidermal cells are usually flattened. Epidermal cells by themselves are not always particularly waterproof, however. The epidermis of leaves, for instance, is overlain by a protective layer of wax that greatly retards evaporation. As we will see, the cells of the leaf epidermis are variously specialized to permit the exchange of atmospheric gases with the leaf tissues while minimizing evaporation from within.

The epidermal cells of the roots may be specialized also, with root hairs that help to increase its surface area for more effective absorption of materials from the soil. But most of the epidermis of stem and root is heavy and often composed of dead cells impregnated with waxes or **suberin,** a gummy, waterproof material.

Figure 17–5 Horse nettle, *Cnidoscolus texanus*. The fuzzy appearance of the leaves of this plant is due to a multitude of hollow hairs, which break off inside the skin of any animal rash enough to brush against it or attempt to eat its seeds. Irritating substances injected into the skin produce a lively stinging sensation. (V. Weinland, Photo Researchers, Inc.)

VASCULAR (CONDUCTIVE) TISSUES

Complex plants have two main types of vascular tissue: xylem and phloem. The **xylem** conducts water and dissolved minerals from the roots to all aerial parts of the plant. The **phloem** transports the products of photosynthesis and also carries hormones and other substances.

Table 17-1
TYPES OF PLANT TISSUE

	Structure	Location	Function
Meristematic	Small, thin-walled cells with small nuclei and few vacuoles	Tips of roots and stems, cambium and stems	Grow, divide, differentiate into other tissues
Permanent (Mature)	Mature, specialized cells		
Surface	Cells usually flattened; may secrete cutin or suberin	Epidermis of leaves; cork layers of stems and roots	Protect; some (like root hairs) absorb water and nutrients
Vascular Xylem	Long tubes composed of tracheids and xylem vessels	Roots, stems, and leaves	Transport water and dissolved salts; support
Phloem	Long tubes composed of sieve tubes and companion cells	Roots, stems, and leaves	Transport nutrients
Fundamental	Simple tissue; usually composed of one type of cell	Bulk of plant body	
Parenchyma	Thin walls, central vacuoles	Roots, stems, and leaves	Photosynthesis, food storage
Chlorenchyma	Thin walls, closely packed; large vacuoles; contain chloroplasts	Leaves, some stems	Photosynthesis
Collenchyma	Cells with walls thickened in the corners	Stems, leaf stalks	Support
Sclerenchyma	Cells with greatly thickened cell walls	Shells of nuts; hard parts of seeds	Support

Phloem transport is not necessarily in a downward direction. When the sap rises in maple trees in early spring, for instance, it travels skyward in the phloem.

When the specialized meristem called **vascular cambium** differentiates into xylem, it forms long, spindle-shaped cells with pointed ends and variously ornamented walls. The cells join end-to-end to form a kind of continuous pipeline, and their contents die and disintegrate so that solutions can move through them almost unimpeded (Fig. 17-6).

Xylem also forms the heartwood (Fig. 17-7) of woody dicots and gymnosperms. Here it is often filled and impregnated with resinous materials and lignin, which converts it into a skeletal tissue capable of supporting crown and branches.

Phloem differs most fundamentally from xylem in being a living tissue, provided with an abundance of cytoplasm but no nuclei. Nuclei are present in adjacent **companion cells.** Also, the ends of the phloem cells do not ever disappear but become perforated to form sievelike partitions between them, called (reasonably enough) **sieve plates.** The total tube thus formed by many adjacent phloem cells is known as a **sieve tube.**

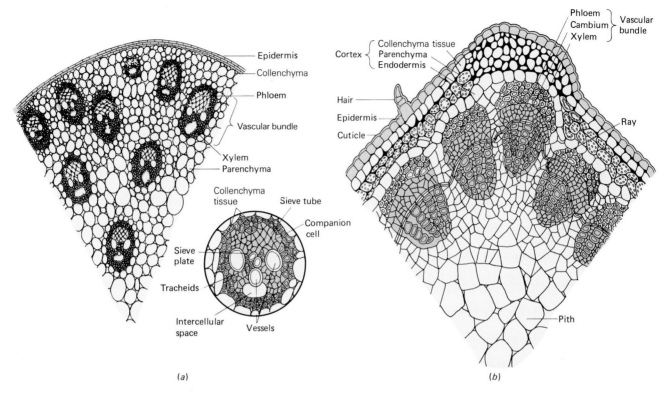

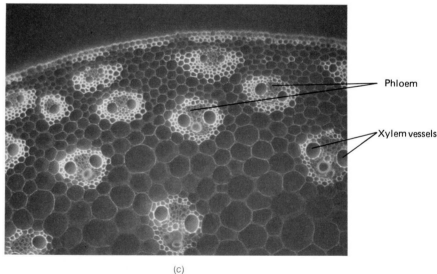

Figure 17–6 (*a*) A sector of a cross section of a stem from a monocot, corn. *Inset*: An enlarged view of a vascular bundle containing phloem (sieve tubes and companion cells) and xylem (tracheids and vessels). (*b*) A sector of a cross section of a stem from a herbaceous dicot, alfalfa. (*c*) Photomicrograph (approximately ×30) of vascular bundles in corn stem prepared with a special fluorescent staining technique. (P. Dayanandan, Photo Researchers, Inc.)

FUNDAMENTAL TISSUES

Most of the body of herbaceous (that is, nonwoody) plants is composed of the mature but relatively undifferentiated pulpy **fundamental tissues.** The green photosynthetic **parenchyma** of the leaf (also called mesophyll) is found in all land plants—at least all that have leaves. The herbaceous plants, annuals usually that attain no very great size, have stems filled with a pulpy parenchyma called **pith.** This tissue offers little mechanical support but is alive. Pith also occurs in the young stems of woody dicots and some may persist in the centers of even large trunks and branches (Fig. 17–8).

There is much parenchyma also in roots, especially those with a pronounced storage function, such as yams or carrots. In these plants the root parenchyma cells are filled with starch grains. Finally, the succulent

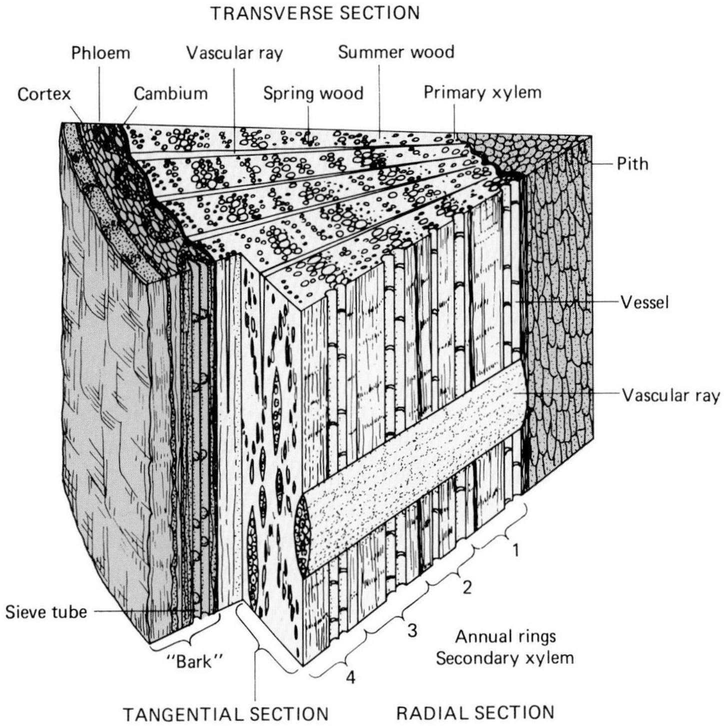

TRANSVERSE SECTION

Phloem   Vascular ray   Summer wood

Cortex   Cambium   Spring wood   Primary xylem

Pith

Vessel

Vascular ray

Sieve tube

1
2
3   Annual rings
     Secondary xylem
4

"Bark"

TANGENTIAL SECTION   RADIAL SECTION

Figure 17–7 Diagram of a four-year-old woody stem, showing transverse, radial, and tangential sections and the annual rings of secondary xylem (heartwood).

tissues of ripe fleshy fruits are really a kind of parenchyma filled with starch, sugar, oil, and sometimes delicious flavor and scent compounds.

THE LEAF

Take a leaf from a tree or shrub and observe it carefully in good, strong light. The stemlike **petiole** attaches the leaf to the stem at a location called the **node.** The angle it forms with the stem is its **axil.** The petiole is continuous with the midrib of the leaf, which usually lies in a central valley that continues to the **tip.** The tip is often shaped so as to facilitate the drip of rainwater from the leaf **blade** (Fig. 17–9). The midrib gives rise to many tiny **veins** in the blade of the leaf, which in dicots branch in several fashions and in monocots run parallel to one another. The two surfaces of the leaf may differ in appearance and somewhat in function as well.

Figure 17–8 Pith cavity of the young branch of a golden rain tree.

Petiole   Veins   Midrib   Blade

Figure 17–9 Parts of a leaf.

Figure 17–10 Many leaves are equipped with hairs that limit the transpiration of water, discourage herbivores, sting, or perform other functions.

Epidermis

The **epidermis,** or outer covering of the leaf, appears to have a smooth and glossy surface. Epidermal cells protect the wet tissues of the leaf from desiccation. Their major means of waterproofing is the layer of waxy, acellular **cuticle** that covers the entire leaf. In some tropical palms this layer of wax is thick enough to be of commercial value and is harvested for use in floor, car, and shoe polish.

Some leaves possess hairs that are outward extensions of epidermal cells. These hairs may have several functions. In some instances they help retard evaporation of water from the leaf by interfering with the free flow of air over the leaf surface; in others they repel grazing mammals or defend the leaf from the attacks of insects (Fig. 17–10). Some even sting. Still others reflect sunlight, helping the leaf to avoid overheating.

Microscopic openings called **stomata** (singular, **stoma**) are usually located on the lower leaf surface. Each stoma is like a tiny mouth surrounded by a pair of liplike **guard cells** (see Fig. 17–11 and Focus on A Sweet Story of Stomata). All these epidermal structures, together with the much more numerous jigsaw puzzle–shaped flat cells that surround them, are the frontier between the semiaquatic inner world of the leaf and the dry and hostile terrestrial environment outside. Except for the guard cells of the stomata, the epidermal cells are not green and contain no chloroplasts.

Mesophyll

Sandwiched between upper and lower epidermis, the **mesophyll** (Fig. 17–12) is the photosynthetic tissue of the leaf. The cells of the mesophyll are not too different from those of many colonial algae, and like algae they must be kept moist (if not wet) in order to live. The loosely packed cells of this tissue are green, a consequence of their abundant chlorophyll content. They resemble the model of a "typical" plant cell shown to students in courses like this one.

Extensive intercellular air spaces ensure the ready access of gases to each and every cell. As a result, oxygen for respiration or carbon dioxide for photosynthesis is made readily available by diffusion alone, as circumstances may require. Given the flattened shape of the leaf, no part of its fleshy mesophyll is very far from a stoma. Its transport system,

Figure 17–11 Guard cells and stomata (approximately ×350) in a *Tradescantia* leaf. (James Bell, Photo Researchers, Inc.)

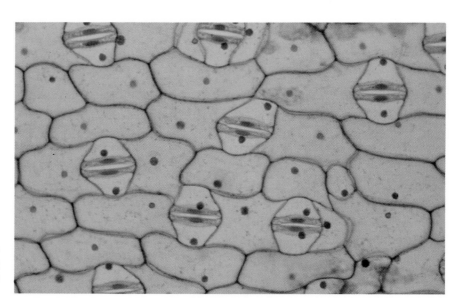

Focus on . . .

A SWEET STORY OF STOMATA

In most species, stomata open in the light and are closed in the dark. Although the mechanisms that control stomatal aperture are not yet fully understood, a widely accepted current explanation is that stomatal opening of most species involves two factors: carbon dioxide and ATP energy. When leaves are exposed to light, photosynthesis reduces CO_2 concentration in the leaf and, as a result, also in guard cells. Because guard cells lack significant levels of enzymes of the carbon fixation cycle, guard cells, in general, *do not* ordinarily carry out the light-independent reactions of photosynthesis. However, it is known that guard cells do contain both photosystems and that guard cell chloroplast photophosphorylation can provide sufficient ATP to drive K^+ intake. Apparently, low guard cell CO_2 content and ATP energy are both needed to activate a K^+ uptake pump. This results in K^+ being actively transported into the guard cells (from surrounding epidermal cells). As these ions accumulate inside the cells, water moves in osmotically. This causes an increased turgor pressure and stomatal opening.

In the dark, when CO_2 concentration in the guard cells becomes high due to respiratory CO_2 production, the K^+ pump is inactivated. The guard cells passively lose most of their K^+, resulting in stomatal closure. Even in full sunlight, guard cell CO_2 content may increase, resulting in stomatal closure. For example, stomata typically begin closing gradually in the afternoon and are fully closed by evening. One explanation for this is that midday water deficits (which occur because transpirational water loss exceeds water absorption, plus high midday temperatures) cause the leaf respiratory rate to exceed its photosynthetic rate. This, in turn, causes an increased guard cell CO_2 concentration and gradual stomatal closure. Under drought conditions, it is known that photosynthetic rates in most species fall to virtually zero, while respiratory rates are elevated. This increases guard cell CO_2 content and induces stomatal closure. However, increased CO_2 content in guard cells during water deficits does not fully account for stomatal closure.

Stomatal closure *due to water deficits* is apparently also mediated by a plant hormone, abscisic acid (ABA). ABA appears to be present in an inactive form in mesophyll cells of well-watered plants. When water deficits in the photosynthetic tissues occur, ABA is converted to an active form and moves into the guard cells. Apparently, the ABA causes stomatal closure by interfering with the K^+ pump. It is not clear whether elevated CO_2 concentration (which also closes stomata) and ABA interact, or whether each acts independently. What is known is that after periods of drought stress, ABA remains active in the guard cells for several days, which delays full stomatal opening during this period. Furthermore, a large amount of ABA produced over a long period of time helps induce formation of an *abscission layer* of waterproof cells at the axil of the leaf. This eventually causes the leaf to drop off, thus protecting the plant against transpirational water loss in a severe drought. Even under less severe water conditions, ABA apparently plays a role in stomatal closure. As the day wears on, as a result of slight dehydration, the mesophyll of the leaf produces increasing amounts of ABA. In response, the stomata close by late afternoon (see figure).

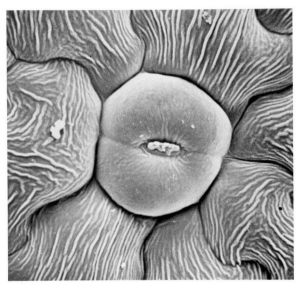

Scanning electron micrograph (approximately ×4000) of closed stoma and guard cells of Loasa volcanica. (Biophoto Associates, Photo Researchers, Inc.)

profoundly different in design from that of almost any animal, does not carry oxygen to its tissues, so diffusion must suffice. The thinness of the leaf also ensures the penetration of light to all chlorophyll-containing cells in amounts adequate for photosynthesis.

In many leaves the mesophyll occurs in two distinguishable layers: the tightly packed **palisade parenchyma,** whose arcades of elongated cells are usually located near the top surface of the leaf, and the somewhat looser **spongy parenchyma** below it. The significance of this arrangement, by no means universal among land plants, is unknown. However, the mesophyll cells by their turgor serve as a hydrostatic skeleton

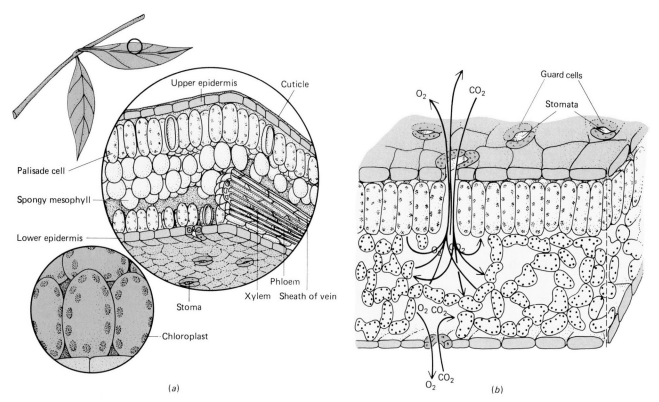

(a)

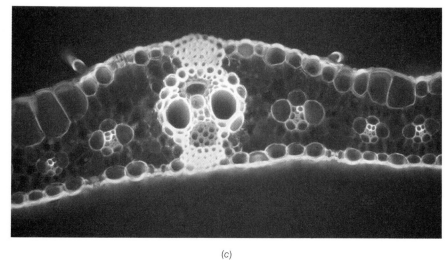

(c)

Figure 17–12 (*a*) Diagram of the microscopic structure of a leaf. Part of a small vein is visible to the right. (*b*) An enlarged cross section of a leaf showing its structure. The diffusion of carbon dioxide through the stomata to the interior of the leaf and the diffusion of oxygen from the photosynthetic cells through the stomata to the exterior are indicated by arrows. (*c*) Photomicrograph (approximately ×60) of a cross section of a corn leaf stained by a special technique that, among other things, causes phloem cells to fluoresce pink. (P. Dayanandan, Photo Researchers, Inc.)

for the leaf, a function for which the distinctive architecture of the mesophyll may be necessary. It has also been suggested that the long, thin palisade cells may function like fiberoptic elements to conduct light efficiently into the center of the leaf. The spongy parenchyma, for its part, may scatter light throughout the leaf by multiple reflections and refractions so that little escapes without capture by chlorophyll.

Veins

The leaf **veins** serve two functions: as a skeleton to support the leaf, and to conduct materials to and from the metabolically active tissues. Although skeletal elements may be lacking in the smaller veins, such structures are prominent in the midrib, whose upper portion may consist entirely of **sclerenchyma,** a tissue of mostly dead cells whose cell walls have become impregnated with hard waxy or gummy materials for strength. Even more important are the vascular tissues, xylem (usually located on the upper side of the vein) and phloem. These carry materials to and from the photosynthetic tissues of the leaf.

Water Loss

The basic function of photosynthesis is served by almost every adaptation of the leaf, from its internal characteristics—such as transparency, shape, and air spaces—to the way the leaf is positioned on the plant. Yet each of these photosynthetic adaptations also promotes the greatest hazard facing the plant: dehydration. The leaves of the vascular land plants represent a design compromise between the conflicting requirements for exposure to atmosphere and light, and protection against water loss. For instance, in the hottest part of the day the leaves of eucalyptus trees orient themselves parallel to the rays of the sun, limiting solar heating of their tissues and minimizing water loss by evaporation. Unfortunately for us, this also has the effect of minimizing the shade they cast just when we would most enjoy it!

A forest interior can be as much as 10° to 15°F cooler than the surrounding countryside not just because of the shade but as a result of the heat consumed by the evaporation of hundreds of gallons of water each day from large trees, each of which is estimated to have the cooling ability of five 10,000-watt air-conditioners. This evaporative water loss, termed **transpiration,** occurs almost entirely through the leaves. One would think this prodigal use would signify an abundance of water, yet, except in a minority of swamp-dwelling plants or aquatic forms, water available to plants is in short supply for much or all of the year in most parts of the world.

Guard Cells and Their Control

Air accounts for much of the volume of even the thinnest leaf. If a leaf is immersed in water and gradually warmed, tiny bubbles of gas can be seen to stream from the stomata as the trapped air expands. The total cellular surface exposed to this internal air may be more than 200 times greater than that of the leaf itself, but evaporation into it is relatively slow, since the internal air is at 100% relative humidity all of the time. Just as the air escapes into the warmed water of our experiment, so water vapor escapes into the air from living plants by the same route.

The role of the stomata is to permit the entry of air, without which growth and eventually life would be impossible for the plant. Yet when the stomata are wide open, they allow water vapor to leave the leaf at a rate only 50% less than would occur if the parenchyma cells were directly exposed to the air. Thus, the open stomata represent a glaring weakness in the water-conserving adaptations of the leaf.

There are some simple plants—the liverworts—whose leaflike structures are equipped only with pores for gas entrance and exit, but these are restricted to wet or very damp habitats. The key stomatal adaptation that allows more complex plants to live even in the desert is the guard cell. These cells enable the plant to close the stomata when the risks of water loss exceed the benefits of continuing photosynthesis (Fig. 17–13). Moreover, guard cells can close the stomata at night when photosynthesis does not normally occur, for oxygen, present in great concentration in the surrounding atmosphere, does diffuse into the leaf through the cuticle in amounts sufficient to support life (see Focus on Photosynthesis in Desert Plants).

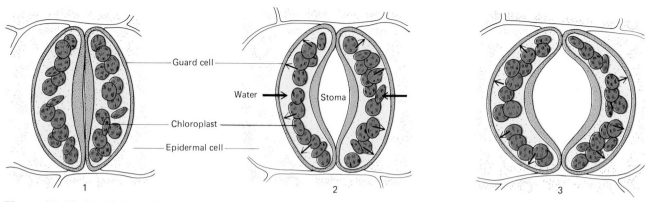

Guard cell

Water

Stoma

Chloroplast

Epidermal cell

1

2

3

Figure 17–13 Regulation of the size of the stoma: (*1*) Nearly closed condition. (*2*) When water enters the guard cells, turgor pressure increases and the guard cells buckle so as to increase the size of the stoma. (*3*) Stoma open.

How the guard cells are able to behave in this manner has been a puzzle for a long time. To understand the current explanation, let us consider the microscopic anatomy of guard cells. Each one is a crescent-shaped semicylinder, and each pair is joined at the ends. One of their most striking features is the chloroplasts, which make them unique among epidermal cells, others of which are not photosynthetic at all.

The cell wall of the guard cell is considerably thicker on the side nearer the stomatal opening. Additionally, fibers of cellulose in the cell wall are arranged in bundles, the **radial micelles,** which converge at the thickened, stomatal side of the cell wall. This arrangement of the cell wall fibers causes the concave surfaces of the two guard cells to pull apart when the volume and pressure of the cell's contents increase, thus opening the stoma. Conversely, when this pair of cells becomes flaccid, it relaxes not uniformly but in such a way that the stoma closes.

THE STEM

The stems of gymnosperms and woody dicots are of tremendous value as building materials. Even the monocot bamboo is pressed into service for scaffolding and fishing rods. The strength and flexibility that give stems their desirable engineering properties are required by the plant to support its leaves above those of its competitors and to hold them up despite every challenge of wind and storm. The stems of even herbaceous plants are the strongest parts of those plants.

The Vascular Tissues

Plants require a transport system much as animals do, and for the same general reasons. The body of any large plant contains so much tissue that diffusion alone could not effectively carry fluids (and the substances dissolved in them) to all the cells. Even the larger algae have some need for a transport system, but surrounded as they are by seawater, minerals and water itself are easily supplied to all their tissues directly. The transport system of a true land plant, however, not only carries the products of photosynthesis (**leaf sap** or **phloem sap**) in a generally downward direction but also conducts water and minerals (**root sap** or **xylem sap**) upward. Before reading further, please turn back to Table 17–1 and review the main plant tissue types.

PHLOEM

Phloem is a vascular tissue specialized to carry the products of photosynthesis from the leaves throughout the body of the plant. This process is known as **translocation.** The principal kind of phloem cell is the **sieve cell.** The sieve cells of phloem are living, and although when mature

Focus on . . .

PHOTOSYNTHESIS IN DESERT PLANTS

Having said that stomata open during the day and close at night, we must now note the exception of many desert plants (including the cacti) that reverse this process, opening their stomata at night and closing them during the day. Such plants are known as **CAM plants,** for crassulacean acid metabolism. (The crassulaceae, for which this metabolic system is named, are an important family of plants known for this adaptation, with many desert-living members.) In at least some CAM species that open their stomata in the dark and close them in the light, the synthesis of organic acids at night from CO_2 is apparently sufficient to result in opening of the stomata without K^+ uptake being required. One can understand the advantage in water conservation, for the heat of the day is just when water will evaporate most from mesophyll or other green tissues. Yet it seems an impossibility as far as carbon dioxide fixation is concerned. How can the plant obtain the raw material of photosynthesis?

Carbon dioxide is as abundant at night as during the day. There is no reason it cannot enter the leaf during the hours of darkness, given open stomata. However, since light reactions are impossible at this time, there is no way that the necessary ATP and protons can be generated for CO_2 fixation via the Calvin cycle. Some other fixation mechanism must be employed.

In the case of wet-leaved succulent desert plants and other CAM plants, the CO_2 is initially fixed by combining with phosphoenolpyruvate (PEP), which you may recall from our discussion of cellular respiration in Chapter 8. The source of the PEP is ultimately starch, from which it is formed by glycolysis. When the PEP accepts CO_2, it is transformed into oxaloacetate. The oxaloacetate is next reduced to malate by NADH. This metabolite is stored in the central vacuoles of the mesophyll cells until daybreak. (Each of these steps is catalyzed by a specific enzyme.)

As dawn approaches and the stomata close, this process starts to operate in reverse. Dehydrogenation of the malate forms oxaloacetate again, followed by decarboxylation, which then yields PEP (which began the process) and CO_2. The PEP can now be made into starch once more, and the CO_2 is available for photosynthesis in the now sealed-off tissues of the parenchyma. It is simply refixed by conventional combination with ribulose bisphosphate, and from then on photosynthesis proceeds more conventionally.

It seems inefficient for CO_2 to be fixed in this roundabout fashion, at what must be a very substantial energy cost. But consider the alternative. Under desert conditions conventional photosynthesis would not be merely inefficient, but fatal. Conventionally photosynthesizing plants can compete with desert plants wher-

Cactus, a typical CAM plant that grows in arid or excessively drained habitats.

ever those conventional plants can survive at all, because under moderate conditions the conventional plants are more productive of carbon compounds. The same thing is true, to a lesser extent, of C_4 plants. Because this process is *less* efficient in CO_2 fixation at low light intensities, C_3 plants actually have an advantage, as a rule, in northern, well-watered locales, and only a few plant members of such communities employ the C_4 process. It is instructive to consider this example of the interaction among biochemistry, plant anatomy, and ecology. We study these things as if they were separate subjects, but to the plant they are all one, with every aspect of its life-style being reflected not only in its individual adaptations but also in the relationships among those adaptations.

most of them contain no nuclei, they do contain a layer of cytoplasm just inside their cell walls. Sieve cells manage without nuclei because they are kept alive by the nucleated **companion cells** that always seem to be adjacent to them. There are profuse cytoplasmic connections between the sieve and the adjacent companion cells. In addition, phloem tissue is strengthened by **fiber cells** adjacent to the sieve and companion cells.

HOW PHLOEM WORKS

Phloem tissue is living and active. Each phloem cell contains a layer next to its cell wall of what is often called **p-protein** or, more traditionally (if less elegantly), **slime.** The currently favored hypothesis of phloem function is the **pressure flow mechanism,** which holds that materials pass through phloem by osmotically facilitated pumping (Fig. 17–14).

Much of what is known about phloem transport of materials has been learned largely from aphids. An aphid has a proboscis so fine that it can be inserted into an individual phloem cell. From this cell the aphid withdraws the food-laden and nourishing leaf sap, substantially free of slime or other contaminants. All the experimenter need do is wait until aphids are feeding on a plant and then snip off their proboscises. This will cause the leaf sap to ooze out of the severed proboscis by internal pressure. Chemical analysis of the phloem sap thus obtained reveals a nutrient content of up to 30% organic matter, containing mostly disaccharide sugars (especially sucrose) but also some amino acids and other substances. Radioactive tracer studies indicate that in intact phloem the leaf sap can flow at rates up to 200 centimeters per hour. This is a high rate of travel, taking into account the microscopically fine diameter of the phloem cells.

Sugars are made rapidly in the daytime by the photosynthetic parenchyma tissue of the leaf. They are then pumped by some cell membrane mechanism[1] of active transport into the sieve cells. This renders the phloem cells hypertonic to the surrounding leaf tissues. The leaf phloem then absorbs water osmotically from the surrounding parenchyma, producing high hydrostatic pressure in the phloem cells of the leaf.

The immensely long tube formed by the interconnected phloem cells connects with sugar-consuming cells somewhere else in the plant, usually in the root. These sugar-consuming tissues actively transport sucrose out of the phloem and then either metabolize the food or store it in the form of starch. (Conversion to starch has the effect of rendering the sugar osmotically inactive.) Thus the sugar-utilizing end of the phloem will become hypotonic to its surroundings and water will flow *out* of it. At one end of the phloem pipeline, therefore, there is high hydrostatic pressure, and at the other, low. There is usually no need for this pressure differential to oppose the force of gravity—the leaf sap flows mainly downhill. Such a pressure difference therefore may easily produce the rapid physical flow of leaf sap that we observe.

Since the direction of phloem transport is really "from" a source of material—usually sucrose—"to" a *sink* for that material (that is, a place where it is consumed or converted to an osmotically inactive form, such as starch), phloem transport is not *necessarily* downhill, although it is "down" an osmotic gradient. In the early spring, for instance, the sap rises in temperate zone trees. What happens is that starch stored in roots and trunk is converted to sugar and transported to the parts of the tree that are now consuming it—the budding leaves. If a shallow hole is bored in the trunk of such a tree, the sweet sap will run out of the layer of phloem that is just beneath the bark.

The sap of maple trees is collected in this way, boiled down, and sold as maple sugar or maple syrup. It is largely sucrose, incidentally, no different from the contents of an ordinary sack of sugar. The difference in taste may be accounted for solely by the impurities maple sugar contains. Not only food materials but also hormones may be distributed to various parts of the plant body by the flow of sap in the phloem.

---

[1] See "Solute transport and the life functions of plants" by Ernest G. Uribe and Ulrich Lüttge. *American Scientist* 72:567–573, Nov./Dec. 1984.

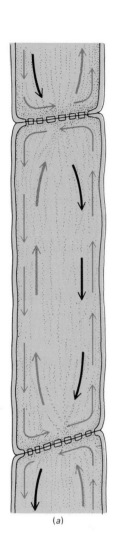

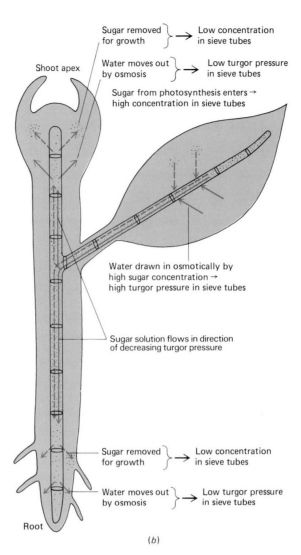

Sugar removed for growth } → Low concentration in sieve tubes

Water moves out by osmosis } → Low turgor pressure in sieve tubes

Sugar from photosynthesis enters → high concentration in sieve tubes

Shoot apex

Water drawn in osmotically by high sugar concentration → high turgor pressure in sieve tubes

Sugar solution flows in direction of decreasing turgor pressure

Sugar removed for growth } → Low concentration in sieve tubes

Water moves out by osmosis } → Low turgor pressure in sieve tubes

Root

(a)　　　　　　　　　(b)

Figure 17–14 (*a*) Cytoplasmic streaming in a phloem tube. The light and dark arrows illustrate that cytoplasmic streaming could account for the simultaneous transport of two different substances in opposite directions. (*b*) Diagram illustrating the pressure-flow theory of phloem transport in plants.

XYLEM

Xylem, in contrast to phloem, consists of hollow, completely dead cells laid end to end. There are several types of xylem cells, the most important being the fibers, vessels, and tracheids. The **fibers,** like sclerenchyma, are skeletal tissues and are vital for support in the stem, especially in plants composed of wood. The vessels and tracheids are conductive elements. The **vessels,** which with a few important exceptions are confined to flowering plants, are the simpler and presumably more efficient of the two types of water-conducting elements. They form from embryonic cells whose walls thicken and become impregnated with impermeable and rigid lignin. Then the cells die, in the process losing their cytoplasm. The walls at the ends of these cells may persist after death in the form of perforated plates somewhat as in phloem, or as ringlike flanges surrounding the lumen, or cavity, of the vessel. Many vessels also have small pits in their walls, which we will consider later in the chapter. The **tracheids** are smaller than the vessels and have acutely slanted end cell walls. Tracheids occur in all tracheophytes—in ferns, in flowering plants, and in gymnosperms. Tracheids are, in fact, the only kind of xylem tissue most gymnosperms possess.

HOW XYLEM WORKS

The tallest trees approach 350 feet in height. A redwood or giant Australian blue gum transpires hundreds of gallons of water between dawn and sunset. Even a medium-sized apple tree transpires as much as 20 liters per day at the height of the growing season. Every ounce of this water must be lifted to the leafy crown at a rate as great as 75 centimeters per minute. It is not as if the tree had ready access to this water, either. The water is usually bound by capillary forces to the soil and must be collected from a large area around the tree base.

How does root sap reach the top of a tree? Mineral ions are absorbed by the root via active transport and move toward the center of the root. As a result of the accumulation of solutes in the root, an osmotic gradient is established in which water moves into the root, creating a hydrostatic pressure called *root pressure*. If the soil contains ample water and an adequate oxygen supply (used in aerobic respiration, which provides the ATP energy for active transport), root pressure can reach a value high enough to force water into the leaves of smaller plants. Leaves possess modified stomata (hydathodes) at their vein endings out of which drips the root sap forced upward by root pressure. The root sap is exuded through the hydathodes, a process known as **guttation** (Fig. 17–15). Root pressure alone, however, cannot account for the rise of root sap to the tops of tall trees, since the root pressures that develop can push sap upward only a few feet. Furthermore, it is well known that during the day, when transpirational water loss is occurring, root sap is under a *negative pressure* or tension, rather than a positive pressure.

Since root pressure cannot account for movement of root sap to the top of a tall tree, is it, perhaps, *pulled* up? Among several explanations that have been proposed, the cohesion hypothesis accounts for most of the facts. Cohesion means "holding together," and that is what the supporters of the cohesion theory believe accounts for the ability of water to ascend to the tops of the tallest plants. It is based upon a behavior of water that is foreign to our macroscopic experience. The highly polar structure of the water molecule (see Fig. 2–6) ensures a significant attraction between the portions of *adjacent* water molecules with opposite partial charges. This produces a regular, somewhat ordered arrangement of these water molecules even in the liquid state. The hydrogen bonds also contribute to the mechanical strength holding the water molecules

Figure 17–15 (a) Guttation from strawberry leaves. (b) A hydathode at the edge of a leaf of *Saxifraga lingulata*. Xylem sap, when under positive pressure, moves through intercellular spaces in the adjacent loose parenchyma tissue and is forced out through the hydathode's pore, a modified stomate. Water is prevented from moving into other leaf tissues by a compact cellular sheath lacking intercellular spaces (a, J.N.A. Lott, McMaster University/ BPS. b, from Ray, Steeves, and Fultz, *Botany*. Saunders College Publishing, 1983)

(a)

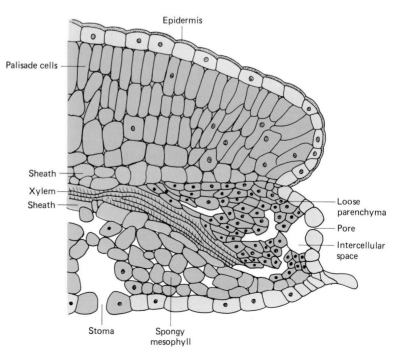

(b)

together. This is reflected, for instance, in an anomalously high boiling point for a liquid composed of such light elements, which might reasonably have been expected to be a gas at room temperature. It also results in the unexpected fact that if a column of water is formed into a filament the diameter of a xylem cell, it has the tensile strength of steel, more than adequate to hang such a filament unbroken from the top of the tallest tree.

The columns of water rushing through the vessels and tracheids are of sufficiently small diameter to hold together indefinitely without forming a cavity. As water evaporates from the mesophyll of the leaf, the solutes in its cells become concentrated. Water then passes by osmosis into the mesophyll from the adjacent xylem in the nearby vascular bundles of the leaf veins. As it leaves the xylem, each molecule of water exerts an infinitesimal pull upon the next in line in an unbroken chain stretching through the xylem of petiole and stem to the vast absorptive root surfaces among the interstices of soil particles below.

The cohesion system of xylem transport that vascular plants employ can function only if there is a continuous filament of water from one end of the xylem to the other. Any break in this liquid thread, however small, would cause the entire mechanism to fail. Though the absence of nuclei on which ice crystals can form does prevent freezing of the water in xylem down to about $-30°C$, in temperate and subarctic climates that filament may be broken when sap in the xylem freezes. Freezing expels any dissolved gas from the sap and produces gaps in the fluid column, resulting from the long bubbles of gas thus released. How can xylem transport resume after thawing, when the cohesion of the water filament has been destroyed?

Let us look again at the anatomy of the xylem cell. Both tracheids and vessels are extensively supplied with tiny pits on their side walls. These pits connect *adjacent* cells with one another so that water can flow around an obstruction by traveling sideways and then upward. It is thought that the distinctive construction of the pits permits them to act as one-way valves, isolating the part of the water filament that has been disrupted by bubble formation and routing root sap around the area.

Monocot Stems

Stems of monocots and dicots are different from one another—sufficiently so to merit separate discussion. The stem of a typical small monocot (like a corn seedling; see Fig. 17–6) is covered with an epidermis and cuticle not too different from that of the leaf. The stem may also have chloroplasts and even stomata. The bulk of its tissue is a pithy parenchyma with mainly skeletal function. In corn, for instance, vascular bundles are distributed rather randomly throughout this parenchyma, most of them in a vaguely defined layer around the periphery. In wheat the center of the stem is actually hollow.

In each vascular bundle one may observe both phloem and xylem, with the xylem located on the interior aspect of the bundle and the phloem on the exterior aspect; that is, they face inward and outward, respectively. If it could be separated from the surrounding pithy parenchyma, each vascular bundle would appear to be a strand reaching from roots to leaves interwoven in complex fashion with other vascular bundles, but in cross section (as it is usually studied) each bundle appears as shown in Figure 17–6. Monocot stems grow by a variety of means but do not usually have any means of secondary vascular growth; that is, once formed, the vascular bundle cannot enlarge further. In general, lateral growth of a monocot stem results from growth of the parenchyma, as can be seen by comparing the closely spaced vascular bundles in a

Figure 17–16 Section through a herbaceous dicot (sunflower) stem showing the arrangement of the vascular bundles. Note that the phloem forms a discontinuous layer around the periphery. (Ed Reschke.)

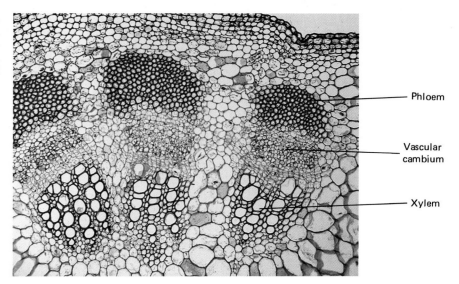

Phloem

Vascular cambium

Xylem

young corn shoot with the widely spaced bundles of a mature stalk. This mode of growth results in a fairly constant stem diameter from the base to the top of the stem, as can be observed in corn, bamboo, or palms, all of which are monocots.

Dicot Stems

Many annual dicots have soft, green, nonwoody stems. Such herbaceous dicot stems resemble those of monocots in exterior appearance, but their interiors are far different (Fig. 17–16). Perennial dicots have a similar stem organization but also contain woody tissues that the short-lived herbaceous dicots do not usually develop. In a typical herbaceous dicot, a skin of cork may cover the stem in place of the simple epidermis. **Cork** is composed of layers of dead cells impregnated with a waterproof substance, suberin, and with the **tannins,** which are commercially extracted for use in leather production. In some ways cork is comparable to the outermost layers of mammalian skin, which are composed of dead cells impregnated with the waterproof keratin protein.

The cork layer is pierced by **lenticels** (Fig. 17–17), openings that permit gaseous exchange between the atmosphere and the respiring tissues of the stem. These lenticels permit gases to diffuse through even the thickest trunk, for in such a trunk only a cardboard-thin layer of tissue beneath the bark is actually alive. Beneath the cork itself is a layer of **cork cambium,** a meristematic tissue whose function is to produce cells that differentiate, mature, and die, becoming converted into cork in the process. Well-developed bark is unusual in herbaceous dicots but is prominent in the woody dicots. It consists of everything external to the vascular cambium, including secondary phloem. As secondary growth of woody dicot stems occurs, the original cork layer thins into insignificance. (The "cork" of commerce, which is not truly cork in the formal botanical sense, is cut from the very thick bark of the drought-resistant species of oak native to the Mediterranean region.)

Beneath the cork cambium lies a layer of parenchyma known as the **cortex.** Embedded within this is a circular array of vascular bundles arranged with the phloem facing outward and the xylem inward, as in monocots, but the bundles are totally confined to this layer, unlike those of monocots. In the center of the stem is an initial core of **pith parenchyma.** Between the phloem and the xylem of the vascular bundles is a layer of vascular cambium. As it grows, the cambium differentiates into a vascular cambium that produces xylem centrally and phloem peripherally.

Figure 17–17 Lenticels on the bark of a young golden rain tree.

This secondary growth considerably increases stem diameter. However, since herbaceous dicots are usually annuals, their stems do not grow very thick as a rule, and it is for this reason that true bark does not ordinarily develop in them.

Most woody dicots and gymnosperms are trees and shrubs. Being perennials, they have ample opportunity to grow in diameter. As they do, the cambium develops into a continuous sheet of cells. This becomes the vascular cambium, which produces a more-or-less continuous layer of phloem and xylem, so the boundaries among the original vascular bundles become a thin layer interior to the phloem. Interior to the cambium is the xylem, which makes up more than 95% of the trunk of a typical tree. As successive seasons pass, ever more **secondary xylem** is laid down in rings, while the phloem remains much the same thickness because old phloem dies and is incorporated into the bark. Thus the bulk of the stem comes to be made up of xylem. The oldest xylem, located near the center of the stem, becomes **heartwood,** so impregnated with resins and other strengthening substances as to be incapable of transporting root sap. No longer a conducting tissue, the heartwood becomes entirely skeletal in function. It may also serve as a repository for the waste products of the tree's metabolism. The best cabinet and construction wood is heartwood.

The younger xylem, just interior to the vascular cambium, is **sapwood,** usually distinguishable by its much lighter color. If the heartwood is destroyed in an old tree by fungus or insect attack, the sapwood continues to sustain the life of the tree until perhaps it is finally blown down and broken. In contrast, a tree cannot survive without sapwood or, in the long run, without phloem. If the bark is removed in a complete circle around the trunk of a tree, the phloem is usually removed with it, and the roots starve for lack of leaf sap. The tree may survive for a season or so and may even produce a bumper crop of fruit (due to the accumulation of sugar in its crown), but it is doomed. Such an attack is called **girdling.** Farmers sometimes clear land in this fashion. Trees can also be killed by the girdling produced by the gnawing of beavers, starving deer, and even by birds, such as sapsuckers.

In temperate and desert climates, and also in some tropical, humid areas, there are often pronounced growing seasons. This results in differences in the amount of xylem produced by trees from season to season. In temperate zone trees there is often a big difference in color and texture between the less compact **spring wood** and the finer-pored **summer wood.** It is these differences that produce the familiar **growth rings** found not only in woody dicots but also in such gymnosperms as pine trees. (Some trees that are not subject to seasonal variations in growth in their natural habitat may lack growth rings entirely.)

ROOTS

All substances the plant requires from its environment (except sunlight, oxygen, and carbon dioxide) must be absorbed through the roots, and the anatomy of most roots is adapted to this function. Roots are by no means static structures; they grow in the soil, extending in directions where richer pockets of water and minerals may occur.

The total mass of a plant's roots is about equal to that of the branches but can be very differently arranged. A small rye plant, for instance, has about 14 million primary and secondary roots, the total length of which is probably much in excess of 300 miles, with a total surface area of about 2,500 square feet, not even counting the root hairs. The taproot of an oak tree may extend 25 feet below ground; the surface-feeding roots of a single prickly pear cactus extend no deeper than 6 inches but occupy a plot of ground 30 square feet.

Figure 17–18 A scanning electron micrograph (approximately ×150) of the root hairs of a radish. (J.N.A. Lott, McMaster University/ BPS.)

Root Structure and Function

Larger root branches much resemble the branches of the trunk of a tree, and like the trunk they are covered with cork provided with lenticels. The smallest roots are covered only with epidermis, many of whose cells are greatly modified to form long **root hairs** through which absorption takes place (Fig. 17–18). If these root hairs are lost during transplantation, plants may be damaged or killed. Thus, when plants are transplanted without accompanying earth most root hairs are lost. Virtually all foliage must also be removed, so as to greatly reduce the loss of water by transpiration until new root hairs can grow.

The collective surface area of the root hairs is many times that of the plain epidermis. Absorption of water and minerals through the root hair is believed to be aided further by a sticky coat of **pectin,** a plant polysaccharide that also occurs in many fruits. (It is the ingredient of jams and jellies that allows them to gel.)

The bulk of most roots consists of the *cortex,* composed mainly of parenchyma (Fig. 17–19). In many roots food is stored in the form of starch grains deposited in the parenchyma cells. The cortex also serves as a pathway for water movement, partially by the infiltration of water through the walls of its cells, but probably mostly by passage from one cell to another via intercellular plasmodesmata. Water and minerals are prevented from passing via cell walls into the central core, or **stele** of the root. This is accomplished by means of the **Casparian strip,** a wax-impregnated area of the walls of the cells that surround the stele (Fig. 17–20). These cells surround the stele in a single layer known as the **endodermis.**

In addition, single-cell root hairs located just behind the growing tip of very young roots absorb water and minerals into their cytoplasm. These substances are then passed through cytoplasmic connections from cell to cell all the way to the center of the root, probably bypassing the Casparian strip when it travels into the root by this route.

If water with its dissolved mineral ions cannot pass directly around and between the cells of the endodermal barrier, it must go *through* them, passing first through their plasma membranes. This permits the endodermal and probably also the stelar cells to regulate the chemical traffic into the plant.

The importance of this arrangement in regulating the composition of a plant's body fluids is dramatically illustrated in the case of mangrove trees. Several tropical shrubs and trees, collectively called **mangroves** even though they belong to several species, live in saline mud

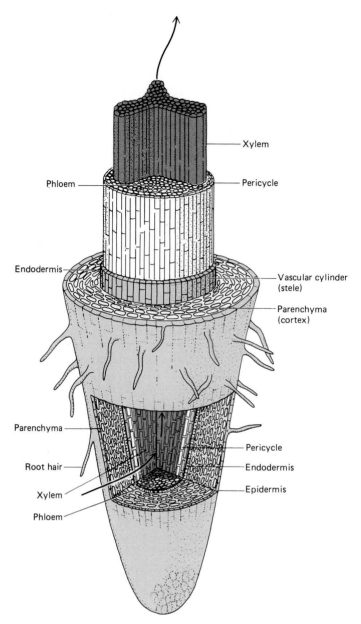

Figure 17–19 Root structure of a typical young dicot root. Water, minerals, and certain organic substances enter the root and cortical cell hairs by diffusion and active transport. These must diffuse through cortex, endodermis, and pericycle in order to reach the xylem. They then travel upward through the xylem. Food materials travel down through the phloem, and some food materials are stored in the form of starch in the cortex parenchyma cells.

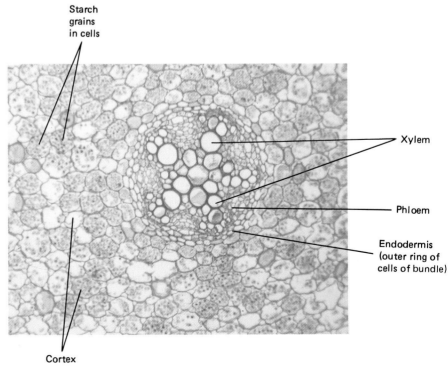

Figure 17–20 Microscopic structure of a typical vascular cylinder (stele) of a dicot root (*Ranunculus*).

371

flats. Yet the sap in the xylem of these trees is far lower in salinity than the water surrounding their roots. The explanation for this is that during transpiration so much pull is exerted on the water filaments in the xylem that water is forced through the membranes of the endodermis by atmospheric pressure. These membranes are differentially permeable, permitting little sodium to pass. The result is that the salt is largely left behind, so the plant functions as a solar-powered, reverse-osmosis desalinization device. What salt does make its way to the leaves is excreted by tiny salt glands, often located in the axils.

It is more usual in typical terrestrial plants, however, for the roots to absorb minerals by active transport systems, causing water to follow by osmosis. This produces a pressure (root pressure) in the root that forces water into the leaves of short plants, sometimes in amounts that threaten to drown the spongy parenchyma cells, or at least to fill the air spaces between them.

Immediately inside the endodermis of the stele lies a layer of parenchyma tissue somewhat resembling cambium. It is called the **pericycle.** The vascular tissues lie within this. In the roots of young dicots, xylem is usually arranged in a star-shaped configuration in the very center, with bundles of phloem occupying the spaces between the arms of the star. The phloem is often separated from the xylem by a layer of cambium, so in perennial dicots the roots as well as the stem are capable of secondary growth. The vascular tissues differentiate from a vascular cambium that originates by a fusion of the existing cambium with the pericycle to make a continuous layer. Branch roots also originate from parts of the pericycle.

Root Respiration

Soils are composed of various mixtures of clay, sand, silt, and the partially decayed organic matter called **humus.** Sandy soils are well oxygenated but hold little water or exchangeable nutrients. Silt and clay soils are poorly oxygenated but retain water and nutrients well. Loam soils are mixtures of these components and often combine their desirable characteristics. Almost any soil can be improved by the addition of humus, which must, however, be continuously renewed because it continuously decays.

A large proportion of the soil is therefore not solid at all but composed of either water or air. Without both, the plant could not exist. Roots respire like any other part of the plant and depend mostly upon diffusion to carry oxygen to their cells. This oxygen must be immediately available to the roots or they will suffocate, as can occur in very wet soils.

Some plants, however, are especially adapted to live in aquatic habitats or those with boggy soil. The water lily is an extreme example. This plant is equipped with floating leaves that have stomata on their upper sides, as might be expected. Its tissues are also more spongy than usual, which aids the flotation of the leaves and also serves to carry air via the stems to the roots in the mud at the bottom. There are even parasitic insects that live on the submerged roots and stems, dependent on the plant not only for food but for air as well.

In certain plants, such as some mangroves, air can reach the roots by special protruding organs called **pneumatophores** that reach above the water surface into the air and can form a small forest around the base of the plant. The protruding "knees" of cypress trees may have a similar function (Fig. 17–21). The roots of some tropical trees also serve to prop up the plant. Logically named **prop roots,** some of these may also serve as respiratory organs, and there is some evidence that respiration may actually be their major function.

(a)

CHAPTER SUMMARY

I. Plant organ systems are less clearly defined than those of animals, but we may distinguish root and shoot as the main divisions of the plant body. The shoot may be further subdivided into leaves, stem, and flowers.

II. Plant tissues are classified as meristematic and permanent tissues.

A. Meristematic tissue is composed of immature dividing cells that differentiate into other types of tissue.

B. Permanent tissues include surface tissues, fundamental tissues, and vascular tissues.

1. Surface tissues make up the outer covering of roots, stems, and leaves.

2. Fundamental tissues make up the great mass of the plant body: the soft parts of the leaf, most of the stem and roots, and the soft parts of flowers and fruits. Fundamental tissues provide support and produce and store food.

3. Two types of conducting tissues are xylem, which conducts water and dissolved salts, and phloem, which conducts dissolved nutrients, such as sugars and amino acids.

III. Within the leaf, the epidermis limits water loss, the mesophyll performs photosynthesis, and vascular elements carry materials to and from the other tissues.

IV. Guard cells limit evaporative water loss from the stomata.

V. Desert plants often absorb carbon dioxide during the night when there is less danger of desiccation than during the day; they then carry out actual photosynthesis during the day with the stomata closed.

VI. Phloem transfers material throughout the plant by an osmotic mechanism; xylem pulls water and dissolved minerals up from the roots by cohesion of the microscopic water columns contained therein.

VII. The vascular tissues are arranged in bundles with the phloem facing outward and the xylem inward.

A. In monocots the vascular bundles are scattered throughout the parenchyma of the stem.

B. In dicots the vascular bundles tend to form a layer near the periphery. If the dicot is woody, extensive secondary growth produces a large heartwood of dead xylem, plus a phloem com-

Figure 17–21 (a) Cypress "knees," possibly respiratory structures that may provide oxygen for the metabolism of the roots, which are buried in anaerobic mud. (b) Pneumatophores of coastal mangrove trees. These serve as a habitat for barnacles, algae, and snails, in addition to fulfilling their principal role in root respiration. (a, W. Harlow, Photo Researchers, Inc.)

ponent of the bark in addition to enlarged areas of functional vascular tissue.

VIII. Water and minerals are absorbed by the root.
A. When the soil solution is absorbed, it may enter through the root epidermis directly or indirectly, for example, via root hairs. It passes through the parenchyma to the stele.
B. The Casparian strip ensures that the endodermis will have the opportunity to regulate the absorption of minerals, usually by active transport mechanisms.

Post-Test

1. The main functions of the plant root are _____ and _____ of water and nutrients.
2. The _____ _____ are the only photosynthetic cells located in the leaf epidermis.
3. The openings between the guard cells are known as the _____ . These close when _____ is in short supply and in response to certain other stimuli.
4. In many leaves, part of the photosynthetic layer consists of the long, cylindrical _____ _____ cells.
5. The turgor pressure within the guard cells changes in response to an influx of _____ into these cells. Water follows by osmosis.
6. Movement of materials in phloem is from some high-sugar _____ to a low-sugar _____ . Most of the time, this gradient produces a flow from the _____ to the lower portions of the plant.
7. Xylem is nonliving. It generally transports _____ _____ and dissolved _____ in response to a combination of transpiration and _____ .
8. A minor function of xylem in some plants is _____ _____ storage. It also performs a skeletal function when it turns into _____ wood, in which resins and other compounds are deposited.
9. Roots generally generate a positive _____ _____ , which can result in the exudation of root sap from specialized stomata in the leaves. This process is known as _____ .

Review Questions

1. Compare the structure and cellular components of xylem and phloem.
2. What are the functions of roots?
3. Describe the processes by which a root absorbs water and salts from the surrounding soil and how these materials enter its vascular tissues.
4. Describe the three-dimensional structure of a woody stem. What are the functions of stems? Can stems and roots always be distinguished reliably from one another? If so, how?
5. Give the functions of the following: (a) cambium, (b) stomata, (c) heartwood, (d) lenticels, (e) cuticle, (f) Casparian strip.
6. Summarize the mechanisms by which guard cells regulate the size of the stomata.
7. What are the functions of leaves? What is the role of transpiration in the plant?
8. What special adaptations do desert plants employ in their photosynthesis, and why are these desirable from the plant's point of view?
9. What are the main differences in stem structure between monocots and dicots?

Readings

Bidwell, R.G.S. *Plant Physiology,* 2nd ed. New York, Macmillan, 1979. A worthwhile general discussion.

Epstein, E. *Mineral Nutrition of Plants: Principles and Perspectives.* New York, John Wiley and Sons, 1972. A clearly written summary of the basis of nutritional plant ecology.

Evert, R.F. "Sieve-Tube Structure in Relation to Function," *Bioscience* March 1982, 789–894. Osmotically driven transport systems in plants would be best served by vascular elements resembling sieve tubes.

Galston, A.W., P.J. Davies, and R.L. Satter. *The Life of the Green Plant,* 3rd ed. Englewood Cliffs, N.J., Prentice-Hall, 1980. A clear and clearly illustrated general plant physiology text.

Heslop-Harrison, Y. "Carnivorous Plants," *Scientific American,* February 1978. The adaptations of carnivorous plants permit them to survive in habitats where few other plants can live.

Marx, J.L. "Exploring Plant Resistance to Insects," *Science,* 216: 722–723 (14 May 1982). How plants fight back against their insect predators.

Marzola, D.L., and D.P. Bartholomew. "Photosynthetic Pathway and Biomass Energy Production," *Science,* 205: 555–558 (10 August 1979). Plant physiology practically applied to agriculture.

Zimmerman, M. "Piping Water to the Treetops," *Natural History* July 1982, 6–13. The micromechanics of xylem transport.

See also readings for Chapter 15.

Chapter 18

REPRODUCTION IN COMPLEX PLANTS

Learning Objectives

After you have studied this chapter you should be able to:

1. Discuss the adaptive advantages and disadvantages of both sexual and asexual reproduction.
2. Describe two methods of asexual reproduction in plants.
3. Compare sexual reproduction in gymnosperms and in angiosperms.
4. Describe a typical flower and identify its parts, using appropriate scientific terminology.
5. Describe the production of pollen and ovules in angiosperms, and summarize the process of fertilization as it occurs in such plants.
6. Outline the adaptive significance of fruits and summarize the processes by which they are formed.
7. Give the main structural features of typical monocot and dicot seeds.
8. Describe germination and seed dormancy.

ike all other living things, plants reproduce themselves. Plants seem to have a special proclivity for asexual modes of reproduction, as every gardener who has rooted a cutting or performed a graft well knows. Asexual reproduction produces genetically identical clones of individuals, and if their genotype is favorable, that exact genotype will be preserved by asexual reproduction to the advantage of all the individuals that have it. Sexual reproduction produces new genetic combinations, some of which may be an improvement. Once established, they too can be asexually propagated. Fortunately, sexual and asexual reproduction are not mutually exclusive options. Reproductive styles can be and are mixed to ensure opportunity for recombination of genes plus reliable propagation of the best combinations. Many plants reproduce asexually as a rule and sexually as an exception.

ASEXUAL PLANT PROPAGATION

A tree covering several acres may seem like a science fiction fantasy, but the East Indian banyan tree is capable of it (Fig. 18–1a). This tropical member of the fig family drops roots from its horizontal branches, which then become trunks in their own right, sending out more branches and yet more roots. If some chance event severs connections among parts of the tree, new individuals may result. For the banyan, asexual reproduction is just an uninhibited form of growth.

(a)

(b)

Figure 18–1 (a) *Ficus indicus*, the banyan tree, a tropical fig that spreads extensively by asexual reproduction via its adventitious roots. (b) Water hyacinths and (c) the Asiatic kudzu vine are obnoxious pests that spread very rapidly by asexual propagation. In each case all the plants visible may be members of a single clone. The kudzu, though it does flower, rarely if ever forms seeds.

(c)

Figure 18–2 Asexual propagation in *Kalanchoe*. Tiny plantlets develop along or at the ends of the branches. When they are mature enough, these propagules drop off the mother plant.

All the examples in Figure 18–1 involve little more than extensions of the original plant body, but some plants produce specialized **propagules**—bits of their bodies specialized for breakaway and dispersal. The *Kalanchoe* house plant buds off complete miniature plants from its leaves (Fig. 18–2), which can be carried off by water or other means and dispersed far from the parent plant. Perhaps the ultimate in propagule production occurs in the common dandelion. As far as is known, the showy yellow flowers produce seed by an asexual process exclusively, so each dandelion seed is genetically identical to its parent plant and, borne by the winds, carries that genotype to far places.

Artificial methods of asexual reproduction are employed by gardeners and agronomists to propagate desirable plant varieties without risking the loss of their desirable combination of traits. Such combinations of traits might be lost by the genetic scrambling that sexual reproduction would necessarily produce. Many people have left a sprig of ivy or Japanese dogwood in a container of water to produce roots; often such plants can be rooted in moist soil just as well. Modern methods of tissue culture carry this process further, sometimes producing an entire plant from a very few cells or even one original cell (Fig. 18–3). Finally, grafting is often employed to propagate fruit trees and other trees asexually. A cutting, called a **scion,** from one plant with desirable characteristics is attached to a root or stem, the **rootstock,** of another plant with different but also desirable characteristics. To take a classic example, in the late 19th century the French wine industry was almost ruined by a root-feeding aphid, *Phylloxera,* that had inadvertently been imported from America. The French saved their precious varieties of grape by grafting grape shoots (scions) onto resistant American rootstocks, whose own grapes would have been of inferior quality.

Figure 18–3 Propagating tobacco by tissue culture. (*a*) A fragment of undifferentiated tissue from the center of a tobacco stem is placed in a culture medium. A complete plant can be formed from the tissue fragment because each cell of the fragment contains all the genetic information of the original organism. Different kinds of hormones in culture medium will produce different growth responses. (*b*) Placed on a callus initiation medium, cells begin to proliferate and undifferentiated tissue called callus forms. (*c*) The callus produces roots on a root initiation medium. (*d*) Roots and shoots form on a root and shoot initiation medium. (Carolina Biological Supply.)

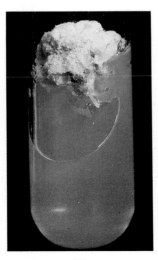
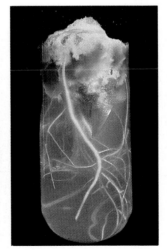
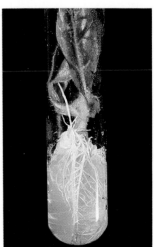

(*a*) (*b*) (*c*) (*d*)

SEXUAL REPRODUCTION IN SEED PLANTS

A **seed** is a sexually produced miniature plant, usually accompanied by a store of food to give the young seedling a vigorous start in life before the leaves are big enough to provide for its needs by photosynthesis. Seeds are almost invariably produced sexually, by the union of an **ovule,** the plant equivalent of an ovum, and a **pollen grain,** which contains or gives rise to the plant sperm. The process differs notably in gymnosperms and angiosperms.

Sexual Reproduction in Gymnosperms

The details of reproduction vary considerably even within the gymnosperms, but space limitations compel us to take the familiar conifers—the pines and spruces—as typical and to bypass the rest. Individual pine cones may be either **staminate** (male) or **pistillate** (female), but both are borne on the same tree (Fig. 18–4). The pistillate cones are what most people think of as pine cones—rather large clusters of scalelike modified leaves attached spiral fashion to a central stemlike structure. Pollen grains are produced in the staminate cones and carried off by the wind (Fig. 18–5). Occasionally, one or more of them will reach the pistillate

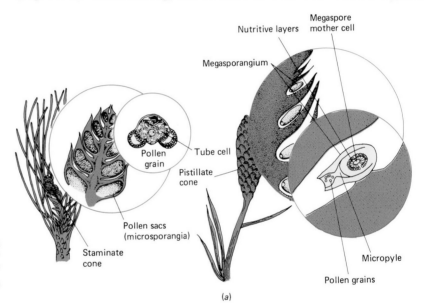

Figure 18–4 (*a*) Reproduction in a yellow pine tree. (*b*) "Male" staminate cones. (*c*) Young "female" pistillate cones. (*d*) Mature pistillate cones.

(*a*)

(*b*)

(*c*)

(*d*)

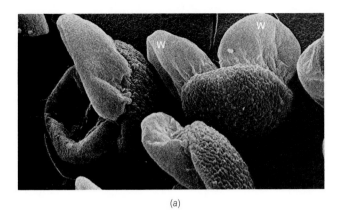

(a)

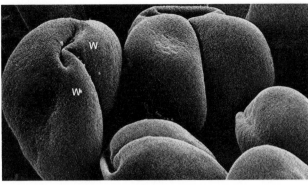

(b)

Figure 18–5 Scanning electron micrographs of pollen grains (approximately ×2250). (a) Each pollen grain of the red pine has two large wings (W) that provide it with greater surface area. This aids in the process of wind dispersal. (b) The blue spruce pollen grains have a different adaptation for wind dispersal. Instead of having wings, this pollen has an outer grain wall in the form of an air bladder (W) that provides extra buoyancy for the pollen. (J.N.A. Lott, McMaster University/BPS.)

cone, whose mature shape is aerodynamically suited to produce currents of air that may carry the pollen between the overlapping, slightly separated scales to the ovules at the bases of the scales.

Staminate cones contain structures called **microsporangia,** pollen sacs, in which pollen grains develop. Each microsporangium contains many **microspore mother cells** that divide by meiosis to produce four haploid **microspores.** Each microspore divides mitotically to form a four-celled **pollen grain.** Sperm develop from one of these cells. Because it contains sperm, the pollen grain may be thought of as a tiny gamete-plant, or **microgametophyte.** Thus, the pollen grain is all that remains of the gametophyte generation that the ancestors of the seed plants may have had in much more fully developed form.

There's a bit more to the female consort of the pollen grain—the **megagametophyte.** The ovule, which contains the megagametophyte, has an opening, the **micropyle,** for the entrance of the pollen grain. A year may elapse before the fortunate pollen grain responds to its surroundings, but when it does, an elongating **pollen tube** grows through an envelope of cells, the **megasporangium,** to the megagametophyte itself. One of the other cells in the pollen grain now divides its nucleus to form two male **gamete nuclei,** considered sperm cells in their own right. Each is surrounded by a delicate cell membrane even though contained in the cytoplasm of the pollen tube (the sperm of cycads and ginkgos is actually ciliated). The sperm-nuclei travel down the pollen tube and eventually reach the female megagametophyte. Only one of the gamete nuclei fertilizes the megagametophyte. The resulting zygote forms the embryo of the pine tree, and the surrounding haploid megagametophyte grows into a mass of nutritive tissue called endosperm. This endosperm is different from the true endosperm of flowering plants. An outer **seed coat** is derived from the tissues of the mother plant, and the seed is complete.

When the cone is mature, its scales separate and dry out, and the seed is released. In some cases the seed is released only under especially favorable conditions, such as the freedom from competition from established trees that might follow a forest fire. In the bristlecone pine, for instance, the seeds are released only from partly burned cones. Though lacking the elaborate adaptations for dispersal found among the seeds of flowering plants, pine seeds have little winglike protuberances that permit the wind to blow them some distance from the parent tree.

Sexual Reproduction in the Flowering Plants

Traditional symbols of romance, flowers, appropriately, are sexual organs. Flowers may vary in the parts that they contain, some having only male parts, some only female. Most, the so-called **perfect flowers,** have

Figure 18–6 The life cycle of an angiosperm.

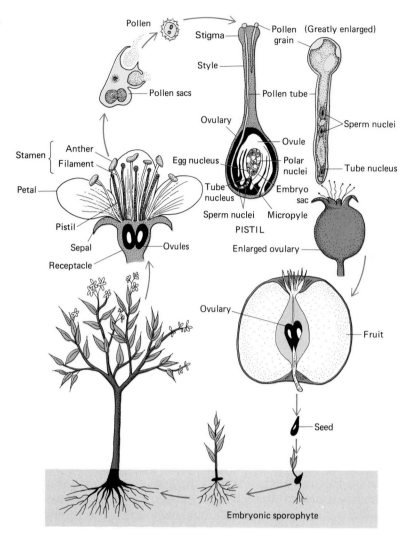

both stamens and pistils. They are so adapted, as a rule, to prevent self-fertilization or to permit it only after ample time has elapsed for fertilization, if possible, by another plant.

A typical flower contains five groups of parts, arranged in a series of approximately concentric circles, or **whorls.** All these are attached to the expanded end of the flower stem, called the **receptacle** (Fig. 18–6). The outermost parts, usually green and leaflike, are the **sepals.** Within the sepals are the brightly colored **petals,** which attract pollinators, such as birds and bees. Collectively, the petals are called the **corolla.**

Just inside the corolla and sometimes attached to it are the **stamens,** the male organs of the flower. A stamen consists of a slender **filament** atop which is an **anther,** consisting of a group of pollen sacs, called **microsporangia,** in which pollen grains develop. At maturity each pollen sac splits along a seam of weakness and turns inside out, exposing the pollen grains for dispersal. Each pollen grain has a distinctive shape, which often permits identification of the plant from which it came sometimes even when the pollen has been fossilized in geological formations.

In the very center of the flower may be one or several **pistils** to which, incidentally, the stamens are sometimes attached. Each pistil consists of a swollen base, the **ovulary,** containing the ovules; a long slender pillar, the **style;** and at the top of the assembly, the sticky **stigma** (Fig. 18–7) for the reception of pollen grains.

(a)

(b)

(c)

Figure 18–7 Some "male" and "female" flower parts. The hibiscus (a) bears the stigma (c) at the end of a stamen tube, which bears stamens and anthers (b) proximally. Each yellow anther carries pollen grains, just visible at this magnification, which stick to the fuzzy surface of the stigma of another flower. (a, E.R. Degginger.)

The stamens and pistils exist for the pollen and ovules that they house, and their function is to bring the two together. Each pollen grain develops from a **pollen mother cell,** which divides by meiosis to form four haploid **microspores.** The microspores give rise to individual pollen grains, each containing *two* mitotically produced nuclei, the larger **tube nucleus** and a smaller **generative nucleus** (Fig. 18–8).

Although some angiosperms (maple trees, for instance, and most grasses) are wind-pollinated like gymnosperms, most employ insects or birds as pollen carriers (Fig. 18–9). For this service they must usually make some payment in the form of **nectar,** secreted usually by special glands situated in the corolla. When a bee, for instance, enters a flower, it must usually brush against both stamens and pistil, transferring pollen from the last flower it visited to the pistil, and picking up pollen grains in its "fur" for delivery to the next. Bees tend to visit only flowers of the same color and species (a behavior called **flower constancy**) and so transfer pollen much more efficiently among these than the promiscuous wind could do. Plants may attract insects to their flowers by using insect pheromones (usually sex attractants), by color, and by perfume. Flowers are usually attractive to humans also, though doubtless by coincidence. This is not true of fly-pollinated flowers, however, which are colored and scented to resemble (in a fly's eye) festering wounds, rotting meat, or even animal anuses (Fig. 18–10).

Once a pollen grain has reached the stigma, by whatever means, it germinates, and a pollen tube grows down through the tissues of the

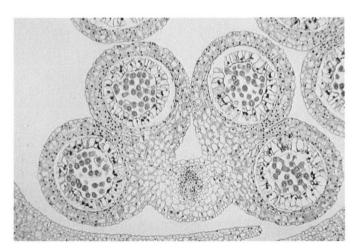

Figure 18–8 A light micrograph of a lily anther, showing mature tetrads, a stage in the meiotic production of pollen nuclei. (J. Robert Waaland, University of Washington/BPS.)

Figure 18–9 Various animal pollinators. (*a*) Honeybee, *Apis mellifera*. (*b*) A hawk moth, *Haemorrhagia thysbe*. (*c*) A greater short-nosed fruit bat, *Cynopterus sphinx*, pollinating a banana plant. The pollen grains on the bat's fur will be carried to the next plant, where cross-pollination will occur. Bats are surprisingly important in the pollination of many tropical fruit plants. (*d*) A ruby-throated hummingbird pollinating a flower from a trumpet vine. (*a*, Stephen Dalton, Photo Researchers, Inc. *b*, R. Humbert/BPS. *c*, Merlin Tuttle, Photo Researchers, Inc. *d*, Steve Maslowski, Photo Researchers, Inc.)

(a)

(b)

(c)

(d)

Figure 18–10 A fly-pollinated flower, skunk cabbage smells as you might expect. To the eye of a fly, as well as to its chemical senses, such oddly patterned and scented flowers may resemble festering wounds. (Alan R. Bleeker, Kings College.)

style till it reaches the female megagametophyte. This latter has developed from an ovule that contained one **megaspore mother cell.** These are diploid but divide by meiosis to form four haploid megaspores, three of which promptly disintegrate. The remaining megaspore then greatly enlarges, and its nucleus divides by mitosis, with each daughter nucleus traveling to opposite ends of the now elongated cell. These nuclei divide twice, converting the cell into a **megagametophyte** or **embryo sac.** It consists of three cells at each end and a large, doubly nucleated cell in the center. Two of the three smaller cells disintegrate; the remaining one becomes the **egg.** Now it is ready for fertilization.

As in gymnosperms, the pollen tube penetrates the megagametophyte. The generative nucleus divides in two, forming two sperm nuclei. These are discharged into the megagametophyte, and one of them finds and fuses with the nucleus of the egg. The other one seeks out the two nuclei of the large cell and fuses with them to form (usually) a triploid zygote that becomes the endosperm, a nutritive tissue found in many angiosperm seeds. The diploid zygote becomes the embryo. The flower parts either wither away or become the **fruit** or part of the fruit (see Focus on Fruit), and when the seeds are mature, they are dispersed by a variety of mechanisms: passed through a digestive tract, carried on an animal's coat, shot through the air like a bullet, blown away by the wind, and still others.

Seeds

If you soak an ordinary dried bean seed from the supermarket and open it (Fig. 18–11), you may easily see the future root stem and leaves already formed in the embryo therein. The embryo nestles between two immensely swollen seed leaves, the **cotyledons,** which in leguminous dicots like peas and beans replace the endosperm as organs of storage. Most dicots have no endosperm, although there are exceptions. The embryo can be divided into two parts on the basis of its attachment to the cotyledons. The **epicotyl** is the part above the attachment, and the **hypocotyl** is the part below.

Figure 18–11 Examples of seed structure. (*a*) Castor bean, a dicot that possesses endosperm. (*b*) Corn, a monocot with endosperm. (*c, d*) Common bean (*Phaseolus vulgaris*), a dicot with no endosperm, stores its food reserves instead in the embryo's thick fleshy cotyledons. *c,* Internal side view with one of the cotyledons removed. *d,* External view, edge on, showing by dotted lines the location of cotyledons and embryo axis within the seed. (*e*) Embryo of shepherd's purse, an herbaceous dicot. (From Ray, Steeves, and Fultz, *Botany,* Saunders College Publishing, 1983.)

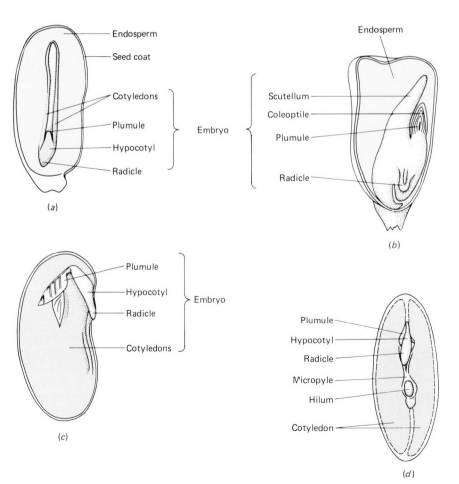

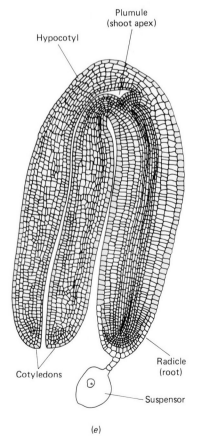

Focus on . . .

FRUIT

In the strict botanical sense of the word, a fruit is a mature ovulary containing seeds—the matured ovules. We usually think of only such sweet, pulpy things as grapes, berries, apples, peaches, and cherries as fruits, but bean and pea pods, corn kernels, tomatoes, cucumbers, and watermelons are also fruits, as are nuts, burrs, and the winged "seeds" of maple trees. A **true fruit** is one developed solely from the ovulary. If the fruit develops from sepals, petals, or the receptacle as well as from the ovulary, it is known as an **accessory fruit.** The apple fruit consists mostly of an enlarged, fleshy receptacle; only the core is derived from the ovulary.

The three types of true and accessory fruits are simple fruits, aggregate fruits, and multiple fruits. **Simple fruits** (e.g., cherries, dates) mature from a flower with a single pistil; **aggregate fruits** (raspberries and blackberries) mature from a flower with several pistils; and **multiple fruits** (pineapples) are derived from a cluster of flowers, all of which contribute structures that unite to form a single fruit. Fruits are also classified as **dry fruits** if the mature fruit is composed of rather hard, dry tissues, or as **fleshy fruits** if the mature fruit is soft and pulpy. Dry fruits represent adaptations for dispersal by the wind or for attachment to animal bodies by hooks. Birds, mammals, and other animals eat fleshy

fruits and their enclosed seeds. The seeds pass through the animal's digestive tract and are dropped with the feces in a new place. Thus fleshy fruits also represent an adaptation for the dispersal of the species.

In one type of dry fruit, termed a **nut,** the wall of the ovulary develops into a hard shell that surrounds the seed. The edible part of a chestnut is the seed within the fruit coat or shell. A Brazil nut is really a seed; there are about 20 such seeds borne within a single fruit. An almond is not a nut at all but the seed or "stone" of a fleshy fruit related to the peach. The outer woody covering of a peach stone or almond is, incidentally, part of the fruit. Only the meat within is the seed. The seeds of apricots, nectarines, and many other members of the rose family are poisonous and should not be eaten, for they contain cyanide.

Grapes, tomatoes, bananas, oranges, and watermelons, although superficially different, are all examples of fleshy fruits. In these the entire wall of the ovulary becomes pulpy. Such fruits are technically called **berries.** Peaches, plums, cherries, and apricots, in contrast, are stone fruits, or **drupes.** In a drupe the outer part of the ovulary wall forms a skin, the middle part becomes fleshy and juicy, and the inner part forms a hard pit or stone around the seed.

There are, then, many kinds of fruit, differing in the

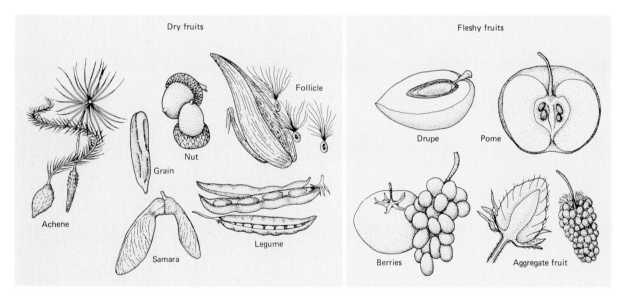

Various types of fruit. Fruits are usually adapted to aid seed dispersal, by being eaten or in some other way, such as flight or carriage on animal fur.

Quite different is a corn (maize) seed, usually studied as a typical monocot. In corn it is not easy to identify any cotyledon at all, but a structure called the **scutellum** that lies between the embryo and the endosperm appears to be homologous with the dicot cotyledon. Another structure, the **coleorhiza,** protects the future root, or **radicle,** and the embryo itself as it grows out of the seed into the surrounding earth. (In corn or

number of seeds, in the part of the flower from which they are derived, as well as in color, shape, consistency, and water and sugar content.

For most of the time that a fruit hangs on a plant it is quite inedible. Before ripening, the fruit is usually green, blending in with the surrounding foliage, unscented, unflavored, and hard. In a word, it is unappetizing. But when the fruit ripens, lovely pigments develop, enticing odors waft into the air, and starch is converted by hydrolytic enzymes into sugar while the flesh becomes deliciously soft. All this is governed by the gaseous plant hormone **ethylene,** which causes ripening to take place. Ripening stimulates the release of more ethylene, which in turn produces more ripening, and so on to completion. The entire process is orchestrated to be complete only when the seed is mature.

It takes a great deal of metabolic energy for the plant to go through all this. One would think that this energy could be better put to use producing more seeds. Yet by itself seed production is not very desirable from the standpoint of the parent plant. If the seeds germinate and grow up next to the parent plant, the two generations could only compete. Many plants, especially desert plants, practice **allelopathy**—they secrete toxic substances into the soil (usually by roots or fallen leaves) that prevent the establishment of seedlings and ensure regular spacing of the plants so as to conserve scarce resources, especially water. So fruit production can be thought of as a kind of toll paid by the plant to animal vectors in return for their services in seed dispersal.

(a) (b) (c)

(d) (e)

Flowering and fruit production in a citrus tree.

grass seeds the epicotyl is often called the **plumule.**) The epicotyl is protected by a special sheath, the **coleoptile,** which has no exact equivalent in the dicot embryo. The coleoptile protects the tightly rolled sheath of long leaves that grows at its *base,* instead of at the tip, as is more usual among plants.

Seed Germination

As a seed develops, its tissues gradually dehydrate even before the seed is released from the parent plant. In many cases, the simple restoration of water from the surrounding moist soil is enough to trigger **germination;** that is, the beginning of growth. However, some seeds have additional and seemingly perverse germination requirements, like prolonged soaking, certain light conditions, scarification of the seed coat or even charring by fire. But each of these requirements is actually an adaptation to the demands of the ecological role that the plant plays. For instance, in a desert annual (Fig. 18–12) it would be a fault indeed if a seed sprouted after any trifling rainfall. It would be best if germination would not occur unless the soil were soaked with enough water to provide for an entire season's growth (for in the desert, rainfall is not frequent). This is accomplished by means of a **germination inhibitor** that requires prolonged soaking to remove it from the seed.

The state of suspended animation of which we have been speaking is called **dormancy,** and breaking dormancy always seems to involve the hormone **gibberellin,** about which there will be more in Chapter 19. When the requirements for germination are met, gibberellin appears and triggers the growth and development of the embryo, together with the production of digestive enzymes that convert the stored starch, fat, and protein of the cotyledons or endosperm into forms usable by the embryo as it becomes a seedling (Fig. 18–13).

CHAPTER SUMMARY

I. Plants reproduce both sexually and asexually. Asexual reproduction preserves advantageous gene combinations, but sexual reproduction permits new combinations to be created.

II. Sexual reproduction in seed plants is the province of a greatly reduced gametophyte generation. The "female" or ovulate representative is the megagametophyte. The "male" or pistillate consort is called the microgametophyte. The microgametophyte is the same as a pollen grain.

   A. In gymnosperms the pollen grain produces a pollen tube within which a male generative nucleus divides mitotically to form two sperm cells. One of these may fertilize the ovum (the other degen-

Figure 18–13 (a) Seedling of bean, *Phaseolus vulgaris*, showing the hypocotyl, storage cotyledon, and the emerging photosynthetic leaves. (b) A coconut germinating on a Pacific coast beach in Costa Rica. Water transport is the usual dispersal mechanism for these huge, ungainly but seaworthy seeds. (a, J.N.A. Lott, McMaster University/BPS. b, L.E. Gilbert, University of Texas at Austin/BPS.)

(a)   (b)

erates). A haploid endosperm develops from macrogametophyte tissue.

B. In angiosperms the pollen grain also produces a pollen tube but the phenomenon of double fertilization produces a triploid or even pentaploid endosperm. The embryo is diploid.

III. Among flowering plants the embryo consists of an epicotyl, which will develop into the above-ground parts of the future plant, and a hypocotyl, which becomes the root and its derivatives. The cotyledons of dicots are usually large and the endosperm small or nonexistent. The opposite is true of monocots.

IV. In both angiosperms and gymnosperms the embryo and nutritive tissue is surrounded by the seed coat. In angiosperms only, the seed is further surrounded by an ovary derivative, the fruit, which usually functions in seed dispersal.

Post-Test

1. In the monocot seed the cotyledon may be reduced to a structure known as a _____ .
2. The portion of the seedling below the cotyledons is known as the _____ .
3. The seeds of desert annuals may contain a _____ _____ _____ , which is washed out by rains heavy enough to ensure sufficient moisture to support a complete life cycle of the plant.
4. The production of _____ by the seed plants largely circumvents the requirement found in ferns, mosses, and the like for _____ as a prerequisite for fertilization.
5. In pines the _____ cones produce pollen.

6. If all goes well, the pollen grain lands in a _____ _____ cone. There it develops a _____ _____ , which conveys two gamete nuclei to the female gametophyte. One survives to fertilize the ovule, producing the sporophyte _____ .
7. In a flower the _____ are usually the most brilliantly colored parts, although other flower parts may sometimes be brightly colored in addition to or instead of them.
8. Perfect flowers must contain both _____ and _____ organs; indeed, they must contain all typical flower parts.

Review Questions

1. What is pollen?
2. Describe fertilization in a pine tree.
3. What is the difference in the food storage tissues of the angiosperm and gymnosperm seeds?
4. Describe factors that influence germination.

5. How do roots manage to branch?
6. Describe secondary growth in dicots.
7. What are the genetic advantages of sexual reproduction? Of asexual?

Readings

Bierzychudek, P. "Jack and Jill in the Pulpit," *Natural History,* March 1982, 23–27. A description of jack-in-the-pulpit fertilization.

Cook, R. "Attractions of the Flesh," *Natural History,* January 1982, 21–24. How fruit serves as an inducement to seed dispersers.

Cook, R. "Reproduction by Duplication," *Natural History,* November 1979, 88–93. The advantages of asexual reproduction, at least in plants.

Echlin, P. "Pollen," *Scientific American,* April 1968. How pollen develops and functions.

Janzen, D.H., and P.S. Martin. "Neotropical Anachronisms: The Fruits the Gomphotheres Ate," *Science,* 215: 19–27 (1 January 1982). A discussion of seed dispersal by fruit-eating animals now extinct, including the interesting suggestion that dinosaurs may have dispersed ginkgo seeds.

Pettit, J., S. Ducker, and B. Knox. "Submarine Pollination," *Scientific American* 224: 135–143 (March 1981). True vascular plants, in fact, angiosperms, live permanently underwater. They reproduce by planktonic pollen.

Rick, C.M. "The Tomato," *Scientific American,* August 1978, 77–87. How we have adapted a flowering—and fruiting—plant to our needs.

Shepard, J.F. "The Regeneration of Potato Plants from Leaf-cell Protoplasts," *Scientific American,* May 1982. A new approach to cloning, or asexual propagation, of plants yields variants that promise future crop improvements.

Taylor, T.N. "Reproductive Biology in Early Seed Plants," *Bioscience,* January 1982, 23–28. Speculations regarding the origin of the seed plants.

Chapter 19

PLANT GROWTH, DEVELOPMENT, AND NUTRITION

Outline

I. Plant development
   A. Meristems
   B. The life of the leaf
   C. Secondary growth
II. Plant responses
III. Plant hormones
   A. Auxin
   B. Other hormones
      1. Gibberellins
      2. Ethylene
      3. Cytokinins
      4. Abscisic acid
      5. Florigen
IV. Plant nutrition
   A. Exchangeable nutrients
   B. Mineralization of organic matter
   C. Mycorrhizae
   D. Nitrogen fixation and root nodules

Learning Objectives

After you have studied this chapter you should be able to:

1. Describe apical (primary) meristems and their derivatives.
2. Describe or diagram a typical leaf bud.
3. Summarize the processes of leaf growth, senescence, and abscission.
4. Summarize the main adaptive features of leaf loss.
5. Describe the principal growth responses in plants, using the proper terminology.
6. Summarize the actions of the major plant hormones: auxins, gibberellins, ethylene, cytokinins, and abscisic acid.
7. Outline the relationship between auxin action and plant tropisms.
8. Summarize plant photoperiodism with respect to (1) phytochromes, (2) the control of flowering.
9. Roughly outline the process of absorption in the root, give the pathway materials take into the plant body after absorption has occurred and their role in plant nutrition.

Among plants, as among animals, the adult does not come into existence full-sized but must grow by a factor of, in some cases, many thousandfold. Growth is not just a matter of enlargement but of differentiation as well. Naturally, much development has already taken place in the seed even before it is released from the parent plant, but since many plants continue to grow throughout life, each branch and root tip (Fig. 19–1) must undergo a continuous process of differentiation similar in principle to what has already occurred in the embryo.

PLANT DEVELOPMENT

Although there are differences in just how the seedling pushes its way out of the soil, perhaps the highest priority of any seedling is the production of a functional root to absorb water and minerals. The developing root tip has a cap of moribund cells that protect it as it pushes through the soil. The cap also serves as a source of growth hormones. Just behind the very apex of the root is a zone of vigorous mitosis but, contrary to what one might expect, of not much growth. The growth takes place just beyond this, in the **zone of elongation.** As its name implies, this consists of elongating cells. The continuous production of new cells and the elongation of those cells produces root growth. However, just behind these growing regions lies a **zone of maturation,** in which tissue differentiation takes place. There, only growth in diameter is possible.

It is in the zone of maturation that root hairs form. These are long, cylindrical extensions of the epidermal cells. In still older regions of the root there are no root hairs and the epidermis becomes converted into a layer of cork. Meanwhile, root growth may help to push the entire epicotyl and cotyledons out of the earth (Fig. 19–2) into the sunlight.

Meristems

The root tip and a somewhat similar **terminal bud** on the epicotyl are the original growing ends of the seedling. The entire body of the future plant will be derived from these **primary (or apical) meristems.** As they grow, however, at certain regularly arranged locations called **nodes** on the stem, lateral buds or lateral roots develop from cambium and pericycle. (These are secondary meristems left behind as more or less peripheral layers in the developing stem and root.) Sometimes these nodes are arranged in such a way that the branches that grow from them spiral around the stem of the older plant; in other cases, they may be arranged

Figure 19–1 Terminal bud, the growing end of a camphor tree branch.

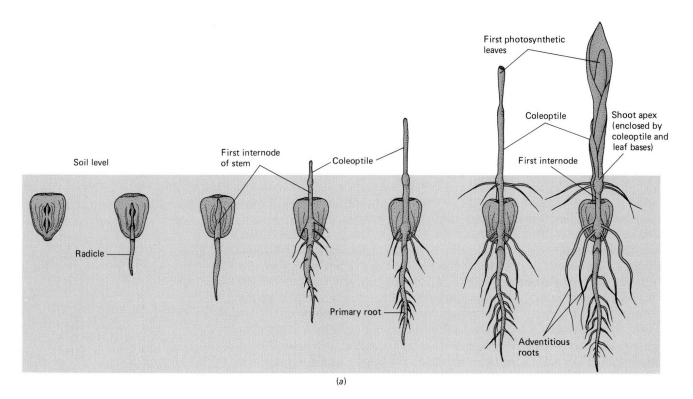

(a)

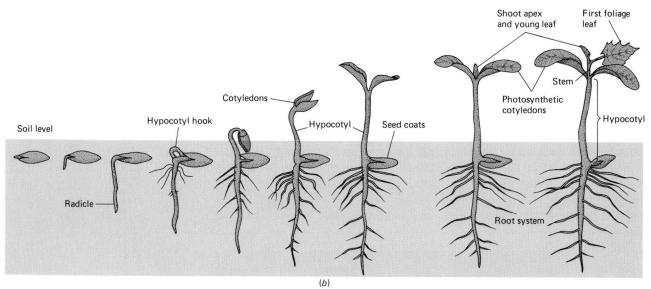

(b)

Figure 19–2 Development. (*a*) Hypogeous germination and early development of a corn plant. In this type of germination the seed remains underneath the surface of the soil. As the shoot grows upward through the soil, the young leaves are protected by the coleoptile that encloses them. Adventitious roots develop from the first-formed node at the base of the coleoptile. (*b*) Epigeous germination in the squash. In this form of germination the seed is forced out of the earth, exposing the cotyledons, which may then function briefly in photosynthesis before they eventually drop off. In the case of the squash (but not of the bean and most other epigeously germinating plants) the seed coat is left beneath the ground. Note the hypocotyl hook, which makes a path as it grows through the soil. Cotyledons are pulled up by growth of the hypocotyl and pass through this path without being broken apart. Above ground, light causes the hook to straighten. (From K. Norstog and R.W. Long.)

at mutual right angles (Fig. 19–3). At the tip of each branch or branch root there is another terminal bud or root tip, indistinguishable from the original terminal bud or root tip, and itself ultimately a product of those earliest meristems.

In dicots the cambium also produces growth in diameter by, for example, producing new vascular tissue, but once a part of the stem has been laid down, it cannot elongate. That is why an inscription of undying love carved in the bark of a tree will stay right there in plain view long after the love has died, instead of being carried upward out of sight, as the carver has perhaps come to wish.

The growth of lateral branches is regulated by the topmost (apical) bud by a mechanism that is still somewhat unclear but probably involves the auxin series of plant hormones. As long as the mechanism is active, the growth of lateral branches is to some degree suppressed, although the exact extent of the suppression varies. This phenomenon, known as apical dominance, accounts for the many different characteristic shapes of trees and shrubs.

The Life of the Leaf

If the season is winter or early spring, you can examine a leafless twig of a deciduous tree. Notice that it bears many leaf scars left by the fall of the last growing season's leaves. If you look closely, you will see a very tiny **axillary bud** just above each leaf scar (Fig. 19–4). Most of those will never develop any further, but the twig you are looking at did develop from one just like them. The branches that may develop from those leaf scars will have the same arrangement as the leaves that left those scars. At the very tip of the twig, though, you will find mature buds that are ready to unfold into leaves and perhaps flowers. Each leaf bud is covered with a shingling of tiny scales that protect it through the winter. These are forced apart by the growing tissues of the swelling bud in the spring (the scales do not occur in summer buds). Within the bud are poorly differentiated embryonic leaves, the **leaf primordia.** In the center of the leaf primordia is an apical meristem.

Each leaf originates on the side of the meristem, as do certain other structures in some kinds of plants (Fig. 19–5). The leaf grows upward and outward as it differentiates into the several kinds of tissue that a

(a)

(b)

Figure 19–3 (a) Spiral arrangement of leaves of a jade plant. (b) Regular arrangement of branches in a young East Indian rosewood tree. This reflects the original arrangement of leaves on a very young stem. Buds and eventually branches developed at each leaf location.

Figure 19–4 Leaf scar and axillary bud in *Cassia.*

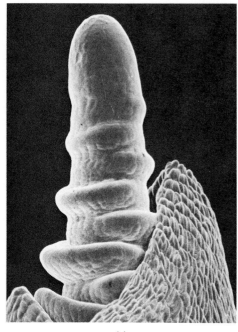

(a)

Figure 19–6 Scanning electron micrograph of the shoot apical meristem of Arawa wheat (approximately ×200). The tip of the apex consists of apical initial cells, which divide to form the shoot, and the leaves are initiated lower down on the side of the apex. The leaves first appear as a ridge, but with successive cell division and expansions, the size of the ridge increases and takes on the more familiar shape of a leaf. The spiral arrangement of the future leaves is already evident in that of the primordia. (*b*) Conventional photomicrograph of the rapidly dividing cells of the apical meristem of *Coleus*. (*a*, from Troughton, J., and L.A. Donaldson. *Probing Plant Structure*. New York, McGraw-Hill, 1972. *b*, Courtesy of Triarch.)

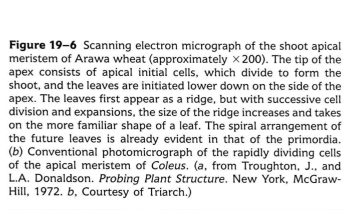

Embryonic leaves

Apical meristem

(b)

mature leaf contains. Once a leaf's growth is well established, new primordia just apical to it begin to differentiate in their turn, and so on in a potentially unending process (Fig. 19–6).

The leaves grow and expand, and as they mature, their content of tannins and other natural insect repellents and pesticides increases. This requires a frantic growth adaptation by the caterpillars that eat these leaves—a kind of race to consume as many new leaves as possible early in the spring before they become inedible. For a week or so the trees may seem to drip caterpillars (Fig. 19–7), and then, suddenly, the insects all but vanish. Recently, it has been shown that trees infested with caterpillars give off some gaseous substance, probably ethylene. This apparently warns other trees of the same species and stimulates them to develop inedible leaves faster than they otherwise would.

Even if unattacked by caterpillars or katydids, leaves die sooner or later. This is true even of evergreen perennials, only their leaves do not all die at once. In deciduous trees, like maples, the leaves all age and die simultaneously, sometimes spectacularly. But why?

Some tropical forests exhibit complete seasonal leaf drop, much as do the temperate deciduous forests. In these tropical forests this seems to be an adaptation to the occurrence of pronounced dry seasons. With the leaves gone, the trees require very little water. (Some tropical trees, however, lose their leaves in the *wet* season. The waterlogged soil inhibits root respiration, and this, in turn, interferes with water absorption.) Physiological drought also occurs in colder climates during the winter. Then almost no water can be absorbed through the roots. One strategy that can be employed in this case is, once again, to lose all leaves and enter a dormant state. (Some broad-leaved evergreens—citrus, for example—become dormant in the winter even without losing their leaves. Broad-leaved evergreens that live in really cold climates, such as *Rhododendron,* have heavy coats of wax on the leaves or other adaptations that tend to minimize water loss.)

As the season progresses, a deciduous tree produces fewer and fewer new leaves. Those that have remained throughout the summer are *senescent,* that is, aging. They are now decidedly tattered in appearance and are performing little photosynthesis. Chemical analysis will disclose a reduction in protein, and often in mineral and sugar content, as vital substances are withdrawn by the main body of the plant to be conserved for next year's use. When the leaf is almost dead, it may begin to change color, especially if the days are sunny and the nights cool. The simultaneous transformation of entire forests produces spectacular autumn scenery in some years, especially in much of North America, Japan, and part of China. The colors result partly from the unmasking, by the removal of green chlorophyll, of xanthophyll and carotene pigments already present and partly from the production of additional pigments. The significance of autumnal coloration is not always clear, but in some plants, like Virginia creeper or bearberry, the bright foliage may attract birds and other seed-dispersing fruit eaters to otherwise inconspicuous berries.

At the base of each petiole a lignified **abscission layer** (Fig. 19–8) develops. The abscission layer is mechanically weak, so the petiole of the leaf can easily break away from the stem. Being impermeable to water, the corky coating of the leaf scar also prevents evaporation from the leaf scar that results. **Abscission** is the process by which plants shed leaves, flowers, and other parts. An abscission layer is an adaptation that permits loss of leaves. When abscission is complete, those few yellow leaves that hung upon the boughs at last drop off to decay and form part of the humus of the soil, their remaining minerals liberated by fun-

Figure 19–7 Many trees are defenseless against herbivores like this *Polyphemus* caterpillar. (Chip Clark.)

(a)

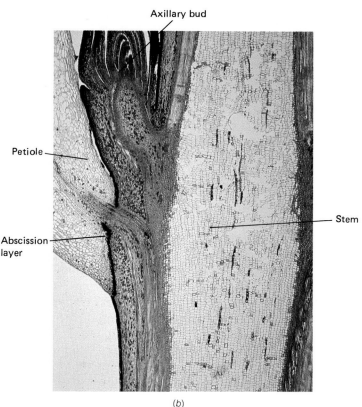

Axillary bud

Petiole

Abscission layer

Stem

(b)

Figure 19–8 (a) Leaf scars resulting from the loss of the giant leaves of *Monstera deliciosa*, a tropical plant. (b) Abscission in a maple tree. The abscission layer is not yet complete but has almost separated the leaf petiole from the stem. Notice that an axillary bud has already formed. This will give rise to an entire new branch, complete with leaves and perhaps flowers of its own. (approximately ×40) (b, Biophoto Associates, Photo Researchers, Inc.)

gal and bacterial action and often reabsorbed by the roots of the very trees that have lost them.

Abscission is not an automatic consequence of leaf death, as can be seen when leaves are killed by an unseasonable frost or other such unusual occurrence. In the absence of an abscission layer, leaves may hang on for months in funereal brown unsightliness. Though plant hormones are surely involved in normal abscission, just how the process is triggered is not fully understood. The decline of auxin hormone production in a senescing leaf may be the main influence, but ethylene may also play a part. Despite its name, the hormone abscisic acid is rarely, if ever, directly involved.

Secondary Growth

In the shoot, as in the root, differentiation begins to occur just behind the apical meristem. In the center of the maturing stem is a **provascular cylinder** that will develop into phloem and xylem and some of the pith. Just outside this lies a layer of **ground tissue,** all of which becomes pith. On the surface a **protoderm** develops, which eventually becomes the epidermis.

Primary growth occurs relatively close to the tips of roots and stems, and results mainly in extending the plant body. **Secondary growth** increases the *thickness* of the roots and stems. At first the developing vascular bundles are widely separated (Fig. 19–9), and in monocots they remain so. But in dicots they enlarge and tend to merge with one another. This is made possible by the continued presence of a layer of meristematic tissue, the **vascular cambium,** between the phloem and the xylem in each vascular bundle. This vascular cambium produces **secondary xylem,** which grows toward the center of the stem, and **secondary**

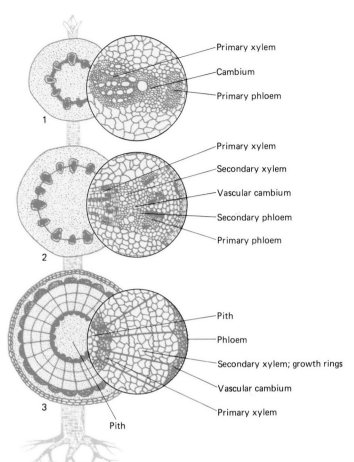

Figure 19–9 Maturation of vascular bundles in a woody dicot. (*1*) Near the growing bud the elements are immature, and vascular bundles are widely separated. (*2*) Farther down the stem the bundles are more mature. Vascular cambium gives rise to secondary tissues. Note that primary xylem is pushed toward the center of the stem. (*3*) Vascular bundles have coalesced in the 3-year-old portion of the stem. (Cork cambium not shown.)

phloem, which grows toward the exterior. The stem of a woody dicot that is a few years old consists mainly of old, hardened, nonfunctional xylem (the heartwood) surrounded by a relatively thin layer of functional xylem and, outside that, a still thinner layer of phloem. In the very center a tiny pith cavity may persist. A young dicot stem is smooth, but secondary growth soon splits the epidermis, which triggers the formation of **cork cambium.** This meristem produces the cork cells of the bark. At intervals nodes are formed, places where an apical meristem produces leaves and may give rise to buds that can develop into branches.

A plant must of course develop roots as well as stems and branches. Which it will be seems to depend on location and more precisely, on the relative proportions of two necessary plant hormones. Experiments have shown that if the hormone cytokinin is present in high concentrations, cultures of plant tissue tend to develop into roots, but if auxin concentration is high, stems develop.

Root development is analogous to stem development. Phloem and xylem differentiate from the provascular tissue in a star-shaped configuration. Usually there is a layer of vascular cambium that can produce secondary vascular tissues and secondary root growth.

Secondary growth in monocots is rare. A palm tree or a cornstalk, for instance, is about the same diameter from bottom to top. Yet the stem of a mature plant is bigger than that of a seedling. In some cases, the parenchyma of the stem retains meristematic properties and grows even after it is first formed. In other cases, as in palm trees, a **primary thickening meristem** occurs just behind the apical meristem. The primary thickening meristem continuously enlarges to produce an immediate increase in diameter just behind the terminal bud. It is the large-diameter primary

Figure 19–10 An unkempt palm tree. There are no branches; all leaves are produced by a single terminal bud. Bears clamber to the tops of cabbage palms to eat this bud, killing the tree in the process. Heart of palm is also prized among humans as a gourmet salad ingredient.

thickening meristem that lays down the vascular and other differentiated tissues. Finally, some monocots—such as bamboo and many grasses that propagate extensively by underground roots or runners—are initially small as seedlings. However, newer, asexually produced individuals are bigger from the outset, since they can draw on the resources of the already established parent plants.[1]

Monocots usually have only one apical meristem, which leaves clusters of leaves behind at the nodes, but no axillary buds and therefore no branches. In palms, which are perennial monocots, old leaves die and hang down the trunk (Fig. 19–10) with only a single cluster of living leaves at the very top producing the distinctive palm appearance. In the center of this leaf cluster is the single bud the tree possesses. It is succulent and tender, somewhat resembling white asparagus in taste and a prized salad ingredient. Yet once this bud has been harvested, the tree can grow no more and soon dies.

PLANT RESPONSES

Plants have no nervous system. Nevertheless, a degree of intercellular electrical transmission does occur in some plants. For example, when a leaf of *Mimosa pudica* is touched, electrochemical impulses occur analogous to the electrochemical changes that occur in nerve and muscle in animals. These impulses are transmitted to a special group of cells, the **pulvinus,** located in the base of each leaflet. The cells abruptly lose turgor pressure and the leaflet collapses (Fig. 19–11). If the stimulus is strong enough, the reaction may spread through the branch or even the entire plant. But this kind of quick irritability is very rare among plants.

Most plant behavior consists of growth responses called **tropisms.** Differential growth, for instance, bends a plant toward a source of light. The stem grows faster on the side *opposite* the light source, which pushes the plant as a whole toward the stimulus. Just the opposite occurs with roots, which grow away from the light. Other growth responses of the same kind account for the curling of vine tendrils around supports, or postural responses to gravity, or the well-known and regrettable attraction of tree roots to the water in sewer pipes.

To form terms describing growth responses, biologists attach specific prefixes to the root *-tropism.* A response to light, then is a **phototropism,** to water a **hydrotropism,** to gravity a **geotropism,** and to touch a **thigmotropism.** To indicate whether the response is "toward" or "away," just add the adjective *positive* or *negative,* respectively. Thus we may call the typical response of a stem to light **positive phototropism,** and that of a root, **negative phototropism.**

PLANT HORMONES

As in animals, the growth and perhaps differentiation of plants is under the control of hormones, but plants have no specific organs comparable to the endocrine glands of animals. Plant hormones are produced by tissues, especially actively growing tissues. The foundations of our still sparse knowledge of plant hormones were first laid between 1910 and 1930 in Europe, but the earliest identifiable investigations of the questions we now know to be explicable in terms of plant hormones were performed by none other than Charles Darwin.

---

[1] Because secondary growth in monocots does not involve cambial layers, some biologists prefer to call such growth **diffuse secondary growth** to distinguish it from the **cambial secondary growth,** which occurs typically in dicots.

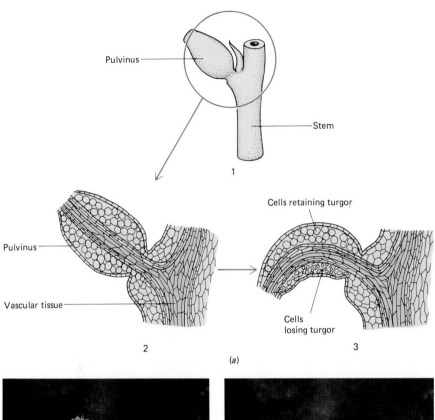

Pulvinus

Stem

1

Pulvinus

Vascular tissue

2

Cells retaining turgor

Cells
losing turgor

3

(a)

Figure 19–11 (a) The mechanism of response in the "sensitive plant," *Mimosa pudica*. (1) The base of the petiole, showing the pulvinus. (2) Section through the pulvinus showing condition of cells when leaf is extended horizontally. (3) Section through the pulvinus showing cells losing turgor to produce folding of the leaves. (b) *Left*, *Mimosa pudica* before being disturbed. *Right*, the plant five seconds after being touched. Note how the leaves have folded and drooped. (b, Richard F. Trump, Photo Researchers, Inc.)

(b)

Darwin worked with the seedlings of a number of kinds of grass (Fig. 19–12). As we have seen, the basal meristems of grass cause their seedling leaves to grow upward from the base, so it is logical, one would think, to look for the controller of growth at the base of the coleoptile, a tightly rolled sheath of young leaves that grows from the basal meristem of the grass seedling. Darwin was amazed to discover that the direction of growth of the coleoptile was influenced by light not on the base but on the *tip*, where growth was *not* taking place. Some kind of message apparently passed between tip and base. But what was the nature of the message?

Figure 19–12 Darwin's experiment with canary grass seedling. *Upper row*, Some plants were uncovered, some were covered only at the tip, and others were completely covered except for the tip. After exposure to light coming from one direction (*lower row*), the uncovered plants and the plants with uncovered tips (*right*) grew in a bent fashion. The plants with covered tips (*center*) grew straight up. Darwin concluded that the tip was light-sensitive, not the base where growth actually took place.

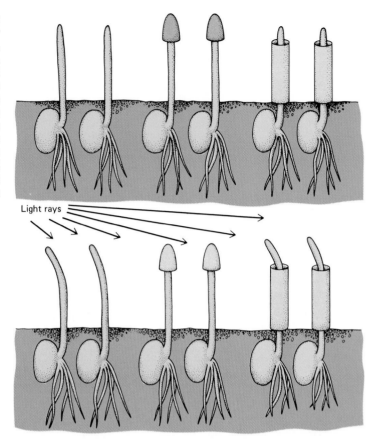

Light rays

Auxin

In this century two researchers, Frits Went and Boysen-Jensen, took up where Darwin had left off many years before. If the tip were indeed responsible for the tropisms of the coleoptile, the tropisms should cease if it were removed. As expected, removal of the tip stopped all growth, tropistic or otherwise. When the tip was replaced, growth resumed. When an inert substance—the mineral mica—was placed between the cut surfaces, growth stopped again (Fig. 19–13). Whatever the nature of the message between tip and base, it was stopped by mica. But that in itself did not disclose its nature. It could, for example, have been electrical or chemical.

The next attempt was made with agar, one of the complex polysaccharides obtained from brown algae (Chapter 15). When mixed with water, agar forms a gel that is mostly water. The investigators placed a block of agar gel—essentially a water bridge—between the cut surfaces of a coleoptile and its severed tip. Growth resumed. Was the growth factor a substance, one that was soluble in water? If so, it should be possible to charge an agar block with it, which should then stimulate growth even in the absence of the tip. The investigators left a block of agar in contact with a tip for some time, then took the agar block alone and placed it on a shaft whose own tip had been removed. Growth resumed. That clinched the matter. The tip secreted a hormone which diffused down into the shaft and there stimulated growth. Went named it **auxin,** from a Greek word that means "growth."

Further studies demonstrated that auxin accumulation on one side of a coleoptile or stem caused differential growth to take place, and that the chemical identity of auxin is, in its most common form, indole acetic acid. The most minute quantities of this terrifically powerful hormone will produce pronounced effects. It is produced from the amino acid tryp-

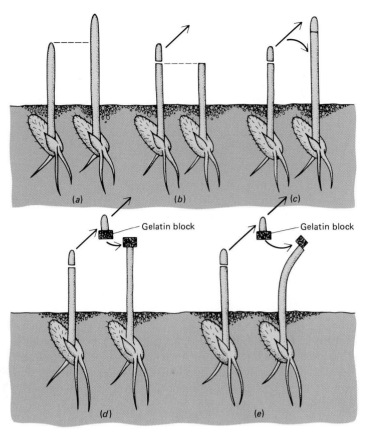

Figure 19–13 A series of experiments that demonstrates the existence and mode of action of plant growth hormones in oat coleoptiles. In each pair of drawings, the figure on the left indicates the experiment performed and the figure on the right the growth after a period of time. (*a*) Control: No operation performed, normal growth. (*b*) If the tip of the coleoptile is removed, no growth occurs. (*c*) If the tip is cut off and then replaced, normal growth ensues. (*d*) If the tip is cut off, placed on a block of gelatin for a time, and then the gelatin block without the coleoptile tip is placed on the seedling, growth occurs. (*e*) If the tip is placed on a gelatin block for a time and then the gelatin block is placed asymmetrically on the seedling, curved growth results.

tophane and actively transported by plant cells. When auxin reaches actively growing plant cells, it stimulates them to reduce their internal pH. This creates optimum conditions for an enzyme that weakens cellulose cross-links. This in turn weakens the cell walls so that they become stretched along their long axes by their turgor pressure.

Roots respond to light in just the opposite fashion from shoots. It is believed that auxin is responsible for the negative phototropism of roots also. It is thought that roots are already so full of auxin that light produces an excess of auxin on their dark sides, and that this additional auxin raises the total concentration to an inhibitory range. Thus, instead of growing faster in response, the dark side of the root grows more slowly. The light side and the root as a whole turn away from the light.

Auxins have many effects on the plant body. We have already mentioned the apical dominance. Auxins are also involved in cambial growth that results in the formation of **reaction wood**—wood that develops in a trunk or branch as a response to stress. Auxins are widely employed in agriculture and horticulture (as an aid to rooting of cuttings for instance), and artificial auxins are widely employed as herbicides.

Other Hormones

Auxin by itself cannot account for all the growth responses of plants, including their flowering. Neither, unfortunately, can the other plant hormones that have been discovered, at least not fully. We will attempt to summarize what is known and a little of what is conjectured about other plant hormones.

GIBBERELLINS

These not very euphoniously named substances were first discovered in Japan, produced by a fungus that caused the "foolish seedling" disease of rice. The young plants grew abnormally tall and died. The causative

Table 19–1 SOME IMPORTANT PLANT HORMONES		
Hormone	**Known targets**	**Action**
Auxin	Roots, young stems	Governs directional growth; stems respond positively, roots negatively; causes elongation of cells
Gibberellin	Stems	Causes elongation
Ethylene	Fruit, leaves	Causes ripening and deposition of toxic substances in leaves under insect attack
Cytokinin	Meristematic tissues, embryos, wounds	Stimulates mitosis
Abscisic acid	Roots, guard cells	Governs dormancy, root growth, and stomatal closure
Florigen (?)	Flowers	Hypothetical—may govern flowering

fungus is named *Gibberella,* hence the hormone name. Now known to be a normally produced plant hormone (though in smaller quantities), gibberellin is deficient in genetically dwarfed varieties of plants and is involved, along with other hormones, in leaf development, flowering, and fruiting.

ETHYLENE

Fruits develop initially from the ovary and adjacent structures in response to auxin hormones from the developing seeds in the ovary or even from the pollen grain. In fact, some fruits, like cultivated bananas and navel oranges, develop without functional seeds at all (and for that reason must be propagated by grafting or other asexual means). Actual ripening, is, as we learned in Chapter 18, stimulated by ethylene. This permits fruits and some vegetables, such as tomatoes, to be picked in a nearly ripened but green, easily shipped state, and then upon arrival at market to be gassed with ethylene so that they ripen. We have also seen that ethylene functions as a **pheromone,** or social hormone, in transmitting information of insect damage from one tree to another.

CYTOKININS

Auxins by themselves do not stimulate mitosis; indeed, they may inhibit it. Yet without young, actively dividing tissue on which to work, the cell elongation produced by auxins would be unable to occur or produce growth. Mitosis is stimulated by the **cytokinins,** which are especially important in the early growth of embryos. In fact, cytokinin action was first observed (though the researchers did not at first appreciate the significance of what they saw) in the action of coconut milk in plant tissue culture. This substance, the liquid endosperm of the coconut, is charged with cytokinin, presumably for the benefit of the developing coconut embryo. It appeared that plant cells would divide in artificial tissue culture only if coconut milk was added to the culture medium. Today, purified cytokinins are employed for this purpose. We have already mentioned the role of cytokinins in root and shoot differentiation, so it should come as no surprise that there is some indication they play a role in the formation of buds and leaves. Cytokinins apparently slow the senescence of plant parts, such as flowers and leaves, although whether this is a normal function outside the laboratory is not known. Cytokinins also promote the healing of wounds.

ABSCISIC ACID

We have discussed the role of abscisic acid in stomatal function. It was originally thought to stimulate leaf abscission, as its name reflects, but this function is now in considerable doubt. Abscisic acid does induce dormancy in many plants and has an important function in root growth control. It accumulates on the lower sides of roots, where it *inhibits* growth so that the root grows downward as a result of the faster growth of its upper-side tissues. This produces the positive geotropism of roots, in addition to whatever action auxin may have on them. It seems that starch grains in the parenchyma cells of the root may sink to the bottom of each cell in response to gravity, thus permitting the plant root to tell down from up. But how this action might determine the distribution of abscisic acid in the root—and even whether it does so in fact—is unknown.

FLORIGEN

There is something amusing about the story of the Maryland Mammoth tobacco variety. In the 1930s this mutant grew to unheard-of size, raising great hopes that were dashed when it proved impossible to get the plant to flower. Without flowers, of course, there can be no seed. To preserve the monstrosity, researchers moved it into a greenhouse where, with perfect perversity, it flowered in the middle of the winter. It turned out that in Maryland Mammoth and in all varieties of tobacco, flowering depended upon day length. A period of light, or **photoperiod,** so short as only to occur naturally in the winter was required to get Maryland Mammoth to flower.

This observation set researchers off on a photoperiodism trail that led to a host of useful and interesting results. It seemed that the flowering of many plants was regulated by day length, and that some of them—**short-day plants**—needed a photoperiod *below* a certain critical length to flower, while others, the **long-day plants** needed a photoperiod *above* a certain critical length. However it was finally discovered that what was important was not the length of the day but that of the night. A short-day plant is also a long-night plant, and vice versa. A short-day plant won't flower, no matter how short the day, if its night is interrupted by even an hour of strong illumination.

We must note that many plants are day-neutral, especially tropical ones (can you think why?), and that factors other than day length also play a part in the control of flowering. Thus one may see a spring-flowering plant confusedly produce blooms in the fall when the day length becomes the same as in spring. Yet there are usually far fewer flowers, and they do not generally set seed.

It seems clear that day length is sensed by a plant pigment **phytochrome,** which is sensitive to light—but not necessarily to visible light. The wavelengths that most affect phytochrome are in the red end of the spectrum. One of these, a red of wavelength 660 nanometers, affects the PR (phytochrome-red) pigment, and the other, a barely visible infrared light of 735 nanometers, affects PFR (phytochrome–far red) pigment. The two pigments are interconvertible; that is, red light converts PR to PFR, and infrared light converts PFR back to PR. But PFR also turns into PR spontaneously in the dark. It is known that flowering is initiated in long-day plants by PFR. Presumably, if the night is too short to permit its

Figure 19 – 14 Hydroponically grown lettuce. There is no soil in the plastic container, which serves only as a float to keep this and other lettuce plants (shown in rows in the background) from sinking into the nutrient solution (kept covered with white plastic film except for holes for the pots). Though soil is not used, these plants yield excellent salad greens. Notice, however, how the roots protrude through the hole in the bottom of the container. Even without soil the roots absorb water and nutrients needed for the plant's growth.

complete conversion back to PR, PFR will accumulate until it reaches a threshold concentration capable of triggering flowering. The reverse would be true in short-day plants.

Yet all this does not tell us how the phytochromes actually exert their control of flowering (and in some cases, seed germination). It has been shown that if just a part of some plants is exposed to the proper photoperiod, flowering can be induced in the whole plant. This would seem to imply the existence of a flowering hormone carried from the exposed part to the unexposed parts. If it is ever shown to exist, such a hormone will be known as **florigen.**

PLANT NUTRITION

What the mouth is to an animal, the root is to a plant; through it the plant must both drink and, after a fashion, eat. To be sure, a plant obtains no energy from the soil in which it grows, but most plants ordinarily have no other source of minerals or water (Fig. 19–14). Water in a soil fills many of the spaces (Fig. 19–15), but some water may flow by gravity to join the groundwater, often out of reach of plant roots. What remains is bound by forces of capillary attraction. Since this water dissolves many mineral ions and holds them in solution, it is called the **soil solution.** When soil contains as much water as it can hold by capillary action, it is said to be at **field capacity.**

The soil solution or its components can enter the root in three ways: by active absorption by the root hairs (Fig. 19–16), by imbibition through the epidermis, and in some cases, by mycorrhizae (which we will presently discuss).

In imbibition, the soil solution is absorbed by the epidermal wall of the root. This is possible over large areas of the root surface. The solution then passes through the cortical cell walls, *without* ever being absorbed by the cells, until it reaches the endodermis. This is known as the **apoplast** route of absorption. There it can go no further by the cell wall route because it is stopped by the Casparian strip (Fig. 19–17). Any of its components that are absorbed must now be taken up by the cells of the endodermis. Therefore, any material entering the vascular tissues of the root must first be absorbed across the membranes of at least one set of cells, even if it is not absorbed by the root hairs.

Exchangeable Nutrients

Some minerals are so tightly bound to soil particles that they do not readily enter the soil solution. One might think them unavailable to plants, but at least in the case of cations they often can be dislodged by exchange with hydrogen ions. Root hairs produce carbon dioxide, which

Figure 19–15 Soil under three conditions of hydration. (*a*) Water-saturated; all pores filled with water. (*b*) Field capacity; macropores air-filled and micropores water-filled. (*c*) Permanent wilting point; pores air-filled, films of hygroscopic water (which roots cannot absorb) around soil particles, small amounts of capillary water at points of particle contact. This is the least amount of soil moisture that will still support some plant activity. (From Ray, P.M., T.A. Steeves, and S.A. Fultz. *Botany.* Philadelphia, Saunders College Publishing, 1983.)

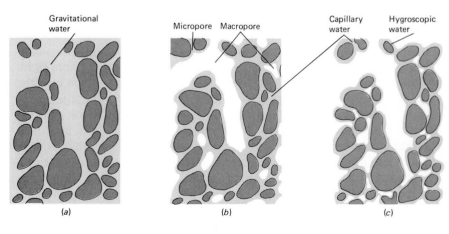

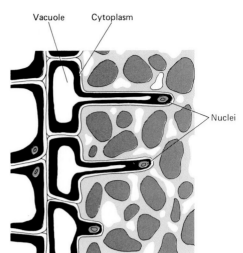

Vacuole   Cytoplasm

Nuclei

Figure 19–16 Root hairs extend into the soil solution, from which not only water but also minerals are absorbed, often by active transport mechanisms that concentrate them greatly. Each root hair is the extension of a single epidermal cell in a young region of the root tip. (From Ray, P.M., T.A. Steeves, and S.A. Fultz. *Botany*. Philadelphia, Saunders College Publishing, 1983.)

Figure 19–17 Absorption of nutrients from the soil. (*a*) Diagram illustrating the symplast theory of ion uptake by roots. Ions diffusing into the cell walls of epidermal or cortical cells can be actively taken up into the cytoplasm through plasma membranes. These ions can then pass without crossing any membranes until the ions reach the endodermis, at which point they must cross the cell membranes of the endodermis to be released into the xylem. Plasmodesmata are represented much larger relative to cell size, and fewer in number, than in actuality. (*b*) Diagrammatic cross section of a root, showing alternative pathways for uptake of water and nutrients. For simplicity the cortex is shown as being only a few cells thick, although in typical roots it is much thicker. The thickness of cell walls of living cells is exaggerated in the diagram. Note how the Casparian strip in cell walls of the endodermis blocks off pathway A all the way around the vascular core of the root. (From Ray, Steeves and Fultz, *Botany*, Philadelphia, Saunders College Publishing, 1983.)

forms carbonic acid. This then dissociates, forming hydrogen ions and bicarbonate ions. An additional quantity of hydrogen ion may be actively secreted by the root hairs. The result is the liberation of such ions as potassium and magnesium and their extraction from the soil in much greater quantities than simple water percolation could ever accomplish (Fig. 19–18).

Mineralization of Organic Matter

Soil is not an inanimate substance but a complex community in its own right, consisting of a bewildering array of inhabitants. Except for the roots of plants, however, all of these creatures are heterotrophs. A few of them eat or are parasitic upon the roots of plants, but most of them are either decomposer organisms (usually fungi) or detritus feeders that consume organic matter from dead or discarded plant parts.

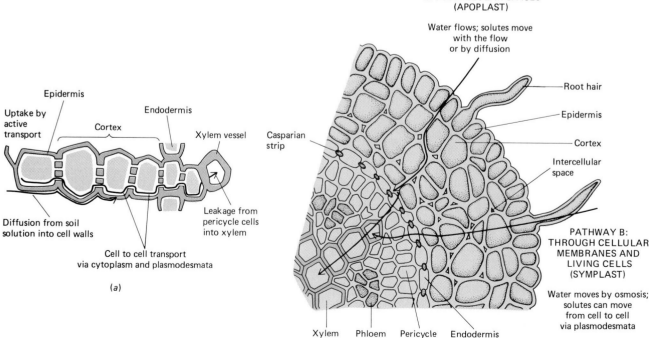

Figure 19–18 Cation exchange with soil particles. Arrows show that cation binding is reversible. (From Ray, P.M., T.A. Steeves, and S.A. Fultz. *Botany.* Philadelphia, Saunders College Publishing, 1983.)

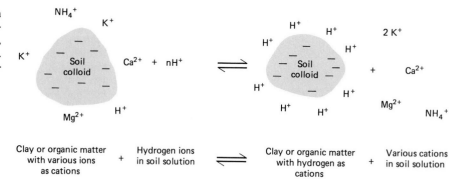

| Clay or organic matter with various ions as cations | + | Hydrogen ions in soil solution | ⇌ | Clay or organic matter with hydrogen as cations | + | Various cations in soil solution |

The organic matter of the soil contains mineral elements that must be liberated from it and converted to inorganic ions before they can be utilized by plants. Some of these ions were incorporated into the organic matter when that was first synthesized by the organisms of which it originally formed a part. In addition, however, organic matter tends to absorb and hold minerals from such sources as inorganic fertilizer, detaining them in the soil and preventing them from being leached out.

Mycorrhizae

Naturalists have long appreciated that some kinds of fungi appear mainly or even only in association with certain species of trees. The extremely poisonous death angel mushroom, *Amanita,* is often found in oak woodlands, for example. These mushrooms are derived from mycelial masses that infest the cortical region of the oak roots but are not parasitic upon them. The growth of oak trees is enhanced by the presence of the fungus rather than retarded by it, as one would expect from a parasitic relationship. Such infested roots have few or no root hairs, which suggests that the fungus might be absorbing minerals from the soil, much as root hairs would. Investigation with radioactive isotopes of phosphorus and other important mineral nutrients has disclosed that this is indeed the case—and that many materials are far better absorbed by fungal hyphae than by ordinary plant root hairs. The fungus, or mycorrhiza, for its part, benefits from the food materials provided to it by the tree, so the association is a classic example of mutualism.

Almost all plants that have been investigated have been shown to be mycorrhizal—to have some symbiotic relationship with a fungus. Many of them are obligately so and will not survive past the seedling stage without their mycorrhizae (Fig. 19–19). Mycorrhizae are particu-

Figure 19–19 Cultivated citrus species show a marked dependency on an endomycorrhizal association for adequate growth. Shown here are orange seedlings after 6 months, with mycorrhizae (left) and without mycorrhizae (right). Applications of calcium nitrate and ammonium nitrate were given to the two sets of plants, but only those with mycorrhizae absorbed these nitrates effectively, aiding in the plants' nutrition and creating a marked difference in growth response. (Dr. J. Menge, University of California, Riverside.)

larly important in tropical ecosystems. The soil in most such communities is deficient in organic matter and usually in available minerals as well. Because they are fungi, the mycorrhizae are able to decompose cellulose and other organic material that roots by themselves do not have access to. Therefore, because of the mycorrhizae, the roots of living plants are able to invade the bodies of freshly dead organisms by proxy and to extract the minerals immediately, before they can be leached away by heavy tropical rainfalls. So effective are these tropical mycorrhizae that the runoff from a rain forest often contains no more minerals than did the rain that fell upon it.

Nitrogen Fixation and Root Nodules

Though all organisms are surrounded by nitrogen, this gas is not readily available to them for incorporation into their tissues because of the tremendous amounts of energy required to bring it into any kind of chemical combination—that is, to accomplish nitrogen fixation. Until this has been done, nitrogen is biologically little more than inert gas (Fig. 19–20).

Most biological nitrogen fixation is carried out by nitrogen-fixing microorganisms, some of which live beneath layers of oxygen-excluding slime on the roots of a number of higher plants. But the most important

Figure 19–20 The nitrogen cycle. Since nitrogen is a gaseous element, it is easily recycled. However, much energy is needed to fix it, that is, to bring it into useful chemical combination. In terrestrial ecosystems solar energy drives photosynthesis, which produces the carbohydrates that in turn are utilized by the nitrogen-fixing microorganisms symbiotic with those plants. Solar energy may drive aquatic nitrogen fixation somewhat more directly, since a great deal of aquatic nitrogen fixation is accomplished by the heterocysts of cyanobacteria.

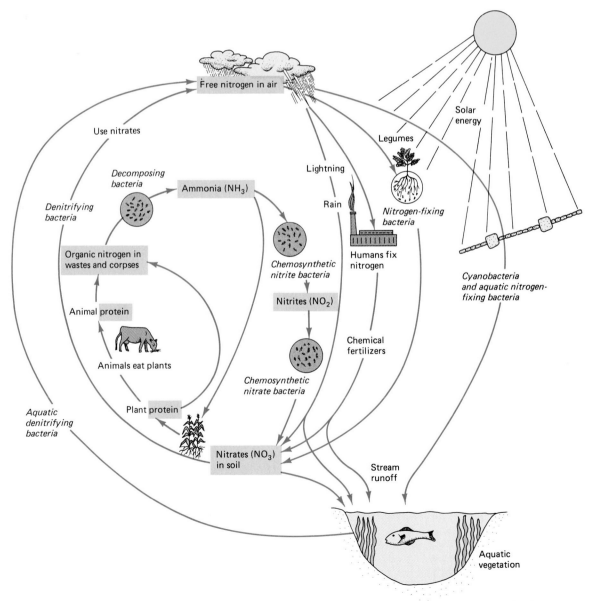

(a)

(b)

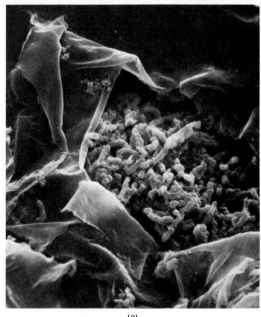

(c)

Figure 19–21 (a) Nitrogen-fixing bacteria nodules on the roots of peas. (b) Nodulated and control soybean plants. The plant on the right was grown in sterilized soil from seed inoculated with specific nodule-forming bacteria; that on the left was grown from uninoculated seed. (c) A look into a plant cell inside a soybean nodule. Note that the cell is packed with *Rhizobium*, symbiotic bacteria capable of nitrogen fixation. (a, Hugh Spencer, Photo Researchers, Inc.; b, USDA; c, courtesy of Dr. Winston Brill.)

of these microorganisms live in special swellings, or **nodules,** on the roots of leguminous plants, such as beans or peas (Fig. 19–21), and certain other woody plants in several families (the actinorhizal plants). It is estimated that in terrestrial environments these mutualistic bacteria can fix nitrogen 100 times faster than other, less vigorous nitrogen-fixing organisms, employing the energy released from carbohydrates provided by the photosynthesis of their host plant.

The reduction of nitrogen gas to ammonia by nitrogenase is a remarkable accomplishment of biological industry. It occurs without the tremendous heat, pressure, and energy consumption of the commercial processes developed by technology. Even so, it takes the consumption of 12 grams of glucose or the equivalent[1] to fix one gram of nitrogen biologically.

In addition to its energy requirements, nitrogenase demands an almost anaerobic environment, which root nodules provide. Oxygen is

---

[1] The energy can also come more directly from photosynthesis in the case of cyanobacteria, whose ability to fix nitrogen is important in many aquatic and swamp ecosystems.

transported to the actively respiring tissues within by **leghemoglobin,** an oxygen carrier very similar to animal hemoglobin. However, the chemical constants of the leghemoglobin are so balanced that little oxygen is available to the nitrogenase enzyme itself; indeed, it functions to maintain the anaerobic conditions that this enzyme requires. Legumes often are planted to introduce ammonia (and ultimately nitrate) into nitrogen-poor soils.

The use of ammonia for fertilizer deserves special mention. It is possible to inject ammonia gas into the ground, from which it is directly absorbed by plant roots. Since ammonia is directly usable in the production of amino acids, the plants need not absorb nitrate, which requires enzymatic reduction to ammonia within the plant before it can be used in amino acid synthesis. The **nitrate reductase** enzyme that plants employ for that purpose disappears from the roots when ammonia is added to the soil. The enzyme is no longer needed. Yet nitrate reductase reappears when ammonia fertilization is discontinued. This is one of the few cases of enzyme induction that has been observed among eukaryotes.

CHAPTER SUMMARY

I. Persistent embryonic tissue in plants is known as meristematic tissue. That of the terminal bud or root tip is primary, or apical, meristem.
   A. All tissue, differentiated or not, is derived from apical meristem.
   B. Secondary meristem, especially cambium, produces secondary growth of differentiated tissue in dicots. Such secondary growth as occurs in monocots results from special thickening meristems or parenchymal growth.
II. Most plant behavior is actually tropistic growth response to stimuli. These tropisms are governed by hormones secreted by various plant tissues but not by special glands, as in animals. The known plant hormones are as follows:
   A. Auxin. Has a general ability to stimulate growth of the above-ground portions of the plant and governs most tropisms.
   B. Gibberellins. Cause stem elongation.
   C. Ethylene. Controls fruit ripening and may act as a plant pheromone.
   D. Cytokinin. Governs shoot-root differentiation and is involved in abscission.
   E. Abscisic acid. Plays a role in guard cell action and perhaps sometimes in abscission.
   F. Florigen. An as-yet hypothetical hormone that may govern flowering. (Flowering and sometimes seed germination are often under the ultimate control of the phytochrome system of photosensitive pigments.)
III. Plants obtain nutrients from the soil solution, and also from clay particles on which they are adsorbed, by ion exchange. In addition, nutrients are liberated from organic material by detritus feeders, decomposer microorganisms, and mycorrhizae.
   A. Mycorrhizae occur in association with most plants. Functionally, they take the place of root hairs, but they are often far more efficient in extracting inorganic nutrients from the organic materials they decompose.
   B. Nitrogen fixation is accomplished by a variety of microorganisms but perhaps with greatest efficiency by those inhabiting the root nodules of legumes.

Post-Test

1. The root hairs form in the zone of _____ of the growing root tip.
2. The growing nodes of embryonic tissue found at the shoot and root tips of a plant are its _____ .
3. The arrangement of secondary branches on a trunk usually reflects the arrangement of the _____ on the parent branch during the preceding season.
4. The basic ecological and physiological function of seasonal leaf loss appears to be that of _____ conservation during times when this substance cannot be readily absorbed or is otherwise unavailable.
5. Secondary growth in monocots is minimal; in some of the treelike monocots, such as palms, there is a _____ _____ meristem located just behind the apical meristem.
6. In the phototropic responses of the above-ground parts of plants, auxins accumulate on the _____ side of the stem.
7. Auxin _____ the pH of elongating cells, which optimizes the attack of enzymes on the _____ _____ of the cell walls.
8. Fruit ripening is accelerated by the gaseous plant hormone _____ .
9. The photoreceptor pigment that controls flowering in long- and short-day plants is _____ , which is sensitive to _____ light.
10. Commonly, nitrogen is absorbed by plants in the form of _____ .
11. As a rule, nitrogen is first liberated from decaying proteins in the form of _____ . Some plants absorb it in this form without change, but it is more usual for them to absorb it in partially oxidized form; that is, as _____ .
12. Most nitrogen fixation from the atmospheric supply of nitrogen is probably accomplished by bacteria and, in aquatic (or even some terrestrial) instances, by _____ .
13. The passage of absorbed materials and water through the cortical cell walls to the vascular tissues of the root is referred to as the _____ route of absorption.
14. The _____ strip prevents the absorption of materials into the stele by the apoplast route.
15. Fungal microorganisms called _____ help the roots of many if not most plants absorb minerals from the surrounding soil.

Review Questions

1. What is an apical meristem?
2. What is the difference between an apical and an axillary bud?
3. How do roots manage to branch?
4. Describe secondary growth in dicots.
5. What are the genetic advantages of sexual reproduction? Of asexual?
6. Construct a table summarizing the known plant hormones, their sources, targets, and actions.
7. What is the relationship between the two forms of phytochrome and the photoperiodism of long- and short-day plants?
8. How do mycorrhizae function?
9. What is the significance of the root nodules of leguminous plants?

Readings

Northcote, D.H. *Differentiation in Higher Plants,* 2nd ed. (booklet). Burlington, N.C., Carolina Biological Supply Company, 1980. Plant tissue differentiation considered from a cellular viewpoint.

Ryan, D.F., and F.H. Bormann. "Nutrient Resorption in Northern Hardwood Forests," *Bioscience,* January 1982, 29–32. Not all nutrient reutilization results from recycling by decomposers. A substantial portion of fixed nitrogen and phosphorus is reabsorbed from aging leaves even before they fall.

Woolhouse, H.W. *Ageing Processes in Higher Plants* (booklet). Burlington, N.C., Carolina Biological Supply Company, 1972. An older but still good discussion of aging in plants.

(See also readings for Chapter 18.)

PART VI
ANIMAL STRUCTURE AND FUNCTION

The leap of a leafhopper. (Animals, Animals, Stephen Dalton.)

Chapter 20

ANIMAL TISSUES, ORGANS, AND ORGAN SYSTEMS

Outline

Learning Objectives

After you have studied this chapter you should be able to:

1. Discuss the advantages of multicellularity.
2. Define tissue, organ, and organ system.
3. Compare the four principal types of animal tissues—epithelial, connective, muscle, and nervous tissues—with respect to general structure and function.
4. List the functions of epithelial tissue, describe the three main shapes of epithelial cells, and describe how these cells can be arranged into tissues.
5. Describe the main types of connective tissue and their functions.
6. Compare the three types of muscle tissue and their functions.
7. Draw and label a typical neuron.
8. Distinguish between benign and malignant neoplasms and contrast the cells of a malignant neoplasm with those of normal tissue.
9. Briefly describe the organ systems of a complex animal.

Animals, like plants and most fungi, are **multicellular,** that is, they are composed of many cells (Fig. 20–1). They are not simply colonies or aggregations of similar cells, but are composed of a number of different types of cells, each with a characteristic size, shape, structure, and function. In most animals, cells are organized into tissues, tissues into organs, and organs into organ systems.

THE ADVANTAGES OF MULTICELLULARITY

Why is a human being or an elephant made up of billions of cells rather than one giant cell? Perhaps the main reason that cells do not get large is that it is inefficient for them to so do. All materials must pass into or out of the cell through the cell membrane, so the size of the membrane in comparison with the rest of the cell is critical. As a cell increases in size its volume increases at a greater rate than its surface. (The surface of a sphere increases in proportion to the square of its radius, while the volume increases as the cube of its radius.) A large cell, then, has proportionately much less cell membrane for its volume than a smaller cell (Fig. 20–2). Perhaps a familiar example will help make this concept clear. Consider potatoes. Would you rather peel one very large potato or four small ones? The amount of mashed potatoes prepared would be the same, but those four small potatoes have a lot more surface to peel.

As a cell grows, its surface (that is, the cell membrane) becomes unable to provide sufficient oxygen and nutrients for all regions of the cell. Wastes produced within the cell must move longer distances to reach the cell membrane and exit from the cell. Too great an increase would threaten the well-being of the cell. This explains why we do not see amebas as large as whales slithering about. Most one-celled organisms are so tiny that even the larger ones are barely visible to the unaided eye. There seems also to be a limit to the amount of cytoplasm that a nucleus can control. In fact, very large cells—including certain species of amebas—may have more than one nucleus.

(a)

(b)

Figure 20–1 All animals are multicellular; a large animal such as the bull elephant from western Kenya (a) is composed of more cells than the much smaller anole lizard (b). (a, R.K. Burnard, Ohio State University/BPS; b, B.J. Miller, Fairfax, VA/BPS.)

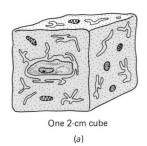 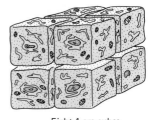

One 2-cm cube   Eight 1-cm cubes
(a)   (b)

Figure 20–2 Eight small cells have a much greater surface (cell membrane) area in relation to their total volume than one large cell.

When its size approaches the limits of efficiency, a cell divides to form two cells. Each new cell is half the size of the mother cell, but the relative size of both the cell and the nuclear membranes is greatly increased in proportion to the volume of the cell.

In unicellular organisms cell division results in the production of two new individuals; in multicellular organisms the two new cells may remain associated to form a *part* of the organism. The number of cells, not their individual size, is responsible for the different sizes of various organisms. The cells of an earthworm and an elephant correspond in size; the elephant is larger because its genes are programmed to provide for a larger number of cells.

Another advantage of multicellularity is specialization. In a one-celled organism the single cell must carry on all life activities. In an organism composed of many cells there can be a division of labor so that each type of cell can perform specific functions. When cells specialize, the organism can become highly proficient at performing a wide variety of activities. For example, some cells can become sensory cells and specialize in receiving information about the environment, while others, specialized to contract, enable the organism to escape a predator or capture food.

TISSUES

Even in the simplest animals, the sponges, there is a division of labor among several cell types. In all other animals, cells are not only specialized but also organized to form tissues. A **tissue** consists of a group of closely associated cells that are adapted to carry out specific functions. In human beings and other complex animals four main groups of specialized tissues are recognized: epithelial, connective, muscle, and nervous tissue. Each kind of tissue is composed of cells with a characteristic size, shape, and arrangement.

Epithelial Tissue

Epithelial tissue consists of cells fitted tightly together, forming a continuous layer or sheet of cells covering a body surface or lining a cavity within the body. One surface of the sheet is attached to the underlying tissue by a **basement membrane** composed of nonliving polysaccharide material (a product of the epithelial cells) and by tiny fibers. The outer layer of skin, the linings of digestive and respiratory tracts, and the lining of kidney tubules are examples of epithelial tissues.

Epithelial tissues may function in protection, absorption, secretion, and sensation. As a covering or lining, epithelial tissue protects the body. The epithelial layer of the skin, called the epidermis, covers the entire body and protects it from mechanical injury, invading bacteria, and excessive water loss. The epithelial tissue lining the digestive tract absorbs nutrients and water into the body. Other epithelial cells may be organized into glands, adapted for the secretion of cell products like hormones, enzymes, or sweat. Sensory epithelium is specialized to receive stimuli. The olfactory epithelium in the lining of the nose, for example, contains cells that respond to the presence of certain chemicals in the air we breathe. These cells are responsible for the sense of smell.

Everything that enters the body or leaves it must cross one or more layers of epithelium. Even food that is taken into the mouth and swallowed is not really "inside" the body until it is absorbed through the epithelial lining of the digestive tract and enters the blood. The permeability properties of the various epithelial tissues regulate to a large extent the exchange of substances between the different parts of the body and between the organism and the external environment.

Many epithelial membranes are subjected to continuous wear and tear. As outer cells are sloughed off, they must be replaced by new ones from below. Such epithelial tissues generally have a rapid rate of cell division so that new cells are continuously produced to take the place of those lost.

Several types of epithelial tissue may be distinguished by the number of cell layers, the shape of the cells, and their arrangement (Table 20–1). Epithelium may be **simple,** consisting of one cell layer, or **stratified,** consisting of many layers (as in the outer layer of the skin). A third arrangement is **pseudostratified** epithelium, in which the cells (falsely) appear to be layered but in fact are not. The epithelial cells may be **squamous** (flattened), **cuboidal** (cube-shaped), or **columnar** (elongated like columns). The free surface of the outer cells of the tissue may have specialized structures like cilia or **microvilli,** tiny cytoplasmic projections that greatly increase the cell's surface area.

A **gland** consists of one or more epithelial cells specialized to produce and secrete a product, such as mucus, sweat, milk, saliva, hormones, or enzymes (Fig. 20–3). The epithelial tissue lining the cavities and passageways of the body typically contains **goblet cells,** unicellular glands specialized to secrete mucus. The mucus lubricates these surfaces and facilitates the movement of materials.

Connective Tissue

Connective tissue joins together and supports the other structures of the body. It protects underlying organs and may be specialized to store or transport materials. When other tissues are injured, connective tissue

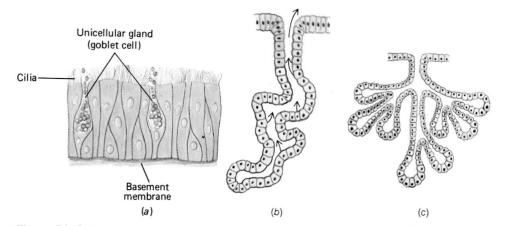

Figure 20–3 A gland consists of one or more epithelial cells. (a) Goblet cells are unicellular glands that secrete mucus. (b) Sweat glands are simple tubular glands with coiled tubes similar to the one shown here. (c) The parotid salivary glands are compound glands like the one shown here.

Table 20–1
EPITHELIAL TISSUES

Tissue name	Main locations	Functions	Description and comments
Simple squamous epithelium Nuclei	Air sacs of lungs, lining of blood vessels	Passage of materials where little or no protection is needed and where diffusion is major form of transport	Cells are flat and arranged as single layer
Simple cuboidal epithelium Lumen of tubule — Nuclei of cuboidal epithelial cells	Lining of kidney tubules, gland ducts	Secretion and absorption	Single layer of cells; from the side each cell looks like short cylinder; sometimes have microvilli for absorption
Simple columnar epithelium Goblet cell — Nuclei of columnar cells	Lining of much of digestive tract, upper part of respiratory tract	Secretion, especially mucus; absorption, protection, movement of mucous layer	Single layer of columnar cells, often with nuclei located in base of each cell almost in row; sometimes with enclosed secretory vesicles (goblet cells), highly developed Golgi complex, and cilia

(Continued)

Table 20–1
EPITHELIAL TISSUES (Continued)

Tissue name	Main locations	Functions	Description and comments
Stratified squamous epithelium	Skin, mouth lining, vaginal lining	Protection only; little or no absorption or transport of materials; outer layer continuously worn away by friction and replaced from below	Several layers of cells, with only lower one columnar and metabolically active; division of lower cells causes older ones to be pushed upward toward surface
Pseudostratified epithelium Approximately ×250 (Ed Reschke)	Some respiratory passages, ducts of many glands, sometimes ciliated	Secretion, protection, movement of mucus	Comparable in many ways to columnar epithelium, except not all cells are of same height; though all cells touch basement membrane, tissue appears stratified; nuclei not in line; may be ciliated, may secrete mucus

acts to repair the damage. In addition, almost every organ in the body has a supporting framework of connective tissue, called **stroma.** The epithelial components of the organ are supported and cushioned by the stroma.

Unlike epithelium, connective tissue consists of relatively few cells separated by large amounts of **intercellular substance.** This intercellular substance consists of threadlike microscopic fibers scattered throughout a **matrix,** a thin gel composed of polysaccharides. The intercellular substance is secreted by connective tissue cells called **fibroblasts** (Fig. 20–4). Three types of fibers found in connective tissue are collagen, elastic,

NEOPLASMS—UNWELCOME TISSUES

A **neoplasm** (new growth), or **tumor,** is an abnormal mass of cells. A **benign** ("kind") tumor tends to grow slowly, and its cells stay together. Because benign tumors form discrete masses, often surrounded by connective tissue capsules, they can usually be removed surgically. Unless a benign tumor develops in a place where it interferes with the function of a vital organ, it is not lethal.

Cancer is the common term for **malignant** ("wicked") neoplasms. Unlike benign neoplasms, cancer invades other tissues, typically spreads to new locations, and does not retain the typical structural features of the cells from which they develop. Neoplasms that develop from connective tissues or muscles are referred to as **sarcomas,** and those that originate in epithelial tissue are **carcinomas.** Common cancers originating in blood or bone marrow are leukemias, lymphomas, and myelomas.

Cancer is thought to be triggered when the DNA of a cell is mutated by radiation, certain chemicals or irritants, or viruses (see the discussion of oncogenes in Chapter 13). When the transformed cell divides, all the cells derived from it bear the identical mutation. The changes in the DNA affect the cell membrane and interfere with the cell's control mechanisms. Membrane proteins that normally help regulate cell division and interaction with other cells are replaced by tumor-specific proteins. Two basic defects in behavior that are typical of cancer cells are wild, often rapid multiplication, and abnormal relations with neighboring cells. Unlike normal cells that respect one another's boundaries and form tissues in an orderly, organized manner, cancer cells grow helter-skelter upon one another and infiltrate normal tissues. Apparently, they are no longer able to receive or respond appropriately to signals from surrounding cells; communication is lacking.

Studies indicate that many neoplasms grow to only a few millimeters in diameter and then enter a dormant stage, which may last for months or even years. At some point, cells of the neoplasm release a chemical substance that stimulates nearby blood vessels to develop new capillaries that grow out toward the neoplasm and invest it. Once a blood supply is ensured, the neoplasm grows rapidly and may soon become life-threatening.

Death from cancer almost always results from **metastasis,** a migration of cancer cells through blood or lymph channels to distant parts of the body. Once there, they multiply, forming new malignant neoplasms; these may interfere with the normal function of the tissues being invaded. Cancer often spreads so rapidly and extensively that surgeons are unable to locate all the malignant masses.

Why some persons are more susceptible to cancer than others remains a mystery. Some researchers think that cancer cells arise daily in everyone, but that in most people the immune system (the system that provides protection from disease organisms and other foreign invaders) is capable of destroying them. According to this theory, cancer is a failure of the immune system. Another suggestion is that people have different levels of tolerance to carcinogens (cancer-producing agents) in the environment. More than 80% of cancer cases are thought to be triggered by carcinogens in the environment.

Cancer is the second-highest cause of death in the United States, and despite advances in its treatment, there is no miracle cure on the horizon. Fewer than 50% of cancer patients survive five years from the time their cancer is first diagnosed. Currently, the key to survival is early diagnosis and treatment with a combination of surgery, radiation therapy, and drugs that suppress mitosis (chemotherapy). Since cancer is actually an entire family of closely related diseases (there are more than 100 distinct varieties), it is probable that there is no single cure. Most investigators agree, however, that a greater understanding of the control mechanisms and communication systems of cells is necessary before effective cures can be developed.

Normal epidermis of skin

cancer tissue

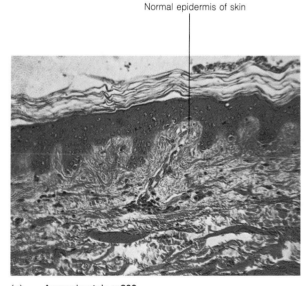 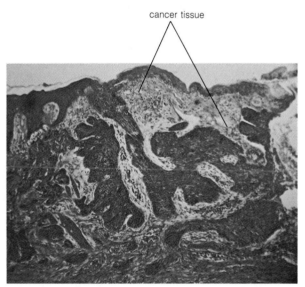

(a)　Approximately ×200

(b)　Approximately ×150

Normal skin (a) compared with cancerous tissue (b). Note the disruption of the normal tissue structure by the invasion of the neoplasm.

Figure 20–4 Loose ordinary connective tissue in the mesentery of a rabbit which had been injected with india ink (approximately ×1000). The macrophages have ingested the ink particles. Fibroblasts (the elongated cells) are also present. Note that the collagen is stained pink and the elastic fibers black. (From Warwick, R., and P.L. Williams, *Gray's Anatomy*, 36th ed. Edinburgh, Churchill Livingstone, 1980.)

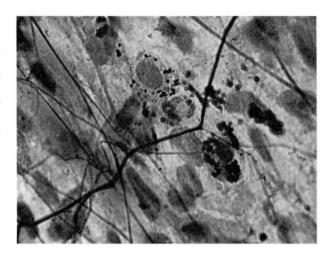

and reticular fibers. Most numerous are the **collagen fibers** (Fig. 20–5), composed of the protein collagen (familiar in its hydrated form as gelatin). The tensile strength of these flexible fibers has been compared to that of steel.

Elastic fibers stretch easily and then, like a rubber band, snap back to their normal length when the stress is removed. These fibers are an important component of structures that must stretch. For example, elastic fibers in the walls of large arteries permit the blood vessel to stretch as it fills with blood. Reticular fibers are very fine, branched fibers that form a supporting network within many tissues and organs.

The main types of connective tissue vary widely in the details of their structure and in the specific functions they perform (Table 20–2).

Two main kinds of **ordinary connective tissue** are loose connective tissue and dense connective tissue. **Loose connective tissue** (also called areolar tissue) is the most widely distributed connective tissue in the body. It is found as a thin filling between body parts and serves as a reservoir for fluid and salts. Nerves, blood vessels, and muscles are wrapped in this tissue. Together with adipose tissue, loose connective tissue forms the subcutaneous (below the skin) layer, the layer that attaches skin to the muscles and other structures beneath. Loose connec-

Figure 20–5 Scanning electron micrograph of collagen fibers taken from the tendon of a human biceps muscle. In transmitting forces generated by muscle to bones, the collagen of tendons acts something like an organic rope and must be flexible, tough, and strong. (Courtesy of L. Jozsa, A. Reffy, and J. Balint, National Institute of Traumatology, Budapest. *Acta Histochemica* 74, 1984. Used by permission.)

tive tissue consists of fibers strewn in all directions through a semifluid matrix. Its flexibility permits the parts it connects to move.

Dense connective tissue is very strong and somewhat less flexible than loose connective tissue. Collagen fibers predominate. In irregular dense connective tissue the collagen fibers are arranged in bundles distributed in all directions through the tissue. This type of tissue is found in the lower layer (dermis) of the skin. In regular dense connective tissue, the collagen bundles are arranged in a definite pattern, making the tissue greatly resistant to stress. Tendons, the cablelike cords that connect muscles to bones, consist of this tissue.

Elastic connective tissue consists mainly of bundles of parallel elastic fibers. It is found in ligaments, the bands of tissue that connect bones to one another. Structures that must expand and then return to their original size, like the walls of large arteries and lung tissue, contain elastic connective tissue. **Reticular connective tissue** is composed mainly of interlacing reticular fibers. It forms a stroma (framework) that supports many organs, including the liver, spleen, and lymph nodes.

Adipose tissue is rich in fat cells, which store fat and release it when fuel is needed for cellular respiration. It is found in the subcutaneous layer and in tissue that cushions internal organs. An immature fat cell is somewhat star-shaped. As fat droplets accumulate within the cytoplasm, the cell assumes a more rounded appearance (Fig. 20–6). Fat droplets eventually merge with one another until finally a single large drop of fat is present. This large drop occupies most of the volume of the mature fat cell. The cytoplasm and organelles are pushed to the cell edges, where a bulge is typically created by the nucleus. A cross section of such a fat cell looks like a ring with a single stone. In a photomicrograph adipose tissue looks somewhat like chicken wire. The "wire" represents the rings of cytoplasm, and the large spaces indicate where fat drops existed before they were dissolved by chemicals used to prepare the tissue. The empty spaces may cause the cells to collapse, resulting in a wrinkled appearance.

The supporting skeleton of vertebrates is composed of cartilage or bone. **Cartilage** is the supporting skeleton in the embryonic stages of all vertebrates, but it is largely replaced by bone in the adult, except in the sharks and rays. In humans the supporting structure of the external ear, the supporting rings in the walls of the respiratory passageways, and the tip of the nose are examples of structures composed of cartilage. Cartilage is firm yet elastic. Cartilage cells called **chondrocytes** secrete this hard, rubbery matrix around themselves and also secrete collagen fibers, which become embedded in the matrix and strengthen it. Chondrocytes eventually come to lie singly or in groups of two or four in small cavities called lacunae in the matrix (Fig. 20–7). The cartilage cells in the matrix remain alive. Cartilage tissue lacks nerves, lymph vessels, and blood vessels. Chondrocytes are nourished by diffusion of nutrients and oxygen through the matrix.

Bone is the principal vertebrate skeletal tissue. It is similar to cartilage in that the **osteocytes** (bone cells) that secrete and maintain the matrix are located in lacunae within the matrix (Fig. 20–8). Unlike cartilage, however, bone is a highly vascular tissue with a substantial blood supply. Diffusion alone would be too slow to provide nourishment for the osteocytes because the matrix is calcified. Thus the osteocytes of bone communicate with one another and with capillaries by tiny channels, **canaliculi,** which contain fine extensions of the cells themselves. In much bone the osteocytes are arranged around central capillaries in concentric layers, called **lamellae,** which form spindle-shaped units known as **osteons.** The capillaries, as well as nerves, run through central microscopic channels in the osteons known as **haversian canals.**

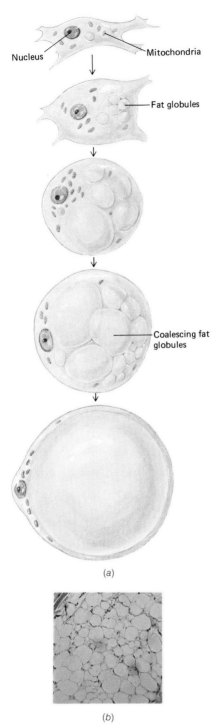

(a)

(b)

Figure 20–6 Storage of fat in a fat cell. (a) As more and more fat droplets accumulate in the cytoplasm, they coalesce to form a very large globule of fat. Such a fat globule may occupy most of the cell, pushing the cytoplasm and the organelles to the periphery. (b) Photomicrograph of adipose tissue (approximately ×100). The fat droplets were dissolved by chemicals used to prepare the tissue, leaving large spaces. Because of these spaces, the cells tend to collapse and no longer appear round.

Table 20–2
CONNECTIVE TISSUES

Tissue name	Main locations	Functions	Description and comments
Loose (areolar) ordinary connective tissue Collagen fibers — Nuclei of fibroblasts — Approximately ×200	Everywhere support must be combined with elasticity, e.g., subcutaneous layer	Support; reservoir for fluid and salts	Fibers produced by fibroblast cells embedded in semifluid matrix and mixed with miscellaneous group of other cells
Dense ordinary connective tissue Approximately ×200	Tendons, strong attachments between organs; dermis of skin	Support; transmission of mechanical forces	Bundles of interwoven collagen fibers interdigitated with rows of fibroblast cells
Elastic connective tissue Approximately ×300	Structures that must both expand and return to their original size, such as lung tissue and large arteries; ligaments	Confers elasticity	Branching elastic fibers interspersed with fibroblast
Reticular connective tissue Approximately ×500	Framework of liver, lymph nodes, spleen	Support	Consists of interlacing reticular fibers

Tissue name		Main locations	Functions	Description and comments
Adipose tissue		Subcutaneous layer; pads around certain internal organs	Food storage, insulation, support of such organs as breast, kidneys	Fat cells star-shaped at first; fat droplets accumulate until typical ring-shaped cells are produced
Cartilage Chondrocytes Lacuna Intercellular substance		Supporting skeleton in sharks, rays, and some other vertebrates; in other vertebrates forms ends of bones; supporting rings in walls of some respiratory tubes; tip of nose; external ear	Flexible support and reduction of friction in bearing surfaces	Chondrocytes separated from one another by gristly intercellular substance, and occupy little spaces in it
Bone Lacunae Haversian canal Matrix Approximately ×150		Most of skeleton in most vertebrates	Support, protection of internal organs, calcium reservoir; skeletal muscles attach to bones	Osteocytes located in lacunae; in compact bone, lacunae arranged in concentric circles about haversian canals
Blood Red blood cells White blood cell Approximately ×1100		Heart and blood vessels of circulatory system	Transports oxygen, nutrients, wastes, other materials	Consists of cells dispersed in fluid intercellular substance

Figure 20–7 Cartilage cells become trapped in small spaces called lacunae. The rubbery matrix contains collagen fibers.

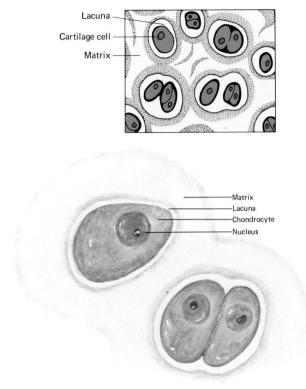

Lacuna
Cartilage cell
Matrix

Matrix
Lacuna
Chondrocyte
Nucleus

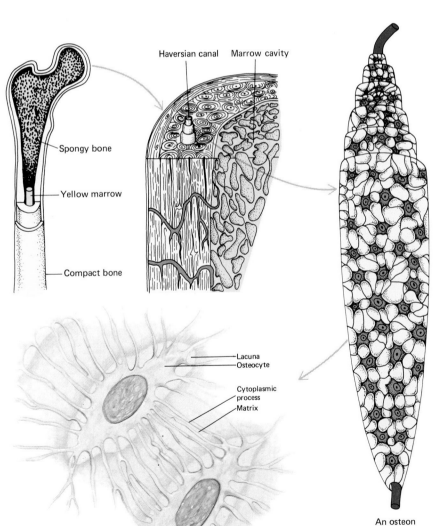

Haversian canal   Marrow cavity

Spongy bone

Yellow marrow

Compact bone

Lacuna
Osteocyte

Cytoplasmic process
Matrix

An osteon

Figure 20–8 Compact bone is made up of units called osteons. Blood vessels and nerves run through the haversian canal within each osteon. In bone the matrix is rigid and hard. Bone cells become trapped within lacunae but communicate with one another by way of cytoplasmic processes that extend through tiny canals.

Bone also contains large, multinucleated cells called **osteoclasts,** which can dissolve and remove the bony substance, as can the osteocytes themselves. The shape and internal architecture of the bone can gradually change in response to normal growth processes and to physical stress. The calcium salts of bone render the matrix very hard, and the collagen prevents the bony matrix from being overly brittle.

Like other connective tissues, **blood** and **lymph** consist of specialized cells dispersed in an intercellular substance. However, these tissues are unique in that the intercellular substance is fluid and this fluid is not secreted by the blood cells. Blood and lymph are circulating tissues that bring various parts of the body into communication. Vertebrate blood consists of a fluid component, called plasma, in which are suspended red blood cells, white blood cells, and platelets. The functions of these various components of blood will be discussed in Chapter 24. Blood cells are produced within another connective tissue, the red bone marrow found within certain bones.

Muscle Tissue

In most animals muscle is the most abundant tissue. It accounts for nearly two-thirds of the body weight in a human being. **Muscle tissue** is specialized for contraction and is the basis for almost all movement in animals. Because they are long and narrow, muscle cells are referred to as fibers. Muscle fibers are usually arranged in layers or bundles surrounded by connective tissue.

There are three types of muscle tissue: skeletal, cardiac, and smooth (Fig. 20–9). **Skeletal muscle,** which is attached to the bones, can

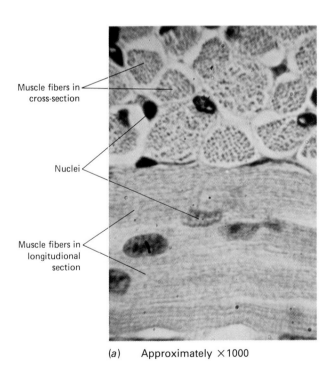

(a)   Approximately ×1000

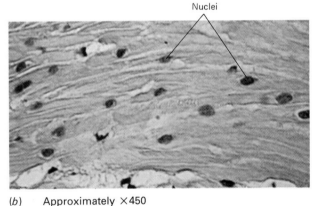

(b)   Approximately ×450

(c)   Approximately ×900

Figure 20–9 Muscle tissue. (a) Skeletal muscle is striated, voluntary muscle. (b) Smooth muscle tissue lacks striations and is involuntary. (c) Cardiac muscle tissue is striated, has branched fibers, and is involuntary. The special junctions between cardiac muscle cells are called intercalated disks.

Table 20–3
THE TYPES OF MUSCLE TISSUES

	Skeletal	Smooth	Cardiac
Location	Attached to skeleton	Walls of stomach, intestines, etc.	Walls of heart
Type of control	Voluntary	Involuntary	Involuntary
Shape of fibers	Elongated, cylindrical, blunt ends	Elongated, spindle-shaped, pointed ends	Elongated, cylindrical fibers that branch and fuse
Striations	Present	Absent	Present
Number of nuclei per fiber	Many	One	One or two
Position of nuclei	Peripheral	Central	Central
Speed of contraction	Most rapid	Slowest	Intermediate
Ability to remain contracted	Least	Greatest	Intermediate

(a) Skeletal muscle fibers   (b) Smooth muscle fibers   (c) Cardiac muscle fibers

be contracted voluntarily. This muscle tissue permits us to walk, run, write, and move the body in other ways. Characterized by a pattern of light and dark stripes, or striations, it is also referred to as **striated muscle.** Each skeletal muscle fiber has several nuclei that lie just under the cell membrane.

Cardiac muscle, the main tissue of the heart, is a kind of striated muscle that is not under voluntary control. The fibers of cardiac muscle are joined end to end and branch and rejoin to form complex networks. One or two nuclei are found within each fiber. A characteristic feature of cardiac muscle tissue is the presence of **intercalated disks,** which mark the junctions between adjoining fibers.

The third type of muscle, **smooth muscle,** lacks striations and is involuntary. Its spindle-shaped cells contain only one nucleus. Found within the walls of many organs, smooth muscle is responsible for such internal movements as moving food through the digestive tract.

Nervous Tissue

Nervous tissue is specialized to receive stimuli and transmit information (nerve impulses) and thereby to control the action of muscles and glands. Though the bulk of nervous tissue is located within the brain and spinal cord, bundles of nerve fibers—that is, the nerves—are dispatched to all parts of the body to pick up information from the sense organs and to return information in the form of decisions from the brain and spinal cord.

Nervous tissue consists of nerve cells, or **neurons,** and supporting cells called **glial cells** (Fig. 20–10). A typical neuron consists of a cell body containing the nucleus, and elongated extensions of the cytoplasm— the dendrites and axon. **Dendrites** are specialized for receiving impulses, and the single **axon** conducts impulses away from the cell body toward

Cell body of neuron   Neurons   Dendrites

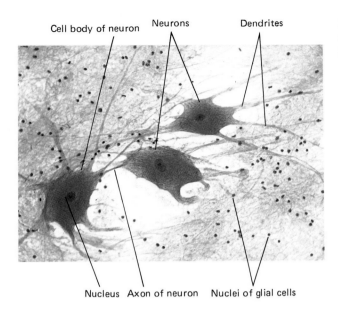

Nucleus   Axon of neuron   Nuclei of glial cells

Figure 20–10 Nervous tissue consists of neurons and glial cells (approximately ×50). (Ed Reschke.)

another neuron or a muscle or gland. (Neurons will be discussed in Chapter 22.)

ORGANS

Different types of tissues cooperating with one another to perform a particular biologic function constitute an **organ.** The brain, heart, stomach, and eye are organs, and though you may not think of the skin as an organ, it is the largest one in the body.

Although an organ may be composed primarily of one type of tissue, other types are needed to provide support, protection, blood supply, and conduction of nerve impulses. For example, the brain is composed mainly of nervous tissue, but epithelial and connective tissues protect it from injury and transport vital nutrients and oxygen to it. The intestine is lined with epithelium that secretes digestive enzymes and absorbs nutrients. Layers of muscle make up the bulk of its wall and contract in waves, moving food along the digestive tube. Nerve tissue places the digestive tube in communication with other parts of the body, such as the brain. Connective tissue supplies the intestine with blood and holds its tissues together, as well as holding the tube in place in the body.

ORGAN SYSTEMS AND THE ORGANISM

Various tissues and organs coordinate their activities to perform a specialized set of functions. Such an organized complex of structures is termed an **organ system.** In the human body, as in other complex animals, we can identify ten organ systems, each responsible for a specific group of activities (Fig. 20–11). Working together, these ten organ systems make up the complex **organism.**

The organ systems of complex animals include the integumentary, skeletal, muscle, nervous, circulatory, digestive, respiratory, urinary, endocrine, and reproductive systems. See Table 20–4 for a summary of their principal organs and functions. As an example of an organ system, consider the digestive system. Organs of the digestive system include the mouth, esophagus, stomach, small and large intestines, liver, pancreas, and salivary glands. This system digests food, reducing it to simple molecular components, and then absorbs these nutrients, enabling them to enter the blood for transport to all the cells of the body.

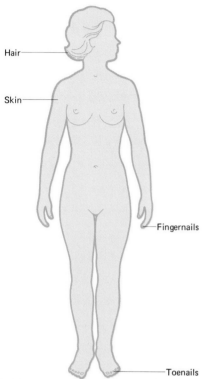

(1) The integumentary system consists of the skin and the structures such as nails and hair that are derived from it. This system protects the body, helps to regulate body temperature, and receives stimuli such as pressure, pain, and temperature.

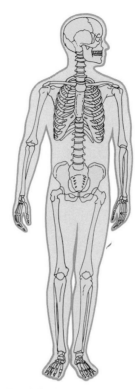

(2) The skeletal system consists of bones and cartilage. This system helps to support and protect the body.

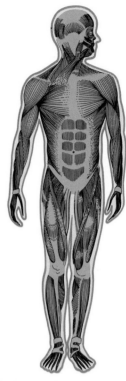

(3) The muscular system consists of the large skeletal muscles that enable us to move, as well as the cardiac muscle of the heart and the smooth muscle of the internal organs.

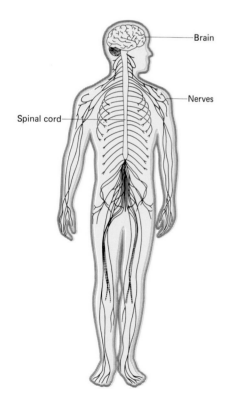

(4) The nervous system consists of the brain, spinal cord, sense organs, and nerves. This is the principal regulatory system.

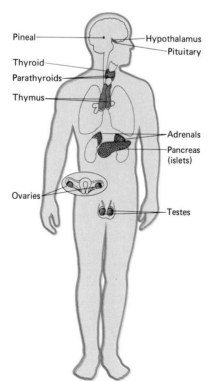

(5) The endocrine system consists of the ductless glands that release hormones. It works with the nervous system in regulating metabolic activities.

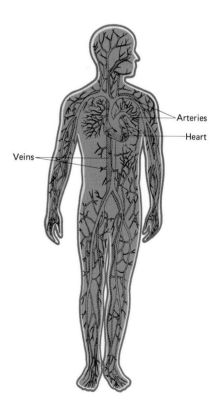

(6a) The circulatory system includes the heart and blood vessels. This system serves as the transportation system of the body.

Figure 20–11 The principal organ systems of the human body.

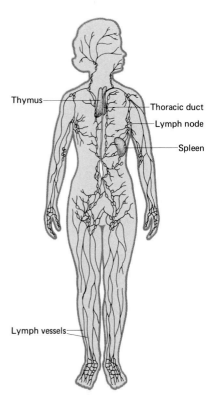

Thymus

Thoracic duct

Lymph node

Spleen

Lymph vessels

(*6b*) The lymphatic system is a subsystem of the circulatory system; it returns excess tissue fluid to the blood and defends the body against disease.

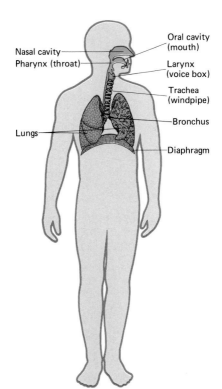

Nasal cavity

Pharynx (throat)

Oral cavity (mouth)

Larynx (voice box)

Trachea (windpipe)

Bronchus

Lungs

Diaphragm

(*7*) The respiratory system. Consisting of the lungs and air passageways, this system supplies oxygen to the blood and excretes carbon dioxide.

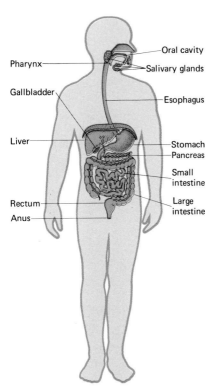

Pharynx

Gallbladder

Liver

Rectum

Anus

Oral cavity

Salivary glands

Esophagus

Stomach

Pancreas

Small intestine

Large intestine

(*8*) The digestive system consists of the digestive tract and glands that secrete digestive juices into the digestive tract. This system mechanically and enzymatically breaks down food and eliminates wastes.

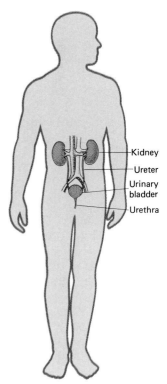

Kidney

Ureter

Urinary bladder

Urethra

(*9*) The urinary system is the main excretory system of the body, and helps to regulate blood chemistry. The kidneys remove wastes and excess materials from the blood and produce urine.

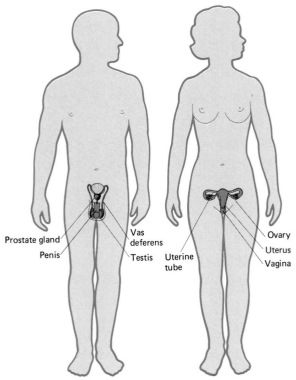

Prostate gland

Penis

Vas deferens

Testis

Uterine tube

Ovary

Uterus

Vagina

(*10*) Male and female reproductive systems. Each reproductive system consists of gonads and associated structures. The reproductive system maintains the sexual characteristics, and perpetuates the species.

Table 20–4
THE ORGAN SYSTEMS OF A MAMMAL AND THEIR FUNCTIONS

System	Components	Functions	Homeostatic ability
Integumentary	Skin, hair, nails, sweat glands	Covers and protects body	Sweat glands help control body temperature; as barrier, the skin helps maintain steady state
Skeletal	Bones, cartilage, ligaments	Supports body, protects, provides for movement and locomotion, calcium depot	Helps maintain constant calcium level in blood
Muscular	Organs mainly of skeletal muscle; cardiac muscle; smooth muscle	Moves parts of skeleton, locomotion; movement of internal materials	Ensures such vital functions as nutrition through body movements; smooth muscle maintains blood pressure; cardiac muscle circulates the blood
Digestive	Mouth, esophagus, stomach, intestines, liver, pancreas	Ingests and digests foods, absorbs them into blood	Maintains adequate supplies of fuel molecules and building materials
Circulatory	Heart, blood vessels, blood; lymph and lymph structures	Transports materials from one part of body to another; defends body against disease	Transports oxygen, nutrients, hormones; removes wastes; maintains water and ionic balance of tissues
Respiratory	Lungs, trachea, and other air passageways	Exchange of gases between blood and external environment	Maintains adequate blood oxygen content and helps regulate blood pH; eliminates carbon dioxide
Urinary	Kidney, bladder, and associated ducts	Eliminates metabolic wastes; removes substances present in excess from blood	Regulates blood chemistry in conjunction with endocrine system
Nervous	Nerves and sense organs, brain and spinal cord	Receives stimuli from external and internal environment, conducts impulses, integrates activities of other systems	Principal regulatory system
Endocrine	Pituitary, adrenal, thyroid, and other ductless glands	Regulates body chemistry and many body functions	In conjunction with nervous system, regulates metabolic activities and blood levels of various substances
Reproductive	Testes, ovaries, and associated structures	Provides for continuation of species	Passes on genetic endowment of individual; maintains secondary sexual characteristics

CHAPTER SUMMARY

I. Multicellular organisms can be much larger than unicellular ones, and their cells can specialize to perform specific functions.

    A. A unicellular organism's size is limited because as a cell approaches a certain size, the ratio of its cell membrane to cell volume becomes inefficient.

    B. When a cell grows to a maximum size determined by its genes, it divides; in multicellular organisms the new cells may remain associated to form tissues.

II. A tissue consists of a group of closely associated cells that are adapted to carry out specific functions.
  A. Epithelial tissue consists of cells fitted tightly together. It covers the body surface and lines its cavities.
    1. Epithelial tissue is specialized to protect, absorb, secrete, or receive stimuli.
    2. Epithelial cells may be squamous, cuboidal, or columnar, and they may be arranged to form simple, stratified, or pseudo-stratified tissue.
  B. Connective tissue supports and binds together structures of the body. It protects underlying structures and may be specialized to store or transport materials.
    1. Connective tissue consists of relatively few cells separated by intercellular substance, which in turn is composed of fibers scattered through a matrix.
    2. The main types of connective tissue are ordinary (loose and dense), elastic, reticular, adipose, cartilage, bone, blood, lymph, and tissues that produce blood cells.
  C. Muscle tissue is specialized to contract, enabling an organism to move. The three types of muscle are skeletal, cardiac, and smooth.
  D. Nervous tissue is specialized to receive and transmit stimuli; neurons control the action of muscles and glands.
III. Cells of a neoplasm differ from cells of a normal tissue in that they divide in an uncontrolled manner and invade normal tissues.
IV. An organ consists of different types of tissue that cooperate to perform a particular biological function. The heart, brain, and stomach are examples of organs.
 V. Organs and tissues work together, forming organ systems. In complex animals about ten principal organ systems work together to constitute the total living organism. Among these are the digestive system, the nervous system, and the skeletal system.

Post-Test

1. A group of closely associated cells adapted to carry out specific functions form a _____ .
2. A _____ consists of epithelial cells specialized to produce and secrete a product.
3. Epithelial cells may be flattened, or _____ ; cube-shaped, or _____ ; or elongated like columns, _____ .

Select the appropriate answer or answers from column B for each entry in column A.

Column A

_____ 4. specialized to protect, absorb, or secrete
_____ 5. specialized to receive stimuli and transmit impulses
_____ 6. specialized to contract
_____ 7. abnormal mass of tissue
_____ 8. contains fibroblasts
_____ 9. may be simple or stratified
_____ 10. bone is an example

Column B

a. muscle tissue
b. epithelial tissue
c. connective tissue
d. nervous tissue
e. neoplasm

_____ 11. forms the subcutaneous layer; the most widely distributed connective tissue
_____ 12. forms a stroma that supports many organs
_____ 13. stores fat
_____ 14. cells found in lacunae
_____ 15. capillaries run through haversian canals
_____ 16. contains chondrocytes

a. bone
b. adipose tissue
c. loose ordinary connective tissue
d. reticular connective tissue
e. cartilage

17. In cardiac muscle _____ _____ mark the junctions between adjoining fibers.
18. Two types of striated muscle are _____ and _____ .
19. Two types of cytoplasmic extensions of neurons are the _____ and _____ .
20. The organ system made up of glands that secrete hormones is the _____ system.
21. The organ system that covers the body is the _____ _____ system.

Review Questions

1. Distinguish among tissues, organs, and organ systems.
2. Imagine that all of the epithelium in a complex animal, such as a human being, suddenly disappeared. What effects would this have on the body and its ability to function?
3. Contrast epithelial tissue with connective tissue.
4. Compare cartilage with bone.
5. What is a gland? Of what tissue is it composed?
6. Compare skeletal, cardiac, and smooth muscle with respect to structure and location in the body.
7. What is a neuron?
8. How do the cells of a malignant neoplasm differ from the cells of a normal tissue? What is metastasis?
9. Locate each of the following: (a) osteocytes, (b) chondrocytes, (c) intercalated disks, (d) fibroblasts, (e) collagen fibers, (f) stratified squamous epithelium.
10. Name the functions of the following: (a) endocrine system, (b) skeletal system, (c) circulatory system, (d) nervous system.

Readings

Ganong, W.F. *Review of Medical Physiology,* 10th ed. Los Altos, Calif., Lange, 1981. A textbook of human physiology.

Gardiner, M.S. *The Biology of Invertebrates.* New York, McGraw-Hill, 1972. An excellent textbook organized around the life processes of invertebrates.

Guyton, A.C. *Textbook of Medical Physiology,* 6th ed. Philadelphia, W.B. Saunders, 1981. A well-written standard reference work in human physiology.

Ralph, C.L. *Introductory Animal Physiology.* New York, McGraw-Hill, 1978. An interesting account of animal physiology.

Solomon, E.P., and P.W. Davis. *Human Anatomy and Physiology.* Philadelphia, Saunders College Publishing, 1983. A very readable presentation of human anatomy and physiology.

(See also the readings for Chapter 16.)

Chapter 21
SKIN, MUSCLE, AND BONE

Learning Objectives

After you have studied this chapter you should be able to:

1. Describe the structure and function of the skin of a typical vertebrate and give its principal derivatives.
2. Summarize and compare different types of skeletal systems that occur in the animal kingdom, emphasizing the hydrostatic skeleton, exoskeleton, and endoskeleton.
3. Contrast cartilage and membrane bone in the human skeleton, using the radius and frontal bones as examples.
4. Describe the gross and microscopic structure of skeletal muscle.
5. List in sequence the events that take place in muscle contraction.
6. Compare the roles of glycogen, creatine phosphate, and ATP in providing energy for muscle contraction.
7. Describe the functional relationship between skeletal and muscular tissues.

ome animals run, some jump, some fly. Others remain rooted to one spot, sweeping their surroundings with tentacles. Many contain internal circulating fluids, pumped by hearts and contained by hollow vessels that maintain their pressure by gentle squeezing. Digestive systems push food along with peristaltic writhings. And in all these cases each action is powered by muscle, a specialized tissue that, however varied its effects, has but *one* action: It can contract. In many animals, the muscle is anchored to the skeleton and, to a lesser extent, the skin, which then serve to transmit forces. In this chapter we will discuss skin, skeleton, and muscle—systems that are closely interrelated in function and significance.

THE VERTEBRATE SKIN

In many fish, in the African ant-eating pangolin (a mammal), and in some reptiles the skin is developed into a set of scales formidable enough to be considered armor. Even in human beings, the skin has considerable strength. Human skin (Fig. 21–1) includes a variety of structures, including fingernails and toenails, hair of various types, sweat glands, oil glands, and several types of sensory receptors responsible for our ability to feel pressure, temperature, and pain. Human skin and the skin of other mammals also contain mammary glands specialized in females for secretion of milk. Human oil glands empty via short ducts into hair follicles (see Fig. 21–3). They secrete a substance called **sebum,** a complex mixture of fats, waxes, and hydrocarbons. In humans these glands are especially numerous on the face and scalp. The oil secreted keeps the hair moist and pliable and prevents the skin from drying and cracking. (It is excessive sebum produced at puberty in response to increased lev-

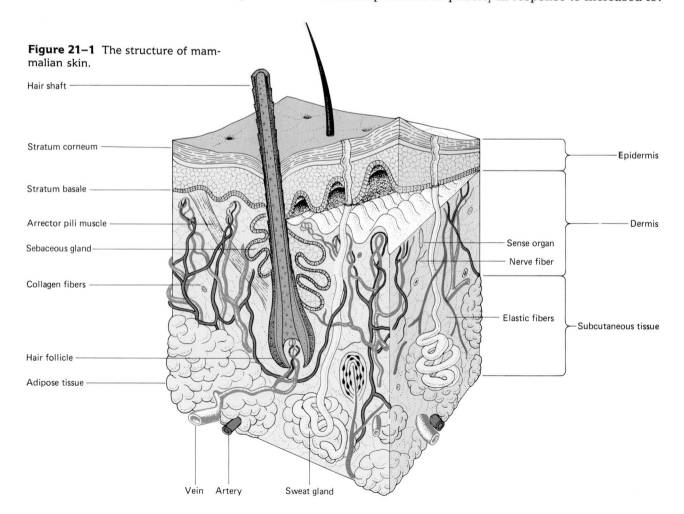

Figure 21–1 The structure of mammalian skin.

Hair shaft

Stratum corneum

Stratum basale

Arrector pili muscle

Sebaceous gland

Collagen fibers

Hair follicle

Adipose tissue

Epidermis

Dermis

Sense organ

Nerve fiber

Elastic fibers

Subcutaneous tissue

Vein   Artery   Sweat gland

els of sex hormones that fills the glands and follicles, producing the too-familiar inflammation called **acne.**)

In a human being the skin functions as a thermostatically controlled radiator, regulating the elimination of heat from the body (Chapter 1). About 2.5 million sweat glands secrete sweat, and its evaporation from the surface of the skin lowers the body temperature.

The skin in some other vertebrates varies considerably from ours (Fig. 21–2). Instead of hairs, birds have feathers, which nevertheless form in a manner comparable to hairs and are even more effective insulation than fur. Among the poikilothermous (cold-blooded) vertebrates one finds epidermal scales (as in reptiles), naked skin covered with mucus (as in many amphibians and fish), and skin with bony or toothlike scales. Some skin, such as that of certain tropical frogs, is even provided with poison glands. Skin and its derivatives are often brilliantly colored in connection with courtship rituals, territorial displays, and various kinds of communication. The human blush pales alongside the spectacular displays of such animals as peacocks.

The Epidermis

The outer layer of skin, the **epidermis,** is the interface between our delicate tissues and the hostile universe. The epidermis consists of several strata, the lowest of which is the **stratum basale,** and the outermost, the **stratum corneum.** In the stratum basale cells continuously divide, and the new cells are forced upward by the pressure of yet other cells being produced below them. As the epidermal cells move upward in the skin, they mature. In almost all vertebrates there are no capillaries in the epidermis, and so the maturing cells are progressively deprived of more and more nourishment and become ever less active metabolically.

As they move upward, epidermal cells manufacture the distinctive skin protein, **keratin,** an elaborately coiled protein that confers on the skin considerable mechanical strength combined with flexibility, for the coils are capable of stretching much like springs. Keratin is quite insoluble and serves as an excellent body surface sealant. When fully mature, epidermal cells are also dead—as dead and as waterproof as shingles. Like shingles, the cells of stratum corneum, the outermost layer, continuously wear off, but unlike shingles, they are continuously replaced.

The Dermis

Underneath the epidermis lies the foundational layer of the skin proper, the **dermis.** The dermis consists of a dense, fibrous connective tissue composed principally of the protein collagen (see Fig. 20–5). The major part of each sweat gland is embedded in the dermis, and the hair follicles

Figure 21–3 (*a*) Photomicrograph of a hair follicle from human scalp (approximately ×25). (*b*) Scanning electron micrograph (approximately ×250) of human skin showing hair follicle. (*a*, Ed Reschke. *b*, courtesy of Dr. Karen A. Holbrook.)

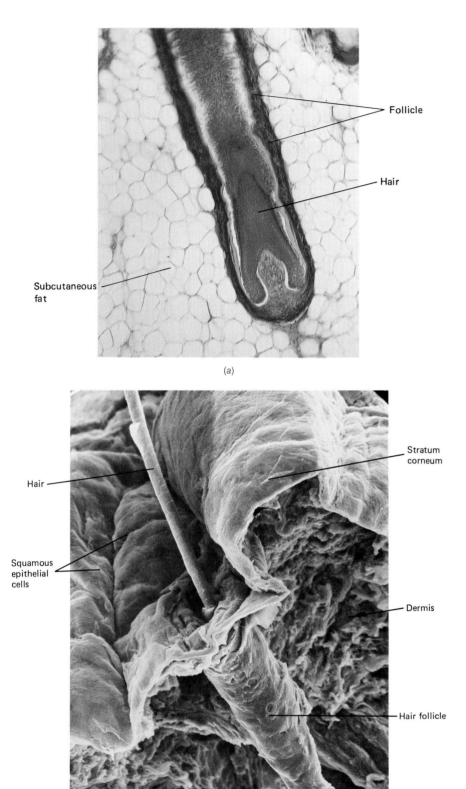

Follicle

Hair

Subcutaneous fat

(*a*)

Hair

Squamous epithelial cells

Stratum corneum

Dermis

Hair follicle

(*b*)

(Fig. 21–3) reach down into it. The dermis also contains blood vessels, which nourish the skin and sense organs concerned with touch.

Mammalian skin rests on a layer of jelly-like subcutaneous tissue, composed mainly of fat that insulates us from unfavorable outside temperature extremes. By localized thickening, it also produces in humans the distinctive body contours that enable us at a glance to tell the difference between the sexes.

SKELETONS

A muscle must have something to act upon, something by means of which its contractions can be transmitted to leg, body, or wing. In some of the simplest animals this is no more than the glutinous, jelly-like substance of the body itself, or perhaps a fluid-filled body cavity. More complex animals, however, require a true skeleton to receive, transmit, and transform the simple movement of their muscular tissues. In a few instances this skeleton is internal—plates or shafts of calcium-impregnated tissue. But in most cases the skeleton is not a living tissue at all but a lifeless deposit atop the epidermis—a shell, or **exoskeleton.**

Hydrostatic Skeletons

Imagine an elongated balloon full of water. If one were to pull on it, it would lengthen, but it would also lengthen if it were squeezed. Conversely, it would shorten if the ends were pushed. In *Hydra* and other cnidarians, cells of the two body layers are capable of contraction. The contractile cells in the outer epidermal layer are arranged longitudinally, whereas the contractile cells of the inner layer (the gastrodermis) are arranged circularly around the central body axis. These two groups of cells work in **antagonistic** fashion. What each can do, the other can undo. When the epidermal, longitudinal layer contracts, the hydra shortens, and because of the fluid present in its gastrovascular cavity, force is transmitted so that it thickens as well. On the other hand, when the endodermal circular layer contracts, the hydra thins, but its fluid contents force it also to lengthen.

The hydra is mechanically little more than a simple bag of fluid (Fig. 21–4). Its fluid interior acts as a hydrostatic skeleton, since it transmits force when the contractile cells contract against it. Hydrostatic skeletons permit only crude mass movements of the body or its appendages. Delicacy is difficult because in a fluid, force tends to be transmitted equally in all directions and hence throughout the entire fluid-filled body of the animal. It is not easy for the hydra to thicken one part of its body, for example, while thinning another.

The annelid worms have a more sophisticated hydrostatic skeleton that permits more versatile body movements than those of *Hydra*. The body of an earthworm, to take a familiar example, may protrude from its burrow on damp evenings to feed on bits of decayed vegetation on the surface. Its posterior end may then protrude in order to defecate the familiar worm castings. This practice is not without its hazards, for if the hungry worm waits too long, an equally hungry early bird is likely to find it. If the bird is quick enough, the story ends right there, but if not, the giant nerve axons of the worm's ventral, solid nerve cord swiftly transmit impulses to the longitudinal and circular muscles of its body that act to inhibit the circular muscles and stimulate the longitudinal. Abruptly the longitudinal muscles contract, pulling the body of the worm toward the safety of its burrow. If the worst occurs and the bird obtains a firm hold, the worm holds on too, with its swollen, contracted anterior now fitting the burrow like a cork in a bottleneck. If the bird releases its hold, the worm will rapidly crawl down the burrow to safety. But how?

If the worm is progressing anterior end first, it must protrude a thinned portion of its body into the burrow ahead. Then, while anchored posteriorly by a thickened portion of itself, the worm must cause its anterior end to swell. Having thus gripped the burrow ahead, the worm releases its posterior grip, and by longitudinal muscle contraction, drags the whole body toward the anchored anterior end. It repeats this process again and again (Fig. 21–5).

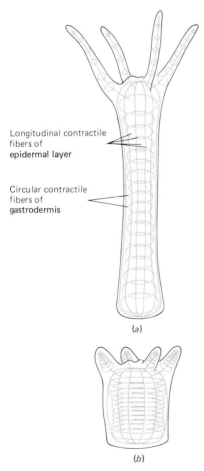

Longitudinal contractile fibers of epidermal layer

Circular contractile fibers of gastrodermis

(a)

(b)

Figure 21–4 Movement in *Hydra.* The longitudinally arranged cells are antagonistic to the cells arranged around the body axis. (*a*) Contraction of the circular muscles elongates the body. (*b*) Contraction of the longitudinal muscles shortens the body.

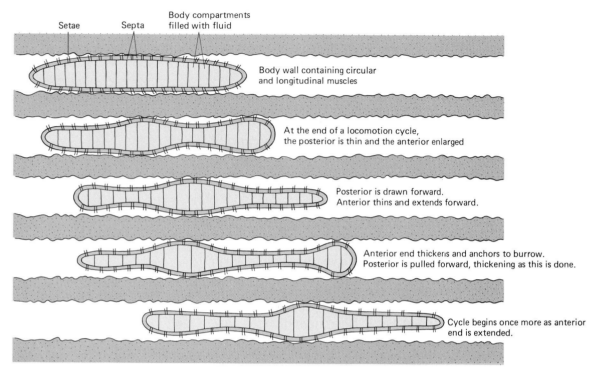

Setae   Septa   Body compartments filled with fluid

Body wall containing circular and longitudinal muscles

At the end of a locomotion cycle, the posterior is thin and the anterior enlarged

Posterior is drawn forward. Anterior thins and extends forward.

Anterior end thickens and anchors to burrow. Posterior is pulled forward, thickening as this is done.

Cycle begins once more as anterior end is extended.

Figure 21–5 Annelid locomotion. Can you infer the segments in which longitudinal or circular muscle is active in each stage? The worm is aided in anchoring itself by bristle-like setae.

All this is made possible by the transverse partitions, or **septa,** that extensively subdivide the body cavity of the worm. These septa isolate portions of the body cavity and its contained fluid, permitting the hydrostatic skeletons of each segment to be largely independent of one another. Thus the contraction of the circular muscle in the elongating anterior end need not interfere with the action of the longitudinal muscle in the segments of the still-anchored posterior.

For animals that do more than drag themselves along on their bellies, the hydrostatic skeleton is insufficient. Yet some examples of it occur to a small extent in the higher invertebrates and even in the vertebrates. Among the mollusks, for example, the feet of bivalves are extended and anchored by a hydrostatic blood pressure mechanism not too different from that used by the earthworm. The multitudinous tube feet of echinoderms, such as the starfish and sea urchin, are moved by an ingenious version of the hydrostatic skeleton, and even in man, the penis becomes erect and stiff because of the turgidity of pressurized blood in its cavernous spaces.

External Skeletons

Although there are others, the two major groups of animals with external skeletons are the mollusks and the arthropods. In both the mollusks and the arthropods the shell is a nonliving product of the cells of the epidermis, but it differs substantially in function between the two phyla. In the mollusks the exoskeleton basically provides protection, with its major muscle attachments serving the skeleton rather than the skeleton serving the muscles. Thus the common clam has a pair of muscles whose major function is to hold the two valves of the shell tightly shut against the onslaughts of seastar and chowder maker.

In the arthropods, however, the skeleton serves not only to protect but also to transmit forces in ways fully comparable to those found in the skeletons of vertebrates. Whereas in mollusks the shell is primarily an emergency retreat, with the bulk of the body nakedly and succulently

Figure 21–6 A cicada molting. This insect requires 13 years to mature. (Chris Simon.)

exposed at other times, in arthropods the exoskeleton covers every bit of the body. It even extends inward as far as the stomach on one end, and for a considerable distance inward past the anus on the other. Though the arthropod exoskeleton is a continuous one-piece sheath, it varies greatly in thickness and flexibility, with large, thick, inflexible plates separated from one another by thin, flexible joints arranged segmentally. Enough joints are provided to make the arthropod's body just as flexible as that of many vertebrates. This exoskeleton is also extensively modified to form specialized tools or weapons, or otherwise adapted to a vast variety of life-styles.

The chief disadvantage of the arthropod exoskeleton is also profound: The rigid exoskeleton prevents growth. To overcome this disadvantage, arthropods must **molt;** that is, cast off their old integument (Fig. 21–6) from time to time to accommodate new growth (see Focus on Molting).

Focus on . . .

MOLTING

To understand molting, we must examine the tissue structure of the arthropod exoskeleton. The living **epidermis** of an arthropod consists of a thin single layer of more or less cuboidal cells interspersed with a variety of glandular and sometimes sensory cells. Directly above this living layer, and in contact with it, is the nonliving **cuticle.** The cuticle consists of a mixture of protein and chitin. **Chitin** is a kind of polysaccharide composed of linked, chemically modified glucose units called glucosamine, which, as the name implies, contain amino groups. In the outer portion of the cuticle cross linkages are established among the chains of chitin by the action of tanning agents secreted by epidermal glands.

The outer portion of the cuticle is the **epicuticle,** composed of proteins, waxes, and sometimes oils. The function of the epicuticle is to retard the evaporation of water, so it is particularly likely to be present and well developed in the terrestrial arthropods, such as insects. If the greasy epicuticle is experimentally removed from a cockroach—little more than thorough wiping is required—the insect dies in a few hours from dehydration. In most aquatic and a few terrestrial arthropods the cuticle is impregnated with calcium salts, producing an exoskeleton almost as hard as a mollusk shell. In the joint membranes these hard layers are reduced or absent for the sake of flexibility.

The developmental stages between moltings are known as **instars.** Somewhere near the end of an instar, an endocrine gland produces a peptide hormone, **ecdysone,** which initiates molting, or **ecdysis.** In response to ecdysone the epidermis secretes both a new cuticle and enzymes that attack the old cuticle at its base. It is not clear why the new cuticle is not digested also, but when the old one has been sufficiently loosened and dissolved, it splits open along predetermined seams and the animal slowly and painstakingly wriggles and pulls every appendage and every other detail of its complicated anatomy out of the old shell. Even the internal linings of mouth and anus must be detached. At last the exhausted animal lies still and recuperates. Eventually it drinks or swallows air to stretch its new suit of hardening chitin in preparation for future tissue growth.

Ecdysis obviously produces a very weak animal, unable to defend itself and almost unable to move, with a soft integument open to attack by any predator. Accordingly, this process is usually carried out in some very sheltered location. During the ensuing hours or days epidermal glands secrete the tanning agents that harden the cuticle, and in some arthropods channels in the cuticle carry other glandular secretions to the surface, which becomes the new epicuticle.

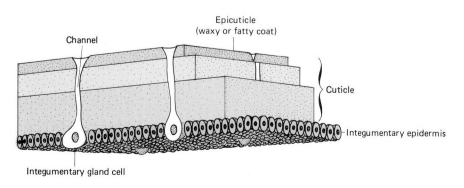

Structure of the arthropod exoskeleton.

Internal Skeletons

Endoskeletons, or internal skeletons, are extensively developed only in the echinoderms and the chordates. The echinoderms have spicules and plates of nonliving calcium salts embedded in the tissues of the body wall. These serve mainly for support and protection, in some cases (such as in sea urchins) forming what amounts to an internal shell. It is the vertebrates that employ the internal skeleton for its full range of potential—for support and for protection, but primarily for the transmission of forces.

Composed of living tissue, the endoskeleton grows in pace with the growth of the animal as a whole, eliminating the need for molting. It also permits the animal to grow, potentially, to great size. Compare the largest land vertebrates—elephants and dinosaurs—to the largest land arthropods—beetles a few inches long. If beetles grew to the size of horses, their external armor would weigh so much that it would probably collapse, or at least prevent the unfortunate animal from moving. So much for the giant insects of the horror movies!

The endoskeleton probably also permits a greater variety of possible motions than does an exoskeleton. In humans, complex motions are produced by an equally complex interaction of many muscles. But there simply is not room for a great many muscles inside the armor of an arthropod limb. Indeed, in some arthropods, especially the spiders, the hydrostatic action of the body fluid is just as important as the intrinsic musculature in producing limb movement.

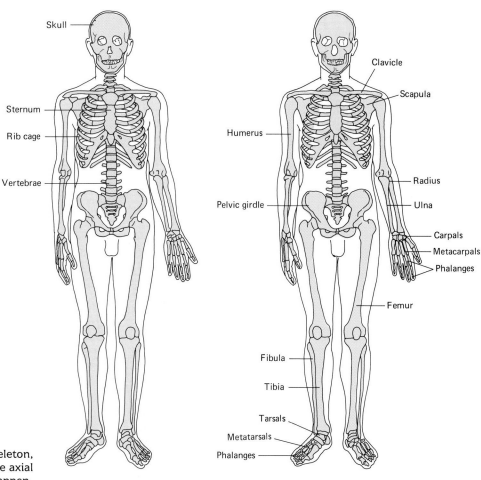

Figure 21–7 The human skeleton, front view. (*a*) The bones of the axial skeleton. (*b*) The bones of the appendicular skeleton.

(*a*)   (*b*)

The human skeleton can be thought of as having two main divisions. The **axial portion** of the skeleton consists of those parts near the skeletal axis—the skull, the vertebral column, the ribs, and the sternum (breastbone). The **appendicular skeleton** consists of the upper and lower extremities, the shoulder girdle, and most of the pelvic bones (Fig. 21–7).

The radius, one of the two bones of the forearm, is a typical long bone (Fig. 21–8). It has numerous muscle attachments, arranged so that the bone operates as a lever that amplifies the motion they generate. The reason is that muscles cannot shorten enough, by themselves, to produce large excursions of the body parts to which they are attached.

At each end of the radius are the **articular cartilages,** which serve as low-friction bearings for the joints (Fig. 21–9) between the radius and

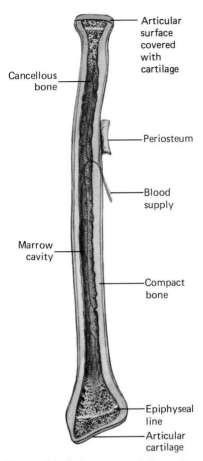

Figure 21–8 Structure of the radius of the arm, a typical long bone. The interior spicules of bone are precisely arranged along the usual lines of greatest stress.

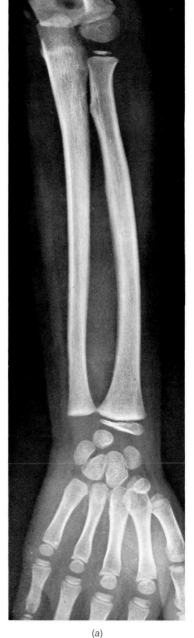

(a)

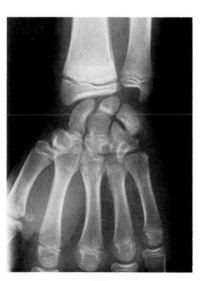

(b)

Figure 21–9 These x-rays illustrate the ends (epiphyses) of immature long bones still separated from the shafts by cartilaginous regions of growth (metaphyses). Greater ossification of the growth region shows that the hand in (b) is older than the one in (a).

the other bones with which it articulates. Damage to the articular carti-lages interferes with the movement of joints and may produce the crip-pling disease **arthritis,** whose causes are still imperfectly understood. The joints are enclosed in **joint capsules** full of a lubricant, the **synovial fluid.**

The articular cartilages are remnants of the original cartilage of which the entire long bone was composed during embryonic life. In the course of development, under the direction of specific developmental proteins, this cartilage is gradually replaced with true bone. Because of this origin the radius, like other long bones, is called a **cartilage bone.** Little cartilage remains, however, in the adult bone (Fig. 21–9).

The radius has a thin, outer shell of hard, **compact bone** consisting of very close-grained osteons (Chapter 20). Interior to the thin shells of

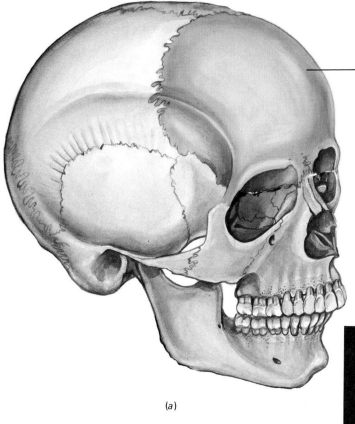

Frontal

(a)

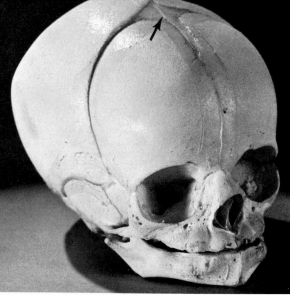

(b)

Figure 21–10 Mature (a) and fetal (b) skulls. Compare these skulls, noting the differences in proportions between the two. Notice that in the fetal specimen the frontal bones have not yet fused to produce the single adult frontal bone. The same is true of the lower jaw. The skull of an infant also has a **fontanel** (*arrow*), or gap, between the skull bones at the superior sur-face of the skull; the fontanel closes in adult life. (b, Carolina Biological Supply Company.)

compact bone is a somewhat spongy filling of **cancellous bone,** which despite its loose structure provides most of the mechanical strength of the bone. The innermost **marrow cavity** contains, in addition to a red or a fatty filling, a good bit of diffuse spongy bone. The radius is enveloped with a tough connective-tissue membrane, the **periosteum,** capable of laying down fresh layers of bone and thus increasing the diameter of the bone.

Quite different in shape and function from the radius, the frontal bone of the skull (Fig. 21–10) is much less important for the transmission of muscular forces than for the dispersal of possibly destructive forces, for it helps to protect that ultimately delicate and unimaginably complex cerebral computer that lies behind it. Shaped almost like a section of eggshell, the frontal bone is, like an eggshell, astonishingly strong for its thickness. This strength is a consequence of its shape, for the frontal bone transforms, redirects, and disperses the mechanical pressure generated by strain and blows, directing them to the sides and away from the soft contents.

The frontal bone is an example of a **membrane bone,** so-called because it originated not from a pre-existing cartilage model but from an embryonic membrane of connective tissue. Splinter-like **spicules** of bone are laid down first in the center of such a bone and then progressively outward until at last the edges are reached. The extreme edges of a baby's skull bones are unossified at birth and are separated from one another by a substantial gap, so some movement of the skull bones is possible. This flexibility allows the baby's head to make its way through the bony confines of the mother's pelvis on the way to the outside world. For some months thereafter the junction of several skull bones at the top of the head (Fig. 21–10b) remains open, though covered with skin, so one may gently feel the soft pulse of the cerebral arteries beneath.

MUSCLE

Locomotion, manipulation, circulation of blood, and the propulsion of food through the digestive tract all require some way of generating mechanical forces and motion. The muscles serve as motors of the body, making possible all these actions and much more. The three types of muscle—skeletal, smooth, and cardiac—each specialized for its particular task, were discussed in Chapter 20.

As in other animals, muscles in humans act antagonistically to one another. Muscles are attached to bones in such a way that the movement produced by one can be reversed by another. The biceps muscle, for example, permits you to flex your arm, whereas the triceps muscle allows you to extend it once again (Fig. 21–11).

Muscle Structure

What we commonly think of as a muscle—the biceps in your arms, for example—consists of thousands of individual cells, each wrapped in connective tissue (Fig. 21–12). Because muscle cells are elongated in shape, they are often referred to as **fibers.**

Each muscle fiber consists of small units of bundled threads, called **myofibrils,** which extend the length of the cell. Cross striations in the myofibrils give skeletal muscle its striped, or striated, appearance. In an electron micrograph it is possible to see that the myofibrils are made up of two types of **myofilaments,** coarse myosin filaments and fine actin filaments. Some of the dark stripes that help to cause the striated appear-

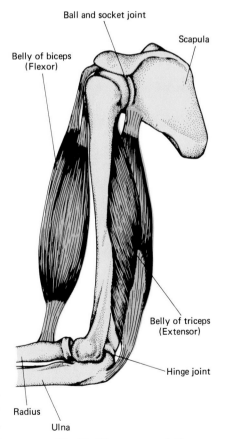

Ball and socket joint

Scapula

Belly of biceps
(Flexor)

Belly of triceps
(Extensor)

Hinge joint

Radius

Ulna

Figure 21–11 The antagonistic arrangement of the biceps and triceps muscles.

Figure 21–12 Cross section of the thigh dissected into smaller and smaller parts. (*a*) Section through the entire thigh. (*b*) A single muscle. (*c*) and (*d*) Microscopic views of the muscle. (*e*) Arrangement of the contractile filaments.

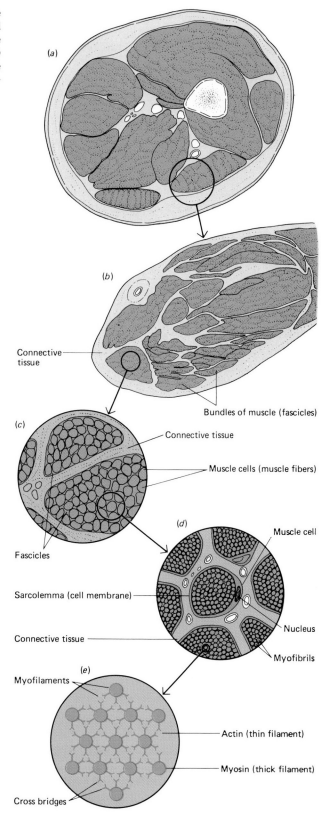

ance of the muscle cells are produced by the optical effect of the overlapping between interdigitating myofilaments (see Fig. 21–12). The typical pull of a muscle results from the shortening of its cells, which in turn results from the myofilaments actively pulling themselves past and between one another (see Fig. 21–13).

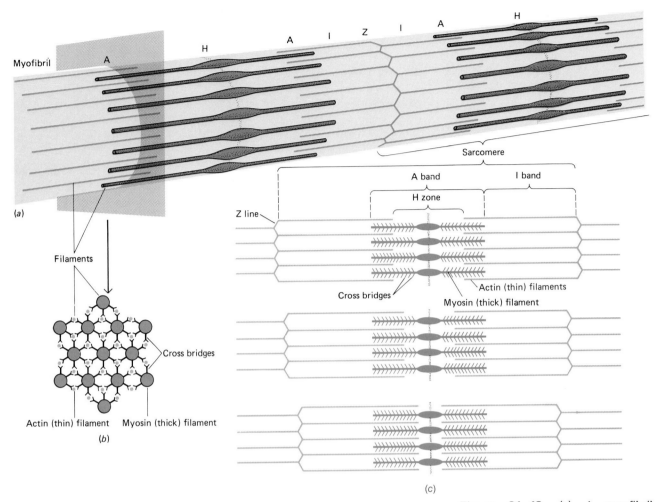

Myofibril

(a)

Filaments

Cross bridges

Actin (thin) filament   Myosin (thick) filament

(b)

Z line

Sarcomere

A band | I band

H zone

Cross bridges
Actin (thin) filaments
Myosin (thick) filament

(c)

The Biochemistry of Muscle Action

The interaction of the actin and myosin filaments produces the contraction of muscle. Each myosin filament consists of about 200 molecules of the protein myosin in a parallel arrangement. A rounded head extends from each rod-shaped myosin molecule. The head of the myosin molecule bears a binding site that is complementary to binding sites on the actin filament. Each actin filament contains 300 to 400 rounded actin molecules arranged in two chains.

A unit of actin and myosin filaments makes up a **sarcomere** (Fig. 21–13). There are many sarcomeres in a skeletal muscle cell, united at their ends by a complex interweaving of filaments called the Z line. Thus, the Z line indicates the boundary of each sarcomere. Each sarcomere is capable of independent contraction. When many sarcomeres contract together, they produce the contraction of the muscle as a whole.

The events that lead up to, trigger, and power contraction begin with a message from a nerve. When a nerve signals a muscle to contract, it releases a substance called **acetylcholine** into the **myoneural cleft,** the space between the nerve ending itself and the muscle cell membrane, or **sarcolemma.** In response, the sarcolemma undergoes an electrical change called **depolarization,** which we will study in more detail as it occurs in nerve cells. Depolarization is unique in muscle cells because it is not confined to the surface membrane of the cell but actually travels *into* the cell along an elaborate system of tubules, which are inward extensions of the cell membrane. These tubules are known collectively as the **T-system** (Fig. 21–14).

Figure 21–13 (a) A myofibril stripped of the accompanying membranes. The Z lines mark the ends of the sarcomeres. (b) Cross section of myofibril shown in a. (c) Filaments slide past each other during contraction. Notice the way the filaments overlap. It is the regular pattern of overlapping filaments that gives rise to the striated appearance of skeletal and cardiac muscle. In the top drawing of c the myofibril is relaxed. In the middle drawing, the filaments have slid toward each other, increasing the amount of overlap and shortening the muscle cell by shortening its sarcomeres. At bottom, maximum contraction has occurred; the sarcomere has shortened considerably. Letters represent zones along the myofibril.

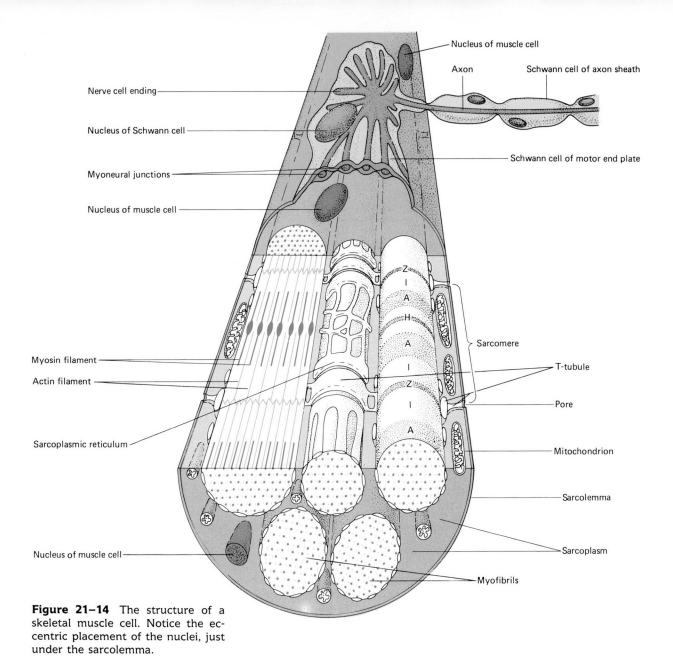

Labels on figure:
- Nucleus of muscle cell
- Axon
- Schwann cell of axon sheath
- Nerve cell ending
- Nucleus of Schwann cell
- Schwann cell of motor end plate
- Myoneural junctions
- Nucleus of muscle cell
- Sarcomere
- Myosin filament
- Actin filament
- T-tubule
- Pore
- Sarcoplasmic reticulum
- Mitochondrion
- Sarcolemma
- Sarcoplasm
- Nucleus of muscle cell
- Myofibrils

(Z I A H A I Z I A letters marked within sarcomere)

Figure 21–14 The structure of a skeletal muscle cell. Notice the eccentric placement of the nuclei, just under the sarcolemma.

The T-system tubules come into contact inside the cell with a specialized endoplasmic reticulum called the **sarcoplasmic reticulum.** As a result of depolarization, calcium is released from specific parts of the sarcoplasmic reticulum. The release of calcium causes a chemical change to take place in the structure of the fine filaments. This change uncovers certain points of anchorage, the myosin binding sites on the actin. Then the binding sites combine temporarily with the ends of the myosin molecules that make up the thick filaments. These ends are called the **cross bridges,** for they bridge the gaps between the thick and thin filaments. The cross bridges now release their hold on the first set of binding sites and reach for the second. The process is repeated with the third, and so on (Fig. 21–15). This series of stepping motions actively pulls the thick and thin filaments past one another with an apparent tendency to combine along as great a length as possible. Relaxation of the muscle occurs when the sarcoplasmic reticulum reabsorbs and binds the calcium once again.

ATP is the immediate source of the energy necessary for muscle contraction (Fig. 21–16). It is necessary both for the pull exerted by the cross bridges and for their release from each active site, as they engage, hand-over-hand fashion, in their tug of war on the thin filaments.

ATP is also necessary to release the grip of the cross bridges. Rigor mortis, the temporary but very marked muscular rigidity that follows death, is due to ATP depletion following the cessation of cellular respiration that occurs at death. Rigor mortis does not persist indefinitely, however, for the entire contractile apparatus of the muscles degenerates eventually, restoring pliability. The phenomenon is temperature-dependent for a number of reasons, so given the prevailing temperature, a police officer can estimate the time of death of a cadaver from its degree of rigor mortis. Perhaps it should be said that rigor mortis is not by itself muscu-

Focus on . . .

FLIGHT MUSCLES OF INSECTS

The only invertebrate phylum that possesses striated muscle to any great extent is the arthropods. In some cases arthropod striated muscle may be more highly differentiated even than that of vertebrates. This is seen best in the **flight muscles** of insects.

In most flying insects the flight muscles are attached not directly to the wings but to the flexible portions of the exoskeleton that articulate with the wings. Each contraction of the muscles produces a dimpling of the exoskeleton in association with a downstroke and, depending on the exact arrangement of the muscles, sometimes on the upstroke as well. When the dimple springs back into its resting position, the muscles attached to it are stretched. The stretching initiates another contraction immediately and the cycle is repeated. The deformation of the cuticle is transmitted as a force to the wings and they beat—so fast that we may perceive the sound as a musical note. In the common blowfly, for instance, the wings may beat at a frequency of 120 cycles *per second*. Yet in that same blowfly, the neurons that innervate those furiously contracting flight muscles are delivering impulses to them at the astonishingly low frequency of three per second. It seems very likely that the mechanical properties of the musculoskeletal arrangement are what provide the stimuli for contraction by stretching the muscle fibers at the resonant frequency of the system. But the nerve impulses are needed to maintain it.

Insect flight muscle in action has a very high metabolic rate, perhaps the highest of any tissue anywhere. Accordingly, it contains more mitochondria than any known variety of muscle, and it is elaborately infiltrated with tiny air-filled tracheae that carry oxygen directly to each cell (see figure). Many insects have special adaptations to rid the body of the excess heat produced by the flight muscles. The rapidly flying sphinx moth, for example, has what amounts to a radiator in its abdomen, a great blood vessel that carries heat from the thorax, where it is generated, and emits it into the cool of the night.

Flight muscles must be kept at operating temperature if they are to function. You have probably noticed the constant twitching of the wings of such insects as wasps even when they are crawling instead of flying. Probably this behavior is necessary to keep the temperature of the flight muscles high for instant combat readiness. You may also have noticed that the bodies of many moths are quite furry. The fur of these insects (more properly called **pilus**) serves the same function as fur in a mammal—to conserve body heat. When the moth awakens and prepares for flight it shivers its flight muscles at a low frequency to warm them up, constricting its abdominal blood vessel to keep the heat in its thorax. Gradually the frequency of the shivering increases until at a critical moment the moth spreads its wings and hums off into the darkness.

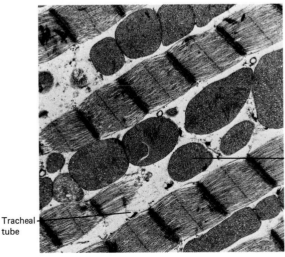

Insect flight muscle, such as that of the bumblebee shown here, may be the most powerful muscle found in any organism. Oxygen is brought directly to the muscle by the tracheal tubes, which convey air into the muscle cell itself. Note the prominent striations and the tremendously convoluted internal membranes of the many mitochondria. (From Heinrich, B.: Bumblebee Economics. Harvard University Press, 1978. Electron micrograph by Mary Ashton. Courtesy of Dr. Heinrich; used by permission.)

Mitochondrion

Tracheal tube

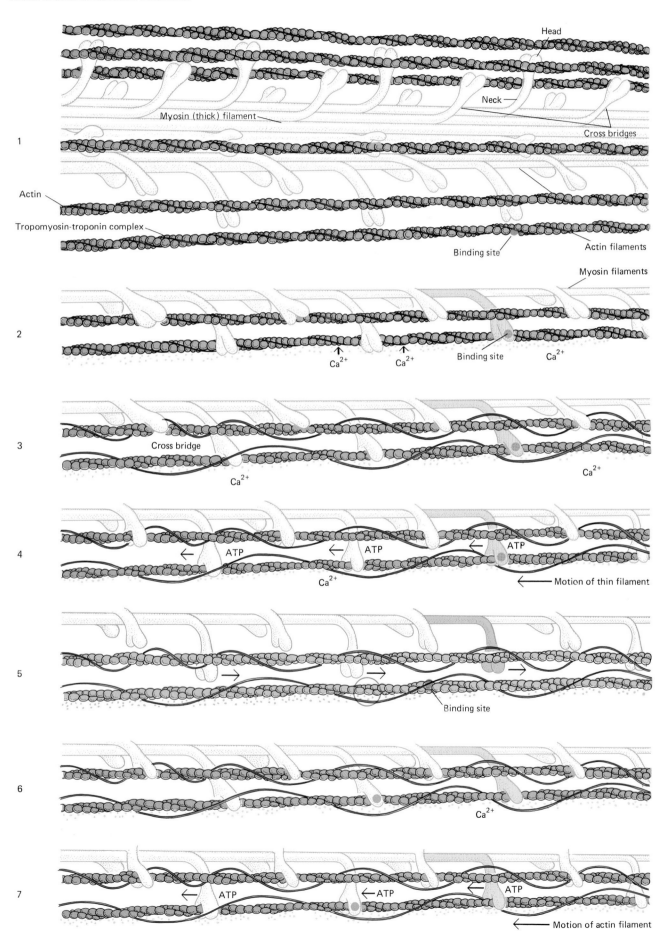

1

Head

Neck

Myosin (thick) filament

Cross bridges

Actin

Tropomyosin-troponin complex

Binding site

Actin filaments

Myosin filaments

2

Ca²⁺   Ca²⁺   Binding site   Ca²⁺

3

Cross bridge

Ca²⁺   Ca²⁺

4

ATP   ATP   ATP

Ca²⁺   ← Motion of thin filament

5

Binding site

6

Ca²⁺

7

ATP   ← ATP   ← ATP

← Motion of actin filament

Figure 21–16 Summary of the events of muscular contraction.

Metabolism of cell produces ATP	Nerve impulse arrives at nerve ending
Energy transferred to creatine phosphate	Acetylcholine released into myoneural cleft
When needed, energy transferred to ATP	Muscle cell membrane depolarized
ATP hydrolysis releases energy	T system depolarized
ADP and phosphate ("spent" ATP)	Ca^{2+} released from sarcoplasmic reticulum

Figure 21–15 Control of the interaction between actin and myosin in muscle contraction. **1.** Thin (actin) filaments are arranged around thick (myosin) in hexagonal fashion. The ends of the myosin molecules are able to interact with the actin molecules. Each such active portion of the myosin molecule is a cross-bridge able to attach to numerous active or binding sites on the actin filaments. When the muscle is not contracted, however, the cross-bridges are prevented from attaching to the binding sites by a tightly wound helix of tropomyosin-troponin protein, which covers the binding sites. **2.** Calcium ions, released from the sarcoplasmic reticulum (not shown), change the shape of the tropomyosin-troponin protein molecule. **3.** The tropomyosin-troponin protein becomes less tightly wound around the actin filament, exposing the binding sites. The cross-bridges begin to attach themselves to the binding sites. We have shown one such in color. **4.** The cross-bridges flex in unison, pulling the thin filaments along as they do so. **5.** If ATP is available, the cross-bridges release the thin filaments. Notice that the binding site formerly grasped by the colored cross-bridge has moved far enough to be grasped by the next cross-bridge. **6.** The next cross-bridge grasps that binding site, and **7.** pulls on it. The same thing is done repeatedly by all the cross-bridges and all binding sites. The muscle contracts.

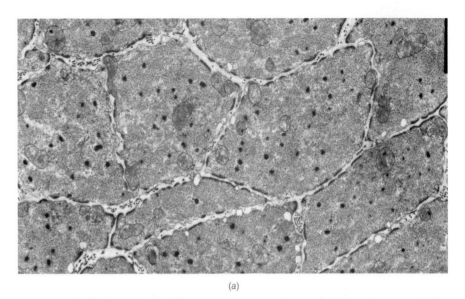

(a)

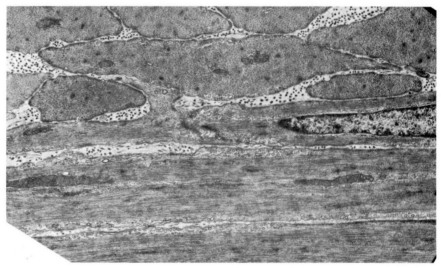

(b)

Figure 21-17 Smooth muscle has fibers, but they do not overlap in the regular fashion of striated and cardiac muscle. These electron micrographs (approximately ×15000) show smooth muscle cells in cross section (a) and longitudinal section (b). The filaments can be easily seen in both views. (Courtesy of Dr. Giorgio Gabella, University College, London.)

lar contraction; it only tends to freeze the corpse in its position at the time of death. Thus, tales of corpses sitting, pointing to their murderers, and otherwise carrying on posthumously may be entertaining but have no factual basis.

As might be imagined, muscle cells require an enormous amount of energy. Fuel is stored in them in the form of muscle **glycogen,** a large polysaccharide molecule formed from hundreds of glucose units. Glycogen is broken down into glucose as needed for fuel in cellular respiration. ATP cannot be stored in appreciable quantity. Instead, muscle cells employ a sort of chemical storage battery that both accepts and gives up its energy to ATP. This is **creatine phosphate (CP),** composed of the organic base creatine plus the energy-rich phosphate donated from ATP. As it is needed, high-energy phosphate is withdrawn from the CP pool and used to recharge the ATP associated with the action of the cross bridges.

The energy conversion of muscular contraction is not very efficient. Only about 30% of the chemical energy of the glucose fuel is actually converted to mechanical work. The remaining energy is accounted for as heat, produced mainly by frictional forces within the muscle cell. This is why we get hot when we work hard physically, and also why we shiver

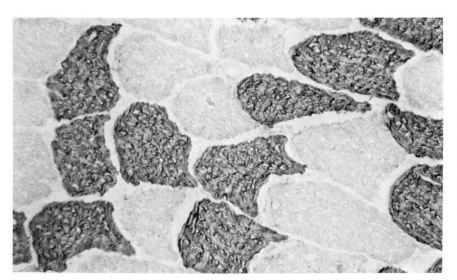

Figure 21–18 Slow (light-colored) and fast (dark) striated muscle fibers in cross section (approximately ×640). The fast fibers have been stained by a technique that identifies a particular kind of quick-acting tropomyosin, a protein that is probably responsible for the characteristic differences in the two kinds of muscle. The white meat of chickens and turkeys is composed of fast-twitch fibers, and the dark, of slow-twitch fibers, but in human beings both kinds of fibers are found in all muscles. (Courtesy of Dr. G.K. Dhoot and G.W. Pearce, *Journal of the Neurological Sciences*, 65, (1984).

when we are cold: The muscle contraction involved in shivering is one way of producing heat to warm the body.

Variations in Muscle Response

The three types of muscle differ in the way they respond. Smooth muscle (Fig. 21–17) often contracts in response to simple stretching, and its contraction tends to be lengthy and sustained. It is well adapted to performing such tasks as the regulation of blood pressure by sustained contraction of the walls of the arterioles. Although smooth muscle contracts slowly, it shortens much more than striated muscle does. Though not well suited for running or flying, smooth muscle squeezes superlatively.

Cardiac muscle contracts abruptly and rhythmically, propelling blood with each contraction. Sustained contraction of cardiac muscle would be disastrous! Skeletal muscle, when stimulated by a single brief stimulus, contracts with a quick, single contraction called a **simple twitch.** Ordinarily, simple twitches do not occur except in laboratory experiments. In the normal animal, skeletal muscle receives a series of separate stimuli very close together. These produce not a series of simple twitches, however, but a single, smooth, sustained contraction called **tetanus.** Depending upon the identity and number of muscle cells tetanically contracting, we thread a needle, haul a rope, or dance a jig.

Not all muscular activities are the same, however. Dancing or, even more so, typing requires quick response rather than the long, sustained effort that might be appropriate in hauling a rope. In many animals entire muscles are specialized for quick or slow responses. In chickens, for instance, the white breast muscles are efficient for quick responses, since flight is an escape mechanism for chickens. On the other hand, chickens walk about on the ground all day, so the dark leg and thigh meat is composed of muscle specialized for more sustained activity.

There is no human equivalent of light and dark meat. However, we do possess individual muscle cells that are specialized for either fast or slow response, which can easily be distinguished microscopically with the appropriate staining (Figure 21–18). The proportions of **slow-twitch** and **fast-twitch** fibers vary from muscle to muscle in the same person and also differ among persons. It has long been believed that the relative proportions of the two determined the kind of athletic activity at which one might have the greatest potential proficiency, and also that this proportion was genetically determined. Recent evidence, however, indicates that the proportions of the two kinds of fibers in human muscle can be changed by appropriate training, at least to some degree.

CHAPTER SUMMARY

I. Human skin includes nails, hair, sweat glands, oil glands, and sensory receptors.
   A. Cells in the stratum basale of the epidermis continuously divide; as they are pushed upward toward the skin surface, these cells mature, produce keratin, and eventually die.
   B. The dermis, which consists of dense, fibrous connective tissue, rests on a layer of subcutaneous tissue composed largely of fat.
II. The skeleton transmits mechanical forces generated by muscle and also supports and protects the body.
   A. Hydra and many other invertebrates have a hydrostatic skeleton in which fluid is used to transmit forces generated by contractile cells or muscle. In *Hydra* the circular and longitudinal layers of contractile cells form an antagonistic relationship.
   B. Exoskeletons are characteristic of mollusks and arthropods. The arthropod skeleton, composed mainly of chitin, is jointed for flexibility. This nonliving skeleton prevents growth, making it necessary for arthropods to molt periodically.
III. Endoskeletons, found in echinoderms and chordates, are composed of living tissue and therefore are capable of growth.
   A. The human skeleton consists of an axial portion and an appendicular portion.
   B. The radius, a typical long bone, consists of a thin outer shell of compact bone surrounding the inner cancellous bone. The radius is referred to as a cartilage bone because in the embryo it consisted of cartilage. During development that cartilage was gradually replaced with bone.
   C. The frontal bone, a membrane bone, functions mainly in protecting the brain.
IV. As muscle contracts (shortens), it moves body parts by pulling on them.
   A. The striations of skeletal muscle fibers reflect the interdigitations of their actin and myosin filaments. A unit of actin and myosin filaments makes up a sarcomere.
   B. Muscle contraction begins when a nerve impulse results in the release of acetylcholine into the myoneural cleft.
   C. This results in depolarization of the sarcolemma (muscle cell membrane) and then of the T-system.
   D. Calcium is released from the sarcoplasmic reticulum and acts to uncover the binding sites of the actin filaments.
   E. Cross bridges of myosin filaments attach to binding sites.
   F. The cross bridges flex and reattach to new binding sites so that the filaments are pulled past one another and the muscle shortens. Bridges are powered by ATP.
   G. ATP is the immediate source of energy for muscle contraction. In muscle tissue, energy is stored in creatine phosphate.

Post-Test

1. The vertebrate skin consists of two main layers, the outer _____ and the inner _____ .
2. The cells of the stratum _____ of the epidermis are dead and almost waterproof.
3. The protein _____ confers mechanical strength, flexibility, and waterproofing on the skin.
4. _____ skeletons have the principal or even sole function of transmitting muscular force.
5. Since an exoskeleton tends to limit size, arthropods must _____ from time to time in order to grow.
6. The internal skeletons of echinoderms and chordates are known as _____ .

7. The radius has a thin outer shell of _____ bone and a spongy filling of _____ bone.
8. Synovial fluid serves as a _____ in _____ .
9. The two types of myofilaments in muscle tissue are _____ filaments and _____ filaments.
10. Unscramble this list of the events of muscle contraction into the correct sequence:
   a. calcium release
   b. T-system depolarization

   c. acetylcholine release
   d. nerve impulse
   e. uncovering of the binding sites of the actin filaments
   f. cross bridges flex
   g. cross bridges release binding sites
11. Creatine phosphate's function is _____ _____ in the muscle cell.
12. Fuel is stored in muscle cells in the form of the polysaccharide _____ ; the immediate source of energy for muscle contraction is _____ .

Review Questions

1. Compare human skin with that of several other types of vertebrates.
2. What properties does keratin confer on human skin?
3. What is a hydrostatic skeleton? Which functions does it perform?
4. What are the disadvantages of an exoskeleton?
5. What are the functions of the human skeleton?
6. What is the difference between cartilage bone and membrane bone?
7. Compare the two types of myofilaments in muscle tissue.
8. What is the role of ATP in muscle contraction? What is the function of creatine phosphate?
9. Outline the sequence of events that causes a muscle cell to contract, beginning with the stimulation of its nerve and including cross-bridge action.

Readings

Austin, P.R., et al. "Chitin: New Facets of Research," *Science* 212: 749–753 (15 May 1981). The practical uses of chitin, the most widely distributed animal skeleton carbohydrate, may eventually rival those of cellulose, the major plant skeletal carbohydrate.

Buller, A.J., and Buller, N.P. *The Contractile Behavior of Skeletal Muscle* (booklet). Burlington, N.C., Carolina Biological Supply Company, 1978. Emphasis is on classical physiology of contraction and stimuli.

Clark, R.B. *Dynamics in Metazoan Evolution: The Origin of the Coelom and Segments.* Oxford, Clarendon Press, 1964. Despite the title, this is basically a discussion of locomotory adaptations in lower animals with emphasis on the hydrostatic skeleton.

Cole, R.P. "Myoglobin Function in Exercising Skeletal Muscle," *Science* 216: 523–525 (30 April 1982). Its function long a mystery, muscle hemoglobin at last yields up some of its secrets. Though the details remain unknown, the substance is shown to be necessary for normal muscular oxygen consumption.

Gray, J. *Animal Locomotion.* London, Weidenfeld and Nicholson, 1968. The hydrostatic skeleton and the role of the musculoskeletal system in higher animal locomotion.

Harrington, W.F. *Muscle Contraction* (booklet). Burlington, N.C., Carolina Biological Supply Company, 1981. Contains hard-to-find and valuable criticisms of the cross bridge theory of muscle contraction.

Heinrich, B., and G.A. Bartholomew. "Temperature Control in Flying Moths," *Scientific American,* June 1972, 87–95. A description of the mechanisms of temperature regulation in moths and of the relationship between temperature and flight.

Huxley, A. *Reflections on Muscle.* Princeton, Princeton University Press, 1980. The originator of the sliding filament theory of muscular contraction discusses the history and prospects of the scientific understanding of contraction.

Lazarides, E., and J.P. Revel. "The Molecular Basis of Cell Movements," *Scientific American,* May 1978, 100–112. The role of microfilaments in cell movement.

Lowenstam, H.A. "Minerals Formed by Organisms," *Science* 211: 1126–1130 (13 March 1981). A fine comparative study of skeletal systems.

Luttgens, K., and K.F. Wells. *Kinesiology: Scientific Basis of Human Motion,* 7th ed. Philadelphia, Saunders College Publishing, 1982. How skeleton, muscles and nervous system interact to permit and produce the multitude of motions of which the human body is capable.

Montagna, W., and P.F. Parakkel. *The Structure and Function of the Skin,* 3rd ed. New York, Academic Press, 1974.

Morey, E.R. "Spaceflight and Bone Turnover," *Bioscience,* 1984, 168–172. Spaceflight may become practical only when the demineralization of bone that it produces is stopped. This will require an extension of our fundamental knowledge of bone mineral turnover mechanisms and their control.

Neville, C. *The Biology of the Arthropod Cuticle* (booklet). Burlington, N.C., Carolina Biological Supply Company. A beautiful short summary.

Chapter 22

RESPONSIVENESS: NEURAL CONTROL

Outline

I. Cells of the nervous system
   A. Glial cells
   B. Neurons
II. Information flow through the nervous system
   A. Neural circuits
   B. Reflex action
III. Transmission of impulses
   A. Transmission along a neuron
   B. Substances that affect excitability
   C. Transmission between neurons
   D. Neurotransmitters
   E. Direction and speed of conduction
IV. Neural integration
V. Organization of neural circuits
Focus on regeneration of an injured neuron

Learning Objectives

1. Describe the functions of glial cells and neurons.
2. Draw a neuron. Label each part and give its function.
3. Trace the flow of information through the nervous system.
4. Draw a reflex pathway consisting of three neurons, label each structure, and indicate the direction of information flow.
5. Describe the mechanism by which an impulse is transmitted along a neuron.
6. Describe the mechanism by which impulses are transmitted from one neuron to another.
7. Describe the factors that determine whether a neuron will transmit an impulse, and summarize their interaction.

The ability of an organism to survive and to maintain its steady state depends largely upon the effectiveness with which it can *respond* to changes in its internal or external environment. Changes within the body or in the outside world that can be detected by an organism are termed **stimuli.** In simple animals with simple nervous systems, the range and types of responses are very limited and stereotyped. In complex animals varied and sophisticated responses to stimuli are possible because responsiveness is controlled by two highly specialized systems—the nervous and the endocrine systems. The nervous system permits very rapid response, while the endocrine system generally provides long-lasting chemical regulation.

CELLS OF THE NERVOUS SYSTEM

Cell types unique to the nervous system are neurons and glial cells. Neurons are specialized to receive and transmit impulses; glial cells play supportive roles.

Glial Cells

There are perhaps ten times as many **glial cells** as neurons in a complex animal, but so far, little is known about the functional roles of these varied and interesting cells. They come in several varieties and are sometimes referred to collectively as the **neuroglia,** which literally means "nerve glue." From a clinical standpoint glial cells are of great interest because they give rise to most of the tumors that develop within the central nervous system (brain and spinal cord).

Some glial cells envelop neurons and form insulating sheaths about them. Others are phagocytic and serve to remove debris from the nervous tissue. A third type of glial cell lines the cavities of the brain and spinal cord. **Schwann cells** (sometimes classified as glial cells) are supporting cells found outside the central nervous system. They form sheaths about some neurons.

Neurons

Highly specialized to receive and transmit messages in the form of neural impulses, the **neuron** is distinguished from all other cells by its long cytoplasmic processes. It will be helpful to examine the structure of the most common variety, a multipolar neuron (Fig. 22–1; several kinds of neurons are shown in Fig. 22–2).

The largest portion of the neuron, the **cell body,** contains the bulk of the cytoplasm, the nucleus, and most of the other organelles. From the cell body project two types of cytoplasmic extensions, the **dendrites** and a long, single **axon.** Dendrites are typically short, highly branched fibers specialized to receive neural impulses and to transmit them to the cell body. The cell body integrates incoming signals and can also receive impulses directly. Although microscopic in diameter, an axon may be 3 feet or more in length. The axon conducts neural messages from the cell body to another neuron or to a muscle or gland. At its end the axon branches and terminates in tiny structures called **synaptic knobs.** These structures release neurotransmitters, chemicals essential to the transmission of impulses from one neuron to another. Along its course an axon can give off branches known as **collaterals.**

Axons of many neurons outside the central nervous system are covered by two sheaths—an outer **cellular sheath,** or **neurilemma,** and an inner **myelin sheath** (Fig. 22–3). Both sheaths are produced by Schwann

Figure 22–1 Structure of a multipolar neuron. The axon of this neuron is myelinated, and so the myelin sheath is shown as well as the cellular sheath.

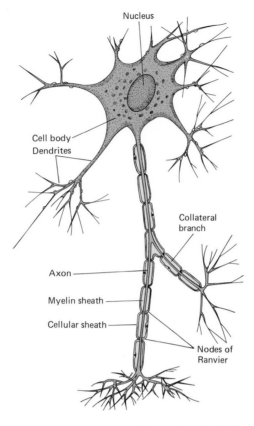

Nucleus

Cell body
Dendrites

Collateral branch

Axon

Myelin sheath

Cellular sheath

Nodes of Ranvier

Figure 22–2 Some representative neurons from a slice of a newborn rat's brain maintained in tissue culture. The differences in their appearance result both from the use of different staining techniques and from the fact that they are different kinds of neurons. (*a*) A pyramidal cell from the hippocampus (approximately × 64). (*b*) A granule cell from the dentate gyrus of the cerebral cortex, (approximately ×64). (*c*) A Purkinje cell from the cerebellum (approximately ×160). (Courtesy of Dr. B.H. Gähwiler, *Experientia*, 40:1984).

cells. To form the cellular sheath, Schwann cells line up along the axon and wrap themselves about it. In producing the myelin sheath, which lies between the axon and the cellular sheath, the Schwann cell winds its cell membrane about the axon several times. **Myelin** is a white, lipid-rich substance that makes up the Schwann-cell membrane. It is an excellent insulator, and its presence influences neural transmission. Between successive Schwann cells, gaps called **nodes of Ranvier** occur in the myelin sheath. At these points the axon is not insulated with myelin.

Almost all axons more than 2 micrometers in diameter are **myelinated,** that is, they possess myelin sheaths. Those of smaller diameter are

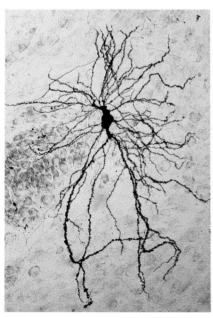

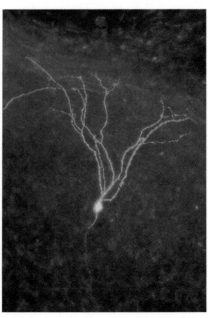

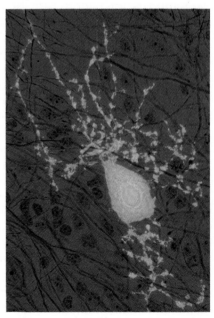

(a)

(b)

(c)

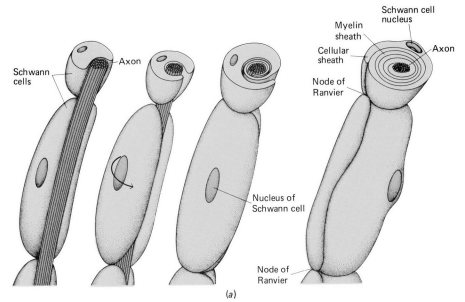

(a)

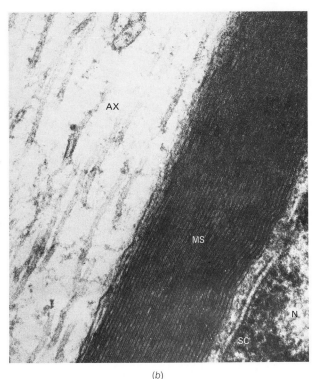

(b)

Figure 22–3 (*a*) Formation of the myelin sheath around the axon of a peripheral neuron. A Schwann cell wraps its cell membrane around the axon many times to form the insulating myelin sheath. The rest of the Schwann cell remains outside the myelin sheath, forming the cellular sheath. (*b*) Electron micrograph of a section through a single myelinated axon (approximately ×173,000). *AX*, axon; *MS*, myelin sheath; *SC*, Schwann cell; *N*, nucleus of Schwann cell. (Courtesy of Dr. Lyle C. Dearden.)

generally unmyelinated. In the central nervous system, myelin sheaths are formed by certain glial cells, but cellular sheaths are not present.

In **multiple sclerosis,** a neurological disease that affects about 300,000 people in the United States alone, patches of myelin deteriorate at irregular intervals along the length of the neurons and are replaced by scar tissue. This damage interferes with conduction of neural impulses, and the victim suffers loss of coordination, tremor, and partial or complete paralysis of parts of the body. The cause of multiple sclerosis has been a medical mystery, but there is some evidence that it is an autoimmune disease, in which the body attacks its own tissue (Chapter 25).

The cellular sheath is important in the regeneration of injured neurons. When an axon is cut, the portion separated from the cell body deteriorates and is digested by surrounding phagocytic cells, but the cellular sheath remains intact. The cut end of the axon regrows slowly through

Focus on . . .

REGENERATION OF AN INJURED NEURON

When an axon is separated from its cell body by a cut, it soon degenerates. A hollow tube of Schwann cells remains, but myelin eventually disappears. As long as the cell body of the neuron has not been injured, it is capable of regenerating a new axon. Sprouting begins within a few days after cutting (see the figure). The growing axon enters the old sheath tube and proceeds along it to its destination in the central nervous system or periphery. Axons can grow in the absence of sheaths if some conduit is provided for them. They can, for example, be made to grow within sections of blood vessels or extremely fine plastic tubes. The time required for regeneration depends on how far the nerve has to grow and may be as long as two years. When cuts occur within the spinal cord or brain, regeneration is very feeble and usually absent. It is thought that growth of new sprouts in the CNS is prevented by scar tissue formed by neuroglial cells at the site of injury. It is remarkable that (if not blocked by scar tissue or other barrier) each regenerating axon of a cut nerve finds its way back to the muscle or gland and functional contact is made.

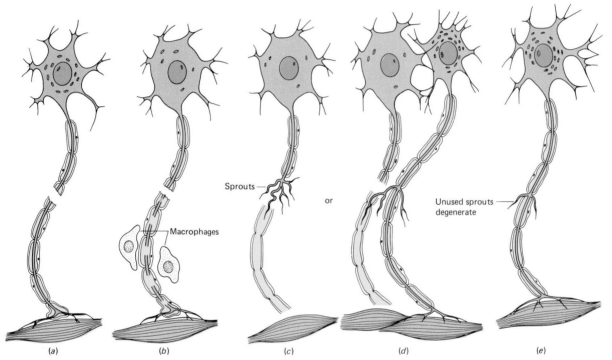

Regeneration of an injured neuron. (a) A neuron is severed. (b) The part of the axon that has been separated from its nucleus degenerates. Its myelin sheath also degenerates, and macrophages phagocytize the debris. (c) The tip of the severed axon begins to sprout, and one or more sprouts may find their way into the empty cellular sheath, which has remained intact. The sprout grows slowly and becomes myelinated. (d) Sometimes an adjacent undamaged neuron may send a collateral sprout into the cellular sheath of the damaged neuron. (e) Eventually the neuron may regenerate completely, so function is fully restored. Unused sprouts degenerate.

the empty cellular sheath, and after a long time at least partial neural function may be restored. (See Focus on Regeneration of an Injured Neuron.)

What we ordinarily think of as a **nerve** is a complex cord consisting of hundreds or even thousands of axons wrapped in connective tissue (Fig. 22–4). We may compare a nerve to a telephone cable. The individual axons correspond to the wires that run through the cable, and the sheaths and connective-tissue coverings correspond to the insulation. If nerves consist of bundles of axons, you might wonder where are the cell bodies that are attached to those axons. These are often grouped together in a mass known as a **ganglion.**

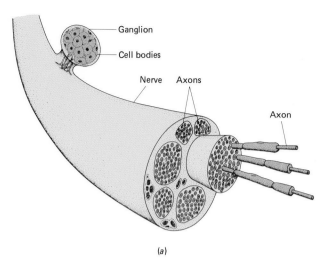

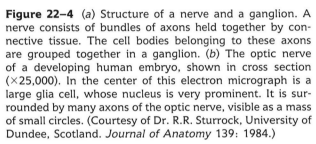

(a)

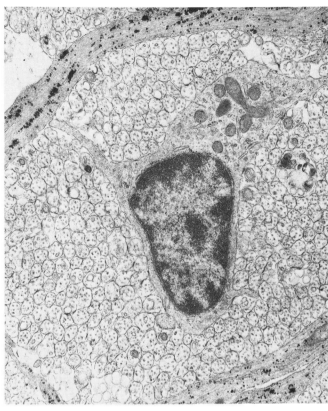

(b)

Figure 22–4 (*a*) Structure of a nerve and a ganglion. A nerve consists of bundles of axons held together by connective tissue. The cell bodies belonging to these axons are grouped together in a ganglion. (*b*) The optic nerve of a developing human embryo, shown in cross section (×25,000). In the center of this electron micrograph is a large glia cell, whose nucleus is very prominent. It is surrounded by many axons of the optic nerve, visible as a mass of small circles. (Courtesy of Dr. R.R. Sturrock, University of Dundee, Scotland. *Journal of Anatomy* 139: 1984.)

INFORMATION FLOW THROUGH THE NERVOUS SYSTEM

An organism is bombarded with thousands of stimuli each day. The job of its nervous system is to receive information, transmit messages, sort out and interpret incoming data, and then issue appropriate commands so that responses will be coordinated and homeostatic. Even the very simple responses to stimuli generally require a sequence of information flow through the nervous system that includes reception, transmission of information, integration, transmission of the "decision," and response (Fig. 22–5).

Reception is the process of detecting or receiving a stimulus; it is the job of specialized sense organs as well as of neurons themselves. Transmission is the process of sending messages along neurons, from one neuron to another, or from a neuron to a muscle or gland. Integration is the process of sorting and integrating incoming information and determining the appropriate mode of response. In complex animals it is primarily the function of the central nervous system. The actual response is carried out by **effectors,** usually muscles and glands.

Neural Circuits

The neurons of the nervous system are organized into millions of sequences called **neural circuits** or **pathways.** Generally, neurons are arranged so that the axon of one neuron in the circuit forms junctions with the dendrites of the next neuron in the circuit. The junction between the two neurons is called a **synapse.** A tiny gap known as the **synaptic cleft** (about 20 or so nanometers, or less than one-millionth of an inch) separates the two neurons. In vertebrates neurons are organized into a central nervous system (CNS), consisting of brain and spinal cord, and a periph-

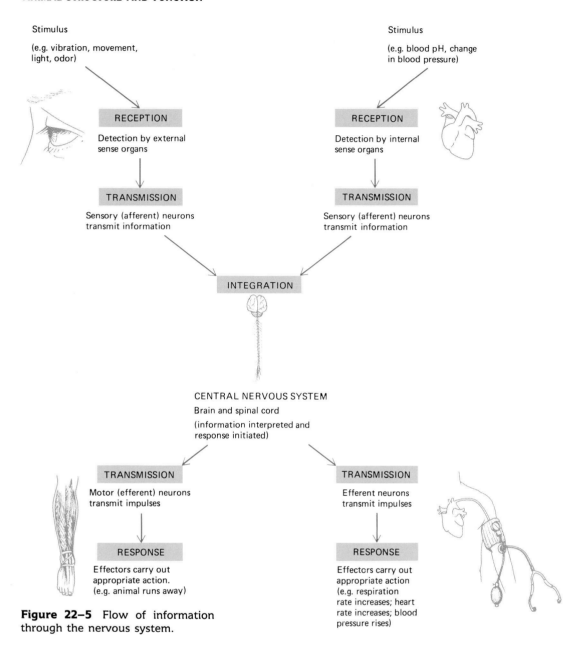

Figure 22–5 Flow of information through the nervous system.

eral nervous system (PNS), which is made up of the sensory receptors and nerves, the communication lines.

Reflex Action

One of the simplest examples of response is the **reflex action,** a relatively fixed reaction pattern to a simple stimulus. The response is predictable and automatic, not requiring conscious thought. Most of the activities of the body are regulated by reflex actions. Though some reflexes, such as those that regulate blood pressure or respiration, are innate, others are learned. The movements involved in walking and many of the movements associated with eating and drinking, for example, become so well established that they are considered reflex actions.

Although most reflex actions are much more complex, let us consider a withdrawal reflex in which a neural circuit consisting of only three neurons is needed to carry out a response to a stimulus (Fig. 22–6). Suppose you touch a hot stove. Almost instantly, and before you are consciously aware of the situation, you jerk your hand away from this

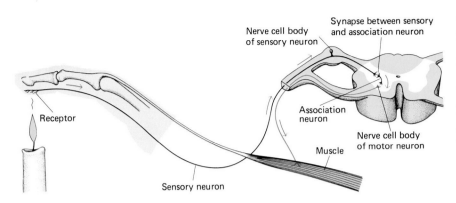

Figure 22–6 The withdrawal reflex involves a chain of three neurons. A sensory neuron transmits the message from the receptor to the central nervous system, where it synapses with an association neuron. Then an appropriate motor neuron (shown in red) transmits an impulse to the muscles that move the hand away from the flame (the response).

unpleasant stimulus. But in this brief instant a message has been carried from pain receptors in the skin to the spinal cord by a **sensory neuron.** In the tissue of the spinal cord the message is transmitted from the sensory neuron to an **association neuron,** then to an appropriate **motor neuron,** which conducts the message to groups of muscles that respond by contracting and pulling the hand from the stove. Actually, many neurons located in sensory, association, and motor nerves participate in such a reaction, and complicated switching is involved. We move our hands *up* from a hot stove but *down* from a hot light bulb. Generally, we are not even consciously aware that all these responding muscles exist.

Quite probably, at the same time that the association neuron sends a message out along a motor neuron, it also sends one up the spinal cord to the conscious areas of the brain. As you withdraw your hand from the hot stove, you become aware of what has happened and feel the pain. This awareness, however, is a feature apart from the reflex response.

TRANSMISSION OF IMPULSES

Once a receptor has been stimulated, the message must be transmitted to the CNS and then back to appropriate effectors. Information must be conducted through a sequence of neurons. How is a neural message transmitted along an individual neuron? And how is it conducted from one neuron to the next in the sequence?

Transmission Along a Neuron

In a resting neuron—that is, one that is not transmitting an impulse—the inner surface of the cell membrane is negatively charged compared with the tissue fluid surrounding it (Fig. 22–7). When electric charges are separated in this way, they have the potential of doing work should

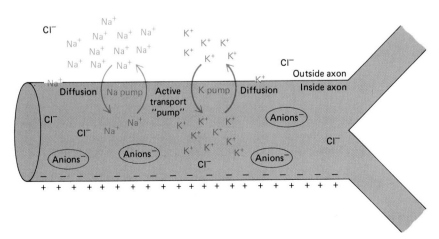

Figure 22–7 Segment of an axon of a resting (nonconducting) neuron. Sodium is actively pumped out of the cell and potassium is pumped in. Sodium is unable to diffuse back to any extent, but potassium does diffuse out along its concentration gradient. Because of the unequal distribution of ions, the inside of the axon is negatively charged compared with the outer tissue fluid. The presence of negatively charged proteins and other large anions in the cell contributes to this polarity.

they be permitted to come together. The amount of work that can be performed may be expressed in volts or, in the case of electrically active cells, in millivolts (a millivolt is one-thousandth of a volt). The **resting potential** of a neuron amounts to about 80 millivolts, expressed as -80 millivolts (because the inner surface of the cell membrane is negatively charged relative to the tissue fluid).

How does the resting potential develop? It results from the presence of a slight excess of negative ions inside the cell membrane compared with a slight excess of positive ions in the tissue fluid immediately surrounding the cell. The membrane of the neuron has very efficient **sodium pumps,** which actively transport sodium out of the cell. At the same time these pumps are thought to transport potassium ions into the cell. However, the membrane is more permeable to potassium than to sodium, so that although only small amounts of sodium are able to leak back in, potassium is able to leak out along a concentration gradient. The end result is that there are more positively charged ions outside the membrane than inside. Negatively charged proteins and some other molecules too large to diffuse out of the cell also contribute to the relative negative charge along the inside of the cell membrane. The resting neuron is said to be electrically polarized (oppositely charged on the inside of the membrane compared with the outside).

Any stimulus (electrical, chemical, or mechanical) that makes the neuron more permeable to sodium may result in the initiation and propagation of a neural impulse, or **action potential.** Some physiologists think that the cell membrane contains specific gates (pores) that are closed when the neuron is at rest but open, admitting sodium, when the neuron is stimulated. Other gates allow potassium to pass through the membrane. When the neuron is stimulated, the sodium gates open so that large numbers of sodium ions diffuse into the cell. This changes the internal electric charge from negative to positive.

When large numbers of sodium ions leak into the cell, making the membrane potential less negative than in a resting neuron, the neuron is said to be depolarized. **Depolarization** is essentially a failure of the cell membrane to keep sodium out. The area of depolarization may spread like a chain reaction down the length of the neuron. Thus a neural impulse is transmitted as a **wave of depolarization,** an electric current that travels down the neuron.

Once a wave of depolarization passes on, resting conditions are quickly re-established. The sodium gates close and the potassium gates open, allowing potassium ions to rush out of the cell. In this way **repolarization** occurs. The sodium pump actively transports excess sodium back out of the cell so that conditions are returned to the resting state. In summary, as the area of depolarization moves down the membrane of the neuron, the normal polarized stage is quickly re-established behind it (Fig. 22–8).

Conduction of an impulse is an active, self-propagating mechanism dependent upon energy expenditure by the neuron. The impulse moves along the axon at a constant velocity (and amplitude) for each type of neuron. Conduction of a neural impulse is somewhat analogous to setting up a path of gunpowder and lighting one end with a match. By igniting the powder particles ahead of it, the flame moves steadily from one end of the trail of gunpowder to the other. The analogy then breaks down: There is no way of restoring the fuse to its original condition after it has burned, but the nerve cell does restore itself.

The membrane of the neuron can depolarize slightly without actually firing, that is, without initiating an impulse. However, when the extent of depolarization reaches about -40 millivolts (the exact amount

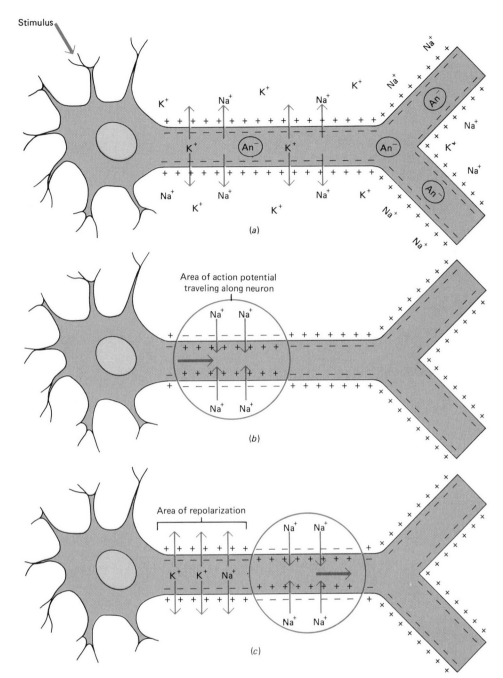

Figure 22–8 Transmission of an impulse along an axon. (*a*) The dendrites (or cell body) of a neuron are stimulated sufficiently to depolarize the membrane to firing level. The axon in (*a*) is shown still in the resting state and has a resting potential. (*b*) and (*c*) An impulse is transmitted as a wave of depolarization that travels down the axon. At the region of depolarization, sodium ions diffuse into the cell. As the impulse passes along from one region to another, polarity is quickly re-established.

varies with each type of neuron), a critical point called the **threshold,** or firing level, is reached (Fig. 22–9). An almost explosive action occurs as the action potential is produced.

Any stimulus weaker than threshold level will not fire the neuron. It merely sets up a local response that fades and dies within a few millimeters from the point of stimulus. A stimulus strong enough to depolarize the neuron to its threshold results in the propagation of an impulse. A stimulus stronger than necessary results in the propagation of an identical impulse. In other words, the neuron either propagates an impulse or it does not. There is no variation in the strength of a single impulse. This is known as the **all-or-none law.**

But how can this be? Sensations do come in different levels of intensity. Certainly, for example, you can tell the difference between a little pain and a big one. How can this apparent inconsistency be explained?

Figure 22–9 An action potential recorded with one electrode inside the cell and one just outside the plasma membrane. When the axon depolarizes to about −55 millivolts, an action potential is generated. (The numerical values are included as representative examples. These values may vary for different nerve cells.)

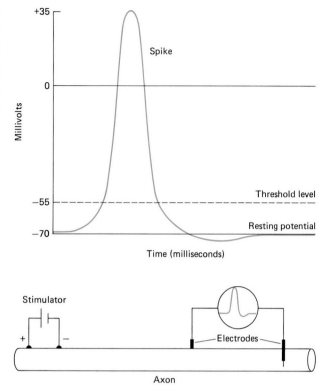

Intensity of sensation depends upon the *number* of neurons stimulated and upon their *frequency of response.* Suppose you burn your hand. The larger the area burned, the more pain receptors will be stimulated and the more neurons will be stimulated. Also, the longer the stimulus persists, the more times per second each neuron will be depolarized.

After an impulse has passed a particular point on the axon, that portion of the axon enters a **refractory period.** During the millisecond or so that it is depolarized, the neuron cannot transmit another impulse. Then, for a few additional milliseconds, while the resting condition is being re-established, it can transmit impulses only when they are more intense than is normally required. Even with the limits imposed by their refractory periods, most neurons can transmit hundreds of impulses per second.

Substances That Affect Excitability

Substances that increase the permeability of the membrane to sodium make the neuron more excitable than normal. Other substances make the neuron less excitable. Calcium balance is essential to normal neural function. When insufficient numbers of calcium ions are present, the sodium gates apparently do not close completely between action potentials, so sodium tends to leak into the cell. This lowers the resting potential, bringing the neuron closer to firing. The neuron thus fires more easily and sometimes even spontaneously. As a result, the muscle (innervated by such neurons) may go into spasm. On the other hand, when calcium ions are too numerous, neurons are less irritable and more difficult to fire.

Local anesthetics such as procaine and cocaine are thought to decrease the permeability of the neuron to sodium. Excitability may be so reduced that the neuron cannot propagate an impulse through the anesthetized region. DDT and related pesticides interfere with the action of the sodium pump. When nerves are poisoned by these substances, they are unable to transmit impulses. Although the human nervous system can also be damaged by these poisons, insects are poisoned by much smaller amounts.

Transmission Between Neurons

A neuron that ends at a specific synapse is referred to as a **presynaptic neuron,** whereas a neuron that begins at a synapse is a **postsynaptic neuron.** The same neuron may be postsynaptic with respect to one synapse and presynaptic relative to another. In some neural pathways neurons may come so close together at the synaptic junction that the impulse is electrically transmitted from presynaptic to postsynaptic neuron. However, this is not the usual case, because most neurons are separated by a wider synaptic cleft. Since depolarization is a property of the neuron membrane, when the impulse reaches the end of the presynaptic axon, it is unable to jump the gap. An entirely different mechanism is needed to conduct the message across the synaptic cleft to the next neuron in the sequence.

When the impulse reaches the synaptic knobs at the end of the axon, it stimulates the release of neurotransmitter into the synaptic cleft. This chemical messenger swiftly diffuses across the gap and may depolarize the dendrites or cell body of postsynaptic neurons so that impulses are initiated in them.

Mitochondria in the synaptic knobs provide the ATP needed for continuous synthesis of the neurotransmitter. Needed enzymes are produced in the cell body and move down the axon to the synaptic knobs. After it is produced, the neurotransmitter is stored in little vesicles (sacs) within the cytoplasm (Fig. 22–10).

Each time an action potential reaches the synaptic knob, calcium gates in the membrane open, permitting calcium ions to pass into the cell. The calcium ions induce several hundred vesicles to fuse with the membrane and release their contents into the synaptic cleft. Neurotransmitter substance diffuses across the synaptic cleft and may be taken up by specific receptors on the dendrites or cell bodies of postsynaptic neurons. Excess neurotransmitter is reabsorbed into the synaptic vesicles or inactivated by enzymes. Transmission of a neural message from a neuron to an effector is also accomplished by the release of neurotransmitter.

Neurotransmitters

About 30 substances are now known (or suspected) neurotransmitters (Table 22–1). Many types of neurons contain two or even three types of neurotransmitter. Furthermore, a postsynaptic neuron can have receptors for more than one type.

The two neurotransmitters that have been investigated most extensively are acetylcholine and norepinephrine (NE). **Acetylcholine** triggers muscle contraction. It is released not only from motor neurons that innervate skeletal muscle but also by some other neurons of the peripheral nervous system and by some neurons in the brain. Cells that release acetylcholine are referred to as **cholinergic neurons.** Acetylcholine has an excitatory effect on skeletal muscle but an inhibitory effect on cardiac muscle. Whether a neurotransmitter excites or inhibits is apparently a property of the postsynaptic receptors with which it combines.

After acetylcholine is released into a synaptic cleft and combines with receptors on the postsynaptic neuron, excess molecules must be removed so that repeated stimulation of the muscle or postsynaptic neuron will not occur. An enzyme called **cholinesterase** catalyzes the breakdown of acetylcholine into its chemical components, choline and acetate.

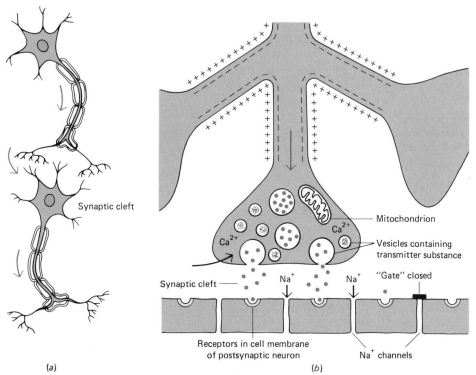

(a)

(b)

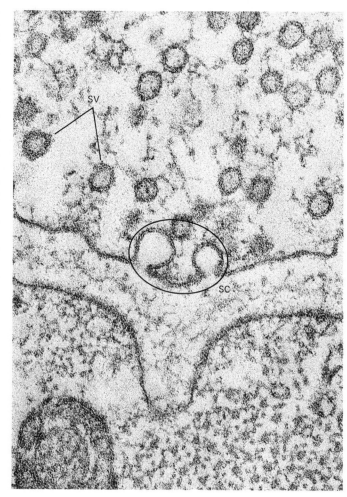

Figure 22–10 Transmission of an impulse between neurons, or from a neuron to an effector. (*a*) In most synapses the wave of depolarization is unable to jump across the synaptic cleft between the two neurons. (*b*) The problem is solved by the release of neurotransmitter from vesicles within the synaptic knobs of the axon. The neurotransmitter diffuses across the synaptic cleft and may combine with receptors in the membrane of the postsynaptic neuron. This may trigger an impulse in the postsynaptic neuron. It is thought that when neurotransmitter combines with the postsynaptic receptors, hypothetical sodium gates open, permitting sodium to rush into the axon. (*c*) Electron micrograph of a synaptic knob and cleft (approximately ×125,000). Two synaptic vesicles (*circled area*) are merging with the cell membrane of knob and discharging neurotransmitter into the cleft. *SV,* synaptic vesicles; *SC,* synaptic cleft; neuron above cleft, muscle below. (Courtesy of Dr. John Heuser.)

(c)

Table 22–1
SOME NEUROTRANSMITTERS

Substance	Locations (where secreted)	Comments
Acetylcholine	Myoneural (muscle-nerve) junctions; autonomic system;[1] parts of brain	Inactivated by cholinesterase
Norepinephrine	Autonomic system; reticular activating system and other areas of brain and spinal cord	Inactivated slowly by monoamine oxidase (MAO); mainly inactivated by reabsorption by vesicles in the synaptic knob; norepinephrine level in brain affects mood
Dopamine	Limbic system; cerebral cortex; basal ganglia; hypothalamus	Thought to affect motor function; may be involved in schizophrenia;[2] amount reduced in Parkinson's disease
Serotonin	Limbic system; hypothalamus; cerebellum; spinal cord	May play role in sleep; LSD antagonizes serotonin; thought to be inhibitory
GABA (gamma-amino butyric acid)	Spinal cord, cerebral cortex, cerebellum	Acts as inhibitor; may play role in pain perception
Endorphins and enkephalins	Many parts of CNS	Group of compounds that affect pain perception and other aspects of behavior

[1] These and other structures listed in this table will be discussed in Chapter 23.
[2] Recent findings suggest that the brains of schizophrenics have more dopamine receptors than those of nonschizophrenics.

Nerve gases and organophosphate-type biocides inactivate cholinesterase. They cause the amount of acetylcholine in the synaptic cleft to increase with successive nerve impulses. This results in repetitive stimulation of the muscle fiber and may lead to potentially life-threatening muscle spasm. Should the muscles of the larynx go into spasm, for example, a person may die of asphyxiation.

Norepinephrine is released by some neurons in the peripheral nervous system as well as by many neurons in the brain and spinal cord. Neurons that release norepinephrine are called **adrenergic neurons.** Norepinephrine and the neurotransmitters epinephrine and dopamine belong to a class of compounds known as **catecholamines.** After release, excess catecholamine molecules are removed mainly by reuptake into the vesicles in the synaptic knobs. Some are degraded by the enzyme monoamine oxidase. Catecholamines affect mood, and many drugs that modify mood do so by altering the levels of these substances in the brain.

Direction and Speed of Conduction

Recent research suggests that within the central nervous system, dendrites may communicate in a variety of ways. For example, the dendrites of one neuron may communicate directly with dendrites of another neuron. However, in the peripheral nervous system, as well as in the longer pathways within the CNS, the usual sequence of transmission is from the axon of one neuron to the dendrite (or cell body) of the next. Because neurotransmitter is found only within the synaptic knobs of the axons (Fig. 22–11), neurons function as one-way streets, transmitting from dendrite to cell body to axon and then across the synaptic cleft to the next neuron in the sequence.

Compared with the speed of an electric current or the speed of light, a nerve impulse travels rather slowly. The speed of a nerve impulse varies from less than 1 meter to more than 120 meters (398 feet) per second, depending upon the type of neuron. What factors affect speed of trans-

Figure 22–11 (a) Neurons in the brain of a squirrel monkey, stained by a recently developed technique. Only a few of the very numerous cells are actually stained, or it would not be possible to make any of them out clearly. (b) Close-up of an axon of one of the neurons shown in (a). Notice the numerous lollipop-shaped structures attached to the side of the axon. Each of these is actually a synapse with another neuron. This should help you to appreciate the tremendous complexity of the interconnections of nerve cells in the central nervous system. (Courtesy of Dr. Fernando E. D'Amelio. *Stain Technology.* 58: 1983.)

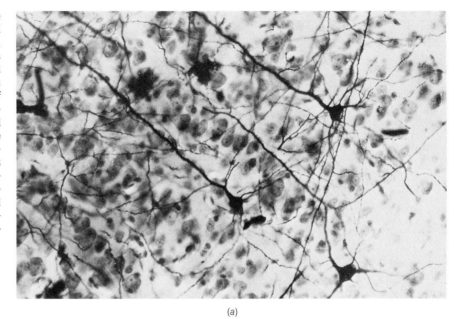

(a)

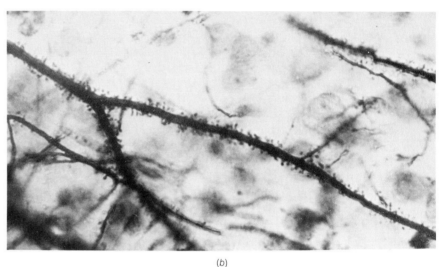

(b)

mission? In general, the greater the diameter of an axon, the greater its speed of conduction. Moreover, the heavier the myelin sheath, the faster the nerve impulse travels. The largest neurons seem also to be the most heavily myelinated.

NEURAL INTEGRATION

Neural integration is the process of adding and subtracting incoming signals and determining an appropriate course of action. More than 90% of the body's neurons are located in the central nervous system, so it is here that most neural integration takes place. The brain and spinal cord are responsible for making the "decisions."

Each neuron may synapse with hundreds of other neurons. Indeed, as much as 40% of a neuron's dendritic surface may be covered by synaptic knobs of neurons communicating with it. It is the job of the dendrites and cell body of every neuron to integrate the hundreds of messages that continually bombard them.

When neurotransmitter combines with a receptor on the surface of a postsynaptic neuron, the effect can be either stimulating (that is, to bring the neuron closer to a state of firing) or inhibiting (to take the neuron further away from the firing state). By a process of adding and subtracting, the membrane of the cell body integrates all the messages converging upon it. This process of integration is quite mechanical and is carried out on the molecular level. Some neurotransmitters raise the threshold, others lower it. Thus, one neurotransmitter may cancel the effect of another neurotransmitter. After these molecular tabulations are carried out, the type and amount of neurotransmitter that predominates determines the result. If sufficient excitatory neurotransmitter is present, the neuron will be stimulated and a message will be transmitted.

ORGANIZATION OF NEURAL CIRCUITS

As illustrated by the reflex pathway discussed earlier in this chapter, neurons are organized into specific pathways, or **circuits.** Within a neural circuit many presynaptic neurons may converge upon a single postsynaptic neuron. In **convergence,** the postsynaptic neuron is controlled by signals from two or more presynaptic neurons (Fig. 22–12). An association neuron in the spinal cord, for instance, may receive converging information from sensory neurons entering the cord, from neurons originating at other levels of the spinal cord, and even from neurons bringing information from the brain. Information from all of these converging neurons must be integrated before an action potential is generated in the association neuron and an appropriate motor neuron stimulated.

In **divergence** a single presynaptic neuron stimulates many postsynaptic neurons (Fig. 22–12). Each presynaptic neuron may synapse with up to 25,000 or more postsynaptic neurons. In **facilitation** the neuron is brought close to threshold level by stimulation from various presynaptic neurons but is not yet at the threshold level. The neuron can be easily excited by further stimulation. Figure 22–13 illustrates facilitation.

The **reverberating circuit** is a neural pathway arranged so that a neuron collateral synapses with an association neuron (Fig. 22–14). The association neuron synapses with a neuron in the sequence that can send new impulses again through the circuit. New impulses can be generated again and again until the synapses fatigue (from depletion of neurotransmitter) or are stopped by some sort of inhibition. Reverberating circuits are thought to be important in rhythmic breathing, in maintaining alertness, and perhaps in short-term memory.

Figure 22–12 Organization of neural circuits. (*a*) Convergence of neural input. Several presynaptic neurons synapse with one postsynaptic neuron. This organization in a neural circuit permits one neuron to receive signals from many sources. (*b*) Divergence of neural output. A single presynaptic neuron synapses with several postsynaptic neurons. This organization allows one neuron to communicate with many others.

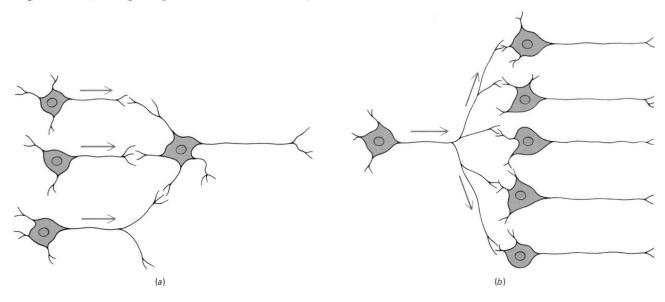

(a)

(b)

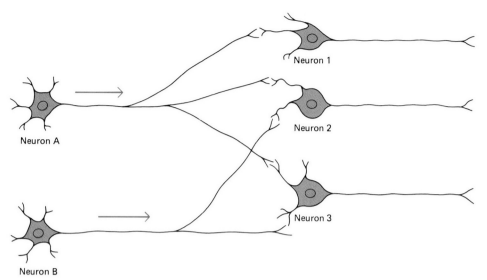

Figure 22–13 Facilitation. Neither neuron *A* nor neuron *B* can itself fire neuron *2* or *3*. However, stimulation by either *A* or *B* does depolarize the neuron toward threshold level (if the stimulation is excitatory). This facilitates the postsynaptic neuron, so if another presynaptic neuron stimulates it, the threshold level may be reached and an action potential generated.

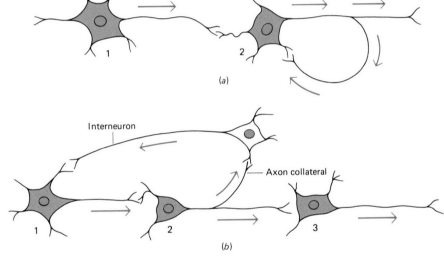

Figure 22–14 Reverberating circuits. (*a*) A simple reverberating circuit in which an axon collateral of the second neuron turns back upon its own dendrites, so the neuron continues to stimulate itself. (*b*) In this neural circuit an axon collateral of the second neuron synapses with an interneuron. The interneuron synapses with the first neuron in the sequence. New impulses are triggered again and again in the first neuron, causing reverberation.

CHAPTER SUMMARY

I. Information flow through the nervous system begins with reception. In a complex nervous system information is then transmitted to the central nervous system via sensory neurons. Integration takes place within the CNS, and appropriate "decisions" are made there. These "decisions" are then transmitted by motor neurons to the effectors that carry out the actual response.

A. Neurons are organized into neural circuits.

B. The junctions between neurons are called synapses.

C. A withdrawal reflex requires a sequence of only three neurons: a sensory neuron, an association neuron, and a motor neuron.

II. Glial cells are supporting cells; neurons are specialized to receive stimuli and transmit impulses.
  A. A typical multipolar neuron consists of a cell body from which project many branched dendrites and a single long axon.
  B. In the peripheral nervous system axons are surrounded by a neurilemma. Many axons are also enveloped in a myelin sheath.
  C. A nerve consists of hundreds of axons wrapped in connective tissue; a ganglion is a mass of cell bodies.
III. Transmission of a neural impulse is an electrochemical mechanism.
  A. A neuron that is not transmitting an impulse has a resting potential.
    1. The inner surface of the plasma membrane is negatively charged compared with the outside.
    2. Sodium pumps continuously transport sodium out of the neuron; potassium pumps transport potassium in.
    3. Potassium ions are able to leak out more readily than sodium ions are able to leak in.
  B. Excitatory stimuli are thought to open sodium gates in the cell membrane. This permits sodium to enter the cell and depolarize the membrane.
  C. When the extent of depolarization reaches threshold level, an action potential may be generated.
    1. The action potential is a wave of depolarization that spreads along the axon.
    2. The action potential obeys an all-or-none law.
    3. As the action potential moves down the axon, repolarization occurs very quickly behind it.
  D. Excitability of a neuron can be affected by calcium balance and by certain substances, such as local anesthetics and pesticides.
  E. Synaptic transmission generally depends upon release of a neurotransmitter from vesicles in the synaptic knobs of the presynaptic neuron.
    1. Neurotransmitter diffuses across the synaptic cleft and combines with receptors on the postsynaptic neuron.
    2. Neurons that release the neurotransmitter acetylcholine are known as cholinergic neurons. Those that release the neurotransmitter norepinephrine or dopamine are adrenergic.
  F. The largest, most heavily myelinated neurons conduct impulses most rapidly.
IV. Neural integration is the process of adding and subtracting incoming signals and determining an appropriate response.
V. Complex neural pathways are possible because of such neuron associations as convergence, divergence, and facilitation.

Post-Test

1. The process of receiving a stimulus is called _____ .
2. The actual response is carried out by effectors, _____ and _____ .
3. The junction between two neurons is called a _____ .
4. A _____ action is a predictable, automatic response to a simple stimulus.
5. Impulses are transmitted from sense organs to the central nervous system by _____ neurons.
6. The supporting cells of the nervous system are called _____ cells; cells specialized to transmit impulses are _____ .
7. The nucleus of a neuron is found within the _____ _____ _____ .
8. The _____ transmits impulses from the cell body to the synapse.
9. The _____ _____ is important in the regeneration of injured neurons.
10. A _____ consists of a mass of cell bodies.
11. Sodium is actively transported out of a resting neuron by _____ _____ .

12. Any stimulus that increases the permeability of the neuron to sodium may result in the transmission of a _____ _____ , also called _____ _____ .

13. The firing level of a neuron is called its _____ _____ .

14. During the _____ _____ a neuron cannot transmit another impulse.

15. Synaptic knobs release _____ .

16. Cholinergic neurons release _____ .

17. In _____ the postsynaptic neuron is controlled by signals from several presynaptic neurons.

18. Label the following diagram.

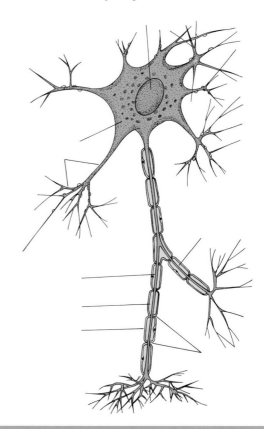

Review Questions

1. Imagine that you are driving down the street when suddenly a child darts in front of your car. What sequence of events must take place within your nervous system before you can slam on the brake?
2. Imagine that you have just burned your finger by touching a hot pot. Draw a diagram to illustrate the reflex action that would occur. Label each structure and indicate the direction of information flow.
3. Contrast the functions of a sensory and a motor neuron.
4. Give the functions of (a) myelin, (b) ganglia, (c) neuroglia, (d) dendrites, (e) an axon.
5. What is meant by the resting potential of a neuron? How does the sodium pump contribute to the resting potential?
6. What is an action potential? What is responsible for it?
7. How does the all-or-none law affect neural action?
8. Contrast a neuron with a nerve.
9. How is neural function affected by the presence of too much calcium? too little calcium?
10. Give the functions of (a) acetylcholine, (b) cholinesterase, (c) norepinephrine.
11. Contrast convergence and divergence.

Readings

Bloom, F.E. "Neuropeptides," *Scientific American,* October 1981. An account of the discovery and actions of neuropeptides, which help regulate bodily activities, in some cases acting as both neurotransmitters and hormones.

Keynes, R.D. "Ion Channels in the Nerve-Cell Membrane," *Scientific American,* March 1979. A discussion of the generation of a nerve impulse by the flow of sodium and potassium ions through channels in the neuron membrane.

Levi-Montalcini and P. Calissano. "The Nerve-Growth Factor," *Scientific American,* June 1979. A description of some important characteristics and effects of nerve-growth factor, a substance that encourages the growth and development of neurons.

Llinas, R.R. "Calcium in Synaptic Transmission," *Scientific American,* October 1982. The role of calcium is studied in the giant synapse of a squid.

Morell, P., and W.T. Norton. "Myelin," *Scientific American,* May 1980. A description of myelin and its functions.

Patterson, P., D. Potter, and E. Furshpan. "The Chemical Differentiation of Nerve Cells," *Scientific American,* July 1978. A discussion of the differentiation of adrenergic and cholinergic nerve cells.

Schwartz, J.H. "The Transport of Substances in Nerve Cells," *Scientific American,* April 1980. A discussion of the movement of substances long distances between the cell body and the neuron endings.

Chapter 23

RESPONSIVENESS: NERVOUS SYSTEMS

Learning Objectives

After you have read this chapter you should be able to:

1. Contrast nerve nets and radial nervous systems with bilateral nervous systems.
2. Describe the process of reception and describe the various types of receptors found in the human body.
3. List the functions of the spinal cord and describe its structure.
4. Locate the following parts of the brain and give the functions for each: medulla, pons, midbrain, thalamus, hypothalamus, cerebellum, and cerebrum.
5. Relate brain-wave patterns to states of consciousness or activity.
6. Compare the reticular activating system with the limbic system.
7. Describe REM sleep.
8. Describe how we perceive sensation, including pain.
9. Summarize current theories of learning and memory.
10. Cite experimental evidence linking environmental stimuli with changes in the brain and with learning ability.
11. Compare the somatic system with the autonomic system and contrast the sympathetic and parasympathetic divisions of the autonomic system.
12. Discuss the biological actions and effects on mood of the following types of drugs: alcohol, barbiturates, antianxiety drugs, antipsychotic drugs, opiates, stimulants, hallucinogens, and marijuana.

An organism's life-style is closely linked with its type of nervous system. The simple, sluggish nervous system of *Hydra* is adequate for an animal that remains rooted in one spot waiting for dinner to brush by its tentacles. With its more sophisticated nervous system, a frog can hop about in search of food and eject its tongue with lightning speed to capture a passing fly. However, neither the hydra nor the frog is able to solve algebra problems or learn about its own physiology. The range of possible responses depends in large part on the number of neurons and how they are organized in an animal's nervous system.

INVERTEBRATE NERVOUS SYSTEMS

There is no nervous system in the sponge. Whatever responses it makes are at the cellular level. Among other invertebrates there are two main types of nervous systems—nerve nets and bilateral nervous systems.

Nerve Nets and Radial Systems

The simplest organized nervous tissue is the **nerve net** found in *Hydra* and other cnidarians (Fig. 23–1). In a nerve net the nerve cells are scattered throughout the body. No central control organ and no definite pathways are present. Sensory cells, specialized to receive stimuli, transmit information to ganglion cells, which are the main cells of the nerve net. From the ganglion cells, information is passed somewhat haphazardly to neurosecretory cells, which apparently send chemical messages to effector cells, such as the cnidocytes (stinging cells). Impulses are transmitted in any direction, becoming less intense as they spread from the region of initial stimulation. If the stimulus is strong, the message will spread to more neurons of the net than if it is weak.

Since it produces responses that involve the body as a whole, or large parts of it at the same time, such a diffuse pattern of transmission is adequate in a radially symmetrical animal with sluggish locomotion. Responses in cnidarians are limited to discharge of nematocysts and contractions that permit the movements associated with locomotion and feeding.

The somewhat more sophisticated nervous system of the echinoderm consists of a circumoral nerve ring that surrounds the mouth from which a large radial nerve extends into each arm. These nerves coordinate movement of the animal. In sea stars a nerve net mediates the responses of the dermal gills to tactile stimulation (touch).

Bilateral Nervous Systems

In bilaterally symmetrical animals the nervous system is usually more complex than in radially symmetrical animals. A bilateral form of symmetry usually reflects a more active way of life, with the need to respond quickly to the environment in a sophisticated manner. The following trends can be identified:

1. Increased number of nerve cells.
2. Concentration of nerve cells to form thick cords or masses of tissue, which become nerves, nerve cords, ganglia, and brain.
3. Specialization of function. For example, transmission of nerve impulses in one direction results in **afferent neurons,** which conduct impulses toward a central nervous system, and **efferent neurons,** which transmit impulses away from the central nervous system and to the effector cells. Also, various parts of the central

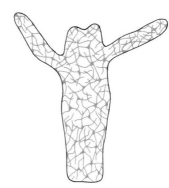

Figure 23–1 The nerve net of *Hydra* and other cnidarians is the simplest organized nervous tissue. No central control organ and no definite neural pathways are present.

nervous system are usually specialized to perform specific functions, so distinct structural and functional regions can be identified.

4. Increased number of association neurons and more complex synaptic contacts (contacts between neurons). This permits much greater integration of incoming messages, provides a greater range of responses, and allows far more precision in responses.

5. Cephalization, or formation of a head. A bilaterally symmetrical animal generally moves in a forward direction. With sense organs concentrated at the front of the body, the animal can detect an enemy quickly enough to escape or sense food in time to capture it. Response can be more rapid if these sense organs are linked by short pathways to decision-making nerve cells nearby. Therefore, nerve cells are also usually concentrated in the head region, constituting a definite brain.

In planarian flatworms there are concentrations of nerve cells in the head region known as **cerebral ganglia** (Fig. 23–2). These serve as a primitive "brain" and exert some measure of control over the rest of the nervous system. Two ventral longitudinal nerve cords extend from the ganglia to the posterior end of the body. Transverse nerves connect the brain with the eyespots and anterior end of the body. This arrangement is called a **ladder-type nervous system.**

In annelids and arthropods there is also typically a pair of ventrally located longitudinal nerve cords (Fig. 23–3). The cell bodies of the nerve cells are massed into pairs of ganglia located in *each* body segment. Afferent and efferent neurons are located in lateral nerves that link the ganglia with muscles and other body structures. In some arthropods specific functional regions have been identified in the cerebral ganglia.

When the earthworm brain is removed, the animal can move almost as well as before, but when it bumps into an obstacle, it persists in futile efforts to move forward instead of turning aside. The brain is therefore necessary for adaptive movements; it enables the earthworm to respond appropriately to environmental change.

In mollusks there are typically at least three pairs of ganglia; each pair has specific functions. In cephalopods, such as the octopus, there is a tendency toward concentration of the nerve cells in a central region. All

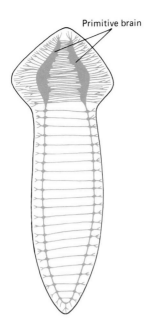

Figure 23–2 Planarian flatworms have a ladder-type nervous system. Cerebral ganglia in the head region serve as a simple brain and, to some extent, control the rest of the nervous system.

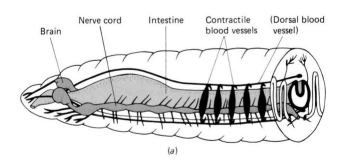

(a)

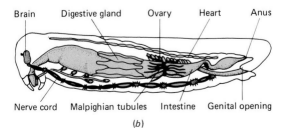

(b)

Figure 23–3 (a) The nervous system of the earthworm is typical of those found in other annelids. The cell bodies of the neurons are located in ganglia found in each body segment. They are connected by the ventral nerve cord. (b) In the insect nervous system the cerebral ganglia serve as a simple brain. Two ventral nerve cords are present.

the ganglia are massed in the **circumesophageal ring,** which contains about 168 million nerve cell bodies. With this complex brain, it is no wonder that the octopus is capable of considerable learning and can be taught quite complex tasks. In fact, the octopus is considered to be among the most intelligent invertebrates.

THE HUMAN NERVOUS SYSTEM

As in other vertebrates, the two main divisions of the human nervous system are the central nervous system (CNS) and the peripheral nervous system (PNS). The CNS consists of a complex tubular brain that is continuous with a single, dorsal, tubular spinal cord. Serving as central control, these organs integrate incoming information and determine appropriate responses. The PNS is made up of the sensory receptors (e.g., touch, auditory, and visual receptors) and the nerves, which are the communication lines. Various parts of the body are linked to the brain by 12 pairs of cranial nerves and to the spinal cord by 31 pairs of spinal nerves. Afferent neurons in these nerves continuously inform the CNS of changing conditions. Then efferent neurons transmit its "decisions" to appropriate muscles and glands, which effect the adjustments needed to preserve homeostasis.

For convenience the PNS may be subdivided into **somatic** and **autonomic** portions. Receptors and nerves concerned with changes in the external environment are somatic; those that regulate the internal environment are autonomic. Both systems have **sensory** (also called **afferent**) **nerves,** which transmit messages from receptors to the CNS, and **motor** (also called **efferent**) nerves, which transmit information back from the nervous system to the structures that must respond. In the autonomic system there are two kinds of efferent pathways—**sympathetic** and **parasympathetic** nerves (see Table 23–1).

THE SENSE ORGANS: RECEPTION

Sense organs link organisms with the outside world and enable them to receive information about the external and internal environment. Most familiar are the complex sense organs located in the head—the eyes, ears, nose, and taste buds. Sense organs are receptors, cells or organs

---

Table 23–1
DIVISIONS OF THE HUMAN NERVOUS SYSTEM

I. **Central nervous system (CNS)**
   A. Brain
   B. Spinal cord
II. **Peripheral nervous system (PNS)**
   A. Somatic portion
      1. Receptors
      2. Afferent (sensory) nerves—transmit information from receptors to CNS
      3. Efferent nerves—transmit information from CNS to glands and involuntary muscle in organs
   B. Autonomic portion
      1. Receptors
      2. Afferent (sensory) nerves—transmit information from receptors in internal organs to CNS
      3. Efferent nerves—transmit information from CNS to internal organs
         a. Sympathetic nerves—generally stimulate activity that results in mobilization of energy (e.g., speeds heartbeat)
         b. Parasympathetic nerves—action results in energy conservation or restoration (e.g., slows heartbeat)

specialized to initiate nerve impulses in response to specific changes in their environment. The sensory cells of the eyes are specialized to react to light waves; cells in the ears are sensitive to sound waves. Dissolved chemicals stimulate the taste buds, and chemicals in the air (odors) stimulate the olfactory receptors in the nose.

Touch Receptors

Thousands of tiny touch receptors are located in the skin. Some, like the **Pacinian corpuscles,** are especially sensitive to pressure, others to light touch. One type of skin receptor is specialized to receive information regarding temperature, whereas the bare endings of sensory neurons react to pain. Specialized receptors located in muscle tissue are sensitive to changes in movement, tension, and position. Their continuous reports to the CNS help ensure that muscle movement will be properly coordinated. Still other types of receptors are located deep within the body and even within the brain itself. These receptors send messages to the CNS concerning changes in the internal environment.

The Eye

In human beings sight is the dominant sense. More than 80% of the incoming information about our environment is received by the eyes. The **eye** forms images as a camera does. The eyelids act as a lens cap. The colored portion of the eye, the **iris,** acts as a diaphragm (Fig. 23–4). In bright light it contracts, narrowing its opening, the **pupil.** In weak light it expands, allowing more light to enter. Light passes through the **lens** and is received by light receptors in the **retina,** which functions as a light-

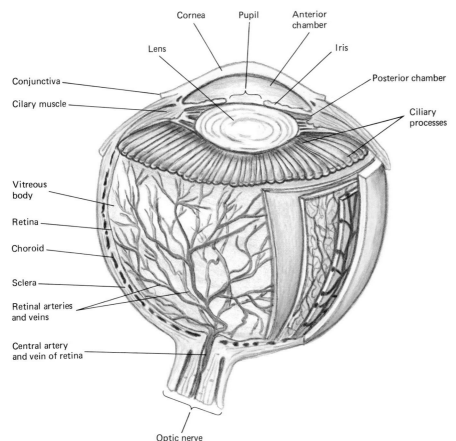

Figure 23–4 The vertebrate eye, partly dissected to show internal structure.

sensitive film. Like a transparent window covering the pupil and iris, the **cornea** keeps dust and other foreign matter out. The cornea is continuous with the **sclera,** a tough covering that envelops the rest of the eyeball. In humans the sclera is visible as the white of the eye. The internal surface of the eye is covered by a black coat, the **choroid,** which functions like the dark interior of a camera to prevent light rays from scattering.

Two types of light receptors in the retina are the **rod cells** and **cone cells.** Rods are sensitive to weak light but not to color. They contain a chemical, **rhodopsin** (or visual purple), that is necessary for the conversion of radiant energy of light into nerve impulses. Vitamin A is required for synthesizing rhodopsin. In the cones a chemical called **iodopsin** functions similarly to rhodopsin. Three types of cone cells are present, each sensitive to one of three colors—red, blue, or green. Some people who are color-blind may lack functional cones. Sensory information is conducted from the retina to the cerebral cortex of the brain by the optic nerve (one of the cranial nerves). In the cerebral cortex the information is perceived as an image.

The Ear

Sound waves travel through the external ear canal and strike the **tympanic membrane** (eardrum), causing it to vibrate (Fig. 23–5). In the middle ear three little bones—the **malleus,** the **incus,** and the **stapes**—transmit vibrations from the tympanic membrane to the **cochlea** of the inner ear. Various types of cells within the cochlea are sensitive to specific frequencies of sound. The brain can distinguish between sounds of different frequency in accordance with the particular cells of the cochlea that have been stimulated and consequently with the neurons that relay the information to the brain. The semicircular canals and vestibule in the inner ear are concerned with dynamic and static balance.

Figure 23–5 The structure of the human ear.

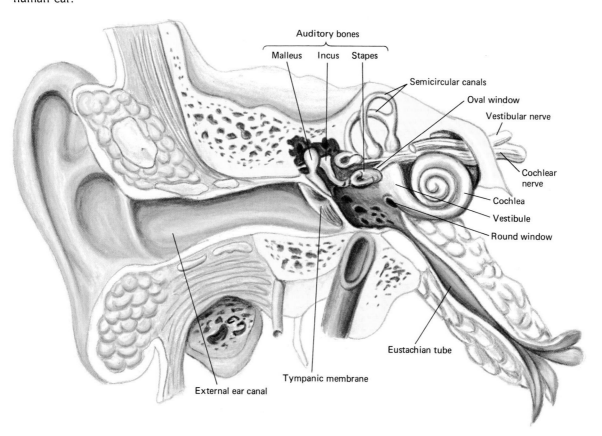

Auditory bones

Malleus   Incus   Stapes

Semicircular canals

Oval window

Vestibular nerve

Cochlear nerve

Cochlea

Vestibule

Round window

Eustachian tube

Tympanic membrane

External ear canal

Taste Buds

The senses of taste and smell depend upon chemoreceptors, receptors that respond to certain chemical stimuli. Reception of taste is the job of little clusters of cells called **taste buds** that are located on the tongue and roof of the mouth. Four basic kinds of taste are sweet, sour, salty, and bitter.

Olfactory Epithelium

In terrestrial vertebrates the sense of smell resides in the epithelium of the nose. In humans the **olfactory epithelium** is located in the roof of the nasal cavity. This epithelium contains specialized olfactory neurons that serve as receptors, as well as conductors, of incoming information regarding odor. These neurons react to as many as 50 odors (chemicals) in the air. A good bit of what we usually think of as taste is actually the sense of smell. When we suffer from a head cold, food does not "taste" nearly so good.

THE HUMAN CENTRAL NERVOUS SYSTEM

The **central nervous system (CNS)** consists of the brain and spinal cord. Both of these soft, fragile organs are well protected. Encased within bone, they are covered by three layers of connective tissue, the **meninges.** A special shock-absorbing fluid, **cerebrospinal fluid,** also cushions brain and spinal cord against mechanical injury.

The Spinal Cord

The **spinal cord** is a hollow cylinder that emerges from the base of the brain and extends downward to about the level of the waist (Fig. 23–6). Its two functions are (1) to control many reflex activities of the body and (2) to transmit messages back and forth to the brain via its **ascending** and **descending nerve tracts.** Each tract is a large bundle of axons. As axons pass down through the brain and spinal cord, some of them cross over from one side of the cord (or brain) to the other. For this reason the right side of the brain mainly controls the left side of the body, and the left side of the brain mainly controls the right side of the body.

When examined in cross section, the spinal cord is seen to have a small central canal surrounded by a butterfly-shaped area of gray matter (Fig. 23–7). Outside the gray matter the cord is composed of white matter. **Gray matter** consists mainly of large masses of cell bodies and dendrites of the neurons present within the cord. **White matter** is composed of the myelinated axons of the tracts within the cord.

The Brain

A soft, wrinkled mass of tissue weighing about 1.4 kilograms (3 pounds), the human **brain** is the most complex organ known. Although computers have been designed along similar principles and have been likened to it, even the most intricate computer does not begin to rival the complexity of the human brain. Each of the brain's 25 billion neurons is functionally connected to as many as 100,000 others, and there may be as many as 10^{14} synapses. No wonder that scientists have barely begun to unravel some of the tangled neural circuits that govern human physiology and behavior!

Brain cells require a continuous supply of oxygen and glucose. Although the brain accounts for only about 2% of the body weight, it re-

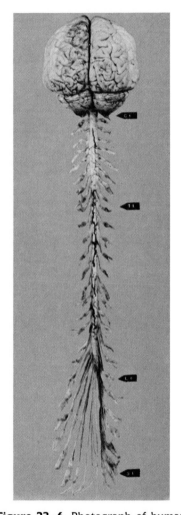

Figure 23–6 Photograph of human brain and spinal cord. The roots of the spinal nerves are still attached. Note the group of nerves that extend caudally from the lower region of the cord. Because they resemble a horse's tail, they are referred to as the cauda equina. These nerves have been left undisturbed on the right but have been fanned out on the left. (Dissection by Dr. M.C.E. Hutchinson, Department of Anatomy, Guy's Hospital Medical School, London. From Williams and Warwick, eds., *Gray's Anatomy.*)

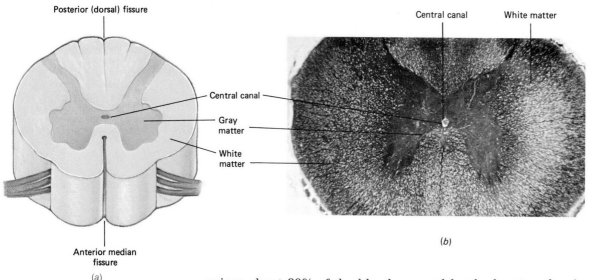

(b)

(a)

Figure 23–7 The spinal cord consists of gray matter and white matter. (a) Cross section through the spinal cord. (b) Photomicrograph of a cross section through the spinal cord (approximately ×25).

ceives about 20% of the blood pumped by the heart each minute, and consumes about 20% of the oxygen used by the body. The brain is so dependent upon its blood supply that when it is deprived of it, consciousness is lost after about five seconds, and irreversible damage occurs within a few minutes. In fact, the most common cause of brain damage is stroke (cerebrovascular accident), in which a portion of the brain is deprived of its blood supply (often because a blood vessel has been blocked by a blood clot).

The brain is an anterior and expanded extension of the spinal cord. Both structures develop from the same tube in the embryo (as will be discussed in Chapter 32). The structure and functions of the main divisions of the brain are given in Table 23–2. The brain is illustrated in Figures 23–8 and 23–9. Also see Focus on Evolution of the Vertebrate Brain.

The largest, most prominent part of the human brain is the **cerebrum.** As in most vertebrates, the human cerebrum is divided into right and left hemispheres. The outer portion of the cerebral tissue consists of gray matter and is called the **cerebral cortex.** The surface area of the cerebral cortex is greatly expanded by numerous folds, called **convolutions** or **gyri.** The furrows between them are called **sulci** when shallow, **fissures** when deep. Functionally, the cerebral cortex is divided into three areas: (1) the **sensory areas,** which receive incoming signals from the sense organs, (2) the **motor areas,** which control voluntary movement, and (3) the **association areas,** which link the sensory and motor areas and are responsible for thought, learning, intelligence, language abilities, memory, judgment, and personality.

ELECTRICAL ACTIVITY OF THE BRAIN

Continuous electrical activity within the brain can be measured by placing electrodes on the surface of the scalp and recording differences in electrical potentials. Patterns of activity called **brain waves** can be traced, producing a record called an **electroencephalogram,** or **EEG** (Fig. 23–10).

Brain waves are produced primarily from the cerebral cortex. The wavelike patterns arise from the synchronized cyclic activity of groups of neurons. Your EEG is as unique as your fingerprints, but the EEG changes with the state of consciousness or emotion.

Often brain waves are irregular, but under some conditions distinct patterns can be recorded. Four main kinds of waves have been distin-

Table 23–2
THE BRAIN

Structure	Description	Function
Brain stem		
Medulla	Continuous with spinal cord; primarily made up of nerves passing from spinal cord to rest of brain	Contains vital centers (clusters of neuron cell bodies) that control heartbeat, respiration, and blood pressure; contains centers that control swallowing, coughing, vomiting
Pons	Forms bulge on anterior surface of brain stem	Connects various parts of brain with one another; contains respiratory center
Midbrain	Just above pons; largest part of brain in lower vertebrates; in human beings most of its functions are assumed by cerebrum	Center for visual and auditory reflexes (e.g., pupil reflex, blinking, adjusting ear to volume of sound)
Thalamus	At top of brain stem	Main sensory relay center for conducting information between spinal cord and cerebrum. Neurons in thalamus sort and interpret all incoming sensory information (except olfaction) before relaying messages to appropriate neurons in cerebrum
Hypothalamus	Just below thalamus; pituitary gland is connected to hypothalamus by stalk of neural tissue	Contains centers for control of body temperature, appetite, fat metabolism, and certain emotions; regulates pituitary gland; link between "mind" (cerebrum) and "body" (physiological mechanisms)
Cerebellum	Second largest division of brain	Reflex center for muscular coordination and refinement of movements; when injured, performance of voluntary movements is uncoordinated and clumsy
Cerebrum	Largest, most prominent part of human brain; more than 70% of brain's cells located here; longitudinal fissure divides cerebrum into right and left hemispheres, each divided by shallow sulci (furrows) into six lobes: frontal, parietal, temporal, insular, occipital, and limbic	Center of intellect, memory, consciousness, and language; also controls sensation and motor functions
Cerebral cortex (outer gray matter)	Arranged into convolutions (folds) that increase surface area; functionally, cerebral cortex is divided into:	
	1. Motor cortex	Controls movement of voluntary muscles
	2. Sensory cortex	Receives incoming information from eyes, ears, pressure and touch receptors, etc.
	3. Association cortex	Site of intellect, memory, language, and emotion; interprets incoming sensory information
White matter	White matter within cortex consists of myelinated axons of neurons that connect various regions of brain; these axons are arranged into bundles (tracts)	Connects: 1. Neurons within same hemisphere 2. Right and left hemispheres 3. Cerebrum with other parts of brain and spinal cord

guished. **Alpha wave** patterns are associated with resting and relaxed activity. **Beta waves** are characteristic of states of heightened mental activity, such as problem solving or information processing. **Delta waves**

Figure 23–8 (a) Photograph of the human brain, lateral view. Note that the cerebrum covers part of the brainstem. (b) Lateral view of the human brain showing the lobes of the cerebrum. Part of the brain has been made transparent so that the underlying insular lobe can be located. (a, from Williams and Warwick, eds., *Gray's Anatomy.*)

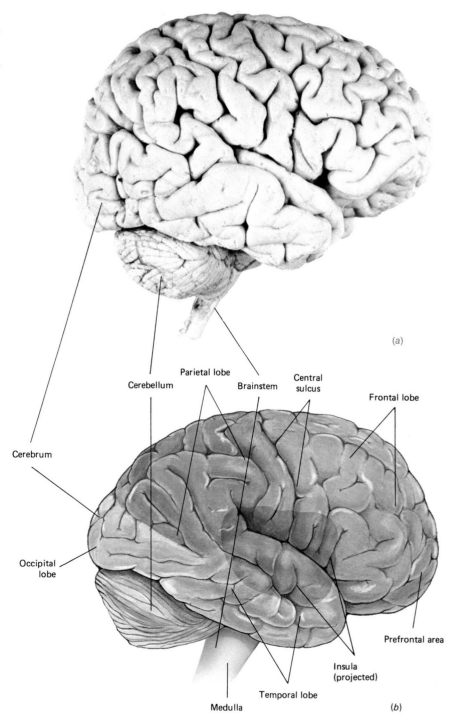

(a)

Cerebellum Parietal lobe Brainstem Central sulcus Frontal lobe

Cerebrum

Occipital lobe

Prefrontal area

Insula (projected)

Medulla Temporal lobe (b)

are slow, large waves associated with normal sleeping. **Theta waves** are recorded mostly in children but are also produced in some adults when they are under emotional stress.

The EEG is used as a diagnostic tool to determine certain disease states, such as epilepsy. Electrical activity of the brain is also used as a criterion of life or death. When electrical activity in the brain ceases, a patient may be pronounced dead.

Can one learn to control brain waves? Some investigators believe that those who tend to be anxious and exhibit beta waves for no good reason might be taught to induce the relaxed alpha state. Subjects have had some success in controlling their brain-wave patterns by the use of biofeedback machines, which permit them to see and listen to their brain

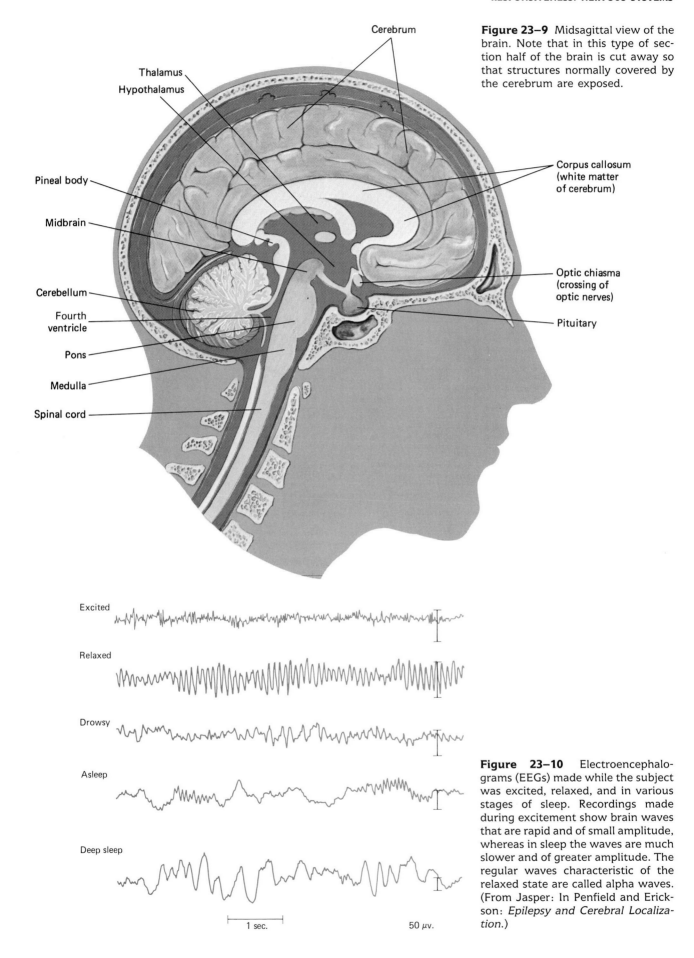

Cerebrum

Thalamus

Hypothalamus

Pineal body

Midbrain

Cerebellum

Fourth ventricle

Pons

Medulla

Spinal cord

Corpus callosum (white matter of cerebrum)

Optic chiasma (crossing of optic nerves)

Pituitary

Figure 23–9 Midsagittal view of the brain. Note that in this type of section half of the brain is cut away so that structures normally covered by the cerebrum are exposed.

Excited

Relaxed

Drowsy

Asleep

Deep sleep

1 sec.

50 μv.

Figure 23–10 Electroencephalograms (EEGs) made while the subject was excited, relaxed, and in various stages of sleep. Recordings made during excitement show brain waves that are rapid and of small amplitude, whereas in sleep the waves are much slower and of greater amplitude. The regular waves characteristic of the relaxed state are called alpha waves. (From Jasper: In Penfield and Erickson: *Epilepsy and Cerebral Localization.*)

Focus on . . .

EVOLUTION OF THE VERTEBRATE BRAIN

All vertebrates, from fish to mammals, have the same basic brain structure (see figure). Certain parts of the brain are specialized to perform specific functions, and some, such as the cerebellum and cerebrum, are vastly more complex in the higher vertebrates.

In fish and amphibians the midbrain is the most prominent part of the brain. In these animals the midbrain is the main association area. It receives incoming sensory information, integrates the information, and sends decisions to appropriate motor neurons. For ex-

ample, the optic lobes, a part of the midbrain, specialize in visual interpretation. In reptiles, birds, and mammals many of the functions of the optic lobes are assumed by the cerebrum.

The size and shape of the cerebellum vary greatly among the vertebrate classes. Development of the cerebellum is correlated roughly with the extent and complexity of muscular activity. In some fish, birds, and mammals the cerebellum is highly developed, whereas it tends to be small in amphibians and reptiles.

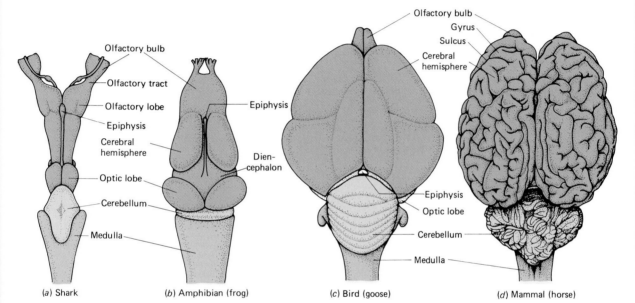

Comparison of the brains of members of four increasingly complex vertebrate classes shows basic similarities and evolutionary trends. Note that different parts of the brain may be specialized in the various groups. For example, the large olfactory lobes in the shark brain (a) are essential to this predator's highly developed sense of smell. During the course of evolution, the cerebrum and cerebellum have become larger and more complex. In the mammal (d) the cerebrum is the most prominent part of the brain.

waves. With such continuous feedback they are able to condition themselves to produce alpha waves. Perhaps someday persons with psychosomatic disorders will be taught to relax in this way and will no longer need tranquilizers and other drugs.

THE RETICULAR ACTIVATING SYSTEM

Sometimes called the arousal system, the **reticular activating system (RAS)** is a complex pathway of neurons in the brain stem. It receives messages from neurons in the spinal cord and from many other parts of the nervous system and communicates with the cerebral cortex by complex circuits. The RAS is ultimately responsible for maintaining consciousness, and the extent of its activity determines the state of alertness. When the RAS bombards the cerebral cortex with stimuli, you feel alert and are able to focus your attention on specific thoughts. When its activity slows, you begin to feel sleepy. Sometimes when you are listening to a boring lecture, the RAS becomes habituated to the monotonous repetition of the professor's voice. As its signals become progressively weaker, the cerebrum may lapse into sleep. Should the RAS be severely damaged, as in an accident, the unfortunate victim may pass into a deep, permanent coma.

THE LIMBIC SYSTEM

The **limbic system** is another action system of the brain. It consists of certain structures of the cerebrum, thalamus, and hypothalamus. The limbic system affects the emotional aspects of behavior, sexual behavior, biological rhythms, autonomic responses, and motivation, including feelings of pleasure and punishment. Stimulation of certain areas of the limbic system in an experimental animal results in increased general activity and may cause fighting behavior or what appears to be extreme rage.

When an electrode is implanted in the so-called reward center of the limbic system, a rat will press a lever that stimulates this area as many as 15,000 times per hour (Fig. 23–11). Stimulation of this area is apparently so rewarding that an animal will forgo food and drink and may continue to press the lever until it drops from exhaustion. When an electrode is implanted in the punishment center of the limbic system, an experimental animal quickly learns to press a lever to *avoid* stimulation. The reward and punishment centers are probably important in influencing motivation and behavior.

SLEEP AND DREAMING

When the RAS slows the process of relaying incoming information to the cerebrum, we lose awareness of stimuli in our surroundings and go to sleep. During sleep electrical activity of the cerebrum decreases and the body relaxes. Some physiologists think that there are sleep centers in the brain stem. When stimulated, their neurons release the neurotransmitter

Focus on . . .

DOPAMINE AND MOTOR FUNCTION

The neurotransmitter dopamine plays an important role in motor function. Its function became known through an interesting series of somewhat unrelated events. During the mid-1950s the drug reserpine became popular as a major tranquilizer used for mental patients. Then, in 1959, investigators noticed that some patients taking reserpine developed extrapyramidal symptoms, such as muscle rigidity and tremor. These symptoms were very similar to those seen in patients with Parkinson's disease, a disorder in which movement is shaky and difficult. Victims of Parkinson's disease suffer from tremors even when they are not attempting to move. This observation led to studies that showed that the drug reserpine greatly reduces the amount of dopamine within the caudate nucleus and putamen (two of the basal ganglia within the white matter of the cerebrum). Investigators then discovered that patients with Parkinson's disease have only about 50% of the normal amount of dopamine in their basal ganglia.

Attempts to administer dopamine to these patients were not successful because dopamine cannot penetrate the blood-brain barrier. However, a substance known as L-dopa, from which dopamine is synthesized in the body, does penetrate the blood-brain barrier and has dramatically relieved the symptoms of Parkinson's disease in most patients.

It has been shown that dopamine is released by neurons that extend from the substantia nigra (an area in the midbrain) to the basal ganglia. The dopamine is thought to inhibit neurons that produce acetylcholine. When dopamine is absent or present in too small a quantity, the acetylcholine causes overactivity of certain neurons, producing the motor symptoms of Parkinson's disease.

Even in healthy persons, the aging process causes changes in motor abilities. Body movements and even reflexes slow, and movement becomes more difficult. Recent studies suggest that these changes may be due to dopamine depletion, and that treatment with L-dopa may be helpful.

Too much dopamine sometimes causes schizophrenic symptoms. In fact, it has been suggested that people suffering from schizophrenia may have too many dopamine receptors in their brains and thus too much dopamine activity.

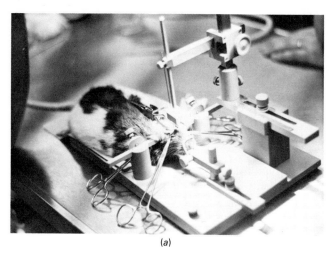

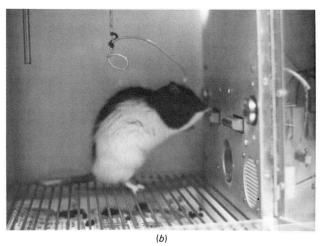

(a)

(b)

Figure 23–11 Electrodes can be implanted in the pleasure center of a rat's brain (a) so when the rat depresses a lever, a stimulating electric current is delivered directly to the pleasure center (b).

serotonin, which inhibits signals passing through the RAS, thus inducing sleep.

Every hour and a half or so during the night the relaxed-sleep pattern shifts to what is called **REM (rapid eye movement)** sleep. In REM sleep the individual dreams, electrical activity greatly increases, and beneath the closed eyelids the eyes move rapidly. Everyone dreams, though some people do not recall their nocturnal adventures.

When experimental subjects are deprived of the normal experiences of REM sleep for several nights by being awakened when these rapid eye movements begin, they become disoriented and eventually exhibit psychotic symptoms.

Mental activity continues during sleep, and many fine discriminations are made within the RAS. People who manage to sleep through the rumbling of passing traffic or the loud noise of a jet flying overhead leap quickly to their feet at the sound of their infant's cry. And the smell of smoke or sound of a burglar's fumbling arouses most individuals from the soundest sleep.

PERCEPTION OF SENSATION

Although you may respond to sensation on a reflex level (when you burn your hand, you jerk it away from the hot pot), for you to be *aware* of sensation, information must be transmitted from the sensory receptors to the brain. Sensory areas in the cerebral cortex continuously receive incoming sensory messages and, with the help of adjacent association areas, analyze them. For example, the visual cortex in the occipital lobe receives continuous messages from the eyes. With the help of its association areas the brain can interpret what it is you are seeing—a monster or a kitten. Damage to the visual cortices in both hemispheres can result in total blindness.

When stimulated, free nerve endings throughout the body signal pain to the CNS via sensory neurons. Such messages are delivered to the brain by neurons within one of the spinal tracts (Fig. 23–12). Pain perception is thought to begin in the thalamus. From there impulses are sent into the parietal lobes of the cerebrum. For example, when you step on a nail, the pain is perceived by the brain and then subjectively projected back to the injured foot, so that you feel pain at the site of puncture. Artificial stimulation of the leg nerves may produce a sensation of pain even though the foot is untouched. In fact, when the foot is amputated, a patient may feel pain from the missing limb. This is called **phantom pain.**

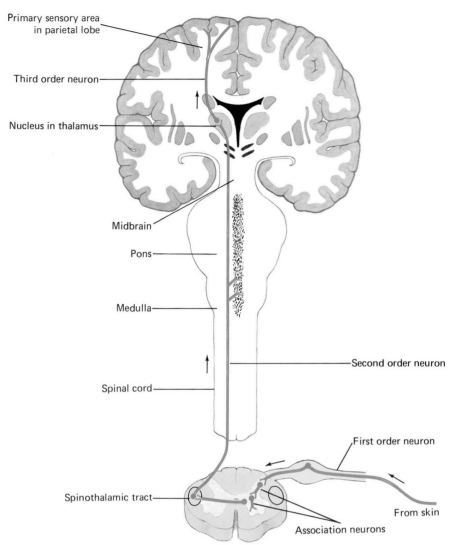

Primary sensory area
in parietal lobe

Third order neuron

Nucleus in thalamus

Midbrain

Pons

Medulla

Spinal cord

Spinothalamic tract

Second order neuron

First order neuron

From skin

Association neurons

Figure 23–12 Sensory pathway for transmission of neural information from pain receptors. The first-order neurons are sensory neurons that synapse with neurons of one of the spinal tracts (the spinothalamic tract) within the spinal cord. The second-order neurons cross to the opposite side of the spinal cord and then pass upward through the brainstem to a nucleus within the thalamus. There they synapse, and the third-order neurons transmit impulses to the sensory areas in the parietal lobe.

Pain from body organs is often poorly localized. In fact, the pain is often *referred* to a superficial area that may be some distance from the organ involved. The area to which the pain is referred generally receives its innervation from the same level of the spinal cord as does the organ involved. A person with angina who feels heart pain in his left arm is experiencing **referred pain.** The pain originates in the heart as a result of ischemia (insufficient blood in the microcirculatory system of the heart muscle). However, the pain is actually felt in the arm. One explanation is that neurons from both the heart and the arm converge upon the same neurons in the central nervous system. The brain interprets the incoming message as coming from the body surface because somatic pain is far more common then visceral (organ) pain; the brain acts on the basis of its past experience. When visceral pain is felt both at the site of the distress and as a referred pain, it may seem to spread, or *radiate,* from the organ to the superficial area.

Pain can be initiated or facilitated at many levels. How intense one's perception of pain is depends upon the particular situation and how one has learned to deal with pain. A child with a bruised knee may emotionally heighten the feeling of pain, whereas a professional fighter may virtually ignore a long series of well-delivered blows.

Recently, it has been discovered that the brain releases peptides that influence pain perception. These substances have been named **endorphins** (for "endogenous morphine-like"). Much more powerful than morphine, the endorphins are currently being investigated as potential analgesic (pain-killing) drugs, as well as to discover their normal role in the nervous system.

For example, some neurobiologists now feel that endorphins may explain the mechanism of action of acupuncture. For thousands of years acupuncture has been used to relieve pain, but how it works has remained a mystery. There is now some evidence that acupuncture needles stimulate nerves deep within the muscles, which in turn stimulate the pituitary gland and parts of the brain to release endorphins. The endorphins may inhibit neurons in the brain that normally discharge in response to pain.

LEARNING AND MEMORY

Despite extensive research, the mechanisms by which the brain thinks, learns, and remembers are still poorly understood. The human brain differs most markedly from the brains of other animals by the remarkable development of its association areas within the cerebral cortex. Neurons within these areas form highly complex pathways. Damage to association areas can prevent a person from thinking logically, even though he may still be able to hear or even read.

To learn, the brain must be able to (1) focus attention on specific stimuli, whether they be the words on this printed page or an angry wasp buzzing overhead, (2) compare incoming sensory stimuli with stimuli it has encountered before, and (3) store information. **Memory** is the ability to recall stored information.

Some forms of **learning** can take place in association areas within lower brain regions, like the thalamus. Even very simple animals that completely lack a cerebral cortex are capable of some types of learning. However, earthworms are unable to solve calculus problems or learn foreign languages. Our ability to do such things results from the greater complexity of our association systems.

Just how the brain stores information and retrieves the memory on command has been the subject of much speculation. According to current theory, there are several levels of memory. Short-term memory involves recalling information for a few seconds or minutes. When you look up a phone number, you usually remember it only long enough to dial. Should you need the same number the next day, you would have to look it up again. One theory of short-term memory suggests that it is based on reverberating neural circuits. In a reverberating circuit a neural message is fed back into the circuit by way of an axon collateral so that a new impulse is sent through the circuit. A memory circuit may continue to

Focus on . . .

CEREBRAL DOMINANCE

How many left-handed people do you know? Probably not very many, for 90% of us are right-handed. The remaining 10% are left-handed or ambidextrous. In right-handed people the left cerebral hemisphere is more highly developed for the motor functions related to handedness. In about 98% of adults (regardless of handedness) the left hemisphere is also dominant for language abilities, including the ability to speak, read, learn mathematics, and perform all other intellectual functions associated with language.

Until recently the left cerebral hemisphere was thought to be dominant in all respects in most people. New research indicates that the two hemispheres actually complement one another and that the right hemisphere specializes in some functions, such as spatiotemporal patterns. The right hemisphere is important in recognizing faces, identifying objects on the basis of shape, and appreciating and recognizing music and form. Some have suggested that creative abilities reside therein.

reverberate for several minutes until it fatigues or until new signals are received that interfere with the old.

When you select a bit of information for long-term storage, the brain is thought to rehearse the material and then store it in association with similar memories. According to one theory, some physical or chemical change takes place in the synaptic knobs or postsynaptic neurons that permanently facilitates the transmission of impulses within a newly established circuit. Perhaps specific neurons become more sensitive to neurotransmitter. Each time a memory is stored, a new neural pathway is facilitated.

Another theory proposes that changes within glial cells may facilitate transmission of impulses through newly formed circuits. Some researchers think that either RNA or protein may serve as memory molecules. A more recent theory suggests that it is not the location of neurons or the establishment of specific pathways that matters but the rhythm at which neurons fire. According to this view, each time something new is learned, cells in many parts of the brain learn a new rhythm of firing. This theory has the advantage of not trying to localize memory in any one area of the brain, something that no one has been able to accomplish.

Researchers have methodically removed portions of the brains of experimental animals without finding specific regions where information is stored. As more cerebral cortex tissue is destroyed, more information is lost, but no *specific* area can be labeled the "memory bank."

Several minutes are required for a memory to become consolidated within long-term memory. Should a person suffer a brain concussion or undergo electroshock therapy, memory of what transpired immediately prior to the incident may be completely lost. When parts of the limbic system are injured or removed, a person can recall information stored in the past but is no longer able to store new information.

Retrieval of information stored in long-term memory is of considerable interest—especially to students. Some researchers believe that once information is deposited in long-term storage, it remains in the brain forever. The only problem is finding the information when we need it. When we seem to forget a particular bit of information, it is because we have not efficiently searched for it.

ENVIRONMENTAL EFFECTS UPON THE BRAIN

In 1925 the behaviorist psychologist John Watson wrote:

> Give me a dozen healthy infants, well-formed, and my own specified world to bring them up in, and I'll guarantee to take any one at random and train him to become any type of specialist I might select—doctor, lawyer, merchant, chief, and yes, even beggarman and thief, regardless of his talents, penchants, tendencies, abilities, vocations, and the race of his ancestors.[1]

Watson's conception of the impact of the environment was considered extreme at that time, but since then his view of the role of the environment in shaping the development and intellectual potential of infants and children has gained wide acceptance. An expanding array of multicolored and multishaped crib mobiles, educational toys for babies, and emphasis on early childhood education are producing more intelligent children and yielding greater opportunity for future academic achievement. Studies indicate that the way a mother relates to her child during the first 18 months of life is critical in establishing personality and academic potential. Mothers who interact warmly with their young children

---

[1] J.B. Watson. *Behaviorism*. New York, Peoples Institute, 1925, p. 82.

and who stimulate their intellectual curiosity from the earliest months of life help to ensure future social and academic success. Brain structure and biochemistry itself may be changed by such environmental enrichment.

Studies on experimental animals confirm that environmental experience may cause physical and chemical changes in brain structure. Several studies have been performed in which one group of rats is provided with a stimulating environment and given the opportunity to learn while other groups are deprived of stimulation and social interaction. After some months or years the animals are sacrificed and their brain structures compared. Those exposed to enriched environments exhibit an increased concentration of synaptic contact. Some investigators have reported that the cerebral cortex actually becomes thicker and heavier. Characteristic biochemical changes also take place. Other experiments have indicated that animals reared in a complex environment may be able to process and remember information more quickly than animals not provided with such advantages.

During early life there are apparently certain critical or sensitive periods of nervous system development that are influenced by environmental stimuli. For example, when the eyes of young mice first open, large numbers of dendritic spines (sites where synapses form) develop on neurons in the visual cortex. When the animals are kept in the dark and deprived of visual stimuli, fewer dendritic spines form. If later in life the mice are exposed to light, some new dendritic spines form, but never as many as develop in a mouse reared in a normal environment.

THE PERIPHERAL NERVOUS SYSTEM

The **peripheral nervous system (PNS)** consists of all the receptors, the nerves that link the receptors with the CNS, and the nerves that link the CNS with the muscles and glands (the effectors). That portion of the PNS that keeps the body in balance with the external environment is the somatic system; the nerves and receptors designed to maintain internal homeostasis make up the autonomic nervous system.

The Somatic System

The **somatic nervous system** includes those receptors that react to changes in the external environment, the sensory neurons that keep the CNS informed of those changes, and the motor neurons that adjust the positions of the skeletal muscles in order to maintain the body's well-being. Twelve pairs of nerves called **cranial nerves** emerge from the brain itself. These nerves transmit information regarding the senses of smell, sight, hearing, and taste from the special sensory receptors, and information from the general sensory receptors, especially in the head region. The cranial nerves also bring orders from the CNS to the voluntary muscles that control movements of the eyes, face, mouth, tongue, pharynx, and larynx.

Thirty-one pairs of spinal nerves emerge from the spinal cord (Fig. 23–13). Named for the general region of the vertebral column from which they originate, there are 8 pairs of cervical spinal nerves, 12 pairs of thoracic, 5 pairs of lumbar, 5 pairs of sacral, and 1 pair of coccygeal spinal nerves.

Each spinal nerve has a **dorsal root** and a **ventral root.** The dorsal root consists of sensory (afferent) fibers, which transmit information from the sensory receptors to the spinal cord. Just before the dorsal root joins with the cord, it is marked by a swelling, the **spinal ganglion,** which consists of the cell bodies of the sensory neurons. The ventral root consists of the motor (efferent) fibers leaving the cord en route to the

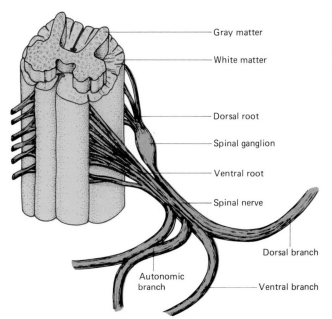

Figure 23–13 Dorsal and ventral roots emerge from the spinal cord and join to form a spinal nerve. The spinal nerve divides into several branches.

muscles and glands. Cell bodies of the motor neurons are located within the gray matter of the cord.

The Autonomic System

The **autonomic system** helps to maintain homeostasis in the internal environment. For instance, it regulates the rate of the heartbeat and maintains a constant body temperature. The autonomic system works automatically and without voluntary input. Its effectors are smooth muscle, cardiac muscle, and glands (Fig. 23–14). Like the somatic system, it

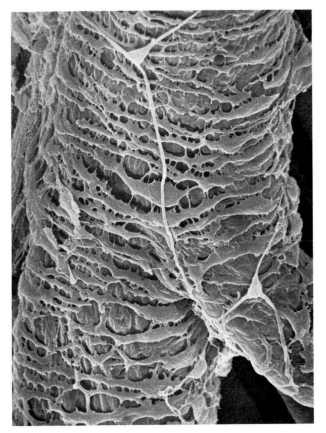

Figure 23–14 Scanning electron micrograph of the autonomic innervation of the smooth muscle cells of a small vein (approximately ×900). The thin white fibers are autonomic nerve axons, and the triangular bodies where these branch are Schwann cells. The neurons are arranged to stimulate large numbers of smooth muscle cells (arranged in rings around the vein) at the same time. (From Desaki, J. "Vascular Autonomic Plexuses and Skeletal Neuromuscular Junctions: A Scanning Electron Microscopic Study," *Biomedical Research* (Supplement), 139–143, 1981.)

is functionally organized into reflex pathways. Receptors within the viscera (internal organs) relay information via afferent nerves to the CNS, the information is integrated at various levels, and the decision is transmitted along efferent nerves to the appropriate muscles or glands.

Information from the various organs is transmitted to the CNS by afferent neurons. Some of these are located within cranial or spinal nerves. These afferent neurons are thought to synapse with association neurons within the CNS. They bring information about blood pressure, respiration, heartbeat, peristalsis, and other visceral activities.

The efferent portion of the autonomic system is subdivided into **sympathetic** and **parasympathetic systems.** In general, the sympathetic nerves operate to stimulate organs and to mobilize energy, especially in response to stress, whereas the parasympathetic nerves influence organs to conserve and restore energy, particularly when one is engaged in quiet, calm activities, such as studying biology. Many organs are innervated by both types of nerves, which act upon the organ in a complementary way (Fig. 23–15). For example, the heart rate is slowed by impulses from its parasympathetic nerve fibers and speeded up by messages from its sympathetic nerve supply.

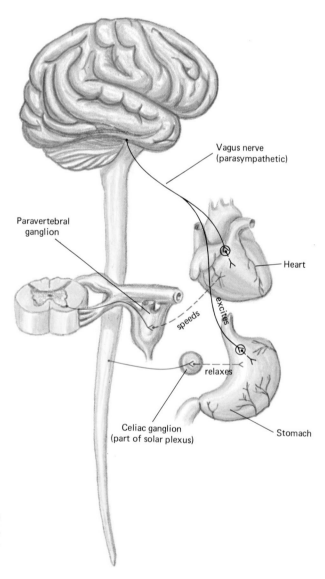

Figure 23–15 Dual innervation of the heart and stomach by sympathetic and parasympathetic nerves. Sympathetic nerves are shown in red.

THE DRUG SCENE

Drugs help us fall asleep, drugs help us stay awake, drugs help us relax, drugs help us forget our problems, drugs help us feel like part of the group, drugs "expand the mind." Some are purchased freely over the counter, some are prescribed by physicians, others are bought furtively from pushers. Sale of prescription drugs amounts to more than $4 billion annually, and nonprescription drugs are a $2.4 billion business. Of all prescribed drugs, some 25% are for psychological conditions. Each year physicians write more than 200 million prescriptions for sedatives, tranquilizers, and pep pills. More and more people are attempting to control their moods with an ever-growing variety of stronger and stronger drugs. We are a pill-popping society and may someday be remembered as "the drugged generation."

Why is this emphasis on drugs undesirable? One reason is that no drug is completely harmless. The body is a delicately balanced machine, and even a mild drug is still a chemical trespasser. Even aspirin has been shown to cause noteworthy side effects such as gastric bleeding. But besides the individual matter of doing harm to one's own body, the use and abuse of drugs have become an important social issue. Drugs affect not only the individual users but also their families, their employers, and even strangers with whom they share the highway or from whom they must steal to support their habit.

In some cases the line between use and abuse of a drug seems ambiguous. Is the anxious middle-class parent who takes Valium several nights each week (as prescribed by a physician) to become calm enough to sleep abusing a drug? Or are social drinkers who drink a few more than usual on New Year's Eve abusing alcohol?

Many mood drugs act by altering the levels of neurotransmitters within the brain. For example, when excessive amounts of norepinephrine are released in the RAS, we feel stimulated and energetic, whereas abnormally low concentrations of this neurotransmitter reduce anxiety.

Habitual use of almost all mood drugs may result in **psychological dependence,** in which the user becomes emotionally dependent upon the drug. When deprived of it, the user craves the feeling of **euphoria** (well-being) that the drug induces. Some drugs induce **tolerance** when they are taken continually for several weeks. This means that increasingly larger amounts are required to obtain the desired effect. Tolerance often occurs because the liver cells produce enzymes that break down the drug more rapidly. Tolerance can also develop if the drug is similar to chemicals naturally produced in the body. In an effort to maintain homeostasis, the body reduces manufacture of those chemicals. Use of some drugs (such as heroin) results in **physical dependence,** or **addiction,** in which physiological changes take place. When the drug is withheld, the addict suffers physical illness and characteristic withdrawal symptoms.

Some commonly used and abused drugs and their effects are listed and described in Table 23–3. Some of the classes of drugs commonly abused are alcohol, stimulants, depressants, opiates, hallucinogens, and marijuana.

CHAPTER SUMMARY

I. Among invertebrates, nerve nets and radial nervous systems are typical of radially symmetrical animals, and bilateral nervous systems are characteristic of bilaterally symmetrical animals.
   A. A nerve net consists of nerve cells scattered throughout the body; no CNS is present. Response in these animals is generally slow and imprecise.

Table 23–3
EFFECTS OF SOME COMMONLY USED DRUGS

Name of drug	Effect on mood	Actions on body	Dangers associated with abuse
Barbiturates (e.g., Nembutal, Seconal)	Sedative-hypnotic[1]; "downers"	Inhibit impulse conduction in RAS: depress CNS, skeletal muscle, and heart; depress respiration; lower blood pressure; cause decrease in REM sleep	Tolerance, physical dependence, death from overdose, especially in combination with alcohol
Methaqualone (e.g., Quaalude, Sopor)	Hypnotic	Depresses CNS; depresses certain polysynaptic spinal reflexes	Tolerance, physical dependence, convulsions, death
Meprobamate (e.g., Equanil, Miltown; "minor tranquilizers")	Antianxiety drug[2]; induces calmness	Causes decrease in REM sleep; relaxes skeletal muscle; depresses CNS	Tolerance, physical dependence; coma and death from overdose
Valium, Librium ("mild tranquilizers")	Reduce anxiety	May reduce rate of impulse firing in limbic system; relax skeletal muscle	Minor EEG abnormalities with chronic use; very large doses cause physical dependence
Phenothiazines (chlorpromazine; "major tranquilizers")	Antipsychotic; highly effective in controlling symptoms of psychotic patients	Affect levels of catecholamines in brain (block dopamine receptors, inhibit uptake of norepinephrine, dopamine, and serotonin); depress neurons in RAS and basal ganglia	Prolonged intake may result in Parkinson-like symptoms
Antidepressant drugs (e.g., Elavil)	Elevate mood; relieve depression	Block uptake of norepinephrine, so more is available to stimulate nervous system	Central and peripheral neurological disturbances; uncoordination; interfere with normal cardiovascular function
Alcohol	Euphoria; relaxation; release of inhibitions	Depresses CNS; impairs vision, coordination, judgment; lengthens reaction time	Physical dependence; damage to pancreas; liver cirrhosis; possible brain damage
Narcotic analgesics (e.g., morphine, heroin)	Euphoria; reduction of pain	Depress CNS; depress reflexes; constrict pupils; impair coordination. Block release of substance P from pain-transmitting neurons.	Tolerance; physical dependence; convulsions; death from overdose
Cocaine	Euphoria; excitation followed by depression	CNS stimulation followed by depression; autonomic stimulation; dilates pupils; local anesthesia; inhibits re-uptake of norepinephrine	Mental impairment; convulsions; hallucinations; unconsciousness; death from overdose

B. Echinoderms typically have a nerve ring and nerves that extend into various parts of the body.

C. In a bilateral nervous system there is a concentration of nerve cells to form nerves, nerve cords, ganglia, and (in complex forms) a brain. There is also an increase in numbers of neurons, especially of the association neurons. This permits greater precision and a wider range of responses.

II. Sense organs receive information about the external and internal

Name of drug	Effect on mood	Actions on body	Dangers associated with abuse
Amphetamines (e.g., Dexedrine)	Euphoria; stimulant; hyperactivity; "uppers," "pep pills"	Stimulate release of dopamine and norepinephrine; block re-uptake of norepinephrine and dopamine into neurons; inhibit monoamine oxidase (MAO); enhance flow of impulses in RAS; increase heart rate; raise blood pressure; dilate pupils	Tolerance; possible physical dependence; hallucinations; death from overdose
Caffeine	Increases mental alertness; decreases fatigue and drowsiness	Acts on cerebral cortex; relaxes smooth muscle; stimulates cardiac and skeletal muscle; increases urine volume (diuretic effect)	Very large doses stimulate centers in the medulla (may slow the heart); toxic doses may cause convulsions
Nicotine	Psychological effect of lessening tension	Stimulates sympathetic nervous system; combines with receptors in postsynaptic neurons of autonomic system; effect similar to that of acetylcholine, but large amounts result in blocking transmission; stimulates synthesis of lipid in arterial wall	Tolerance; physical dependence; stimulates development of atherosclerosis
LSD (lysergic acid diethylamide)	Overexcitation; sensory distortions; hallucinations	Alters levels of transmitters in brain (may inhibit serotonin and increase norepinephrine); potent CNS stimulator; dilates pupils sometimes unequally; increases heart rate; raises blood pressure	Irrational behavior
Marijuana	Euphoria	Impairs coordination; impairs depth perception and alters sense of timing; inflames eyes; causes peripheral vasodilation; exact mode of action unknown	In large doses, sensory distortions, hallucinations. Evidence of lowered sperm counts and testosterone (male hormone) levels

[1] Sedatives reduce anxiety; hypnotics induce sleep.
[2] Antianxiety drugs reduce anxiety but are less likely to cause drowsiness than the more potent sedative-hypnotics.

environment. Sense organs include the eyes, ear, taste buds, olfactory epithelium, and many types of touch receptors within the skin, as well as receptors in muscles and other internal organs.

    A. The retina of the eye contains two types of light receptors: rods, sensitive to weak light, and cones, sensitive to color.

    B. The cochlea of the inner ear contains cells sensitive to specific frequencies of sound.

III. The human central nervous system consists of brain and spinal cord. These organs are protected by bone and by three meninges, and they are cushioned by cerebrospinal fluid.

A. The spinal cord consists of ascending tracts, which transmit information to the brain, and descending tracts, which transmit information from the brain. Its gray matter consists of many nuclei, which serve as reflex centers.

B. The human brain consists of the cerebrum, cerebellum, and brain stem. The brain stem includes the medulla, pons, midbrain, thalamus, and hypothalamus.

C. The cerebral cortex contains motor areas, which control voluntary movement; sensory areas, which receive incoming sensory information; and association areas, which link sensory and motor areas and are also responsible for learning, language, thought, and judgment.

   1. Alpha wave patterns are characteristic of relaxed states, beta wave patterns of heightened mental activity, and delta waves of non-REM sleep.

   2. The reticular activating system is responsible for maintaining consciousness.

   3. The limbic system affects the emotional aspects of behavior, motivation, sexual behavior, autonomic responses, and biological rhythms.

   4. Metabolic rate slows during non-REM sleep. REM sleep is characterized by dreaming.

   5. Association areas in the cerebral cortex are responsible for the perception of sensation. Endorphins affect pain perception.

   6. Short-term memory may depend upon reverberating circuits in the brain. Mechanisms of long-term memory are not understood.

   7. Environmental experience can cause physical and chemical changes in the brain.

IV. The peripheral nervous system consists of sensory receptors and nerves, including the cranial and spinal nerves and their branches.

V. The autonomic system regulates the internal activities of the body.

A. The sympathetic system enables the body to respond to stressful situations.

B. The parasympathetic system influences organs to conserve and restore energy.

VI. Many drugs alter mood by increasing or decreasing the concentrations of specific neurotransmitters within the brain.

Post-Test

1. The simplest organized nervous tissue is the _____ _____ _____ which is found in *Hydra.*

2. Afferent nerves conduct impulses toward a _____ _____ _____ _____ .

3. In a planarian flatworm the _____ _____ serve as a primitive brain.

4. The human central nervous system consists of the _____ and _____ _____ _____ .

5. The light-sensitive layer of the eye is the _____ _____ ; it contains two types of light receptors, _____ and _____ .

6. The sense of _____ and _____ depend upon chemoreceptors.

7. Pacinian corpuscles detect _____ .

8. The function of the cerebrospinal fluid is to _____ .

9. The largest, most prominent part of the brain is the _____ .

10. Voluntary movement is controlled by _____ in the cerebral cortex.

11. As you take this post-test, your brain should be emitting _____ _____ .

12. The reticular activating system (RAS) is responsible for _____ _____ .

Match the following:

13. Coordinates and refines muscular movement
14. Contains center that controls heartbeat
15. Regulates pituitary gland
16. Receives incoming signals from sense organs
17. Interprets incoming sensory information

a. hypothalamus
b. sensory cortex
c. cerebellum
d. medulla
e. none of the above

18. A person with angina who feels heart pain in his left arm is experiencing _____ _____ .
19. The efferent portion of the autonomic system is subdivided into the _____ and _____ systems.
20. An example of _____ is the need for a user to increase the dose of a drug to obtain the desired effect; this often occurs after several weeks of taking the same drug.
21. Label the diagram at the right. For correct labeling, see Figure 23–9.

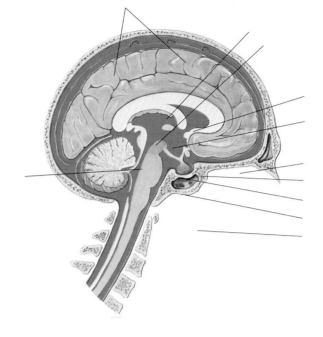

Review Questions

1. Compare the nervous system of a hydra with that of a planarian flatworm.
2. Compare the flatworm nervous system with that of a human.
3. What are the functions of sense organs? Identify the receptors in (a) the eye, (b) the ear.
4. What structures protect the brain and spinal cord?
5. What are the functions of the spinal cord?
6. Identify the part of the brain most closely associated with each of the following: (a) regulation of body temperature, (b) regulation of heart rate, (c) link between nervous and endocrine systems, (d) interpretation of incoming sensory messages, (e) coordination of movements.
7. Imagine that you have just become a parent. What kind of things could you do to ensure development of your child's academic abilities?
8. What is the RAS? How does it function?
9. In what way does electrical activity of the brain reflect a person's state of consciousness? Of what use might it be to learn to control one's brain wave patterns?
10. Contrast somatic and autonomic systems.
11. Contrast sympathetic and parasympathetic systems.
12. Describe how these drugs affect the CNS: (a) alcohol, (b) Dexedrine or other amphetamines, (c) barbiturates, (d) antipsychotic drugs, (e) hallucinogens.

Readings

Constantine-Paton, M., and M. Law. "The Development of Maps and Stripes in the Brain," *Scientific American*, December 1982, 62–70. Vol. 247, No. 6. An account of studies of the visual system and its organization.

Hudspeth, A.J. "The Hair Cells of the Inner Ear," *Scientific American*, Vol. 248, No. 1, January 1983, 54–64. A description of the mechanism by which hair cells in the inner ear respond to convey information about acoustic tones and acceleration. Some of this author's conclusions are controversial.

Kandel, E.R., and J.H. Schwartz. "Molecular Biology of Learning: Modulation of Transmitter Release." *Science*, Vol. 218, (29 October 1982). A review that focuses on the biochemistry of learning in a marine mollusk.

Levine, J.S., and E.F. MacNichol, Jr. "Color Vision in Fishes," *Scientific American*, February 1982, 140–149.

A discussion of retinal pigments in fishes and their relationship to the evolution of the eye.

Miller, J.A. "Colorful Views of Vision," *Science News*, Vol. 120, Oct. 3, 1981. An account of a technique for mapping components of the visual system.

Morrison, A.R. "A Window on the Sleeping Brain," *Scientific American*, Vol. 248, No. 4, April, 1983, 94–101. A study of REM sleep and accompanying paralysis.

Newman, E.A., and P.H. Hartline. "The Infrared 'Vision' of Snakes," *Scientific American*, March 1982, pp. 116–124. Snakes of two families can detect and localize sources of infrared radiation.

Wurtman, R.J. "Nutrients That Modify Brain Function," *Scientific American*, April 1982, 50–59. An exciting discussion of the effect of such nutrients as choline, tryptophane, and tyrosine on brain function.

Chapter 24
INTERNAL TRANSPORT

Outline

Learning Objectives

After you have read this chapter you should be able to:

1. Compare internal transport in different types of organisms, list the functions of the circulatory system, and explain each function.
2. Describe the principal components of the blood and give their functions.
3. Compare the structure and function of the different types of blood vessels, including arteries, arterioles, capillaries, and veins.
4. Describe the structure and function of the parts of the heart and label them on a diagram.
5. Describe the heartbeat and its neural regulation and describe the sounds produced by the heart.
6. Give the basis of arterial pulse and tell how pulse is measured.
7. Compare blood pressure in different types of blood vessels and summarize how arterial blood pressure is regulated.
8. Trace a drop of blood through the pulmonary and systemic circulations, naming in proper sequence each structure through which it passes.
9. Describe the risk factors of atherosclerosis, trace the progress of the disorder, and summarize its possible complications (including angina pectoris and myocardial infarction).

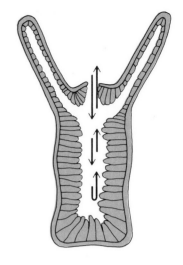

ach cell of a living system requires a constant supply of oxygen and nutrients and some means of getting rid of metabolic wastes. Very small organisms have few cells and each is in close contact with the surrounding environment. Gas exchange in such organisms depends upon simple diffusion, and each cell is able to take care of its own nutritional needs. However, diffusion of gases occurs more than 10,000 times faster through air than through tissues. As long as an organism is less than 1 millimeter thick, diffusion is adequate, but if an organism is to grow any larger, it must possess some other mechanism for transporting materials to all its cells.

In complex animals internal transport is accomplished by specialized structures that make up the **circulatory system.** Most circulatory systems consist of (1) blood, a fluid connective tissue consisting of cells and cell fragments dispersed in fluid, (2) a pumping device, usually called a heart, and (3) a system of blood vessels through which the blood circulates.

(a) Hydra

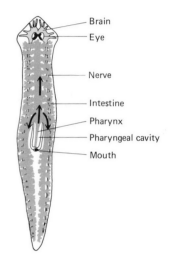

(b) Planarian flatworm

INVERTEBRATE CIRCULATORY SYSTEMS

Cnidarians and many flatworms combine the digestive and circulatory functions in a gastrovascular cavity (Fig. 24–1). Movement of the animal's body, as it stretches and contracts, stirs up the contents of the cavity and aids circulation.

In the arthropods and most mollusks, there is an **open circulatory** ✗ **system,** in which the heart pumps blood into vessels that have open ends (Fig. 24–2a). Blood spills out of them, filling large spaces that make up the **hemocoel** (blood cavity) and bathing the cells of the body. Blood re-enters the circulatory system through openings in the heart or through open-ended vessels that lead to the gills. The blood of many of these animals contains a pigment (hemocyanin) that transports oxygen. Insects employ blood mainly to distribute nutrients and hormones, since oxygen is delivered directly to the cells by a tracheal system of air tubes.

Circulation of blood through an open system is not as rapid or efficient as through a **closed circulatory system,** ✗ in which blood flows through a continuous circuit of blood vessels. Earthworms and other annelids have a complex closed circulatory system (Fig. 24–2b). Although they have no heart, contractions of certain blood vessels, as well as contractions of the body wall muscles, circulate the blood. Earth-

Figure 24–1 (*a*) In *Hydra* and other cnidarians the gastrovascular cavity serves a circulatory function, permitting nutrients to reach the body cells. (*b*) In planarian flatworms the branched intestine conducts food to all regions of the body.

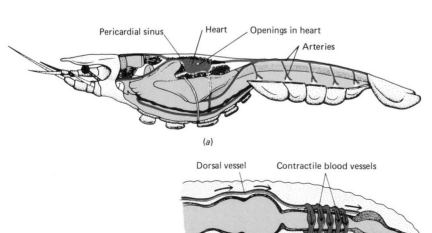

Pericardial sinus   Heart   Openings in heart

Arteries

(a)

Dorsal vessel   Contractile blood vessels

Ventral vessel   Subneural vessel

(b)

Figure 24–2 (*a*) Open circulatory system of the crayfish. (*b*) Closed circulatory system of the earthworm.

Figure 24–3 The circulatory system of the cat is a typical vertebrate system, in which a ventral heart pumps blood into a system of arteries. The arteries branch into smaller and smaller vessels until blood flows through the thin-walled capillaries, where materials are exchanged between blood and body cells. Blood returns to the heart through a system of veins. (*R*, right; *L*, left; *a*, artery; *v*, vein.)

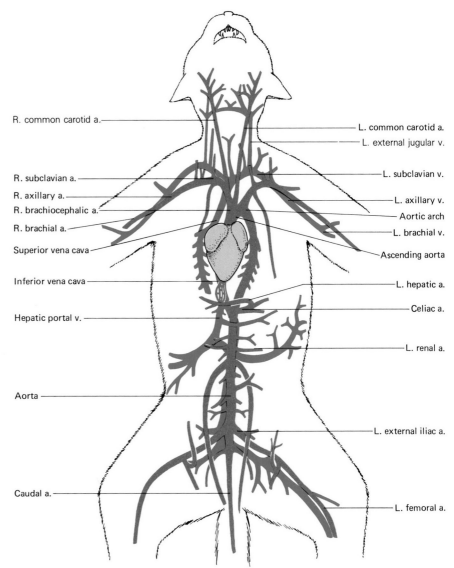

R. common carotid a.

L. common carotid a.

L. external jugular v.

R. subclavian a.

L. subclavian v.

R. axillary a.

L. axillary v.

R. brachiocephalic a.

Aortic arch

R. brachial a.

L. brachial v.

Superior vena cava

Ascending aorta

Inferior vena cava

L. hepatic a.

Celiac a.

Hepatic portal v.

L. renal a.

Aorta

L. external iliac a.

Caudal a.

L. femoral a.

worms possess hemoglobin, the same red pigment that transports oxygen in vertebrate blood. In earthworms, the hemoglobin is not within red blood cells but is dissolved in the blood plasma.

Although other mollusks have an open circulatory system, the fast-moving cephalopods (squid, octopus) require a more efficient means of internal transport. They have a closed system made even more effective by the presence of extra "hearts" at the base of the gills that speed the passage of blood through the gills.

The circulatory system of the sea cucumbers is the most highly developed system of any of the echinoderms. Its vessels parallel the tubes of the water vascular system, and it appears to transport both nutrients and oxygen. Invertebrate chordates typically have a closed circulatory system, with a ventral heart.

THE VERTEBRATE CIRCULATORY SYSTEM

In vertebrates the circulatory system is closed and consists of a ventral heart, blood vessels, blood, lymph, lymph vessels, and associated organs, such as the thymus, spleen, and liver (Fig. 24–3). Some of the vital functions it performs are the following:

1. Transports nutrients from the digestive system and from storage depots to each cell of the body.
2. Transports oxygen from gills or lungs to the cells of the body.
3. Transports metabolic wastes from each cell to excretory organs.
4. Transports hormones from endocrine glands to target tissues.
5. Helps maintain fluid balance.
6. Defends the body against invading microorganisms.
7. Helps to distribute metabolic heat within the body and to maintain normal body temperature in homeothermic (warm-blooded) animals.

THE BLOOD

An adult man weighing about 70 kilograms (154 pounds) possesses about 5.6 liters (6 quarts) of blood. Whole **blood** consists of a straw-colored fluid known as **plasma** in which red blood cells, white blood cells, and platelets are suspended (Fig. 24–4).

Plasma

Plasma consists of water (92%), plasma proteins, a sprinkling of salts, and a vast array of materials being transported, such as nutrients, dissolved gases, metabolic wastes, hormones, and even contaminants, such as drugs. Plasma proteins may be divided into three groups, or fractions—the albumins, globulins, and fibrinogen. One of the main functions of these plasma proteins is to maintain blood volume. As blood flows through the capillaries, the smallest blood vessels, some of the plasma seeps through the capillary walls and passes into the tissues. However, because the large protein molecules have difficulty passing through the capillary walls, most of them remain in the blood. There they exert an osmotic force, called **colloid osmotic pressure,** which helps pull plasma back into the blood.

Certain **albumin** proteins transport fatty acids, and certain other albumins carry specific hormones, keeping them bound in the blood

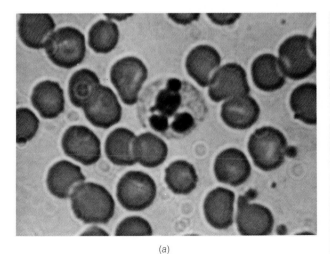

(a)

(b)

Figure 24–4 (a) Photomicrograph of blood (approximately ×1200). All the cells shown are red blood cells except for a single neutrophil. (b) A scanning electron micrograph of a red blood cell. (b, from Bernstein, E. *Science* 173, 1971. Copyright 1971, the American Association for the Advancement of Science.)

until needed. Many of the **globulins** are antibodies, substances which provide immunity against invading disease organisms. **Fibrinogen** and several other plasma proteins are involved in the clotting process.

When the proteins involved in blood clotting have been removed from the plasma, the remaining liquid is called **serum.**

Red Blood Cells where produced?

Each of us has about 30 trillion (3×10^{13}) **red blood cells,** more formally known as **erythrocytes,** suspended in our plasma. These cells are so tiny that about 3000 of them lined up end-to-end would measure only 1 inch. Red blood cells transport oxygen.

Red blood cells develop inside certain bones in special tissue called **red bone marrow.** As each cell differentiates, it manufactures great quantities of **hemoglobin,** the pigment which gives blood its characteristic red color. Each hemoglobin molecule contains four atoms of iron, each of which can combine with one molecule of oxygen. Thus, each hemoglobin molecule can combine chemically with four molecules of oxygen to form the compound **oxyhemoglobin.** It is in this molecular form that oxygen is transported throughout the body.

The mature red blood cell is a tiny sac of hemoglobin, shaped with a considerable surface area for gas exchange (Fig. 24–4). In mammals the mature red blood cell lacks a nucleus and is severely limited in its ability to carry on normal cellular functions, It has a short life span— about 120 days. As blood circulates through the liver and spleen, worn-out red blood cells are removed from circulation and destroyed. Their hemoglobin molecules are taken apart so that some of the components, such as iron, can be sent back to the red bone marrow for reuse. About 2.4 million red cells are destroyed each second; these are immediately replaced by new red blood cells from the bone marrow.

A deficiency of hemoglobin, usually accompanied by a reduced number of red blood cells, is called **anemia.** With less hemoglobin, oxygen transport is reduced, so the body cells do not receive sufficient amounts of oxygen. An anemic person complains of never having enough energy—the "tired-blood" syndrome.

Three general causes of anemia are (1) loss of blood due to hemorrhage or internal bleeding, (2) decreased production of hemoglobin or red blood cells as in iron deficiency anemia, and (3) increased rate of red blood cell destruction (the **hemolytic anemias** such as sickle cell anemia).

White Blood Cells

White blood cells, known as **leukocytes,** defend the body against invading bacteria and other foreign substances. These cells are able to leave the blood, passing out through the walls of the capillaries. Since they are capable of independent locomotion similar to that of an amoeba, they wander through the tissues of the body.

There are several types of leukocytes, but they all have their origin in the red bone marrow. Two kinds of **agranular leukocytes** (that is, those lacking specific granules) are lymphocytes and monocytes (Figs. 24–5 and 24–6). **Monocytes** are thought to be immature cells that leave the circulation and complete their development in the tissues. There, they increase to about five times their original size and become **macrophages,** the giant scavenger cells of the body. Macrophages have voracious appetites for bacteria, dead cells, and other matter littering the tissues. The role of lymphocytes in immune responses will be discussed in Chapter 25.

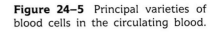

Figure 24–5 Principal varieties of blood cells in the circulating blood.

7 μm

ERYTHROCYTES

1 to 2 μm

THROMBOCYTES
(PLATELETS)

LEUKOCYTES

Granular leukocytes

Agranular leukocytes

10 to 14 μm

Neutrophil

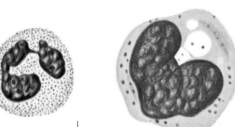

15 to 20 μm

Monocyte

10 to 14 μm

Eosinophil

8 to 10 μm

Lymphocyte

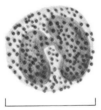

10 to 14 μm

Basophil

———Whole Blood———

Plasma

1. Fibrinogen
2. Serum
   a. Albumins }
   b. Globulins } Plasma proteins
   c. Glucose and other nutrients
   d. Salts
   e. Dissolved gases
   f. Hormones
   g. Water
   h. Wastes

Cellular Components

1. Red blood cells
2. White blood cells
   a. Granular leukocytes
      (1) Neutrophils
      (2) Basophils
      (3) Eosinophils
   b. Agranular leukocytes
      (1) Lymphocytes
      (2) Monocytes
3. Platelets

Figure 24–6 Components of whole blood.

Three types of **granular leukocytes** are the neutrophils, eosinophils, and basophils. **Neutrophils** are especially adept at seeking out and ingesting bacteria. **Eosinophils** are thought to play a part in allergic reactions. **Basophils** contain large amounts of **histamine,** a chemical released in injured or infected tissues and in allergic reactions. Basophils are also thought to contain **heparin,** an anticlotting chemical that may be important in preventing inappropriate clotting within the blood vessels.

Normally, there are about 7000 white blood cells per cubic millimeter of blood. In contrast, there may be as many as 5 million red blood cells per cubic millimeter, so there are almost 700 red cells to every white cell. Measuring the number of white blood cells is a useful diagnostic tool because the white blood cell count becomes elevated in bacterial infections and in certain other disorders. On the other hand, lowered white cell counts often accompany viral infections, rheumatoid arthritis, and a few other conditions. This explains why physicians often perform white blood cell counts before prescribing antibiotics, for these drugs are effective against bacteria but not against viruses.

Leukemia is a form of cancer in which any one of the kinds of white cells multiply rapidly within the bone marrow. Many of these cells do not mature, and their large numbers crowd out developing red blood cells and platelets, leading to anemia and impaired clotting. A common cause of death from leukemia is internal hemorrhaging, especially in the brain. Another frequent cause of death is infection because, although there may be a dramatic rise in the white cell count, the cells are immature and abnormal and unable to defend the body against disease organisms. Although no cure for leukemia has been discovered, radiation treatment and therapy with antimitotic drugs can induce partial or complete remissions lasting as long as 15 years in some patients.

Platelets

Platelets are small fragments of cytoplasm that separate from certain large cells in the bone marrow. When the wall of a blood vessel is injured, as when you cut your finger, platelets seal the break by adhering to the wall in large numbers. This process is aided by a complex series of chemical reactions that produce tiny fibers. The fibers reinforce the platelets, forming a strong clot. This process may be summarized as follows:

$$\text{Prothrombin} \xrightarrow[\text{compounds released from platelets}]{\text{Several clotting factors, Calcium}} \text{thrombin}$$

(a plasma
protein)　　　　　　　　　　　　　　　　　　　　　(active form
of prothrombin)

$$\text{Fibrinogen} \xrightarrow{\text{thrombin}} \text{fibrin}$$

Prothrombin is a globulin manufactured in the liver. Vitamin K is necessary for its production. Fibrin consists of tiny threads that serve to reinforce the platelet clot and to entrap red and white blood cells. (The red blood cell in Figure 24–4b is enmeshed in the fibrin of a developing clot.) In people with hemophilia ("bleeder's disease") one of the clotting factors is absent because of a genetic mutation.

BLOOD VESSELS

Three main types of blood vessels are arteries, capillaries, and veins (Fig. 24–7). An **artery** carries blood away from the heart, toward other tissues. When an artery enters an organ, it divides into many smaller branches called **arterioles.** The arterioles deliver blood into the micro-

Figure 24–7 Types of blood vessels and their relationship to one another. Lymphatic vessels return excess interstitial fluid to the blood by way of ducts that lead into large veins in the shoulder region.

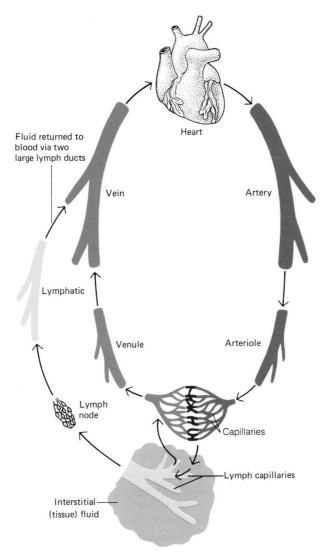

Heart

Fluid returned to blood via two large lymph ducts

Vein

Artery

Lymphatic

Venule

Arteriole

Lymph node

Capillaries

Lymph capillaries

Interstitial (tissue) fluid

scopic **capillaries.** After coursing through an organ, capillaries eventually merge to form larger and larger vessels that transport the blood back toward the heart. These vessels are known as **veins.**

The walls of arteries and veins are thick, which prevents gases and nutrients from passing through them. The exchange of materials between the blood and individual cells of the body tissues takes place through the capillary walls, which are only one cell thick (Fig. 24–8). Capillary networks in the body are so extensive that at least one of these tiny vessels is located close to almost every cell in the body. It has been estimated that the total length of all capillaries in the body is more than 60,000 miles.

Smooth muscle in the arteriole wall can constrict **(vasoconstriction)** or relax **(vasodilatation),** changing the radius of the arteriole. Such changes help maintain appropriate blood pressure and determine the volume of blood passing to a particular tissue. Changes in blood flow are regulated by the nervous system in response to the metabolic needs of the tissue, as well as by the demands of the body as a whole. For example, when a tissue is metabolizing rapidly, it needs a greater supply of blood to bring it nutrients and oxygen. During exercise arterioles within the muscles dilate, increasing the amount of blood flowing to the muscle cells by more than tenfold.

If all the blood vessels were dilated at the same time, there would not be sufficient blood to fill them completely. Normally, the liver, kidneys, and brain receive the lion's share of the blood. However, if sud-

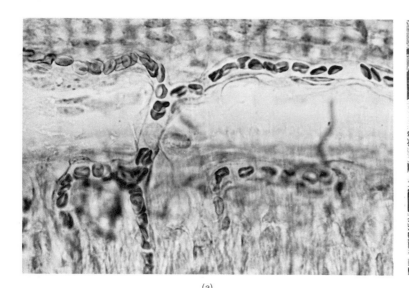

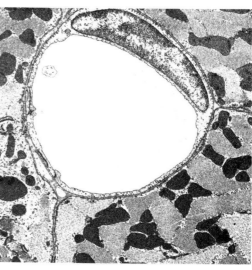

(a)

(b)

Figure 24–8 (a) Red blood cells must pass through capillaries in almost single file. (b) Cross section of capillary in the papillary muscle of a rat heart. Notice that the wall of the capillary is composed of a single cell formed into a tube. A series of such cells makes up the entire capillary. The light-and-dark pattern surrounding the capillary consists of cross sections of cardiac muscle cells. The heart is much too thick and active to be supplied by the blood it contains; it therefore requires its own system of capillaries to nourish it and provide it with oxygen. (b, courtesy of Torsten Mattfeldt and Gerhard Mall, University of Heidelberg, *Cardiovascular Research* 17, 1983.)

denly a monster began to chase you, your blood would be rerouted quickly in favor of your heart and muscles. This would enable you to escape effectively. At such a time the digestive system and kidneys can do with a bit less blood, for they are not critical in your race for survival.

THE HEART

Not much bigger than a fist and weighing less than a pound, the human **heart** (Fig. 24–9) is a remarkable organ that beats about 2.5 billion times in an average lifetime, pumping about 300 million liters (80 million gallons) of blood. It can vary its output from 5 to 35 liters (5 to 37 quarts) of blood per minute in accordance with the body's need (see Focus on the Evolution of the Vertebrate Heart).

Located in the chest cavity, the heart is a hollow, muscular organ consisting of four chambers (Fig. 24–10). Each side of the heart has an **atrium** (which receives blood) and a **ventricle** (which pumps blood out into certain blood vessels). The wall of the heart is composed mainly of cardiac muscle covered by a tough connective-tissue membrane. Another membrane, the **pericardium,** surrounds the entire heart but is separated from it by a space known as the **pericardial cavity.** Normally, a thin film of lubricating fluid within the pericardial cavity moistens the contacting surfaces, facilitating smooth movement of the heart as it contracts and relaxes.

To prevent backflow of blood, a flaplike structure called a **valve** guards the entrance and the exit of each ventricle (Fig. 24–10). When the ventricles contract, the **atrioventricular (AV) valves** between the atria and the ventricles close so that blood does not flow backward into the atria. The AV valve between the left atrium and ventricle is known as the **mitral valve.** As blood leaves the ventricles and is forced into the great arteries, the **semilunar valves** close, preventing backflow into the ventricles.

The Heartbeat

You may have watched horror films in which a heart separated from the body of its owner continues to beat spookily. Script writers of such films may have actually rooted their fantasies in fact, for when removed from the body, the heart will continue to beat for many hours if bathed in an

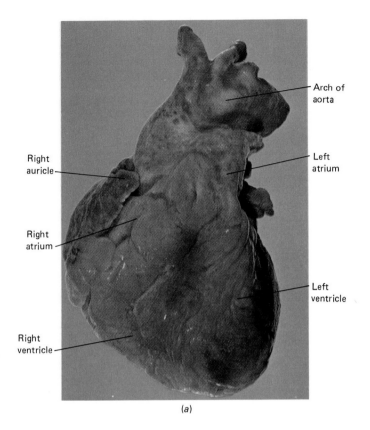

Arch of
aorta

Right
auricle

Left
atrium

Right
atrium

Left
ventricle

Right
ventricle

(a)

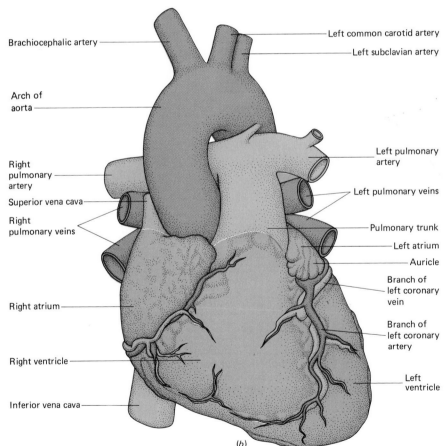

Brachiocephalic artery

Left common carotid artery

Left subclavian artery

Arch of
aorta

Left pulmonary
artery

Right
pulmonary
artery

Left pulmonary veins

Superior vena cava

Right
pulmonary veins

Pulmonary trunk

Left atrium

Auricle

Branch of
left coronary
vein

Right atrium

Branch of
left coronary
artery

Right ventricle

Left
ventricle

Inferior vena cava

(b)

Figure 24–9 (a) Photograph of a human heart, anterior external view. (Courtesy of Phil Horne, Stanford University School of Medicine.) (b) Diagram of the heart, anterior view. Note the coronary blood vessels that bring blood to and from the heart muscle itself.

Focus on . . .

THE EVOLUTION OF THE VERTEBRATE HEART

A comparison of the development of the heart and circulation in different classes of vertebrates reveals a progression from the relatively simple heart and single-circuit circulation of the fish to the complex heart and double-circuit circulation of the bird and mammal. Because it has only one atrium and one ventricle, the fish heart is usually described as a two-chambered heart. Actually, two accessory chambers are present. A thin-walled **sinus venosus** receives blood returning from the tissues and pumps it into the atrium. The atrium then contracts, sending blood into the ventricle. The ventricle in turn pumps the blood into an elastic **conus arteriosus,** which does not contract. These four compartments are separated by valves, which prevent blood from flowing backward. From the conus, blood flows into a large artery, the ventral aorta, which branches to distribute blood to the gills. Because blood must pass through the capillaries of the gills before flowing to the other tissues of the body, blood pressure is low through most of the system. This single-circuit, low-pressure circulatory system permits only a low rate of metabolism in the fish and helps explain its inability to maintain a constant, high body temperature.

The three-chambered amphibian heart consists of two atria and a ventricle. A thin-walled sinus venosus collects blood returning from the veins and pumps it into the right atrium. Blood returning from the lungs passes directly into the left atrium. Both atria pump into the single ventricle. In the frog heart, oxygenated and deoxygenated blood are kept somewhat separate. Deoxygenated blood is pumped out of the ventricle

first and passes into the tubular conus arteriosus, which has a spiral fold that helps keep the blood separate. Much of the deoxygenated blood is directed to the lungs and skin, where it can be charged with oxygen. Oxygenated blood is sent into arteries that conduct it to the various tissues of the body.

In reptiles the heart consists of two atria and two ventricles. In all reptiles except the crocodiles, however, the wall between the ventricles is incomplete, so some mixing of oxygenated and deoxygenated blood occurs. Mixing is minimized by the timing of contractions of the left and right side of the heart and by pressure differences.

The hearts of birds and mammals have completely separate right and left sides. The wall between the ventricles is complete, preventing the mixture of oxygenated blood in the left side with deoxygenated blood in the right side. The conus has split and become the base of the aorta and pulmonary artery. No sinus venosus is present as a separate chamber, although a vestige remains as the sinoatrial node (the pacemaker).

Complete separation of right and left hearts makes it necessary for blood to pass through the heart twice each time it makes a tour of the body. As a result, blood in the aorta of birds and mammals contains more oxygen than that in the aorta of other vertebrates. Thus, the tissues of the body receive more oxygen, a higher metabolic rate can be maintained, and the homeothermic (warm-blooded) condition is possible. Birds and mammals can maintain a constant, high body temperature even in cold surroundings.

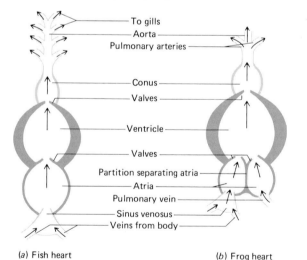 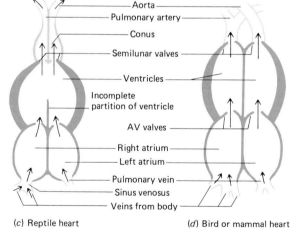

(a) Fish heart   (b) Frog heart   (c) Reptile heart   (d) Bird or mammal heart

The evolution of the vertebrate heart. (a) In the fish heart there is one atrium and one ventricle. (b) The amphibian heart consists of two atria and one ventricle. (c) The reptilian heart has two atria and two ventricles, but the wall separating the ventricles is incomplete, so blood from the right and left chambers mixes to some extent. (d) Birds and mammals have two atria and two ventricles, and oxygenated blood is kept completely separate from deoxygenated blood.

appropriate nutritive fluid. This is possible because the heart has its own specialized conduction system and can beat independently of its nerve supply.

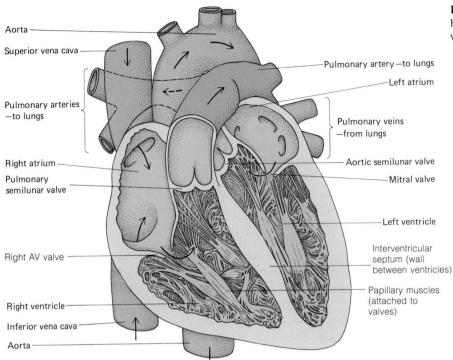

Figure 24–10 Section through the human heart showing chambers, valves, and connecting blood vessels.

Aorta

Superior vena cava

Pulmonary arteries —to lungs

Right atrium

Pulmonary semilunar valve

Right AV valve

Right ventricle

Inferior vena cava

Aorta

Pulmonary artery—to lungs

Left atrium

Pulmonary veins —from lungs

Aortic semilunar valve

Mitral valve

Left ventricle

Interventricular septum (wall between ventricles)

Papillary muscles (attached to valves)

Each **heartbeat** begins in a specialized node of tissue called the **pacemaker** (sinoatrial node, or simply SA node), which is located in the posterior wall of the right atrium. From the pacemaker, impulses are transmitted through the muscle fibers of the atria, causing them to contract. One group of atrial muscle fibers transmits the muscle impulse directly to a second node (the atrioventricular or AV node), located in the wall between the atria. From this node impulses sweep through specialized fibers to all parts of the ventricles, causing them to contract.

Cardiac muscle fibers are separated at their ends by dense bands called **intercalated discs** (Fig. 24–11). Each intercalated disc is actually

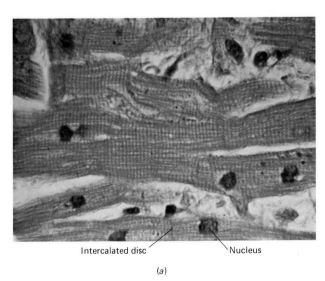

Intercalated disc   Nucleus

(a)

Intercalated discs

(b)

Figure 24–11 (a) Cardiac muscle as seen with the light microscope (approximately ×400). (b) An electron micrograph of cardiac muscle. (b, courtesy of Lyle C. Dearden.)

a tight junction between two cells. This junction is of great significance because it offers very little resistance to the passage of a muscle impulse. It allows an impulse to pass across the disc, so the entire mass of atrial or ventricular muscle tends to contract as if it were one giant cell.

Regulation

Although the heart is capable of beating rhythmically on its own, it cannot by itself effectively change the strength and rate of contraction to meet the changing needs of the body. This kind of control is the function of the autonomic nervous system. Under conditions of stress sympathetic nerves increase the strength of contraction as much as 100%. Under more placid conditions, the vagus nerve, a parasympathetic nerve, slows the heart. It is the balance between the two that determines heart rate, and this balance is determined by the brain.

The endocrine system also plays a part in regulating heartbeat. When the body is under stress, the hormone epinephrine, released from the adrenal glands, stimulates the force of the heartbeat.

The normal heart rate is about 70 beats per minute, and the **cardiac output,** the volume of blood pumped by one ventricle, is about 5 liters (5 quarts) per minute. This amount is equivalent to the total volume of blood in the body. Cardiac output depends primarily on the volume of blood delivered to the heart by the veins **(venous return).** During vigorous exercise the heart may beat as many as 200 times per minute and its output may increase to about 28 liters (30 quarts) per minute. In a trained athlete the heart actually enlarges (in extreme cases up to 50%) and is capable of pumping a greater quantity of blood per beat. An athlete's heart is thus more efficient and does not have to beat as often to distribute the same quantity of blood as does the heart of a person who is not in good physical condition.

Heart Sounds

A physician listening to the heartbeat with a stethoscope can distinguish two principal sounds, which occur in repeating rhythm. These sounds, usually described as a "lub-dup," are produced each time the heart valves close. The first sound, the "lub," is caused by the closing of the AV valves, and marks the beginning of ventricular **systole,** the phase of the heart's cycle when the ventricles contract. The "dup" sound is heard as a quick snap and is caused by the closing of the semilunar valves. This marks the beginning of ventricular **diastole,** the phase of the heart's cycle when the ventricles relax.

A **heart murmur** is a common type of abnormal sound that sometimes indicates a valve disorder. When a valve does not close properly, some blood may flow backward, creating a hissing sound. Characteristic murmurs may also be heard when a valve is enlarged with scar tissue and is rough, so the passageway is narrowed.

The Electrocardiogram

As each wave of contraction spreads through the heart, electrical currents spread into the tissues surrounding the heart and onto the body surface. This electrical activity can be recorded by placing electrodes on the body surface on opposite sides of the heart. The **electrocardiograph** is the machine used to amplify and record the electrical activity, and the record produced is called an **electrocardiogram** (ECG or EKG). In intensive-care units and in operating rooms an oscilloscope is often used in-

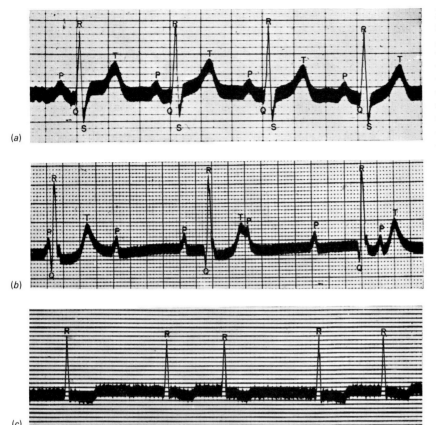

Figure 24–12 Electrocardiograms. (a) Tracing from a normal heart. The P wave corresponds to the contraction of the atria, the QRS complex to the contraction of the ventricle, and the T wave to the relaxation of the ventricle. (b) Tracing from a patient with a complete block of the atrioventricular node. The atria and ventricles beat independently, each at their own rate. Note that the P waves appear at regular intervals and QRS and T waves appear at regular but longer intervals, but that there is no relation between the P and QRS waves. (c) Tracing from a patient with atrial fibrillation. The individual muscle fibers of the atrium twitch rapidly and independently. There is no regular atrial contraction and no P wave. The ventricles beat independently and irregularly, causing the QRS wave to appear at irregular intervals. (Courtesy of Dr. Lewis Dexter and the Peter Bent Brigham Hospital, Boston, Mass.)

stead of an electrocardiograph. The oscilloscope continuously monitors the heart, displaying a moving beam of electrons on a screen.

An ECG begins with a **P wave,** which represents the spread of an impulse through the atria just before atrial contraction (Fig. 24–12). Then a **QRS complex** appears, reflecting the spread of an impulse through the ventricles just before they contract. As the ventricles recover, currents generated are reflected upon the graph as a **T wave.**

Abnormalities in the ECG indicate disorders in the heart or its rhythm. One class of disorders which can be diagnosed with the help of the ECG is **heart block.** In this condition transmission of an impulse is delayed or blocked at some point in the conduction system. Artificial pacemakers are now implanted in patients with severe heart block. A pacemaker is implanted beneath the skin, and its electrodes are connected to the heart. This device provides continuous rhythmic impulses that drive the heart.

PULSE AND BLOOD PRESSURE

Each time the left ventricle pumps blood into the aorta (the great blood vessel conducting blood away from the heart), the elastic wall of the aorta expands. This expansion moves down the aorta and its branches in a wave. As soon as the wave has passed, the elastic arterial wall snaps back to its normal size. This alternate expansion and recoil of an artery constitute the arterial **pulse.**

When you place your finger over the radial artery in the wrist or over the carotid artery in the neck region, you can feel the pulse. The number of pulsations counted per minute indicates the number of heartbeats per minute, since every time the heart contracts, a pulse wave begins.

Blood pressure is the force exerted by the blood against the inner walls of the blood vessels. It is determined by the blood flow and the

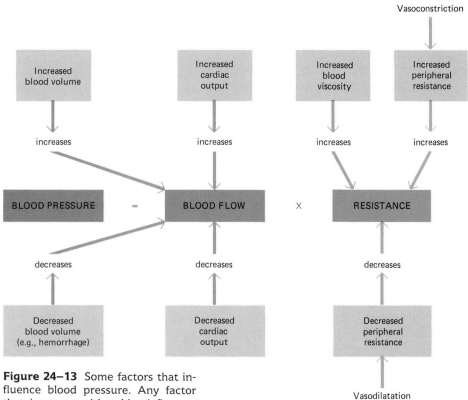

Figure 24–13 Some factors that influence blood pressure. Any factor that increases either blood flow or resistance increases blood pressure.

resistance to that flow (Fig. 24–13). Blood flow depends directly upon the pumping action of the heart. When cardiac output increases, blood flow increases, causing a rise in blood pressure. When cardiac output decreases, blood flow decreases, causing a fall in blood pressure. The volume of blood flowing through the system also affects blood pressure. When blood volume is reduced by hemorrhage or by chronic bleeding, the blood pressure drops. On the other hand, an increase in blood volume results in an increase in blood pressure. For example, a high dietary salt intake causes water retention. This results in an increase in blood volume and leads to higher blood pressure.

Blood flow is impeded by resistance; when the resistance to flow increases, blood pressure rises. **Peripheral resistance** is the resistance to blood flow caused by the viscosity of the blood and by friction between the blood and the wall of the blood vessel. The length and diameter of a blood vessel determine the surface area of the vessel in contact with the blood. The length of a blood vessel does not change, but the diameter, especially of an arteriole, does. Even a small change in the diameter of a blood vessel causes a big change in blood pressure. For example, if the radius of a blood vessel is doubled, the resistance is reduced to one sixteenth of its former value, and the blood flow increases 16-fold.

Blood pressure in arteries rises during systole and falls during diastole. A blood pressure reading is expressed as systolic pressure over diastolic pressure. Normal blood pressure of a young adult would approximate 120/80 (measured in millimeters of mercury, or mm Hg). When the diastolic pressure consistently reads over 95 mm Hg, the patient may be suffering from **hypertension** (high blood pressure). Hypertension is one of the most common cardiovascular disorders in the United States. Although the cause of hypertension is not known, heredity, obesity, stress, and high dietary salt intake are thought to be important factors in its development. This disorder places a heavy burden upon the heart, which must pump against greater blood pressure.

As you might imagine, blood pressure is greatest in the arteries and lessens as blood flows through the capillaries. By the time blood reaches the veins, its pressure is very low. When the body is in an upright position, gravity offers a great deal of resistance to blood flow through the veins. It is really quite remarkable that blood in the feet manages to make its way back up to the heart. Much of the success of this journey may be attributed to flaplike **valves** within the veins. These prevent the blood from flowing (or falling) backwards. Blood is pushed along through the veins by the pressure of the blood behind it and by compression of veins by muscular movement that occurs when we move about.

In people whose jobs require that they stand for long periods each day, blood accumulates in the veins of the legs. Excessive pooling of the blood stretches the veins, so that the cusps of their valves no longer meet. This may lead to **varicose veins,** especially in those who are obese or who have inherited weak vein walls. A varicose vein is dilated, tortuous, and elongated. Varicose veins occur most often in the superficial veins (those close to the surface), especially those in the legs. These veins have the least external support and are subjected to the greatest increases in pressure. **Hemorrhoids,** which are varicose veins in the anal region, occur when venous pressure in that region is constantly elevated, as in chronic constipation (because of straining) and during pregnancy (because of pressure of the enlarged uterus).

When one stands perfectly still for a long period of time (as when a soldier stands at attention), blood pools in the veins. Within a few moments pressure increases in the capillaries because the veins are not able to accept more blood from them. This causes large amounts of plasma to leave the circulation through the thin walls of the capillaries. Within a few minutes so much blood volume can be lost from the circulation that arterial blood pressure may fall drastically. Blood supply to the brain is diminished, and sometimes fainting occurs. Less dramatic changes in blood pressure occur each time you get up from a horizontal position.

Several complex homeostatic mechanisms interact to maintain normal blood pressure. Tiny pressure receptors called **baroreceptors,** located in certain arteries, are sensitive to changes in blood pressure and, when stimulated, send messages to centers in the medulla of the brain. Nerves then signal the heart to slow down or speed up, and other nerves send messages to arterioles and veins, causing them to dilate or constrict. For example, when blood pressure decreases as you get out of bed in the morning, the heart rate increases slightly and blood vessels are constricted, so blood pressure increases. These neural reflexes act continuously to maintain a steady state of blood pressure. Hormones are also involved in regulation of blood pressure. The **angiotensins** are a group of hormones that are powerful vasoconstrictors. When blood pressure is low, the kidneys release the hormone **renin,** which stimulates the formation of angiotensins from a plasma protein. The kidneys also help maintain blood pressure indirectly by influencing blood volume. In response to hormones, the kidneys vary the rate of excretion of salts and water.

THE PATTERN OF CIRCULATION

Blood flows through a continuous network of blood vessels that form a double circuit—the **pulmonary circulation,** connecting heart and lungs, and the **systemic circulation,** connecting the heart with all the tissues of the body. This general pattern of circulation is illustrated in Figure 24–14.

Figure 24–14 Highly simplified diagram showing the pattern of circulation through the systemic and pulmonary circuits. Red represents oxygenated blood; blue represents deoxygenated blood.

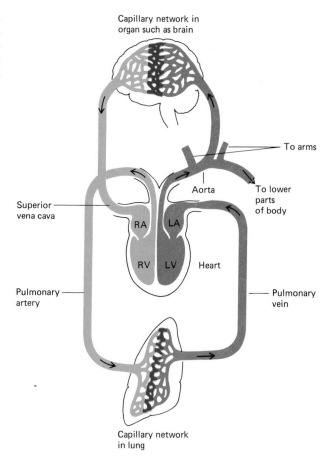

Capillary network in organ such as brain

To arms

Aorta

To lower parts of body

Superior vena cava

RA | LA

RV | LV | Heart

Pulmonary artery

Pulmonary vein

Capillary network in lung

Pulmonary Circulation

Blood from the tissues returns to the right atrium of the heart depleted of its oxygen supply, or **deoxygenated,** but loaded with carbon dioxide wastes. This deoxygenated blood must be sent to the lungs to obtain more oxygen and to get rid of its carbon dioxide wastes. Accordingly, it is pumped from the right ventricle into the pulmonary circulation (Fig. 24–14). The **pulmonary arteries** carry blood to the lungs. These are the only arteries in the body that carry deoxygenated blood. In the lungs the pulmonary arteries branch into smaller and smaller vessels and finally into an extensive network of **pulmonary capillaries** that course throughout the lung tissue. As blood circulates through this capillary network, gases are exchanged. Then **pulmonary veins,** the only veins in the body to carry oxygenated blood, return the oxygen-rich blood to the left atrium of the heart.

In summary, blood flows through the vessels of the pulmonary circulation in the following sequence: Right atrium → right ventricle → pulmonary artery → pulmonary capillaries (in lung) → pulmonary vein → left atrium.

Systemic Circulation

Blood returning from the pulmonary circulation enters the left atrium of the heart, then passes into the left ventricle. From there it is pumped into the largest artery of the body, the **aorta.** The aorta divides into arterial branches that carry blood to all regions of the body, including the heart muscle itself. Some of the principal branches include the **carotid** arteries to the brain, the **subclavian** arteries to the shoulder and arm region, the **mesenteric** arteries to the intestine, the **renal** arteries to the kidneys, and the **iliac** arteries to the legs (Figure 24–15). Each of these branches into smaller and smaller arteries that bring blood to the capillary networks within each tissue and organ.

Blood returning from the brain is carried back toward the heart by the **jugular** veins. Blood from the shoulders and arms drains into the **subclavian** veins. These veins and others bringing blood from the upper portion of the body merge to form the **superior vena cava**, a very large vein that empties blood into the right atrium. **Renal** veins from the kidneys, **iliac** veins from the lower limbs, **hepatic** veins from the liver, and other veins returning blood from the lower regions of the body empty

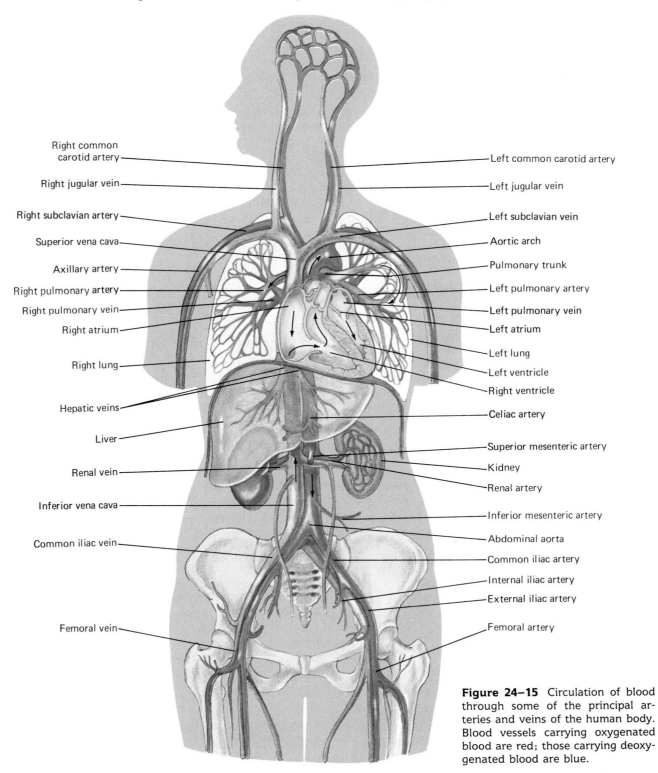

Right common carotid artery

Right jugular vein

Right subclavian artery

Superior vena cava

Axillary artery

Right pulmonary artery

Right pulmonary vein

Right atrium

Right lung

Hepatic veins

Liver

Renal vein

Inferior vena cava

Common iliac vein

Femoral vein

Left common carotid artery

Left jugular vein

Left subclavian vein

Aortic arch

Pulmonary trunk

Left pulmonary artery

Left pulmonary vein

Left atrium

Left lung

Left ventricle

Right ventricle

Celiac artery

Superior mesenteric artery

Kidney

Renal artery

Inferior mesenteric artery

Abdominal aorta

Common iliac artery

Internal iliac artery

External iliac artery

Femoral artery

Figure 24–15 Circulation of blood through some of the principal arteries and veins of the human body. Blood vessels carrying oxygenated blood are red; those carrying deoxygenated blood are blue.

Focus on . . .

CARDIOVASCULAR DISEASE

Cardiovascular disease is the number one cause of death in the United States and in most other industrial societies. Most often death results from some complication of **atherosclerosis**[1] (hardening of the arteries as a result of lipid deposition). Although atherosclerosis can affect almost any artery, the disease most often develops in the aorta and in the coronary and cerebral arteries. When it occurs in the cerebral arteries, it can lead to a **cerebrovascular accident (CVA),** commonly referred to as a stroke.

Although there is apparently no single cause of atherosclerosis, several major risk factors have been identified:

1. Elevated levels of cholesterol in the blood, often associated with diets rich in total calories, total fats, saturated fats, and cholesterol.
2. Hypertension. The higher the blood pressure, the greater the risk.
3. Cigarette smoking. The risk of developing atherosclerosis is two to six times greater in smokers than in nonsmokers and is directly proportional to the number of cigarettes smoked daily.
4. Diabetes mellitus, an endocrine disorder in which glucose is not metabolized normally.

The risk of developing atherosclerosis also increases with age. Estrogen hormones are thought to offer some protection in women until after menopause, when the concentration of these hormones decreases. Other suggested risk factors that are currently being studied are obesity, hereditary predisposition, lack of exercise, stress and behavior patterns, and dietary factors, such as excessive intake of salt or refined sugar.

In atherosclerosis, lipids are deposited in the smooth muscle cells of the arterial wall. Cells in the arterial wall proliferate and the inner lining thickens. More lipid, especially cholesterol from low-density lipoproteins, accumulates in the wall. Eventually calcium is deposited there, contributing to the slow formation of hard plaque. As the plaque develops, arteries lose their ability to stretch when they fill with blood, and they be-

come progressively occluded (blocked), as shown in the figure. As the artery narrows, less blood can pass through to reach the tissues served by that vessel and the tissue may become **ischemic** (lacking in blood). Under these conditions the tissue is deprived of an adequate oxygen supply.

When a coronary artery becomes narrowed, **ischemic heart disease** can occur. Sufficient oxygen may reach the heart tissue during normal activity, but the increased need for oxygen during exercise or emotional stress results in the pain known as **angina pectoris.** People with this condition often carry nitroglycerin pills with them for use during an attack. This drug dilates veins so the venous return is reduced. Cardiac output is lowered, so the heart is not working so hard and requires less oxygen. Nitroglycerin also dilates the coronary arteries slightly, allowing more blood to reach the heart muscle.

Myocardial infarction (MI) is the very serious, often fatal, form of ischemic heart disease that often results from a sudden decrease in coronary blood supply. The portion of cardiac muscle deprived of oxygen dies within a few minutes and is then referred to as an **infarct.** The term myocardial infarction is used as a synonym for heart attack. MI is the leading cause of death and disability in the United States. Just what triggers the sudden decrease in blood supply that causes MI is a matter of some debate. It is thought that in some cases an episode of ischemia triggers a fatal arrhythmia, such as **ventricular fibrillation,** a condition in which the ventricles contract very rapidly without actually pumping blood. In other cases, a **thrombus** (clot) may form in a diseased coronary artery. Because the arterial wall is roughened, platelets may adhere to it and initiate clotting. If the thrombus blocks a sizable branch of a coronary artery, blood flow to a portion of heart muscle is impeded or completely halted. This condition is referred to as a **coronary occlusion.** If the coronary occlusion prevents blood flow to a large region of cardiac muscle, the heart may stop beating—that is, **cardiac arrest** may occur—and death can follow within moments. If only a small region of the heart is affected, however, the heart may continue to function. Cells in the region deprived of oxygen die and are replaced by scar tissue.

[1]Atherosclerosis is the most common form of arteriosclerosis, any disorder in which arteries lose their elasticity.

---

blood into the **inferior vena cava,** which returns blood to the right atrium.

As an example of blood circulation through the systemic system, let us trace a drop of blood from the heart to the brain and back: Left atrium → left ventricle → aorta → carotid artery → capillaries in brain → jugular vein → superior vena cava → right atrium → right ventricle → into pulmonary circulation.

CORONARY CIRCULATION

The heart is an actively metabolizing organ that requires a large and continuous supply of nutrients and oxygen. Blood flowing through its chambers cannot serve these needs because the heart wall is far too thick to permit effective diffusion. Fortunately, the heart is equipped with its own complex of blood vessels, the **coronary circulation** (see Fig. 24–9b).

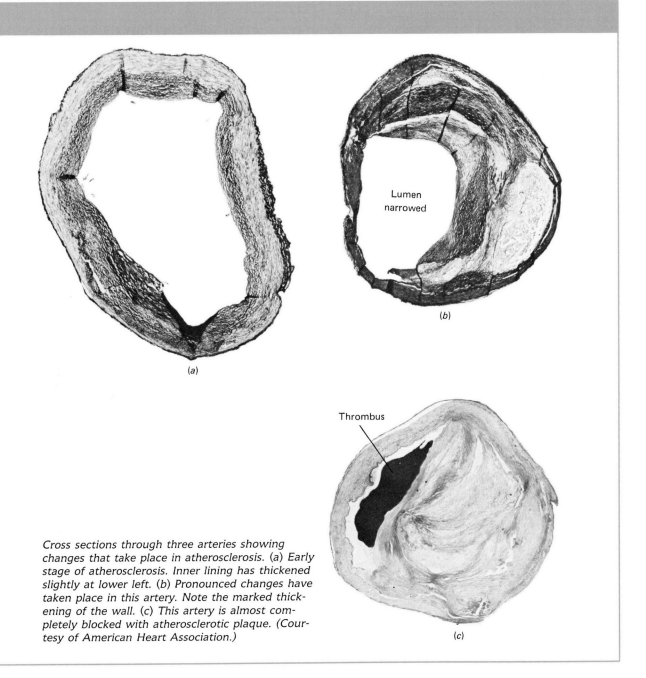

(a)

(b)

Lumen
narrowed

Thrombus

(c)

Cross sections through three arteries showing changes that take place in atherosclerosis. (a) Early stage of atherosclerosis. Inner lining has thickened slightly at lower left. (b) Pronounced changes have taken place in this artery. Note the marked thickening of the wall. (c) This artery is almost completely blocked with atherosclerotic plaque. (Courtesy of American Heart Association.)

Two **coronary arteries** branch off from the aorta just as it leaves the heart. These arteries give rise to an extensive system of blood vessels within the heart tissue. Most of the coronary capillaries empty into veins that join to form a large vein, the **coronary sinus,** which empties into the right atrium. Blockage of the coronary arteries is a principal cause of heart disease (see Focus on Cardiovascular Disease).

HEPATIC PORTAL SYSTEM

Generally, blood travels from artery to capillary to vein. An exception to this sequence is the **hepatic portal system,** which transports nutrients from the intestine to the liver. Blood reaches the intestines via the mesenteric artery and enters capillaries in the intestinal villi, where it receives nutrients. Capillaries from the intestinal villi merge to form the **superior mesenteric vein,** which empties into the **hepatic portal vein.** This vein goes not to the heart but to the liver. In the liver the hepatic portal vein branches into a vast network of tiny vessels called sinusoids. (Sinusoids are exchange vessels similar to capillaries.) As the blood courses through these sinusoids, the liver cells absorb and process the excess nutrients present. Liver capillaries eventually merge to form the **hepatic veins,** which in turn empty into the inferior vena cava. Thus, in the hepatic portal system we find this unusual sequence of blood vessels: Capillaries → vein → sinusoids → vein.

Thus there is an extra set of exchange vessels. The liver is also supplied with freshly oxygenated blood by a hepatic artery, for the oxygen content of the blood in the hepatic portal vein would not be sufficient to sustain the liver cells.

THE LYMPHATIC SYSTEM

The lymphatic system is an accessory circulatory system that feeds into the blood circulation. Its three principal functions are (1) to collect and return excessive tissue fluid to the blood, (2) to defend the body against disease organisms, and (3) to absorb lipids from the digestive system and transport them to the blood. Here we will focus upon the first function; the other two will be discussed in later chapters.

Design of the Lymphatic System

The lymphatic system has tiny "dead-end" capillaries that extend into almost all tissues of the body (Fig. 24–16). Excessive tissue fluid enters the lymph capillaries, where it is then known as **lymph.** Lymph capillar-

Focus on . . .

CARDIOPULMONARY RESUSCITATION (CPR)

Cardiopulmonary resuscitation, or **CPR,** is a method for aiding victims of accidents or heart attacks who have suffered cardiac arrest and respiratory arrest. It should not be used if the victim has a pulse or is able to breathe. It must be started immediately, because irreversible brain damage may occur within about 3 minutes of respiratory arrest. Here are its ABCs:

Airway Clear airway by extending victim's neck. This is sometimes sufficient to permit breathing to begin again.
Breathing Use mouth-to-mouth resuscitation.
Circulation Attempt to restore circulation by using external cardiac compression.

The procedure for CPR may be summarized as follows.

I. Establish unresponsiveness of victim.
II. Procedure for mouth-to-mouth resuscitation:
  1. Place victim on his or her back on firm surface.
  2. Clear throat and mouth and tilt head back so that chin points outward. Make sure that the tongue is not blocking airway. Pull tongue forward if necessary.
  3. Pinch nostrils shut and forcefully exhale into victim's mouth. Be careful, especially in children, not to overinflate the lungs.
  4. Remove your mouth and listen for air rushing out of the lungs.

  5. Repeat about 12 times per minute. Do not interrupt for more than 5 seconds.
III. Procedure for external cardiac compression:
  1. Place heel of hand on lower third of breastbone. Keep your fingertips lifted off the chest. (In infants, two fingers should be used for cardiac compression; in children, use only the heel of the hand.)
  2. Place heel of the other hand at a right angle to and on top of the first hand.
  3. Apply firm pressure downward so that the breastbone moves about 4 to 5 cm (1.6 to 2 in.) toward the spine. Downward pressure must be about 5.4 to 9 kg (12 to 20 lb) with adults (less with children). Excessive pressure can fracture the sternum or ribs, resulting in punctured lungs or a lacerated liver. This rhythmic pressure can often keep blood moving through the heart and great vessels of the thoracic cavity in sufficient quantities to sustain life.
  4. Relax hands between compressions to allow chest to expand.
  5. Repeat at the rate of at least 60 compressions per minute. (For infants or young children, 80 to 100 compressions per minute are appropriate.) If there is only one rescuer, 15 compressions should be applied, then two breaths, in a ratio of 15:2. If there are two rescuers, the ratio should be 5:1.

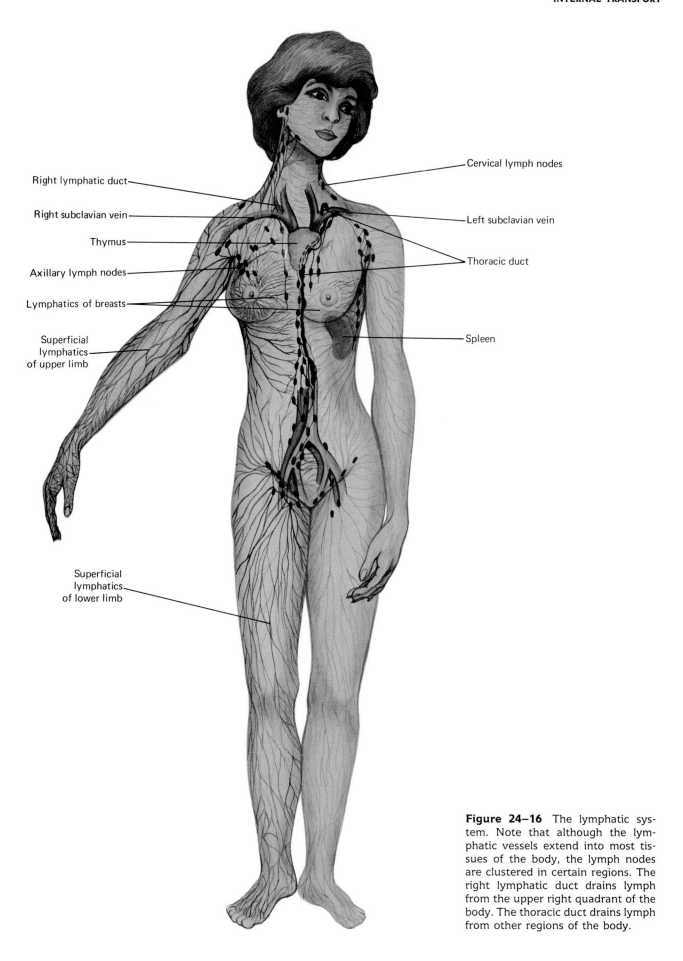

Cervical lymph nodes

Right lymphatic duct

Right subclavian vein

Left subclavian vein

Thymus

Axillary lymph nodes

Thoracic duct

Lymphatics of breasts

Superficial lymphatics of upper limb

Spleen

Superficial lymphatics of lower limb

Figure 24–16 The lymphatic system. Note that although the lymphatic vessels extend into most tissues of the body, the lymph nodes are clustered in certain regions. The right lymphatic duct drains lymph from the upper right quadrant of the body. The thoracic duct drains lymph from other regions of the body.

ies convey the lymph to larger vessels called **lymph veins** (or **lymphatics**).

At strategic locations lymph veins enter **lymph nodes,** small organized masses of lymph tissue. Lymph nodes have two main functions: (1) they filter the lymph as it slowly passes through, and (2) they produce lymphocytes, white blood cells concerned with immune responses. Lymph nodes (sometimes called **lymph glands**) are most numerous in the neck region, under the arms, in the groin region, and in the chest and abdomen. Lymph nodes in an infected area enlarge conspicuously and may be felt as hard little knots below the skin.

Lymph veins that leave the lymph nodes conduct lymph toward the shoulder region. Eventually lymph veins empty their contents into the subclavian veins by way of the **thoracic** and **right lymphatic ducts.**

Tonsils are masses of lymph tissue under the lining of the oral cavity and throat. (The pharyngeal tonsils in back of the nose are called **adenoids** when they are enlarged.) Tonsils help protect the respiratory system from infection by destroying bacteria and other foreign matter that enter the body through the mouth or nose. Unfortunately, tonsils are sometimes overcome by invading germs, become the site of frequent infection themselves, and then become prime targets for surgical removal.

Some nonmammalian vertebrates, such as the frog, have lymph "hearts" that pulsate and squeeze lymph along. In mammals, however, the walls of the lymph vessels themselves pulsate, pushing the lymph through the vessels. When muscles contract or arteries pulsate, pressure on the lymph vessels enhances lymph flow. Valves within the lymph vessels prevent backflow.

Fluid Balance

As blood enters a capillary network, it is under high enough pressure that some of the plasma is forced out through the capillary wall (Fig. 24–17). Plasma that has escaped from the blood circulation is known as **tissue fluid,** or **interstitial fluid.** It contains few red cells or platelets and only about 25% as much protein as the circulating plasma. Rich in nutrients and oxygen, the tissue fluid bathes the cells in a rich sea of needed materials. It also receives wastes from the cells, and eventually delivers them back into the blood.

At the venous ends of the capillaries some tissue fluid moves back into the blood. This is because blood pressure is much lower there, and

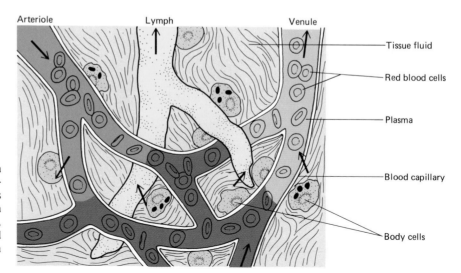

Figure 24–17 The relation of lymph capillaries to blood capillaries and tissue cells. Note that blood capillaries are connected to vessels at both ends, whereas lymph capillaries, shown in yellow, are dead-end streets. The arrows indicate direction of flow.

osmotic pressure of the blood tends to pull fluid back into the vessels. However, not as much fluid returns to the blood as escaped, and the plasma protein does not effectively re-enter the blood through the capillary walls. These problems would cause a serious fluid imbalance if it were not for the lymphatic system. This system makes a vital contribution to fluid homeostasis by collecting the excess fluid (amounting to about 10% of the tissue fluid) and protein and returning them to the blood.

When for any reason lymph vessels become blocked, tissue fluid accumulates in the affected area, causing **edema** (swelling). Obstruction of lymph flow may result from injury, inflammation, parasitic infection, or surgery (Fig. 24–18). For example, when a breast is removed (radical mastectomy) because of cancer, lymph nodes in the region under the arm are surgically removed in an effort to prevent the spread of cancer cells. The patient's arm may swell tremendously because of the disrupted lymph circulation. However, new lymph vessels usually develop within a few months, and the swelling slowly subsides.

Figure 24–18 Lymphatic drainage is blocked in the limbs of this individual because of a parasitic infection known as filariasis. The condition characterized by such swollen limbs is elephantiasis. (From Markell, E.K., and M. Voge. *Medical Parasitology*, 5th ed. Philadelphia, W.B. Saunders, 1981.)

CHAPTER SUMMARY

  I. Small, simple invertebrates, such as sponges, cnidarians, and flatworms, depend upon diffusion for internal transport. More complex invertebrates require a specialized circulatory system.
   A. Arthropods and most mollusks have an open circulatory system, in which blood flows into a hemocoel, bathing the tissues directly.
   B. Other invertebrates have a closed circulatory system, in which blood flows through a continuous circuit of blood vessels.
 II. The vertebrate circulatory system is a closed system that transports nutrients, oxygen, wastes, and hormones; it helps maintain fluid balance and body temperature; and through its subsystem, the lymphatic system, it is responsible for immune responses.
III. The blood consists of the straw-colored liquid plasma in which are suspended red blood cells, white blood cells, and platelets.
   A. Plasma consists of water, salts, substances in transport, and three fractions of proteins: albumins, globulins, and fibrinogen.
   B. Red blood cells transport oxygen.
   C. Lymphocytes and monocytes are agranular white blood cells; neutrophils, eosinophils, and basophils are granular white cells. White blood cells can leave the blood and wander through the body's tissues.
   D. Platelets patch damaged blood vessels and release substances essential for blood clotting.
 IV. Arteries carry blood away from the heart, whereas veins bring blood back toward the heart. Capillaries are tiny, thin-walled vessels through which materials are exchanged between blood and tissues.
  V. The heart is a muscular pump consisting of a right atrium and ventricle and a left atrium and ventricle. Entrances and exits of the ventricles are guarded by valves.
   A. Although the heart has its own conduction system and can beat independently of its nerve supply, the beat is regulated by sympathetic and parasympathetic nerves.
   B. Heart sounds can be heard with a stethoscope, and electrical activity of the heart may be measured with an ECG.
 VI. Blood pressure is greatest in arteries and least in veins.

A. Neural and hormonal mechanisms act continuously to maintain normal blood pressure.

B. Arterial pulse is the alternate expansion and recoil of an artery that occurs each time the left ventricle pumps blood into the aorta.

VII. Blood flows through a double circuit—the pulmonary and systemic circulations.

A. The right atrium receives deoxygenated blood from the tissues of the body and pumps it into the right ventricle, which sends it into the pulmonary circulation.

B. Blood from the lungs returns via the pulmonary veins to the left atrium and is pumped into the aorta by the left ventricle. The aorta sends arterial branches into all parts of the systemic circulation.

1. In the coronary circulation blood is transported to the cells of the heart itself.

2. In the hepatic portal system a vein gives rise to an extensive network of exchange vessels within the tissues of the liver.

VIII. Atherosclerosis is the basis of most cardiovascular disease. It can lead to ischemic heart disease or cerebrovascular accidents.

IX. The lymphatic system, a subsystem of the circulatory system, collects interstitial fluid and returns it to the blood.

Post-Test

1. In an _____ circulatory system the heart pumps blood into a hemocoel.

2. Small invertebrates, such as sponges, depend upon _____ for internal transport.

3. The fluid component of blood is _____ .

Select the most appropriate answers from column B for each item in column A.

Column A

_____ 4. Transport oxygen
_____ 5. Seek out and ingest bacteria
_____ 6. Become macrophages
_____ 7. Initiate clotting
_____ 8. Contain hemoglobin
_____ 9. Agranular leukocyte

Column B

a. platelets
b. monocytes
c. red blood cells
d. neutrophils
e. eosinophils

10. A deficiency in hemoglobin is referred to as _____ .

11. During clotting, fibrinogen is converted to fibrin by the action of _____ .

Select the most appropriate answers from column B for each item in column A.

Column A

_____ 12. Conduct blood toward heart
_____ 13. Help regulate blood pressure
_____ 14. Exchange vessels
_____ 15. Largest blood vessel
_____ 16. Equipped with valves

Column B

a. capillaries
b. arteries
c. veins
d. arterioles
e. aorta

17. The force exerted by the blood against the inner walls of the blood vessels is known as _____ .

18. The pulmonary vein delivers blood to the _____ _____ _____ .

19. The hepatic portal vein delivers blood to the _____ _____ .

20. Blood pressure is sensed by _____ within certain arteries.

21. The _____ valve is located between the left atrium and ventricle.

22. The angiotensins are a group of powerful _____ _____ .

23. When a tissue is ischemic, it lacks sufficient _____ _____ .

24. In _____ the arterial wall thickens and may block the passage of blood.

25. Lymph is produced when _____ _____ _____ enters _____ vessels.

26. Label the figure at the top of the next page. For correct labeling, see Figure 24–10.

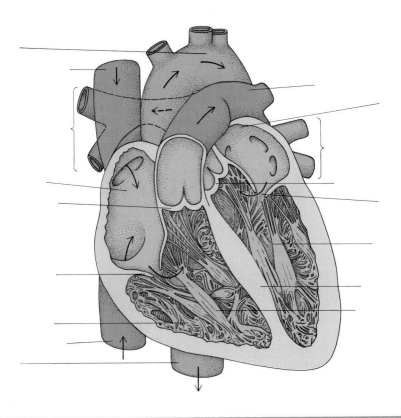

Review Questions

1. Give the functions of (a) red blood cells, (b) plasma proteins, (c) platelets, (d) monocytes, (e) macrophages, (f) neutrophils.
2. Where do red blood cells originate? How are they destroyed?
3. What are three general causes of anemia?
4. What are the functions of arterioles? of capillaries?
5. Draw a diagram of the heart and label its chambers, valves, and the principal blood vessels that enter and exit from it.
6. How is the heartbeat initiated? regulated?
7. How does blood manage to flow against gravity through veins in the legs on its route back to the heart?
8. Give an example of a normal blood pressure reading and of a blood pressure reading from an individual with hypertension.
9. Trace a drop of blood from (a) superior vena cava to aorta, (b) brain to kidney, (c) intestine to lung.
10. How does the lymphatic system help maintain fluid balance?
11. What is the relationship among plasma, interstitial (tissue) fluid, and lymph?

Readings

Bodde, T. "Coping in Space: The Body's Answer to Zero Gravity," *Bioscience,* Vol. 32 No. 4, April 1982. An interesting discussion of the physiological changes experienced by astronauts.

Doolittle, R.F. "Fibrinogen and Fibrin" *Scientific American,* December, 1981, pp. 126–135. A description of the process of blood clot formation and of clot breakdown.

Levy, R.I., and J. Moskowitz. "Cardiovascular Research: Decades of Progress, a Decade of Promise," *Science,* 217: 121–126 (9 July 1982). A review of current techniques for dealing with cardiovascular disease and a discussion of the decrease in mortality due to these diseases.

Stallones, R.A. "The Rise and Fall of Ischemic Heart Disease," *Scientific American,* November 1980, pp. 53–59. An interesting statistical analysis of the decline of U.S. death rates from heart disease since the 1960's.

Chapter 25

INTERNAL DEFENSE: IMMUNITY

Outline

Learning Objectives

After you have studied this chapter you should be able to:

1. Compare in general terms the types of immune responses in invertebrates and vertebrates.
2. Distinguish between specific and nonspecific immune responses.
3. Describe nonspecific defense mechanisms, such as inflammation and phagocytosis, and summarize their role in the defense of the body.
4. Contrast T and B lymphocytes with respect to life cycle and function.
5. Describe the mechanisms of cell-mediated immunity, including development of memory cells.
6. Define the terms antigen and antibody and describe how antigens stimulate immune responses.
7. Describe the mechanisms of antibody-mediated immunity, including the effects of antigen-antibody complexes upon pathogens. (Include a discussion of the complement system.)
8. Describe the functions of the thymus in immune mechanisms.
9. Contrast active and passive immunity and give examples of each.
10. Explain the theory of immunosurveillance and tell how the body destroys cancer cells.
11. Explain the immunological basis of allergy and briefly describe the events that occur during a hay fever response and during systemic anaphylaxis.
12. Review the major problems encountered in organ transplantation and the attempts being made to solve them.

All animals have defense mechanisms that provide protection against disease-causing organisms, or **pathogens.** When pathogens gain entrance to the body, they are likely to be greeted by a barrage of defenses designed to destroy them. Defense mechanisms against pathogens are collectively referred to as **immunity** or **immune responses.** The term immunity is derived from a Latin word meaning *safe.* **Immunology,** the study of these defense mechanisms, is one of the most exciting fields of medical research today.

Defense mechanisms can be nonspecific or highly specific. **Nonspecific defense mechanisms** deter a variety of pathogens. They prevent entrance of pathogens and destroy them quickly if they penetrate the body's outer covering (Fig. 25–1). Phagocytosis of invading bacteria is an example. **Specific defense mechanisms** are more sophisticated; they are tailor-made to the particular type of pathogen that infects the body. The production of antibodies (highly specific proteins that help destroy pathogens) is one of the body's most important specific defense mechanisms.

SELF AND NONSELF

Immune responses depend upon the ability of an organism to distinguish between itself and foreign matter. Such recognition is possible because organisms are biochemically unique. Many cell types have surface macromolecules (proteins or large carbohydrates) that are slightly different from the surface macromolecules on the cells of other species or even other organisms of the same species. An organism "knows" its own macromolecules and "recognizes" those of other organisms as foreign.

A single bacterium may have from 10 to more than 1,000 distinct macromolecules on its surface. When a bacterium invades another organism, these macromolecules stimulate the organism to launch an immune response. A substance capable of stimulating an immune response is called an **antigen,** or **immunogen.**

IMMUNITY IN INVERTEBRATES

All invertebrate species that have been studied demonstrate the ability to distinguish between self and nonself. However, most invertebrates are

Figure 25–1 Summary of the body's nonspecific and specific defense mechanisms.

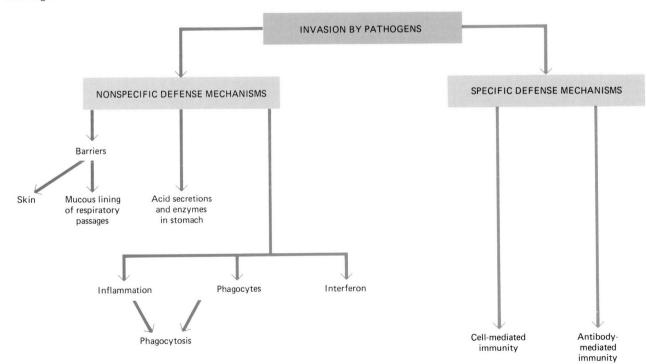

able to make only nonspecific immune responses. Members of many invertebrate phyla are capable of making two important nonspecific responses—phagocytosis and the inflammatory response.

Sponge cells possess specific glycoproteins on their surfaces that enable them to distinguish between self and nonself. When experimentally separated cells of two species of sponges are mixed together, they reaggregate according to species. Cnidarians (jellyfish, corals, and their relatives) also have this ability. They can, in fact, reject grafted tissue and cause the death of foreign tissue.

In complex invertebrates, **coelomocytes** (wandering ameboid phagocytes) very effectively phagocytize bacteria and other foreign matter. Any particle too large to be phagocytized is walled off or encapsulated. Complex invertebrates also possess nonspecific substances that kill bacteria, inactivate cilia in some pathogens, and cause the agglutination (clumping) of some foreign cells.

Echinoderms and invertebrate chordates (tunicates) are the simplest animals known to have specialized white blood cells that perform immune functions. Among the invertebrates, only certain annelids (earthworms, for example) and cnidarians (such as corals) are thought to possess specific immune mechanisms and immunological memories. In them and in some echinoderms and simple chordates, the body appears to remember antigens for a short period of time and can respond to them more effectively in a second encounter.

IMMUNE MECHANISMS IN VERTEBRATES

All vertebrates can launch both nonspecific and specific immune responses. Although vertebrates possess many of the basic mechanisms seen in invertebrates, they also possess many more sophisticated defense mechanisms, which are made possible by the development of a specialized lymphatic system. In the discussion that follows we will focus on the human immune system, with references to those of other vertebrates.

Nonspecific Defense Mechanisms

The outer covering of an animal, the first line of defense against pathogens, is often more than just a mechanical barrier. The human skin, for example, is populated by millions of harmless microorganisms that appear to inhibit the multiplication of potentially harmful organisms that happen to land on it. In addition, sweat and sebum contain chemicals that destroy certain kinds of bacteria.

Microorganisms that enter with food are usually destroyed by the acid secretions and enzymes of the stomach. Pathogens that enter the body with inhaled air may be filtered out by hairs in the nose or trapped in the sticky mucous lining of the respiratory passageways. They are then destroyed by phagocytes.

Should pathogens invade the tissues, other nonspecific defense mechanisms are activated. Certain kinds of cells, for instance, when infected by viruses or other intracellular parasites (some types of bacteria, fungi, and protozoa), respond by secreting proteins called **interferons.** This group of proteins stimulates other cells to produce antiviral proteins, which prevent the cell from manufacturing macromolecules required by the virus. The virus particles produced by cells exposed to interferon are not very effective at infecting cells. Drug companies have invested millions of dollars trying to develop an inexpensive, effective method of producing human interferon. Such a drug might be useful in

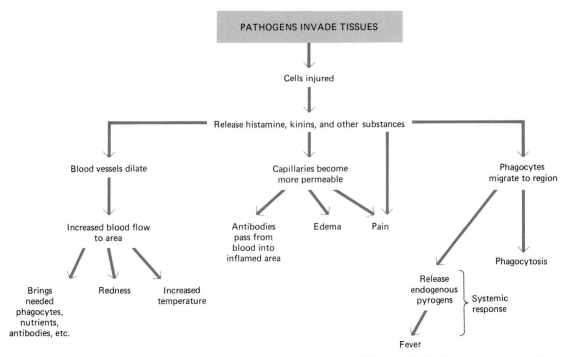

Figure 25–2 The physiology of inflammation.

preventing and treating some viral infections and might even be helpful in treating certain forms of cancer.

Inflammation and phagocytosis are two very important nonspecific defense mechanisms. When pathogens invade tissues, they trigger an **inflammatory response** (Fig. 25–2). Blood vessels in the affected area dilate, increasing blood flow to the infected region. The increased blood flow makes the skin look red and feel warm. Capillaries in the inflamed area become more permeable, allowing more fluid to leave the circulation and enter the tissues. As the volume of interstitial fluid increases, **edema** (swelling) occurs. The edema (and also certain substances released by the injured cells) cause the pain that is characteristic of inflammation.

Although inflammation is often a local response, sometimes the entire body is involved. **Fever** is a common clinical symptom of widespread inflammatory response. Proteins called **endogenous pyrogens** are released by neutrophils and macrophages and somehow reset the body's thermostat in the hypothalamus. Substances known as prostaglandins are also involved in this resetting process. Fever interferes with bacterial iron uptake and places the invaders at a metabolic disadvantage.

One of the main functions of inflammation appears to be increased phagocytosis (Fig. 25–3). The increased blood flow to the injured area brings large numbers of phagocytes (neutrophils and monocytes; see Chapter 24). As it ingests a bacterium, a phagocyte wraps it within membrane pinched off from the cell membrane. The vesicle containing the bacterium is called a **phagosome.** One or more lysosomes adhere to the phagosome membrane and fuse with it. The lysosome releases potent digestive enzymes onto the captured bacterium, and the phagosome membrane releases hydrogen peroxide onto the invader. These substances destroy the bacterium and break down its macromolecules to small, harmless compounds that can be released or utilized by the phagocyte.

After a neutrophil phagocytizes 20 or so bacteria, it becomes inactivated (perhaps by leaking lysosomal enzymes) and dies. A macrophage can phagocytize about 100 bacteria during its lifespan. Can bacteria

Figure 25–3 Scanning electron micrographs of macrophages recovered from the peritoneal cavity of mice. The upper photograph (*a*, approximately ×14,400) is of an unstimulated macrophage. The macrophage in the lower picture (*b*, approximately ×16,200) was taken from a mouse that had received an injection of mineral oil in the peritoneal cavity a few days earlier. The mineral oil acts as an irritant, producing "angry" macrophages with greatly increased metabolic rates, folding of the plasma membrane and increased ability to ingest bacteria. The irregular surface and many lateral microprojections of the "angry" macrophage are evident. The macrophages were grown in cover slips in tissue culture medium for a brief period before photography. (From Albrecht *et al. Experimental Cell Research,* 70: 230–232, 1971.)

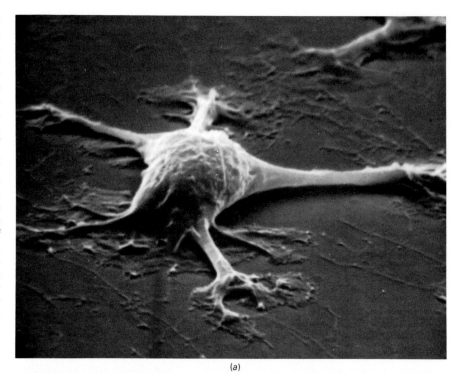

(*a*)

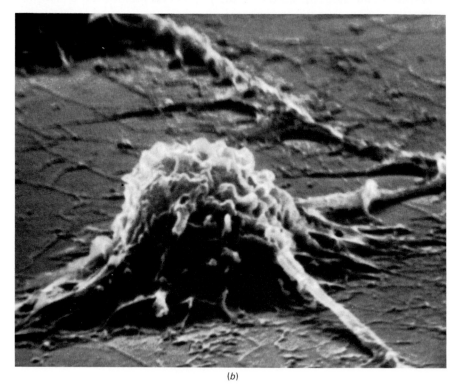

(*b*)

counteract the body's attack? Certain bacteria are able to release enzymes that destroy the membranes of the lysosomes. The powerful lysosomal enzymes then spill out into the cytoplasm and may destroy the phagocyte.

Specific Defense Mechanisms

Nonspecific defense mechanisms destroy pathogens and prevent the spread of infection while the specific defense mechanisms are being mobilized. Several days are required to activate specific immune responses, but once in gear, these mechanisms are extremely effective. There are two main types of specific immunity: **cell-mediated immunity,** in which lymphocytes attack the invading pathogen directly, and **antibody-mediated immunity,** in which lymphocytes produce specific antibodies designed to destroy the pathogen.

T AND B LYMPHOCYTES

The principal warriors in specific immune responses are the trillion or so lymphocytes stationed strategically in the lymphatic tissue throughout the body. In amphibians and more complex vertebrates there are two types of lymphocytes, T lymphocytes and B lymphocytes. Both types are thought to originate either in the bone marrow or in the embryonic liver (Fig. 25–4). On their way to the lymph tissues, the **T lymphocytes (T cells)** stop off in the thymus gland for processing. (The T in T lymphocytes refers to thymus-derived.) Somehow the thymus gland influences the differentiation of lymphocytes so that they become capable of immunological response. T lymphocytes are responsible for cellular immunity.

B lymphocytes (B cells) are responsible for antibody-mediated immunity. In birds they are processed in a lymphatic organ called the **bursa of Fabricius.** (The B in B lymphocytes refers to bursa-derived.) Other vertebrates do not have a bursa and an equivalent organ has not yet been

(b)

Figure 25–4 (a) Origin and functions of T and B lymphocytes. (b) Electron micrograph (approximately ×8000) of a B lymphocyte. (Courtesy of Robert A. Good, Sloan-Kettering Cancer Center.)

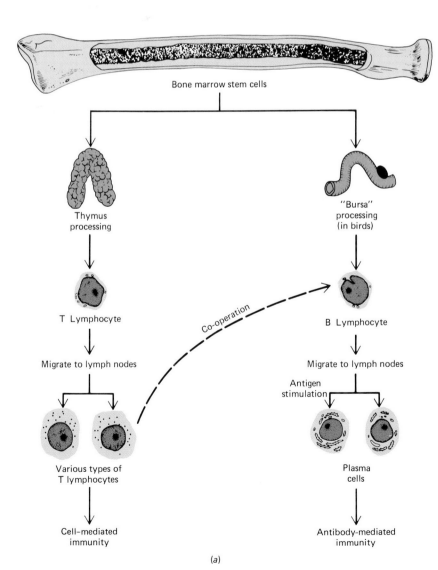

Bone marrow stem cells

Thymus processing

"Bursa" processing (in birds)

T Lymphocyte

Co-operation

B Lymphocyte

Migrate to lymph nodes

Migrate to lymph nodes

Antigen stimulation

Various types of T lymphocytes

Plasma cells

Cell-mediated immunity

Antibody-mediated immunity

(a)

identified. In mammals, B lymphocytes may be processed as they form in the bone marrow, or they may be processed in the fetal liver or spleen.

CELL-MEDIATED IMMUNITY

Cell-mediated immunity is the responsibility of the T cells and macrophages. These cells destroy viruses or foreign cells that enter the body. Most lymphocytes are usually in an inactive state. These are called small lymphocytes. There are thousands of varieties of small lymphocytes, each capable of responding to a specific type of antigen. When an antigen invades the body, macrophages engulf it and bring the antigen to the lymphocytes. The variety of lymphocyte able to react to that antigen—that is, the **competent lymphocyte**—becomes activated, or **sensitized.** Once stimulated, the lymphocytes increase in size. Then they divide by mitosis, each giving rise to a sizable clone of cells identical to itself (Fig. 25–5). These T lymphocytes then differentiate to become killer T lymphocytes, helper T lymphocytes, suppressor T lymphocytes, or memory cells. Members of this cellular infantry then leave the lymph nodes and make their way to the infected area.

Killer T lymphocytes combine with the antigen on the surface of the invading cell and then release powerful enzymes directly into the attacked cell. These enzymes destroy the foreign or malignant cell by disrupting its cell membrane. **Helper T lymphocytes,** the most numerous type of T cell, secrete substances known as lymphokines that enhance the response of killer T cells, suppressor T cells, and B cells. One kind of lymphokine inhibits macrophages from leaving the infection site and stimulates them to be more effective at phagocytosis so that they attack and destroy more invading pathogens. Such stimulated macrophages are sometimes called "angry macrophages." **Suppressor T lymphocytes** inhibit immune defenses several weeks after an infection activates them.

T lymphocytes are especially effective in attacking viruses, fungi, and the types of bacteria that live within host cells. How do the T cells know which cells to attack? Once a pathogen invades a body cell, the host cell's macromolecules may be altered. The immune system then regards that cell as foreign, and T lymphocytes destroy it. Killer T cells also destroy cancer cells and, unfortunately, the cells of transplanted organs.

When T lymphocytes are activated and give rise to a clone, not all of the sensitized cells produced leave the lymph tissue. Some remain behind as **memory cells.** Such cells, or their progeny, live on for many years. Should the invading pathogen ever attack again, the memory cells launch a far more rapid response than was possible during the first invasion. In this secondary immune response, the pathogens are usually destroyed before they have time to establish themselves and cause symptoms of the disease. This is why you usually do not suffer from the same disease several times. Most people get measles or chicken pox, for example, only once.

However, if this is true, how can a person get flu or a cold more than once? Unfortunately, there are many varieties of the common cold and of flu, each caused by a slightly different virus with slightly different antigens. To make matters worse, viruses mutate often (a survival mechanism for them), which may change their surface antigens. Even a slight change may prevent recognition by memory cells. Each "different" antigen presents the body with a new immunological challenge.

ANTIBODY-MEDIATED IMMUNITY

The B lymphocytes are responsible for antibody-mediated immunity. Just as with the T lymphocytes, there are thousands of varieties of B lymphocytes, each capable of responding to a specific type of antigen.

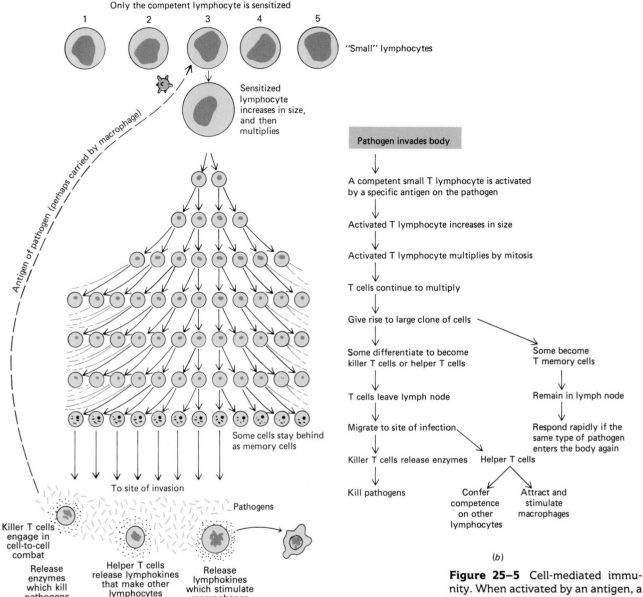

Only the competent lymphocyte is sensitized

1 2 3 4 5

"Small" lymphocytes

Sensitized lymphocyte increases in size, and then multiplies

Antigen of pathogen (perhaps carried by macrophage)

Some cells stay behind as memory cells

To site of invasion

Pathogens

Killer T cells engage in cell-to-cell combat

Release enzymes which kill pathogens

Helper T cells release lymphokines that make other lymphocytes competent to help

Release lymphokines which stimulate macrophages

(a)

Pathogen invades body

A competent small T lymphocyte is activated by a specific antigen on the pathogen

Activated T lymphocyte increases in size

Activated T lymphocyte multiplies by mitosis

T cells continue to multiply

Give rise to large clone of cells

Some differentiate to become killer T cells or helper T cells

Some become T memory cells

T cells leave lymph node

Remain in lymph node

Migrate to site of infection

Respond rapidly if the same type of pathogen enters the body again

Killer T cells release enzymes   Helper T cells

Kill pathogens

Confer competence on other lymphocytes

Attract and stimulate macrophages

(b)

Figure 25–5 Cell-mediated immunity. When activated by an antigen, a T lymphocyte gives rise to a large clone of cells. Many of these differentiate to become killer T cells, which migrate to the site of infection and attempt to destroy the invading pathogens.

Activated B lymphocytes divide to produce large clones of immunologically identical lymphocytes (Fig. 25–6). Most of these cells increase in size and differentiate into **plasma cells,** which may be thought of as graduate lymphocytes. Plasma cells have an extensive, highly developed rough endoplasmic reticulum for the synthesis of proteins, for they are the cells that produce antibodies. Plasma cells do not leave the lymph nodes, as do T cells. Only the antibodies they secrete leave the lymph tissues and make their way via the lymph and blood to the infected area.

Some activated B lymphocytes do not differentiate into plasma cells but instead become memory cells, similar in function to T lymphocyte memory cells. B memory cells continue to produce small amounts of antibody long after an infection has been overcome. This antibody, part of the gamma globulin fraction of the plasma, becomes part of the body's arsenal of chemical weapons. Should the same pathogen enter the body again, this circulating antibody is present to destroy it. At the same time memory cells quickly divide to produce new clones of plasma cells.

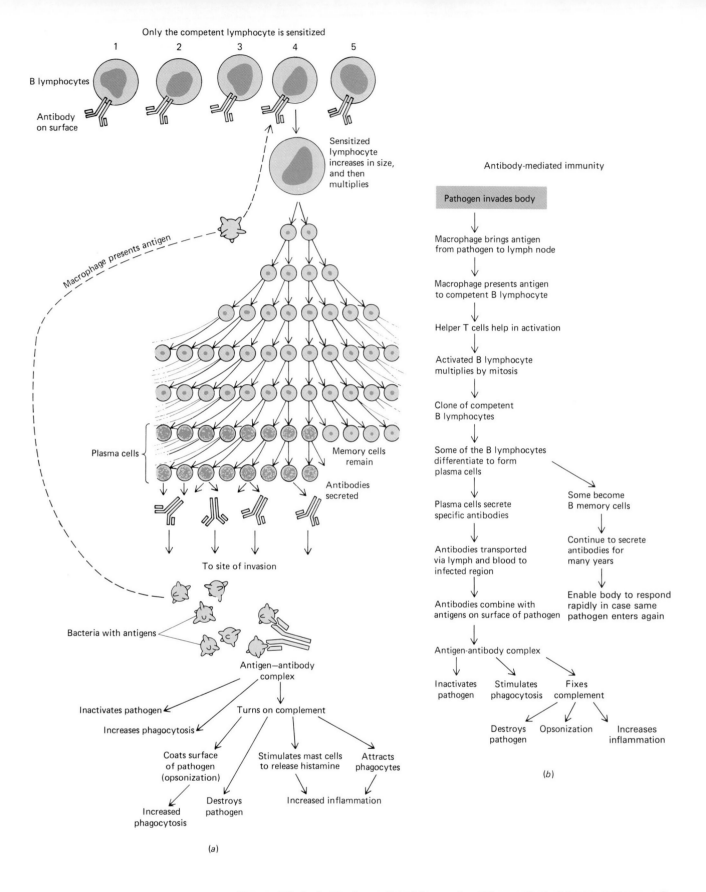

Only the competent lymphocyte is sensitized

1 2 3 4 5

B lymphocytes

Antibody on surface

Sensitized lymphocyte increases in size, and then multiplies

Macrophage presents antigen

Plasma cells

Memory cells remain

Antibodies secreted

To site of invasion

Bacteria with antigens

Antigen–antibody complex

Inactivates pathogen

Increases phagocytosis

Turns on complement

Coats surface of pathogen (opsonization)

Stimulates mast cells to release histamine

Attracts phagocytes

Increased phagocytosis

Destroys pathogen

Increased inflammation

(a)

Antibody-mediated immunity

Pathogen invades body

Macrophage brings antigen from pathogen to lymph node

Macrophage presents antigen to competent B lymphocyte

Helper T cells help in activation

Activated B lymphocyte multiplies by mitosis

Clone of competent B lymphocytes

Some of the B lymphocytes differentiate to form plasma cells

Plasma cells secrete specific antibodies

Antibodies transported via lymph and blood to infected region

Antibodies combine with antigens on surface of pathogen

Antigen-antibody complex

Inactivates pathogen

Stimulates phagocytosis

Fixes complement

Destroys pathogen

Opsonization

Increases inflammation

Some become B memory cells

Continue to secrete antibodies for many years

Enable body to respond rapidly in case same pathogen enters again

(b)

Figure 25–6 Antibody-mediated immunity. When activated by an antigen, a B lymphocyte multiplies, producing a large clone of cells. Many of these differentiate and become plasma cells, which secrete antibodies. The plasma cells remain in the lymph tissues, but the antibodies are transported to the site of infection by the blood. Antigen-antibody complexes form that directly inactivate some pathogens and also turn on the complement system. Some of the B lymphocytes become memory cells, which continue to secrete small amounts of antibody for years after the infection is over.

ANTIBODIES AND THEIR STRUCTURE. Antibodies, also called **immunoglobulins,** are highly specific proteins that may be produced in response to specific antigens. They are among the body's most powerful chemical weapons.

A typical immunoglobulin consists of four polypeptide chains: two identical heavy chains and two identical light chains (Fig. 25–7). Each chain has a constant region and a variable region. The constant region may be thought of as the standard part of a key, the handle that one holds. At its variable regions the antibody folds three-dimensionally, assuming a shape that enables it to recognize and combine with a specific antigen. The variable region is the part of the key that is unique for a specific antigen. When they meet, antigen and antibody fit together *somewhat* like a lock and key, and they *must* fit in just the right way for the antibody to be effective. However, the fit is not as precise as with an enzyme and its substrate. A given antigen can bind with different strengths, or affinities, with different antibodies. In the course of an immune response better, stronger (higher-affinity) antibodies are generated. A typical antibody is a Y-shaped molecule that contains two binding sites (Fig. 25–7), enabling the antibody to combine with two antigens. This permits formation of antigen-antibody complexes.

How does an antibody "recognize" a particular antigen? In a protein antigen there are amino acids that constitute an **antigenic determinant.** These amino acids give part of the antigen molecule a specific configuration that can be "recognized" by an antibody or cell receptor. However, the mechanism is even more complicated. Usually, an antigen has five to ten antigenic determinants on its surface. Some have 200 or

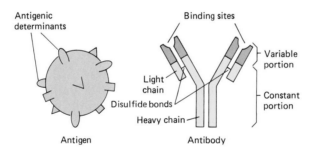

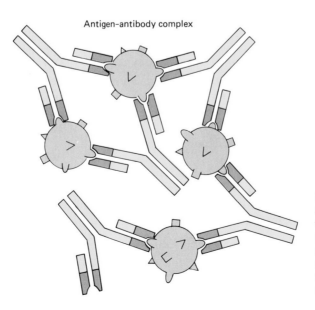

Figure 25–7 Antigen, antibody, and antigen-antibody complex. Note that the antibody molecule is composed of two light chains and two heavy chains, which are joined together by disulfide bonds. The constant and variable regions of the chains are indicated.

even more. These antigenic determinants may differ from one another, so several kinds of antibodies can combine with such a complex antigen.

Some substances found in dust and certain drugs are too small to be antigenic, yet they do stimulate immune responses. These substances, called **haptens,** become antigenic by combining with a larger molecule, usually a protein.

CLASSES OF ANTIBODIES. Immunoglobulins are grouped in five classes according to their structure. At its constant end, the heavy chains of antibody have an amino acid sequence characteristic of the particular antibody class. Using the abbreviation Ig for immunoglobulin, the classes are designated IgG, IgM, IgA, IgD, and IgE. In the simpler vertebrates only IgM is present. In the amphibians IgM and IgG are characteristic. These two classes and also IgA are found in mammals, and all five classes are produced in humans. The classes of antibodies have different functions. Perhaps as animals became more complex, it was advantageous to have a variety of antibody classes with specialized functions.

In humans about 75% of the antibodies in the blood belong to the **IgG** group; these are part of the gamma globulin fraction of the plasma. IgG is thought to contribute to immunity against many blood-borne pathogens, including bacteria, viruses, and some fungi. **IgA** immunoglobulins are the principal antibodies found in body secretions, such as mucous secretions of the nose, respiratory passageways, and digestive tract, and tears, saliva, and vaginal secretions. IgA antibodies are thought to be important in protecting the body from infections by inhaled or ingested pathogens.

HOW ANTIBODIES WORK. Antibodies identify a pathogen as foreign by combining with an antigen on its surface. Often several antibodies combine with several such antigens, creating a mass of clumped antigen-antibody complex (Fig. 25–7). The combination of antigen and antibody activates several defense mechanisms.

1. The antigen-antibody complex may inactivate the pathogen or its toxin. For example, when an antibody attaches to the surface of a virus, the virus loses its ability to attach to a host cell.
2. The antigen-antibody complex stimulates phagocytosis of the pathogen by macrophages and neutrophils.
3. The antigen-antibody complex stimulates a series of reactions that activate the **complement system.** This system consists of about 11 proteins present in plasma and other body fluids. Normally, complement proteins are inactive, but the antibody is said to **fix complement.** Proteins of the complement system then work to destroy pathogens. Some complement proteins digest portions of the pathogen cell. Others coat the pathogens, a process called **opsonization.** This seems to make the pathogens "tastier" so that the macrophages and neutrophils rush to phagocytize them. Complement proteins also increase the extent of inflammation. Antibodies of the IgG and IgM groups work mainly through this **complement** system.

Complement proteins are not specific. They act against any antigen, provided they are activated by antigen-antibody complex. Antibodies identify the pathogen very specifically; then complement proteins complement their action by destroying the pathogens.

ANTIBODY DIVERSITY. One of the most puzzling problems in immunology has been accounting for the tremendous diversity and numbers of antibodies. The immune system has the potential to produce millions of different antibodies, each programmed to respond to a different antigenic determinant. Recent research indicates that the ability to make many different antibodies is inherited, but that diversity is probably increased by mutation, as well as by recombination. According to current theory, there is only one gene that codes for the constant region portion of each type of antibody. However, one of several hundred genes can code for the variable region of the light chain, and one of several hundred can specify the variable region of the heavy chain. There are also four or five other genes that code for joining segments that lie between the constant and variable portions of each chain.

The formation of an active antibody involves the recombination of one of many possible variable region sequences with joining segments. These are then combined with a constant region segment. This process occurs in the production of both the light and heavy chains. Then the two chains associate to form the completed antibody. During the course of development there is also a very high mutation rate in the variable region DNA. This, of course, contributes to the diversity in antibody structure. However, even without these mutations, it has been estimated that from about 300 separate genes, about 18 billion possible antibodies could be produced by the recombination of the various segments of protein that make up the antibody.

FUNCTION OF THE THYMUS

Present in all vertebrates, the thymus gland is thought to have at least two functions. First, in some unknown way, the thymus confers immunological competence upon T lymphocytes. Within the thymus these cells develop the ability to differentiate into cells that can respond to specific antigens. This "instruction" within the thymus is thought to take place just before birth and during the first few months of postnatal life. When the thymus is removed from an experimental animal before this processing takes place, the animal is not able to develop cellular immunity. If the thymus is removed after that time, cellular immunity is not seriously impaired.

The second function of the thymus is that of an endocrine gland. It secretes several hormones, including one known as **thymosin.** Though not much is known about these hormones, thymosin is thought to affect T cells after they leave the thymus, stimulating them to complete their differentiation and to become immunologically active. Thymosin has been used clinically in patients who have poorly developed thymus glands. It is also being tested as a modifier of biological response in patients with certain types of cancer. By stimulating their cellular immunity, it may help prevent the spread of the disease.

Active and Passive Immunity

We have been discussing **active immunity,** immunity developed from exposure to antigens. If you had measles as a young child, you developed memory cells and immunity that have kept you from contracting measles again. Active immunity can be *naturally* or *artificially induced* (Table 25–1). If someone with measles sneezes in your direction and you contract the disease, you develop the immunity naturally. However, such immunity can also be artificially induced by **immunization,** that is, by injection of a vaccine. In this case, the body launches an immune response against the antigens contained in the measles vaccine and develops memory cells so that future encounters with the same pathogen will be dealt with swiftly.

Focus on . . .

HOW THE BODY DEFENDS ITSELF AGAINST CANCER

Some immunologists think that cancer cells form every day in each of us in response to viruses, hormones, radiation, or carcinogens in the environment. Because they are abnormal cells, some of their surface proteins are different from those of normal body cells. Such proteins act as antigens, stimulating an immune response. According to the **theory of immunosurveillance,** the body's immune system destroys these abnormal cells whenever they arise. Only when the mechanism fails do these abnormal cells divide rapidly, causing cancer.

Killer T cells and macrophages attack cancer cells (see figure). Another cell type, the **natural killer cell,** is now thought to be important in destroying cancer cells. These natural killer cells are capable of killing tumor cells or virally infected cells upon first exposure to the foreign antigen. Patients with advanced cancer are thought to have less natural killer cell activity than normal people.

What prevents killer T cells, macrophages, and natural killer cells from effectively destroying cancer cells in some people? In some cases the immune system cells may fail to recognize the cancer cells as foreign. In other cases they may recognize them but be unable to destroy them. Sometimes the presence of cancer cells stimulates B cells to produce IgG antibodies that combine with antigens on the surfaces of the cancer cells. These **blocking antibodies** may block the T cells so that they are unable to adhere to the surface of the

cancer cells and destroy them. For some unknown reason the blocking antibodies are not able to turn on the complement system, which would destroy the cancer cells.

An exciting new approach in cancer research involves the production of **monoclonal antibodies.** In this procedure, mice are injected with antigens from human cancer cells. After the mice have produced antibodies to the cancer cells, their spleens are removed and cells containing the antibodies are extracted from this tissue. These cells are fused with cancer cells from other mice. Because of the apparently unlimited ability of cancer cells to divide, these fused hybrid cells will continue to divide indefinitely. Researchers select hybrid cells that are manufacturing the antibodies needed and then clone them in a separate tissue culture. Cells of this clone produce large amounts of the specific antibodies needed; hence the name monoclonal antibodies.

Such antibodies can be injected into the very cancer patients whose cancer cells were used to stimulate their production, and they are highly specific for destroying the cancer cells. (Monoclonal antibodies specific for a single antigenic determinant can now be produced.) These antibodies could also be tagged with toxic drugs, which would theoretically be delivered specifically to the cancer cells.

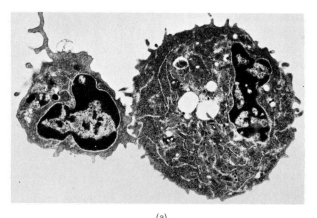
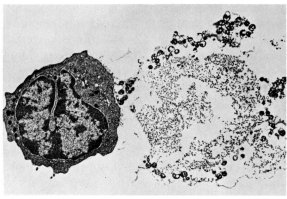

(a) (b)

(a) An electron micrograph of a killer T cell (the smaller of the two cells) approaching a cancer cell. (b) After about two hours of contact, a killer T cell has destroyed a cancer cell. (Courtesy of Prof. Daniel Zagury.)

Effective vaccines can be prepared in a number of ways. A virus may be weakened (attenuated) by successively passaging it through cells of nonhuman hosts. Mutations occur that adapt the pathogen to the nonhuman host and it can no longer cause disease in humans. This is how polio, smallpox, and measles vaccines are produced. Whooping cough and typhoid fever vaccines are made from killed pathogens that still have the necessary antigens to stimulate an immune response. Tetanus and

Table 25–1
ACTIVE AND PASSIVE IMMUNITY

Type of immunity	When developed	Memory cells	Duration of immunity
Active			
Naturally induced	Pathogens enter the body through natural encounter, e.g., infected person sneezes	Yes	Many years
Artificially induced	After immunization	Yes	Many years
Passive			
Naturally induced	After transfer of antibodies from mother to developing baby	No	Few months
Artificially induced	After injection with gamma globulin	No	Few months

botulism vaccines are made from toxins secreted by the pathogens. The toxin is altered so that it can no longer destroy tissues, but its antigens are still intact. When any of these vaccines is introduced into the body, the immune system actively develops clones, produces antibodies, and develops memory cells.

In **passive immunity** an individual is given antibodies actively produced by another organism. The serum or gamma globulin containing these antibodies can be obtained from humans or animals. Animal sera are less desirable because nonhuman proteins can give rise to an immune response that may cause clinical illness (serum sickness).

Passive immunity is borrowed immunity, and its effects are not lasting. It is used to boost the body's defense temporarily against a particular disease. For example, during the Vietnam War, in areas where hepatitis was widespread, soldiers were injected with gamma globulin containing antibodies to the hepatitis pathogen. Such injections of gamma globulin offer protection for only a few months. Because the body has not actively launched an immune response, it has no memory cells and cannot produce antibodies to the pathogen. Once the injected antibodies wear out, the immunity disappears.

Pregnant women confer natural passive immunity upon their developing babies by manufacturing antibodies for them. These maternal antibodies (of the IgG class) pass through the placenta (the organ of exchange between mother and developing child) and provide the fetus and newborn infant with a defense system until its own immune system matures. Babies who are breast-fed continue to receive immunoglobulins (particularly IgA) in their milk. These immunoglobulins provide considerable immunity to the pathogens responsible for gastrointestinal infection, and perhaps to other pathogens as well.

Hypersensitivity

The immune system normally functions to defend the body against pathogens and to preserve homeostasis, but sometimes the system malfunctions. **Hypersensitivity** is a state of altered immune response that is harmful to the body. Two familiar types of hypersensitivity are allergic reactions and autoimmune disease.

ALLERGIC REACTIONS

About 15% of the U.S. population is plagued by a major allergic disorder, such as allergic asthma or hay fever. There appears to be an inherited tendency to these disorders. Allergic people have a tendency to manufacture antibodies against mild antigens, called **allergens,** which do not stimulate a response in nonallergic individuals. In many kinds of allergic reaction, distinctive IgE immunoglobulins called **reagins** are produced.

Let us examine a common allergic reaction—a hay fever response to ragweed pollen (Fig. 25–8). When an allergic person inhales the microscopic pollen, allergens stimulate the release of reagin from sensitized plasma cells in the nasal passages. The reagin attaches to receptors on the membranes of mast cells, large connective tissue cells filled with distinctive granules. Each mast cell has thousands of receptors to which the reagin molecules may attach. Each reagin molecule attaches to a mast cell receptor by its constant end, leaving the variable end of the reagin free to combine with the ragweed pollen allergen.

When the allergen combines with IgE antibody, the mast cell secretes histamine, serotonin, and other chemicals that cause inflammation. These substances cause dilation of blood vessels, and increased capillary permeability, leading to edema and redness. Such physiological responses cause the victims' nasal passages to become swollen and irritated. Their noses run, they sneeze, their eyes water, and they feel generally uncomfortable.

Certain foods or drugs act as allergens in some persons, causing the swollen red welts known as hives. Here the allergen-reagin reaction takes place in the skin, with the histamine released by mast cells causing the hives. In **allergic asthma** an allergen-reagin response occurs in the bronchioles of the lungs. Mast cells release SRS-A (slow-reacting substance), which causes smooth muscle to constrict. The airways in the lungs can constrict for several hours, making breathing difficult.

Systemic anaphylaxis is a dangerous kind of allergic reaction that can occur when a person develops an allergy to a specific drug, such as

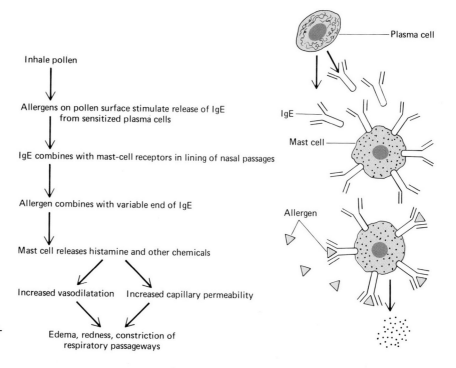

Inhale pollen

↓

Allergens on pollen surface stimulate release of IgE from sensitized plasma cells

↓

IgE combines with mast-cell receptors in lining of nasal passages

↓

Allergen combines with variable end of IgE

↓

Mast cell releases histamine and other chemicals

↙ ↘

Increased vasodilatation   Increased capillary permeability

↘ ↙

Edema, redness, constriction of respiratory passageways

Plasma cell

IgE

Mast cell

Allergen

Figure 25–8 A common type of allergic response.

penicillin, or to insect poison, such as that in a bee sting. Within minutes after the substance enters the body, a widespread allergic reaction takes place. Large amounts of histamine are released into the circulation, causing extreme vasodilatation and permeability. So much plasma may be lost from the blood that circulatory shock and death can occur within a few minutes.

The symptoms of allergic reactions are often treated with **antihistamines,** drugs that block the effects of histamines. These drugs compete for the same receptor sites on cells targeted by histamine. When the antihistamine combines with the receptor, it prevents the histamine from combining, and thus prevents its harmful effects. Antihistamines are useful clinically in relieving the symptoms of hives and hay fever. They are not completely effective, however, because besides releasing histamine, mast cells release other substances that cause allergic symptoms.

In serious allergic disorders patients are sometimes given **desensitization therapy.** Very small amounts of the antigen to which they are allergic are either injected or administered in the form of drops daily over a period of months or years. This stimulates production of IgG antibodies against the antigen. When the patient encounters the allergen, the IgG immunoglobulins combine with the allergen, blocking its receptors so that the IgE cannot combine with it. In this way a less harmful immune response is substituted for the allergic reaction.

AUTOIMMUNE DISEASE

Sometimes regulatory mechanisms malfunction and the body reacts immunologically against its own tissues, causing an **autoimmune disease.** Some of the diseases that result from such failures in self-tolerance are rheumatoid arthritis, multiple sclerosis, myasthenia gravis, systemic lupus erythematosus (SLE), and perhaps infectious mononucleosis.

Rebuilding the Body—Transplanting Organs

Hundreds of kidneys, hearts, and other organs have been transplanted from human donors to recipients during the past several years, but only with limited success. Unfortunately, organ and tissue transplants bring with them a host of antigens that the immune system views as foreign.

GRAFT REJECTION

Skin can be successfully transplanted from one part of the body to another. However, when skin is taken from one individual and grafted onto the body of another, the graft is rejected and it sloughs off. Why?

Each of us has several groups of antigens called **histocompatibility antigens.** In humans the most important group seems to be the **HLA** (human leukocyte antigen) **group,** which is determined by five different linked genes. These genes are multiple alleles at each locus. Tissues from the same individual or from identical twins have the same histocompatibility alleles and thus the same histocompatibility antigens. Such tissues are compatible. Tissue transplanted from one location to another in the same individual is called an **autograft.**

Because the histocompatibility genes are multiple alleles, it is difficult to find identical matches among strangers. If a tissue or organ is taken from a donor and transplanted to the body of an unrelated host, several of the HLA antigens are likely to be different. Such a graft made between members of the same species but of different genetic makeup is called a **homograft.** The host's immune system regards the graft as for-

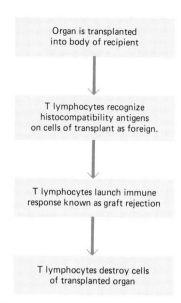

Figure 25–9 Graft rejection.

eign and launches an effective immune response called **graft rejection.** T lymphocytes attack the transplanted tissue and can destroy it within a few days (Fig. 25–9).

Before transplants are performed, tissues from the patient and from potential donors must be typed and matched as well as possible. However, not all of the HLA antigens can be readily typed by serological means. (Sera typing is somewhat similar to blood typing but is more complex, in part because there are more possible alleles.) In fact, one of the loci takes about five days to identify, since an in vitro lymphocyte proliferation test must be performed. Therefore the results of a tissue match may not be known until after the organ has been transplanted. The information is still useful, however, because it gives the physician an idea of how serious the graft rejection may be and how to treat it. If all five of the HLA group of antigens are matched, the graft has about a 95% chance of surviving the first year.

Perfect matches are hard to come by, since not many people are fortunate enough to have an identical twin to supply spare parts. Furthermore, some parts, such as the heart, cannot be spared. Most transplanted organs, therefore, are removed from unrelated donors, often from dying patients or from those who have just died.

To try to prevent graft rejection in less compatible matches, drugs and x-rays are used to kill T lymphocytes. These methods do not kill T lymphocytes selectively, however, so all types of lymphocytes are indiscriminately destroyed. Unfortunately, lymphocyte destruction suppresses not only graft rejection but other immune responses as well, so many transplant patients succumb to pneumonia or other infections. In immunosuppressed patients there is also an increased incidence of certain types of tumor growths. An antibiotic called cyclosporin A appears to suppress T cells that have been activated by antigens on the graft, but it has little effect on B cells. Now being used experimentally, this drug may prove very valuable.

NEW PARTS FOR OLD

With the increasing use of organ transplants comes a variety of medical, social, and ethical problems. A few structures, such as pacemakers, valves, and joints, have been constructed, but organs of flesh are usually required. Many researchers have been trying to devise artificial organs, particularly hearts. However, although the heart is one of the less complex organs (since its function is restricted to pumping blood), even the heart has not proved easy to duplicate. For one thing, blood tends to clot and red blood cells tend to break when they come in contact with anything but a genuine blood-vessel wall. Most organs are complex almost beyond imagination, and researchers cannot even begin to replace them with organs fashioned from nonliving materials.

Another possible solution to the spare-part problem for human beings might be to use healthy organs from other animals. But use of animal parts has not been very successful because animal tissues contain many foreign proteins that stimulate graft rejection. Until the problem of graft rejection has been solved, the best source for spare human organs is other human organs.

Unfortunately, many patients wait for months until a suitable donor is found. Kidneys can be spared, since we are equipped with two and can survive with only one. However, taking organs from living donors presents a variety of moral and psychological problems. A patient in need of a new heart must, of course, wait for a suitable donor to die, and often there are legal complications, as well as the unpleasant task of dealing with mourning relatives. To help solve some of these latter problems the

Focus on . . .

AIDS

The acquired immunodeficiency syndrome (AIDS) is a disorder which results in irreversible defects in cellular immunity. Victims of AIDS have a deficiency of T lymphocytes. They also have an abnormally high ratio of suppressor T cells to helper T cells. (Suppressor T cells inhibit secretion of harmful substances by killer T cells and inhibit the development of B cells into antibody-secreting plasma cells. Helper T cells help trigger B cells to make antibodies.) The ability to resist infection is severely depressed, and AIDS victims die of pneumonia, cancer, or other infections.

AIDS is thought to result from infection with a retrovirus. Current evidence indicates that the disease is transmitted by body secretions during very intimate contact or by direct exposure to blood. AIDS is not spread by casual contact. Those most at risk are homosexual and bisexual men (about 75% of cases) and in-

travenous drug users (20% of cases). Very effective blood-screening procedures have been developed to safeguard blood bank supplies so that hemophiliacs and others requiring blood transfusions are no longer at risk.

Research laboratories throughout the world are searching for drugs that will successfully combat the AIDS virus. Some drugs being tested strengthen the immune system of AIDS victims; others are antiviral drugs. One antiviral drug (azidothymidine, or AZT) currently being tested blocks the action of reverse transcriptase, the enzyme needed by the retrovirus for incorporation into the host cell's DNA. Because the AIDS virus often infects the central nervous system, an effective antiviral drug must cross the blood-brain barrier. Current research is also directed at developing a vaccine against this virus.

Uniform Anatomical Gift Act was passed in 1973 allowing adults to assign their organs for later use as transplants; many people now carry cards in their wallets indicating that in the event of their death they want to donate their organs to those in need.

IMMUNOLOGICALLY PRIVILEGED SITES

Certain locations in the body are said to be **immunologically privileged.** This means that foreign tissue placed at those sites is not subjected to immunologic attack. Cornea transplants, for example, are highly successful: Since there are almost no blood or lymphatic vessels associated with the cornea, it is out of reach of lymphocytes. Also, there is little chance that the antigens in a corneal graft would find their way to a blood vessel and reach the lymph system.

The uterus appears to be another immunologically privileged site. There the human fetus is able to develop its own biochemical identity in safety. Just how this is arranged is not known.

CHAPTER SUMMARY

I. Immune responses depend upon the ability of an organism to distinguish between self and nonself.
II. Most invertebrates are capable only of nonspecific immune responses, such as phagocytosis and the inflammatory response.
III. All vertebrates can launch both nonspecific and specific immune responses.
   A. Nonspecific defense mechanisms that prevent entrance of pathogens include the skin, acid secretions in the stomach, and the mucous lining of the respiratory passageways.
   B. Should pathogens succeed in breaking through the first line of defense, other nonspecific defense mechanisms, including inflammation and phagocytosis, are activated to destroy the invading pathogens.
   C. Specific immune responses include cell-mediated immunity and antibody-mediated immunity.
      1. In cell-mediated immunity, specific T lymphocytes are acti-

vated by the presence of specific antigens. Activated T lymphocytes multiply, giving rise to a clone of cells.
   a. Some T cells differentiate to become killer T cells. Some of these then migrate to the site of infection and destroy pathogens.
   b. Some sensitized T cells remain in the lymph nodes as memory cells; others become helper T cells or suppressor T cells.
   2. In antibody-mediated immunity, specific B lymphocytes are activated by the presence of specific antigens. Activated B lymphocytes multiply, giving rise to clones of cells that differentiate to become plasma cells. These plasma cells secrete specific antibodies.
   a. Antibodies, also called immunoglobulins, are highly specific proteins produced in response to specific antigens. They are grouped in five classes according to their structure.
   b. Antibodies diffuse into the lymph and are transported to the site of infection by lymph and blood.
   c. An antibody combines with a specific antigen to form an antigen-antibody complex that may inactivate the pathogen, stimulate phagocytosis, or activate the complement system. The complement system increases the inflammatory response and phagocytosis. Some complement proteins digest portions of the pathogen cell.
D. Active immunity develops as a result of exposure to antigens. It may occur naturally or may be artificially induced by immunization. Passive immunity develops when an individual is injected with antibodies produced from another person or animal. Passive immunity is temporary.
E. According to the theory of immunosurveillance, the immune system destroys abnormal cells whenever they arise. Such diseases as cancer develop when this immune mechanism fails to operate effectively.
F. Hypersensitivity is a state of altered immune response that is harmful to the body.
   1. In an allergic response, an allergen can stimulate production of IgE antibody, which combines with the receptors on mast cells. When the allergen combines with the antibody, the mast cells release histamine and other substances, causing inflammation and other symptoms of allergy.
   2. In autoimmune diseases the body reacts immunologically against its own tissues.
G. Transplanted tissues possess antigens that stimulate graft rejection, an immune response launched mainly by T cells, which destroys the transplant.

Post-Test

1. Substances capable of stimulating an immune response are called _____ ; specific proteins produced by the body during an immune response are called _____ .
2. On their way to the lymph tissues, T lymphocytes are processed in the _____ .
3. Enzymes that destroy foreign cells directly are released by _____ _____ .
4. Differentiated B lymphocytes are known as _____ _____ _____ .
5. In _____ _____ an individual is given antibodies actively produced by another organism.

Select the most appropriate answer from column B for each item in column A.

Column A

_____ 6. Become antigenic by combining with a larger molecule

_____ 7. Group of proteins in plasma activated by antigen-antibody complex

_____ 8. Nonspecific defense mechanism

_____ 9. Gives specific configuration to molecule that can be recognized by an antibody

_____ 10. Affects T cells, perhaps stimulating them to complete their differentiation

Column B

a. antigenic determinant
b. thymosin
c. haptens
d. complement
e. inflammation

11. The process by which neutrophils and macrophages engulf pathogens is termed _____ .

12. _____ _____ is a type of allergic response in which a great deal of plasma is lost from the blood causing circulatory shock.

13. A homograft is one made between _____ _____ .

14. The immune response directed at transplanted tissues is _____ _____ .

15. An example of an immunologically privileged site is the _____ .

Review Questions

1. How does the body distinguish between self and nonself? Are invertebrates capable of making this distinction?
2. Contrast specific and nonspecific defense mechanisms. Which type confronts invading pathogens immediately?
3. How does inflammation help to restore homeostasis?
4. Give two specific ways in which cell-mediated and antibody-mediated immune responses are similar, and give three ways in which they are different.
5. Describe three ways in which antibodies work to destroy pathogens.
6. John is immunized against measles. Jack contracts measles from a playmate in nursery school before his mother gets around to having him immunized. Compare the immune responses in each child. Five years later, John and Jack are playing together when Judy, who is coming down with measles, sneezes on both of them. Compare the responses in Jack and John.
7. Why is passive immunity temporary?
8. What is immunological tolerance?
9. What is graft rejection? What is the immunological basis for it?
10. List the immunological events that take place in a common type of allergic reaction, such as hay fever.
11. Explain the theory of immunosurveillance. What happens when immunosurveillance fails?
12. What is an autoimmune disease? Give two examples.

Readings

Bolotin, Carol. "Drug as Hero," *Science 85,* June 1985, 68–71. The story of cyclosporin, a drug that selectively inhibits the rejection of transplanted organs.

Buisseret, P.D. "Allergy," *Scientific American,* August 1982, 86–95. A discussion of the cellular and biochemical changes that occur during an allergic response.

Edelson, R.L., and J.M. Fink. "The immunologic function of skin," *Scientific American,* Vol. 252, No. 6, June 1985, 46–53. Specialized cells in the skin play interacting roles in the response to foreign invaders.

Laurence, J. "The Immune System in AIDS," *Scientific American,* Vol. 253, No. 6, December 1985, 84–93. An excellent account of the mechanisms by which the AIDS virus alters the immune system.

Lerner, R.A. "Synthetic Vaccines," *Scientific American,* Vol. 48, No. 2, February 1983, 66–74. A report on experiments on the preparation of synthetic vaccines.

Marx, J.L. "Antibodies: Getting Their Genes Together," *Science,* 212: 1015–1017 (29 May 1981). A brief review of studies on the rearrangement of genes to provide diverse antibodies.

Marx, J.L. "Monoclonal Antibodies in Cancer," *Science,* 216: 283–285 (16 April 1982). An interesting account of the clinical use of monoclonal antibodies.

Chapter 26
GAS EXCHANGE

Outline

Learning Objectives

After you have studied this chapter you should be able to:

1. Compare various adaptations for gas exchange, including tracheal tubes, gills, lungs, and the body surface.
2. Compare air and water as sources of oxygen.
3. Trace the route traveled by a breath of air through the human respiratory system from nose to air sacs and, finally, to recipient cells.
4. Describe breathing and its regulation.
5. Compare the composition of exhaled air with that of inhaled air, and describe the exchange of oxygen and carbon dioxide in the lungs and tissues.
6. Summarize the mechanisms by which oxygen and carbon dioxide are transported in the blood.
7. Describe the effects of breathing polluted air on the respiratory system.

Gas exchange between the organism and the environment is known as **respiration.** During **organismic respiration** oxygen from the environment is taken up by the organism and delivered to its individual cells, and carbon dioxide is excreted into the environment. **Cellular respiration** is the complex series of reactions (discussed in Chapter 8) generally requiring oxygen, by which cells break down fuel molecules, releasing carbon dioxide and energy.

ADAPTATIONS FOR GAS EXCHANGE

Gases move in and out of cells by diffusion. The air or water supplying the oxygen must be continuously renewed so that as soon as oxygen is used up, more will be available. For this reason animals carry on **ventilation,** that is, they actively move their air or water supply over their cells or over the surfaces of specialized respiratory structures. Sponges do this by setting up a current of water through the channels of their bodies by means of flagella; most fish gulp water; we breathe air.

Gas exchange is a fairly simple process in small, aquatic organisms, such as sponges, hydras, and flatworms. Dissolved oxygen from the surrounding water diffuses into the cells, and carbon dioxide diffuses out of the cells and into the water. In large, complex animals, cells deep within the body cannot efficiently exchange gases directly with the environment because oxygen cannot diffuse rapidly enough through tissues to reach all the cells. Specialized respiratory structures are required.

Specialized respiratory structures must have thin walls so that diffusion can easily occur. They must be kept moist so that oxygen and carbon dioxide can be dissolved in water. And they are generally richly supplied with blood vessels to ensure transport of respiratory gases. Four main types of respiratory structures used by animals are the body surface, tracheal tubes, gills, and lungs.

The Body Surface

All gas exchange occurs through the entire body surface in some animals, including nudibranch mollusks, most annelids, small arthropods, and a few vertebrates (Fig. 26–1). Not only are such animals usually small, but they also have a relatively low metabolic rate, so only small quantities of oxygen are needed. An earthworm, for example, has gland cells in the epidermis that secrete mucus, which keeps the body surface moist. Oxygen from tiny air pockets in the loose soil that the earthworm inhabits dissolves in the mucus and then diffuses through the body wall. The oxygen diffuses into blood circulating in a network of capillaries just beneath the outer cell layer.

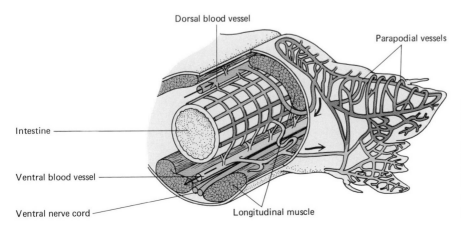

Dorsal blood vessel

Parapodial vessels

Intestine

Ventral blood vessel

Ventral nerve cord

Longitudinal muscle

Figure 26–1 Gas exchange across the body surface. Vascular system within a segment of the clam worm, *Nereis virens.* Arrows indicate the direction of blood flow. The limb-like parapodium acts as an extension of the body wall in gas exchange with the surrounding water. (From Barnes after Nicoll.)

(a)

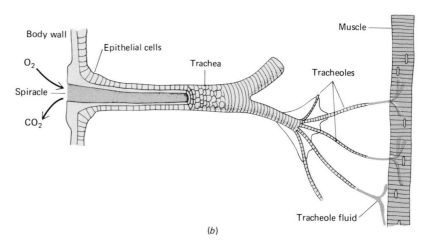

(b)

Figure 26–2 Tracheal tubes. (*a*) A scanning electron micrograph of a mole cricket trachea (approximately ×1300). The corrugations are a very long spiral tube, which may strengthen the tracheal wall somewhat as does the spring within the plastic hoses of many vacuum cleaners and hair dryers. The tracheal wall is composed of chitin. (*b*) Each tracheal tube and its branches conduct oxygen to the body cells of the insect. (*a*, courtesy of Dr. James L. Nation and *Stain Technology*, vol. 58, 1983.)

Tracheal Tubes

In insects and some other arthropods and even in some snails the respiratory system consists of a network of **tracheal tubes.** Air enters the tracheal tubes through a series of tiny openings called **spiracles** along the body surface (Fig. 26–2). The maximum number of spiracles in an insect is 20—two thoracic pairs and eight abdominal pairs—but the number and position vary in different species. In large or active insects air moves in and out of the spiracles by movements of the body or by rhythmic movements of the tracheal tubes. For example, the grasshopper draws air into its body through the first four pairs of spiracles when the abdomen expands. Then the abdomen contracts, forcing air out through the last six pairs of spiracles.

Once inside the body, the air passes through the branching tracheal tubes, which extend to all parts of the body. Gas exchange by diffusion is thought to occur throughout the tracheal system. The tracheal tubes terminate in microscopic, fluid-filled tracheoles. Gases are also exchanged between the body cells and these tracheoles, some of which actually penetrate the cells. (See the figure in Focus on Insect Flight Muscle, Chapter 21.)

Gills

Gills are respiratory structures found mainly in aquatic animals. They are moist, thin structures that extend from the body surface. In many animals the outer surface of the gills is exposed to water, and the inner side is in close contact with networks of blood vessels.

Sea stars and sea urchins have simple **dermal gills,** which project from the body wall. Cilia of the epidermal cells ventilate the gills by beating a stream of water over them. Gases are exchanged through the gills between the water and the coelomic fluid inside the body.

Mollusks' gills are folded, providing a large surface for respiration. In bivalve mollusks and in simple chordates, the gills are adapted for trapping and sorting food. Rhythmic beating of cilia draws water over the gill area, and food is filtered out of the water as gases are exchanged. In mollusks gas exchange also takes place through the mantle.

In chordates the gills are usually internal (Fig. 26–3). A series of slits perforate the pharynx, and the gills are located along the edges of these gill slits. In bony fish, the fragile gills are protected by an external bony plate, the **operculum.** Movements of the operculum help to pump oxygenated water in through the mouth. The water flows over the gills and then leaves via the gill slits.

Each gill in the bony fish consists of many **filaments,** which provide an extensive surface for gas exchange. A capillary network delivers blood to the gill membranes, facilitating diffusion of oxygen and carbon dioxide between blood and water. The direction of blood flow increases the efficiency of this system. Blood flows in a direction opposite to that of the water. This arrangement is referred to as a **countercurrent exchange system.** Water entering the gills is oxygen-rich. Blood, on the other hand, is oxygen-poor when it enters the gill circulation. Blood entering the gills picks up some oxygen from the water passing over the gills, but it can hold more oxygen. As it leaves the gills, the blood comes in contact with oxygen-rich water entering the gills, and diffusion of additional oxygen into the blood further saturates the blood with this needed gas. As the water passes over the gill surface, it contains less and less oxygen. How-

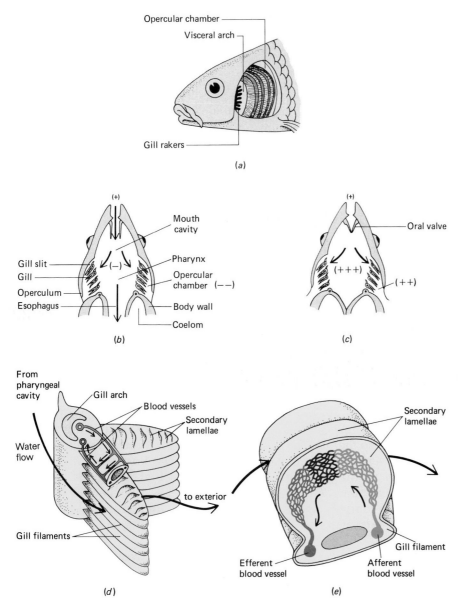

Figure 26–3 Function of the fish gill. (*a*) A fish's gills are located under a plate, the operculum, which has been removed in this side view. They occupy the opercular chamber, and form the lateral wall of the pharyngeal cavity. (*b*) The fish respires by movements of its mouth and pharyngeal cavity. Initially, in inhalation, it enlarges its pharyngeal cavity, seen here from the top. This produces a negative (−) pressure in the pharyngeal cavity and water enters through the mouth. (*c*) When the fish exhales, it closes the mouth and the oral valve behind the lips so that water is forced between the gills and out. Muscular action produces the high positive (+++) pressure in the pharyngeal cavity that is necessary to force the water between the gills. (*d*) Each gill consists of a cartilaginous gill arch to which two rows of leaflike gill filaments are attached. Blood circulates through the gill filaments as water passes among them. (*e*) Each gill filament has many even smaller extensions called secondary lamellae. These contain capillaries full of initially deoxygenated blood. The blood flows through the capillaries in a direction *opposite* to that taken by the water. In this example of countercurrent exchange, the blood is charged with oxygen very efficiently.

ever, the water is coming in contact with blood containing less and less oxygen, so oxygen continues to diffuse into the blood. This very effective mechanism ensures that a great deal of the oxygen in the water will diffuse into the blood.

Oxygen and carbon dioxide do not interfere with one another's diffusion, and they diffuse in opposite directions at the same time. This is because oxygen is more concentrated outside the gills than within, but carbon dioxide is more concentrated inside the gills than outside. Thus the same countercurrent exchange mechanism that ensures efficient influx of oxygen works in reverse fashion to ensure equally efficient outgo of carbon dioxide.

Lungs

Lungs are broadly defined as respiratory structures that develop as ingrowths of the body surface or from the wall of a body cavity, such as the pharynx. Arachnids and some small mollusks (particularly terrestrial and some amphibious snails and slugs) have lungs that depend almost entirely on diffusion for gas exchange. The **book lungs** of some spiders are enclosed in an inpocketing of the abdominal wall. These lungs consist of a series of parallel, thin plates filled with blood. The plates are separated by air spaces that are connected to the outside environment through a spiracle. Larger mollusks and vertebrates with lungs have some means of forcefully moving air across the lung surface, that is, of ventilating the lung.

Not all fish breathe exclusively by gills. Although modern fish lack lungs, most do possess homologous **swim bladders.** These are hydrostatic organs that may also store oxygen. In some fish the swim bladder, which is connected to the pharynx, can serve as an accessory respiratory organ. When the oxygen level in the pond or lake is low the fish may come to the surface and gulp air into its pharynx, which then enters the swim bladder. From there, oxygen passes into the blood circulating through the swim bladder wall.

The lungs of salamanders are not much more elaborate than swim bladders (Fig. 26–4). They are two long, simple sacs, covered on the outside by capillaries. Frogs and toads have ridges containing connective tissue on the inside of the lung, which increase the respiratory surface somewhat. Though there are exceptions, the lungs of modern reptiles are not much more complex than those of amphibians. In some lizards and in turtles and crocodiles, the lungs do have many subdivisions that increase the surface area for gas exchange.

Birds are very active animals with high metabolic rates. They require large amounts of oxygen to sustain their activities, and they have highly effective respiratory systems. In them the lungs have developed several extensions (usually nine) called **air sacs,** which reach into all parts of the body, and even penetrate into some of the bones, where their main function may be simply to lighten the body. The respiratory system is arranged so that there is a one-way flow of air through the lungs, and the air is renewed during each inspiration. The lungs have tiny, thin-walled ducts, the **parabronchi,** which are open at both ends. Gas exchange takes place across the walls of these ducts. The direction of blood flow in the lungs is opposite that of air flow through the parabronchi. This countercurrent flow increases the amount of oxygen that enters the lung.

When the bird is at rest, a forward and upward movement of the ribs and a forward and downward movement of the sternum expand the trunk volume, drawing air into the body. When the bird is flying, the chest wall must be held rigid to form an anchor for the flight muscles. However, air

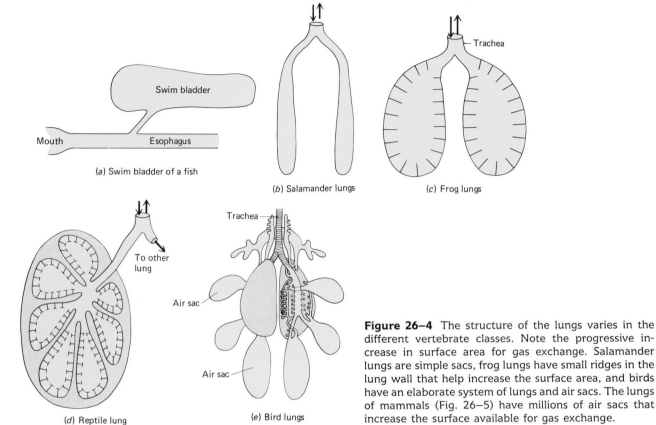

Figure 26-4 The structure of the lungs varies in the different vertebrate classes. Note the progressive increase in surface area for gas exchange. Salamander lungs are simple sacs, frog lungs have small ridges in the lung wall that help increase the surface area, and birds have an elaborate system of lungs and air sacs. The lungs of mammals (Fig. 26-5) have millions of air sacs that increase the surface available for gas exchange.

sacs lying between certain flight muscles are squeezed and relaxed on each stroke of the wing. These act as bellows to move air in and out of the lungs. The faster the bird flies, the more rapidly air circulates through the lungs.

The lungs of mammals are very complex and have an enormous surface area. We will examine the human respiratory system, a typical mammalian respiratory system, in some detail in a later section.

AIR VERSUS WATER

Organisms must be specifically adapted to breathe either air or water. Some respiratory structures, like tracheal tubes and lungs, seem best adapted for gas exchange in air, whereas others, like gills, function best in water. Gas exchange with air is more effective than gas exchange with water because air contains far more oxygen than water, and oxygen diffuses more rapidly through air than through water.

Because water has a much greater density and viscosity (resistance to flow) than air, an animal must expend more energy to ventilate its respiratory surface with water than with air. A fish uses up to 20% of its total energy expenditure to perform the muscular work needed to ventilate its gills. An air breather expends much less energy, only 1% or 2% of the total, to ventilate its lungs. Moreover, air is not salty, so air breathers do not have to cope with the diffusion of ions into their body fluids along with their oxygen. Water carries off body heat much faster than air does. Homeothermic aquatic animals frequently require special insulation to compensate for this, such as thick subcutaneous fat or special air-trapping fur or feathers.

Air breathers must have adaptations to help them avoid desiccation. In addition to having fairly impermeable integuments, air-breathing vertebrates possess lungs located deep within the body, preventing excessive loss of water from the respiratory surface. Air must pass through a long sequence of passageways before reaching the blood-rich, wet respiratory surfaces of the lung, and expired air must again pass through these passageways before leaving the body. The lungs are thus partially protected from the drying effects of air.

THE HUMAN RESPIRATORY SYSTEM

The respiratory system in humans and other air-breathing vertebrates consists for the most part of a series of tubes through which air passes on its journey from the nostrils to the air sacs of the lungs and back. A breath of air enters the body through the nose, flows through the twin compartments of the nasal cavity to the pharynx, through the larynx, and into the trachea, or windpipe (Fig. 26–5). From the trachea the stream of air divides, first into two streams as the trachea branches into the bronchi (one bronchus enters each lung), then into the many bronchioles of the lung. As these, too, divide again and then again, the breath becomes a thousand tiny breezes that finally reach the microscopic air sacs. From

Figure 26–5 The human respiratory system. The paired lungs are located in the thoracic cavity. The muscular diaphragm forms the floor of the thoracic cavity, separating it from the abdominal cavity below. An internal view of one lung illustrates its extensive system of air passageways. The microscopic alveoli are shown in later figures.

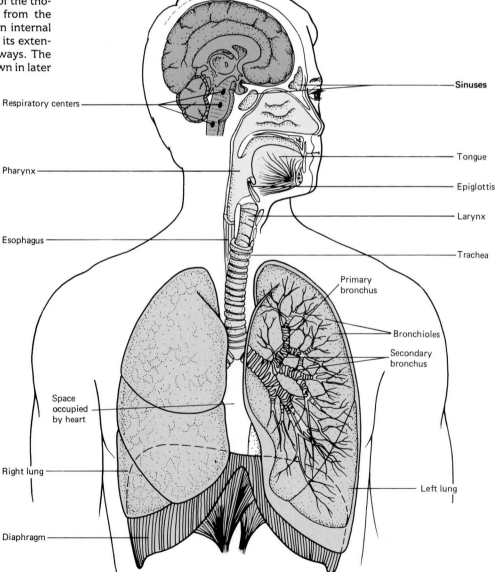

them oxygen diffuses into the blood of the multitude of capillaries enveloping each air sac.

The Respiratory Passageways

Air finds its way into the pharynx whether one breathes through the nose or mouth. Nose breathing is more desirable, however, because as air passes through the nose, it is filtered, moistened, and brought to body temperature. The nostrils are fringed with coarse hair that helps to prevent the entrance of insects and other large foreign matter.

The nostrils open into the **nasal cavities,** which are lined with moist, ciliated epithelium. The lining of the nose has a rich blood supply that warms and humidifies the incoming air. Mucous cells within the epithelium produce more than a pint of mucus a day. Inhaled dirt and other foreign particles are trapped in the layer of mucus and pushed along with the stream of mucus toward the throat by the cilia. In this way foreign particles are delivered to the digestive system, which is far more capable of disposing of such materials than the delicate lower part of the respiratory system. A person normally swallows more than a pint of nasal mucus each day, more if he has an allergy or infection.

The back of the nasal cavities is continuous with the throat region, or **pharynx.** An opening in the floor of the pharynx leads into the **larynx,** sometimes called the "Adam's apple." Because it contains the vocal cords, the larynx is also referred to as the "voice box." Cartilage embedded in its wall prevents the larynx from collapsing and makes it hard to the touch when felt through the neck.

During swallowing, a flap of tissue, the **epiglottis,** automatically closes off the larynx so that neither food nor liquid can enter the lower airway. Should this defense mechanism fail and foreign matter come in contact with the sensitive larynx, a **cough reflex** is initiated to expel the material from the respiratory system.

From the larynx air passes into the **trachea.** Like the larynx, the trachea is kept from collapsing by rings of cartilage in its wall. The trachea divides into two branches, the **bronchi** (singular, **bronchus**), one going to each lung. Both trachea and bronchi are lined by a mucous membrane containing ciliated cells. Many medium-sized particles that have escaped the cleansing mechanisms of nose and larynx are trapped here. Mucus containing these particles is constantly beaten upward by the cilia to the pharynx, where it is periodically swallowed. This mechanism keeps foreign material out of the lungs, functioning as a cilia-propelled mucus elevator.

The Lungs

The lungs are large, paired, spongy organs occupying the thoracic (chest) cavity. The right lung is divided into three lobes, the left lung into two lobes. Each lung is covered with a membrane, the **pleural membrane,** which forms a continuous sac enclosing the lung and continuing as the lining of the chest cavity. The space between the pleural membranes covering the lung and the pleural membrane lining the chest cavity is called the **pleural cavity.** A film of fluid in the pleural cavity provides lubrication between the lungs and the chest wall.

Inside the lungs the bronchi branch into smaller and smaller airways, the **bronchioles.** There are more than a million tiny bronchioles in each lung, and each leads into a cluster of tiny air sacs, the **alveoli** (singular, **alveolus**) (Fig. 26–6). These tiny air sacs are lined by an extremely

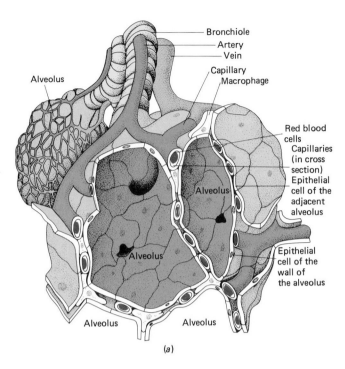

Bronchiole
Artery
Vein
Capillary
Macrophage
Alveolus
Red blood cells
Capillaries (in cross section)
Epithelial cell of the adjacent alveolus
Alveolus
Alveolus
Epithelial cell of the wall of the alveolus
Alveolus
Alveolus

(a)

Figure 26–6 Gas exchange takes place across the thin wall of the alveolus. (a) Structure of the alveolus. Note that the alveolar wall consists of extremely thin squamous epithelium. The alveoli share their walls, and between the walls of the alveoli lie extensive capillary networks. (b) View through the terminal portion of a bronchiole, the smallest respiratory duct, into a cluster of alveoli (approximately ×370). (c) The wall of a single alveolus together with its capillaries (approximately ×2000), which in this scanning electron micrograph somewhat resemble a network of hoses. (d) A cross section through one of these capillaries (approximately ×11,450). The nucleus of the endothelial cell that makes up the capillary wall is visible at the bottom of the photograph. Large, dark structures within the capillary are red blood cells. (e) An enlargement (approximately ×48,100) of a portion of a capillary similar to that shown in (d). The dark structure extending through the capillary is a portion of a red blood cell. The wall of the alveolus is visible just above the wall of the capillary. Notice the very short distance oxygen need travel to get from the air within the alveolus to the red blood cell in which it is transported to the body tissues. (b–e), courtesy of Drs. Peter Gehr, Marianne Bachofen and Ewald R. Wiebel. (b, originally published in *Respiration Physiology*, vol. 32, 1978.)

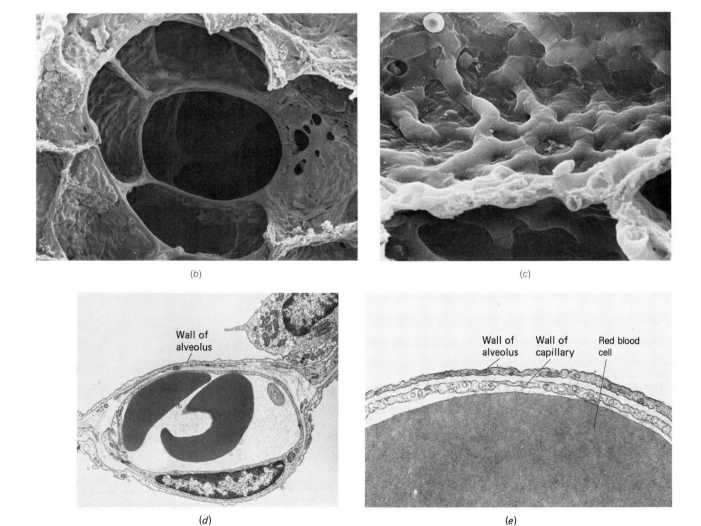

(b)

(c)

Wall of alveolus

(d)

Wall of alveolus   Wall of capillary   Red blood cell

(e)

thin single layer of epithelial cells, through which gases can diffuse freely to the capillary blood carried within their walls. Alveoli are covered by a thin film of lipoprotein, known as surfactant, which prevents them from collapsing so that gases do not have far to travel between air sac and blood. Because the lung consists largely of air tubes and elastic tissue, it is a spongy, elastic organ with a very large surface area for gas exchange. In normal adults the surface area of the lung is estimated as approximately that of a tennis court.

Breathing

Breathing is the mechanical process of moving air from the environment into the lungs, or inspiration, and of expelling air from the lungs, or expiration. A resting adult breathes about 12 times each minute.

MECHANICS OF BREATHING

The chest cavity is closed so that no air enters. It possesses a muscular floor, the **diaphragm,** and is divided into two lateral pleural cavities, each containing a lung, and a central compartment, the **mediastinum,** containing the heart, esophagus, trachea, and other structures.

During inspiration, or breathing in, the chest cavity is expanded by the downward contraction of the diaphragm and by the upward contraction of the rib muscles, which act to increase the circumference of the chest cavity (Fig. 26–7). Since the chest cavity is closed, when it expands, the film of pleural fluid holds the lung surfaces in contact with the chest wall, causing the membranous walls of the lungs to be pulled outward along with the chest walls. This increases the space within each lung. The air in the lungs at first tends to expand to fill this increased space, but then the pressure of the air in the lungs falls below the pressure of the air outside the body. As a result, air from the outside rushes in through the respiratory passageways and fills the lungs until the two pressures are again equal.

Expiration, or breathing out, occurs when the diaphragm and rib muscles relax, allowing the elasticity of the structures in the chest cavity to make it smaller. Pressure increases in the lung, and the pushing of its elastic fibers against the air forces air to rush out of the lung. The millions of tiny alveoli deflate and the lung is ready for another change of air. More active expiration, such as might occur during running, singing, or even speaking, requires the coordinated actions of several sets of muscles, including the diaphragm, the abdominal muscles, and possibly the internal intercostal muscles between the ribs. By pushing on the abdominal or chest contents, all three actively force air out of the lungs.

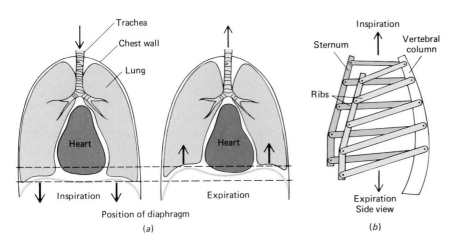

Figure 26–7 The mechanics of breathing. (*a*) Changes in the position of the diaphragm in inspiration and expiration result in changes in the volume of the chest cavity. (*b*) Changes in position of the rib cage in expiration and inspiration. The elevation of the front ends of the ribs by the chest muscles causes an increase in the front-to-back dimension of the chest, and a corresponding increase in the volume of the chest cavity. When the volume of the chest cavity increases, a corresponding amount of air moves into the lungs.

REGULATION OF BREATHING

The amount of oxygen used by the body varies with different levels of activity. When you are engaged in a strenuous basketball game, for example, you require more oxygen than when reading quietly. Breathing is controlled by respiratory centers in the brain that are indirectly sensitive to increases in the amount of carbon dioxide in the blood (see Fig. 26–5). Nerves from the respiratory centers stimulate the contraction of chest muscles. During exercise greater amounts of carbon dioxide are produced. The carbon dioxide stimulates the respiratory centers to produce more rapid and more forceful breathing. In this way sufficient oxygen is provided to meet the body's increased need.

Gas Exchange and Transport

The respiratory system delivers oxygen to the air sacs, but if oxygen were to remain in the lungs, all the other body cells would soon die. The vital link between air sac and body cell is the circulatory system. Each air sac serves as a tiny depot from which oxygen is loaded into blood brought close to the alveolar air by capillaries (Fig. 26–8).

Oxygen molecules diffuse from the air sacs into the blood because the air sacs contain a greater concentration of oxygen than does blood entering the pulmonary capillaries. Similarly, carbon dioxide moves from the blood, where it is more concentrated, to the air sacs, where it is less concentrated. Each gas diffuses through the cells lining the alveoli and the cells lining the capillaries.

Table 26–1 shows the percentages of oxygen and carbon dioxide present in exhaled air compared with inhaled air. Because carbon dioxide is produced during cellular respiration, there is more of this gas—100 times as much—entering the alveoli from the blood than there is in air inhaled from the environment.

The movement of gases between air sacs and blood is not completely efficient. Not every molecule of inhaled oxygen actually finds its way into the blood, and not every molecule of carbon dioxide is removed from the blood. Part of the reason for this is that newly inhaled air must mix with air from the last inhalation, so the carbon dioxide content of alveolar air is always higher than that of the atmosphere, and its oxygen content is lower. The amount of exchange that takes place, however, is obviously sufficient to support the metabolic well-being of the body.

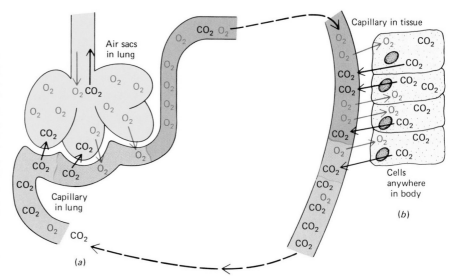

Figure 26–8 Gas exchange. (*a*) Exchange of gases between air sacs and capillary in the lung. The concentration of oxygen is greater in the air sacs than in the capillaries, so oxygen moves from the air sacs into the blood. Carbon dioxide is more concentrated in the blood than in the air sacs, so it moves out of the capillaries and into the air sacs. (*b*) Exchange of gases between capillary and body cells. Here, oxygen is more concentrated in the blood, so it moves out of the capillary and into the cells. Carbon dioxide is more concentrated in the cells, and so it diffuses out of the cells and moves into the blood.

Table 26–1
COMPOSITION OF INHALED AIR COMPARED WITH THAT OF EXHALED AIR

	% oxygen (O_2)	% carbon dioxide (CO_2)	% nitrogen (N_2)
Inhaled air (atmospheric air)	20.9	0.04	79
Exhaled air (alveolar air)	14.0	5.60	79

Note: As indicated, the body uses up about one-third of the inhaled oxygen. The amount of CO_2 increases more than 100-fold because it is produced during cellular respiration.

When oxygen diffuses into the blood, it forms a weak chemical bond with hemoglobin molecules in the red blood cells, forming **oxyhemoglobin.** Each hemoglobin molecule can combine with, and thus transport, four molecules of oxygen. Since the chemical bond formed between the oxygen and the hemoglobin is weak, the reaction is readily reversible. In the pulmonary capillaries:

Hemoglobin + Oxygen $\longrightarrow$ Oxyhemoglobin

When oxyhemoglobin reaches body cells low in oxygen, the reverse reaction occurs:

Oxyhemoglobin $\longrightarrow$ Oxygen + Hemoglobin

The oxygen released diffuses from the capillaries into the cells.

As blood flows through the capillary networks of an organ, carbon dioxide moves out of the cells where it has accumulated and into the blood, where it is less concentrated. Carbon dioxide is transported in the blood in three ways. About 20% is carried on the hemoglobin molecule, 10% is transported in the plasma as carbon dioxide itself, and the remainder is dissolved in the plasma after being converted into **bicarbonate ions** (HCO_3^-), a portion of which is present in the form of carbonic acid.

When it reaches the lungs, carbon dioxide diffuses out of the blood and into the alveoli. Most but by no means all the transported carbon dioxide leaves the blood and eventually the respiratory system. The human body is adjusted to function at an internal pH of about 7.4 that results from a balance of substances in the blood, one of which is a moderate amount of carbon dioxide. If too much is retained because of respiratory insufficiency, the blood pH becomes abnormally low **(acidosis)** due to excess carbonic acid formation. If hyperventilation blows off too much carbon dioxide, an equally undesirable state, **alkalosis,** results from abnormally low concentrations of carbonic acid in the blood.

Physiological adaptation to pressure change takes time. Divers who return to the surface too quickly, or pilots who ascend to over 35,000 feet too rapidly, may suffer from **decompression sickness** ("the bends"). While submerged, a diver breathes gases under high pressure, and because of this pressure, excessive amounts of nitrogen gas dissolve in his blood and tissues. As he ascends to a lower pressure, the dissolved nitrogen is liberated. If he surfaces too rapidly, tiny nitrogen bubbles form in the blood and tissues and may block capillaries and cause other damage. These bubbles cause the symptoms of decompression sickness: pain, paralysis, even death.

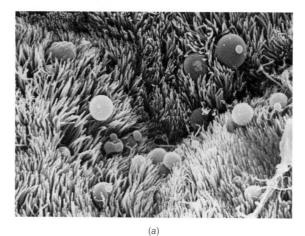

(a)

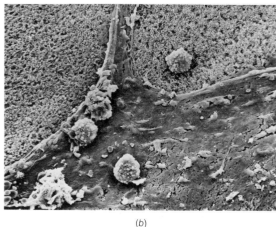

(b)

Figure 26–9 (a) Scanning electron micrograph showing a carpet of cilia in the lining of a small bronchus. Note the globules of mucus that have emerged from goblet (gland) cells in the epithelial lining. (b) View of irritated bronchial lining at a lower magnification. Excess mucus has formed a pool (bottom of photograph) overlaying part of the ciliary carpet. Four macrophages (rounded structures) are present. Macrophages engulf particles and then carry them onto the cilia-mucus elevator, which may sweep them out of the respiratory system. (Courtesy of Dr. R.J. Pack, University of London.)

Effects of Smoking and Air Pollution

A human being breathes about 20,000 times each day, inhaling about 35 pounds of air—six times the amount of food and drink consumed. Most of us breathe dirty urban air laden with particulates, carbon monoxide, sulfur oxides, and other harmful substances that cannot fail to affect the health and function of the respiratory system.

Mucus cells in the epithelial lining of the respiratory passageways react to the irritation of inhaled pollutants by secreting increased amounts of mucus (Fig. 26–9). The ciliated cells, damaged by the pollutants, cannot effectively clear the mucus and trapped particles from the airways. The body attempts to clear the airways by coughing. (Smoker's cough is a well-known example.) The coughing causes increased irritation, and the respiratory system becomes susceptible to viral, bacterial, and fungal infections.

One of the body's fastest responses to inhaling dirty air is **bronchial constriction.** The bronchial tubes constrict, providing increased opportunity for the inhaled particles to be trapped along the sticky mucous lining of the airway. Unfortunately, the narrowed bronchial tubes allow less air to pass through to the lungs, so the oxygen available to the body cells is decreased. People who smoke or breathe heavily polluted air may have chronically constricted bronchioles. Chronic bronchial constriction resulting from unknown causes or from allergy is known as **asthma.** It is greatly aggravated by air pollution, including pollution of indoor air by nearby smokers.

Neither mucus nor ciliated cells are found in the terminal bronchioles or air sacs. Any foreign particles that find their way into the air sacs may remain lodged there indefinitely or be engulfed by macrophages (Fig. 26–9). Often the macrophages make their way to lymph nodes within the lung tissue and accumulate there.

Chronic bronchitis and emphysema are chronic obstructive pulmonary diseases that have been linked to smoking and air pollution. More than 75% of patients with chronic bronchitis have a history of heavy cigarette smoking (see Focus on Smoking Facts). In this disease, the bronchioles secrete too much mucus and are constricted and inflamed. Respiratory cilia cannot clear the passages of the mucus and particles that partly clog them. Affected persons are short of breath and often cough up mucus. Persons with chronic bronchitis often develop emphysema.

The alveoli are damaged in the very serious and irreversible disease **pulmonary emphysema** (Fig. 26–10). In this disorder, by far most common in smokers, neutrophils are attracted in vast numbers to the alveoli

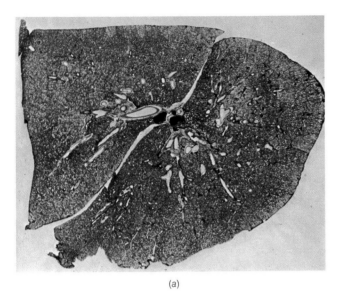

(a)

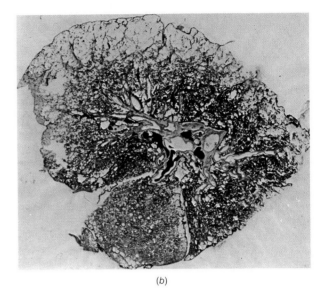

(b)

Figure 26–10 Freeze-dried lungs. (a) A normal lung without emphysema. (b) A lung with advanced emphysema, in which alveoli have been destroyed and have run together. Alveolar walls have also been compressed, hampering circulation. (From Turk, et al: *Environmental Science*, 2nd ed. W. B. Saunders, 1978.)

that have been inflamed by inhaled irritants. The released contents of their lysosomes damages the connective and other tissues of the alveoli, so many of the interalveolar walls are destroyed. Adjacent alveoli enlarge, merge and can lose so much surface area that gas exchange becomes seriously impaired. In most people the damage is limited somewhat by a protein, **antitrypsin,** that interferes with the ability of the lysosomal enzymes to digest protein. In smokers with hereditary deficiencies of this antienzyme, emphysema is much more severe and is much earlier in its onset.

Focus on . . .

SMOKING FACTS

- The life of a 30-year-old who smokes 15 cigarettes a day is shortened by more than five years.
- If you smoke more than one pack per day, you are about 20 times more likely to develop lung cancer than a nonsmoker. According to the American Cancer Society, cigarette smoking causes more than 75% of all lung cancer deaths.
- If you smoke, you are more likely than a nonsmoker to develop atherosclerosis, and you double your chances of dying from cardiovascular disease.
- If you smoke, you are 20 times more likely to develop chronic bronchitis and emphysema.
- If you smoke, you are seven times more likely to develop peptic ulcers (especially malignant ulcers).
- If you smoke, you have about 5% less oxygen circulating in your blood.
- If you smoke when you are pregnant, your baby will weigh about 6 ounces less at birth, and there is double the risk of miscarriage, stillbirth, and infant death.
- Workers who smoke one or more packs of cigarettes per day are absent from their jobs because of illness 33% more often.
- Risks increase with the number of cigarettes smoked, the number and depth of inhalations per cigarette,

and smoking down to a short stub. Cigar and pipe smokers have lower risks than cigarette smokers because they do not inhale as much smoke. Cigarette smokers who switch to cigars and continue to inhale actually increase their risks.
- Nonsmokers confined in living rooms, offices, automobiles, or other places with smokers are adversely affected by the smoke. For example, when parents of infants smoke, the infant has double the risk of contracting pneumonia or bronchitis in its first year of life.
- When smokers quit smoking, their risk of dying from chronic pulmonary disease, cardiovascular disease, or cancer decreases. (Precise changes in risk figures depend upon the number of years the person smoked, the number of cigarettes smoked per day, the age of starting to smoke, and the number of years since quitting.)
- Switching to low-tar, low-nicotine or filter brands does not help as much as one might think. Most smokers tend to compensate by increasing their consumption of cigarettes.
- If everyone in the United States stopped smoking, more than 300,000 lives would be saved each year.

CHAPTER SUMMARY

I. In very small organisms gas exchange depends upon diffusion. Larger animals have specific adaptations to ensure that oxygen reaches each cell of the body. These include the body surface, gills, tracheal tubes, and lungs.

   A. In insects and some other arthropods, the respiratory system consists of a network of tracheal tubes that extend to all parts of the body.

   B. Gills are moist, thin extensions of the body surface that occur mainly in aquatic animals.

   C. Large mollusks and terrestrial vertebrates have lungs, respiratory structures that develop as ingrowths of the body surface or from the wall of the pharynx.

II. Gas exchange in air is more efficient than in water because air contains more oxygen than water, and oxygen diffuses more rapidly through air than through water.

III. In humans respiration is accomplished by a system of air passageways that branch into smaller and smaller tubes, ending finally in the alveoli within the lungs.

   A. In the nose air is filtered, brought to body temperature, and humidified.

   B. From the nasal cavities air passes through the pharynx and into the larynx. The larynx helps prevent the entrance of foreign material into the lungs by initiating a cough reflex when touched by foreign matter.

   C. From the larynx inhaled air passes into the trachea and then into the right or left bronchus.

   D. Within the lungs the bronchi branch into an extensive system of bronchioles, which eventually terminate in the millions of tiny alveoli, through which gas exchange takes place with the blood.

   E. Breathing is the mechanical process of moving air back and forth between the environment and the air sacs of the lungs.

      1. When the diaphragm and rib muscles contract, expanding the chest, air rushes into the lungs. When these muscles relax, pressure in the lung increases and air is expired.

      2. Breathing is normally regulated by respiratory centers in the brain, which are sensitive to the amount of carbon dioxide in the blood.

   F. Oxygen diffuses from the air sacs into the blood and is transported to the body cells in the form of oxyhemoglobin. As oxygen is needed by the cells, the oxyhemoglobin dissociates, and oxygen diffuses from the blood into the cells.

   G. Carbon dioxide is transported in several forms by the blood to the alveoli from which it is expired.

   H. Inhaling polluted air can cause serious damage to the respiratory system.

      1. The cilia mucus elevators in the trachea and bronchi continuously move a stream of mucus, in which harmful particles and microorganisms are trapped, toward the throat.

      2. Bronchial constriction and macrophage action are other defense mechanisms of the lung.

Post-Test

1. Specialized respiratory structures must have _____ walls so that _____ can easily occur; they must be _____ so that gases can be dissolved; and they must be richly supplied with _____

_____ _____ to ensure transport of gases.

2. In insects air enters a network of _____ tubes through openings called _____ .

3. Respiratory structures that develop from the wall of a body cavity, such as the pharynx, or as ingrowths of the body surface are called _____ .

4. In birds the lungs have several extensions referred to as _____ _____ .

5. In the mammalian respiratory system, inhaled air passing through the larynx would next enter the _____ _____ .

6. In the mammalian respiratory system, gas exchange takes place through the thin walls of the _____ _____ .

7. In mammals the floor of the thoracic cavity is formed by the _____ .

Select the most appropriate answer in column B for each description in column A.

Column A   *Column B*

_____ 8. Seals off larynx during   a. sinuses
           swallowing   b. larynx
_____ 9. Cavities in bones of skull   c. alveoli
_____ 10. Initiates cough reflex   d. pleural
_____ 11. Covers lung   membrane
_____ 12. Coated with thin film of   e. epiglottis
            lipoprotein

13. _____ _____ can result when a diver surfaces too rapidly and nitrogen bubbles form in the blood and tissues.

14. Bronchial constriction is one of the body's most rapid responses to _____ .

15. In _____ , the alveolar walls break down so that several air sacs join to form larger, less elastic alveoli.

16. The main cause of lung cancer is _____ _____ .

17. Label the following diagram. For correct labeling, see Figure 26–5.

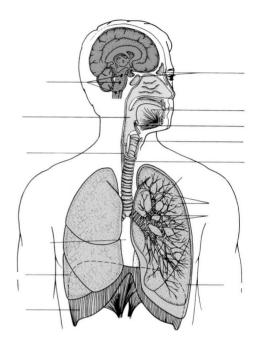

Review Questions

1. What adaptations for gas exchange are found in fish? in insects? in terrestrial vertebrates? Why cannot large animals depend solely upon diffusion for gas exchange?

2. Why are lungs more suited for an air-breathing vertebrate and gills more effective in a fish?

3. Trace a breath of inhaled air from nose to alveoli, listing each structure through which it must pass.

4. What are the advantages of having millions of alveoli rather than a pair of simple, balloon-like lungs?

5. Describe the protective mechanisms of the respiratory system, including the cilia-mucus elevator and bronchial constriction.

6. Describe the processes of inspiration and expiration.

7. How is breathing regulated?

8. What role does diffusion play in gas exchange in humans?

9. How is oxygen transported in the blood of humans?

10. In what way does the composition of inhaled air differ from that of exhaled air? Why?

11. Summarize the health effects of smoking.

12. What is decompression sickness?

Readings

Cholander, P.F. "The Master Switch of Life," *Scientific American,* Dec. 1963, 85–94.

Kolata, G. "Cell Biology Yields Clues to Lung Cancer," *Science,* 218: 38–39 (1 October 1982). By studying lung cancer cells in culture, researchers are gaining insights into how the cells multiply and how they can be prevented from multiplying.

Kooyman, G.L. "The Weddell Seal," *Scientific American,* August 1969, 101–106. An account of the diving behavior of the Weddell seal.

Chapter 27
PROCESSING FOOD

Learning Objectives

After you have read this chapter you should be able to:

1. Describe in general terms the following steps in processing food: ingestion, digestion, absorption, and elimination.
2. Compare adaptations that herbivores, carnivores, and omnivores possess for their particular mode of nutrition.
3. Compare the nutritional life-styles of parasites, commensals, and mutualistic partners.
4. Contrast the digestive systems of cnidarians, flatworms, and earthworms.
5. Identify on a diagram or model each of the structures of the human digestive system described in this chapter, and give the function of each structure.
6. Trace the pathway traveled by an ingested meal, describing each of the changes that takes place en route. (Your instructor may ask you to describe in sequence the step-by-step digestion of carbohydrates, lipids, and proteins.)
7. Describe the protective mechanisms that prevent gastric juice from digesting the stomach wall, and explain what happens when these mechanisms fail.
8. List the functions of the liver and pancreas.
9. Draw and label a diagram of an intestinal villus, and describe its function.

Nutrients are the substances present in food that are needed by an organism as an energy source to run the machinery of the body, as ingredients to make compounds for metabolic processes, and as building blocks to ensure growth and repair of tissues. Obtaining food is of such vital importance that both individual organisms and ecosystems are designed around the central theme of **nutrition,** the process by which an organism takes in and assimilates food. An organism's body plan, as well as its life-style, is adapted to its particular mode of procuring food.

All animals are heterotrophs, organisms that must obtain their energy and nourishment from the organic molecules manufactured by other organisms. Most animals have a **digestive system** that processes the food they eat. Food processing may be divided into several steps: ingestion, digestion, absorption, and elimination.

After foods are selected and obtained, they are **ingested,** that is, taken into the body. In humans and other vertebrates, ingestion includes taking the food into the mouth and swallowing it. Because animals eat the macromolecules tailor-made by and for other organisms, they must break down these molecules and refashion them for their own needs. We cannot incorporate the proteins in steak directly into our own muscles, for example. The body must **digest** the steak, mechanically breaking down the large bites of meat into smaller ones and then enzymatically degrading the proteins into their component amino acids. The amino acids can then be absorbed and transported to the muscle cells, which must arrange these components into human muscle proteins.

In organisms with intracellular digestion, **absorption** is merely the passage of nutrients from the food vacuole into the cytoplasm. In animals equipped with digestive tracts, absorption is the passage of nutrients through the cells lining the digestive tube and into the blood. Nutrients are then transported and distributed to all parts of the body and used for metabolic activities within each cell. Food that is not digested and absorbed is discharged from the body, a process referred to as **egestion** in simple animals and **elimination** in more complex animals.

ALL KINDS OF DINNER JACKETS

Consider how any organism processes food and you will see that it has a built-in dinner jacket. Consider the sharp teeth and claws of the lion, for example, as well as its long, quick legs. These structural adaptations enable it to hunt and kill other animals. Or analyze the hydra's body plan. This little animal is exquisitely designed for obtaining dinner. Its long tentacles, equipped with stinging cells, permit the hydra to capture prey without ever having to move from its perch. The hydra's body is little more than a double-layered digestive sac with a single opening at the top that serves as both mouth and anus.

Adaptations of Herbivores

Some animals are **herbivores,** or **primary consumers,** which eat only plant materials. Terrestrial plants contain a great deal of supporting material, including cell walls made of cellulose. Animals cannot digest cellulose, so it is somewhat of a problem for them to obtain nutrients from the plant material they eat. The adaptations utilized to exploit plant food sources are strikingly varied (Fig. 27–1). Some insects have piercing and sucking mouthparts so that they can pierce through the tough cell walls and suck the sap or nectar within the plant cells. Other herbivores simply eat great quantities of food. Grasshoppers, locusts, elephants, and cattle, for example, all spend a major part of their lives eat-

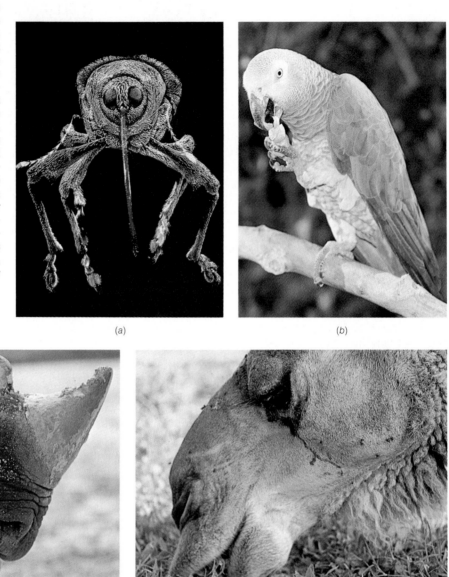

Figure 27–1 Adaptations of herbivores. (*a*) An acorn weevil. The impressively long "snout" of this little beetle is used both for feeding and to make a hole in the acorn through which an egg is deposited. When it has hatched, the larva feeds on the contents of the acorn seed. (*b*) The parrot has a powerful beak for cracking nuts and seeds. Unlike most birds, the parrot can use one foot to manipulate its food and feed itself. (*c*) The rhinoceros can use its horn to uproot and overturn small trees and bushes; it then eats the leaves. Members of some species use their lips to break off grass. (*d*) Camels live on seeds, dried leaves, and whatever desert plants they can find. This animal can eat sharp cactus thorns without injury because the lining of its mouth is very tough. (*a*, Darwin Dale, Photo Researchers Inc. *b*–*d*, courtesy of Busch Gardens.)

(a)

(b)

(c)

(d)

ing. Most of what they eat is not efficiently digested but moves out of the body as waste, almost unchanged. However, by eating large enough quantities, they digest and absorb sufficient material to provide the nourishment necessary to sustain their life processes.

Many herbivores are equipped with jaws and teeth or toothlike structures for ingesting food. The teeth of herbivorous mammals include wide molars for grinding plant food, which often have enamel specially adapted to heavy wear. Herbivores also have long, elaborate digestive tracts so that food is retained for digestion for a long time.

A common and very interesting adaptation of herbivores is an intimate (symbiotic) relationship with microorganisms in their digestive tracts that can digest cellulose for them. Vertebrate herbivores generally have a specialized section of the digestive tract in which live bacteria capable of digesting cellulose.

Adaptations of Carnivores

Herbivores are eaten by the flesh-eating **carnivores.** Most carnivores are predators, so their first problem is to find and capture their prey. Even simple invertebrate carnivores have adaptations for this purpose. For

(a)

(b)

(c)

Figure 27–2 Adaptations of carnivores. (a) The long-nose butterfly fish, *Forcipiger longirostris*, has a mouth adapted for picking small worms and crustaceans from tight spots in coral reefs. (b) With lightning speed the Burmese python strikes at its prey, then suffocates it before consuming it whole. (c) With its wide field of vision and fast reflexes, the California mantid is a very able carnivore. (a, Charles Seaborn. b, courtesy of Mical Solomon and Trudi Segal. c, P.J. Bryant, UC-Irvine/BPS.)

example, *Hydra* and its relatives have long tentacles equipped with stinging cells called cnidocytes.

Vertebrate carnivores have many interesting adaptations for catching prey (Fig. 27–2). The fast-moving tongue of the frog captures many a fly, and the long, quick legs, and sharp teeth, of the cheetah enable it to catch and kill gazelles. All carnivorous mammals have well-developed canine teeth for stabbing during combat. Their molars are modified for shredding meat into small chunks that can be swallowed easily. The digestive juices of the stomach break down proteins, and because meat is more easily digested than plant food, their digestive tracts are shorter than those of herbivores.

Adaptations of Omnivores

Omnivores, such as bears and humans, include both plant material and meat in their diet. Some aquatic herbivores are filter-feeding organisms that ingest both tiny plants and animals. Earthworms take in large amounts of soil containing both animal and plant material. They use the organic material for food and egest the rest. Omnivores generally possess adaptations that permit them to distinguish among a wide range of smells and tastes. This ability enables them to select various foods.

Symbionts

A symbiont is an organism that lives in intimate association with a member of another species. One or both of the organisms usually derive nutritional benefit from the association. Three types of symbionts are parasites, commensals, and mutualistic partners.

(a)

(b)

Figure 27–3 Parasitic symbionts. (a) The flea is a well-known ectoparasite. Its body shape is adapted for slipping through fur, and its long hind legs are built for jumping. Note the claws, which help the flea hold on to hairs on the host's body. (b) A most interesting form of predation is practiced by the brachonid wasp. The adult wasp lays her eggs under the skin of an insect like this saddleback caterpillar of the flannel moth. When the eggs hatch, the larvae feed on the body fluids of the caterpillar until ready to spin cocoons. This caterpillar is almost covered by cocoons and has little chance of developing into an adult moth. Note that wasps are beginning to emerge from their cocoons. (b, Luci Giglio.)

A **parasite** lives in or on the body of a living plant or animal, the **host** species, and obtains its nourishment from the host (Fig. 27–3). Since the parasite's environment is its host, an effective parasite does not actually kill its host. For example, a tapeworm might live within the intestine of its host for many years. **Ectoparasites,** such as fleas and ticks, live outside the host's body; **endoparasites,** such as tapeworms and hookworms, live inside the host. Whether the parasite nourishes itself from food ingested by its host or by sucking the host's blood, it is strictly a freeloader.

A **commensal** is an organism that derives benefit from its host without either harming or benefiting the host. Commensalism is especially common in the ocean. Practically every worm burrow and hermit crab shell contains some uninvited guests that take advantage of the shelter and abundant food supplied by the host.

Mutualistic partners are two species of organisms that live together for their mutual benefit. They may be unable to survive separately. For example, the cellulose-digesting bacteria that inhabit the digestive tracts of herbivores help provide nourishment for their hosts in exchange for a place to live. A classic example of mutualism is the flagellate protozoan that lives in the intestine of the termite. The termite eats wood but, like most animals, does not produce the enzymes necessary to digest it. Its mutualistic partner, the flagellate, cannot chew wood and cannot survive outside the termite's gut, but it does produce the enzymes necessary to digest cellulose. Termites cannot survive without these intestinal inhabitants, and newly hatched termites lick the anus of another termite to obtain a supply of flagellates. This need requires these insects to form social groups.

INVERTEBRATE DIGESTIVE SYSTEMS

Sponges, the simplest animals, are filter feeders. They obtain food by filtering microscopic plants and animals out of seawater or pond water. Individual cells phagocytize the food particles. Digestion is intracellular within food vacuoles and depends upon the action of lysosomes. Wastes are egested into the water that continuously circulates through the sponge body.

Cnidarians, such as hydras and jellyfish, capture small aquatic animals with the help of their cnidocytes and tentacles (Fig. 27–4). The prey is pushed into the mouth by the tentacles. The mouth opens into a large gastrovascular cavity lined by cells that secrete enzymes to break down proteins. During digestion within the gastrovascular cavity, proteins are enzymatically cleaved to polypeptides. Digestion continues intracellularly within food vacuoles, and diffusion circulates the digested nutrients. Undigested food is ejected through the mouth by contraction of the body.

Freeliving flatworms begin to digest their prey even before ingesting it. They extend the pharynx through the mouth and secrete digestive enzymes onto the prey. When ingested the food enters the branched intestine. Extracellular digestion proceeds as intestinal cells secrete digestive enzymes. Partly digested food fragments are then phagocytized by cells of the intestinal lining, and digestion is completed intracellularly within food vacuoles. Like cnidarians, flatworms have no anal opening, so undigested wastes are eliminated through the mouth.

In most other invertebrates, and in all vertebrates, there is a **complete digestive system**—the digestive tract is a complete tube with *two* openings. Food enters through the mouth, and undigested food is eliminated through the anus. Waves of muscular contraction, known as **peri-**

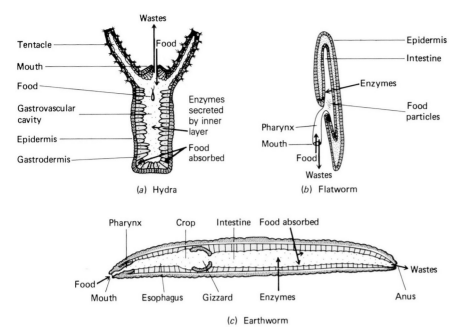

staltic contractions, push the food in one direction, so that more food can be taken in while previously eaten food is being digested and absorbed farther down the tract. Various parts of the tube are specialized to perform specific functions.

THE VERTEBRATE DIGESTIVE SYSTEM

Vertebrates have a complete digestive system. Food successively passes through these parts of the digestive tract: mouth, pharynx (throat), esophagus, stomach, small intestine, large intestine, and anus. All vertebrates have accessory glands that secrete digestive juices into the digestive tract. These include the liver, the pancreas, and in terrestrial vertebrates, the salivary glands. In the discussion that follows we will focus upon the human digestive system (Fig. 27–5).

Inside the Mouth

Imagine that you have just taken a big bite of a hamburger. Mechanical digestion begins as you bite and chew the meat and bun with your teeth. As the food is moistened by saliva, some of its molecules dissolve, enabling you to taste it. (Taste buds, which are located on the tongue and other surfaces of the mouth, were discussed in Chapter 23.) Three pairs of salivary glands secrete about a liter of saliva into the mouth cavity each day. Saliva contains an enzyme, called **salivary amylase,** which initiates the digestion of carbohydrates. You can demonstrate this for yourself by chewing a piece of bread for a few minutes. As the starch in the bread is reduced to maltose, a disaccharide sugar, you begin to notice a sweet taste.

Through the Pharynx and Esophagus

After the bite of food has been chewed and fashioned into a lump called a **bolus,** it is swallowed, that is, moved through the pharynx and into the **esophagus.** The **pharynx,** or throat, is a muscular tube that serves as the hallway of the respiratory system as well as the digestive system. During swallowing, the opening to the airway is closed by a small flap of tissue,

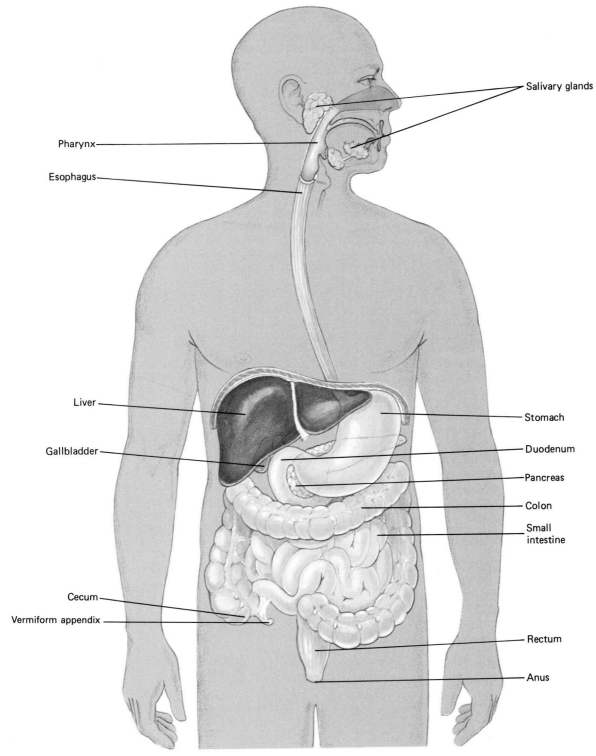

Figure 27–5 The human digestive system. Note the complete digestive tract, a long, coiled tube extending from mouth to anus. Locate the three types of accessory glands.

the **epiglottis.** A peristaltic contraction sweeps the bolus through the pharynx and esophagus toward the stomach.

Movement of food through the digestive tract by peristaltic contractions is illustrated in Figure 27–6. Circular muscle fibers in the wall of the esophagus contract around the top of the bolus, pushing it downward. Almost at the same time, longitudinal muscles around the bottom of the bolus and below it contract, shortening the tube.

When the body is in an upright position gravity helps to move the food through the esophagus, which is about 25 centimeters (10 inches)

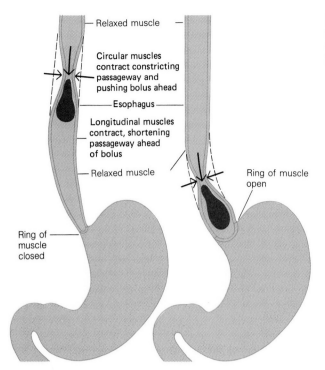

Figure 27–6 Peristalsis. Food is moved through the digestive tract by waves of muscular contraction known as peristalsis.

Relaxed muscle

Circular muscles contract constricting passageway and pushing bolus ahead

Esophagus

Longitudinal muscles contract, shortening passageway ahead of bolus

Relaxed muscle

Ring of muscle open

Ring of muscle closed

long, but gravity is not necessary. Astronauts are able to eat in its absence, and even if you are standing on your head, food will reach the stomach.

Inside the Stomach

The entrance to the stomach is normally closed by a ring of muscle at the lower end of the esophagus. When a peristaltic wave passes down the esophagus, the muscle relaxes, permitting the bolus to enter the **stomach,** a large muscular organ (Fig. 27–7). When empty, the stomach

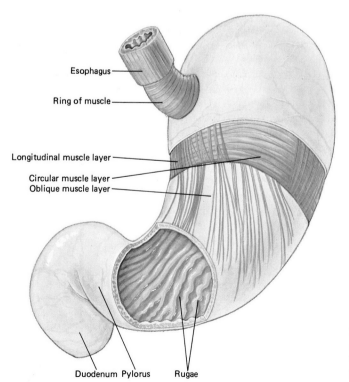

Esophagus

Ring of muscle

Longitudinal muscle layer

Circular muscle layer
Oblique muscle layer

Duodenum   Pylorus   Rugae

Figure 27–7 Structure of the stomach. From the esophagus, food enters the stomach, where it is mechanically and enzymatically digested.

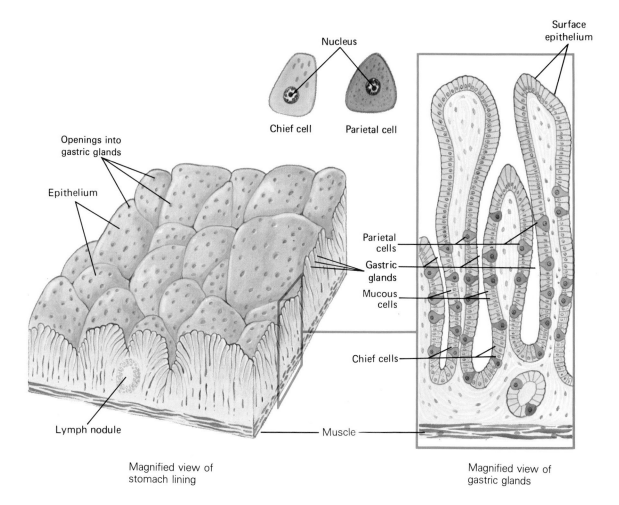

Nucleus

Chief cell   Parietal cell

Surface epithelium

Openings into gastric glands

Epithelium

Parietal cells

Gastric glands

Mucous cells

Chief cells

Lymph nodule

Muscle

Magnified view of stomach lining

Magnified view of gastric glands

Figure 27–8 The stomach lining and gastric glands.

is collapsed and shaped almost like a hot dog. Folds of the stomach wall called **rugae** give the inner lining a wrinkled appearance. As more and more food enters the stomach, the rugae gradually iron out, stretching the capacity of the stomach to more than a quart (about a liter).

The stomach is lined with a simple columnar epithelium that secretes large amounts of mucus. Tiny pits mark the entrances to the millions of **gastric glands,** which extend deep into the stomach wall (Fig. 27–8). **Chief cells** in the gastric glands secrete **pepsinogen,** the inactive form of the enzyme pepsin. When pepsinogen comes in contact with the acidic gastric juice it is converted to **pepsin,** the main digestive enzyme of the stomach. Pepsin hydrolyzes proteins, reducing them to polypeptides. (Also see Focus on Peptic Ulcers.) Other cells in the gastric glands, called **parietal cells,** secrete hydrochloric acid and a substance known as **intrinsic factor,** which is needed for adequate absorption of vitamin B_{12}.

What changes have occurred in our bite of hamburger during its stay in the stomach? The stomach has mashed and churned the food so that it has assumed the consistency of a thick soup which is known as **chyme.** Protein digestion has proceeded, so that much of the hamburger protein has been degraded to polypeptides.

Inside the Small Intestine

After three or four hours of digestion in the stomach, peristaltic waves propel a few milliliters of chyme at a time through the stomach exit, the **pylorus,** and into the **duodenum,** the first portion of the small intestine. Most chemical digestion takes place in the duodenum, not in the stomach, as is commonly believed. Bile from the liver and enzymes from the

Focus on . . .

PEPTIC ULCERS

One of the marvels of physiology is that the gastric juice does not digest the walls of the stomach. The acidity and protein-digesting activity of the gastric juice is so great that if you dipped your finger in a cup of it, your skin would be burned and digested. Several mechanisms protect the stomach from that fate. By releasing pepsin in an inactive form, the gastric glands protect themselves from being digested as the enzyme oozes out. The cells of the epithelial lining of the stomach are fitted tightly together so that gastric juice cannot leak into the intercellular spaces. The cells themselves are protected by the thick layer of mucus, which they secrete, and by their rapid turnover. About a half million of these cells are shed and replaced every minute. Since each epithelial cell of the stomach lining has a life span of only about three days, any damaged ones are soon replaced by healthy cells.

Still, occasionally something goes wrong and a small bit of the stomach wall is digested, leaving an open sore called a **peptic ulcer** (see figure). Such ulcers are thought to occur when the protective mechanisms of the stomach fail or become inadequate. Such substances as alcohol or aspirin are known to reduce the resistance of the stomach to digestion by gastric juice. Ulcers are even more common in the duodenum than in the stomach. Duodenal ulcers appear to be related to excessive secretion of gastric juice, especially between meals. Stress is thought to be a factor in the development of ulcers, and both gastric and duodenal ulcers are significantly more common in cigarette smokers than in nonsmokers.

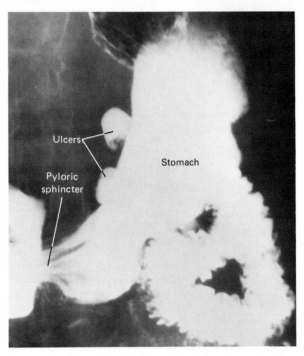

An x-ray of an ulcer in the wall of the stomach. The stomach and intestine have been filled with a contrast medium, making them appear white. This fluid also fills the cavities of the ulcers. (Courtesy of Dr. Jon Ehringer.)

pancreas are released into the duodenum to act upon the chyme. Then enzymes produced by the epithelial cells lining the duodenum catalyze the final steps in the digestion of the major types of nutrients (Table 27–1).

The lining of the small intestine is velvety because of millions of tiny finger-like projections in the lining called the intestinal **villi** (Figs. 27–9 and 27–10). The villi (singular, villus) serve to increase the surface area of the small intestine for digestion and absorption of nutrients. The intestinal surface is further expanded by thousands of **microvilli,** folds of cytoplasm on the exposed borders of the epithelial cells. About 600 microvilli protrude from the surface of each cell, giving the epithelial lining a fuzzy appearance when viewed with the electron microscope. This fuzzy surface is referred to as a **brush border.**

If the intestinal lining were smooth like the inside of a water pipe, food would zip right through the intestine and many valuable nutrients would be wasted. Folds in the wall of the intestine, the villi, and microvilli together increase the surface area of the small intestine by about 600 times. If we could unfold and spread out the lining of the small intestine, its surface would approximate the size of a tennis court.

The Liver

Just under the diaphragm lies the **liver,** the largest and also one of the most complex organs in the body (Fig. 27–11). A single liver cell can

Table 27–1
SUMMARY OF DIGESTION

Location	Source of enzyme	Digestive process*

Carbohydrate digestion

Mouth — Salivary glands — Polysaccharides (e.g., starch) $\xrightarrow{\text{salivary amylase}}$ Maltose + Small polysaccharides

Stomach — — Action continues until salivary amylase is inactivated by acidic pH

Small intestine — Pancreas — Undigested polysaccharides and small polysaccharides $\xrightarrow{\text{pancreatic amylase}}$ Maltose

Intestine — Disaccharides hydrolyzed to monosaccharides as follows:

Maltose (malt sugar) $\xrightarrow{\text{maltase}}$ Glucose + Glucose

Sucrose (table sugar) $\xrightarrow{\text{sucrase}}$ Glucose + Fructose

Lactose (milk sugar) $\xrightarrow{\text{lactase}}$ Glucose + Galactose

Protein digestion

Stomach — Stomach (gastric glands) — Protein $\xrightarrow{\text{pepsin}}$ Polypeptides

Small intestine — Pancreas — Polypeptides $\xrightarrow[\text{chymotrypsin}]{\text{trypsin,}}$ Tripeptides + Dipeptides

A—A—A—A—A   A—A—A   A—A
A—A—A—A—A

Dipeptides $\xrightarrow{\text{carboxypeptidase}}$ Free amino acids
A—A

Small intestine — Tripeptides + Dipeptides $\xrightarrow{\text{peptidases}}$ Free amino acids
A—A—A   A—A

Lipid digestion

Small intestine — Liver — Glob of fat $\xrightarrow{\text{bile salts}}$ Emulsified fat (individual triacylglycerols)

Pancreas — Triacylglycerol $\xrightarrow{\text{lipase}}$ Fatty acids + Glycerol

---

* ⌀ = monosaccharide; = triacylglycerol; ⌶ = glycerol; ⌇ = fatty acid; A = amino acid units or, when standing alone, a free amino acid.

569
PROCESSING FOOD

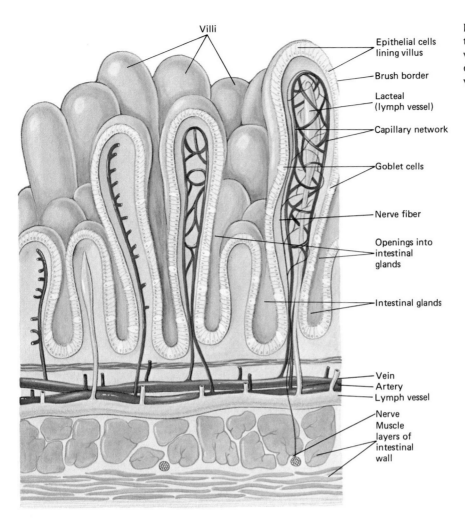

Villi

Epithelial cells
lining villus

Brush border

Lacteal
(lymph vessel)

Capillary network

Goblet cells

Nerve fiber

Openings into
intestinal
glands

Intestinal glands

Vein
Artery
Lymph vessel

Nerve
Muscle
layers of
intestinal
wall

Figure 27–9 Diagram of the wall of the human intestine showing the villi. Some of the villi have been opened to show the blood and lymph vessels within.

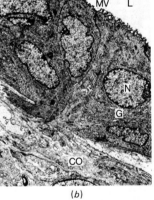

(a)

(b)

carry on more than 500 separate metabolic activities. Here is a partial list of the complex and diverse actions of the liver. The liver:

1. Secretes **bile,** which is important in the mechanical digestion of fats.
2. Inventories nutrients absorbed from the intestine, removing excess amounts of each from the blood.

Figure 27–10 (a) Photomicrograph (×100) of the intestinal lining, showing the villi. V, villus; L, lumen of the intestine. A close-up of the circled area is shown in b. (b) Electron micrograph (approximately ×8000) of epithelial cells lining the small intestine, showing the microvilli (MV). L, lumen; N, nucleus of an epithelial cell; G, Golgi complex; CO, collagen. (c) Scanning electron micrograph (approximately ×14,000) of the surface of an epithelial cell from the lining of the small intestine showing microvilli. The epithelium has been cut vertically to allow the microvilli to be viewed from the side as well as from above. (b, courtesy of Dr. Lyle C. Dearden. c, courtesy of J.D. Hoskins, W.G. Henk, and Y.Z. Abdelbaki, from *American Journal of Veterinary Research,* Vol. 43, No. 10.)

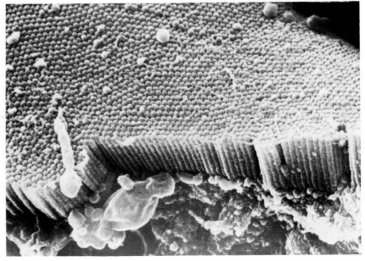

(c)

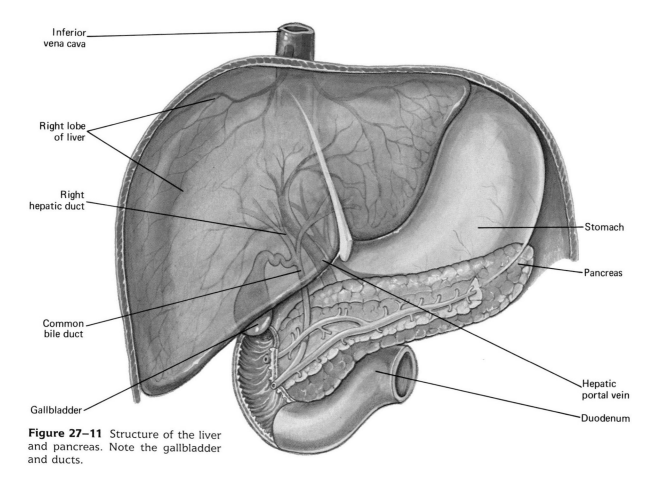

Inferior vena cava

Right lobe of liver

Right hepatic duct

Common bile duct

Gallbladder

Stomach

Pancreas

Hepatic portal vein

Duodenum

Figure 27–11 Structure of the liver and pancreas. Note the gallbladder and ducts.

3. Converts excess glucose to glycogen and stores it.
4. Stores iron and certain vitamins.
5. Converts excess amino acids to fatty acids and urea.
6. Manufactures many proteins that are components of the blood.
7. Detoxifies many drugs and poisons that enter the body.
8. Removes bacteria and worn-out red blood cells from the circulation.

Bile produced in the liver is stored in the pear-shaped **gallbladder,** then released into the duodenum as needed. Bile mechanically digests fats by a detergent-like action in which it decreases the surface tension of fat particles. This permits them to break apart into smaller particles as they are pushed about by the mixing movements of the small intestine. This mechanical digestion of fats by bile is called **emulsification.** Bile does not digest any food material chemically, for it contains *no digestive enzymes.*

The Pancreas

The **pancreas** is a large gland that secretes both digestive enzymes and hormones that help regulate the level of glucose in the blood. Among its enzymes are (1) **trypsin** and **chymotrypsin,** which digest polypeptides to dipeptides, (2) **pancreatic lipase,** which degrades neutral fats, (3) **pancreatic amylase,** which breaks down almost all types of carbohydrates, except cellulose, to disaccharides, and (4) **ribonuclease** and **deoxyribonuclease,** which split the nucleic acids RNA and DNA to free nucleotides. Digestion of carbohydrates, protein, and lipid is summarized in Table 27–1.

Table 27–2
REGULATION OF DIGESTION BY HORMONES AND NERVES

Name	Where secreted	Target tissue	Actions	Factors that stimulate release
Hormone				
Gastrin	Stomach	Stomach (gastric glands)	Stimulates gastric glands to secrete	Distention of stomach by food; certain substances, such as partially digested proteins and caffeine
Secretin	Duodenum (mucous membrane)	Pancreas	Stimulates release of alkaline component of pancreatic juice	Acidic chyme acting on duodenum
		Liver	Increases rate of bile secretion	
Cholecystokinin (CCK)	Duodenum	Pancreas	Stimulates release of digestive enzymes	Presence of fatty acids and partially digested proteins in duodenum
		Gallbladder	Stimulates contraction and emptying	
Gastric inhibitory peptide	Duodenum	Stomach	Decreases motor activity of stomach and thus slows stomach's emptying	Presence of fat or carbohydrate in duodenum
Nerves				
Parasympathetic branches of vagus nerve		Most of digestive system	Stimulates general activity, such as peristalsis; increases stomach secretion	Calm, restful state
Sympathetic nerves		Most of digestive system	Inhibits general activity	Stress, fear, rage
Neural plexuses (networks) within wall of intestine		Portions of digestive system	Stimulates contraction and secretion	Local stretch receptors stimulated by distention with chyme

Regulation of Digestion

Activities of the digestive system are regulated by both nerves and hormones (Table 27–2). As an example, consider the secretion of gastric juice. Seeing, smelling, tasting, or even thinking about food causes the brain to send neural messages to the glands in the stomach, stimulating them to secrete. In addition, when food distends the stomach, it stimulates the stomach wall to release a hormone called **gastrin.** Gastrin is absorbed into the blood and transported to the gastric glands, where it stimulates release of gastric juice.

Absorption

Only a few substances—water, simple sugars, salts, alcohol, and certain drugs—have small enough molecules to be absorbed through the wall of the stomach. Absorption of nutrients is primarily the job of the intestinal villi. As illustrated in Figure 27–9, the wall of a villus consists of a single layer of epithelial cells. Inside each villus is a network of tiny blood vessels (capillaries) and a central lymph vessel called a **lacteal.** To reach the blood (or lymph), a nutrient molecule must pass through an epithelial cell of the lining and through a cell lining the vessel. Glucose and amino acids are thought to be transported into the blood by an active transport mechanism. Lipids enter the lymph system and are transported to the upper trunk region, where the lymph fluid and its contents enter the blood circulatory system.

Elimination

Indigestible material, such as the cellulose of plant foods, along with unabsorbed chyme, passes into the large intestine. This organ, though only about 1.3 meters long (about 4 feet), is called "large" because its diameter is greater than that of the small intestine. The small intestine joins the large intestine about 7 centimeters (2.8 inches) from the end of the large intestine, thereby creating a blind pouch, the **cecum.** The **vermiform appendix** projects from the end of the cecum. (**Appendicitis** is an inflammation of the appendix.) The functions of the cecum and appendix in human beings are not known, and they are generally considered vestigial organs, perhaps important in the vegetarian past of the human species. Herbivores such as rabbits have a large, functional cecum containing bacteria that digest cellulose.

From the cecum to the rectum (the last portion of the large intestine) the large intestine is known as the **colon.** As the chyme passes slowly through the large intestine, water and sodium are absorbed from it, and it gradually assumes the consistency of normal feces. Bacteria inhabiting the large intestine enjoy the last remnants of the meal and return the favor by producing vitamin K and certain B vitamins that can be absorbed and utilized.

When chyme passes through the intestine too rapidly, defecation (expulsion of feces) becomes more frequent and the feces are watery. This condition, called **diarrhea,** may be caused by certain disease organisms that irritate the intestinal lining, by emotional tension, or by certain foods. Prolonged diarrhea results in loss of water and salts, leading to dehydration, a serious condition, especially in infants.

Constipation refers to slow movement of feces through the large intestine. Because more water than usual is removed from the chyme, the feces may be hard and dry. Constipation is often caused by a diet containing insufficient fiber.

Cancer of the colon is one of the most common causes of cancer deaths in the United States. Research indicates that this type of cancer may be related to diet, for the disease is more common in people whose diets are very low in fiber. It has been suggested that less fiber results in less frequent defecation, allowing prolonged contact between the mucous membrane of the colon and such carcinogens as nitrites (used as preservatives) in foods.

CHAPTER SUMMARY

I. An organism's body plan and life-style are adapted to its mode of nutrition.
   A. Among the adaptations of herbivores are molars for grinding plant food, long and elaborate digestive tracts, and symbiotic relationships with microorganisms that digest cellulose.
   B. Carnivores have adaptations for capturing their prey. Carnivorous mammals have canine teeth for stabbing and modified molars for shredding meat.
   C. Omnivores can distinguish among a wide range of smells and tastes.
   D. In a symbiotic relationship one or both organisms derive nutritional benefit from the association.

II. In the simplest invertebrates, the sponges, there is no digestive system; digestion is carried on intracellularly.

   A. Cnidarians and flatworms have digestive systems with only one opening, which serves as both mouth and anus.

   B. In more complex invertebrates the digestive tract is a complete tube with an opening at each end.

III. The vertebrate digestive system is a complete tube leading from mouth to anus.

   A. Mechanical digestion and chemical digestion of carbohydrates begin in the mouth.

   B. As food is swallowed, it is propelled through the pharynx and esophagus. A bolus of food is moved along through the esophagus and through the rest of the digestive tract by peristaltic action.

   C. In the stomach, food is mechanically digested by vigorous churning and proteins are chemically digested by the action of pepsin in the gastric juice.

   D. Most chemical digestion takes place in the duodenum, which receives secretions from the liver and pancreas and produces several digestive enzymes of its own.

   E. The liver produces bile, which mechanically digests fats.

   F. The pancreas releases enzymes that chemically digest protein, lipid, and carbohydrate, as well as RNA and DNA.

   G. Activities of the digestive system are regulated by both nerves and hormones.

   H. Most nutrients are absorbed through the thin walls of the intestinal villi.

   I. The large intestine is responsible for the elimination of undigested wastes. It also incubates bacteria that produce vitamin K and certain B vitamins.

Post-Test

1. The process of taking food into the body is called _____ .

2. _____ consists of mechanically and enzymatically breaking down food into molecules small enough to be absorbed.

3. _____ is the process of getting rid of undigested and unabsorbed food.

4. An animal with wide molars for grinding plant food and symbiotic microorganisms that digest cellulose is a _____ .

5. Carnivorous mammals have well-developed _____ teeth.

6. An organism that lives in intimate association with a member of another species is a _____ .

7. The most characteristic feature shared by the cnidarian and flatworm digestive systems is that _____ .

8. In a complete digestive system the digestive tract has _____ .

9. Saliva contains an enzyme called _____ , which initiates the digestion of _____ .

Select the most appropriate term in column B for each description in column A.

Column A

_____ 10. Protein digestion begins here

_____ 11. Incubates bacteria

_____ 12. Receives secretions from pancreas

_____ 13. Secretes bile

_____ 14. Converts food to chyme

_____ 15. Conducts food to stomach

Column B

a. duodenum
b. stomach
c. liver
d. large intestine
e. none of the above

16. The surface area of the stomach is increased by the presence of folds called _____ .

17. Absorption takes place through the finger-like projections in the lining of the small intestine called _____ .

18. Food leaving the stomach next enters the _____ .

19. An open sore in the wall of the stomach or duodenum is called a _____ _____ .
20. The function of the gallbladder is to _____ _____ .
21. Label the diagram at the right. For correct labeling, see Figure 27–5.

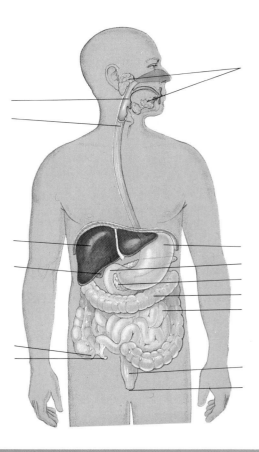

Review Questions

1. Describe adaptations in herbivores and carnivores that help each group succeed in its particular nutritional life-style.
2. Give examples of parasites, commensals, and mutualistic partners, and describe how each relates to its host.
3. Compare food processing in sponges, hydras, flatworms, and earthworms.
4. Why must food be digested?
5. Trace a bite of food through the digestive tract, listing each structure through which it passes.
6. What mechanisms prevent gastric juice from digesting the wall of the stomach? What happens when they fail?
7. List three types of accessory glands that secrete digestive substances into the digestive tract. Identify their secretions.
8. Summarize the step-by-step digestion of (a) carbohydrates, (b) lipids, (c) proteins.
9. What happens to ingested cellulose? Why?
10. Draw and label an intestinal villus.

Readings

Kessel, R.G., and R.H. Kardon. *Tissue and Organs: A Text-Atlas of Scanning Electron Microscopy.* San Francisco, Freeman, 1979. Chapter 9, which focuses on the digestive system, includes many fascinating electron micrographs.

Moog, F. "The Lining of the Small Intestine," *Scientific American,* November 1981, 154–176. The cells lining the small intestine are covered by a membrane that actively digests foods and speeds nutrients into the blood.

Solomon, E.P., and P.W. Davis. *Human Anatomy and Physiology.* Philadelphia, Saunders College Publishing, 1983. Chapter 21 presents a readable discussion of the structure and function of the digestive system.

Chapter 28
NUTRITION

Learning Objectives

After you have read this chapter you should be able to:

1. Identify each of the basic nutrients, give its role in the body, and describe the effect of its deficiency in the diet.
2. Give three reasons why it is difficult to obtain adequate amounts of amino acids in a vegetarian diet, and tell how a nutritious vegetarian diet could be planned.
3. Describe the fate of glucose after its absorption and define the following terms: glucogenesis, glycogenolysis, and gluconeogenesis.
4. Describe the fate of amino acids after they are absorbed from the intestine.
5. Describe the utilization of lipids after they are absorbed from the intestine.
6. Distinguish between basal metabolic rate and total metabolic rate.
7. Write the basic energy equation for maintaining body weight and tell what happens when it is altered in either direction.
8. Describe the effects of obesity on health, its causes, and its cure. Basing your answer on energy considerations, describe the most effective reducing diet.
9. In general terms, describe the problem of world food supply relative to world population, and describe the effects of malnutrition.
10. Discuss the solutions to problems of inadequate food supply proposed in this chapter, citing advantages, disadvantages, and examples of each.

I n the last chapter we focused upon the digestive system and how it processes food. In this chapter we will discuss the basic nutrients and how they are used metabolically by the cells of the body.

THE BASIC NUTRIENTS

From about 20 chemical elements, which are absorbed in the form of simple compounds and salts, plants are able to produce all the different kinds of organic molecules that they need. Animals are not such sophisticated chemists. Although they require approximately the same 20 chemical elements, these elements must already be chemically combined in the form of about 40 chemical substances, many of them organic compounds. These essential nutrients are provided by a balanced diet consisting of proteins, carbohydrates, lipids, vitamins, minerals, and water (Tables 28–1 and 28–2). With only slight variation, all animals require the same basic nutrients. Our discussion will primarily center on human nutritional needs.

Water

A daily dietary intake of about 2.5 quarts of water (about 2.4 liters) is required by an average adult. Because plant and animal tissues are composed largely of water, much of this daily requirement is met by eating food. For example, a raw apple is composed of almost 85% water; a slice of roast beef is 60% water. We drink the remaining quart or so directly as water or other beverages.

The human body is about 65% water by weight, but the amount in specific organs varies. Skin is about 70% water, bone is only 25%. Most surprising, perhaps, is that the human brain is about 80% water. The functions performed by water in the body have been discussed in Chapter 2.

Minerals

Minerals are inorganic nutrients generally ingested as salts dissolved in food and water. Essential minerals include sodium, chlorine, potassium, magnesium, calcium, sulfur, phosphorus, iron, iodine and fluorine (Table 28–1). Several others, such as cobalt and selenium, are called **trace elements** because they are required in minute quantities.

Minerals are needed as components of all body tissues and fluids. Salt content (about 0.9%) is vital in maintaining the fluid balance of the body, and since salts are lost from the body daily in sweat, urine, and feces, they must be replaced by dietary intake. Sodium chloride (common table salt) is the salt needed in largest quantity in blood and other body fluids. A deficiency of sodium chloride results in dehydration.

Phosphorus and calcium are important structural components of bones and teeth. Phosphorus is also a component of nucleic acids, the high-energy molecule ATP (adenosine triphosphate), and several other important organic compounds. Calcium is required for biochemical reactions essential to blood clotting, muscle contraction, and transmission of nerve impulses.

Iron, the mineral most likely to be deficient in the diet, is an essential part of the cytochrome components of the electron transport system and of the hemoglobin molecule. When iron intake is inadequate, the body cannot synthesize enough hemoglobin, and as a result, the capacity of the blood to transport oxygen to the cells is reduced. People with this condition, known as **iron-deficiency anemia,** lack energy and are easily fatigued.

Table 28–1
SOME IMPORTANT MINERALS AND THEIR FUNCTIONS

Mineral	Functions	Comments
Calcium	Component of bones and teeth; essential for normal blood clotting; needed for normal muscle and nerve function	Bones serve as calcium reservoir; sources: milk and other dairy products, green leafy vegetables
Phosphorus	As calcium phosphate, an important structural component of bone; essential in energy transfer and storage (component of ATP) and in many other metabolic processes; component of DNA and RNA	Performs more functions than any other mineral; antacids can impair absorption
Sulfur	As component of many proteins (e.g., insulin), essential for normal metabolic activity	Sources: high-protein foods, such as meat, fish, legumes, nuts
Potassium	Principal positive ion within cells; influences muscle contraction and nerve excitability	Occurs in many foods
Sodium	Principal positive ion (cation) in interstitial fluid; important in fluid balance; essential for conduction of nerve impulses	Occurs naturally in foods; sodium chloride (table salt) added as seasoning; too much is ingested in average American diet; excessive amounts may lead to high blood pressure
Chlorine	Principal negative ion (anion) of interstitial fluid; important in fluid balance and acid-base balance	Occurs naturally in foods; ingested as sodium chloride
Copper	Component of enzyme needed for melanin synthesis; component of many other enzymes; essential for hemoglobin synthesis	Sources: liver, eggs, fish, whole wheat flour, beans
Iodine	Component of thyroid hormones (hormones that stimulate metabolic rate)	Sources: seafoods, iodized salt, vegetables grown in iodine-rich soil; deficiency results in goiter (abnormal enlargement of thyroid gland)
Cobalt	As component of vitamin B_{12}, essential for red blood cell production	Sources: meat, dairy products; strict vegetarians may become deficient in this mineral
Manganese	Necessary to activate argininase, an enzyme essential for urea formation; activates many other enzymes	Sources: whole-grain cereals, egg yolks, green vegetables; poorly absorbed from intestine
Magnesium	Appropriate balance between magnesium and calcium ions needed for normal muscle and nerve function; component of many coenzymes	Occurs in many foods
Iron	Component of hemoglobin, myoglobin, important respiratory enzymes (cytochromes), and other enzymes essential to oxygen transport and cellular respiration	Mineral most likely to be deficient in diet; deficiency results in anemia; sources: meat (especially liver), nuts, egg yolk, legumes
Fluorine	Component of bones and teeth; makes teeth resistant to decay	Where it does not occur naturally, fluorine may be added to municipal water supplies (fluoridation); excess causes tooth mottling
Zinc	Component of at least 70 enzymes; component of some peptidases, and thus important in protein digestion; may be important in wound healing	Occurs in many foods

Table 28–2
THE VITAMINS

Vitamins and U.S. RDA*	Actions	Effect of deficiency
Fat-soluble		
A 5000 IU†	Component of retinal pigments, essential for normal vision; essential for normal growth and integrity of epithelial tissue; promotes normal growth of bones and teeth by regulating activity of bone cells	Failure of growth; night blindness; atrophy of epithelium; epithelium subject to infection; scaly skin
D 400 IU	Promotes calcium absorption from digestive tract; essential to normal growth and maintenance of bone	Bone deformities; rickets in children; osteomalacia in adults
E 30 IU	Inhibits oxidation of unsaturated fatty acids and vitamin A that help form cell and organelle membranes; precise biochemical role not known	Increased catabolism of unsaturated fatty acids, so not enough are available for maintenance of cell membranes and other membranous organelles; prevents normal growth
K probably about 1 mg	Essential for blood clotting	Prolonged blood-clotting time
Water-soluble		
C (ascorbic acid) 60 mg	Needed for synthesis of collagen and other intercellular substances; formation of bone matrix and tooth dentin, intercellular cement; needed for metabolism of several amino acids; may help body withstand injury from burns and bacterial toxins	Scurvy (wounds heal very slowly and scars become weak and split open; capillaries become fragile; bone does not grow or heal properly)
B-complex vitamins		
Thiamine (B$_1$) 1.5 mg	Acts as coenzyme in many enzyme systems; important in carbohydrate and amino acid metabolism	Beriberi (weakened heart muscle, enlarged right side of heart, nervous system and digestive tract disorders)
Riboflavin (B$_2$) 1.7 mg	Used to make coenzymes (e.g., FAD) essential in cellular respiration	Dermatitis, inflammation and cracking at corners of mouth; mental depression
Niacin (nicotinic acid) 20 mg	Component of important coenzymes (NAD and NADP) essential to cellular respiration	Pellagra (dermatitis, diarrhea, mental symptoms, muscular weakness, fatigue)
Pyridoxine (B$_6$) 2 mg	Coenzyme needed for amino acid synthesis and protein metabolism	Dermatitis, digestive tract disturbances, convulsions
Pantothenic acid 10 mg	Constituent of coenzyme A (important in cellular metabolism)	Deficiency extremely rare
Folic acid 0.4 mg	Coenzyme needed for reactions involved in nucleic acid synthesis and for maturation of red blood cells	A type of anemia
Biotin 0.3 mg	Coenzyme needed for carbon dioxide fixation	Deficiency unknown
B$_{12}$ 6 mg	Coenzyme important in nucleic acid metabolism	Pernicious anemia

*RDA: the recommended dietary allowance, established by the Food and Nutrition Board of the National Research Council, to maintain good nutrition for healthy persons.
†International Unit: the amount that produces a specific biological effect and is internationally accepted as a measure of the activity of the substance.

Sources	Comments
Liver, fish-liver oils, egg, yellow and green vegetables	Can be formed from provitamin carotene (a yellow or red pigment); sometimes called anti-infection vitamin because it helps maintain epithelial membranes; excessive amounts harmful
Liver, fish-liver oils, egg yolk, fortified milk, butter, margarine	Two types: D_2 (calciferol), a synthetic form, and D_3, formed by action of ultraviolet rays from sun upon a cholesterol compound in the skin; excessive amounts harmful
Oils made from cereals, seeds, liver, eggs, fish	
Normally supplied by intestinal bacteria; green leafy vegetables	Antibiotics may kill bacteria, then supplements are needed in surgical patients
Citrus fruits, strawberries, tomatoes	Possible role in preventing common cold or in the development of acquired immunity(?); harmful in very excessive dose
Liver, yeast, cereals, meat, green leafy vegetables	Deficiency common in alcoholics
Liver, cheese, milk, eggs, green leafy vegetables	
Liver, meat, fish, cereals, legumes, whole-grain and enriched breads	
Liver, meat, cereals, legumes	
Widespread in foods	
Produced by intestinal bacteria; liver, cereals, dark-green leafy vegetables	
Produced by intestinal bacteria; liver, chocolate, egg yolk	
Liver, meat, fish	Contains cobalt; intrinsic factor secreted by gastric mucosa needed for absorption

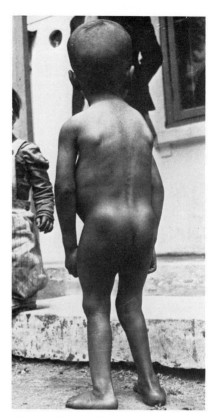

Figure 28–1 Child with rickets, a condition that results from deficiency of vitamin D during childhood. This deficiency decreases the body's ability to absorb and use calcium and phosphorus and produces soft, malformed bones. Note the bowed legs. (United Nations, Food and Agricultural Organization photo.)

Vitamins

Vitamins are organic compounds required by the body for biochemical processes. Some function as *coenzymes* (see Chapter 6). Very small amounts are required in comparison with other dietary constituents. Vitamins may be divided into two main groups. **Fat-soluble vitamins** are those that can be dissolved in fat and include vitamins A, D, E, and K. **Water-soluble vitamins** are the B and C vitamins. Table 28–2 gives the sources, functions, and consequences of deficiency for most of the vitamins (also see Fig. 28–1).

While millions of people gulp down their daily vitamin pills with great fervor, two opposing schools of thought debate the need for vitamin supplements. One group, which includes many physicians, contends that people who eat a nutritionally balanced diet have no need for vitamin pills. The other group argues that most of us do not eat a balanced diet and therefore are likely to suffer from vitamin deficiencies. Debates also rage over the advisability of taking large amounts of certain specific vitamins, such as vitamin C for colds or vitamin E to protect against vascular disease. To date there is no compelling evidence to support claims that large quantities of any vitamin are beneficial, and in fact more conclusive evidence is needed before we will fully understand all the biochemical roles played by vitamins or the interactions between various vitamins and other nutrients. Meanwhile, the vitamin supplement industry is doing a booming business.

Some people think that if one vitamin capsule daily is healthy, four or five might be even better. Actually, such a daily overdose may be harmful. *Moderate* overdoses of the B and C vitamins are excreted in the urine, but surpluses of the fat-soluble vitamins are not easily excreted. An excess of vitamin D can cause weight loss, nausea, diarrhea, and eventually, mineral loss from the bones and calcification of soft tissues, including heart, blood vessels, and kidney tubules. One common result is kidney disease. In children 2000 units or more a day result in growth retardation, and high doses of vitamin D taken by pregnant women have been linked to mental retardation in the developing child. Overdose of vitamin A results in skin ailments, slow growth, enlargement of liver and spleen, and painful swelling of the long bones.

Carbohydrates

Carbohydrates are the body's principal fuel. Our cells "burn" them to obtain energy for metabolic activities, growth, repair, and physical activity. In the average American diet carbohydrates provide about 50% of the Calories (Cal) ingested daily. Nutritionists measure the energy value of food in Calories per gram of food. (A **Calorie,** spelled with a capital C, is actually a kilocalorie. It is defined as the amount of heat required to raise the temperature of a kilogram of water from 15° to 16°C. The calorie—spelled with a lower case c—used by chemists is 1000 times smaller.)

Foods rich in carbohydrates include rice, potatoes, corn, and other cereal grains. These are the least expensive foods, and for this reason the proportion of carbohydrate in a family's diet may reflect economic status. Very poor people may subsist on diets that are almost exclusively carbohydrate, while the more affluent enjoy the more expensive protein-rich foods, such as meat and dairy products.

Most carbohydrates are ingested in the form of starch and cellulose, both polysaccharides. (You may want to review the discussion of carbohydrates in Chapter 3.) In affluent societies about 25% of the carbohydrate intake (more in children) is in the form of the disaccharide sucrose—cane or beet sugar. Sucrose is the so-called refined sugar put in coffee and desserts. During digestion each molecule of sucrose is degraded to one molecule of glucose and one of fructose. Another important disaccharide is lactose, the sugar in milk. Each molecule of lactose is digested into one of glucose and one of galactose.

Lipids

Cells use ingested lipid nutrients as fuel, as components of cell membranes, and to make complex lipid compounds, such as steroid hormones and bile salts. Lipid accounts for about 40% of the Calories in the average American diet. In poor countries this percentage falls to less than 10%, because most lipid-rich foods—meats, eggs, and dairy products—are relatively expensive.

About 98% of lipids in the diet are ingested in the form of triacylglycerols. (You should recall from Chapter 3 that a triacylglycerol is a glycerol molecule chemically combined with three fatty acids; see Figure 3–8.) Triacylglycerols may be saturated, that is, fully loaded with hydrogens, mono-unsaturated (containing one double bond in the carbon chain of a fatty acid, so two more hydrogen atoms can be added), or polyunsaturated (containing two or more double bonds, so four or more hydrogen atoms can be added). As a rule, animal foods are rich in both saturated fats and cholesterol, and plant foods contain unsaturated fats and no cholesterol. Commonly used polyunsaturated vegetable oils are corn, soya, cottonseed, and safflower oils. Olive and peanut oils contain large amounts of mono-unsaturated fats. Butter contains mainly saturated fats.

The average American diet provides about 700 milligrams of cholesterol each day, whereas only about 300 milligrams is recommended. Cholesterol sources are egg yolks, butter, and meat. However, the body is not dependent upon dietary sources because it is able to synthesize cholesterol from other nutrients.

Lipids have been the focus of much research because of their role in atherosclerosis, a progressive disease in which the arteries become occluded with fatty material. As was discussed in Chapter 24, atherosclerosis leads to circulatory impairment and heart disease. A diet high in saturated fats and cholesterol raises the blood-cholesterol level by as much as 25% and thus greatly increases the chances of developing atherosclerosis. On the other hand, ingestion of polyunsaturated fats tends to decrease the blood cholesterol level, thus probably affording some protection against atherosclerosis. For these reasons many people now cook with vegetable oils rather than butter and lard, drink skim milk rather than whole milk, eat ice milk instead of ice cream, and use margarine instead of butter. (It should be pointed out that in the production of margarines, oils are partially hydrogenated, a process that reduces their degree of polyunsaturation.)

Proteins

Protein consumption is an index of a country's (or an individual's) economic status, because high-quality protein is the most expensive and least available of all the nutrients. Protein poverty is one of the world's most pressing health problems; millions of human beings suffer from poor health, disease, and even death as a consequence of protein malnutrition.

Proteins are critical as nutrients because they are essential building blocks of cells. Indeed, 75% of body solids consists of protein. These nutrients also serve as enzymes and are used to make many needed substances in the body.

Ingested proteins are degraded in the digestive tract to their molecular subunits—amino acids. These are absorbed and used by the cells to make the types of proteins needed. Of the 20 or so amino acids important in nutrition, the body is able to make several by rearranging the atoms of certain organic acids. About eight of the amino acids (nine in children) cannot be synthesized by the body cells at all, or at least not in sufficient quantity to meet the body's needs. These, which must be provided in the diet, are called the **essential amino acids.**

Not all proteins contain the same kinds or quantities of amino acids, and many proteins lack some of the essential amino acids. Highest-quality proteins, those that contain the most appropriate distribution of amino acids for human nutrition, are found in eggs, milk, meat, and fish. Some foods, such as gelatin or soybeans, contain a high proportion of protein but do not contain all the essential amino acids, or they do not contain them in nutritional proportions. Most plant proteins are deficient in one or more essential amino acids, usually lysine, tryptophan, or threonine.

The recommended daily amount of protein is about 56 grams—only about an eighth of a pound (half a quarter-pound burger). In the United States and other developed countries most people eat far more protein than required. It has been estimated that the average American eats about 300 pounds of meat and dairy products per year. In some underdeveloped countries an average of only 2 pounds per person per year is consumed.

Most human beings depend upon cereal grains as their staple food—usually rice, wheat, or corn. None of these foods provide an adequate proportion of total amino acids or distribution of essential amino acids, especially not for growing children. In some underdeveloped countries starchy crops, such as sweet potatoes or cassava, are the principal food. Total protein content of these foods is less than 2%, far below minimum needs.

METABOLISM

Once absorbed, nutrients in the blood are transported to the large hepatic portal vein, which conducts them to the liver. Inside the liver the hepatic portal vein divides into a vast network of capillary-like sinusoids, which enable material to be exchanged between blood and liver cells. As the nutrient-rich blood courses slowly through these sinusoids, liver cells take inventory. Surplus nutrients are absorbed into the liver cells, where they are either stored or converted into other materials. Under normal circumstances blood leaves the liver carrying sufficient nutrients to meet the requirements of all the cells of the body. The blood has been appropriately described as a traveling smorgasbord from which each cell selects whatever nutrients it needs.

Fate of Carbohydrates

One of the most important jobs of the liver is to help regulate the blood sugar level. Cells of the body are extremely dependent upon a constant supply of glucose delivered by the blood. Brain cells are especially dependent because they are unable to store glucose themselves. If deprived of this vital source of energy for even a few minutes, they cease to function. After a meal or rich dessert, when there is an excess of glucose in the blood, the liver cells remove and store it. Between meals, when the glucose level begins to fall, the liver cells release glucose back into the blood. In this way the liver maintains a rather steady glucose level in the

Focus on . . .

THE VEGETARIAN DIET

Most of the world's population depends almost exclusively upon plant foods for proteins and other nutrients. However, besides being deficient in some of the essential amino acids, plant foods have a lower percentage of protein than animal foods. Meat contains about 25% protein; the new high-yield grains contain 5% to 13%. Another problem is that plant protein is less digestible than animal protein. Because we cannot digest the cellulose cell walls, much of the protein encased within the cells passes on through the digestive tract as part of the bulk.

Yet another problem with the vegetarian diet is that the body has no mechanism for storing amino acids. Cells cannot make a protein storage compound comparable to glycogen, and they have no protein depot, in the sense that fat is stored in adipose tissue. One might almost say that all the essential amino acids must be ingested in the same meal. For example, if corn is eaten for lunch and beans for dinner, the body will not have all the essential amino acids at the same time and will not be able to manufacture needed proteins.

Does this mean, then, that vegetarian diets are always unhealthful? Not at all. Given a variety of plant foods and some knowledge of nutrition,[1] a vegetarian can plan a diet that provides all the needed amino acids. The main task is to select foods that complement one another. For example, if beans and corn or beans and rice are eaten together in the proper proportions, they will provide the required essential amino acids. When dairy products are available, they should be eaten with the plant foods. For example, when cereal is eaten with milk, the meal has a much greater nutritional value. Macaroni and cheese is another example.

Soybeans, peanuts, and other legumes have more than twice the protein content of the cereal grains. Unfortunately, yields per acre of these crops are much lower than yields per acre of cereal grains. Also, most of the legumes produced are used as livestock feed, rather than for human consumption. In fact, in the United States 91% of the cereal, legume, and other vegetable protein suitable for human use is fed to livestock. This represents a tremendous and serious loss of protein to human beings, because for every 5 kilograms (11 pounds) of vegetable and fish protein fed to livestock in addition to their forage intake, only 1 kilogram (2.2 pounds) of animal protein is produced. Meat is expensive economically because it is ecologically expensive. As the human population continues to expand, more and more of us will have to turn to vegetarian diets.

[1]See the following for more information: Ewald, E.B. *Recipes for a Small Planet*, New York, Ballantine, 1973. Lappé, F.M. *Diet for a Small Planet*, New York, Ballantine, 1975.

blood. The normal blood glucose content while fasting is about 90 milligrams per 100 milliliters blood. After a carbohydrate-rich meal the level may increase briefly to about 140 milligrams per 100 milliliters. If the liver did not remove the excess, the level would rise to more than three times normal after a carbohydrate-rich meal, and then fall disastrously between meals or during the night.

Once taken up by the liver cells, a molecule of glucose is chemically joined to many other molecules of glucose. This process is called **glycogenesis** (meaning "production of glycogen"). Individual glucose molecules would disturb the cells' osmotic balance, for thousands of small, soluble molecules in a cell would exert an osmotic pressure, drawing water into the cell. Hundreds, even thousands, of glucose molecules can be conveniently packaged as glycogen, a large, stable molecule that usually precipitates out of solution and forms little granules. Although most cells store some glycogen, liver and muscle cells are able to store large amounts.

When the blood glucose level begins to fall between meals, glycogen is slowly disassembled, a process called **glycogenolysis** (meaning "glycogen breakdown"), and the glucose units are released into the blood (Fig. 28–2).

The amount of glycogen stored in the liver is sufficient to maintain the blood glucose level for several hours. After the glycogen runs out, liver cells begin to convert amino acids and the glycerol portions of fat into glucose (Fig. 28–2). This process is appropriately named **gluconeogenesis** ("new glucose production"). Several hormones influence the various processes by which the liver regulates blood glucose

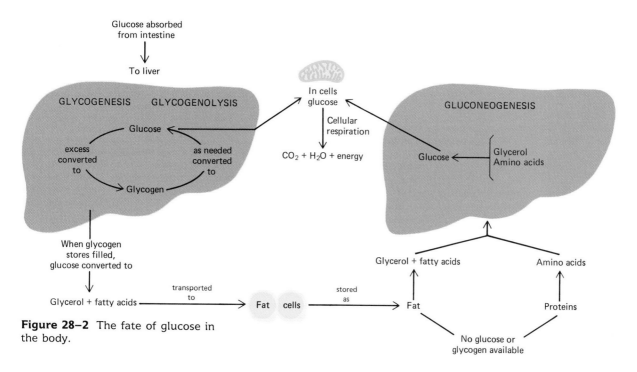

Figure 28–2 The fate of glucose in the body.

level; these will be discussed in Chapter 30. When too much carbohydrate-rich food is eaten, the liver cells may become fully packed with glycogen and still have excess glucose to manage. In this situation liver cells convert excess glucose to fatty acids and glycerol. These compounds are then synthesized into triacylglycerols and sent to the fat depots of the body for storage.

Fate of Amino Acids

Most excess amino acids are removed from circulation by the liver. In the liver cells these are **deaminated;** that is, the amine group is removed. Amine groups are waste products and are converted to **urea** for excretion from the body. The remaining carbon chain of the amino acid (called a **keto acid**) may be converted into carbohydrate or lipid and used as fuel or stored (Fig. 28–3). Thus, even a person who eats protein almost exclusively can become fat if he eats too much. Amino acids circulating in the blood are removed as needed by individual body cells and used primarily for synthesis of proteins.

Fate of Lipids

Although they take a circular route through the lymphatic system, absorbed fats eventually enter the blood. From the blood, fat is taken up by the adipose tissues and stored. When needed, stored fat is hydrolyzed to fatty acids and released into the blood. Before these fatty acids can be used by cells as fuel, they must be broken down into smaller compounds and combined with coenzyme A to form molecules of acetyl coenzyme A (Fig. 28–4). This transformation is accomplished in the liver by a process known as **β (beta) oxidation.** For transport to the cells, acetyl coenzyme A is converted into one of three types of **ketone bodies.** Normally, the level of ketone bodies in the blood is low, but in certain abnormal conditions, such as starvation or diabetes mellitus, fat metabolism is tremendously increased. Ketone bodies are then produced so rapidly that their level in the blood becomes excessive and causes the blood to become too acidic. Such disruption of normal pH balance can lead to

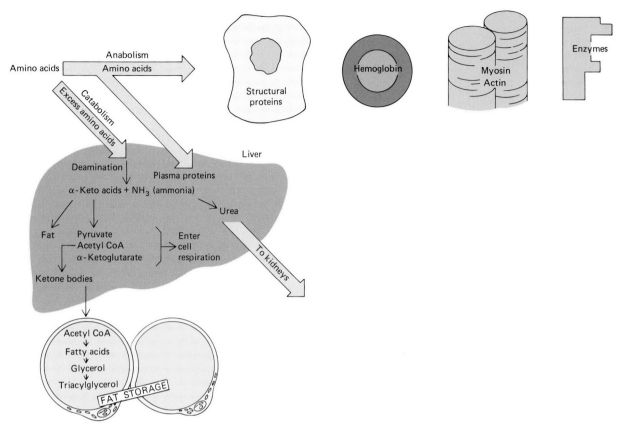

Figure 28–3 Overview of protein metabolism.

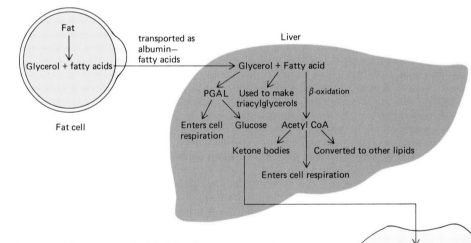

death. (How Eskimos, who live on diets extremely high in fat, manage to maintain acid-base homeostasis is somewhat of a mystery.) Figure 28–5 summarizes the interrelationships of carbohydrate, protein, and lipid metabolism.

Energy Balance

As you know, the term *metabolism* refers to all the chemical reactions that take place in the body. The amount of energy (heat) liberated by the body during metabolism is a measure of the metabolic rate; much of the energy expended by the body is ultimately converted to heat. Metabolic rate may be expressed either in Calories of heat energy expended per hour per day, or as a percentage above or below a standard normal level.

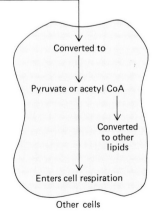

Figure 28–4 Overview of lipid metabolism.

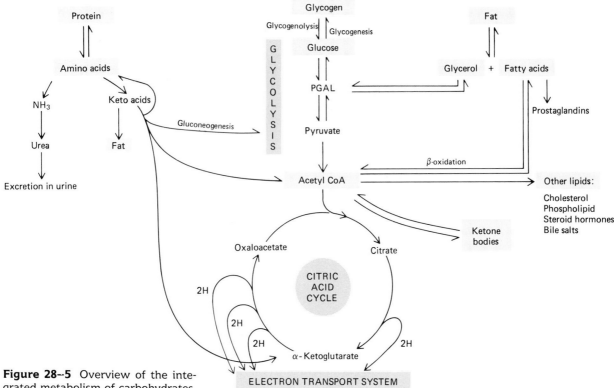

Figure 28–5 Overview of the integrated metabolism of carbohydrates, fats, and proteins. This diagram is greatly simplified and illustrates only a few of the principal pathways.

The **basal metabolic rate (BMR)** is the body's basic cost of metabolic living, that is, the rate of energy used during resting conditions. An individual's **total metabolic rate** is the sum of his or her BMR and the energy used to carry on all daily activities. An athlete or a laborer has a greater total metabolic rate than a teacher or executive who does not exercise regularly.

An average-sized man who does not engage in any exercise program and who sits at a desk all day expends about 2000 Calories daily. If the food he eats contains about 2000 Calories, he will be in a state of energy balance; that is, his energy input will equal his energy output. This is an extremely important concept, for when

Energy (Calorie) input = energy output,

body weight remains constant. When energy output is greater than energy input, stored fat is burned and body weight decreases. On the other hand, people gain weight when they take in more energy (Calories) in food than they expend in daily activity—in other words, when

Energy (Calorie) input > energy output.

THE PROBLEM OF OBESITY

Some 40 million Americans eat too much. **Obesity,** the excess accumulation of body fat, is a serious form of malnutrition, and in our affluent society it has become a problem of epidemic proportions. An overweight person places an extra burden upon his or her heart and, being more susceptible to heart disease and other ailments, tends to die at a younger age than people of normal weight. According to insurance statistics, men who are 20% or more overweight bear a 43% greater risk of dying from heart disease, a 53% greater risk of dying from cerebral hemorrhage, and a 133% greater risk of dying as a result of diabetes, compared with men of normal weight. A man who is 20% overweight is 30% more likely to die before retirement age than if his weight were normal. Yet one-third of our working population is 25% or more overweight.

Why are so many people overweight? Recent studies show that in some cases the problem stems from early childhood. The number of fat cells in the body is apparently determined early in life. When babies or small children are overfed, abnormally large numbers of fat cells are formed. Later in life these fat cells may be fully stocked with excess lipids or may be shrunken, but they are always there. People with such increased numbers of fat cells are thought to be more susceptible to obesity than those with normal numbers. Recent research also indicates that there may be individual differences in the efficiency with which we metabolize our food. In some, food may be very efficiently processed, so a high percentage of nutrients ingested is actually absorbed. In others, food may be less efficiently processed, so more is wasted. These fortunate individuals can perhaps eat more because a lower percentage of Calories is actually absorbed.

Whatever the underlying causes, overeating is the only way to become obese. Although water retention does increase body weight, it does not affect fat storage; water excesses can be diminished faster and more easily than fat excess. It has been estimated that for every 9.3 Calories of excess food taken into the body, 1 gram of fat is stored. (An excess of about 140 Calories per day for a month will result in gaining 1 pound.)

A 15-year-old male athlete requires a much larger food intake to support his growth and physical activity than he does ten years later as a sedentary accountant. Those who continue eating at the same level even though their activities require far less energy are destined to become overweight.

Because so many people are overweight, dieting has generated a multimillion-dollar industry embracing diet foods, formulas, pills, books, clubs, and slenderizing devices. Despite this, the only cure for obesity is to shift the energy balance so that intake is less than output. Then the body will have to draw on its fat stores for the missing calories, and as the fat is mobilized and burned, body weight decreases. This can best be accomplished by a combination of increased exercise and decreased total caloric intake. Most nutritionists agree that the best reducing diet is a well-balanced one containing a normal proportion of fats, carbohydrates, and proteins. In other words, eat everything but in smaller quantity.

THE PROBLEM OF WORLD FOOD SUPPLY

If all food in the world were equally distributed and each person received identical quantities, we would all be malnourished. If the entire world's food supply were parceled out at the dietary level of the United States, it would feed only about one-third of the human population. We have managed to increase world food supply greatly by developing scientific methods of agriculture. But human population has multiplied even faster than the food needed to nourish it, and nutrition is today one of the world's most serious problems. More than 50% of the 4 billion human beings on earth suffer from malnutrition. Some 15 million, most of them children, die each year as a result of malnutrition. Even in the United States hunger is a daily problem for more than 14 million people, nearly half of them children.

Millions of people suffering from mild malnutrition do not exhibit obvious disease symptoms. However, they cannot function efficiently

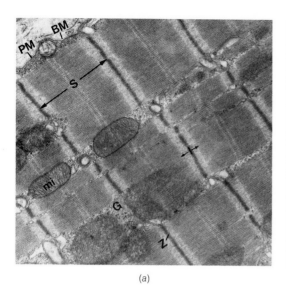

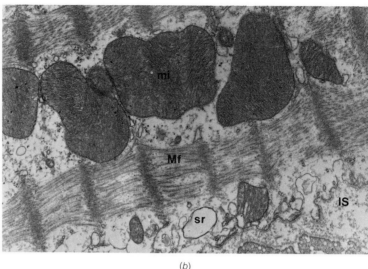

(a)

(b)

Figure 28–6 Protein-calorie malnutrition in rats over a six-week period resulted in changes in the heart muscle. (a) Cardiac muscle from a control rat (approximately ×14,900). Note the very regular arrangement of the contractile elements. *PM*, plasma membrane; *BM*, basement membrane; *mi*, mitochondria; *G*, glycogen granule; *Z*, Z line; *S*, sarcomere; *I*, I band. (b) Cardiac muscle from a protein- and calorie-malnourished rat (approximately ×20,000). Note the disruption of the contractile elements due to rupture and loss of myofilaments (*Mf*). The endoplasmic reticulum (*sr*) is dilated. *IS*, interstitial space. (Courtesy of Dr. M.A. Rossi and S. Zucoloto.)

because they are weak, easily fatigued, and highly susceptible to infection. Iron, calcium, and vitamin A are commonly deficient nutrients. An estimated quarter of a million children become permanently blind every year as a result of inadequate intake of vitamin A.

Protein Poverty

Of all the required nutrients, essential amino acids are the ones most often deficient in the diet. Millions of people suffer from poor health and a lowered resistance to disease because of protein deficiency (Figs. 28–6, 28–7). Children's physical and mental development may be retarded when the essential building blocks of cells are not provided in the diet. Common childhood diseases, such as measles, whooping cough, and chicken pox, are often fatal in the malnourished.

When a pregnant woman subsists on a diet lacking sufficient protein, the development of the fetus is jeopardized. Perhaps the greatest toll is taken upon the brain. Studies of the brains of babies who died of malnutrition showed that they had as few as 40% of the normal number of brain cells. Brain development is most critical before birth and during the first two years of life. A child deprived of needed nutrients during that time may never make up the lost growth and development. Even moderate protein malnutrition manifests itself in clumsiness, reduced manual skills, retarded language development, and lower intelligence.

Figure 28–7 A child who has experienced three periods of severe protein deficiency. The three discolored, light strips in her hair correspond to the periods of protein deficiency, when her body was unable to produce the pigment melanin, which is a protein. (FAO photo by Marcel Ganzin.)

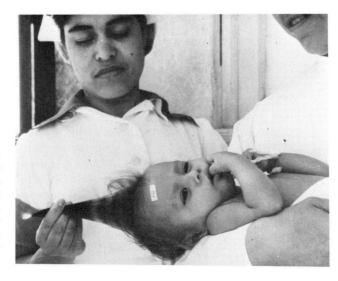

Severe protein malnutrition results in the condition known as **kwashiorkor.** The term, an African word that means "first-second," refers to the situation in which a first child is displaced from its mother's breast when a younger sibling is born. The older child is placed on a diet of starchy cereal or cassava that is deficient in protein. Growth becomes stunted, muscles are wasted, edema develops (as displayed by a swollen belly), the child becomes apathetic and anemic, and metabolism is impaired (Fig. 28–8). Without essential amino acids the digestive enzymes themselves cannot be manufactured, so that eventually what little protein is ingested cannot be digested. Dehydration and diarrhea develop, often leading to death.

Population and Food Supply

For the current human population to be adequately fed, food production must be substantially increased. Meanwhile the population continues to increase at a frightening rate. Figure 28–9 illustrates the quickening pace of human population expansion. This graph shows that it took from the time of the first human beings on earth until the year 1830 for the human population to reach 1 billion. But in only 100 years, from 1830 to 1930, the population doubled to 2 billion. In only 30 more years (by 1960) the third billion was added. By 1975 the human population probably exceeded 4 billion. The doubling time for world population is now only 35 years. By the year 2000 about 7 billion people will be crowded onto our planet.

It has been estimated that by the year 2000 the United States will have only 84% of the land necessary to produce food for its own citizens (more than 300 million). The rest of the land will be severely eroded or covered by housing and shopping centers. Each year more than 2 million acres of United States agricultural land is paved over for new housing developments, highways, and parking lots.

Figure 28–8 Child suffering from kwashiorkor, a disease caused by severe protein deficiency. Note the characteristic swollen belly, which results from fluid imbalance. (United Nations, Food and Agricultural Organization photo by P. Pittet.)

What Are the Solutions?

Is it possible to multiply our food supply sufficiently to meet the demands of a hungry world? We will discuss three possibilities: (1) improving conventional agriculture, (2) farming the sea, and (3) developing exotic food sources.

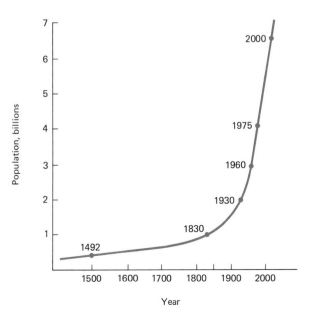

Figure 28–9 Human population growth.

IMPROVING CONVENTIONAL AGRICULTURE

Food production could be increased either by cultivating more land or by improving crop yields on land already being farmed.

CULTIVATING MORE LAND. Most land that can be cultivated, from an economically feasible standpoint, is now under cultivation. Indeed, such land is diminishing as a result of erosion and urbanization. In many regions the soil is so poor that it is not practical to attempt to cultivate it. Some areas possess fertile soil but little rainfall. In these regions farming is dependent upon artificial irrigation. Unfortunately, the water demands of an expanding population, as well as the consequent pollution of waterways, are steadily reducing the amount of fresh water available for irrigation.

IMPROVING CROP YIELDS—THE GREEN REVOLUTION. The most practical solution, and the one with the most potential for alleviating the hunger crisis, is the application of new technological advances for improving crop yields on land already under cultivation. More intensive and sophisticated methods of agriculture have contributed to recent significant increases in crop yields. Some agriculturists are so hopeful of continued progress in multiplying food supply by these methods that they have suggested that we are in the midst of a Green Revolution. This agricultural revolution is based primarily upon the development of new "miracle grains" of rice and wheat, more intensive use of fertilizers, pesticides, and irrigation, and recent breakthroughs in the development of high-protein strains of cereals. However, using these techniques requires government subsidies and sophistication and know-how. Manufacture and use of fertilizers, pesticides, and farm machinery require a great deal of energy input, and are therefore very expensive.

What are "miracle grains"? These are new genetic varieties of wheat and rice that have been especially bred to produce greater yields. They are short-stemmed to permit a heavier head of grain, they are adaptable to a wide range of climates, and they mature so quickly that two or three crops can be grown per year. Their chief disadvantage is that their success depends upon large quantities of fertilizer. Most countries do not have the natural resources, including fuel, needed to produce fertilizer, so it must be imported at great cost. Moreover, intensive use of chemical fertilizers destroys the bacteria necessary for normal cycling of nitrogen, while favoring bacterial denitrification (so large quantities of the inorganic nitrogen added to the fertilizer become useless to the plants). The result is that the soil soon becomes unable to maintain itself, becomes "addicted" to fertilizer, and fertilizers must be added continually. Nature requires about 500 years to produce one inch of good topsoil, but humans can destroy it in just a few years.

Scientists have managed to breed strains of grain that are resistant to various plant pests and diseases. Although such disease-resistant crops may appear to be an answer to the farmer's prayers, they are not without genetic hazard. If a few supergrains with the same disease-resistant characteristics are planted worldwide, a new plant disease or insect could destroy large portions of the world food supply with a single disastrous blow. And new varieties of insects and disease organisms do appear in nature constantly. Genetic variability is the key to survival. When we impose an artificial homogeneity upon the genetic complements of our crops, we make them vulnerable to extinction. Food production could be devastated before breeders had time to develop new strains.

At best the Green Revolution can buy only a few more years. Our central problem is the population problem, and if we go on producing more human beings than we are able to feed, the Green Revolution will not be able to bail us out.

FARMING THE SEA

There is a popular myth that the seas are filled with food resources just waiting to be harvested. The truth is that overfishing has already caused the decline of many marine species. Because most fish are near the top of long food chains, they are relatively few in number. Fish are actually most plentiful not in the open sea but in coastal areas and estuaries where an abundance of minerals supports food chains. Unfortunately, these are also the areas closest to dense human populations, and therefore they are becoming increasingly polluted by human wastes. If we were to harvest algae on a large scale, we would also deplete fish populations by depriving them of their food-chain base. Furthermore, technology is not yet available for cultivating and harvesting algae economically, or for converting it into edible food once it is harvested.

Several research projects are in progress in the area of fish culture. Salmon, catfish, carp, and trout are among the many species whose suitability for fish farming is being tested. One of the problems is that expensive high-protein food is used to feed the fish being cultured.

EXOTIC FOOD SOURCES

Bacteria, yeast, and algae can be cultivated on sewage. These single-celled organisms produce great quantities of protein and can be harvested for animal feeds. Development of such animal feeds would free millions of tons of grain and legumes (now being fed to animals) for human use.

One promising exotic food source is the culture of yeast and bacteria on petroleum. These organisms synthesize protein several thousands times faster than domestic animals. Such single-celled organisms can be harvested, dried, and purified. The resulting white powder can be added to various foods. So far, protein produced in this way is being used only for livestock feed, but eventually it will be processed for human consumption. Although protein from petroleum offers some promise, it is not a long-term solution because the amount of petroleum on earth is limited and will eventually run out.

CHAPTER SUMMARY

I. For a balanced diet human beings and other animals require proteins, carbohydrates, lipids, vitamins, minerals, and water.
   A. Vitamins, which serve as components of coenzymes, may be divided into the fat-soluble group (A, D, E, K) and the water-soluble group (B complex and C).
   B. Most carbohydrates are ingested in the form of polysaccharides, starch, and cellulose. Carbohydrates are used primarily as fuel.
   C. Most lipids are ingested in the form of triacylglycerols. Lipids are used as fuel and as components of cell membranes. They are also used to synthesize steroid hormones and other lipid substances.
   D. The best distribution of essential amino acids is found in the high-quality proteins of animal foods.
II. After absorption, monosaccharides and amino acids are transported to the liver, where they are inventoried. Lipids are absorbed into the lymph system but eventually find their way into the blood.
   A. Excess glucose is removed from the blood by the liver cells, con-

verted to glycogen (glycogenesis), and stored until needed. When the glycogen stores of the liver (and muscle) cells are filled, excess glucose is converted to fatty acids and glycerol and stored as fat.

B. Excess amino acids are deaminated by the liver cells. Amine groups are excreted in urine, and the remaining keto acids are converted to carbohydrate and used as fuel, or converted to lipid and stored in fat cells.

C. Fatty acids are converted to molecules of acetyl-coenzyme A (by beta oxidation) and used as fuel. Excess fatty acids are stored as fat.

D. Basal metabolic rate is the body's cost of metabolic living. Total metabolic rate is the BMR plus the energy used to carry on daily activities. When energy (Calorie) input equals energy output, body weight remains constant.

III. Obesity is a serious health problem.

A. A person gains weight by taking in more energy, in the form of Calories, than he expends in activity.

B. Weight may be lost by expending more energy than is taken in. The needed energy is obtained by mobilizing fat and using it as fuel.

IV. More than half the people on earth suffer from malnutrition.

A. Essential amino acids are the nutrients most often deficient in the diet.

B. The only long-term solution to the problem of world food supply is slowing population growth.

C. Food supply can be increased by application of new technology for increasing yields on land already under cultivation (the Green Revolution).

Post-Test

1. Inorganic nutrients generally ingested as dissolved salts are called _____ .
2. Vitamins function mainly as _____ .

Match the nutrients in column B with the descriptions in column A.

Column A

_____ 3. Needed for hemoglobin synthesis
_____ 4. Used as fuel molecule
_____ 5. Deficiency results in goiter
_____ 6. Water-soluble vitamin
_____ 7. Deficiency results in rickets

Column B

a. carbohydrate
b. vitamin D
c. iron
d. iodine
e. none of the above

8. Most lipids are ingested in the form of _____ _____ .
9. In the digestive tract proteins are degraded to _____ _____ _____ ; most carbohydrates are degraded to _____ .
10. In glycogenesis glucose is chemically _____ .
11. In the liver cells excess _____ _____ _____ are deaminated; the amine groups are converted to _____ .
12. The _____ _____ _____ is the body's rate of energy use during resting conditions.
13. When energy input is greater than energy output, _____ .
14. Kwashiorkor is a disease caused by extreme _____ deficiency.

Review Questions

1. List the nutrients that must be included in a balanced diet.
2. Why, specifically, are each of the following essential? (a) iron, (b) calcium, (c) iodine, (d) vitamin A, (e) vitamin K, (f) essential amino acids, (g) pantothenic acid
3. Draw a diagram to illustrate the fate of carbohydrates after absorption.

4. Describe the fate of absorbed amino acids.
5. Describe the fate of absorbed fat.
6. Write an equation to describe energy balance and tell what happens when the equation is altered in either direction.
7. What is the best "cure" for obesity?
8. What is the Green Revolution? What are some of its drawbacks?

9. According to your text, what is the only logical long-term solution to the problem of insufficient world food supply?
10. Describe two means of increasing food supply that could be used in this century.

Readings

Beddington, J.R., and R.M. May. "The Harvesting of Interacting Species in a Natural Ecosystem," *Scientific American,* November, 1982, 62–69. Harvesting a biological resource such as krill affects the whales and other animals that normally feed upon it.

Brown, M.S., and J.L. Goldstein. "How LDL Receptors Influence Cholesterol and Atherosclerosis," *Scientific American,* November 1984, 58–66. Many Americans have too few LDL receptors, which normally remove particles carrying cholesterol from the circulation. Absence of these receptors puts individuals at high risk for atherosclerosis and heart attacks.

Hartbarger, J.C. and N.J. Hartbarger. *Eating for the Eighties: A Complete Guide to Vegetarian Nutrition.* Philadelphia, W.B. Saunders, 1981.

Krause, M.B., and L.K. Mahan. *Food, Nutrition, and Diet Therapy,* 7th ed., Philadelphia, W.B. Saunders Company, 1984. A comprehensive discussion of the science of nutrition and its application to the maintenance of health.

Kretchmer, N., and W. van B. Robertson. *Human Nutrition.* San Francisco, W.H. Freeman Company, 1978. An interesting collection of articles from *Scientific American* covering all levels of nutrition.

Swaminathan, M.S. "Rice," *Scientific American,* Vol. 250, no. 1, 80–93. This member of the grass family is one of three on which the human species largely subsists.

Chapter 29

DISPOSAL OF METABOLIC WASTES

Learning Objectives

After you have read this chapter you should be able to:

1. Describe the functions of the excretory system, including excretion, osmoregulation, and maintaining homeostasis, and identify principal waste products.
2. Summarize the mechanisms for waste disposal used by various types of animals, and describe how freshwater and marine fish deal with problems of osmoregulation.
3. Identify the organs that participate in waste disposal in human beings and other vertebrates.
4. Describe the role of the liver in processing wastes.
5. Draw, label, and describe the organs of the urinary system, giving the function of each.
6. Draw and label the principal parts of the nephron and give the function of the following structures: Bowman's capsule, glomerulus, collecting duct, afferent arteriole, efferent arteriole.
7. Describe the process of urine formation and the composition of urine.
8. Define renal threshold and describe its relationship to the disorder diabetes mellitus.
9. Describe the effects of ADH and variations in fluid intake on the volume and composition of urine, and give the physiological basis for diabetes insipidus.
10. Summarize the functions of the kidneys in maintaining homeostasis.

As cells carry on metabolic activities, waste products are generated. If allowed to accumulate, metabolic wastes would eventually reach toxic concentrations and threaten the homeostasis of the organism. For that reason, they must be removed from the body in a process known as **excretion.** Single-cell organisms can dispose of their wastes by simple diffusion, but larger animals may require elaborate excretory systems.

Excretory systems do more than remove metabolic wastes from the body. They also remove desirable substances that have reached undesirable concentrations. After you consume a large soft drink and a bag of pretzels, for example, your circulatory system contains excess water and salt that must be excreted. If a particular substance is in short supply, however, the excretory system will actively conserve it. Excretory systems are designed to maintain homeostasis by selectively adjusting the concentration of salts and other substances in blood and other body fluids.

Fluid balance is also largely regulated by the excretory system. The ability of an organism to regulate its fluid content, known as **osmoregulation,** is just as important in land-dwelling organisms as in aquatic. The body cells of higher organisms depend on the excretory system to preserve an optimum fluid environment around them.

WASTE PRODUCTS

The principal metabolic waste products in most animals are water, carbon dioxide, and nitrogenous (nitrogen-containing) wastes. Carbon dioxide is mainly excreted by respiratory structures; excretory organs, such as kidneys, excrete water and nitrogenous wastes.

The nitrogenous wastes are the special province of animal excretory systems. You should recall that amino acids and nucleic acids contain nitrogen. During the breakdown of excess amino acids the nitrogen-containing amino group is removed (deamination) and converted to **ammonia.** However, ammonia is highly toxic. Some organisms excrete it into the surrounding water before it can build up to toxic concentrations in their tissues, and a few even vent it directly to the air. But in many organisms, ourselves included, ammonia is quickly converted to some less toxic nitrogenous waste, such as **urea** or **uric acid.** When nucleotides from nucleic acids are broken down, uric acid is formed.

METABOLIC WASTE DISPOSAL AND OSMOREGULATION IN INVERTEBRATES

Sponges and cnidarians have no specialized excretory structures. Their wastes pass by diffusion from the intracellular fluid to the external environment. They save much of the energy that they would otherwise have to expend in respiration and excretion by taking advantage of the free energy provided by the water currents that sweep by or even through them. For this reason certain environments, such as coral reefs, are especially prone to damage when changes in water currents or stagnation occur. The vast majority of sponges and cnidarians live in a marine environment that is isotonic to their cells. They have no special problems of excess water intake.

Nephridial Organs

Nephridial organs are a common type of excretory organ in invertebrates. They consist of simple or branching tubes that usually open to the outside of the body through nephridial pores. Flatworms are the simplest

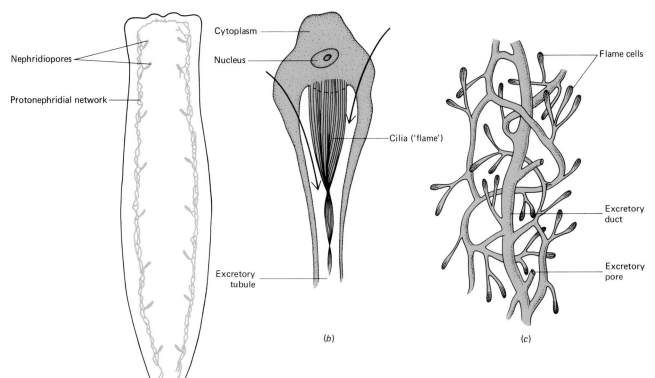

Nephridiopores

Protonephridial network

Cytoplasm

Nucleus

Cilia ('flame')

Excretory
tubule

Flame cells

Excretory
duct

Excretory
pore

(a)

(b)

(c)

Figure 29–1 (a) The excretory system of a typical flatworm. (b) Diagram of a flatworm flame cell. (c) Wastes collected by flame cells pass through excretory ducts and out of the animal through excretory pores.

animals with specialized excretory organs. Although metabolic wastes pass by diffusion through their body surface, these animals also have osmoregulatory nephridial organs, which consist of tubules with enlarged blind ends containing cilia. These organs, known as **protonephridia,** have **flame cells,** which possess brushes of cilia whose constant motion reminded early biologists of flickering flames (Fig. 29–1). A system of branching excretory ducts connects the protonephridia with the outside. The flame cells lie in the fluid that bathes the body cells, and wastes diffuse into the flame cells and from there into the excretory ducts. The beating of the cilia then expels the wastes through the excretory pores.

Annelids have nephridial organs called **metanephridia** in each segment of their bodies. The metanephridium is a tubule open at both ends through which drains coelomic fluid and all that it contains (Fig. 29–2). The inner end opens into the coelom as a ciliated funnel. Around each tubule is a network of capillaries, which removes wastes from the blood and also reabsorbs needed materials, such as water or glucose, from the coelomic fluid passing through the tubule. Since other materials are reabsorbed, wastes are concentrated before they pass out of the body. About 60% of an earthworm's total body weight is lost each day in the form of urine, so the animal would quickly dehydrate in surroundings where it could not absorb water to replace this.

Antennal Glands

Antennal glands, also called **green glands,** are the principal excretory organs of crustaceans (Fig. 29–2). A pair of these structures is located in the head, often at the base of the antennae. Each gland consists of a coelomic sac, a greenish glandular chamber with folded walls, an excretory tubule, and an exit duct that is enlarged to form a bladder in some species.

Fluid from the blood is filtered into the coelomic sac, and its composition is adjusted as it passes through the excretory organ. Needed materials are reabsorbed into the blood. Wastes can also be actively secreted from the blood into the filtrate within the excretory organ.

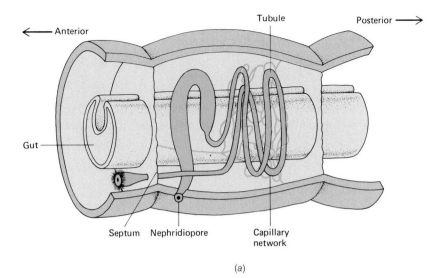

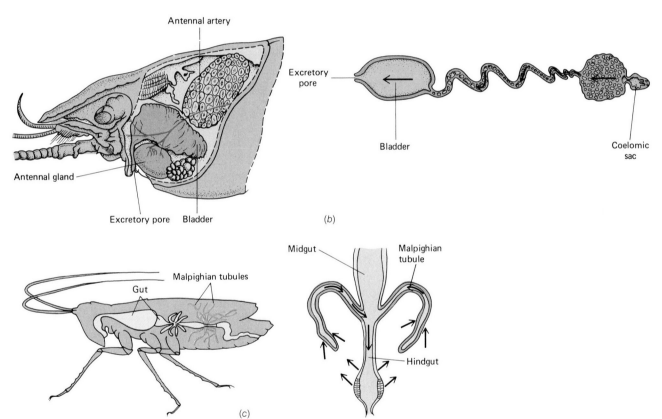

Figure 29–2 Representative invertebrate excretory systems. (*a*) The earthworm possesses metanephridia, which receive fluid from the coelom and an associated network of capillaries that reabsorb needed materials before final elimination. (*b*) Many crustaceans, such as the crayfish, have antennal glands, located at the base of the antennae, which serve the function of kidneys. (*c*) Insects possess Malpighian tubules that concentrate waste materials from blood in the hemocoel and convey them into the hindgut for elimination.

Malpighian Tubules

The excretory system of insects consists of organs called **Malpighian tubules** (Fig. 29–2). There may be two to several hundred tubules, depending upon the species. Malpighian tubules have blind ends that lie in the hemocoel, bathed in blood. Their cells transfer wastes by diffusion or active transport from the blood to the cavity of the tubule. Each tubule has a muscular wall and writhes slowly to help move wastes to the gut cavity. The Malpighian tubules empty into the intestine between the midgut and the hindgut. Water is reabsorbed from the contents by the walls of the Malpighian tubules and returned to the hemocoel. Uric acid, the major waste product, is excreted as a semidry paste with a minimum of water. This adaptation helps to conserve the insect's body fluids.

METABOLIC WASTE DISPOSAL AND OSMOREGULATION IN VERTEBRATES

In most vertebrates, not only the urinary system but also the skin, lungs or gills, and digestive system function to some extent in metabolic waste disposal and fluid balance. Some vertebrates have special salt glands that excrete excess salt. By removing excess water, salts, and other potentially toxic waste materials, all these organs help to maintain homeostasis.

The Vertebrate Kidney

The main excretory organs in vertebrates are kidneys, which excrete nitrogenous wastes, water, a variety of salts, and some other substances in the form of urine. The vertebrate kidney is composed of functional units called nephrons, which will be described in the section on the human urinary system. Most vertebrate kidneys function by filtration and reabsorption. Blood is filtered nonselectively, and the initial filtrate that enters the nephron contains all the substances present in the blood except large compounds, such as proteins. (Blood cells and platelets are, of course, too large to be filtered out of the blood.) As the filtrate passes through the coiled tubules of the nephron, needed materials, such as glucose, amino acids, salts, and water, are selectively reabsorbed into the blood. Thus the composition of the filtrate is slowly adjusted, and the urine that is finally excreted consists of metabolic waste products and excess materials such as salts.

Osmoregulation

Aquatic vertebrates have a continuous problem of osmoregulation (Fig. 29–3). Because marine bony fishes have blood and body fluids that are hypotonic to sea water, they tend to lose water osmotically even though they live in water. Many marine bony fish compensate by drinking sea water constantly. They retain the water and excrete salt by the action of specialized cells in their gills. These fish have small, sometimes almost vestigial kidneys that excrete very little urine.

Marine chondrichthyes (sharks and rays) have a different set of osmoregulatory features that allows them to tolerate the high salt concentrations of their environment. They are able to accumulate, retain, and tolerate urea in concentrations high enough to be somewhat hypertonic to sea water. This results in a net inflow of water into their bodies and necessitates large kidneys to get rid of the excess. The system evidently does result in some imbalance, which is compensated for by a salt-excreting rectal gland in many species.

Freshwater fish are hypertonic to their surroundings and are therefore in continuous danger of becoming waterlogged, given the osmotic

Figure 29–3 Problems of osmoregulation in marine and freshwater fish. (a) Marine fish continuously lose water to their surroundings. They compensate by drinking large amounts of seawater, producing only a small amount of urine and excreting excess salt from the gills. (b) Sharks retain urea and gain water by osmosis; their large kidneys excrete the excess water. (c) In freshwater fish, water continuously enters the body by osmosis. To compensate, they produce large amounts of urine. A freshwater fish has the opposite salt problem of a marine fish. It must absorb salt through the gills from the surrounding water.

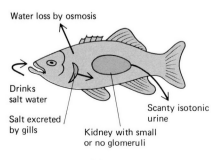

(a)

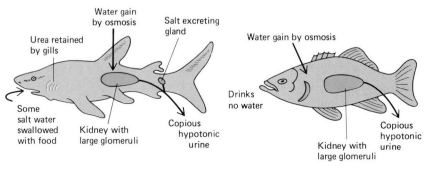

(b)

(c)

inward flow of water. Although they are covered with a mucous secretion that retards the passage of water into the body, water does enter through the gills. The kidneys of these fish are adapted to filter out excess water, and they excrete a copious dilute urine. Water entry, though, is only part of the problem of osmoregulation in freshwater fish. These animals also tend to lose salts to the surrounding fresh water. To compensate, special cells in the gills actively transport salt from the water into the body.

Most amphibians are at least semiaquatic, and their mechanisms of osmoregulation are similar to those of freshwater fish. They produce a large amount of dilute urine. A frog can lose through its urine and skin an amount of water equivalent to one-third of its body weight in a day. Loss of salt, both through the skin and in the urine, is compensated for by active transport of salt by special cells in the skin.

Whales, dolphins, and other marine mammals ingest sea water along with their food. Their kidneys produce a very concentrated urine, much more salty than sea water. This is an important physiological adaptation, especially for marine carnivores. The high-protein diet of these animals results in production of large amounts of urea, which must be excreted in the urine or, in some cases, by special accessory salt glands. Desert-dwelling mammals must also conserve water. The kangaroo rat has such efficient water recycling in its kidneys that it would even be able to drink sea water without harm and usually subsists on the water content of its food plus whatever water it generates metabolically.

THE MAMMALIAN URINARY SYSTEM

The mammalian **urinary system** consists of the kidneys, the urinary bladder, and associated ducts.

Most of the deamination of amino acids takes place in the liver, which is also the site of production of both urea and uric acid (Fig. 29–4). In addition, most of the bile pigments produced by the breakdown of red blood cells are normally excreted by the liver into the intestine and then pass out of the body with the feces.

Figure 29–4 (*a*) In terrestrial animals the kidney conserves water by reabsorbing it. (*b*) Disposal of metabolic wastes in humans and other terrestrial animals. To conserve water, a small amount of hypertonic urine is produced. Nitrogenous wastes are produced by the liver and transported to the kidneys. All cells produce carbon dioxide and some water during cellular respiration. The kidneys, lungs, skin, and digestive system all participate in the disposal of metabolic wastes.

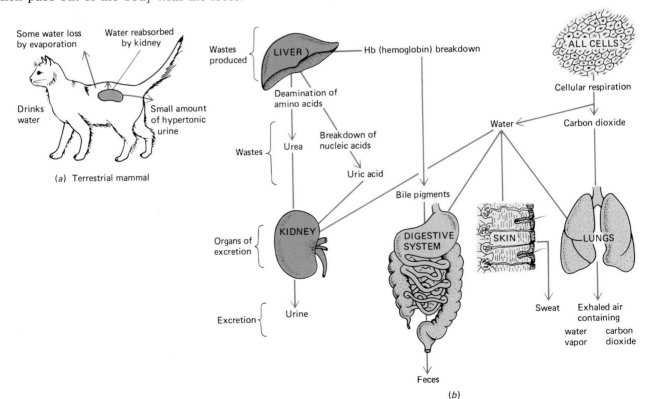

(*a*) Terrestrial mammal

(*b*)

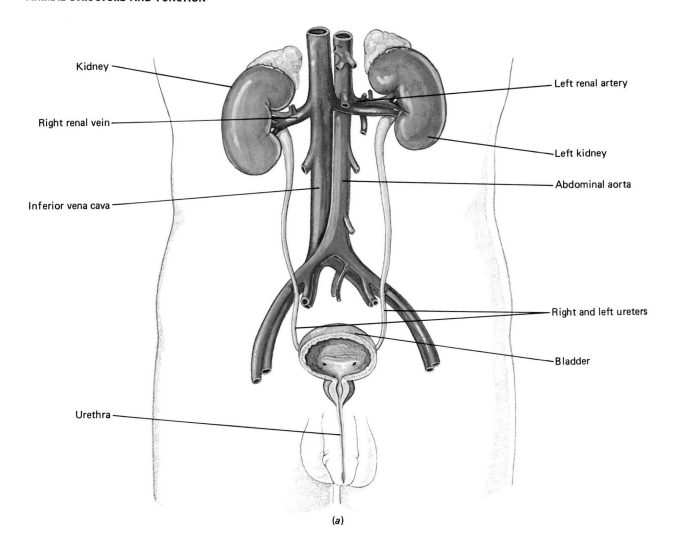

Kidney

Right renal vein

Inferior vena cava

Urethra

Left renal artery

Left kidney

Abdominal aorta

Right and left ureters

Bladder

(a)

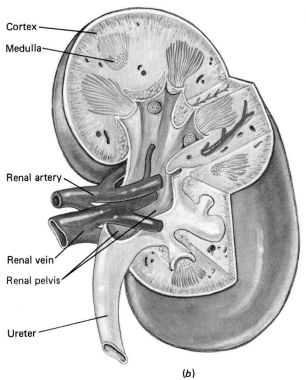

Cortex

Medulla

Renal artery

Renal vein

Renal pelvis

Ureter

(b)

Figure 29–5 (a) The human male urinary system. (b) The human kidney.

Design of the Urinary System

The overall structure of a typical mammalian urinary system (that of the human being) is shown in Figure 29–5. Located just below the diaphragm, the kidneys look like a pair of giant, dark-red lima beans, each about the size of a fist. The outer portion of the kidney is called the **cortex;** the inner portion, the **medulla;** and the portion that connects to the ureters, the **renal pelvis.** The job of the kidneys is to produce urine from the excess water, wastes, and other materials that they filter from the blood. By adjusting the amount of water and salts excreted, the kidneys play a vital role in maintaining the internal chemical balance of the body.

Urine passes from the kidneys through the paired **ureters,** ducts about 25 cm (10 inches) long, which connect the kidneys with the urinary bladder. The **urinary bladder** is a remarkable organ capable of holding (with practice) up to 800 ml (about a pint and a half) of urine. Emptying the bladder changes it in a moment from the size of a melon to that of a pecan. This remarkable feat is made possible by the smooth muscle and special epithelium of the bladder wall, which is capable of great shrinkage and stretching (Fig. 29–6).

As urine leaves the bladder, it flows through the **urethra,** a duct leading to the outside of the body (Fig. 29–7). In the male the urethra is lengthy and passes through the penis. Semen as well as urine is transported through the male urethra. In the female the urethra is short and transports only urine. Its opening to the outside is just above the opening of the vagina. The length of the male urethra discourages bacterial invasions of the bladder, and thus such infections are more common in females than in males.

The urethra has two sphincter valves. These muscles open by reflex when the volume of urine in the bladder reaches about 350 ml (10.5 ounces), at which time **urination** (release of urine from the bladder) takes place. The bladder is not under voluntary nervous control in the sense that our skeletal muscles are voluntary; what we call bladder control depends upon the ability to facilitate or inhibit this reflex voluntarily. For example, one can voluntarily empty the bladder at a convenient time even before it is full. On the other hand even when the volume of urine in the bladder has exceeded 350 ml, one can inhibit urination for some time

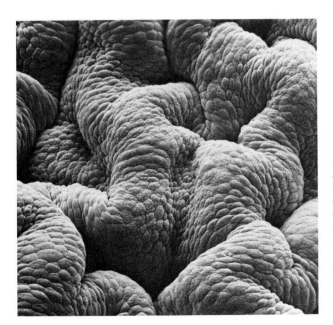

Figure 29–6 Scanning electron micrograph of the lining of the urinary bladder illustrating the special transitional epithelium and folds resulting from the contraction of the underlying smooth muscle. As urine accumulates in the bladder, the folds stretch out. (From Kessel, R.G., and R.H. Kardon. *Tissues and Organs: A Text-Atlas of Scanning Electron Microscopy,* San Francisco, W.H. Freeman, 1979.)

Figure 29–7 The epithelium of the urethra. This scanning electron micrograph (approximately ×2400) shows a stratified squamous epithelial cell that has broken from its comrades and is about to be flushed away in the urine stream. (From Colleen, S. "A Histochemical and Ultrastructural Study of Human Urethral Uroepithelium," *Acta. Path. Microbiol. Immunol. Scand.* Sect A, 90:103–111, 1982.)

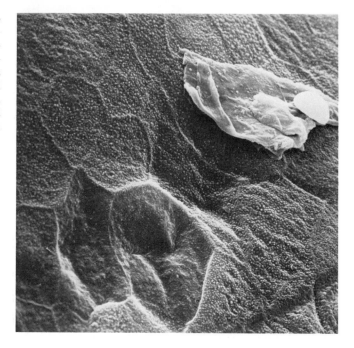

until convenient. Such voluntary control cannot be exerted by an immature nervous system, so that most babies are unable to develop urinary control until about age 2—no matter how hard anxious parents try to teach them.

The emptying of the bladder constitutes one of the few normal occurrences of positive feedback in biology. One might think that when urination begins, the pressure within the bladder would swiftly fall below the threshold necessary to trigger the urination reflex so that urination would promptly stop, leaving the bladder mostly full. Yet the healthy bladder empties completely. The explanation is that the bladder contracts so strongly that the pressure within it actually rises once urination has begun, which stimulates the urination reflex even more strongly. That is why it is so difficult to stop urination once it has begun.

The Nephron

Each kidney consists of more than a million functional units called **nephrons.** Each nephron (Fig. 29–8) consists of a cuplike **Bowman's capsule** connected to a long, partially coiled **renal tubule.** Positioned within the cup-shaped capsule is a cluster of capillaries known as the **glomerulus.**

Urine Formation

Blood circulating through the glomerulus is filtered by Bowman's capsule, whose walls are made up of cells called **podocytes.** These cells keep most of the protein of the blood out of the filtrate but otherwise do not greatly influence its content, so the filtrate chemically resembles blood plasma minus albumins, globulins, and of course blood cells. The filtrate flows into the renal tubule, where its contents are inventoried. Useful substances are returned to the blood, but waste materials pass on into a larger **collecting duct** that eventually leads to the ureter. Thus, urine formation consists of two important phases: filtration through Bowman's capsule and reabsorption by the tubules. A third mechanism, secretion, involves only a few substances, such as certain drugs.

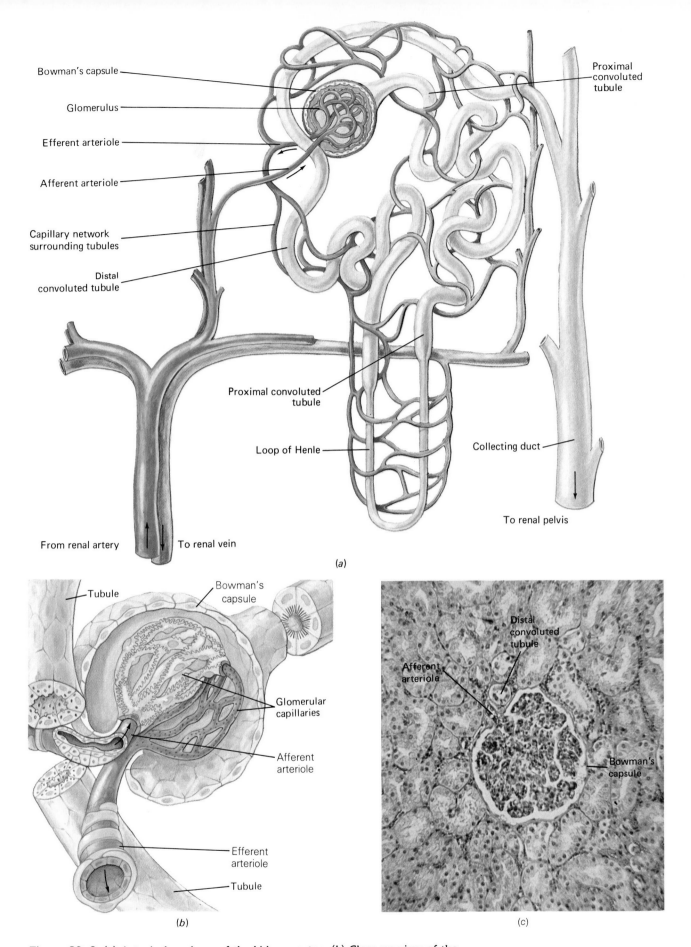

Figure 29–8 (*a*) A typical nephron of the kidney cortex. (*b*) Close-up view of the glomerulus and Bowman's capsule. (*c*) Appearance of Bowman's capsule under a light microscope.

603

FILTRATION

Blood is delivered to the kidneys by the **renal arteries.** Branches of the renal arteries give rise to **afferent arterioles.** The term *afferent* signifies that they conduct blood *into* the glomerulus. Then blood flows into the capillaries that make up the glomerulus. Because it flows through the capillaries under high pressure, more than 10% of the plasma is forced out of the capillaries and into Bowman's capsule. This process of **filtration** is somewhat similar to the mechanism whereby tissue fluid is formed as blood flows through other capillary networks in the body (see Chapter 24). The fluid that passes into Bowman's capsule is known as the **glomerular filtrate.**

Almost 25% of the cardiac output is delivered to the kidneys each minute, so every 4 minutes the kidneys receive a volume of blood equal to the total volume of blood in the body. Every 24 hours about 180 liters (about 45 gallons) of filtrate are produced. Common sense tells us that no one could excrete urine at the rate of 45 gallons per day: Within a few minutes dehydration would become a life-threatening problem.

REABSORPTION

Fortunately, about 99% of the filtrate is reabsorbed into the blood, leaving only about 1.5 liters to be excreted as urine. **Reabsorption** is the function of the renal tubules (Fig. 29–9). Reabsorption not only reduces the volume of the filtrate but also returns useful substances to the blood.

You may recall that in the usual circulatory pattern capillaries deliver blood into veins. Circulation in the kidneys is an exception in that blood flowing from the glomerular capillaries next passes into an **efferent arteriole,** so called because it conducts blood *away* from the glomerulus. The efferent arteriole delivers blood to a second capillary network, which surrounds the renal tubule. (Find these structures in Figure 29–8.) The second set of capillaries delivers blood into small veins, which eventually merge to form the large renal vein draining each kidney. The first set of capillaries, those of the glomerulus, provides the blood to be filtered; the second set receives materials returned to the blood by the tubule.

Filtration is not a selective process. Although the filtrate contains very little protein, it does contain such vital materials as glucose, amino acids, and salts, as well as large amounts of water. Reabsorption, on the other hand, *is* highly selective. Wastes, surplus salts, and water are re-

Figure 29–9 Scanning electron micrograph of part of a nephron located in the kidney medulla. Notice that the tubule shown here has a wall composed of hexagonal cells (called "cuboidal" because of the way they look when a conventionally prepared slide is viewed under a light microscope). The speckled appearance of some of the cells results from the presence of hundreds of microvilli on each cell. (Courtesy of Drs. J.L. Ojeda, J.A. Garcia-Porrero, and M.A. Ros, University of Santander, Spain. *Stain Technology* Vol. 59, 1984. Used by permission of the Williams and Wilkins Company.)

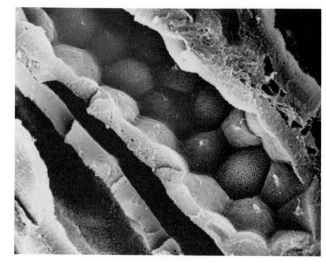

tained by the tubule, but glucose, amino acids, and other useful materials are reabsorbed into the blood. This is accomplished by a combination of diffusion, osmosis, and active transport.

An important example of this involves the transport of sodium and water. Sodium is actively transported out of the renal tubule, and water follows by osmosis, but more sodium than water is transported into the intercellular fluid (interstitium) of the kidney. Consequently, a hypertonic state develops in the medullary portion of the kidney through which the collecting ducts must pass on their way to the pelvis. A large part of the reabsorption of water occurs not in the nephron but in the larger collecting ducts that receive the output of several nephrons. The tissue fluid of the medulla of the kidney is made hypertonic by a countercurrent exchange mechanism. Water passes out through the walls of the collecting ducts by osmosis and is ultimately carried away from the kidney by the blood.

Normally, substances that are useful to the body are completely reabsorbed from the tubules. However, if a large excess of a particular substance is present in the blood, the tubules may not be able to return all of it. The maximum concentration of a specific substance in the blood at which complete reabsorption can take place is termed the **renal threshold** for that substance. When a substance exceeds its renal threshold, the portion not reabsorbed is excreted in the urine. Some substances, such as urea, have very low thresholds, so even when present in small concentrations, not much is reabsorbed. Other substances, such as glucose, amino acids, and hormones, have high renal thresholds and are normally completely reabsorbed. The threshold values of all these substances are so arranged that the tubules not only largely cleanse the blood of wastes but also regulate each component of the internal chemical environment of the body. Every day the tubules reabsorb more than 40 gallons of water, 2.5 pounds of salt, and about 0.5 pound of glucose. Most of this is reabsorbed many times over.

What happens if a substance in the blood exceeds its threshold value? An important example of this occurs in the condition **diabetes mellitus.** Because of an insufficiency of the hormone insulin, a diabetic suffers from impaired carbohydrate metabolism. Glucose accumulates in the blood instead of being efficiently absorbed and utilized by the cells. The concentration of glucose filtered into the nephron exceeds the renal threshold, and the balance of the glucose is excreted in the urine. Presence of glucose in the urine is evidence of this disorder, as is an increased total production of urine.

SECRETION

A few substances are actively moved from the capillaries surrounding the tubules into the tubules. This mechanism is called **secretion.** Just how important a role secretion plays in human urine formation is not known. Potassium is one substance known to be both reabsorbed and secreted by the tubules. Certain drugs, such as penicillin, appear to be excreted from the body by tubular secretion.

Composition of Urine

By the time the filtrate reaches the ureter, its composition has been precisely adjusted. Useful materials have been returned to the blood by reabsorption. Wastes and excess materials that entered by filtration or secretion have been retained by the tubules. The adjusted filtrate, called urine, is composed of about 96% water, 2.5% nitrogenous wastes (primarily urea), 1.5% salts, and traces of other substances, such as bile pigments, which may contribute to the characteristic color and odor. Healthy urine is sterile and has been used to wash battlefield wounds where clean water is not available. However, urine swiftly decomposes when exposed to bacterial action, forming ammonia and other products. It is the ammonia that produces the diaper rash of infants.

Regulation of Urine Volume

The amount of urine produced depends upon the body's need for retaining or ridding itself of water. When one drinks a great deal of water, a correspondingly large amount of urine is produced. Excess water absorbed from the digestive tract into the blood is removed by the kidneys, so the steady volume and composition of blood are maintained.

The kidney receives its information regarding the current state of the blood by a rather circuitous route (Fig. 29–10). When fluid intake is low, the body begins to dehydrate. The concentration of salts dissolved in the blood becomes greater, causing an increase in the osmotic pressure of the blood. Specialized receptors in the brain and in large blood vessels are sensitive to such change. The posterior lobe of the pituitary gland responds by releasing **antidiuretic hormone (ADH),** which serves as a chemical messenger carrying information from the brain to the collecting ducts of the kidney. There it causes the walls of the ducts to become more permeable so that water is more efficiently reabsorbed into the blood. In this way more water is conserved for the body and the blood volume is increased, restoring conditions to normal. A small amount of

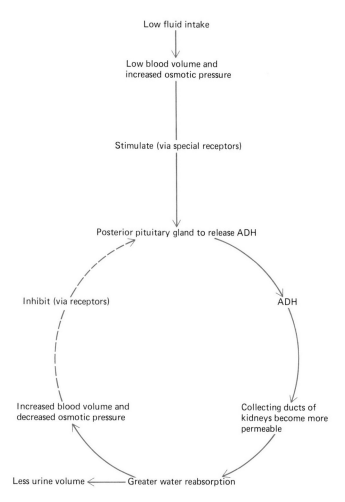

Figure 29–10 Regulation of urine volume reflects the blood volume and osmotic pressure.

concentrated urine is produced. A thirst center in the brain also responds to dehydration, stimulating an increase in fluid intake.

On the other hand, when a lot of fluid is consumed, the blood becomes diluted and its osmotic pressure falls. Release of ADH by the pituitary gland decreases, lessening the amount of water reabsorbed from the collecting ducts. A large volume of dilute urine is thus produced.

Occasionally, the pituitary gland malfunctions and does not produce sufficient ADH. The resulting condition is termed **diabetes insipidus** (not to be confused with the more common disorder, diabetes mellitus). This condition can also result from a developed insensitivity of the kidney to ADH. In sufferers from diabetes insipidus, water is not efficiently reabsorbed from the ducts, and therefore a large volume of urine is produced. A person with severe diabetes insipidus may excrete up to 25 quarts of urine each day, a serious loss of water to the body. The affected individual becomes dehydrated and is constantly thirsty. He must drink almost continually to offset fluid loss. Diabetes insipidus can often be controlled by injections of ADH or with an ADH nasal spray.

ADH regulates only the rate at which water is excreted by the kidney. Salt excretion is controlled by hormones secreted by the adrenal glands.

CHAPTER SUMMARY

I. The principal waste products of animal metabolism are water, carbon dioxide, and nitrogenous wastes (e.g., ammonia, urea, and uric acid).

II. The mechanisms of waste disposal are diverse and are adapted to the body plan and life-style of an organism.
   A. Sponges and cnidarians have no known excretory organs.
   B. Flatworms have nephridial organs characterized by protonephridia with flame cells.
   C. Annelids possess coelomic metanephridia.
   D. Among arthropods, crustaceans employ antennal glands, and insects, Malpighian tubules.
   E. Vertebrates have kidneys which are vital in maintaining homeostasis of body fluids, as well as in disposal of metabolic wastes.

III. The urinary system is the principal excretory system in humans and other vertebrates.
   A. Nitrogenous wastes are produced principally in the liver, then transported by the blood to the kidneys.
   B. The kidneys produce urine, which then passes through the ureters to the urinary bladder for storage. During urination the urine passes through the urethra to the outside of the body.
   C. Each nephron consists of Bowman's capsule, a cluster of capillaries called a glomerulus, and a long, coiled renal tubule.
   D. Urine formation is accomplished by filtration of plasma, reabsorption of needed materials, and secretion of a few substances into the renal tubule.
      1. Plasma filters out of the glomerular capillaries and into Bowman's capsule. Because filtration is mostly a nonselective process, both needed materials, such as glucose, and wastes become part of the filtrate.
      2. About 99% of the filtrate is reabsorbed from the renal tubules into the blood; this is a highly selective process that permits needed materials to be returned to the blood but leaves wastes

and excessive quantities of other substances to be excreted from the body.

3. In secretion certain substances and drugs are actively transported into the renal tubule to become part of the urine.

4. Urine finally consists of water, nitrogenous wastes, salts, and excesses of other substances.

E. Urine volume is regulated by the hormone ADH, which is released at appropriate times by the posterior lobe of the pituitary glands. ADH governs the permeability of the kidney's collecting ducts.

Post-Test

1. The process of removing metabolic wastes from the body is called _____ .
2. The principal nitrogenous waste product of insects and birds is _____ _____ .
3. The principal nitrogenous waste product of amphibians and mammals is _____ .
4. Flatworms have excretory structures called _____, which are characterized by _____ cells.
5. Earthworms have _____ in each of their body segments.
6. The principal excretory organs of crustaceans are _____ _____ .
7. The excretory system of insects consists of organs called _____ _____ .
8. The vertebrate kidney consists of functional units called _____ .

Select the most appropriate answer from column B for each description in column A.

Column A

_____ 9. outer portion of human kidney
_____ 10. delivers urine to bladder
_____ 11. part of kidney that receives urine from collecting ducts and delivers it to ureters

_____ 12. site of filtration
_____ 13. site of reabsorption
_____ 14. delivers urine to outside of body

Column B

a. cortex
b. medulla
c. ureter
d. urethra
e. renal pelvis

a. urethra
b. renal tubules
c. renal pelvis
d. Bowman's capsule
e. ureter

15. The glomerulus consists of a tuft of _____, which project into _____ _____ .
16. Blood is delivered to the glomerulus by the

_____ _____ and leaves the glomerulus in the _____ _____ .

17. Once the fluid that has left the glomerular capillaries enters Bowman's capsule, it is known as _____ _____ .

18. When a substance exceeds its renal threshold concentration, the portion not reabsorbed is _____ _____ .

19. _____ urine is low in water and so its production conserves water.

20. Antidiuretic hormone (ADH) increases the permeability of the _____ _____ so that more water is _____ and the volume of urine is _____ (increased or decreased).

21. Label the following diagram. (For correct labeling, see Figure 29–8.)

Review Questions

1. Describe the mechanisms of waste disposal used in (1) flatworms, (2) insects.
2. What is osmoregulation? What type of osmoregulatory problem is faced by marine fish? By freshwater fish?

How are these problems met in each case? Do human beings ever have osmoregulatory problems?

3. Name the human structure that is associated with each of the following: (1) urea formation, (2) urine forma-

tion, (3) temporary storage of urine, (4) conduction of urine out of the body.

4. Draw a diagram of a nephron and label its parts.
5. Which part of the nephron is associated with the following: (1) filtration, (2) reabsorption, (3) secretion.
6. What is the principal difference between reabsorption and secretion?
7. List the sequence of blood vessels through which a drop of blood passes as it makes its way from renal artery to renal vein.
8. Why is glucose normally not present in urine? Why is it present in the urine of people with diabetes mellitus? Why do you suppose diabetics experience an increased output of urine?
9. How is urine volume regulated? Explain.

Readings

Dantzler, W.H. "Renal Adaptations of Desert Vertebrates," *Bioscience,* Vol. 32, No. 2, 108–112. Adaptations in renal physiology are integrated with other physiological and behavioral mechanisms for desert survival.

Heatwole, H. "Adaptations of Marine Snakes," *American Scientist,* Sept–Oct 1978, pp 594–604. Among the adaptations discussed are those that permit several groups of snakes to inhabit the marine environment by maintaining fluid balance.

Pollie, R. "Comprehending Kidney Disease," *Science News,* 2 October 1982, p 218. A simple theory is offered to explain a complex syndrome.

Solomon, E.P., and P.W. Davis. *Human Anatomy and Physiology,* Philadelphia, Saunders College Publishing, 1983. Chapters 23 and 24 focus on the human urinary system and fluid balance.

Chapter 30
ENDOCRINE REGULATION

Outline

I. How hormones work
   A. Mechanisms of hormone action
   B. Regulation of hormone secretion
II. Invertebrate hormones
III. Vertebrate hormones
   A. Endocrine malfunctions
   B. Hypothalamus and pituitary gland
      1. Posterior lobe of the pituitary gland
      2. Anterior lobe of the pituitary gland
         a. Growth
         b. Growth problems
   C. Thyroid gland
      1. Regulation of secretion
      2. Disorders of the thyroid gland
   D. Parathyroid glands
   E. Islets of the pancreas
      1. Actions of insulin
      2. Actions of glucagon
      3. Regulation of insulin and glucagon secretion
      4. Diabetes mellitus
      5. Hypoglycemia
   F. Adrenal glands
      1. Adrenal medulla
      2. Adrenal cortex
      3. The adrenal glands and stress
   G. Other hormones

Learning Objectives

After you have read this chapter you should be able to:

1. Define the terms *hormone* and *endocrine gland* and distinguish between endocrine and exocrine glands.
2. Describe the mechanisms of hormone action, including the role of second messengers, such as cyclic AMP.
3. Summarize the role of hormones in invertebrates and describe the interaction of hormones that control development in insects.
4. Identify the principal vertebrate endocrine glands and locate them in the body. (Consult Figure 30–2.)
5. Summarize the regulation of endocrine glands by negative-feedback mechanisms and relate the concept of negative feedback to specific hormones discussed.
6. Justify the description of the hypothalamus as the link between nervous and endocrine systems and describe the mechanisms by which the hypothalamus exerts its control.
7. Identify the hormones released by the posterior and anterior lobes of the pituitary, give their origin, and describe their actions.
8. Describe the actions of growth hormone on growth and metabolism and describe the consequences of hyposecretion and hypersecretion.
9. Give the actions of the thyroid hormones, describe their regulation, and describe the thyroid disorders discussed in this chapter.
10. Contrast the actions of insulin and glucagon and describe the disorders associated with malfunction of the islets of the pancreas.
11. Describe the actions of the mineralocorticoid and glucocorticoid hormones and give the effects of malfunction of the adrenal cortex.
12. Summarize the role of the adrenal glands in helping the body to adapt to stress.

T he **endocrine system** is a diverse collection of glands and tissues that secrete chemical messengers responsible for the remote control of many body processes. These chemical messengers, or **hormones,** have specific regulatory effects on the activities of other tissues. The term *hormone* is derived from a Greek word meaning "to excite." Hormones do indeed excite their target tissues, usually by stimulating a change in some metabolic activity. The study of endocrine activity, **endocrinology,** is an exciting field of medical research.

Endocrine glands do not have ducts through which they release their secretions. Instead, hormones are secreted into the surrounding tissue fluid and diffuse into capillaries. Hormones are then transported by the blood to **target tissues,** where they exert their influence. The target tissue may be another endocrine gland or it may be an entirely different type of organ, such as a bone. Often the target tissue is located far from the endocrine gland. Figure 30–1 illustrates the difference between endocrine glands and **exocrine glands** (such as sweat glands and gastric glands), which release their secretions into ducts.

In recent years it has been discovered that specialized cells in the digestive tract and in some other organs, such as the kidneys, also release hormones. The scope of endocrinology has been broadened to include the study of chemical messengers produced by cells that are widely distributed in the body, rather than by single, discrete organs. Chemically, hormones either are lipids or belong to the protein family (proteins, peptides, and derivatives of amino acids).

HOW HORMONES WORK

Most endocrine glands secrete at least small amounts of their hormones continuously, so that at any moment 30 or 40 hormones may be present in the blood. Most are there in minute amounts, some in concentrations as low as 1/1,000,000 of a milligram (1 picogram) per milliliter.

A hormone may pass through many tissues seemingly "unnoticed" until it reaches its target tissue. How does the target tissue "recognize" its hormone? Specific receptor proteins in the target tissues bind the hormone. This is a highly specific process. The receptor site is like a lock, and the hormones are like different keys. Only the hormone that fits the lock can influence the metabolic machinery of the cell.

Mechanisms of Hormone Action

Once a hormone is taken up by a particular tissue, how does it influence the activity of the cells? Many protein-type hormones combine with re-

Figure 30–1 Comparison of (*a*) an exocrine with (*b*) an endocrine gland. The secretion of an exocrine gland passes through a duct to reach its final destination. For example, sweat passes through the duct of a sweat gland to reach the surface of the skin. The hormone of an endocrine gland is released into the surrounding tissue fluid and diffuses into the blood, which transports it to its target tissue.

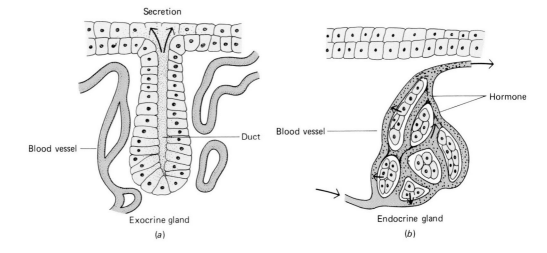

Secretion

Blood vessel — — Duct

Exocrine gland
(*a*)

Blood vessel — — Hormone

Endocrine gland
(*b*)

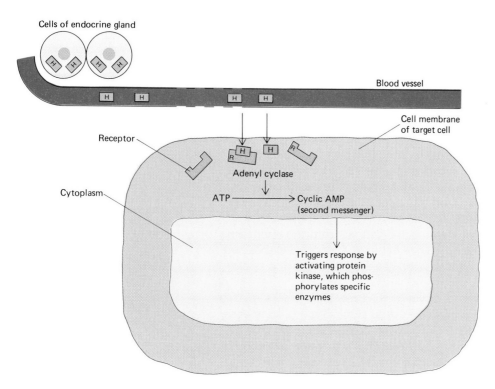

Figure 30–2 Second-messenger mechanism of hormone action. Many hormones combine with receptors in the cell membrane of target cells. The hormone-receptor complex stimulates an enzyme, adenyl cyclase, which catalyzes the conversion of ATP to cyclic AMP, a second messenger. Cyclic AMP then triggers the response.

ceptors on the cell membrane of a target cell. Then the actual hormonal message is relayed to the appropriate site within the cell by a second messenger. One second messenger, known as **cyclic AMP** (adenosine monophosphate), has been extensively studied. When a hormone combines with a receptor on the target cell, it increases the level of cyclic AMP within the cell. Here is how it happens (Fig. 30–2). First, the enzyme **adenyl cyclase,** which is attached to the cell membrane of almost all cells in the body, is activated. It catalyzes the conversion of ATP to cyclic AMP, as follows:

$$ATP \xrightarrow{\text{Adenyl Cyclase}} \text{Cyclic AMP}$$

The cyclic AMP then triggers the chain of reactions that leads to the metabolic effect.

The particular action initiated by cyclic AMP depends upon the specific kinds of enzyme systems present in the cell. This explains how the same hormone can promote different responses in different cell types. In some cases cyclic AMP is thought to affect the activity of specific genes, so that specific proteins are synthesized. When the target tissue is another endocrine gland, cyclic AMP regulates the release of its hormones.

Steroid hormones are relatively small, lipid-soluble molecules that easily pass through the cell membrane into the cytoplasm. Instead of having receptors for steroid hormones on the cell membranes, target cells have them within the cytoplasm (Fig. 30–3). When a hormone combines with a receptor, the hormone-receptor complex moves into the nucleus and activates a particular gene. The result is synthesis of some type of protein.

Several types of hormones may be involved in regulating the metabolic activities of a particular type of cell. In fact, most hormones produce a synergistic effect: The presence of one hormone may enhance the effects of another.

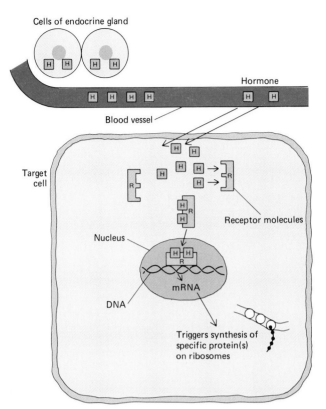

Cells of endocrine gland

Hormone

Blood vessel

Target cell

Receptor molecules

Nucleus

DNA

mRNA

Triggers synthesis of specific protein(s) on ribosomes

Figure 30–3 Activation of genes by steroid hormones. Steroid hormones are small lipid-soluble molecules that pass freely through the cell membrane. Some steroid hormones combine with receptors within target cells. The steroid hormone-receptor complex moves into the nucleus and combines with a protein associated with the DNA. This activates specific genes, leading to the synthesis of mRNA coding for specific proteins. The proteins cause the response recognized as the hormone's action.

Regulation of Hormone Secretion

How does an endocrine gland "know" how much hormone to release at any given moment? Hormone secretion is self-regulated by **negative-feedback control mechanisms.** Information regarding the hormone level or its effect is fed back to the gland, which then responds in a homeostatic manner. The parathyroid glands of the neck, which regulate calcium level in the blood, provide a good example. The parathyroid hormone causes the level of calcium in the blood to rise. A low level of calcium in the blood signals the parathyroid glands to release more hormone (Fig. 30–4). But when the calcium level rises beyond normal limits, the parathyroid glands are inhibited and slow their output of hormone. Both responses are negative-feedback mechanisms, since in both cases the effects are opposite (negative) to the stimulus.

Negative feedback forms the basis of hormone regulation. As you will see, variations of this theme abound, many involving the hypothalamus and pituitary gland.

INVERTEBRATE HORMONES

Among many invertebrates hormones are secreted mainly by neurons rather than by endocrine glands. These **neurohormones** help regulate such processes as regeneration in hydras, flatworms, and annelids; molting and metamorphosis in insects; color changes in crustaceans; and gamete production, reproductive behavior, and metabolism in other groups.

Insects have both endocrine glands and neurons that secrete hormones. Their various hormones interact with one another to regulate growth and development, including molting and morphogenesis. Hormones also help regulate insect metabolism and reproduction.

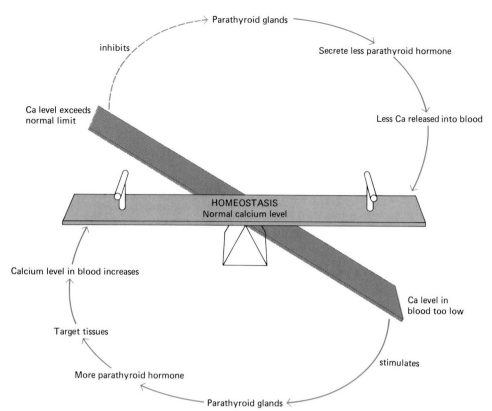

Figure 30–4 Regulation of hormone secretion by negative feedback. When the calcium level in the blood falls below normal, the parathyroid glands are stimulated to secrete more parathyroid hormone. This hormone acts to increase the calcium level in the blood, thus restoring homeostasis. Should the calcium level exceed normal, the parathyroid glands are inhibited and slow their secretion of hormone. This diagram has been greatly simplified; calcitonin, a hormone secreted by the thyroid gland, works antagonistically to parathyroid hormone and is important in lowering blood calcium levels.

Hormonal control of development in insects is complex and varies from species to species. Generally, some environmental factor (temperature change, for instance) affects certain neurons in the brain. Once activated, these cells produce a hormone referred to as **BH** (brain hormone), which their axons discharge into organs called the **corpora cardiaca.** BH is stored in the corpora cardiaca. When released, it stimulates the **prothoracic glands,** endocrine glands in the prothorax, to produce **molting hormone,** also called ecdysone. Molting hormone stimulates growth and molting.

In the immature insect endocrine glands called **corpora allata** secrete **juvenile hormone.** This hormone suppresses metamorphosis at each larval molt so that the insect retains its immature state (Fig. 30–5). After the molt the insect is still in a larval stage. When the concentration of juvenile hormone is lower, metamorphosis occurs and the insect is transformed into a pupa. In the absence of the juvenile hormone the pupa molts to become an adult. The secretory activity of the corpora allata is regulated by the nervous system, and the amount of juvenile hormone decreases with successive molts.

VERTEBRATE HORMONES

In vertebrates there are ten or so discrete endocrine glands located throughout the body as well as many specialized tissues that secrete hormones. Vertebrate hormones help regulate such diverse activities as growth, metabolic rate, utilization of nutrients by cells, and reproduction. They are largely responsible for regulating fluid balance and blood homeostasis, and they help the body cope with stress. The principal human endocrine glands are illustrated in Figure 30–6. Most vertebrates possess similar endocrine glands. Table 30–1 gives the physiological actions and sources of some of the major vertebrate hormones.

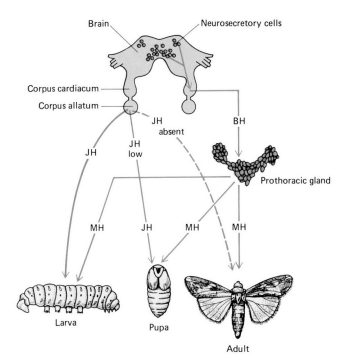

Figure 30–5 The neural and endocrine control of growth and molting in a moth. Neurons in the brain secrete a hormone, BH, which stimulates the prothoracic glands to secrete molting hormone (MH). In the immature insect, the corpora allata secrete juvenile hormone (JH), which suppresses metamorphosis at each larval molt. Metamorphosis to the adult form occurs when MH acts in the absence of JH.

Endocrine Malfunctions

In such a complex system there are many opportunities for things to go wrong. When a disorder affects an endocrine gland, the rate of secretion often becomes abnormal. The gland may decrease its hormone output,

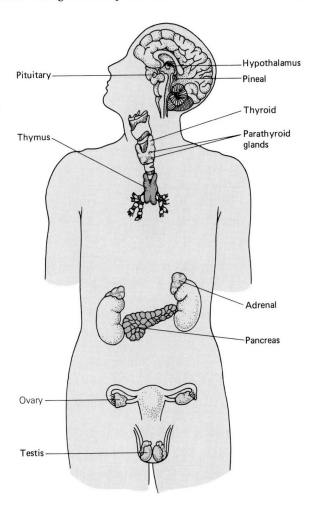

Figure 30–6 Location of the principal endocrine glands of the human male and female.

Table 30–1
PRINCIPAL ENDOCRINE GLANDS AND THEIR HORMONES*

Endocrine gland and hormone	Target tissue	Principal actions
Hypothalamus		
Releasing and inhibiting hormones	Anterior lobe of pituitary gland	Stimulates or inhibits secretion of specific hormones
Hypothalamus (production) and posterior lobe of pituitary (storage and release)		
Oxytocin	Uterus	Stimulates contraction
	Mammary glands	Stimulates ejection of milk into ducts
Antidiuretic hormone (ADH)	Kidneys (collecting ducts)	Stimulates reabsorption of water; conserves water
Anterior lobe of pituitary		
Growth hormone (GH)	General	Stimulates growth by promoting protein synthesis
Prolactin	Mammary glands	Stimulates milk secretion
Thyroid-stimulating hormone (TSH)	Thyroid gland	Stimulates secretion of thyroid hormones; stimulates increase in size of thyroid gland
Adrenocorticotropic hormone (ACTH)	Adrenal cortex	Stimulates secretion of adrenal cortical hormones
Gonadotropic hormones (FSH, LH)	Gonads	Stimulate gonad function
Thyroid gland		
Thyroxine (T_4) and triiodothyronine (T_3)	General	Stimulate metabolic rate; essential to normal growth and development
Calcitonin	Bone	Lowers blood calcium level by inhibiting removal of calcium from bone
Parathyroid glands		
Parathyroid hormone	Bone, kidneys, digestive tract	Increases blood calcium level by stimulating bone breakdown; stimulates calcium reabsorption in kidneys; activates vitamin D
Islets of Pancreas		
Insulin	General	Lowers blood glucose level by facilitating glucose uptake and utilization by cells; stimulates glycogenesis; stimulates fat storage and protein synthesis
Glucagon	Liver, adipose tissue	Raises blood glucose level by stimulating glycogenolysis and gluconeogenesis; mobilizes fat

hyposecretion, depriving target cells of needed stimulation, or it may increase output to abnormal levels, **hypersecretion,** causing imbalance in the opposite direction. In some endocrine disorders an appropriate amount of hormone may be secreted, but target cells may not be able to take it up and use it. There may be insufficient numbers of receptors or the receptors may not function properly. Any of these malfunctions leads to predictable metabolic abnormalities, with accompanying clinical symptoms (Table 30–2). Some of these will be described as specific endocrine glands are discussed.

Hypothalamus and Pituitary Gland

Nervous and endocrine systems are linked by the **hypothalamus,** which regulates the activity of the pituitary gland. Because it controls the activity of several other endocrine glands, the **pituitary gland** has been

Endocrine gland and hormone	Target tissue	Principal actions
Adrenal medulla		
Epinephrine and norepinephrine	Skeletal muscle, cardiac muscle, blood vessels, liver, adipose tissue	Help body cope with stress; increase heart rate, blood pressure, metabolic rate; reroute blood; mobilize fat; raise blood sugar level
Adrenal cortex		
Mineralocorticoids (Aldosterone)	Kidney tubules	Maintain sodium and phosphate balance
Glucocorticoids (Cortisol)	General	Help body adapt to long-term stress; raise blood glucose level; mobilize fat
Ovary†		
Estrogens	General	Stimulate development of secondary sex characteristics
	Reproductive structures	Stimulate growth of sex organs at puberty; prompt monthly preparation of uterus for pregnancy
Progesterone	Uterus	Completes preparation of uterus for pregnancy
	Breasts	Stimulates development
Testis		
Testosterone	General	Stimulates development of secondary sex characteristics and growth spurt at puberty
	Reproductive structures	Stimulates development of sex organs; stimulates spermatogenesis
Pineal gland		
Melatonin	Gonads, pigment cells, other tissues	Influences reproductive processes in hamsters and other animals; pigmentation in some vertebrates; may control biorhythms in some animals; may help control onset of puberty in humans

*The digestive hormones are described in Chapter 27.
†The reproductive hormones will be discussed in Chapter 31.

dubbed the master gland of the body. Truly a biological marvel, the pituitary gland is only the size of a large pea and weighs only about 0.5 gram (0.02 ounce), yet it secretes at least nine distinct hormones that exert far-reaching influence over body activities. Connected to the hypothalamus by a stalk of nervous tissue, the pituitary gland consists of two main lobes, the **anterior** and **posterior lobes.**

POSTERIOR LOBE OF THE PITUITARY GLAND

Two peptide hormones, oxytocin and antidiuretic hormone (ADH), are secreted by the **posterior lobe of the pituitary gland.** These hormones are actually produced by specialized nerve cells in the hypothalamus. They reach the posterior lobe of the pituitary by flowing through axons that connect the hypothalamus with the pituitary (Fig. 30–7). Enclosed within tiny vesicles, the hormones pass slowly down the axons of these

Table 30–2
CONSEQUENCES OF ENDOCRINE MALFUNCTION

Hormone	Hyposecretion	Hypersecretion
Growth hormone	Pituitary dwarf	Gigantism if malfunction occurs in childhood; acromegaly in adult
Thyroid hormones	Cretinism (in children); myxedema, a condition of pronounced adult hypothyroidism (BMR is reduced by about 40%; patient feels tired all of the time and may be mentally slow); goiter, enlargement of the thyroid gland (see Fig. 30–12)	Grave's disease; goiter
Parathyroid hormone	Spontaneous discharge of nerves; spasms; tetany; death	Weak, brittle bones; kidney stones
Insulin	Diabetes mellitus	Hypoglycemia
Adrenocortical hormones	Addison's disease (body cannot synthesize sufficient glucose by gluconeogenesis; patient is unable to cope with stress; sodium loss in urine may lead to shock)	Cushing's disease (edema gives face a full-moon appearance; fat is deposited about trunk; blood glucose level rises; immune responses are depressed)

nerve cells. The axons extend through the pituitary stalk and into the posterior lobe. Hormone accumulates in the axon endings until the neuron is stimulated; then it is released and diffuses into surrounding capillaries.

Toward the end of pregnancy, **oxytocin** levels rise, stimulating the strong contractions of the uterus needed to expel the baby. Oxytocin is sometimes administered clinically (under the name Pitocin) to initiate or

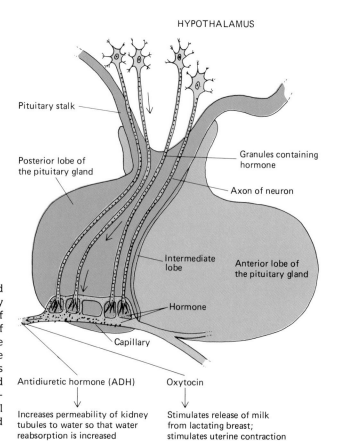

HYPOTHALAMUS

Pituitary stalk

Posterior lobe of the pituitary gland

Granules containing hormone

Axon of neuron

Intermediate lobe

Anterior lobe of the pituitary gland

Hormone

Capillary

Antidiuretic hormone (ADH)

Oxytocin

Increases permeability of kidney tubules to water so that water reabsorption is increased

Stimulates release of milk from lactating breast; stimulates uterine contraction

Figure 30–7 The hormones secreted by the posterior lobe of the pituitary are actually manufactured in cells of the hypothalamus. The axons of these neurons extend down into the posterior lobe of the pituitary. The hormones are packaged in granules that flow through these axons and are stored in their ends. The hormone is secreted into the interstitial fluid as needed and then transported by the blood.

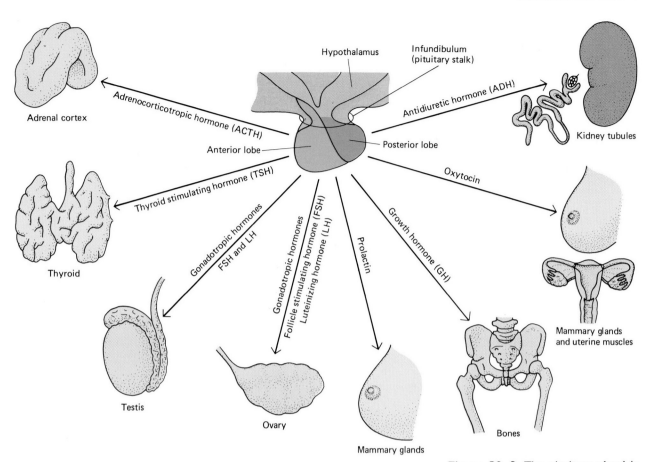

Figure 30–8 The pituitary gland is suspended from the hypothalamus by a stalk of neural tissue. The hormones secreted by the anterior and posterior lobes of the pituitary gland and the target tissues they act upon are shown.

speed labor. When an infant sucks at its mother's breast, sensory neurons are stimulated and signal the hypothalamus to release oxytocin. The hormone stimulates contraction of cells surrounding the milk glands so that milk is let down into the ducts, from which it can be sucked. Because the oxytocin also stimulates the uterus to contract, breast feeding promotes rapid recovery of the uterus to nonpregnant size. Males have about the same amount of oxytocin circulating in their blood as females, but its function in them is unknown.

ANTERIOR LOBE OF THE PITUITARY GLAND

The **anterior lobe of the pituitary** secretes growth hormone, prolactin, and several tropic hormones, which stimulate other endocrine glands (Fig. 30–8). **Prolactin** is the hormone that stimulates the cells of the mammary glands to secrete milk during lactation. Each of the anterior pituitary hormones is in some way regulated by a separate releasing hormone, and sometimes also by an inhibiting hormone produced in the hypothalamus. At appropriate times these **neurohormones** (hormones secreted by neural tissue) are released by the hypothalamus. They diffuse into capillaries and are transported to the hypothalamus by portal veins. (A portal vein connects two sets of exchange vessels.) Within the anterior pituitary the portal veins divide into a second set of capillaries, from which the hormone diffuses into the tissue of the anterior lobe of the pituitary (Fig. 30–9). When we speak of the pituitary as being stimulated or inhibited, it should be understood that certain receptors in the hypothalamus are generally affected first. They in turn control the pituitary.

GROWTH. Small children measure themselves periodically against their parents, eagerly awaiting that time when they, too, will be "big." Whether

Figure 30–9 The hypothalamus secretes several specific releasing and inhibiting hormones, which reach the anterior lobe of the pituitary gland by way of portal veins. Each releasing hormone stimulates the synthesis of a particular hormone by the cells of the anterior lobe. (*R*, releasing hormone; *H*, hormone.)

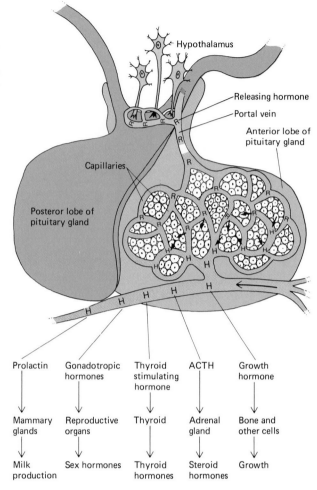

one will be tall or short depends upon many factors, including genes, diet, and hormonal balance.

Growth hormone (also called somatotropin) stimulates body growth mainly by promoting protein synthesis. It acts by increasing uptake of amino acids by the cells and by stimulating protein synthesis. Growth hormone promotes mobilization of fat from adipose tissues, raising the level of free fatty acids in the blood. Fatty acids become available for cells to use as fuel, a protein-sparing action. How does this help to promote growth? Fat mobilization by growth hormone is also important during fasting or when a person is under prolonged stress, situations in which the blood sugar level is low. Can you explain why?

Secretion of growth hormone is regulated both by a growth-hormone–releasing hormone and a growth-hormone–inhibiting hormone (also called somatostatin) released by the hypothalamus. A high level of growth hormone in the blood signals the hypothalamus to secrete the inhibiting hormone. This results in decreased release of growth hormone by the pituitary. A low level of growth hormone in the blood causes the hypothalamus to secrete the releasing hormone. This hormone stimulates the pituitary gland to release more growth hormone. Many other factors influence secretion. It is increased by low blood sugar level, increased amino acid concentration in the blood, and stress.

Your parents probably used to say you must get plenty of sleep and exercise to grow properly. These age-old notions are supported by recent studies. Secretion of growth hormone does increase during exercise, probably because rapid metabolism by muscle cells lowers blood sugar level. Growth hormone secretion also increases during non-REM sleep.

Children who get lots of loving also have an advantage. Surprisingly, growth may be retarded in children who are deprived of emotional attention and support even when their physical needs for food and shelter are amply met. Cuddling, playing, and other forms of nurture are apparently essential to normal development. Some emotionally deprived children exhibit psychosocial dwarfism, together with abnormal sleep patterns, which may be the basis for decreased secretion of growth hormone.

Other hormones also influence growth. Thyroid hormones appear to be necessary for normal growth-hormone secretion and function. Sex hormone must be present for the growth spurt associated with puberty to occur. However, the presence of sex hormone eventually causes the growth centers within the long bones to fuse to the shafts, so further increase in height is impossible even when growth hormone is present.

GROWTH PROBLEMS. Have you ever wondered why midgets failed to grow normally? They are probably **pituitary dwarfs**—individuals whose pituitary glands did not produce sufficient growth hormone during childhood. Though miniature, a pituitary dwarf has normal intelligence and is usually well proportioned. If the growth centers in the long bones are still active when this condition is diagnosed, it can be treated clinically by injection with growth hormone, which can now be synthesized commercially. Growth problems may also result from the malfunction of other mechanisms, such as the regulating hormones from the hypothalamus. Abnormally tall individuals develop when the anterior pituitary secretes excessive amounts of growth hormone during childhood. This condition is referred to as **gigantism.**

If pituitary malfunction leads to hypersecretion of growth hormone during adulthood, the individual cannot grow taller. Connective tissue proliferates, and bones in the hands, feet, and face may increase in diameter. This condition is known as **acromegaly,** meaning "large extremities" (Fig. 30–10).

Thyroid Gland

The **thyroid gland** is located in the neck region, in front of the trachea and below the larynx (see Figure 30–6). Its principal hormones are thyroxine and calcitonin. **Thyroxine,** also known as T_4, is synthesized from the amino acid tyrosine. Its most distinctive chemical feature is the four iodine atoms attached to each molecule. Thyroxine is essential to normal growth and development and is important in stimulating the rate of metabolism (probably by stimulating the electron transport system). It is also necessary for cellular differentiation. Tadpoles cannot develop into adult frogs without thyroxine; the hormone appears to regulate selectively the synthesis of needed proteins.

REGULATION OF SECRETION

The principal regulation of thyroid hormone secretion depends upon a feedback system between the anterior pituitary and the thyroid gland (Fig. 30–11). When concentration of thyroid hormone in the blood rises above normal, the anterior pituitary is inhibited. When the level falls, the pituitary secretes more **TSH (thyroid-stimulating hormone).** The thyroid-stimulating hormone acts by way of cyclic AMP to stimulate increased iodine uptake, synthesis, and secretion of thyroid hormones, and also increased size of the gland itself.

Too much thyroid hormone in the blood also affects the hypothalamus, inhibiting secretion of TSH-releasing hormone. However, the hypo-

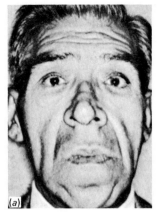
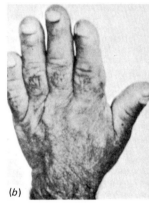

Figure 30–10 A case of acromegaly. (a) Note enlarged nose and ears and prominent jaw and cheekbones. (b) Dorsal aspect of hand of same patient. Note broadened fingers and "meaty" appearance due to an increase in connective tissue (From Wyngaarden and Smith, eds., *Cecil Textbook of Medicine*, 17th ed. Philadelphia, W.B. Saunders, 1985.)

Figure 30–11 Regulation of thyroid hormone secretion. Dotted lines indicate inhibition.

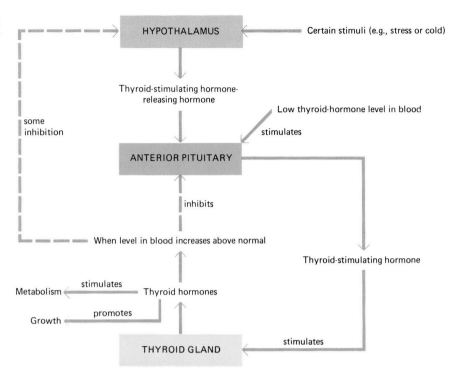

thalamus is thought to exert its regulatory effects primarily in certain stressful situations, such as extreme weather change. Exposure to very cold weather may stimulate the hypothalamus to increase secretion of TSH-releasing hormone, thereby raising body temperature through increased metabolic heat production.

DISORDERS OF THE THYROID GLAND

Extreme hypothyroidism during infancy and childhood results in low metabolic rate and retarded mental and physical development, a condition called **cretinism.** When hypothyroidism is treated early enough, cretinism can be prevented.

An adult who feels like sleeping all the time, has little energy, and is mentally slow or even confused may be suffering from hypothyroidism. When there is almost no thyroid function, the basal metabolic rate is reduced by about 40% and the patient develops the condition called **myxedema,** characterized by a slowing down of physical and mental activity. Hypothyroidism can be treated by using thyroid pills to replace missing hormones.

Hyperthyroidism does not cause abnormal growth but does increase metabolic rate by 60% or even more. This increase in metabolism results in swift utilization of nutrients, causing the individual to be hungry and to increase food intake. But this is not sufficient to meet the demands of the rapidly metabolizing cells, so individuals with this condition often lose weight. They also tend to be nervous, irritable, and emotionally unstable.

Any abnormal enlargement of the thyroid gland is termed a **goiter** and may be associated with either hyposecretion or hypersecretion (Fig. 30–12). One cause is dietary iodine deficiency. Without iodine the gland cannot make thyroid hormones, so their concentration in the blood decreases. In compensation the anterior pituitary secretes large amounts of TSH. The thyroid gland enlarges, sometimes to gigantic proportions. Since the problem is lack of iodine, enlargement of the gland cannot

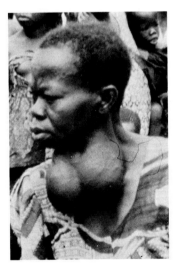

Figure 30–12 Goiter resulting from iodine deficiency. (Courtesy of United Nations Food and Agricultural Organization.)

increase production of the hormones, for the needed ingredient is still missing. Thanks to iodized salt, goiter is no longer common in the United States. In other parts of the world, however, hundreds of thousands still suffer from this easily preventable disorder.

Parathyroid Glands

The **parathyroid glands** are embedded in the connective tissue surrounding the thyroid gland. The parathyroid glands secrete **parathyroid hormone,** which regulates the calcium level of the blood and tissue fluid. Parathyroid hormone stimulates release of calcium from bones and calcium reabsorption from the kidney tubules. It also activates vitamin D, which then increases the amount of calcium absorbed from the intestine. **Calcitonin,** secreted by the thyroid gland, works antagonistically to parathyroid hormone. When calcium levels become too high, calcitonin is released and acts rapidly to inhibit removal of calcium from bone.

Islets of the Pancreas

Besides secreting digestive enzymes (Chapter 27), the pancreas serves as an important endocrine gland. Its hormones, insulin and glucagon, are secreted by cells that form little clusters, or islets, dispersed throughout the pancreas. These islets, first described by the German histologist Paul Langerhans, are often referred to as the **islets of Langerhans** (Fig. 30–13). About a million islets are present in the human pancreas. They are composed of beta cells, which secrete insulin, and alpha cells, which secrete glucagon.

ACTIONS OF INSULIN

Insulin stimulates cells, especially muscle cells, to take up glucose from the body by enabling them to transport it across the cell membrane. Once glucose enters muscle cells, it is either used immediately as fuel or stored as glycogen, a process called **glycogenesis.** Insulin activity results in lowering the glucose level in the blood. It also influences fat and

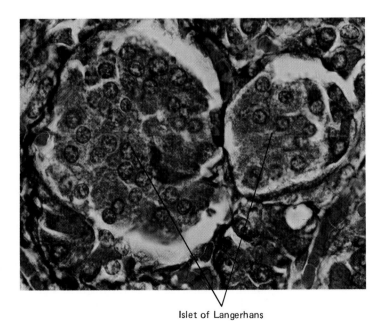

Islet of Langerhans

Figure 30–13 Photomicrograph showing the islets of Langerhans.

protein metabolism. Insulin reduces the use of fatty acids as fuel and instead stimulates their storage in adipose tissue. In a similar manner it inhibits the use of amino acids as fuel, thus promoting protein synthesis.

ACTIONS OF GLUCAGON

Glucagon acts antagonistically to insulin. Its principal effect is to raise blood sugar level. It does this by stimulating liver cells to convert glycogen to glucose, a process called **glycogenolysis,** and by stimulating liver cells to make glucose from other metabolites, **gluconeogenesis.** Note that these actions are opposite to those of insulin. Glucagon is thought to be secreted also by certain cells in the wall of the stomach and duodenum.

REGULATION OF INSULIN AND GLUCAGON SECRETION

Secretion of insulin and glucagon is directly controlled by the blood sugar level (Fig. 30–14). After a meal, when the blood glucose level rises as a result of intestinal absorption, beta cells are stimulated to increase insulin secretion. Then, as the cells remove glucose from the blood, decreasing its concentration, insulin secretion decreases accordingly.

When one has not eaten for several hours, the blood sugar level begins to fall. When it falls from its normal fasting level of about 90 milligrams per 100 milliliters to about 70 milligrams, the alpha cells of the islets secrete large amounts of glucagon. Glucose is mobilized from the liver cells' stores, and blood sugar level returns to normal. The alpha cells actually react to the glucose concentration within their own cytoplasm, which is a reflection of the blood sugar level. When blood sugar level is high, there is generally a high level of glucose within the alpha cells, and glucagon secretion is inhibited.

It should be clear that insulin and glucagon work oppositely to keep blood sugar level within normal limits. When glucose level rises, insulin

Figure 30–14 Regulation of blood sugar level by insulin and glucagon.

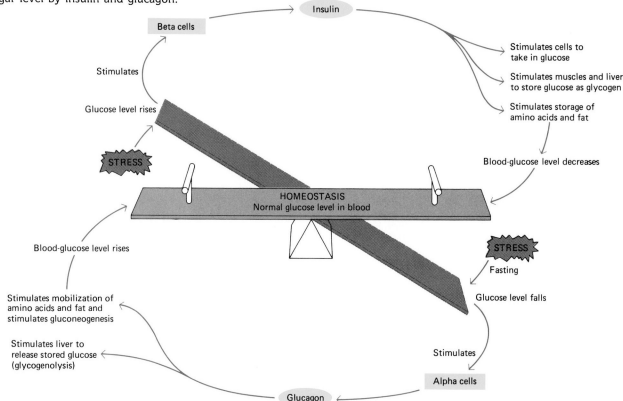

release brings it back to normal; when it falls, glucagon acts to raise it again. The insulin-glucagon system is a powerful, fast-acting mechanism for keeping blood sugar level normal. Can you think of reasons why it is important to maintain a constant blood sugar level? Perhaps the most important one is that brain cells are completely dependent upon a continuous supply of glucose, since they ordinarily are unable to utilize other nutrients as fuel. As you will see, several other hormones also affect blood sugar level.

DIABETES MELLITUS

The principal disorder associated with pancreatic hormones is **diabetes mellitus.** There are an estimated 10 million diabetics in the United States alone, and about 40,000 die each year as a result of this disorder, making it the third most common cause of death. Diabetes is a leading cause of blindness, kidney disorders, disease of small blood vessels, gangrene of the limbs, and various other malfunctions.

The symptoms of diabetes mellitus have been associated with insulin deficiency, but evidence indicates that some diabetics may have enough insulin, and the problem may be that receptors on target cells cannot bind it. Two distinct clinical types of diabetes are juvenile-onset diabetes, which occurs before age 15, and maturity-onset diabetes, which develops gradually, usually after age 40. Patients with juvenile-onset diabetes have a short life expectancy because of atherosclerotic disease, which develops as a result of impaired lipid metabolism.

A tendency for developing diabetes is inherited as a recessive trait, but the mechanisms that actually trigger the disorder are not completely understood. A recent study suggests that diabetes may be triggered in genetically susceptible persons by a common poliomyelitis-related virus. If this proves true, a vaccine might be developed to prevent infection.

In experimental animals diabetes has been induced by administering large amounts of certain hormones, such as growth hormone, or by diets excessively high in carbohydrates. When the blood sugar level is kept very high by these methods, it is thought that the beta cells eventually become exhausted or "burned out." It has been suggested that diabetes can be triggered in genetically susceptible persons by excessive ingestion of sweets or other carbohydrates.

Most of the metabolic disturbances associated with diabetes can be traced to three major effects of insulin deficiency: (1) decreased utilization of glucose, (2) increased fat mobilization, and (3) increased protein utilization.

1. In diabetes, cells dependent upon insulin can take in only about 25% of the glucose they require for fuel. Glucose remains in the blood, so the blood sugar level rises **(hyperglycemia).** Instead of the normal fasting level of 90 milligrams per 100 milliliters, the diabetic may have from 300 to more than 1000 milligrams. Blood glucose level is so high that it exceeds the renal threshold: The tubules in the kidneys are unable to return all the glucose in the filtrate to the blood. Glucose therefore is excreted in the urine. The presence of glucose in the urine is a simple diagnostic test for diabetes.

2. Despite the large quantities of glucose in the blood, most cells cannot utilize it and must turn to other sources of fuel. The absence of insulin promotes the mobilization of fat stores, providing nutrients for cellular respiration. But unfortunately, the blood lipid level may reach five times the normal level, leading to development of atherosclerosis. Also, the increased fat metabo-

lism increases the formation of ketone bodies. These build up in the blood, causing **ketosis,** a condition which results in the body fluids and blood becoming too acidic. If sufficiently marked, ketosis can lead to coma and death. When ketone level in the blood rises, ketones appear in the urine, another clinical indication of diabetes. Because of osmotic pressure, when ketone bodies and glucose are excreted in the urine, they take water with them, so that urine volume increases. The resulting dehydration causes the diabetic to be continually thirsty.

3. Lack of insulin also results in increased protein breakdown relative to protein synthesis, so the untreated diabetic becomes thin and emaciated.

Treatment of diabetes involves a carefully managed, balanced diet designed to reduce carbohydrate intake and to maintain an appropriate body weight. In serious cases daily injections of insulin are required. Because insulin is a protein, it cannot be ingested orally, for it would be digested by enzymes in the intestine. Mild cases of diabetes are sometimes treated with drugs, such as tolbutamide, which stimulate the islets to produce insulin and may be taken orally. However, studies indicate that heart disease may be a serious side effect of these drugs, so there has been a shift away from their use.

HYPOGLYCEMIA

Hypoglycemia (low blood sugar level) is sometimes seen in people who later develop diabetes. It may be an overreaction by the islets to glucose challenge. Too much insulin is secreted in response to carbohydrate ingestion. About 3 hours after a meal the blood sugar level falls below normal, making the individual feel very drowsy. If this reaction is severe enough, the patient may become uncoordinated or even unconscious.

Serious hypoglycemia can develop if diabetics inject themselves with too much insulin or if too much is secreted by the islets because of a tumor. The blood sugar level may then fall drastically, depriving the brain cells of their needed supply of fuel. Insulin shock may result, a condition in which the patient may appear to be drunk or may become unconscious, suffer convulsions, or even die.

Adrenal Glands

The paired **adrenal glands** are small yellow masses of tissue that lie in contact with the upper ends of the kidneys. Each gland consists of a central portion, the adrenal medulla, and a larger outer section, the adrenal cortex. Although wedded anatomically, the adrenal medulla and cortex develop from different types of tissue in the embryo and function as distinct glands. Both secrete hormones that help to regulate metabolism, and both help the body deal with stress.

ADRENAL MEDULLA

The **adrenal medulla** develops from neural tissue, and its secretion is controlled by sympathetic nerves. Two hormones, **epinephrine** (sometimes called **adrenaline**) and **norepinephrine** (noradrenaline), are secreted by the adrenal medulla. Chemically, these hormones are very similar; they belong to the chemical group known as **catecholamines** (derived from amino acids). Norepinephrine is the same substance secreted as a neurotransmitter by sympathetic neurons and by some neurons in the central nervous system. Its effects on the body are similar but last about ten times longer because the hormone is removed from the blood slowly. About 80% of the hormone output of the adrenal medulla is epinephrine.

Often referred to as the emergency gland of the body, the adrenal medulla prepares you physiologically to deal with threatening situations. Were a monster suddenly to appear before you, hormone secretion from this gland would initiate an alarm reaction enabling you to think more quickly, fight harder, or run faster than normally. Metabolic rate would increase as much as 100%.

The adrenal medullary hormones cause blood to be rerouted in favor of those organs essential for emergency action. Blood vessels to the skin and kidneys are constricted, while those going to the brain, muscles, and heart are dilated. Constriction of blood to the skin has the added advantage of decreasing blood loss in case of hemorrhage. (It explains the sudden paling that comes with fear or rage.) At the same time, the heart beats faster and thresholds in the reticular activating system of the brain are lowered, so you become more alert. Strength of muscle contraction increases. The adrenal medullary hormones also raise fatty acid and glucose levels in the blood, assuring needed fuel for extra energy.

Under normal conditions both epinephrine and norepinephrine are secreted continuously in small amounts. Their secretion is under nervous control. When anxiety is aroused, messages are sent from the brain through sympathetic nerves to the adrenal medulla. Acetylcholine released by these neurons triggers release of the hormones.

ADRENAL CORTEX

All the hormones of the **adrenal cortex** are steroids synthesized from cholesterol, which is made from acetyl coenzyme A. (You should recall that steroids are a chemical group classified with the lipids.) Three types of hormones are produced by the adrenal cortex: (1) sex hormones, (2) mineralocorticoids, and (3) glucocorticoids.

Very small amounts of both **androgens** (hormones that have masculinizing effects) and **estrogens** (the female hormones) are secreted by the adrenal cortex in both sexes. Normally, the amounts of these hormones released are so small that they have little physiological effect, although adrenal tumors may secrete large quantities, especially of the androgens. Some of the "bearded ladies" of circus sideshows owe their claim to fame to adrenal tumors.

Mineralocorticoids help regulate salt balance. The principal mineralocorticoid is **aldosterone.** This hormone increases the rate at which sodium is reabsorbed by kidney tubules, thereby helping to maintain sodium balance in the body and an appropriate blood pressure. Aldosterone also helps regulate the phosphate concentration in the body. When the adrenal glands do not function, large amounts of sodium are excreted in the urine. Water leaves the body with the sodium (because of osmotic pressure), and the blood volume may be so markedly reduced that the patient dies of low blood pressure.

Cortisol, also called hydrocortisone, accounts for about 95% of the **glucocorticoid** activity of the adrenal cortex. The principal action of cortisol is to promote gluconeogenesis in the liver. Cortisol helps provide nutrients for gluconeogenesis by stimulating transport of amino acids into liver cells and by promoting fat mobilization so that fatty acids are available for gluconeogenesis. Large amounts of glucose and glycogen are produced in the liver, and the blood glucose level rises. Cortisol helps ensure adequate fuel supplies for the cells when the body is under stress. Thus the adrenal cortex provides an important backup system for the adrenal medulla.

Glucocorticoids are used clinically to reduce inflammation in allergic reactions, infections, arthritis, and certain types of cancer. These hormones help stabilize lysosome membranes so that they do not destroy tissues with their potent enzymes. Glucocorticoids also reduce inflammation by decreasing the permeability of capillary membranes, thereby reducing swelling, and they reduce the effects of histamine.

When used in large amounts over long periods of time, glucocorticoids can cause serious side effects. They decrease the number of lymphocytes in the body and can cause atrophy of lymph tissue, reducing the patient's ability to fight infections. Other side effects include ulcers, hypertension, diabetes mellitus, and atherosclerosis.

Glucocorticoid secretion (like aldosterone secretion) is regulated by **ACTH** (adrenocorticotropic hormone) from the pituitary. In turn ACTH secretion is controlled by a releasing hormone from the hypothalamus. Almost any type of stress is reported to the hypothalamus, which activates the system so that large amounts of cortisol are rapidly released. When the body is not under stress, high levels of cortisol in the blood inhibit both the hypothalamus and the pituitary. Abnormally large amounts of glucocorticoids, whether due to disease or drugs, result in the condition called **Cushing's disease.** Fat is mobilized from the lower part of the body and deposited about the trunk. Edema gives the patient's face a full-moon appearance. Blood sugar level rises to as much as 50% above normal, causing adrenal diabetes. If this condition persists for several months, the beta cells in the pancreas can "burn out" from trying to secrete excessive insulin. This can result in permanent diabetes mellitus. Reduction in protein synthesis causes weakness and decreases immune responses, so the patient often dies of infection.

Destruction of the adrenal cortex and the resulting decrease in aldosterone and cortisol secretion cause **Addison's disease.** Reduction in cortisol prevents the body from regulating blood sugar levels because it cannot synthesize enough glucose by gluconeogenesis. The patient also loses the ability to cope with stress. If cortisol levels are significantly depressed, even the stress of mild infections can cause death.

THE ADRENAL GLANDS AND STRESS

Stressors, whether in the form of noise, infection, or even the anxiety of taking a test for which one is not fully prepared, arouse the adrenal glands to action. The brain signals the adrenal medulla rapidly via neural connections to release epinephrine and norepinephrine, hormones that prepare the body for fight or flight. The hypothalamus also signals the anterior pituitary hormonally to secrete ACTH, which increases cortisol secretion, thereby adjusting metabolism to meet the increased demands of the stressful situation (Fig. 30–15).

Some forms of stress are short-lived. One reacts to the situation and quickly resolves it. Other stressors may last for days, weeks, or even years—a chronic disease, for example, or an unhappy marriage or job situation. General anxiety and tension are examples of nonspecific stress.

Chronic stress is harmful because of the side effects of long-term elevated levels of hormones such as cortisol. Though glucocorticoids help reduce inflammation, they also interfere with normal immune responses. Chronic high blood pressure may contribute to heart disease, and increased levels of fat in the blood may promote atherosclerosis. When experimental animals are injected with large amounts of glucocor-

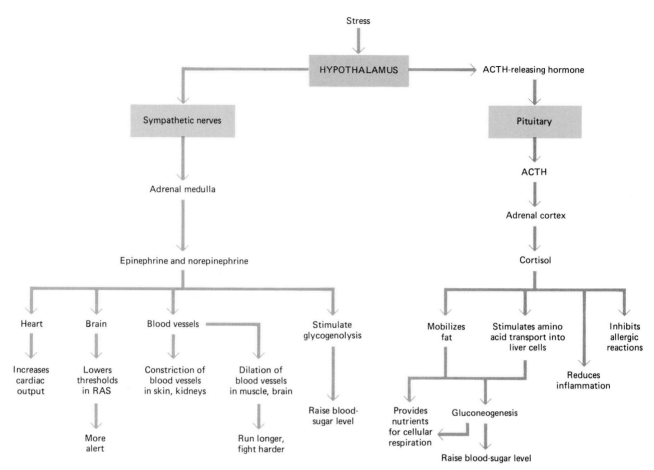

Figure 30–15 Some effects of stress.

ticoids, such disease states are induced, and similar effects are seen when large doses are administered clinically to human patients. Among the diseases linked to excessive amounts of adrenocortical hormones are ulcers, high blood pressure, atherosclerosis, and arthritis.

Other Hormones

Many other tissues of the body secrete hormones. In Chapter 27 we described the hormones released by the digestive tract. The **thymus gland** releases a hormone (thymosin) that plays a role in immune responses, and the kidneys release hormones, one of which (renin) helps regulate blood pressure. The **pineal gland,** located in the brain, produces a hormone called **melatonin,** which may affect reproduction. In Chapter 31 we will discuss the principal reproductive hormones.

Prostaglandins are a group of hormones released by many different tissues in the body, including the lungs, liver, and digestive tract, which influence a variety of metabolic activities. They dilate the bronchial passageways, inhibit gastric secretion, stimulate contraction of the uterus, affect nerve function, affect blood pressure, and influence metabolism. Those synthesized in the temperature-regulating center of the hypothalamus can cause fever, and it is now known that the ability of aspirin to reduce fever (long a mystery) depends upon inhibiting prostaglandin synthesis.

Prostaglandins mimic many of the actions of cyclic AMP, and depending upon the specific tissue type, they stimulate or inhibit cyclic AMP formation. Thus, they may help regulate other hormones. Prostaglandins are used clinically to initiate labor and induce abortion, and their use as a birth-control drug is being investigated.

CHAPTER SUMMARY

I. The endocrine system consists of endocrine glands and tissues that secrete hormones; this system helps regulate many aspects of metabolism, growth, and reproduction.

II. Hormones are transported to their target tissues via the blood.

 A. Some hormones combine with receptors on the cell membrane of their target cells and act by way of a second messenger, such as cyclic AMP.

 B. Steroid hormones combine with receptor proteins within the cytoplasm of the target cell, and the hormone-receptor complex may stimulate a particular gene to initiate protein synthesis.

 C. Hormone secretion is self-regulated by negative-feedback control mechanisms.

III. In invertebrates hormones influence growth, reproduction, molting, morphogenesis, and pigmentation. The interaction of three hormones is responsible for insect development.

IV. In vertebrates hormones help regulate growth, reproduction, salt and fluid balance, and many aspects of metabolism.

V. Nervous and endocrine systems are linked by the hypothalamus, which regulates the activity of the pituitary gland.

 A. The hormones of the posterior lobe of the pituitary are actually produced by the hypothalamus.

 B. Secretion of anterior pituitary hormones is regulated by releasing and inhibiting hormones secreted by the hypothalamus.

 1. Growth hormone stimulates body growth by promoting protein synthesis.

 2. Malfunctions in growth hormone secretion can lead to pituitary dwarfism and gigantism.

VI. Thyroxine from the thyroid gland stimulates the rate of metabolism.

 A. Regulation of thyroxine secretion depends mainly upon a feedback system between the anterior pituitary and the thyroid gland.

 B. Hyposecretion of thyroxine during childhood may lead to cretinism; during adulthood it may result in myxedema. Goiter may develop from hyposecretion or hypersecretion.

VII. The islets of the pancreas secrete insulin and glucagon.

 A. Insulin stimulates cells to take up glucose from the blood and so lowers blood sugar level.

 B. Glucagon raises blood glucose level by stimulating glycogenolysis and gluconeogenesis.

 C. Insulin and glucagon secretion are regulated directly by blood glucose levels.

 D. In diabetes mellitus insulin deficiency results in decreased utilization of glucose, increased fat mobilization, and increased protein utilization.

VIII. The adrenal glands secrete hormones that help the body cope with stress.

 A. The adrenal medulla, sometimes considered the emergency gland, secretes epinephrine and norepinephrine.

 B. The adrenal cortex secretes sex hormones; mineralocorticoids (such as aldosterone, which increases the rate of sodium reabsorption); and glucocorticoids, such as cortisol, which promotes gluconeogenesis.

 C. The hormones of the adrenal medulla help the body respond to stress by increasing heart rate, metabolic rate, and strength of muscle contraction, and causing blood to be rerouted to those organs needed for fight or flight. The adrenal cortex acts as a backup system, ensuring adequate fuel supplies for the rapidly metabolizing cells.

Post-Test

1. Endocrine glands lack _____ ; they release _____ .
2. A chemical messenger that is produced by an endocrine gland or some special cell type and has a specific regulatory effect on the activity of another cell is called a _____ .
3. A second messenger important in the action of many hormones is _____ _____ .
4. The _____ serves as a link between the nervous and endocrine systems.
5. _____ hormone suppresses metamorphosis in insects.

Select the most appropriate term in column B for each description in column A.

Column A	Column B
_____ 6. located in neck region	a. anterior lobe of pituitary
_____ 7. secrete insulin	b. adrenal medulla
_____ 8. sometimes called emergency gland	c. thyroid
_____ 9. regulates other endocrine glands via tropic hormones	d. islets of pancreas
_____ 10. secretes glucocorticoids	e. none of the above

Column A	Column B
_____ 11. stimulates rate of metabolism	a. thyroxine
_____ 12. stimulates release of calcium from bone	b. glucagon
_____ 13. helps maintain sodium balance	c. aldosterone
_____ 14. raise blood sugar level	d. epinephrine
_____ 15. causes heart to beat faster and blood to be rerouted	e. none of the above

16. Cretinism is caused by _____ secretion of _____ during childhood.
17. An abnormal enlargement of the thyroid gland is a _____ .
18. The principal action of _____ is to promote gluconeogenesis.
19. Calcitonin is secreted by the _____ gland.
20. In untreated _____ _____ glucose utilization is decreased and ketosis occurs.

Review Questions

1. What is a hormone? What are some of the important functions of hormones?
2. How are hormones transported? How do they "recognize" their target tissues? What is the role of cyclic AMP in hormone action?
3. How do hormones interact to regulate development in insects?
4. Why is the hypothalamus considered the link between the nervous and the endocrine systems? Explain.
5. Describe the actions of (1) prolactin, (2) oxytocin, (3) thyroid-stimulating hormone.
6. Draw a diagram to illustrate the regulation of thyroxine secretion by the anterior pituitary gland.
7. Explain the hormonal basis for (1) acromegaly, (2) pituitary dwarfism, (3) cretinism, (4) hypoglycemia, (5) Cushing's disease.
8. Explain the antagonistic actions of insulin and glucagon in regulating blood glucose level. What other hormones studied in this chapter affect blood glucose level?
9. Describe several physiological disturbances that result from diabetes mellitus.
10. What are the actions of epinephrine and norepinephrine?
11. How is the adrenal medulla regulated?
12. What three types of hormones are released by the adrenal cortex, and what are the actions of each type?
13. Explain how the adrenal glands help the body deal with stress.
14. What are prostaglandins?

Readings

Bloom, F. "Neuropeptides," *Scientific American,* October 1981. Neuropeptides are sometimes considered hormones, sometimes neurotransmitters.

Crews, D. "The Hormonal Control of Behavior in a Lizard," *Scientific American,* August 1979. Sexual behavior is controlled by the interaction of the brain and hormones from the gonads.

Guillemin, R., and R. Burges. "Hormones of the Hypothalamus," *Scientific American,* November 1972. Hormones released by the hypothalamus control the pituitary.

Chapter 31

REPRODUCTION: PERPETUATION OF THE SPECIES

Outline

I. Asexual reproduction
II. Sexual reproduction
III. Human reproduction
   A. The male
      1. Production of sperm
      2. Sperm transport
      3. Semen production
      4. The penis
      5. Male hormones
   B. The female
      1. Production of ova and hormones
      2. Ovum transport—the uterine tubes
      3. Incubating the embryo—the uterus
      4. The vagina
      5. External genital structures
      6. Breasts
         a. Lactation
         b. Breast cancer
      7. The menstrual cycle
         a. Interaction of hormones
         b. Menopause
   C. Physiology of sexual response
   D. Fertilization
IV. Birth control
   A. Oral contraceptives—the Pill
   B. Intrauterine devices (IUDs)
   C. Other common methods
   D. New methods
   E. Sterilization
      1. Male sterilization
      2. Female sterilization
   F. Abortion
V. Sexually transmitted diseases
Focus on fertilization

Learning Objectives

After you have read this chapter you should be able to:

1. Compare asexual with sexual reproduction, giving two specific examples of asexual reproduction.
2. Trace the development of sperm cells and their passage through the male reproductive system, labeling on a diagram each male structure and giving its function.
3. Give the actions of testosterone and of gonadotropic hormones in the male.
4. Describe or label on a diagram each structure of the female reproductive system and give its function.
5. Trace the development and fate of the ovum.
6. Describe the hormonal control of the menstrual cycle and identify the timing of important events, such as ovulation and menstruation.
7. Describe the process of fertilization and list its functions. Identify the time of the menstrual cycle at which sexual intercourse is most likely to result in pregnancy.
8. Describe the mode of action and give the advantages and disadvantages of each of the methods of contraception described: the Pill, IUD, spermicides, condom, diaphragm, rhythm, douche, withdrawal, and sterilization.
9. Describe common methods of induced abortion and summarize the current legal status of this practice.
10. Discuss sexually transmitted diseases, describing the symptoms, effects, and treatment of gonorrhea, syphilis, and genital herpes.

Perhaps the most unique feature of a living organism is its ability to reproduce and perpetuate its species. The survival of each species depends upon the success of its members in replacing those individuals that die.

ASEXUAL REPRODUCTION

In **asexual reproduction** a single parent splits, buds, or fragments, giving rise to two or more offspring that are identical copies of the parent. The parent endows the offspring with a set of genes exactly like its own, so the new individual becomes a carbon copy of the parent. Mutation provides the only means of varying the genetic composition of members of the species.

Sponges and hydras can reproduce by **budding,** in which a new individual grows as an extension of the parent's body. It may remain attached, becoming part of a colony of closely associated but independent organisms, or it may separate completely from the parent (Fig. 31–1).

Some flatworms can reproduce by **fragmentation,** in which the body of the parent breaks into several pieces and each piece develops into a whole new animal. Oyster farmers learned long ago that when they tried to kill sea stars by chopping them in half and throwing the pieces back into the sea, the number of sea stars preying on their oyster beds doubled. A sea star can regenerate an entire new individual from a single arm.

SEXUAL REPRODUCTION

Sexual reproduction requires the fusion of two types of **gametes,** or sex cells, to form a **zygote,** or fertilized egg. The genetic composition of each gamete is different, so the zygote contains genetic information that is a unique combination of the genes contributed by each. In species that reproduce sexually there usually are two types of individuals—males, which produce motile sperm cells, and females, which produce egg cells. However, some invertebrate species (earthworms and certain flatworms, for example) are **hermaphroditic,** meaning that a single individual produces both eggs and sperm. Some hermaphroditic animals, such as the parasitic tapeworms, are capable of self-fertilization. In other species, such as the earthworm, each organism produces eggs and sperm but does not fertilize its own gametes. Instead, two individuals come together in copulation and exchange sperm cells, a kind of double reproductive act.

Fertilization, the fusion of sperm and egg, may be **internal** and take place inside the female body, or be **external** and occur outside (Fig. 31–2). Most aquatic animals practice external fertilization, with mating partners simultaneously releasing eggs and sperm into the water. Dispersion frequently results in loss of many gametes, but so many are released that a sufficient number of sperm and egg cells meet and unite to ensure propagation of the species.

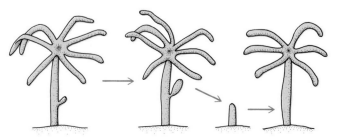

Figure 31–1 Hydras and many other organisms reproduce asexually by budding. A part of the body grows outward, then separates and develops into a new individual. The portion of the parent body that buds is not specialized exclusively for performing a reproductive function.

(a)

(b)

(c)

Figure 31–2 External fertilization (a) and internal fertilization (b). (a) Spawning pickerel frogs. Most amphibians must return to water for mating. The female lays a mass of eggs, while the male mounts her and simultaneously deposits his sperm in the water. (b) Desert tortoises mating. (c) A butterfly, *Perhybris pyrra*, has just finished ovipositing (depositing eggs) on the leaf of a plant. (a, E.R. Degginger; b, Paul Kuhn, Tom Stack and Associates; c, L.E. Gilbert, University of Texas at Austin/BPS.)

In internal fertilization matters are left less to chance. The male generally delivers sperm cells directly into the body of the female. Most terrestrial animals, as well as a few fish and other aquatic animals, practice the internal method of fertilization.

Although sex is not universally essential to reproduction, it does have certain biological advantages. Species that reproduce sexually exhibit a great *variety* among their individual members, and variety has been described as nature's grand tactic of survival. In sexual reproduction the offspring is the product of the genes contributed by both parents, rather than a genetic copy of a single parent. What is more, the genes are thoroughly shuffled during meiosis (the production of sex cells) so that no two offspring even of the same parents are likely to be very similar.

In sexual reproduction only the tiny gamete of the parent lives on in the new generation. The rest of the parental body ages and eventually dies. Although the individual does not attain immortality, it enjoys the advantages of specialization, not possible in asexual forms. Only a small part of the organism is occupied with the reproductive function; other cells are free to specialize in other tasks, such as thinking and locomotion.

HUMAN REPRODUCTION

In human beings and other mammals reproductive processes include formation of gametes (eggs and sperm), preparation of the female body for pregnancy, sexual intercourse, fertilization, pregnancy, and lactation (producing milk for nourishment of the young). These events are precisely coordinated by the interaction of hormones secreted by the anterior lobe of the pituitary gland and by the **gonads** (sex glands).

The Male

The reproductive responsibility of the male is to produce sperm cells and to deliver them into the female reproductive tract. When a sperm combines with an egg, it contributes half the genetic endowment of the offspring and determines its sex. The male reproductive system is illustrated in Figure 31–3. Male structures include the testes (which produce sperm and the male hormone testosterone), the scrotum (which contains the testes), conducting tubes (which lead from the testes to the outside of the body), accessory glands (which produce semen), and the penis (which serves as a copulatory organ).

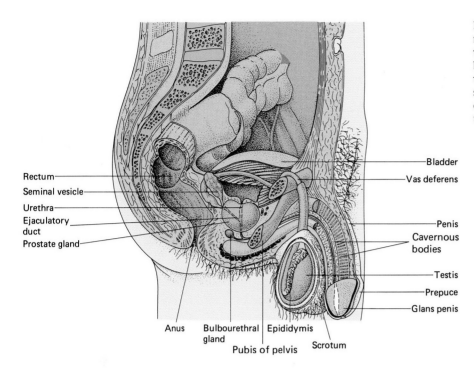

Figure 31–3 Anatomy of the human male reproductive system. The scrotum, penis, and pelvic region have been cut sagittally to show their internal structures. Identify the accessory glands and try to trace the conducting tubes from the testis to the urethra.

Rectum
Seminal vesicle
Urethra
Ejaculatory duct
Prostate gland

Bladder
Vas deferens
Penis
Cavernous bodies
Testis
Prepuce
Glans penis

Anus
Bulbourethral gland
Pubis of pelvis
Epididymis
Scrotum

PRODUCTION OF SPERM

In the adult male millions of sperm are produced each day by the paired male gonads, the **testes.** Each testis is a small oval organ packed with an extensive system of tiny coiled **seminiferous tubules** (Fig. 31–4). These tubules are the sperm cell factories. About 1000 seminiferous tubules can be found in each testis, and if uncoiled and placed end to end, these threadlike microscopic tubules would span almost one-third of a mile.

Sperm develop from stem cells called **spermatogonia** found next to the wall of the tubules. These cells continuously multiply by mitosis in order to maintain a large population of stem cells. About half of them differentiate to become **primary spermatocytes** and then undergo meiosis. During their development sperm cells are intimately associated with large supporting cells, called **Sertoli cells,** which may help nourish them. The process of sperm cell production, called **spermatogenesis,** requires about 2 months.

As sperm cells arise from meiosis, they move farther and farther from the tubule wall toward the center of the tubule (Fig. 31–5). The immature sperm cells (**spermatids**) that are produced from meiosis differentiate to become mature sperm cells. Although highly specialized, the mature sperm cell is perhaps the tiniest cell in the human body. Its structure is described in Chapter 9.

In the embryo the testes develop in the abdominal cavity, but about 2 months before birth they descend into the **scrotum,** a skin-covered bag suspended from the groin. As they descend, the testes move through the **inguinal canals,** passageways connecting the cavities of the pelvis and scrotum. The inguinal region remains a weak place in the abdominal wall and therefore a common site of hernia in the male. Straining the abdominal muscles by lifting a heavy object sometimes results in a tear in the muscles and connective tissues of the abdominal wall, through which a loop of intestine can bulge into the scrotum. This condition is called an **inguinal hernia.**

Sperm cells cannot develop at body temperature. The scrotum serves as a cooling unit, maintaining them about 2°C (about 3.5°F) below body temperature. An abundance of both sweat glands and blood vessels

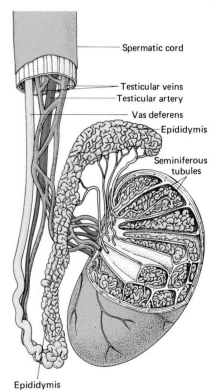

Spermatic cord

Testicular veins
Testicular artery
Vas deferens
Epididymis

Seminiferous tubules

Epididymis

Figure 31–4 Structure of the testis and epididymis. The testis is shown in sagittal section to illustrate the arrangement of the seminiferous tubules.

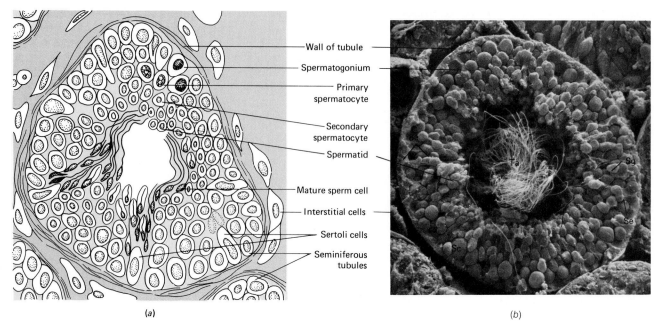

Labels for (a):
Wall of tubule
Spermatogonium
Primary spermatocyte
Secondary spermatocyte
Spermatid
Mature sperm cell
Interstitial cells
Sertoli cells
Seminiferous tubules

(a)

(b)

Figure 31–5 Structure of a seminiferous tubule showing developing sperm cells in various stages of spermatogenesis. (a) Identify the sequence of sperm cell differentiation. Note the Sertoli cells and the interstitial cells. (b) A scanning electron micrograph of a transverse section through a seminiferous tubule (approximately ×580). *Se,* Sertoli cell; *Sc,* primary spermatocyte; *Sg,* spermatogonium; *Ta,* tails of spermatozoa. (Kessel, R.G., and R.H. Kardon. *Tissues and Organs: A Text-Atlas of Scanning Electron Microscopy.* San Francisco, W.H. Freeman & Company, 1979.)

in the wall of the scrotum permits heat loss and so helps to maintain the cool temperature. In hot weather involuntary muscles in the wall of the scrotum relax, positioning the testes away from the body so that the sperm cells remain cool. In cold weather these muscles contract, drawing the testes up close to the abdominal wall where they are kept warm.

Although the testes ordinarily descend into the scrotum before birth, occasionally this fails to happen, and then viable sperm cells are not produced. This condition can be corrected with hormone therapy, which stimulates the testes to descend, or by surgery. If not treated, the seminiferous tubules eventually degenerate and the male becomes **sterile,** that is, unable to father offspring. However, the masculinity of such a person is not affected: The interstitial cells, located between the seminiferous tubules, are not damaged by abdominal temperatures and continue to produce the male hormone testosterone.

SPERM TRANSPORT

Sperm cells leave the seminiferous tubules of the testes through a series of small tubules that empty into a larger tube, the **epididymis** (see Fig. 31–4). The epididymis of each testis is a much-coiled tube in which sperm cells complete their maturation and are stored. The epididymis continues as a straight tube, the **vas deferens** (plural, vasa deferentia), or sperm duct, which passes from the scrotum through the inguinal canal and into the abdominal cavity. Each vas deferens empties into a very short **ejaculatory duct,** which passes through the prostate gland (see below) and then opens into the urethra. The single **urethra** (which conducts both urine and semen) passes through the penis to the outside of the body.

SEMEN PRODUCTION

As they are transported through the conducting tubes, sperm are mixed with secretions from the accessory glands. About 3.5 milliliters of **semen** is ejaculated during sexual climax. Semen consists of about 400 million sperm cells suspended in the secretions of the seminal vesicles, prostate, bulbourethral glands, and small glands in the walls of the ducts. Sperm cells are so tiny that they account for very little of the semen volume.

The paired **seminal vesicles** are saclike glands that empty into the vasa deferentia. Secretions of the seminal vesicles account for about 60% of semen volume. The mucus-like fluid secreted by the seminal vesicles contains the sugar fructose, plus other nutrients that nourish and provide fuel for the active sperm cells.

The single **prostate gland** contributes a thin, milky, alkaline secretion important in neutralizing the acidity of the other fluids in the semen. In older men the prostate gland often enlarges and exerts pressure on the urethra, making urination difficult. When necessary, the prostate can be removed surgically, often with no adverse effect on sexual performance. Cancer is another common affliction of this gland. Most commonly, prostate cancer grows slowly and may be confined to the gland for a long time.

Upon sexual arousal the **bulbourethral glands** release a few drops of a clear, sticky, alkaline fluid that may neutralize the acidity of the urethra in preparation for ejaculation.

Men with fewer than 40 million sperm per milliliter of semen usually are sterile. When a couple's attempts to produce a child are unsuccessful, sperm counts and analyses may be undertaken in clinical laboratories. Sometimes semen is found to contain large numbers of abnormal sperm or, occasionally, none at all. Fever or infection of the testes may cause temporary sterility. In about one-fourth of mumps cases in adult males, the testes become inflamed, and in some of these cases the spermatogonia deteriorate, resulting in sterility.

THE PENIS

The **penis** is an erectile copulatory organ designed to deliver sperm into the female reproductive tract. It consists of a long **shaft,** which enlarges to form an expanded tip, the **glans** (Fig. 31–6). Part of the loose-fitting skin of the penis folds down and covers the proximal portion of the glans, forming a cuff called the **prepuce** or **foreskin.** This cuff of skin is removed during **circumcision.** Although the pros and cons of circumcision are a matter of debate, this simple surgical procedure is routinely performed on male infants in many hospitals. (Circumcision is also performed as a religious rite by Jews and Moslems.)

Under the skin the penis consists of three parallel cylinders of erectile tissue called the **cavernous bodies** (corpora cavernosa). One of these columns of erectile tissue surrounds the portion of the urethra that passes through the penis. The erectile tissue consists of large blood vessels called venous sinusoids. When the male is sexually stimulated, nerve impulses cause the arteries of the penis to dilate. Blood rushes into the sinusoids of the erectile tissue. As the erectile tissue fills with blood, it swells, compressing veins that conduct blood away from the penis and hindering the outflow of blood through these veins. Much more blood enters the penis than can leave, causing the erectile tissue to become engorged with blood. The penis thus becomes **erect,** that is, longer, larger in circumference, and firm.

MALE HORMONES

Puberty, the period of sexual maturation, usually begins at about age 12 or 13. With its onset the hypothalamus begins to secrete releasing hormones, which stimulate the anterior pituitary to secrete gonadotropic hormones. These hormones are **follicle-stimulating hormone (FSH)** and **luteinizing hormone (LH).** (Discovered first in the female, these hormones were named for their functions in the female body. In the male LH is often referred to as *interstitial cell–stimulating hormone, or ICSH.*)

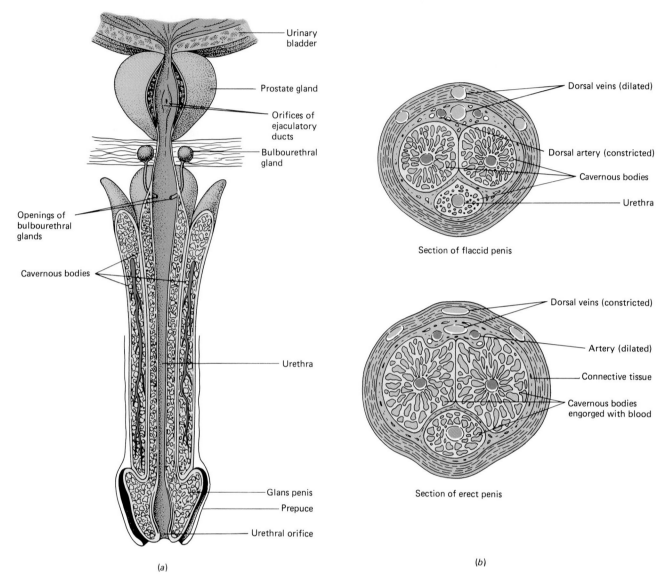

(a)

(b)

Figure 31–6 Internal structure of the penis. (a) Longitudinal section through the prostate gland and penis. (b) Cross section through flaccid and erect penis. Note that the erectile tissues of the cavernous bodies are engorged with blood in the erect penis.

FSH stimulates development of the seminiferous tubules in the testes and promotes sperm production (Table 31–1). LH stimulates the **interstitial cells** in the testes to secrete the male hormone **testosterone.** Small clusters of these endocrine cells are found between the seminiferous tubules of the testes (see Fig. 31–5).

Secretion of these hormones is regulated by feedback systems (Fig. 31–7). High levels of testosterone in the blood inhibit the hypothalamus, so LH secretion by the pituitary is slowed. This feedback system maintains a steady level of testosterone. Testosterone may also have a slight inhibiting effect on FSH secretion, but it is thought that there may be yet another hormone produced during sperm production that is important in regulating FSH secretion.

One of the main actions of testosterone is to promote growth. As large quantities of this hormone begin to circulate in the blood during puberty, the male experiences an adolescent growth spurt. Testosterone is responsible for the development of both primary and secondary sexual characteristics in the male. The penis and scrotum enlarge and the internal reproductive structures increase in size and become active. Secondary characteristics include deepening of the voice, muscle development, and growth of pubic, facial, and underarm hair. Testosterone also stimulates the rate of secretion of the oil glands in the skin, predisposing the adolescent to acne.

Table 31–1
PRINCIPAL MALE REPRODUCTIVE HORMONES

Endocrine gland and hormones	Principal site of action	Principal actions
Anterior pituitary		
Follicle-stimulating hormone (FSH)	Testes	Stimulates development of seminiferous tubules; may stimulate spermatogenesis
Luteinizing hormone (LH); also called interstitial cell–stimulating hormone (ICSH)	Testes	Stimulates interstitial cells to secrete testosterone
Testes		
Testosterone	General	Before birth, stimulates development of primary sex organs and descent of testes into scrotum; at puberty, responsible for growth spurt, stimulates development of reproductive structures and secondary sex characteristics (male body build, growth of beard, deep voice, etc.); in adult, responsible for maintaining secondary sex characteristics and may stimulate spermatogenesis

What happens when testosterone is absent? When a male is **castrated** (i.e., his testes are removed) before puberty, he becomes a **eunuch.** He retains childlike sex organs and does not develop secondary

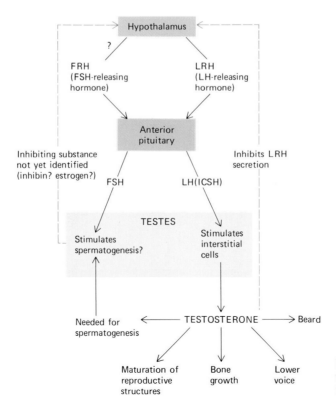

Figure 31–7 Known and suspected reproductive hormone relationships in the male.

sex characteristics. If castration occurs after puberty, increased secretion of male hormone by the adrenal glands helps to maintain masculinity.

Males normally possess small amounts of female sex hormones, and females have small amounts of male hormone. These hormones are secreted partly by the adrenal glands and partly by the gonads. Sexuality is determined not so much by the presence or absence of male or female hormones as by the balance that exists between them.

The Female

The female reproductive system is designed to produce **ova** (eggs), to receive the penis and the sperm released from it during sexual intercourse, to house and nourish the embryo during its prenatal development, and to nourish the infant. Because it must perform all these diverse functions, the physiology of the female system is more complex than that of the male. Much of its activity is regulated by hormonal changes that take place during the **menstrual cycle**, the monthly preparation for possible pregnancy.

Principal organs of the female reproductive system (Figs. 31–8 and 31–9) are the ovaries (which produce ova and female hormones), the uterine tubes (where fertilization takes place), the uterus (incubator for the developing child), the vagina (which receives the penis and serves as part of the birth canal), the vulva (external genital structures), and the breasts.

PRODUCTION OF OVA AND HORMONES

The paired **ovaries,** the female gonads, produce ova and the female sex hormones, **estrogen** and **progesterone.** About the size and shape of large almonds, the ovaries are located close to the lateral walls of the pelvic cavity. Each ovary is covered with a single layer of epithelium. Its internal structure consists mainly of connective tissue called **stroma,** through which are scattered ova in various stages of maturation (Fig. 31–10).

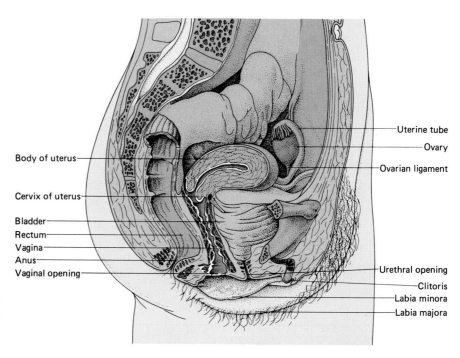

Figure 31–8 Midsagittal section of female pelvis showing reproductive organs. Note the position of the uterus relative to the vagina.

Body of uterus

Cervix of uterus

Bladder
Rectum
Vagina
Anus
Vaginal opening

Uterine tube
Ovary
Ovarian ligament

Urethral opening
Clitoris
Labia minora
Labia majora

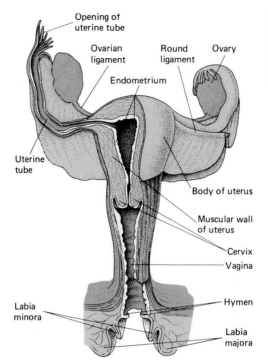

Opening of
uterine tube

Ovarian
ligament

Round
ligament

Ovary

Endometrium

Uterine
tube

Body of uterus

Muscular wall
of uterus

Cervix

Vagina

Labia
minora

Hymen

Labia
majora

Figure 31–9 Anterior view of female reproductive system. Some organs have been cut open to expose
the internal structure. The ligaments
help to hold the reproductive organs
in place.

Before birth hundreds of thousands of **oogonia** (cells that give rise
to ova) may be seen in the ovaries. Each is surrounded by a cluster of
cells. Together the ovum and its surrounding cells constitute a **follicle.**
The female's entire lifetime complement of gametes is established during embryonic development; no new oogonia arise after birth.

During prenatal development the oogonia increase in size to become **primary oocytes.** By the time of birth they are in the prophase of the
first meiotic division. At this stage they enter a resting phase, which lasts
throughout childhood and into adult life.

With the onset of puberty a few of the follicles develop each month
in response to FSH secreted by the anterior pituitary gland. As the follicle grows, the primary oocyte completes its first meiotic division, producing two cells that are markedly disproportionate in size. The smaller
one, called the **first polar body,** later divides to form two polar bodies.
Both of these eventually disintegrate. The larger cell, a **secondary oocyte,**
may continue to develop. The second meiotic division, however, is not
completed until after fertilization. The development of a diploid oogonium to a haploid ovum is called **oogenesis** (see Fig. 9–13).

As an oocyte develops, it becomes separated from its follicle by a
thick membrane, the **zona pellucida.** The follicle increases in size as its
cells multiply and as specialized connective-tissue cells of the ovary
grow around it. An outer layer of connective tissue called **theca** surrounds the follicle cells. Cells of the theca secrete estrogen. As the follicle develops, its cells secrete fluid, which collects in a cavity (the an-
trum) created between them.

As the follicle matures it moves closer to the surface of the ovary,
eventually coming to resemble a fluid-filled blister on the ovarian surface. Mature follicles are known as **Graafian follicles.** Normally, only one
follicle matures each month. Several others develop for about a week,
then deteriorate.

In response to FSH and LH from the anterior pituitary gland, the
single mature follicle ruptures after about 2 weeks of development. During this process, called **ovulation,** the ovum is ejected through the wall of

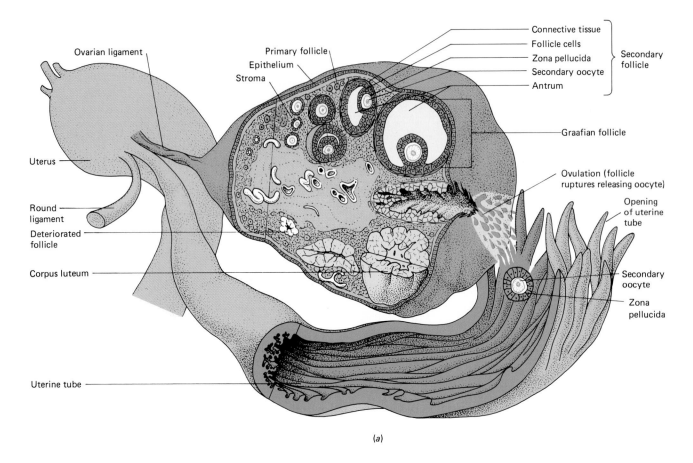

(a)

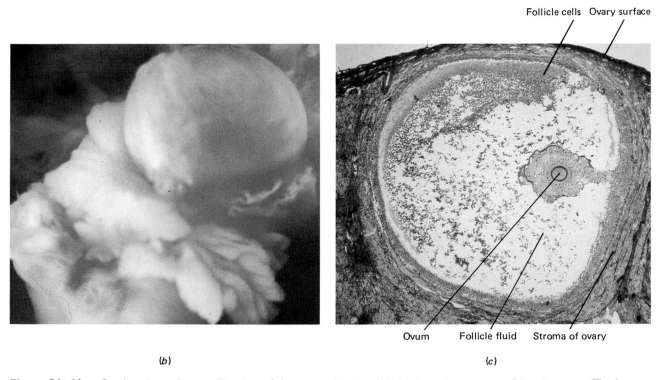

(b)

(c)

Figure 31–10 *a*, Section through a small region of the ovary showing follicles in various stages of development. The largest is the most mature. (*b*) The mature follicle forms just under the surface of the ovary, producing a tightly stretched fluid-filled blister (shown here from the outside) that will eventually burst, releasing the ovum and the follicle cells surrounding it. The remaining follicle cells then form the corpus luteum. (*c*) This stained microscope slide shows the ovum surrounded by a layer of follicle cells that will be released along with it. These follicle cells become the corona radiata, a layer that acts as a barrier to sperm cells and may help to ensure that the egg is fertilized by only one of the many sperm that crowd around it. (*b*, Petit Format, Photo Researchers, Inc.; *c*, Biophoto Associates, Photo Researchers, Inc.)

the ovary and into the body cavity. The portion of the follicle that remains behind in the ovary develops into an important endocrine structure, the **corpus luteum** (literally, "yellow body"). This transformation is directed by the hormone LH released by the anterior pituitary. The corpus luteum secretes progesterone and estrogen, female hormones that help prepare the uterus for possible pregnancy.

OVUM TRANSPORT—THE UTERINE TUBES

At ovulation the ovum is released into the pelvic cavity. The free end of the **uterine tube** (also called fallopian tube) is strategically designed and located so that the ovum enters into it almost immediately (Fig. 31–10). Peristaltic contractions of the muscular wall of the uterine tube and beating of cilia in its lining help move the ovum along toward the uterus. Fertilization takes place within the upper portion of the uterine tube. If fertilization does not occur, the ovum degenerates there.

Inflammation of the uterine tubes sometimes results in scarring that may partially constrict the tubes. If passage of the ovum is blocked, sterility may result. Sometimes partial tubal constriction results in **tubal pregnancy,** in which the embryo begins to develop in the wall of the uterine tube because it cannot progress to the uterus. (Such pregnancies must be diagnosed early so that the tube can be surgically removed before it ruptures. Uterine tubes are not designed to bear the burden of a developing baby.)

INCUBATING THE EMBRYO—THE UTERUS

Each uterine tube joins the hollow, muscular **uterus,** or womb. Normally about the size of a fist, the uterus is a pear-shaped organ that occupies a central position in the pelvic cavity. The uterus has thick walls of smooth muscle and is lined by a mucous membrane, the **endometrium.** The endometrium thickens each month in preparation for possible pregnancy. If an egg is fertilized, the tiny embryo finds its way into the uterus and implants itself in the thick lining. Here it grows and develops, sustained by the nutrients and oxygen delivered by surrounding maternal blood vessels. It is one of nature's wonders that the uterus, ordinarily only about 8 centimeters (3 inches) long, can expand to accommodate a fully developed fetus. At the time of birth contractions of the thick muscular walls of the uterus help expel the baby. If fertilization does not occur during the monthly cycle, the mucous lining sloughs off and is discharged in the process of **menstruation.**

The lower portion of the uterus, called the **cervix,** projects slightly into the vagina. Cancer of the cervix is one of the most common types of cancer in women, accounting for about 13,000 deaths per year in the United States alone. About 50% of cervical cancer is now detected at very early stages, when the patient can be cured. Detection can be accomplished by the routine **Papanicolaou test** (Pap smear), in which a few cells are scraped from the cervix during a gynecological examination and are studied microscopically.

THE VAGINA

The **vagina,** an elastic, muscular tube capable of considerable distension, extends from the cervix to the external genital structures. The vagina receives the penis during sexual intercourse. It also serves as an exit through which the discarded uterine lining is discharged during menstruation, and as the lower part of the birth canal. The vagina is internal and should not be confused with the external genital structures.

Figure 31–11 The external female genital structures.

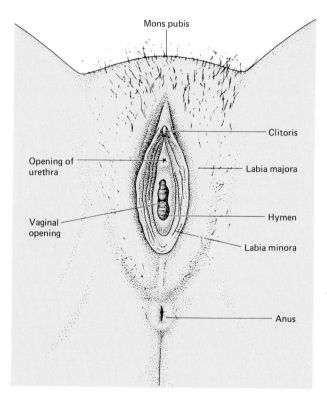

EXTERNAL GENITAL STRUCTURES

The external female genital structures are referred to collectively as the **vulva** (Fig. 31–11). Two pairs of liplike folds surround the opening to the vagina. The delicate inner folds of skin are the **labia minora.** External to these are the heavier, hairy **labia majora.** In the embryo the labia majora develop from the same structures that form the scrotum in the male.

Anteriorly, the labia minora merge to form the foreskin of the **clitoris,** a very small erectile structure comparable to the male glans penis. Rich in nerve endings, the clitoris serves as a center of sexual sensation in the female. During sexual excitement it becomes enlarged and stiff.

The **hymen** is a thin ring of tissue that may partially block the entrance to the vagina. Through the ages it has been considered the symbol of virginity because it is often ruptured during a woman's first sexual intercourse. Not all the attention focused upon the hymen is merited, however, since its absence is not a reliable indicator of previous sexual relations. In many women the hymen is destroyed by strenuous physical exercise during childhood, or by use of tampons inserted to absorb the menstrual flow. In a few women the hymen persists after sexual relations or even after childbirth.

Just anterior to the vaginal opening is the opening of the urethra. In the female the urinary system is entirely separated from the reproductive system.

BREASTS

The **breasts,** containing the mammary glands, overlie the pectoral muscles and are attached to them by connective tissue. Fibrous bands of tissue called **ligaments of Cooper** connect the breasts firmly to the skin. The basic function of the breasts is **lactation,** production of milk for nourishment of the young. Each breast is composed of about 20 lobes of glandular tissue (Fig. 31–12). A duct drains milk from each lobe and

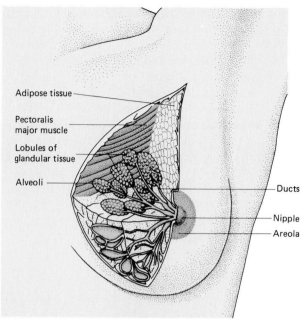

Figure 31–12 The mature female
breast.

Adipose tissue

Pectoralis
major muscle

Lobules of
glandular tissue

Alveoli

Ducts

Nipple

Areola

opens onto the surface of the nipple. Thus there are about 20 tiny openings on the surface of each nipple. (Some types of baby bottle nipples emulate this natural arrangement by having several small openings in the nipple rather than a single one, as in old-fashioned bottle nipples.) The amount of adipose tissue around the lobes of the glandular tissue determines the size of the breasts and accounts for their soft consistency. The size of the breasts does not affect their capacity to produce milk.

The nipple consists of smooth muscle, which can contract to make the nipple erect in response to sexual or certain other stimuli. In the pinkish **areola** surrounding the nipple are several rudimentary milk glands.

In childhood the breasts contain only rudimentary glands. At puberty estrogen and progesterone (in the presence of growth hormone and prolactin) stimulate development of the glands and ducts and the deposition of fatty tissue characteristic of the adult breast.

LACTATION. During pregnancy high concentrations of estrogen and progesterone produced by the corpus luteum and by the placenta stimulate the glands and ducts to develop, resulting in increased breast size. For the first few days after childbirth the mammary glands produce a fluid called **colostrum,** which contains protein and lactose but little fat. After birth *prolactin,* secreted by the anterior pituitary, stimulates milk production, and by the third day after delivery milk itself is produced. When the infant suckles at the breast, a reflex action results in release of prolactin and oxytocin from the pituitary gland. Oxytocin permits release of milk from the glands and ejection from the breasts.

Breast feeding a baby offers many advantages besides promoting a close bond between mother and child. Milk content is tailored to the nutritional needs of a human infant. Babies nourished with cow's milk are more likely to develop allergies to dairy products. Antibodies in the colostrum and mother's milk are thought to play a protective role, for breast-fed babies have a lower incidence of respiratory and digestive tract infection during the second 6 months of life. Breast feeding also enhances recovery of the uterus, since the oxytocin released stimulates it to contract to nonpregnant size.

BREAST CANCER. The breasts are the most common site of cancer in women. Incidence has increased in recent years and now stands at more

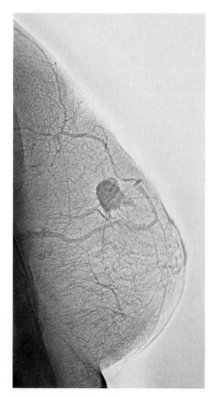

Figure 31–13 Xeromammogram showing cancer of the breast.

than 70 per 100,000 women. About 50% of breast cancers begin in the upper, outer quadrant of the breast. As a malignant tumor grows, it may adhere to the deep connective tissue of the chest wall. Sometimes it extends to the skin, causing dimpling. Eventually the cancer spreads to the lymphatic system. About two-thirds of breast cancers have spread to the lymph nodes by the time they are diagnosed.

Mastectomy (removal of the breast), radiation treatment, and chemotherapy are considered the most reliable methods of treating breast cancer. When diagnosis and treatment begin early, 80% of patients survive for 5 years, and 62% for 10 years or longer. Untreated patients have only a 20% 5-year survival rate.

Since early detection of these cancers greatly increases the chances of cure and survival, self-examination is important. Xeromammography, a soft-tissue radiological study of the breast, helps detect very small lesions (Fig. 31–13).

THE MENSTRUAL CYCLE

As a female approaches puberty, the anterior pituitary gland secretes the gonadotropic hormones FSH and LH, which signal the ovaries to begin functioning. Interaction of FSH and LH with estrogen and progesterone from the ovaries regulates the menstrual cycle, which runs its course every month from puberty until **menopause,** the end of a woman's reproductive life. The menstrual cycle stimulates production of an ovum each month and prepares the uterus for pregnancy.

Although there is wide variation, a "typical" menstrual cycle is 28 days long. The first day of menstruation (bleeding) marks the first day of the cycle. Ovulation occurs about 14 days before the next cycle begins. In a 28-day cycle this would correspond to about the 14th day (Fig. 31–14).

During menstruation, which lasts about 5 days, the thickened endometrium of the uterus is sloughed off. Total blood loss is about 35 milliliters (1 fluid ounce), but an additional 35 milliliters of fluids of the uterine glands is also discharged. During this phase of the menstrual cycle FSH is the principal hormone released by the pituitary gland. It stimulates a group of follicles to develop in the ovary.

During the *preovulatory phase* of the menstrual cycle, estrogen released from the theca of the developing follicle in the ovary stimulates the growth of the endometrium once again. Its blood vessels and glands begin to develop anew. At midcycle estrogen is thought to stimulate the hypothalamus to secrete a releasing hormone that results in release of LH from the pituitary. LH is necessary for final maturation of the follicle, for ovulation, and later for development of the corpus luteum.

After ovulation the *postovulatory phase* begins. The corpus luteum releases progesterone as well as estrogen, and these hormones stimulate continued thickening of the endometrium. Progesterone especially affects the little glands in the endometrium, stimulating them to secrete a fluid rich in nutrients. Should fertilization occur (Figure 31–15), this fluid nourishes the embryo when it first arrives in the uterus on about the fourth day of development. On about the seventh day after fertilization the embryo begins to implant itself into the thick endometrium of the uterus. The placenta begins to develop and secretes a hormone called **human chorionic gonadotropin (HCG),** which signals the corpus luteum to continue to function.

If pregnancy does not occur, the corpus luteum begins to degenerate, and progesterone and estrogen levels in the blood fall markedly. Spiral arteries in the uterine wall constrict, and the part of the endometrium they supply becomes ischemic (deprived of blood and thus of oxy-

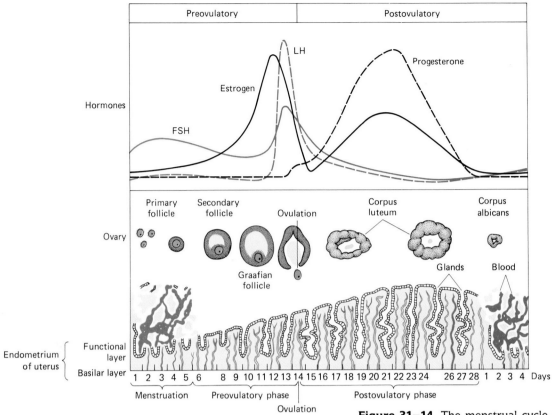

Figure 31–14 The menstrual cycle. The events that take place within the pituitary, ovary, and uterus are precisely synchronized. When fertilization does not occur, the cycle repeats itself about every 28 days. Compare this illustration with Figure 31–15.

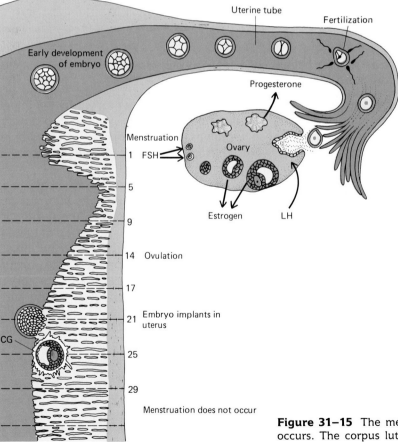

Figure 31–15 The menstrual cycle is interrupted when pregnancy occurs. The corpus luteum does not degenerate, and menstruation does not take place. The wall of the uterus remains thickened, providing the embryo with an ideal environment for development.

Figure 31–16 Known and suspected reproductive hormone relationships in the female. Blue arrows indicate inhibition.

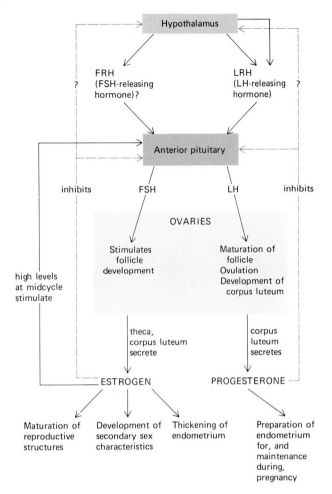

gen). Menstruation begins once again as cells start to die and damaged arteries rupture and bleed.

INTERACTION OF HORMONES. The menstrual cycle is regulated by the interaction of releasing factors from the hypothalamus, along with pituitary and ovarian hormones. These hormones signal the organs involved by a system of feedback control (Fig. 31–16). Table 31–2 gives the actions of the pituitary and ovarian female hormones.

Estrogen is a general term that is used here to refer collectively to a group of closely related steroid hormones. Like testosterone in the male, estrogen is responsible for the growth of the sex organs at puberty and for the development of the secondary sex characteristics: initiation of breast development, broadening of the pelvis, and characteristic distribution of fat and muscle. During the menstrual cycle estrogen promotes the growth of follicles, stimulates growth of the endometrium, and makes the cervical mucus thinner and more alkaline, changes favorable to sperm survival.

Progesterone is a steroid secreted by the corpus luteum and, during pregnancy, by the placenta. It stimulates the endometrium to complete its preparation for pregnancy. Under its influence glands of the endometrium secrete glycogen, which nourishes the early embryo.

MENOPAUSE. At about age 50 the ovaries become less responsive to gonadotropic hormones, and the amount of estrogen and progesterone secreted diminishes. Perhaps there are not enough follicles left to develop and secrete hormones. The ovaries begin to degenerate, and the menstrual cycle becomes irregular and eventually halts. Hot flashes some-

Table 31–2
PRINCIPAL FEMALE REPRODUCTIVE HORMONES

Endocrine gland and hormones	Principal target tissue	Principal actions
Anterior pituitary		
Follicle-stimulating hormone (FSH)	Ovary	Stimulates development of follicles; with LH, stimulates secretion of estrogen and ovulation
Luteinizing hormone (LH)	Ovary	Stimulates ovulation and development of corpus luteum
Prolactin	Breast	Stimulates milk production (after breast has been prepared by estrogen and progesterone)
Ovary		
Estrogen	General	Growth of sex organs at puberty; development of secondary sex characteristics (breast development, broadening of pelvis, distribution of fat and muscle)
	Reproductive structures	Maturation; monthly preparation of the endometrium for pregnancy; makes cervical mucus thinner and more alkaline
Progesterone	Uterus	Completes preparation of endometrium for pregnancy
	Breast	Stimulates development

times occur, probably because of the effect of decreased estrogen on the temperature-regulating center in the hypothalamus. Estrogen deficiency may also contribute to feelings of depression and headaches experienced by some women. The vaginal lining thins, and the breasts and vulva begin to atrophy.

Despite these physical changes, menopause does not usually affect a woman's interest in sex or her sexual performance. Although replacing missing hormones has been used clinically to alleviate many of the symptoms of menopause, this is no longer widely practiced because continued use of estrogen has been linked to increased risk of cancer.

Physiology of Sexual Response

Sexual stimulation causes two basic physiologic responses: *vasocongestion* and *increased muscle tension.* Vasocongestion occurs as blood flow is increased to the genital structures and to certain other tissues, such as the breasts, skin, and earlobes. These structures become engorged with blood, and the penis and clitoris become temporarily erect.

In describing physiological changes, it is helpful to divide sexual response into four phases—*excitement, plateau, orgasm,* and *resolution.* To function in sexual intercourse, or **coitus,** the penis must be erect. During the **excitement phase,** psychological or physical stimulation, usually both, provides the needed arousal. In the female, vaginal lubrication is the first response to effective sexual stimulation. The wall of the vagina lacks glands; the fluid produced is a product of the vasocongestion that occurs in the vaginal wall. During the excitement phase the vagina lengthens and expands in preparation for receiving the penis. The clitoris and breasts become vasocongested, and the nipples become erect. In both sexes the heart rate increases and blood pressure is elevated.

If the erotic stimuli continue, sexual excitement heightens to the **plateau phase.** During this phase sexual excitement intensifies and sexual climax is approached. During coitus the penis is moved back and forth in the vagina by movements known as pelvic thrusts. Physical and psychical sensations resulting from this friction and from the entire intimate experience between the partners lead to **orgasm,** the climax of sexual excitement. Though lasting only a few seconds, orgasm is the achievement of maximal sexual tension and its release.

In the male, orgasm is marked by **ejaculation** of the semen. Contraction of the vas deferens propels sperm into the ejaculatory duct, while the accessory glands contract, adding their secretions. Contractions of the ejaculatory ducts, certain muscles of the pelvic floor, and urethra eject the semen from the penis. After ejaculation the urethra ducts, accessory glands, and muscles surrounding the root of the penis continue to contract at 0.8-second intervals. After the first several contractions their intensity decreases and they become less regular and less frequent.

In the female, orgasm is marked by rhythmic contractions of the pelvic muscles and the vagina, starting at 0.8-second intervals and recurring 5 to 12 times. Orgasm in the female is not marked by fluid ejaculation. Stimulation of the clitoris appears to be important in heightening the sexual excitement that leads to orgasm. In both sexes heart rate and respiration more than double, and blood pressure rises to a peak during orgasm. In the **resolution phase,** relaxation and detumescence (subsiding of swelling) restore the body to its normal state.

Erection of the penis is necessary for effective coitus. Chronic inability to sustain an erection is called **erectile dysfunction** (impotence). Although erectile dysfunction is experienced from time to time by almost all men, when chronic this condition is often associated with psychological problems. Erectile dysfunction should not be confused with sterility, although both may result in failure to produce offspring.

Fertilization

During coitus sperm are released in the vicinity of the cervix. The female reproductive tract is a hostile environment for sperm during most of the menstrual cycle. The acidic nature of the vagina is spermicidal, and a thick plug of mucus blocks the cervix. As the time of ovulation approaches, however, conditions begin to change. The vagina becomes slightly alkaline and the cervical mucus thins, permitting sperm to pass into the uterus. Glycogen content of the cervical mucus increases greatly, providing nourishment for sperm cells.

Only one sperm actually fertilizes the ovum, yet millions are required. Apparently, many die or lose their way, because only about 2000 reach the "correct" upper uterine tube. (Remember that only one ovum is released each month, and it moves into the uterine tube nearer the ovary that produced it.)

As soon as one sperm penetrates the ovum, no other sperm is able to get through. As the fertilizing sperm enters the ovum, it usually loses its middle piece and its tail. Sperm entry stimulates the ovum to complete its second meiotic division. The head of the sperm swells to form the **male pronucleus,** and the nucleus of the ovum becomes the **female pronucleus.** Then, sperm and ovum pronuclei become one, completing the process of fertilization (see Focus on Fertilization).

Fertilization serves several functions: (1) the diploid number of chromosomes is restored, as the sperm contributes a new set of chromosomes bearing a unique complement of genes, (2) the sex of the offspring is determined by the sperm cell, and (3) fertilization initiates the developmental events leading to the birth of a new individual.

After ejaculation sperm remain viable for only about 48 hours. The ovum remains fertile for about 24 hours after ovulation. Thus, there are only a few days during each menstrual cycle (perhaps days 12 to 15 in a regular 28-day cycle) when sexual intercourse is likely to result in conception.

BIRTH CONTROL

Most couples agree that it is best to have babies by choice, not by chance, but unfortunately the majority of couples who engage in sexual intercourse have only vague notions of how to prevent conception. In underdeveloped countries an estimated 88% of women lack the means to limit family size. Studies indicate that many of these women would use birth control methods if they were available, and if someone showed them how.

More than 1 million teenagers become pregnant every year, and thousands of girls aged 14 or younger are having babies each year. Yet only one in five sexually active teen-agers consistently use contraception. People old enough to produce babies should know where they come from and how to prevent conception. Because parents, for a host of reasons, often neglect to provide such information, it is important that sex education be included in the public school curriculum. One may hope that, in addition, the family and other social and religious institutions will increasingly accept the responsibility of teaching not only about sex itself but also the social and ethical dimensions of sexuality.

When a sexually active woman uses no form of birth control, her chances of becoming pregnant during the course of a year are about 90%. Any method for deliberately separating sexual intercourse from production of babies is considered **contraception** (literally, "against conception"). Since ancient times humans have searched for effective contraceptive methods. Modern science has developed a variety of contraceptives with a high percentage of reliability, but the ideal contraceptive has not yet been devised. Some of the more common methods of birth control are described below and in Table 31–3. (Also see Fig. 31–17.) Note that IUDs as well as some types of oral contraceptives may not actually prevent fertilization; they probably destroy the tiny embryo or prevent its implantation in the wall of the uterus.

Oral Contraceptives—The Pill

More than 10 million women in the United States alone use **oral contraceptives,** and worldwide the figure is estimated at more than 80 million. The most common preparations are composed of a combination of progestin (synthetic progesterone) and synthetic estrogen. (Natural hormones are destroyed by the liver almost immediately, but synthetic ones resist destruction.) When taken correctly, these pills are about 99.9% effective in preventing pregnancy.

Most oral contraceptives prevent pregnancy by preventing ovulation. By maintaining critical concentrations of ovarian hormones in the

Focus on . . .

FERTILIZATION

The production of a new individual, genetically as well as physically distinct from its parents, begins dramatically at the instant of fertilization, when sperm and egg are united. It is now known that the egg plays a very active part in the process of fertilization. Perhaps the reason that this role of the egg has so long been obscured is the more obvious and frenetic activity of the sperm. The sperm cell's extreme functionalism is apparent from its anatomy. It is a missile of information, driven through body fluids by the lashing of its flagellum. It is entirely by chance that one sperm penetrates the outer coats of the egg cell and happens to contact the plasma membrane of the egg.

The egg responds almost instantly. A cone of microvilli reaches up from the surface of the egg to embrace the head (and sometimes the tail) of the sperm and draw it into the cytoplasm. At the same time the egg's plasma membrane undergoes an electrical change, like that of a depolarized muscle or nerve cell, which renders it temporarily impermeable to other sperm. Shortly thereafter (about 30 seconds) the egg erects an extracellular coat and develops a heavy cytoskeleton of microfilaments just under the plasma membrane. These permanently prevent the entrance of other sperm.

These events are triggered by the sperm's acrosome, a tiny organelle located in the tip of the sperm's head. Seemingly a modified Golgi complex, the acrosome has long been an object of biological curiosity. Its

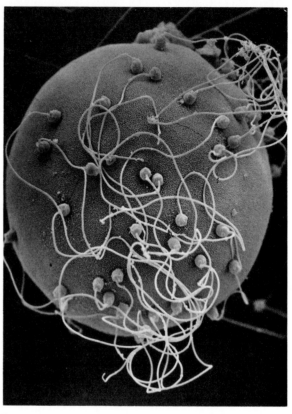

Sperm swarm around an egg of a surf clam in this scanning electron micrograph (approximately ×2100). (David M. Phillips, The Population Council.)

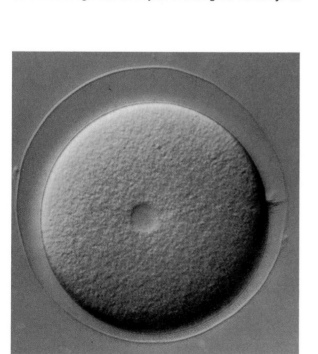

The vast egg of a sea urchin awaits the arrival of a sperm. The egg is almost in a state of suspended animation. A human egg can last for 50 years in this state, while it resides in the ovary. The large nucleus can be seen in the center. (Courtesy of Drs. Gerald and Heide Schatten.)

most important function now appears to be the activation of the egg's complex response to sperm penetration. The acrosome secretes into the egg factors that act like hormones, and it also produces a bundle of microfilaments, which is shot into the egg. This submicroscopic harpoon is made of a kind of actin, similar to that of muscle cells. Perhaps by interacting with myosin in the egg, the acrosomal microfilaments help hold the sperm to the egg and may also help pull it inside. Once the head is inside, other substances, especially calcium, are released by the sperm head into the egg's cytoplasm to initiate further changes.

The journey of the sperm is not yet complete. It is still separated from the egg's nucleus (with which it must fuse) by a considerable distance. How can the two nuclei find one another in this great sphere?

At this point, the egg assumes the responsibility of transportation. A system of microtubules forms, which guides the *egg's* nucleus to the sperm's. When they meet, the sperm nucleus is largely decondensed and expanded, its chromosomes ready to mingle with those of the egg to form the homologous pairs of the new individual.

The zygote, or fertilized egg, does not long remain its original bloated size. Though the egg does not seem

(a)

(b)

(a) *When the first sperm makes contact with the egg's plasma membrane, a cone of microvilli reaches for it and draws it into the cytoplasm of the egg.*
(b) *This computer reconstruction shows the fusion of sperm and egg cell membranes that marks the instant of fertilization. (a, David M. Phillips, The Population Council; b, courtesy of Drs. Gerald and Heide Schatten.)*

to possess centrioles, the sperm does. These form the foci of a mitotic spindle, which separate the already duplicated sets of double chromosomes into two identical groups. With cytokinesis a two-cell embryo is formed. As many as a hundred cellular generations later, a larva, hatchling, or newborn organism is ready to confront the world. Each of its cells contains every bit of the volumes of genetic information contained in the original set of chromosomes donated mutually by egg and sperm. In due course it, too, may produce sperm or eggs and re-enact the dramatic events of its own beginning.

Sperm head

Fertilization
cone

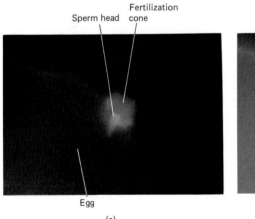

Egg

(a)

Egg   Sperm nucleus

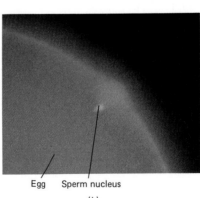

(b)

Chromosomes

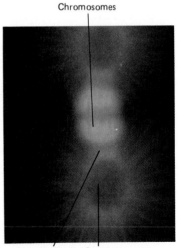

Spindle fibers   Centriole

(a) *This light micrograph, obtained using special fluorescent staining techniques, shows the sperm nucleus (blue) as it first enters the egg, surrounded by a corona of egg microfilaments (red). (b) Now well on its way into the egg, the sperm nucleus no longer requires the microfilaments, which have already begun to disperse. A dense network of microfilaments has formed just under the plasma membrane to prevent the penetration of more sperm cells. (Courtesy of Drs. Gerald and Heide Schatten.)*

The first mitotic division of the new organism. The centrioles are derived from the sperm, but the microtubules (in green) of the mitotic spindle are part of the egg's dowry. The brilliantly blue-fluorescing chromosomes were provided by both. (Courtesy of Drs. Gerald and Heide Schatten.)

**Table 31–3
CONTRACEPTIVE METHODS**

Method	Failure rate* (%)	Mode of action	Advantages	Disadvantages
Oral contraceptives	0.3; 5	Prevent ovulation; may also affect endometrium and cervical mucus and prevent implantation	Highly effective; sexual freedom; regular menstrual cycle	Minor discomfort in some women; possible thromboembolism; hypertension, heart disease in some users
Intrauterine device (IUD)	1; 5	Unknown; probably sets up minor inflammation	Provides continuous protection; highly effective	Cramps; increased menstrual flow; spontaneous expulsion
Spermicides (foams, jellies, creams)†	3; 22	Chemically kill sperm	No known side effects	Unreliable; recent epidemiological evidence suggests that when used at or around the time of conception, spermicides may cause birth defects (e.g., Down's syndrome and limb malformations) or miscarriages
Contraceptive diaphragm (with jelly)	3; 13	Diaphragm mechanically blocks entrance to cervix; jelly is spermicidal	No side effects	Must be prescribed (and fitted) by physician; must be inserted prior to coitus
Condom	2.6; 10	Mechanical; prevents sperm from entering vagina	No side effects; some protection against STD	Interruption of foreplay to fit; slightly decreased sensation for male
Rhythm‡	35, but varies greatly	Abstinence during fertile period	No side effects	Not very reliable
Douche	40	Flushes semen from vagina	No side effects	Unreliable; sperm beyond reach of douche within seconds
Withdrawal (coitus interruptus)	20?	Male withdraws penis from vagina prior to ejaculation	No side effects	Unreliable; contrary to powerful drives present as orgasm is approached; sperm present in fluid secreted before ejaculation may be sufficient for conception
Sterilization				
Tubal ligation	0.04	Prevents ovum from leaving uterine tube	Most reliable method	Usually not reversible
Vasectomy	0.15	Prevents sperm from leaving scrotum	Most reliable method	Usually not reversible
Chance (no contraception)	About 90			

*The first figure is the failure of the method; the second figure includes method failure plus failure of the user to utilize the method correctly. Both are based on number of failures per 100 women who use the method per year in the United States.
†Failure rate is lower when used together with spermicidal foam.
‡There are several variations of the rhythm method. For those who use the calendar method alone, the failure rate is about 35. However, if a woman takes her temperature daily and keeps careful records (since temperature rises after ovulation), the failure rate can be reduced. Also, by keeping a daily record of the type of vaginal secretions, she can note changes in cervical mucus and use them to determine time of ovulation. This type of rhythm contraception is also slightly more effective. When women use the temperature or mucus method and wait more than 48 hours after ovulation to have intercourse, the failure rate can be reduced to about 7%.

Oral contraceptives

Condom

Figure 31–17 Some commonly used contraceptive devices.

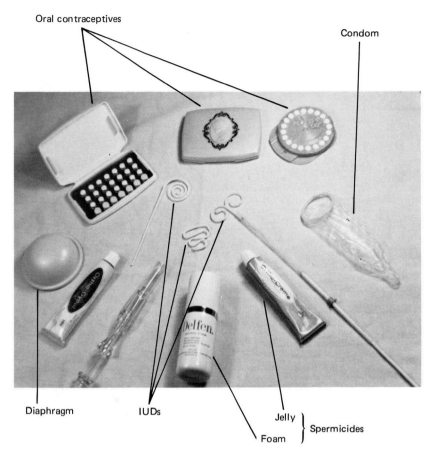

Diaphragm

IUDs

Jelly } Spermicides

Foam

blood, these substances "deceive" the hypothalamus so that it does not secrete the LH-releasing hormone necessary for the mid-cycle surge of LH. Thus, ovulation does not occur. The hypothalamus, of course, has no way of "knowing" that the hormones are synthetic rather than the product of a fully functioning ovary.

Oral contraceptives that utilize very low doses of hormones probably do not prevent ovulation. They are thought to accelerate passage of the embryo (should there be one) through the uterine tube so that it arrives in the uterus too soon to implant itself. They also affect the endometrium so that it is not prepared to receive an embryo.

The chief advantage of oral contraceptives is their high rate of effectiveness. Many women enjoy the sexual freedom they afford, as no special preparations must be made for birth control prior to each coitus. Another advantage is that because of their precise hormonal control they maintain a highly regular menstrual cycle.

A direct relationship between the estrogen component of the Pill and hypertension has been demonstrated in about 18% of users. Studies have suggested that women over age 40, especially those who smoke, should not take the Pill because of a possible link with heart attacks. Serious side effects of oral contraceptives such as thromboembolic disease (in which blood clots develop within the blood vessels and may lead to death) are rare. Oral contraceptives result in death in about 3 per 100,000 users. This compares favorably with the death rate of about 25 per 100,000 pregnancies (Table 31–4).

Intrauterine Devices (IUDs)

The **intrauterine device,** or **IUD,** is a small plastic loop or coil, which must be inserted into the uterus by a medical professional. Once in place, some types of IUDs may be left in the uterus for up to three years or

Table 31–4
DEATHS IN THE UNITED STATES FROM PREGNANCY AND CHILDBIRTH AND FROM VARIOUS BIRTH CONTROL METHODS

	Death rate per 100,000
Pregnancy and childbirth	20
Oral contraception	4
IUD	0.5
Legal abortions—first trimester	<1
Legal abortions—after first trimester	12.5*
Illegal abortion performed by medically untrained individuals	About 100

*Figure based on saline and prostaglandin injection. The mortality rate for hysterectomy is 41.3 per 100,000 procedures.

until the woman wishes to conceive. When used properly newer types of IUDs are about 99% effective. The IUD is thought to prevent implantation of the ovum within the uterus. As a foreign body, the IUD sets up a minor local inflammation in the uterus, attracting macrophages that destroy the embryo, as well as sperm cells. Increased levels of antibodies in those who use IUDs support the idea that an immunological mechanism may be involved.

Women who suffer no side effects say that the IUD is an excellent contraceptive because, once inserted, it provides continuous protection. There is no need to remember to take a pill each day or to do anything specific before coitus.

Some women, especially those who have never borne children, suffer pain when the IUD is inserted. Others experience side effects, such as cramps, bleeding, or increased menstrual flow. During the first year about 20% of IUD users either have them removed for these reasons or find that they have been spontaneously expelled.

Other Common Methods

Other common contraceptive methods are listed and compared in Figure 31–17 and Table 31–3. The contraceptive diaphragm and the condom are enjoying a revival in popularity. Many who fear the side effects of contraceptive pills and IUDs are turning to these methods because of their safety.

New Methods

The main focus of contraception research is making hormonal contraceptives safer and more convenient. Currently available hormonal contraceptives are taken orally, but new methods for introducing long-acting hormones into the body are being tested. One is a silicone rod that can be inserted under the skin in the buttock or forearm. The rod is impregnated with a hormone that is released gradually over a period of months. Another is a vaginal ring that is inserted like a contraceptive diaphragm and slowly releases synthetic progesterone. IUDs containing hormones are also being tested.

Prostaglandin tampons may eventually be developed which will prevent implantation of an embryo. Or perhaps a woman will insert one each month just before she expects menstruation to occur. If by chance she were pregnant the prostaglandin would stimulate a miniabortion.

Research continues on prostaglandin contraceptives; so far the problem has been that they cause unpleasant side effects, such as nausea.

Short of sterilization, the only male method of contraception is the condom. Recent attempts to suppress sperm production by interfering with secretion of FSH or releasing hormones have not been successful, but research continues in this area.

Sterilization

The only foolproof method of contraception (besides abstinence) is sterilization. About 75% of sterilization operations are currently performed on males.

MALE STERILIZATION

An estimated 1 million **vasectomies** are performed each year in the United States. Using a local anesthetic a small incision is made on each side of the scrotum. Then each vas deferens is cut and its ends are tied or clipped so that they cannot grow back together (Fig. 31–18). Since testosterone secretion and transport are not affected, a vasectomy in no way affects masculinity. Sperm continue to be produced, though at a much slower rate, and are destroyed by macrophages in the testes. No change in the amount of semen ejaculated is noticed, since sperm account for very little of the semen volume.

In a study of more than 1000 men who had vasectomies, 99% said they had no regrets, and 73% claimed an increase in sexual pleasure, probably because anxiety about pregnancy was erased. By surgically reuniting the ends of the vasa deferentia, surgeons have reversed sterilization in about 30% of attempts made. The low success rate may be because some sterilized men eventually develop antibodies against their own sperm, making them nonviable.

An alternative to reversible vasectomy is the storage of frozen sperm in sperm banks. If the male should decide to father another child after he has been sterilized, he simply "withdraws" his sperm for use in artificially inseminating his wife. Sperm banks are currently being established throughout the United States. Not much is known yet about the effects of long-term sperm storage, but there may be an increased risk of genetic defects.

Figure 31–18 Sterilization. (a) In vasectomy the vas deferens (sperm duct) on each side is cut and tied. (b) In tubal ligation each uterine tube is cut and tied so that ovum and sperm can no longer meet.

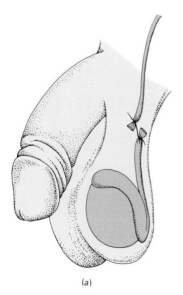

(a)

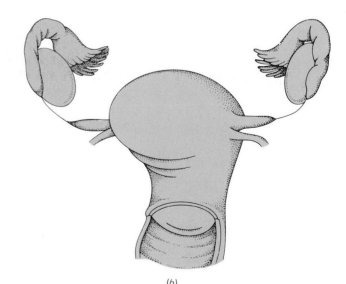

(b)

FEMALE STERILIZATION

Several techniques are in current use for prevention of ova transport. Most of them involve **tubal ligation,** cutting and tying the uterine tubes (Fig. 31–18). This can be done through the vagina but is usually performed through an abdominal incision, requiring a general anesthetic. Female sterilization carries with it an estimated 25 deaths per 100,000 procedures performed, whereas there is almost no risk of death in a vasectomy procedure. New techniques being developed will make tubal ligation a simpler, safer procedure and also improve chances of reversing it. As in the male, hormone balance and sexual performance are not affected.

Abortion

More than 1.5 million abortions are performed each year in the United States (and an estimated 40 million worldwide). There are three kinds of abortion. **Spontaneous abortions,** popularly known as miscarriages, occur without intervention and often are nature's way of destroying a defective embryo. More than 50% of all pregnancies end in spontaneous abortion, although in many cases the woman does not even know that she has been pregnant. **Therapeutic abortions** are induced to preserve the health of the mother or when there is reason to believe that the embryo is grossly abnormal. The third type of abortion—the kind performed as a means of birth control—is the most controversial. All societies employ abortion as a means of preventing unwanted births, and abortion is probably the most widely used single method of birth control in the world.

Most first-trimester abortions (those done in the first 3 months of pregnancy) and some later ones are performed using a suction method. After the cervix is dilated, a suction aspirator is inserted into the uterus and the products of conception are quickly evacuated.

Later in pregnancy, abortions are often performed using saline injections. Amniotic fluid surrounding the embryo is removed with a needle and replaced with a salt solution. The fetus dies within 1 to 2 hours, and labor begins several hours later. Prostaglandins are also used to induce abortions, especially during the second trimester. They appear to be a safe, nonsurgical method of chemically terminating pregnancy.

What of the safety of abortions? When performed by skilled medical personnel, the risk of death is about 1.9 per 100,000 abortions performed. The death rate from illegal abortions performed by untrained individuals is about 100 per 100,000 (Table 31–4).

The entire issue of abortion is a matter of current controversy. Abortions have always been available to those who could afford to travel to places where they were legal. The poor woman took her chances with the unqualified, often unsanitary, but local and less expensive, practitioner. Thousands of women have died as a result of such illegal abortions.

In January 1973 the U.S. Supreme Court ruled that during the first 3 months of pregnancy the decision to have an abortion rests entirely with the pregnant woman and her physician, and that the state cannot interfere. After the first 3 months the state may regulate abortion procedure in order to protect maternal health. A state can prohibit abortion during the last 10 weeks of pregnancy, when the fetus can survive outside the uterus. The Supreme Court's decision came after several states had liberalized restrictive abortion laws.

A concerted drive by anti-abortion groups, often called "right-to-life" societies, has been mounted in reaction to the relaxation of legislative guidelines regarding abortion. Such groups see abortion as a moral issue affecting the embryo, which they consider to be a human being with legal rights. In 1977 the Supreme Court ruled that states may deny Medicaid payments for abortions for women who depend upon public assistance for medical care. It is possible that all abortions may become illegal again.

SEXUALLY TRANSMITTED DISEASES

Sexually transmitted diseases (STD), also called venereal diseases (VD), are, next to the common cold, the most prevalent communicable diseases in the world. The World Health Organization has estimated that more than 250 million people are infected each year with gonorrhea and more than 50 million with syphilis. An estimated 3 million persons are infected with gonorrhea and 400,000 with syphilis each year in the United States. The most common sexually transmitted diseases are listed and described in Table 31–5 on the following page.

CHAPTER SUMMARY

I. In asexual reproduction a single parent endows its offspring with a set of genes identical to its own. In sexual reproduction each of two parents contributes half of the offspring's genetic endowment.

II. Human reproductive processes include gamete formation, preparation of the female for pregnancy, sexual intercourse, fertilization, pregnancy, and lactation.

   A. The reproductive role of the male is to produce sperm cells and to deliver them into the female reproductive tract.

     1. Sperm are produced in the seminiferous tubules of the testes.

     2. From the tubules of the testes, sperm pass through an epididymis, vas deferens, ejaculatory duct, and then through the urethra, which passes through the penis.

     3. Semen, produced by the seminal vesicles, prostate gland, and bulbourethral glands, consists of about 400 million sperm cells suspended in about 3.5 milliliters of fluid.

     4. The penis consists of three columns of erectile tissue. When the erectile tissue becomes engorged with blood, the penis becomes erect.

     5. The gonadotropic hormones FSH and LH stimulate sperm production and testosterone secretion. Testosterone is responsible for both primary and secondary male characteristics.

   B. The reproductive role of the female includes production of ova, reception of sperm, incubation and nourishment of the developing embryo, and lactation.

     1. Ova develop in the ovaries as part of follicles. At ovulation an ovum is ejected from the ovary into the pelvic cavity.

     2. After ovulation the ovum finds its way into the uterine tube and is transported to the uterus. Fertilization generally takes place within the uterine tube.

     3. The uterus serves as an incubator for the developing embryo.

     4. The vagina is the lower part of the birth canal. It receives the penis during sexual intercourse and serves as an outlet for menstrual discharge.

     5. The clitoris is a small erectile structure that serves as a center of sexual sensation in the female.

     6. Lactation depends on the hormones prolactin and oxytocin.

     7. The first day of menstrual bleeding marks the first day of the menstrual cycle. Ovulation occurs at about day 14 in a typical

Table 31–5
SOME COMMON SEXUALLY TRANSMITTED DISEASES

Disease and causative organism	Course of disease	Treatment
Gonorrhea (Gonococcus bacterium)	Infection by sexual contact. Bacterial toxin may produce redness and swelling at infection site. Symptoms in males: painful urination and discharge of pus from penis. In about 60% of infected women no symptoms occur in initial stages. Can spread to epididymis (in males) or uterine tubes and ovaries (in females), causing sterility. Can cause widespread pelvic or other infection, plus damage to heart valves, meninges (outer coverings of brain and spinal cord), and joints.	Penicillin, or other antibiotic if penicillin-resistant strain involved.
Syphilis (*Treponema pallidum*, a spirochete bacterium)	Bacteria enter body through defect in skin near site of infection. Spread throughout body by lymphatic and circulatory routes. Primary chancre (a small, painless ulcer) forms at site of initial infection; heals in about a month. Highly infectious at this stage. Secondary stage follows, in which a widespread rash and influenza-like symptoms may occur. Scaly lesions may occur that teem with bacteria and are highly infectious. Latent stage that follows can last 20 years. Eventually, lesions called gummae may occur, consuming parts of the body surface or damaging liver, bone, or spleen. Serious brain damage may occur. Death results in 5% to 10% of cases.	Penicillin. Sensitive blood tests can detect antibodies and hence infection. About one-third of cases recover spontaneously.
Genital herpes (herpes simplex type 2 virus)	Tiny, painful blisters appear on genitals; may develop into ulcers. Influenza-like symptoms may occur. Recurs periodically. Threat to fetus or newborn infant. May predispose to cervical cancer in females.	No effective cure. Some drugs may shorten outbreaks or reduce severity of symptoms.
Pelvic inflammatory disease (PID) (usually chlamydial bacteria)	Generalized infection of reproductive organs and pelvic cavity; usually chronic and difficult to treat. May lead to sterility (more than 15% of cases). PID now most common STD in the U.S.	Antibiotics, surgical removal of affected organs.
Trichomoniasis (a protozoan)	Symptoms include itching, discharge, soreness. Can be contracted from dirty toilet seats and towels.	Drugs.
Yeast infections (genital candidiasis) (yeasts)	Irritation, soreness, discharge; especially common in females.	Drugs.

28-day cycle. Events of the menstrual cycle are coordinated by the interaction of gonadotropic and ovarian hormones.

III. The cycle of sexual response comprises the excitement stage, plateau stage, orgasm, and resolution. Vasocongestion and increased muscle tension are two basic physiological responses to sexual stimulation.

IV. Fertilization restores the diploid chromosome number, determines the sex of the offspring, and triggers development.

V. Effective methods of birth control are currently available, but the ideal contraceptive has not yet been developed.

A. Oral contraceptives work hormonally, inhibiting the production of FSH and LH, so ovulation does not occur.

B. The IUD is thought to set up a local inflammation in the uterus, so the embryo is destroyed by macrophages.

C. The condom and diaphragm are mechanical methods that prevent sperm and ovum from meeting. Contraceptive foams, jellies, and creams are spermicides.

D. Sterilization is the only foolproof method of birth control; it does not affect hormone balance and so does not interfere with sexuality.

VI. Gonorrhea, syphilis, genital herpes, and pelvic inflammatory disease are common types of sexually transmitted disease.

Post-Test

1. From the epididymis sperm pass into the _____ _____ .

2. The cavernous bodies are columns of _____ tissue in the _____ .

Find the correct term in column B for each description in column A.

Column A

_____ 3. Sperm produced here
_____ 4. Severed during vasectomy
_____ 5. Passes through the penis
_____ 6. Contributes alkaline secretion to semen
_____ 7. Testosterone produced here
_____ 8. Removal referred to as castration

Column B

a. prostate gland
b. testis
c. vas deferens
d. urethra
e. none of the above

9. The period of sexual maturation is called _____ .

10. The two ovarian hormones are _____ and _____ .

Find the correct term in column B for each description in column A.

Column A

_____ 11. Produces female gametes
_____ 12. Lining is endometrium
_____ 13. External female genital structures
_____ 14. Fertilization takes place here
_____ 15. Projects into vagina
_____ 16. Embryo implants here

Column B

a. uterine tube
b. ovary
c. cervix
d. vulva
e. uterus

17. Ejection of the ovum from the follicle is called _____ _____ .

18. In the female FSH is released by the _____ _____ and stimulates development of _____ .

19. In a typical 28-day menstrual cycle, ovulation takes place on about the _____ day.

20. Most oral contraceptives prevent pregnancy by preventing _____ .

21. Label the following diagram. For correct labeling, see Figure 31–9.

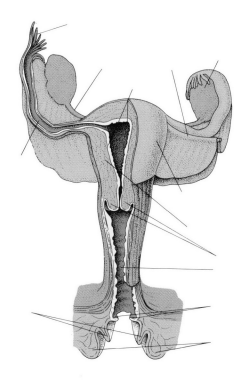

Review Questions

1. Compare asexual with sexual reproduction. Give examples of asexual reproduction.

2. Compare sperm production with ovum production.

3. Trace the path traveled by a sperm cell from the tubules of the testes until it is released from the body.

4. What is sterility? What is castration?

5. What are the actions of testosterone?
6. What are the actions of the gonadotropic hormones in the male?
7. Trace the development of the ovum from its origin to fertilization.
8. What is the function of the corpus luteum? Which hormone is necessary for its development?
9. What are the actions of FSH and LH in the female?
10. What are the actions of estrogen and progesterone?
11. Draw a diagram illustrating the menstrual cycle and indicate important events such as menstruation and ovulation.
12. What are the advantages of breast-feeding an infant?

13. Which methods of birth control are most effective? least effective?
14. What is the mode of action of oral contraceptives? of the condom? of the IUD?
15. Is masculinity affected by vasectomy? Why or why not?
16. What is puberty? What is menopause?
17. What are the symptoms of syphilis? of gonorrhea? of genital herpes?
18. What are the arguments for and against abortion as a means of birth control? How might the number of women seeking abortions be reduced?

Readings

Cates, W. "Legal Abortion: The Public Health Record," *Science,* Vol. 215 (26 March 1982). An interesting analysis of the effects of the increasing availability and use of legal abortion in the United States during the 1970s.

Herbert, W. "Premenstrual Changes," *Science News,* Vol. 122, No. 24 (11 December 1982). A discussion of premenstrual syndrome.

Langone, J. "The Quest for the Male Pill," *Discover,* October 1982.

Schatten, G. "Motility During Fertilization," *Endeavor* (New Series), Vol. 7, No. 4, 1983.

Small, S.A. "Birth Control from Amulets to the Pill," *MD,* March 1983.

Chapter 32

DEVELOPMENT: THE ORIGIN OF THE ORGANISM

Learning Objectives

After you have read this chapter you should be able to:

1. Describe the preformation theory and the theory of epigenesis, and relate these theories to current concepts of development.
2. Define growth, morphogenesis, and cellular differentiation, and describe the role of each process in the development of an organism.
3. Describe the principal events and characteristics of each of the early stages of development: zygote, cleavage, blastocyst, gastrula, and nervous system development.
4. Describe the implantation of the zygote and the development of the placenta, and give the functions of the placenta and the amnion.
5. Describe the general course of the later development of the human being, from 1 month after conception until the time of birth.
6. Describe the birth process, distinguishing among the three stages of labor.
7. Describe specific steps that the pregnant woman can take to promote the well-being of her developing child, and tell how the embryo can be affected by nutrients, drugs, oxygen deprivation, pathogens, and ionizing radiation.
8. Trace the stages of the human life cycle.
9. Identify changes that occur with aging and discuss current theories of aging.

Figure 32–1 The preformed "little man" within the sperm as visualized by 17th-century scientists. (After Hartseeker's drawing from *Essay de Dioptrique*, Paris, 1694.)

Many scientists of the 17th century thought that either the sperm or the egg cell contained a completely formed, though miniature, human being (Fig. 32–1). They believed that all the parts were already present, so the embryo had only to grow in size. This concept is known as the **preformation theory.**

An opposing view, the **theory of epigenesis,** gained experimental support as better techniques for investigation were developed. This theory held that the embryo develops from an undifferentiated zygote and that the structures of the body emerge in an orderly sequence, developing their characteristic forms only as they emerge.

Today we know that development is largely epigenetic. No invisible "little man" waits preformed in either gamete. Development proceeds from one cell to billions, from a formless mass of cells to an intricate, highly specialized, and organized organism. However, a spark of truth can be found in the preformationist view. Although the "little man" himself is not to be found within the zygote, his blueprint is there, precisely encoded in the form of chemical specifications within the DNA of the genes.

THE PATTERN OF DEVELOPMENT

How does the microscopic, unspecialized zygote give rise to the blood, bones, brain, and all the other complex structures of the completed organism? As we shall see, development is a balanced combination of three processes: growth, morphogenesis, and cellular differentiation.

Growth of the embryo includes both cellular growth and mitosis. An orderly pattern of growth provides the cellular building blocks of the organism. But growth alone would produce only a formless heap of cells. Cells must arrange themselves into specific structures and appropriate body forms. The precise and complicated cellular movements that bring about the form of a multicellular organism with its intricate pattern of tissues and organs constitute **morphogenesis.**

Not only must cells be arranged into specific structures, but they must also be specialized and organized to perform specific functions. To function in different fashions, body structures must be composed of different components. During early development cells begin to differentiate from one another, specializing biochemically and structurally to perform specific tasks. The process by which cells become specialized is known as **cellular differentiation.**

As you read the following section on development, bear in mind that growth, morphogenesis, and cellular differentiation are intimately interrelated. Although our discussion will center upon human development, the pattern of early development is basically similar for all vertebrates.

Cleavage—From One Cell to Many

Shortly after fertilization the zygote undergoes a series of mitoses, collectively referred to as the **cleavage stage.** About 24 hours after fertilization the human zygote has completed the first mitotic division and become a 2-cell **embryo** (Fig. 32–2). Each of the cells of the 2-cell embryo undergoes mitosis, and repeated divisions increase the number of cells making up the embryo. At the 16-cell stage the embryo consists of a tiny cluster of cells called the **morula.** As cleavage takes place, the embryo is pushed along the uterine tube by ciliary action and muscular contraction. By the time the embryo reaches the uterus on about the 5th day of development, it is in the morula stage (Fig. 32–3).

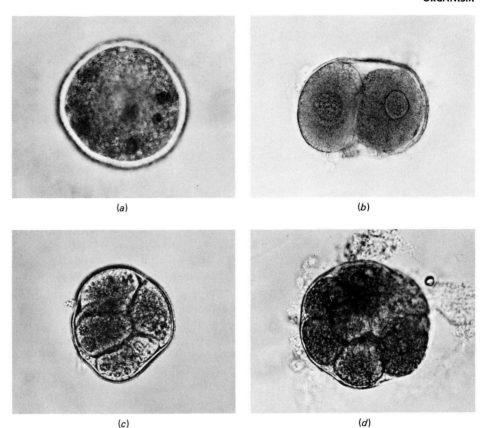

Figure 32–2 Early human development (approximately ×250). (*a*) Human zygote. This single cell contains the genetic instructions for producing a complete human being. (*b*) Two-cell stage. (*c*) Eight-cell stage. (*d*) Cleavage continues, giving rise to a cluster of cells called the morula. (Courtesy of the Carnegie Institution of Washington.)

Early Development in the Uterus

When the embryo enters the uterus, it is bathed in a nutritive fluid secreted by the glands of the uterus. Nourished in this manner, the embryo continues its development for 2 or 3 days, floating freely in the uterine cavity. During this period its cells arrange themselves into a hollow ball called a **blastocyst,** or **blastula** in many other animal groups (Figs. 32–4 and 32–5). The outer layer of cells, the **trophoblast,** eventually forms the protective and nutritive membranes (the **chorion** and **placenta**) which surround the embryo. A little cluster of cells, the **inner cell mass,** which projects into the cavity of the blastocyst, gives rise to the embryo itself.

Occasionally, the inner cell mass subdivides to form two independent groups of cells, and each develops into a complete organism. Since

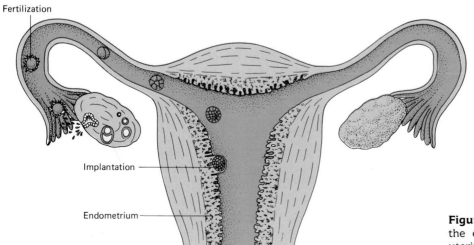

Figure 32–3 Cleavage takes place as the embryo is moved through the uterine tube to the uterus.

Figure 32–4 Implanted blastocyst, 12 days after fertilization. (Carnegie Institution of Washington.)

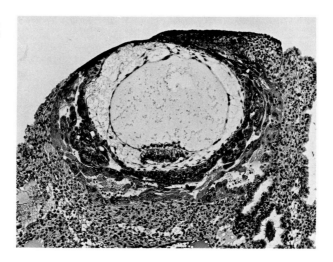

each cell has an identical set of genes, the individuals formed are exactly alike—**identical twins.** Very rarely, the two inner cell masses are not completely separated and give rise to **conjoined** (Siamese) **twins. Fraternal twins** develop when a woman ovulates two eggs and each is fertilized by a separate sperm. Each zygote has its own distinctive genetic endowment, so the individuals produced are not identical. Similarly, triplets (and other multiple births) may be either identical or fraternal. In the United States twins are born once in about 88 births, triplets once in 88 squared, and quadruplets once in 88 cubed.

On about the 7th day of development the embryo begins to **implant** itself in the wall of the uterus. The trophoblast cells in contact with the uterine lining secrete enzymes that erode an area of the uterine wall just large enough to accommodate the tiny embryo. Slowly the embryo works its way down into the underlying connective tissues. All further development of the embryo takes place within the wall of the uterus (Figs. 32–4 and 32–5).

Extraembryonic Membranes and Placenta

Extraembryonic membranes protect the embryo and help in obtaining food and oxygen and eliminating wastes. Since they are not part of the embryo proper, however, they are discarded at birth. Reptiles, birds, and mammals have four extraembryonic membranes—the *amnion, yolk sac, chorion,* and *allantois* (Fig. 32–5). In human beings and other mammals these are often called **fetal membranes.**

Even though there is no yolk in the human ovum, a **yolk sac** forms as an outpocketing of the developing gut and is quite prominent between the 2nd and 6th weeks of development. Its walls serve as an important temporary center for the formation of blood cells.

The **chorion** develops from the trophoblast; the **allantois** grows outward from the primitive gut. The allantois, like the yolk sac, appears to be a vestigial structure in the human embryo.

AMNION

The **amnion** begins to develop even before the first structures of the embryo proper take form, and eventually it expands to surround the entire embryo. The amniotic cavity (between the amnion and embryo) is filled with a clear amniotic fluid that bathes the embryo and in effect provides it with its own private swimming pool. Besides helping to keep the embryo moist, the amniotic fluid serves as an effective shock absorber, protecting it from mechanical injury.

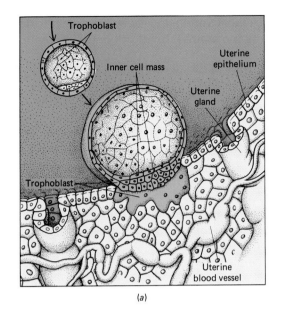

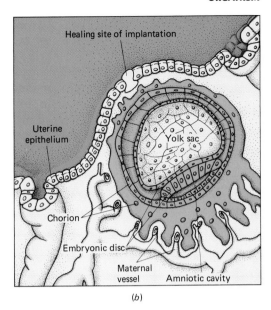

(a)

(b)

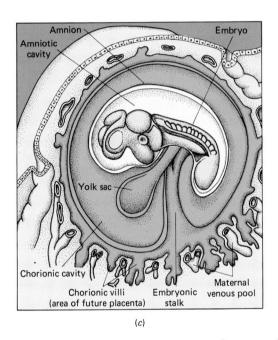

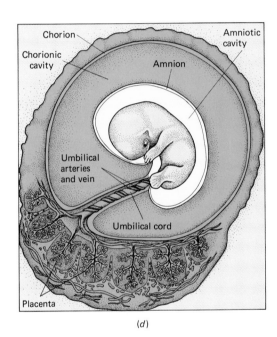

(c)

(d)

Figure 32–5 Formation of the fetal membranes and development of the early embryo. (a) About 7 days after fertilization the blastocyst drifts to an appropriate site along the uterine wall and begins to implant itself. The cells of the trophoblast proliferate and invade the endometrium. (b) About 10 days after fertilization. The chorion has formed from the trophoblast. The inner cell mass has differentiated into the two-layered embryonic disc, from which the embryo will develop. Extensions of the upper endodermal layer form the wall of the developing yolk sac. Extensions of the lower epidermal layer give rise to the developing amnion. (c) By 25 days intimate relationships have been established between the embryo and the maternal blood vessels. Oxygen and nutrients from the maternal blood are now satisfying the embryo's needs. Note the specialized region of the chorion that will soon become the placenta. The embryonic stalk will become the umbilical cord. (d) At about 45 days the embryo and its membranes together are about the size of a Ping-Pong ball, and the mother still may be unaware of her pregnancy. The amnion filled with amniotic fluid surrounds and cushions the embryo. The yolk sac has been incorporated into the umbilical cord. Blood circulation has been established through the umbilical cord to the placenta.

PLACENTA

In placental mammals the **placenta** is the organ of exchange between the blood of the mother and that of the embryo. It provides nutrients and oxygen for the embryo and removes wastes from the developing organism for excretion by the mother. In addition, the placenta serves as an endocrine organ.

The placenta develops from both the embryonic chorion and the maternal uterine tissue. After implantation the chorion continues to grow rapidly, invading the endometrium and forming villi, which become vascularized as the embryonic circulation develops (Fig. 32–5).

An umbilical cord develops, connecting the embryo with the chorion. Two umbilical arteries pass through the umbilical cord and merge into a vast network of capillaries developing within the villi. The placenta actually consists of the portion of the chorion that has developed villi together with the uterine tissue between the villi, which contains maternal capillaries and small pools of maternal blood. Blood from the villi returns to the embryo through the umbilical vein, which passes through the umbilical cord.

The placenta brings maternal blood close to the blood of the embryo, *but the two circulatory systems are completely separate.* Oxygen and nutrients pass from the maternal blood through the placental tissue and diffuse into the embryo's blood, which then transports these materials to developing tissues of the embryonic body. Wastes pass from the embryonic blood through the placental tissue and into the maternal blood, which transports them to the mother's kidneys for disposal.

Several hormones are produced by the placenta. From the time the embryo first begins to implant itself, its trophoblast cells release **human chorionic gonadotropin (HCG),** which signals the corpus luteum that pregnancy has begun. In response, the corpus luteum increases in size and releases large amounts of progesterone and estrogen, which in turn stimulate continued development of the endometrium and placenta. Without HCG the corpus luteum would degenerate, and the embryo would be aborted and flushed out with the menstrual flow. The woman might not even know that she had been temporarily pregnant. After about the 11th week of pregnancy the placenta itself produces sufficient amounts of progesterone and estrogen to maintain pregnancy.

Development of the Germ Layers

While the fetal membranes and placenta develop, important changes are also taking place within the inner cell mass. The cells arrange themselves in a two-layered disc. Then cells of the lower level emerge to line an inner cavity, the primitive gut, which eventually develops into the digestive tract. These cells (which become flattened) are known as **endoderm;** the columnar cells that remain to cover the embryo and become its outermost layer are the **ectoderm.** A third layer of cells, the **mesoderm,** grows between the ectoderm and endoderm. Ectoderm, mesoderm, and endoderm are known as the three **germ layers,** or embryonic tissue layers. Each gives rise to specific structures in all vertebrate embryos (Table 32–1).

The process by which the blastocyst develops into a three-layered embryo is known as **gastrulation,** and during this period the embryo is called a **gastrula.** Gastrulation can be most easily observed in a simple organism, such as the sea star. (See Focus on Comparative Development.)

Development of the Nervous System

The brain and spinal cord are among the first organs to develop in the early embryo. Late in the gastrula stage the ectoderm thickens in the midline of the embryo to form the **neural plate.** The neural plate develops

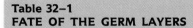

Table 32–1
FATE OF THE GERM LAYERS

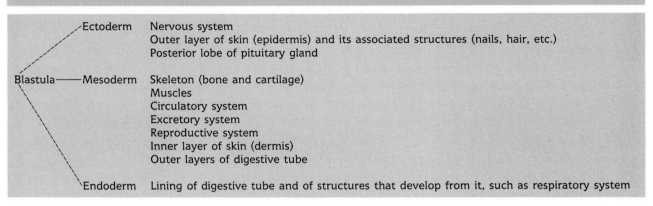

Blastula	Ectoderm	Nervous system Outer layer of skin (epidermis) and its associated structures (nails, hair, etc.) Posterior lobe of pituitary gland
	Mesoderm	Skeleton (bone and cartilage) Muscles Circulatory system Excretory system Reproductive system Inner layer of skin (dermis) Outer layers of digestive tube
	Endoderm	Lining of digestive tube and of structures that develop from it, such as respiratory system

just above a rod of tissue called the **notochord,** which serves as a flexible skeletal axis in all chordate embryos. (In vertebrate embryos it is eventually replaced by the vertebral column.) The notochord actually stimulates the tissue above it to differentiate into nervous tissue, a process known as induction. Figure 32–6 illustrates the development of the neural tube from the neural plate. (Also see Fig. 32–7.) The anterior portion of the neural tube grows and differentiates to form the brain, and the remainder of the tube develops into the spinal cord. While the nervous system is forming, the heart, blood vessels, digestive tube, and many other structures also begin their development.

Later Development

Development proceeds as an orderly, predictable sequence of events. Normally, one can predict with startling precision which structures will begin to develop, or begin to function, on a particular day or even at a specific hour of development (Table 32–2).

By 4 weeks the embryo has already developed the rudiments of many of its organs (Fig. 32–8). A simple but functional circulatory sys-

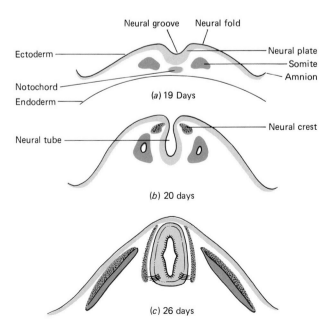

Figure 32–6 Cross sections of the ectoderm of human embryos at successively later stages illustrating the early development of the nervous system. The neural crest cells form the spinal ganglia and the sympathetic nerve ganglia. (*a*) Approximately 19 days. The neural plate is surrounded by neural folds so that it now forms a shallow groove. (*b*) Approximately 20 days. The neural folds approach one another to form the neural tube. The neural crest cells will give rise to nerves. (*c*) Approximately 26 days. The neural tube has now formed and will give rise to the brain in the anterior end of the embryo and the spinal cord posteriorly.

Focus on . . .

COMPARATIVE DEVELOPMENT

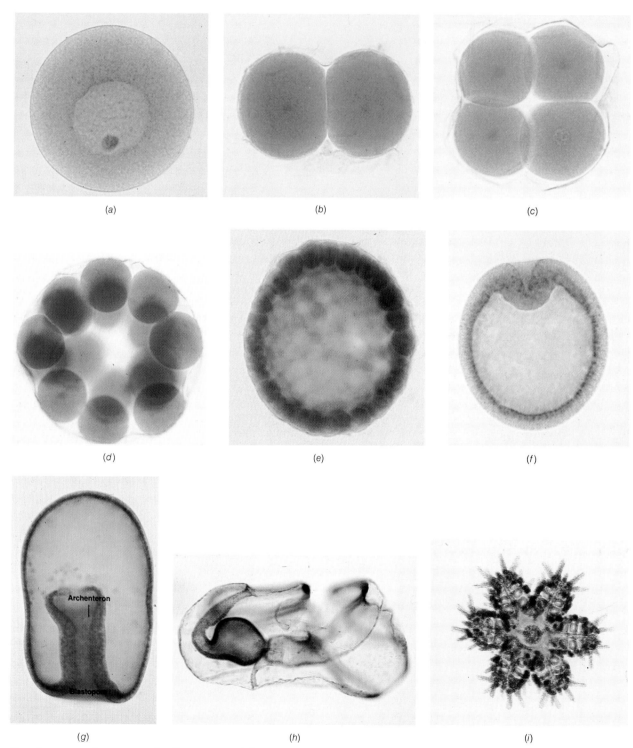

Development of a seastar. (a) Seastar egg. (b) Two-cell stage. (c) Four-cell stage. (d) Sixteen-cell stage. (e) Cross section through 64-cell blastula. (f) Section through early gastrula. (g) Section through middle gastrula. (h) Seastar larva. (i) Young seastar. In this type of cleavage the whole egg becomes partitioned into cells. The blastopore is the opening into the inner cavity, the archenteron. (Carolina Biological Supply Company)

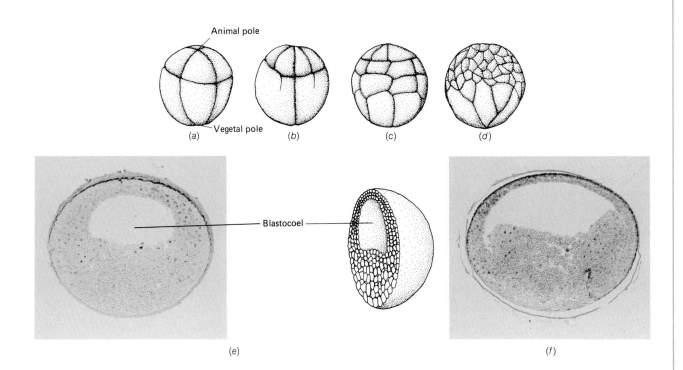

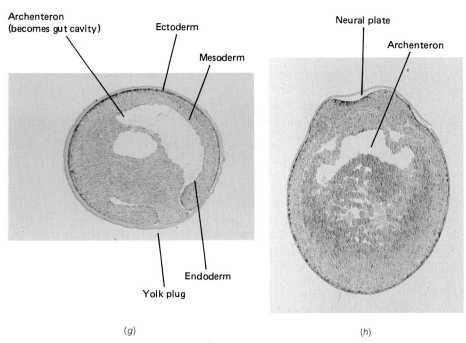

Successive stages in cleavage and gastrulation in the frog. Parts a to d are viewed from the side; e to h have been cut in half so that you can view the inside of the embryo. (e) Late blastula. (f) Early gastrula. (g) Middle gastrula. (h) Late gastrula. Note that in the frog, cleavage is incomplete. The presence of a large amount of yolk, concentrated at the vegetal pole of the egg, inhibits cleavage at one end of the egg.

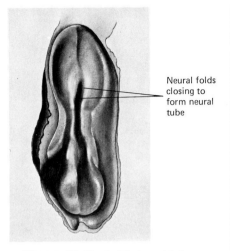

Neural folds closing to form neural tube

Figure 32–7 A 20-day-old human embryo illustrating the early development of the nervous system. (Courtesy of the Carnegie Institution of Washington.)

Table 32–2
SOME IMPORTANT DEVELOPMENTAL EVENTS

Time from fertilization	Event
24 hours	Embryo reaches two-cell stage
3 days	Morula reaches uterus
7 days	Blastocyst begins to implant
2.5 weeks	Notochord and neural plate form; tissue that will give rise to heart differentiates; blood cells forming in yolk sac and chorion
3.5 weeks	Neural tube forming; primordial eye and ear visible; pharyngeal pouches forming; liver bud differentiating; respiratory system and thyroid gland just beginning to develop; heart tubes fuse, bend, and begin to beat; blood vessels laid down
4 weeks	Limb buds appear; three primary vesicles of brain formed
2 months	Muscles differentiating; embryo capable of movement; gonads distinguishable as testes or ovaries; bones begin to ossify; cerebral cortex differentiating; principal blood vessels assume final positions
3 months	Sex can be determined by external inspection; notochord degenerates; lymph glands develop
4 months	Face begins to look human; lobes of cerebrum differentiate; eye, ear, and nose look more "normal"
6–9 months (Third trimester)	Lanugo appears (later shed); neurons become myelinated; tremendous growth of body
266 days	Birth

tem is in operation. At this stage the heart is an S-shaped tube that beats about 60 times each minute.

After the 2nd month (Fig. 32–9) the human embryo is often referred to as a **fetus.** By this time the basic forms of all the organs have been laid down; the fetus becomes recognizable as human.

The last 3 months (last trimester) of development are a period of rapid growth and final differentiation of tissue and organs (Fig. 32–10). If born prematurely at this age, the fetus is able to move about, cry, and try to breathe, but often it dies because its brain is not sufficiently developed to sustain such vital functions as rhythmic breathing and regulation of body temperature.

During the 7th month the cerebrum grows rapidly, and its convolutions become differentiated. The grasp and sucking reflexes are apparent, and the fetus may suck its thumb. Most of the body is covered with a downy hair called **lanugo,** which is usually shed before birth. At birth the full-term baby weighs about 3000 grams (7 pounds) and measures about 52 centimeters (20 inches) in length.

BIRTH

From the time of conception about 266 days are required for the baby to complete its prenatal development. Late in pregnancy the placenta begins to degenerate. As the fetus begins to outgrow its home, the uterus is distended to its fullest capacity. Several days before birth the fetus usually assumes an upside-down position so that its head will enter the birth canal first. The factors that actually initiate the birth process are not well understood.

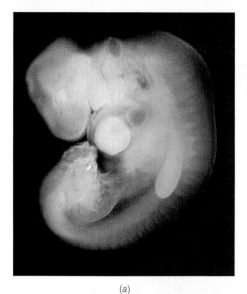

(a)

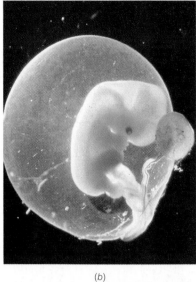

(b)

Figure 32–8 (a) Human embryo at 29 days, about 7 mm (0.3 inch) long. Note the slender tail and developing limb buds. The heart can be seen below the head near the mouth of the fetus. The "double chins" are known as branchial arches (gills). (b) Human embryo at 5½ weeks, 1 cm (0.4 inch) long. Limb buds have lengthened, and eye has become prominent. Note the thin tail, which regresses during later development. (c) In its seventh week of development, the embryo is 2 cm (0.8 inch) long. The dark red object inside the embryo is the liver. (a, c, Lennart Nilsson, *A Child Is Born.* 1977. New York: Dell Publishing Co., Inc.; b, Guigoz, Petit Format, Photo Researchers, Inc.)

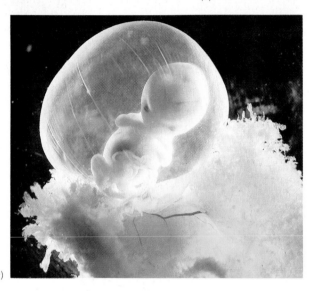

(c)

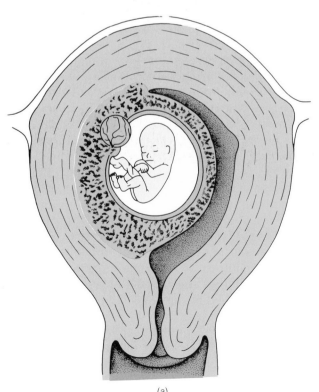

(a)

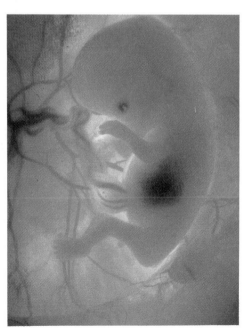

(b)

Figure 32–9 Human embryo at 10 weeks, 6 cm (2.4 inches) long. (Nestle, Petit Format, Photo Researchers, Inc.)

Figure 32–10 Human fetus at 16 weeks, 16 cm (6.4 inches). (Lennart Nilsson, *A Child Is Born.* 1977. New York: Dell Publishing Co., Inc.)

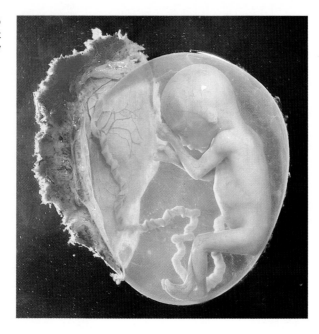

Childbirth begins with a long series of involuntary contractions of the uterus, experienced as **labor** contractions. Labor may be divided into three stages. During the first stage the mother becomes aware of strong uterine contractions, which at first may occur every 30 minutes or so but then become more intense, rhythmic, and frequent, occurring as often as every minute (or even less) later in labor. The sensation accompanying each contraction usually begins in the lower-back area and extends around to the ventral portion of the abdomen. As the first stage of labor progresses, the cervix becomes *dilated* to about 10 centimeters (4 inches) and *effaced* (that is, continuous with the uterine wall, so it cannot be distinguished from the adjoining portion of the uterus), allowing passage of the fetal head. The first stage of labor is the longest, often lasting 8 to 24 hours in a first pregnancy.

During the second stage of labor, the baby is expelled through the vagina and is delivered (Fig. 32–11). By contracting her abdominal muscles the mother can help push the baby along through the vagina. Often the amnion ruptures during the second stage of labor, releasing the amniotic fluid in a gush through the vagina. Popularly referred to as "breaking of the bag of waters," this event sometimes occurs earlier in labor or may even precede it. However, it is not uncommon for the physician to "break the water" with an instrument during preparation for delivery.

As the baby moves through the vagina, its head rotates, facilitating passage to the outside world. Just before birth the physician usually makes a surgical incision called an **episiotomy**, extending from the vagina toward the anus. This facilitates expulsion of the baby and prevents tearing of the maternal tissue. After the delivery the incision is neatly sutured and usually heals within a few weeks.

The infant's head emerges slowly from the vagina, and then two or three additional contractions eject the entire body into the waiting hands of the physician. At this time the umbilical cord still connects the infant with the placenta. Most physicians clamp and cut the cord immediately after the infant has been delivered.

During the third stage of labor the placenta separates from the uterus and is expelled. Generally this occurs within 10 to 20 minutes after the birth of the baby. Now referred to as the *afterbirth,* the placenta is inspected for abnormalities and later discarded.

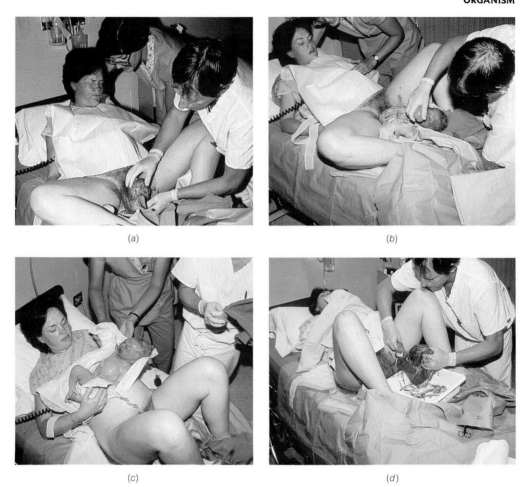

(a)

(b)

(c)

(d)

ADJUSTING TO LIFE AFTER BIRTH

During prenatal life the embryo exists as a parasite, taking from its mother all the oxygen and nutrients it needs, even at the expense of her homeostasis. For example, if the mother's dietary intake of calcium is insufficient, calcium will be taken from her bones to supply the needs of the rapidly growing skeletal system of the fetus.

At birth the infant suddenly must begin to live independently. Normally the **neonate** (newborn infant) begins to breathe within a few seconds of birth and cries within half a minute. However, if anesthetics have been given to the mother, the fetus may also be anesthetized, and breathing and other activities may be depressed. In some cases respiration may not begin for several minutes. This is one of the principal reasons for the current trend toward more natural childbirths, with the use of as little medication as possible.

Soon after birth important changes take place within the circulatory system. Since the lungs cannot function during prenatal life, an extensive circulation through them is not necessary. Accordingly, blood is routed differently in the fetus. Most of the blood entering the right atrium of the fetal heart is routed away from the pulmonary circulation through an opening in the wall between right and left atria, called the **foramen ovale.** The blood passes into the left atrium and into the systemic circulation.

At birth the umbilical blood vessels close and circulation through the placenta comes to an abrupt halt. Because of pressure changes the

Figure 32–11 Birth of a baby. In about 95% of all human births the baby descends through the cervix and vagina in the head-down position. (a) The mother bears down hard with her abdominal muscles, helping to push the baby out. When the head fully appears, the physician or midwife can gently grasp it and guide the baby's emergence into the outside world. (b) Once the head has emerged, the rest of the body usually follows readily. The physician gently aspirates the mouth and pharynx to clear the upper airway of any amniotic fluid, mucus, or blood. At this time the neonate usually takes its first breath. (c) The baby, still attached to the placenta by its umbilical cord, is presented to its mother. (d) During the third stage of labor the placenta is delivered. (Courtesy of Dan Atchison.)

Figure 32–12 Thalidomide administered to the marmoset (*Callithrix jacchus*) produces a pattern of developmental defects similar to those found in humans. (*a*) Control marmoset fetus obtained from an untreated mother on day 125 of gestation. (*b*) Fetus (same age as control) of a marmoset treated with 25 mg/kg thalidomide from days 38 to 52 of gestation. The drug suppresses limb formation, perhaps by interfering with the function of cholinergic nerves. (Courtesy of Dr. W.G. McBride and P.H. Vardy, Foundation 41; from *Development, Growth and Differentiation*, 25 [4]: 361–373, 1983.)

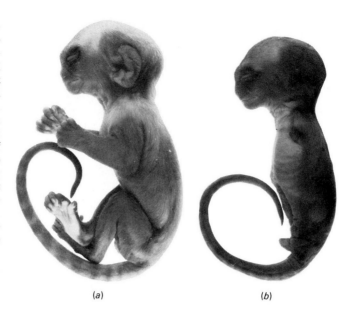

(*a*)   (*b*)

direction of blood flow through the hole in the heart wall reverses, and a small flap of tissue is pressed into place, repairing the wall.

ENVIRONMENTAL INFLUENCES ON THE EMBRYO

We all know that a baby's growth and development are influenced by the food we give him, the air he breathes, the disease organisms that infect him, the chemicals or drugs he receives, and other factors. But how many of us stop to think that a baby's life begins at the moment of conception, and that *prenatal* development is also affected by these environmental influences? In fact, life before birth is even more sensitive to environmental changes than it is in the fully formed baby.

About 7% of all babies born alive in the United States, or 175,000 babies per year, arrive with a defect of clinical significance. Such birth defects account for about 10% of deaths among newborns. Birth defects may result from genetic or environmental factors or a combination of the two (Fig. 32–12). In this section we shall discuss the environmental conditions that affect the well-being of the embryo. (Genetic defects were discussed in Chapter 10.)

Since most structures form during the first 3 months of embryonic life, the developing baby is most susceptible to environmental conditions during this early period. The mother may not even realize that she is pregnant and may therefore take no special precautions to minimize potentially dangerous influences. Table 32–3 describes some of the environmental influences upon development.

Recent advances in medicine have enabled physicians to diagnose some defects while the embryo is in the uterus. In some cases treatment is possible before birth. In **amniocentesis** a sample of amniotic fluid is withdrawn through the abdominal wall and used in diagnosing certain genetic disorders. If it is determined that the embryo will develop into a grossly deformed baby, the parents can elect to abort the pregnancy. Figure 32–13 is a **sonogram,** a photograph taken of the embryo by using ultrasound. The exact age of the embryo, in terms of weeks, can be determined by measuring the head and bones of the extremities. Such previews are helpful in diagnosing defects, and also in determining the position of the fetus and whether a multiple birth is pending. By the 5th month, depending on the position of the embryo, its sex can sometimes be determined by a sonogram.

Table 32–3
ENVIRONMENTAL INFLUENCES ON THE EMBRYO

Factor	Example and effect	Comment
Nutrition	Severe protein malnutrition doubles number of defects; fewer brain cells are produced, and learning ability may be permanently affected	Growth rate mainly determined by rate of net protein synthesis by embryo's cells
Excessive amounts of vitamins	Vitamin D essential, but excessive amounts may result in form of mental retardation; too much vitamins A and K may also be harmful	Vitamin supplements are normally prescribed for pregnant women, but some women mistakenly reason that if one vitamin pill is beneficial, four or five might be even better
Drugs	Many drugs affect development of fetus: Even aspirin has been shown to inhibit growth of human fetal cells (especially kidney cells) cultured in laboratory; it may also inhibit prostaglandins, which are concentrated in growing tissue	Common prescription drugs are generally taken in amounts based on body weight of mother, which may be hundreds or thousands of times too much for tiny embryo
Alcohol	When woman drinks heavily during pregnancy, baby may be born with fetal alcohol syndrome—that is, deformed and mentally and physically retarded; low birth weight and structural abnormalities have been associated with as little as two drinks a day; some cases of hyperactivity and learning disabilities may be caused by alcohol intake of pregnant mother	Fetal alcohol syndrome thought to be one of leading causes of mental retardation in United States
Heroin	Heroin results in high mortality rate and high prematurity rate	Infants that survive are born addicted and must be treated for weeks or months
Methadone	Methadone results in fetal addiction	
Thalidomide	Thalidomide, marketed as mild sedative, was responsible for more than 7000 grossly deformed babies born in the late 1950s in 20 countries. Principal defect was **phocomelia,** condition in which babies are born with extremely short limbs, often with no fingers or toes	This drug interferes with cellular metabolism; most hazardous when taken during fourth to sixth weeks, when limbs are developing
Cigarette smoking	Cigarette smoking reduces amount of oxygen available to fetus because some of maternal hemoglobin is combined with carbon monoxide; may slow growth and can cause subtle forms of damage; in extreme form carbon monoxide poisoning causes such gross defects as hydrocephaly	Mothers who smoke deliver babies with lower-than-average birth weights and have higher incidence of spontaneous abortions, stillbirths, and neonatal deaths; studies also indicate possible link between maternal smoking and slower intellectual development in offspring
Pathogens	Rubella (German measles) virus crosses placenta and infects embryo; interferes with normal metabolism and cell movements; causes syndrome that involves blinding cataracts, deafness, heart malformations, and mental retardation; risk is greatest (about 50%) when rubella is contracted during first month of pregnancy; risk declines with each succeeding month	Rubella epidemic in the United States in 1963–65 resulted in about 20,000 fetal deaths and 30,000 infants born with gross defects
	Syphilis is transmitted to fetus in about 40% of infected women; fetus may die or be born with defects and congenital syphilis	Pregnant women are routinely tested for syphilis during prenatal examinations
Ionizing radiation	When mother is subjected to x-rays or other forms of radiation during pregnancy, infant has higher risk of birth defects and leukemia	Radiation was one of earliest teratogens to be recognized

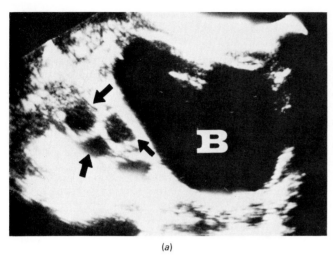

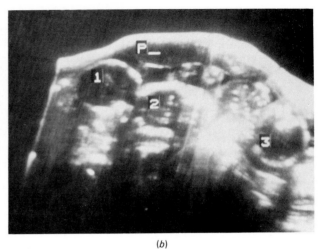

(a)   (b)

Figure 32–13 Ultrasonic techniques can be used to monitor follicle maturation and ovulation, as well as to give the physician information about the fetus. (a) Sonogram showing three follicles of equal maturity in the left ovary of a human. (b) Triplets in the same patient at 16 weeks of pregnancy. P, placenta. Such previews are valuable to the physician in diagnosing defects and predicting multiple births. (Courtesy of Biserka Funduk-Kurjak and Asim Kurjak, from *Acta Obstetricia et Gynecologica Scandinavica*, 61, 1982.)

THE HUMAN LIFE CYCLE

Development begins at conception and continues through the stages of the human life cycle until death (Table 32–4). We have examined briefly the development of the embryo and fetus, the birth process, and the adjustments it requires of the neonate. The neonatal period is usually considered to extend from birth to the end of the first month of extrauterine life. **Infancy** follows the neonatal period and lasts until the rapidly developing infant can assume an erect posture and walk, usually between 10 and 14 months of age. Some regard infancy as extending to the end of the first 2 years. **Childhood,** also a period of rapid growth and development, continues from infancy until adolescence.

Adolescence is the time of development between puberty and adulthood. During adolescence a young person experiences the physical and physiological changes that result in physical and reproductive maturity (Fig. 32–14). This is also a time of profound psychological development, as young people make adjustments that help prepare them to assume the responsibility of adulthood.

Young adulthood extends from adolescence until about age 40. Middle age is usually considered to be the period between ages 40 and 65. Old age begins after age 65.

THE AGING PROCESS

Since development in its broadest sense includes any biological change with time, it also includes those changes that result in the decreased functional capacities of the mature organism, the changes commonly called **aging.** The declining capacities of the various systems in the human body, though most apparent in the elderly, may begin much earlier in life, during childhood, or even during prenatal life. The newborn female has only 400,000 oocytes remaining of the 4,000,000 she had three months earlier in fetal life.

The aging process is far from uniform in various parts of the body. Various systems of the body may begin their decline at quite different times. A 75-year-old man, for example, has lost 64% of his taste buds, 44% of the renal glomeruli, and 37% of the axons in his spinal nerves that he had at age 30. His nerve impulses are propagated at a rate 10% slower, the blood supply to his brain is 20% less, and the vital capacity of his lungs has declined 44%. The aging process is also marked by a progressive decrease in the body's homeostatic ability to respond to stress.

Several theories have been advanced regarding the cause of the aging process—hormonal changes; the development of *autoimmune re-*

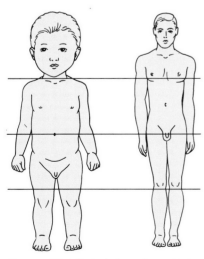

Figure 32–14 Relative sizes of the head, arms, and legs in a young child and in an adult.

Table 32–4
STAGES IN THE HUMAN LIFE CYCLE

Stage	Time period	Characteristics
Embryo	Conception to end of 8th week of prenatal development	Development proceeds from single-celled zygote to embryo that is about 30 mm long, weighs 1 g, and has rudiments of all its organs
Fetus	Beginning of 9th week of prenatal development to birth	Period of rapid growth, morphogenesis, and cellular differentiation, changing tiny parasite to physiologically independent organism
Neonate	Birth to 4 weeks of age	Neonate must make vital physiological adjustments to independent life: It must now process its own food, excrete its wastes, obtain oxygen, and make appropriate circulatory changes
Infant	End of 4th week to 2 years of age (sometimes, ability to walk is considered end of infancy)	Rapid growth; deciduous teeth begin to erupt; nervous system develops (myelinization), making coordinated activities possible; language skills begin to develop
Child	Two years to puberty	Rapid growth; deciduous teeth erupt, are slowly shed and replaced by permanent teeth; development of muscular coordination; development of language skills and other intellectual abilities
Adolescent	Puberty (approximately ages 11–14) to adult	Growth spurt; primary and secondary sexual characteristics develop; development of motor skills; development of intellectual abilities; psychological changes as adolescent approaches adulthood
Young adult	End of adolescence (approximately age 20) to about age 40	Peak of physical development reached; individual assumes adult responsibilities that may include marriage, fulfilling reproductive potential, and establishing career. After age 30, physiological changes associated with aging begin
Middle-aged adult	Age 40 to about age 65	Physiological aging continues, leading to menopause in women and physical changes associated with aging in both sexes (e.g., graying hair, decline in athletic abilities, wrinkling skin). This is period of adjustment for many as they begin to face their own mortality
Old adult	Age 65 to death	Period of senescence (growing old); physiological aging continues; maintaining homeostasis more difficult when body is challenged by stress; death often results from failure of cardiovascular or immune system

actions (allergies against certain components of the organism's own body that result in destruction of those components by antibodies); the accumulation of specific waste products within the cell (the "clinker" theory); changes in the molecular structure of macromolecules, such as collagen (an increased cross-linkage between the helical chains); a decrease in the elastic properties of connective tissues owing to an accumulation of calcium, which results in stiffening of the joints and hardening of the arteries; the peroxidation of certain lipids by free radicals; or the destruction of cells by hydrolases released by the breaking of lysosomes.

Other current theories suggest that aging involves the accumulation of mutations caused by continued exposure to cosmic radiation and

Focus on . . .

NOVEL ORIGINS

About 10,000 children born each year are products of **artificial insemination.** Usually this procedure is sought when the male partner of a couple desiring a child is sterile or carries a genetic defect. Although the sperm donor remains anonymous to the couple involved, his genetic qualifications are screened by physicians.

In vitro fertilization is a technique by which an ovum is removed from a woman's ovary, fertilized in a test tube, and then reimplanted in her uterus. Such a procedure may be attempted if a woman's fallopian tubes are blocked, or if they have been surgically removed. With the help of this technique a healthy baby was born in England in 1978 to a couple who had tried unsuccessfully for several years to have a child. Since that time, many other children have been conceived within laboratory glassware.

Another novel procedure is **host mothering.** In this procedure, a tiny embryo is removed from its natural mother and implanted into a female substitute. The foster mother can support the developing embryo either until birth or temporarily until it is implanted again into the original mother or into another host. This technique has already proved useful to animal breeders. For example, embryos from prize sheep can be temporarily implanted into rabbits for easy shipping by air, and then reimplanted into a foster mother sheep, perhaps of inferior quality. Host mothering also has the advantage of allowing an animal of superior quality to produce more offspring than would be naturally possible. In one recent series of experiments mouse embryos were frozen for up to 8 days and then successfully transplanted into host mothers. Host mothering may someday be popular with women who can produce embryos but are unable to carry them to term.

Someday society may have to deal with **cloning** (not yet a reality in humans). In this process the nucleus would be removed from an ovum and replaced with the nucleus of a cell from a person who wished to produce a human copy of himself. Theoretically, any cell nucleus could be used, even a white blood cell nucleus. The fertilized ovum would then be placed into a human uterus for incubation; the resulting baby would be an identical, though younger, twin to the individual whose nucleus was used.

---

x-radiation, mutations that decrease the ability of the cell to carry out its normal functions at the normal rate. In all likelihood, aging is a part of and due to the same kinds of developmental processes that bring about the increasing functional capacities of the various systems of the body during earlier development. The processes may be part of the program of timed development built into the genes. Like other developmental processes, aging may be accelerated by certain environmental influences and may occur at different rates in different individuals because of inherited differences. For example, there is some experimental evidence that aging, at least in rats, can be delayed by dietary means, by caloric restriction: Thin rats generally live longer than fat rats. Genetic predisposition may, however, be the best guarantee of a long life.

CHAPTER SUMMARY

I. Development is largely epigenetic, although the blueprints for the organism are preformed in the DNA of the zygote. Development proceeds as a balanced combination of growth, morphogenesis, and cellular differentiation.
   A. Cleavage takes place and a morula develops as the human embryo passes along the uterine tube to the uterus.
   B. The blastocyst forms after the embryo arrives in the uterus, and on about the 7th day of development the embryo implants itself in the thick uterine wall.
   C. Extraembryonic membranes protect the embryo and help in its nourishment. All vertebrates have an amnion, yolk sac, chorion, and allantois.

1. The amnion serves as a shock absorber and a fluid medium in which the embryo develops.
   2. In placental mammals the placenta is the organ of exchange between mother and embryo.
D. During gastrulation, the ectoderm, mesoderm, and endoderm are formed, each of which gives rise to specific tissues.
II. Labor can be divided into three stages, with the actual delivery of the baby occurring during the second stage.
III. At birth the neonate must make the physiological adjustments that permit him to function as an independent organism.
IV. By controlling environmental factors, such as nutrition, vitamin and drug intake, cigarette smoking, and disease-causing organisms, a pregnant woman can help ensure the well-being of her unborn child.
V. The human life cycle can be divided into the following stages: embryo, fetus, neonate, infant, child, adolescent, young adult, middle-aged adult, and old adult.
VI. The aging process is marked by a progressive decrease in the body's homeostatic abilities to respond to stress.

Post-Test

1. The term morphogenesis refers to the process by which _____ .
2. The process by which cells become specialized is called _____ _____ .

Find the most appropriate term in column B for each description in column A.

Column A

_____ 3. Series of mitoses during early development
_____ 4. Human embryo after second month
_____ 5. Acts as shock absorber
_____ 6. Organ of exchange between mother and embryo
_____ 7. Hollow ball of cells that gives rise to gastrula

Column B

a. blastocyst
b. cleavage
c. amnion
d. placenta
e. fetus

8. The inner cell mass gives rise to the _____ ___ .
9. Human chorionic gonadotropin (HCG) signals the _____ _____ that _____ _____ .
10. The _____ is the germ layer that gives rise to the nervous system; the _____ lines the digestive tract.
11. The baby is delivered during the _____ stage of labor.
12. The foramen ovale is _____ _____ .
13. A neonate is a _____ _____ .
14. The stage in the human life cycle between puberty and adulthood is _____ .

Review Questions

1. Contrast the preformation theory with the theory of epigenesis, and relate these theories to current concepts of development.
2. Explain how growth, development, and morphogenesis are all necessary for a developmental process like formation of the neural tube.
3. Trace the development of the embryo from zygote to gastrula. Draw and label diagrams to illustrate your description.
4. What are the functions of the amnion? of the placenta?
5. What is implantation? Describe.
6. When during development do most of the organs form? Which organs begin to develop first?
7. What are the germ layers? What adult structures develop from ectoderm?
8. What happens during each stage of labor?
9. What are some nongenetic factors that influence development of the embryo? What effects do they have?
10. What are two specific pathogens that affect the development of the embryo? What effects do they have?
11. How can the following affect development? cigarette smoking; drugs; ionizing radiation.
12. How would you compare the human cycle with the life cycle of a dog? an insect?

Readings

Beaconsfield, P., G. Birdwood, and R. Beaconsfield. "The Placenta," *Scientific American,* August 1980. A summary of the development of the placenta and its functions.

De Robertis, E.M., and J.B. Gurdon. "Gene Transplantation and the Analysis of Development," *Scientific American,* December 1979. The amphibian oocyte serves as a living test tube for studying the biochemistry of gene regulation in development.

Garcia-Bellido, A., P.A. Lawrence, and G. Morata. "Compartments in Animal Development," *Scientific American,* July 1979.

Hayflick, L. "The Cell Biology of Human Aging," *Scientific American,* January 1980. A very interesting theory of the aging of cells and of the relationship of aging cells to the human life span.

Johnson, E.M. "Screening for Teratogenic Hazards," *Annual Review of Pharmacological Toxicology,* Vol. 21, 1981.

Jones, M.D., Jr., et al. "Oxygen Delivery to the Brain Before and After Birth," *Science,* 16 April 1982.

PART VII
EVOLUTION, BEHAVIOR, AND ECOLOGY

Brown pelican, *Pelecanus occidentalis,* and its prey. Pelicans, fish, and the microscopic plankton of the ocean are parts of an intricate network of interwoven food chains. (P. and W. Ward, Animals, Animals.)

Chapter 33

VARIABILITY AND EVOLUTION

Outline

Learning Objectives

After you have studied this chapter you should be able to:

1. Discuss the history of evolutionary thought (with emphasis on Charles Darwin), listing the most important contributors and summarizing their ideas.
2. State the Hardy-Weinberg law and discuss its significance and consequences in terms of population genetics and evolution.
3. Define *organic evolution* in genetic terms and distinguish between microevolution and macroevolution.
4. Discuss the relationship between the Hardy-Weinberg law and evolution, and describe the dependence of evolution upon genetics.
5. Summarize the "modern" concept of evolution (neo-Darwinism), discuss its mechanisms (mutation, genetic recombination, genetic drift, natural selection, reproductive isolation) and the adaptive results, and cite specific examples and illustrative experiments.
6. Discuss the distinctive features of the punctuated equilibrium theory of evolution.

Figure 33–1 Living "stones." These unusual desert plants have leaves so greatly thickened that they resemble spheres. This shape approaches the minimal surface area that any solid object may have, which helps to minimize their potential water loss under the hot, dry conditions of their arid habitat. The adaptations of organisms usually fit them for the demands of their environment and lifestyle with great precision. The complexity and effectiveness of these adaptations have always been a source of wonder. How do such design features originate?

Among all human cultures there is some explanation for the origin of the universe, of the world, of humans , and of other organisms. Our curiosity about our origins has demanded explanations, and these explanations continue to be modified today.

THE MEANING OF EVOLUTION

The general meaning of the word evolution is simply gradual change. In biology, **evolution** refers to **organic evolution,** the theory that living things, and the populations of which they form a part, change gradually over the course of their history, and that all living things are, to some extent, genetically related to one another.

Most biologists use the term **microevolution** to refer to changes in gene frequencies that occur within populations, so ordinarily, microevolution refers to changes that take place within a species. **Macroevolution** generally refers to evolution above the species level that might give rise to new genera or higher-level categories of organisms. In this latter sense, evolution is the idea that all forms of life developed gradually from very different and often much simpler ancestors, and that all lines of their descent can be traced back to a common ancestral organism.

DEVELOPMENT OF EVOLUTIONARY THINKING

The name most commonly associated with theories of evolution is Charles Darwin, the father of almost all modern evolutionary thinking (Fig. 33–2). Darwin, however, did not inhabit an intellectual vacuum. His predecessors and contemporaries made important contributions to his thinking.

Darwin's most important predecessor was probably the Frenchman Lamarck. All too often Lamarck is remembered only as someone who proposed an unsatisfactory theory of how evolution proceeded, but he was also the first to tackle the question seriously. Those writing before him toyed with some evolutionary ideas, merely proposing that evolution did take place. However, Lamarck proposed a theory of *how* evolution might take place—by use and disuse of organs combined with the inheritance of acquired characteristics. He held, for example, that by persistently stretching their necks in search of food, ancestral giraffes had, over the course of many generations, produced our present long-necked giraffes. Disuse of the eyes over a long period might similarly produce blind cave fish. Thus Lamarck thought that an individual organism's adaptive response to the demands of the environment was somehow

Figure 33–2 Charles Darwin (1809–1882). Darwin's studies in natural history had made him well known even before the publication of *The Origin of Species* in 1859. His role as the leader of an intellectual revolution was accepted reluctantly. He confronted his critics only in his letters and books. Most of his adult years were spent in Down, England, where he continued to do investigations in botany, ecology, and animal behavior until his death. This photograph was taken in 1881. (Mary Evans Picture Library, Photo Researchers, Inc.)

Figure 33–3 *Archaeopteryx* (left) and a modern bird (right). Known only from fossils, *Archaeopteryx* seems physically intermediate between certain small dinosaurs and modern birds. Assuming that birds are in fact descended from ancestors similar to *Archaeopteryx*, how did the changes necessary to produce birds take place? Teeth would have to be lost, as would the tail. A host of rearrangements of feather patterns would be required, together with numerous changes in behavior. Lamarck would have seen these adaptational changes as being purposive in that they result from the striving of the organism to meet the demands of its environment. Any changes would become a part of the organism's heredity, be passed on to its offspring, and serve as a baseline from which further changes could take place. Darwin, on the other hand, stressed the role of natural selection—only chance governed variation in body form, but only favorable variations would survive (or at any rate, would survive better than alternative traits).

transmitted to succeeding generations. As we will discuss later, however, there is no evidence that the environment *directly* influences the hereditary traits of organisms in an adaptive fashion (Fig. 33–3).

The first and, to date, only plausible alternative to Lamarck's explanation of evolutionary change was proposed by Charles Darwin, an Englishman born in 1809. The grandson of Erasmus Darwin, a well-known physician and naturalist, and the son of a successful physician, Darwin was sent to the best schools and universities. After deciding that he did not have the stomach for the crude surgical techniques of his time, Charles Darwin abandoned a proposed career in medicine and studied instead for the clergy. He earned a degree but showed more interest in the considerable biological knowledge that he absorbed from his scholarly environment. He put his hobby of natural history to good use when, in 1831, he joined (at his own expense) a British naval oceanographic expedition as a naturalist. His voyage on the H.M.S. *Beagle* lasted almost five years and took him to many parts of the world (Fig. 33–4). The observations made and specimens collected on this trip suggested the theory associated with his name today.

Perhaps the most important port of call on the *Beagle* voyage—at least for the history of biology—was the Galapagos Islands. There Darwin observed a number of species of a bird called the finch, each special-

Figure 33–4 The voyage of H.M.S. *Beagle*.

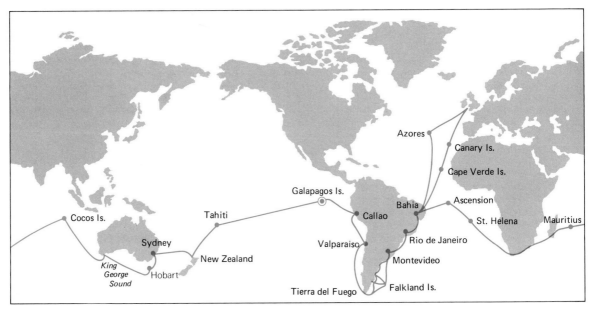

(a)

(b)

Figure 33–5 Two species of Darwin's famous Galapagos Islands finches. Although these drab, unremarkable birds differ considerably in their adaptations, there seems no doubt that they have a common ancestry. Darwin observed that the various species of Galapagos finches occupied ecological niches filled elsewhere by birds of different families. The continental equivalents of the various Galapagos finches are presumably better adapted to these lifestyles than almost all of the finches are, but as chance had it, were never afforded the opportunity to colonize the Galapagos Islands. The likely derivation of such different birds from a common ancestry helped suggest to Darwin that species were not unchanging and that they originated by natural selection. (a) Cactus finch, *Geospiza scandens.* The cactus finch may be close to the relatively unspecialized form of the original finch colonists. (b) A large ground finch, *Geospiza magnirostra.* This bird has an extremely heavy, nutcracker-type bill adapted for eating heavy-walled seeds. (a, Jeanne White, Photo Researchers, Inc., b, Miguel Castro, Photo Researchers, Inc.)

ized for a particular life-style (Fig. 33–5). All were apparently related, for they were highly similar in anatomy, coloration, and the like. Darwin felt sure that all had had a common ancestor, perhaps blown to the islands during a severe storm. The several modern species must have come into existence as modifications of the ancestral one. The conclusion seemed inescapable—new species originate from old ones. Since Darwin considered this view to be in direct opposition to the doctrine of special creation, the dominant religious belief and also the view widely held among scientists of the day, he felt that expressing his new opinion was almost "like confessing to a murder."

THE MECHANISM OF EVOLUTION: DARWIN'S PROPOSAL

In 1858 Darwin (with a fellow naturalist, A.R. Wallace, who had independently reached similar conclusions) published a paper summarizing his view of evolution by natural selection. In 1859 there followed Darwin's epochal book, *On the Origin of Species by Means of Natural Selection.* Darwin and Wallace's proposal involved four main points: (1) the overproduction of offspring by all organisms, (2) variation among those offspring, (3) inheritability of that variation, and (4) natural selection of the variants by the demands of the environment and each organism's lifestyle. The two scholars felt certain that these mechanisms were enough to produce the evolution that they took to be a demonstrable fact.

Today we realize that the competition implied by Darwin is by no means the only basis of selection. Organisms are stressed naturally by changes in climate and many other dimensions of the environment that may not directly involve competition with members of their own species or even with members of other species. Moreover, Darwin did not consider his ideas and Lamarck's mutually exclusive. The alleged inheritance of acquired characteristics did find a minor place in his thinking, especially in his later years. Still, in its broad outlines, Darwin's proposal has survived as the dominant theory of evolution to this day.

Darwin retired to a country home and spent most of the remainder of his life there, insulated from the controversy his new theory was generating. The cudgels of debate were taken up by others, most notably T.H. Huxley. Darwin continued to conduct research and to publish, for the most part on evolutionary topics. By the time of his death in 1882 his name had become a household word.

GENETICS AND MICROEVOLUTION

The modern concept of evolution is a logical extension and development of Darwin's ideas. For this reason it is known as Neo-Darwinism. In its modern form, evolution depends heavily on the science of genetics, which was unknown to Darwin. (Gregor Mendel's work was just then being published, but it never attracted Darwin's attention.) The development of modern evolutionary theory had to await the rediscovery of genetics in the early 1900s. The modern genetic principles most essential to our present understanding of evolution are those of the gene pool and the Hardy-Weinberg law.

Gene Pool and the Hardy-Weinberg Law

A population of organisms in which genes combine essentially at random is said to be **panmictic.** Even human populations are panmictic for certain genes; that is, mating is at random with respect to them. People are not greatly concerned about blood types, for example, and therefore usually marry without taking them into account. Consequently, we probably are panmictic with respect to blood types. However, mating among human beings is not panmictic with respect to some genes—those governing skin color, for example.

A panmictic population may be considered a **pool of allelic genes,** any of which has a chance of combining with any other when borne by an individual of the opposite sex. That chance will depend upon the frequencies of the genes involved, and upon no other factor. This was not always appreciated. At the beginning of this century, when Mendel's genetic principles were rediscovered, scientists believed that dominating genes would eventually crowd out all recessive alleles in a population simply because they *were* dominant. This is obviously not the case. The explanation of why they do not involves some mathematics, which we have confined to the accompanying box (see Focus on the Hardy-Weinberg Law). To summarize the concept, one can demonstrate that, regardless of dominance or recessiveness, the relative frequencies of allelic genes do not change from generation to generation if they are left undisturbed. This principle is known, after its discoverers, as the **Hardy-Weinberg law.** It applies, however, only in a panmictic population and in the *absence* of certain perturbing forces, including mutation, genetic drift, selective migration, and any kind of consistent selection for or against any genetically determined phenotype.

Changes in Gene Frequency

It is true, as far as is known, that without those perturbing forces (soon to be discussed) genetic frequencies in a freely interbreeding population will not change from generation to generation. Yet those forces, or at least some of them, are always present. In reality, the Hardy-Weinberg law does little more than establish a base line from which evolutionary departures must take place.

MUTATION

Clearly, genetic frequencies *would* change if one kind of gene changed into another form of the same allele by mutation. This in fact does happen in all known populations. Yet it happens at a very low rate, and the most radical of such changes are swiftly removed by natural selection.

Focus on . . .

THE HARDY-WEINBERG LAW

At the beginning of our century Hardy, Weinberg, and other scholars showed that in a freely interbreeding population that mates randomly with respect to two allelic genes, the genotypes of the population may be described by the following expression:

$$p^2 + 2pq + q^2$$

where

p = the frequency of the dominant allele
q = the frequency of the recessive allele
p^2 = the frequency of the homozygous dominant genotype
2pq = the frequency of the heterozygote
q^2 = the frequency of the recessive homozygous genotype

Imagine a population of organisms that is panmictic for two allelic genes: *A*, whose frequency is p, and *a*, whose frequency is q. Their combined frequency is obviously 1, so we may write:

$$p + q = 1 \qquad (1)$$

so

p = 1 − q

q = 1 − p

A gamete bearing gene *A* may combine with a similar one to form an *AA* zygote. Similarly, $A \times a \rightarrow Aa$; $a \times A \rightarrow aA$, and $a \times a \rightarrow aa$. The frequency of each combination is the product of the frequencies of its component genes, as follows:

	p(*A*)	q(*a*)
p(*A*)	p^2(*AA*)	pq(*aA*)
q(*a*)	pq(*Aa*)	q^2(*aa*)

When we add these, the total of all genotypes is as follows:

$$p^2 + 2pq + q^2 = 1 \qquad (2)$$

where

p^2 = frequency of *AA*
2pq = frequency of *Aa*
q^2 = frequency of *aa*

In the next generation the present *Aa* organisms will be able to produce *either A* or *a* gametes; that is, half the gametes will be *A* and half *a* from this source. The *AA* organisms will, of course, yield only *A* gametes; the *aa*, only *a* gametes. Thus the total frequency of all *A* and *a* may be expressed by the following equations, where p′ stands for the new frequency of *A* and q′ for the new frequency of *a*.

$$
\begin{aligned}
p' &= p^2 + \tfrac{1}{2}(2pq) \qquad (3)\\
   &= p^2 + pq\\
   &= p^2 + p(1 - p)^*\\
   &= p^2 + p - p^2 = p
\end{aligned}
$$

$$
\begin{aligned}
q' &= q^2 + \tfrac{1}{2}(2pq) \qquad (4)\\
   &= q^2 + pq\\
   &= q^2 + q(1 - q)^*\\
   &= q^2 + q - q^2 = q
\end{aligned}
$$

Therefore, if left undisturbed, gene frequencies in a panmictic population do not change from generation to generation, regardless of dominance or recessivity.

*See equation 1.

Although mutations do provide evolution's raw material, they do not determine the direction of evolutionary change.

Rather, as unfavorable alleles are continuously weeded out by selective pressures of the environment, the production of new mutations simply keeps up a genetic variability within the population (Fig. 33–6). The new mutations, in turn, are usually weeded out and replaced with other variants. Evolutionary changes are based on the ability of a population to produce recombinations (that is, to have sufficient genetic variability) that allow enough individuals to survive under the stress of a changed environment. Thus, evolutionary change does not rely on a *high* rate of mutation.

The question arises whether enough mutations are likely to occur to produce the necessary genetic variability. Some researchers who studied rates of mutation occurring within selected species in recent years have estimated that at least 500 million mutations that are reproduced in offspring occur during the evolutionary life of a species (based on 50,000 generations in a population of 100 million individuals). Most of these would be duplications, or decidedly unfavorable. However, according to

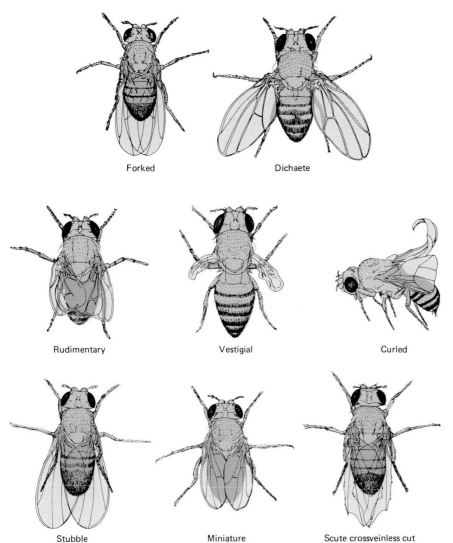

Forked   Dichaete

Rudimentary   Vestigial   Curled

Stubble   Miniature   Scute crossveinless cut

Figure 33–6 Some wing and bristle mutations in the fruit fly *Drosophila melanogaster*. Hundreds of mutations are known to occur in these little fruit flies, which breed rapidly enough to allow us to easily follow the transmission of such altered traits from generation to generation. In addition to those shown here, mutations affecting eye color, behavior, and many other traits have been described and studied. Since mutations are random changes in genetic material, most of them are bound to be harmful to the organism, as is obviously the case in the *vestigial* mutation shown here. Yet for island dwelling insects, fully developed wings might well be more of a disadvantage than an advantage, permitting the insect to be too easily blown away from land. Perhaps for this reason, flies and other insects that dwell on small oceanic islands frequently have reduced wings or are entirely wingless. (After Wallace, Sturtevant, and Beadle.)

some estimates, perhaps as few as 500 would be sufficient to provide the genetic basis for evolution, provided that they were indeed useful for the organism.

Chromosomal mechanisms can also supply genetic variability. For instance, more than one-fourth of the species of flowering plants are **polyploids** of their closest relatives. This means they arose as a result of defects in meiosis that produced multiples of the normal number of chromosomes—4*n*, 8*n*, and so on. A polyploid human being, if one could exist, would have 92 chromosomes or some even larger number instead of the normal 46.

That polyploidy is a mechanism of plant evolution has been demonstrated experimentally. One researcher synthesized a species of hemp nettle by interbreeding two species, each with 8 pairs of chromosomes. The offspring contained 16 pairs of chromosomes and turned out to be identical with a naturally occurring species that also contained 16 pairs. The synthesized species would not interbreed with either of the species from which it had originated. Rather, it reproduced true to its kind, producing plants with 16 chromosome pairs, and it was capable of crossing with the natural 16-chromosome species. This is an obvious, directly observable instance of evolution at the present time. There is also evidence that a few animal species, such as some within the catastomid family of fish, arose by means of polyploidy.

NATURAL SELECTION

Another influence that tends to perturb the genetic equilibrium of populations is natural selection. Darwin saw the demands of the environment as exercising a selective effect upon the organisms present in that environment. You may recall, as an example, the case of sickle-cell anemia from Chapter 10. In this disorder an abnormal form of hemoglobin occurs in the red blood cells of those homozygous for the trait. If not treated, the condition is almost invariably fatal. Yet a heterozygote suffers from no more than a mild, harmless anemia. In East Africa, where the trait probably originated, the heterozygote enjoys some protection against the malaria parasite. So prevalent is malaria in the Congo Basin and similar areas (Fig. 33–7) that someone heterozygous for sickling apparently has a considerable advantage over a person of normal genotype. But not everyone with sickle-cell anemia lives in the Congo Basin; the trait also persists in many persons of African origin in the United States, now nearly malaria-free. Obviously, the sickle-cell heterozygote has no advantage in Chicago. Natural selection may eventually be expected to remove this gene from the American population, given enough time in which to do so, for here it constitutes a disadvantage. (Modern methods of medical treatment may, however, preserve the gene despite the workings of natural selection.)

SELECTIVE IMMIGRATION OR EMIGRATION

Theoretically, selective immigration or emigration can change genetic frequencies, since it represents the arrival or departure of organisms possessing one genotype at a greater rate than others in the population, and thus changes their proportional representation in the gene pool. Probably this is not an important mechanism of genetic change in the history of most populations.

GENETIC DRIFT

Genetic frequencies can remain constant only if a population is fairly large. Otherwise, random events will tend to cause changes. If a population consists of only a few individuals, predators could destroy the only representatives of a particular genotype and miss the others purely at random. Such an event would be most unlikely in a large population. Then, too, it is a matter of chance whether a particular genotype will be represented in those gametes that do manage to unite in fertilization—a chance that depends upon the random distribution of maternal and pa-

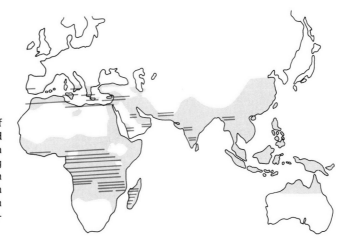

Figure 33–7 The distribution of sickle-cell anemia (*bars*) compared with the distribution of malaria (*shaded region*). Probably originating in East Africa, this mutation is an advantage wherever there is a high incidence of malaria, but it is an unmitigated disadvantage elsewhere.

ternal chromosomes in meiosis. The production of changes in gene frequency by random events is known as **genetic drift.**

Even if a gene were of no particular advantage or disadvantage to an organism, its frequency could change by genetic drift, and it can be demonstrated that this process will continue until eventually only one of the alternative alleles is present in the population. (This is called **fixation.**) Genetic drift is a hazard in small populations of organisms with limited gene pools that are in danger of extinction. It results in a loss of genetic variability, which tends to reduce the versatility of organisms, so they may become more susceptible to stressful changes in their environment.

FOUNDERS AND BOTTLENECKS

If a new habitat is colonized by a small number of organisms, the only genes that will be represented among their descendants will be those few that the founders chanced to possess. Thus, isolated populations may have very different gene frequencies than those characteristic of the species elsewhere. The disproportionate effect exerted upon a population by a limited number of ancestors is termed the **founder effect.** Like genetic drift, it can produce great changes in gene frequency even in the absence of natural selection (Fig. 33–8).

In some species, very few individuals survive some critical stage of their life cycle. For example, among houseflies in northern areas, only a few survive the winter, and they give rise to most of the summer population. In principle this is similar to the founder effect. Only a few individuals, which are perhaps not truly representative of the genetics of the population from which they came, will give rise to the entire future population. They will by chance exert a disproportionate influence over its prospective genetic frequencies. Since we can think of this phenomenon as a periodic squeezing out of some of the genes in a gene pool in random fashion, it is termed the **bottleneck effect.**

Like genetic drift and the founder effect, the bottleneck effect can change gene frequencies even in the absence of natural selection, usually by greatly reducing the genetic variability of a population. It has been

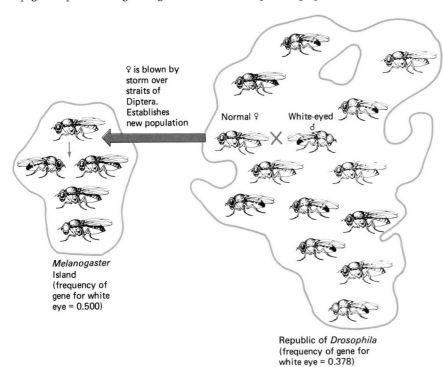

♀ is blown by storm over straits of Diptera. Establishes new population

Normal ♀   White-eyed ♂

Melanogaster Island (frequency of gene for white eye = 0.500)

Republic of *Drosophila* (frequency of gene for white eye = 0.378)

Figure 33–8 Founder effect. In this instance, a single female fruit fly is blown to an island. Even though not typical of the population of fruit flies as a whole, her genes and those of her mate will serve as the foundation for the entire gene pool of all fruit flies living on that island. Their gene frequencies will not be the same as those of the continental fruit flies. In such cases, there is usually a greater range of variation among the ancestral population than among the island group.

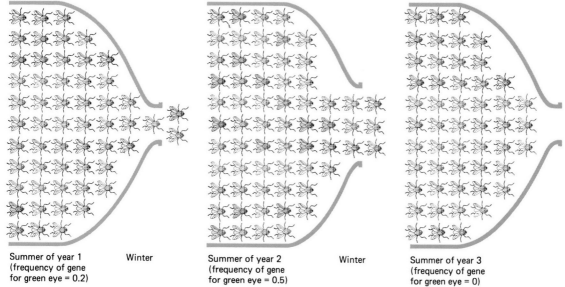

Summer of year 1
(frequency of gene
for green eye = 0.2)

Winter

Summer of year 2
(frequency of gene
for green eye = 0.5)

Winter

Summer of year 3
(frequency of gene
for green eye = 0)

Figure 33–9 The bottleneck effect. An occasional or periodic crisis in the history of an organism that greatly reduces its population may have results similar to those of the founder effect. Again, descendant populations will have gene frequencies typical of their ancestors, but these may not be typical of the larger populations from which their ancestors came.

shown, for example, that the genetic variability of African cheetahs is very low. This is taken to indicate that the species came close to extinction in fairly recent times, perhaps 3,000 years ago (Fig. 33–9).

To sum up, despite the Hardy-Weinberg law, gene frequencies are always changing. Panmixis seldom prevails, mutation and natural selection never rest, and many collections of organisms in nature are divided into small, more or less isolated gene pools separated by geographical barriers. In most instances, then, microevolutionary changes in genetic frequencies are inevitable in natural populations of organisms. Like Darwin himself, Darwin's modern followers emphasize the role of natural selection in bringing about these changes.

Examples of Selection

In recent years many examples of natural selection have come to light in the laboratory or in the field, affording excellent opportunities to study this evolutionary mechanism in action. Among these are the development of DDT resistance in insect populations exposed to this pesticide, industrial melanism in moths (described in Chapter 1), and the development of resistance to drugs among disease bacteria. (See Focus on Natural Selection in Bacteria.)

PROTECTIVE COLORATION

A subtle example of natural selection is the development of **protective coloration,** coloration that permits an organism to blend with its surroundings. This protects it from its predators or, in the case of a predator, keeps the victims from noticing it until too late.

Many examples of protective form and coloration readily come to mind (Fig. 33–10). Stick insects resemble sticks so closely that you would never guess that they are animals—until they start to walk. The chicks of ground-nesting birds are usually colored to blend in with the surrounding weeds and earth so that they cannot be discerned from a distance. Some leafhoppers resemble leaves not only in color but in the pattern of veins in their wings. There are praying mantids that resemble flowers—pity the poor bee that comes to visit! Crab spiders are often colored like the flowers they inhabit; in fact, they seek out flowers whose

Focus on . . .

NATURAL SELECTION IN BACTERIA

Does evolution proceed by the direct and directive effect of the environment upon genes? If so, the genes should change in all bacteria subjected to an environmental stress in such a way as to adapt them to that stress. To test this prediction, a series of experiments of a type first developed by Joshua and Esther Lederberg may be performed.

1. A dish is filled with jelly-like culture medium, and bacteria are grown upon it. The carpet of bacteria may contain a few clones of individuals that are already suited by mutation of the appropriate genes to resist a given antibiotic. The vast majority, however, are not resistant.

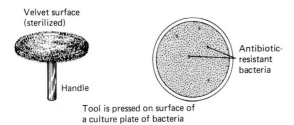

Velvet surface
(sterilized)

Handle

Tool is pressed on surface of
a culture plate of bacteria

Antibiotic-
resistant
bacteria

2. Since any such mutants cannot be detected by inspection, it is necessary to discover them by means of their antibiotic resistance. A sterilized velvet carrier is placed in contact with the original plate, and bacteria of all kinds adhere to it.

Pressed on
fresh medium

3. When the carrier is placed on a series of fresh plates *containing antibiotic*, only representatives of the original antibiotic-resistant clones will survive. These were present in the original culture plate, and did not develop in the antibiotic plates, since the resistant colonies always occur in the same position in all the replica plates. This can only reflect their *original* distribution in the *antibiotic-free* original plate. Due to natural selection these antibiotic-resistant bacteria survive and reproduce, generating whole populations of bacteria resistant to the antibiotic.

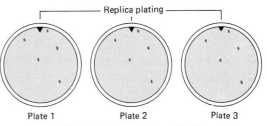

Replica plating

Plate 1   Plate 2   Plate 3

All replica plates contain antibiotic

4. If the antibiotic *stimulated* bacteria to become resistant to it, one would find a random and probably much denser pattern of colonies on all replica plates. Since this result is not observed, Lamarckian evolution would seem not to occur in these bacteria.

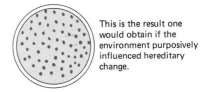

This is the result one
would obtain if the
environment purposively
influenced hereditary
change.

---

colors are similar to their own. Evidently, such protective coloration has been preserved and accentuated by means of natural selection. How might these adaptations originate? (See Chapter 1, pages 10 and 11.)

MIMICRY

Some organisms closely imitate the appearance of models to which they are unrelated. This resemblance is termed **mimicry.** One type, **Batesian mimicry,** is the resemblance by a harmless or palatable species to one that is dangerous, obnoxious, or poisonous (Figs. 33–11 and 33–12). A harmless moth may resemble a bee or wasp so closely that even a biologist would hesitate to pick it up. Likewise, many butterflies mimic the monarch butterfly, which birds avoid because it is toxic, having fed as a caterpillar on the poisonous milkweed plant. Its imitators resemble it in color but are nonpoisonous to birds.

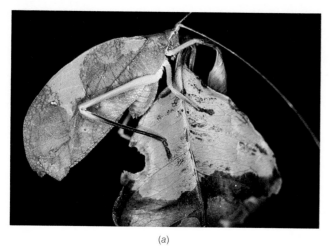

(a)

(b)

Figure 33–10 Camouflage conferred by protective coloration. (a) Leaf-mimicking katydid from Peru. The coloring of this insect matches the mottling of the partly dead leaf it rests on. Protectively colored insects tend not to rest in surroundings that do not resemble their color patterns. (b) As with many ground-nesting birds, in their natural surroundings the chicks of nighthawks are almost invisible both to us and to sight-hunting predators. (a, James L. Castner.)

Apparently, natural selection has maintained a resemblance that gives the mimic almost as much protection as the model, for as soon as predators learn to associate the distinctive markings of the model with its undesirable characteristics, they tend to avoid all similarly marked animals.

Genetic Recombination

Almost all members of a species have some means of transmitting genes to one another, or from one generation to another, in such a way that new combinations of genes can result. This is true even of viruses and bacteria. Without genetic recombination offspring would resemble their parents precisely, and the only means of genetic change from one generation to another would be mutation. Even favorable genes would come to pre-

(a)

(b)

(c)

Figure 33–11 Batesian mimicry. (a) At the right is Jordan's salamander, *Plethodon jordani*, a distasteful species. To its left is *Desmognathus imitator*, a palatable species (at least to its predators). (b) Few people would want to get close enough to this insect to discover that it is actually a moth (note the mothlike antennae). (c) A genuinely noxious insect, the golden paper wasp. (a, E.D. Brodie, Jr., Adelphi University/BPS. b, L.E. Gilbert, University of Texas at Austin/BPS. c, P.J. Bryant, UC–Irvine/BPS.)

Figure 33–12 Mullerian mimicry. These various lepidopterans are all unpalatable. Their similar coloration works to their mutual advantage since it makes it easier for potential predators to learn their common warning coloration than if each species had its own distinctive pattern. (L.E. Gilbert, University of Texas at Austin/BPS.)

dominate in a population of such organisms much more slowly than if they could reproduce sexually. Often new *combinations* of genes are more favorable than old, resulting in phenotypes that no one of the members of the combination could produce by itself. When two genetically uniform inbred strains of organisms are crossed, their offspring display **hybrid vigor;** that is, they are usually more vigorous than the parents.

THE EVOLUTION OF SPECIES

Clearly, it is not enough to speak of mutations and the change of gene frequencies as accounting for all the major differences among organisms. By themselves, these processes tell us little about the mode of formation of new species and still less about the production of higher taxonomic categories.

Speciation

A **species** is a reproductively isolated population of organisms that have a common ancestry and, in nature, breed only with each other. The key to the development of a species, or **speciation,** is the development of its reproductive isolation so that it cannot interbreed with other species. It is thought that there are two main ways in which such a situation may arise—through allopatric or sympatric speciation.

    Allopatric speciation is the development of new species through the chance effects of long physical isolation and independent development. For example, geographical barriers may separate populations isolated on islands or mountaintops or in streams belonging to separate watersheds. Recent studies indicate that most species are composed of populations that are more or less isolated from one another and among which there is little or no gene exchange. Allopatric speciation could take place easily among such populations, producing several species from one, especially where geographical barriers accentuate the separation. It is easy to see that the forces of natural selection and genetic drift could not be the same in any two locales. Given enough time, sufficient genetic differences may develop so that interbreeding cannot take place even if the two populations should come into contact with one another again.

Figure 33-13 Blood lust among fire-flies. The firefly on the right is a female of the genus *Photuris;* being tightly grasped is a male of the genus *Photinus.* This embrace will be his last, for she is gnawing through his neck, having lured him from afar by imitating the light signals of a female of his own species. Courtship rituals have evolved among many animals as a mechanism of reproductive isolation; that is, they help ensure that mating takes place only among members of the same species, preventing genetic exchange among species. (James E. Lloyd.)

In **sympatric speciation** two populations occupy the same territory. Sometimes sympatric species can be detected only because members will not interbreed. How might this come about?

Imagine two diverse habitats in the same location. A versatile organism is able to live in both. However, certain mutants are capable of living in one place—call it habitat A—better than the normal members of the population. It is obviously to their advantage if no genes from the parent population find their way into the gene pool of this new group, which we will call population A. Meanwhile, a population B is developing, specialized for life in habitat B. It is to their mutual advantage that the two populations not interbreed. Natural selection can be expected to favor those genotypes that make interbreeding impossible, so adaptations that prevent or minimize it will develop. It is thought that this does not happen very often.

Isolating Mechanisms

Whether as the cause or the result of speciation, differing but related species of organisms have various mechanisms whereby they achieve reproductive isolation from one another. These are known as **isolating mechanisms.**

Isolating mechanisms that interfere with mating are known as **pre-zygotic isolating mechanisms** because they act *before* the zygote can be formed and in fact prevent its formation. For instance, fruit flies exhibit a definite courting behavior that is species-specific. Unless both partners behave in just the proper instinctive fashion, they will not mate successfully. Part of the behavior has been shown to be a love song—a series of buzzes of just the right pitch and rhythm performed by the male. If he does not have the right beat, the encounter ends right there. These differences in song seem to be all that keeps some *Drosophila* species apart. Even geneticists often cannot distinguish some species in any other way, and apparently the flies cannot, either. Similar though more elaborate mechanisms isolate various species of crickets, birds, and many other animals that have instinctive, highly stereotyped patterns of courtship behavior that must be performed precisely for mating to take place (Fig. 33-13).

Probably the most obvious prezygotic isolating mechanisms are physical barriers to mating—gross inappropriateness of structure, such that reproductive parts simply will not fit each other. But even among species that thus differ physically, an attempt at mating rarely actually takes place. The effort, often a considerable and even dangerous effort, would be wasted.

Currently, two kinds of **postzygotic isolating mechanisms** are known. One is embryonic lethality, in which the zygote or embryo does not develop properly and is aborted. The other, hybrid sterility, is a more subtle mechanism. In this case two species can mate and the offspring may be healthy or even unusually vigorous but sterile, and in the long run genetic mixture between the two species does not truly take place. One example of hybrid sterility will occur to you—that of the mule, the sterile offspring of a horse and a donkey.

It is clearly most advantageous to the organisms concerned to employ prezygotic isolating mechanisms, operating as early in the courtship, mating, and reproduction sequence as possible. Courtship tends to consume time and energy and can even be dangerous. Pregnancy, nest building, and territorial defense may require a large portion of the organism's life span. It is adaptive to prevent the initiation of any of these processes if gene propagation cannot result from them. Probably for this reason most animals belonging to differing species hardly give one another a second glance, since the stimuli that release courtship behavior are likely to be quite different. The brilliantly colored buttocks of the estrous female chimpanzee, for instance, are highly attractive to the male chimpanzee but hardly appealing to the human male.

Cases are known (for example, lion-tiger hybrids) in which the offspring of an interspecies cross are fertile. Presumably, one might argue that the two species are not "really" separate, but common sense is against that. It seems more reasonable to view them as truly separate species, kept genetically isolated *in a state of nature* by a combination of behavioral and ecological mechanisms that can be induced to break down in captivity.

Stabilizing Selection

Toward the end of the 19th century an ornithologist named Hermon Bumpus collected a group of sparrows killed in an exceptionally severe snowstorm. He compared these dead birds with those that survived the storm, using nine characteristics, such as wingspread and body weight, as the bases of his comparison.

Bumpus discovered that the dead birds tended to be abnormal; that is, they represented the extreme ends of the normal range of variation in a sparrow population. Bumpus concluded that there is a more or less standard body build suitable for a bird with the life-style of a sparrow. Although extreme deviants from this standard may do well when conditions are not rigorous, unusual stresses, such as a blizzard, periodically tend to weed out the unsuitable phenotypes, with the genotypes that produce them.

Though Bumpus's data have been subjected to several reanalyses (one cannot duplicate the observations until a comparable blizzard bird-kill recurs), his basic thesis seems to stand. The phenomenon is known as **stabilizing selection.** It is, in a sense, antievolutionary: It tends to maintain a standard phenotype in a population of organisms. In other words, stabilizing selection is genetically homeostatic as long as the environment of an organism does not undergo long-term change, for ordinarily, natural selection will tend to stabilize the genetic composition of populations. Should the environment change, however, or should the organism find itself able to expand its range into a new kind of environment or ecological niche, then (and only then) natural selection might produce evolutionary changes.

CHAPTER SUMMARY

I. Organic evolution is the theory that species change gradually over the course of their history and that they are genetically related to one another. Microevolution refers to the change of gene frequencies within populations. Macroevolution refers to the concept that all life commonly descended from one or a few ancestors and is therefore genetically related.

II. Recognizable evolutionary theories in the modern sense were first propounded in the 18th and 19th centuries. The principal theory that has survived to the present day was proposed by Darwin and Wallace.

A. Darwin and Wallace did not originate the concept of evolution but proposed a plausible mechanism whereby it might take place.

Focus on . . .

PUNCTUATED EQUILIBRIUM

In his 1940 book, *The Material Basis of Evolution,* Goldschmidt proposed that extreme mutations, **macromutations,** might take place from time to time. These would produce nothing so trivial as a changed number of bristles on a fruit fly's posterior or an altered enzyme with a new temperature optimum. Goldschmidt pointed out that just one or a few changes in genes governing early development could produce proportionately large results. Were he alive today, Goldschmidt might point to recent studies showing that we and chimpanzees have more than 90% of our protein amino acid sequences in common (probably 99%, in fact), and since genes make proteins, we must have most of our genes in common. The differences between the two species must rest on a few key genes—**regulatory genes.** Such genes presumably would regulate such things as the relative rate of growth of brain parts, limb bones, and the forehead. In principle, then, genetic engineering might make something virtually human out of a chimpanzee with rather little effort. Goldschmidt thought that something like this may have taken place naturally, and he called such a major evolutionary leap **saltation.**

However, even today no serious evolutionist would argue, as Goldschmidt did, that the first bird hatched from a reptilian egg. Nevertheless, a view somewhat similar to Goldschmidt's saltation theory has gained some credence of late; it is called **punctuated equilibrium.** This theory suggests that long periods of no change in populations are punctuated by periods of rapid speciation. According to the promoters of this view, the major events of evolution, and many minor ones as well, occurred in isolated, out-of-the-way habitats, such as islands or glacially isolated terrain. These changes resulted from the action of gradual natural selection on the process of genetic drift, the founder effect, and similar mechanisms fostered by intense inbreeding in isolated populations.

Other theorists blend aspects of gradualism—classical Darwinism—with a form of modern catastrophism. A sudden change in environment, whether of terrestrial or extraterrestrial origin, could decimate the populations of otherwise successful species while affording surviving organisms an opportunity to radiate into newly vacated niches. Such is the scenario whereby mammals might have replaced reptiles as the dominant large land animals. Such sudden changes might have occurred during other periods of time for which evolutionary change was rapid and the evidence of transitional forms is poor.

Natural selection, according to the different views we have discussed, would come into play mainly when isolating barriers were broken down or existing niches were left open to exploitation. At that point the evolutionary novelties would be loosed on the world. If their bizarre new combinations of traits rendered them adaptively inferior to their more normally endowed cousins in the outside world, nothing more would be heard of them. If they were *superior,* however, they would swiftly take over any available ecologic niche. Thus, punctuated equilibrium *predicts* the very gaps in the fossil record that have continued to bother scientists since Darwin's time. If the major changes took place with unusual rapidity in small isolated areas, then only a very few if any transitional forms would be represented in the fossil record.

B. Their view comprised four main points: overproduction of offspring, variation among those offspring, inheritability of that variation, and natural selection of the variants.

III. The modern concept of evolution continues to be based upon the idea of natural selection but emphasizes changes in gene frequencies within populations.

A. Such changes will not take place, according to the Hardy-Weinberg law, if mating is at random (panmixis), if there is no selection against a genotype, if there is no selective immigration or emigration, and if the population is relatively large.

B. Since these conditions are rarely met, changes in gene frequency usually do take place in most populations of organisms over time.

IV. An intrafertile population of organisms that does not exchange genes with other populations is termed a species. The development of this genetic isolation is called speciation.

A. Allopatric speciation takes place when populations are separated by geographical barriers.

B. Sympatric speciation occurs when two populations of an organism occupy the same territory and become specialized for different ecological niches.

V. Stabilizing selection tends to fix genetic frequencies within a population and to preserve a standard type of organism if the environmental conditions remain the same. If environmental conditions change systematically, however, genetic frequencies will tend to change also.

Post-Test

1. The basic meaning of the term evolution is simply _____ , but _____ evolution specifically implies the gradual development of complex organisms from simple beginnings.
2. Lamarck proposed a theory of evolution based upon the supposed inheritance of _____ traits.
3. Darwin served as a _____ aboard the oceanographic ship *Beagle.* On that voyage he observed that certain species of birds called _____ _____ in the Galapagos Islands appeared to be related.
4. _____ _____ is the production of changes in allele frequencies by random events.
5. In the bottleneck effect, genetic diversity is _____

_____ by temporary but extreme population _____ _____ , such as near-extinction.
6. The process of _____ formation necessarily involves the development of _____ isolation so that gene flow between populations becomes restricted.
7. Species that coexist geographically are said to be _____ .
8. The commonest and most efficient isolating mechanisms are _____ . Usually they prevent even the initiation of _____ .
9. The _____ _____ view of evolution suggests that evolution often proceeds by abrupt, major steps.

Review Questions

1. Contrast Lamarck's and Darwin's evolutionary theories.
2. What is meant by *panmixis?* What relation does the concept of panmixis bear to the concept of a species?
3. What bearing do dominance and recessiveness have upon change in gene frequencies? List the factors that can be expected to change the frequencies of genes within a population of panmictic organisms.

4. What mechanisms of speciation have been proposed?
5. Propose a mechanism whereby DDT resistance in mosquitoes could develop.
6. What is punctuated equilibrium? What evolutionary problems is it intended to solve?
7. What is genetic drift? Give an example.
8. Why is it that only inherited changes are important in the evolutionary process?

Readings

Bonnell, M., and R.K. Selander. "Elephant Seals: Genetic Variation and Near Extinction," *Science,* 24 May 1974. A possible example of the bottleneck effect.

Carson, H.L., P.S. Nair, and F.M. Sene. "*Drosophila* Hybrids in Nature: Proof of Gene Exchange Between Sympatric Species," *Science,* 5 September 1975. Speciation in progress?

Dobzhansky, T., F.G. Ayala, and G.L. Stebbins. *Evolution.* San Francisco, W.H. Freeman & Company, 1977. One of the very best summaries of the action of the genetic mechanisms of evolution.

Gould, S.J. "Darwinism and the Expansion of Evolutionary Theory," *Science,* 23 April 1982. A more complete but also somewhat more difficult discussion of the recent changes in evolutionary thinking.

Karp, L.E. "The Immortality of a Cancer Victim Dead Since 1951," *Smithsonian,* March 1976. The ubiquitous HeLa cell line, ideally adapted to laboratory conditions, has tended to contaminate cultures of other cells and displace them, so the most stringent measures must be taken to protect tissue cultures from these useful alien invaders. A bizarre example of natural selection in an unnatural environment.

Mayr, E. "Darwin and Natural Selection," *American Scientist,* May–June 1977. How did the concept of natural selection ever occur to Darwin?

Nijhout, H.F. "The Color Patterns of Butterflies and Moths," *Scientific American,* November 1981. A study of the development of the more than 100,000 wing patterns of butterflies and moths.

Savage, J.M. *Evolution,* 3rd ed. New York, Holt, Rinehart & Winston, 1977. A brief college-level paperback.

Stanley, S. *Macroevolution: Pattern and Process.* San Francisco, W.H. Freeman & Company, 1979. Does speciation and evolution of higher categories proceed in fits and starts? An exposition of punctuated equilibrium.

Wiens, J.A. "Competition or Peaceful Coexistence?" *Natural History,* March 1983. Sometimes there may be less competition between species than evolutionists believe.

Chapter 34

EVOLUTION: ORIGINS AND EVIDENCE

Outline

I. The origin of life
   A. Origin of organic molecules
   B. Origin of cells
   C. Origin of eukaryotes
   D. Human evolution
II. Evolutionary evidence
   A. Microevolution
   B. Morphological resemblances
   C. Biochemistry
   D. Genetics
   E. Biogeography and distribution
   F. Structural parallelism
   G. Adaptive radiation
   H. Dating methods and the fossil record

Learning Objectives

After you have read this chapter you should be able to:

1. Summarize the general features of the current theory of the spontaneous origin of macromolecules and other organic compounds necessary for life's beginning.
2. Summarize the endosymbiotic theory of the origin of eukaryotic cells.
3. List and critically discuss the various types of evidence bearing upon evolution, specifically (1) microevolution, (2) comparative anatomy, with emphasis on homology, (3) biochemistry, (4) genetics, (5) distribution, and (6) the fossil record.
4. Summarize the events in the history of life that are inferred from geological evidence.
5. Summarize the relationships among the members of the human family tree, based on the fossil record.

We now leave the discussion of evolutionary mechanisms to consider the broader question of the history of life on our planet. Here we enter a realm more of inference than of direct observation, a realm in which there has always been more room for debate, discussion, and even controversy.

THE ORIGIN OF LIFE

As discussed in Chapter 4, the principle that all organisms arise from living parents is firmly established in biology today. Yet according to evolutionary theory, life was generated from nonliving molecules, that is, by **abiogenesis.** This apparent paradox is usually explained by the assertion that conditions on earth were far different billions of years ago, when life first began to evolve. Then, once they came into being, living organisms changed the conditions of their environment, so abiogenesis is no longer probable on most parts of the earth.

Origin of Organic Molecules

Most theorists believe that the earth's primitive atmosphere consisted of a mixture of nitrogen, carbon dioxide, methane, ammonia, hydrogen, and water vapor. These constituents could have reacted with one another in the presence of volcanic heat, lightning, or ultraviolet light (energy sources) to form rather complex organic compounds—amino acids, carbohydrates, and even nucleic acids. Investigators have constructed experimental models of the supposed conditions of the primitive earth, (Fig. 34–1), and when the inorganic substances mentioned above are permitted to react under these conditions, a surprising array of organic

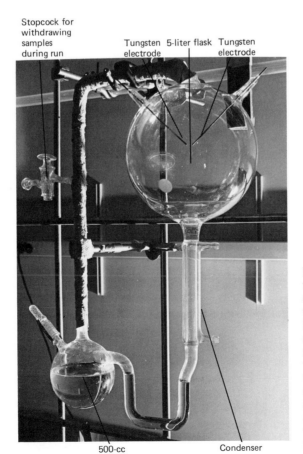

Stopcock for withdrawing samples during run

Tungsten electrode 5-liter flask Tungsten electrode

500-cc flask   Condenser

Figure 34–1 Apparatus similar to that used by Stanley Miller and Harold Urey in an attempt to replicate what they thought conditions might have been in the earth's primitive atmosphere. The products of the reactions produced by the sparking of the electrodes in the flask on the upper right accumulated in the smaller flask on the lower left. When analyzed, some of the organic chemicals that are characteristic of living things were found in that mixture. (Stanley L. Miller, courtesy of University of California, San Diego.)

compounds form, including some macromolecules—that is, giant molecules, such as protein fragments—together with an assortment of materials that play no part in any known living thing.

The composition of the atmosphere of the early earth is not known with any certainty, and there is some controversy concerning even such important points as how much oxygen, if any, may have been present. Some theorists even believe that the initial biogenetic events took place extraterrestrially, in clouds (such as those now covering the planet Venus), or in subterranean volcanic chambers. The majority, however, side with Darwin, who thought that life probably began in something like a warm little pond.

Experiments constructed in accordance with these different assumptions yield varying results, as might be expected. But although the chemicals produced by these experiments may vary greatly from what we find in life today,[1] the very fact that the raw materials of life can come into existence without deliberate plan indicates that deliberate organic synthesis is not necessarily required for the production of complex biochemicals. The real problem (not yet solved) lies in the assembly of *some* of those biochemicals into a living cell, however primitive.

Origin of Cells

If organic molecules such as those we have been discussing could form spontaneously under present-day conditions, they would quickly be consumed by bacteria or slowly oxidized by oxygen. But neither bacteria nor oxygen is thought to have existed on the earth at that time. Macromolecules could have arranged themselves to form tiny bubble-like structures called **coacervates,** which have been observed to form in laboratory experiments. These coacervates possess catalytic activity and share some properties of living cells. Nucleic acids (especially RNA) might have duplicated themselves, and eventually by chance some could have developed the ability to synthesize needed enzymes. Natural selection would, of course, have favored those aggregates that developed such abilities.

The main difficulty with this proposal is that if structures resembling living cells came about by chance, they could not reproduce themselves unless they possessed nucleic acids containing the necessary genetic information to be passed on to their descendants, some way of replicating those nucleic acids, and some way of expressing them. In modern forms of life, all these processes require the presence of very specific enzymes, enzymes that themselves require information deposited in the nucleic acids for their production. We are thus left with a neat chicken-and-egg dilemma, for it would seem that the enzymes could not have existed without the nucleic acids, nor could the nucleic acids have existed without the enzymes. One proposed solution is that the nucleic acid RNA can possess some catalytic enzyme-like properties and might not have needed the accessory enzymes that modern forms of RNA require. In the beginning, in other words, was RNA; only later did DNA and the accessory proteins evolve. This still leaves many questions unanswered.

The first living things must have been heterotrophs, nourishing themselves on the pre-existing organic molecules around them. Once the supply of organic molecules (especially ATP) began to diminish, however, survival would depend on the ability to capture energy from organic

---

[1] For instance, modern living things employ only the left-handed (1 −) versions of amino acids, and employ only 20 of them. Most experiments of the type described here produce equal mixtures of left- and right-handed amino acids, and many more kinds of them than occur in modern living things.

molecules (glycolysis) and finally, from sunlight (photosynthesis). Once photosynthetic producers evolved, only then could oxygen eventually be produced, making possible the development of the modern aerobically respiring class of consumer organisms.

Origin of Eukaryotes

Evolutionists think the earliest cells must have been the simplest, and the simplest known cells today are prokaryotes. You will recall from Chapter 4 (see also Chapter 14) that prokaryotic cells lack nuclear membranes as well as other membranous organelles, such as mitochondria, ER, and chloroplasts. But how might these structures have developed in the ancient prokaryotic ancestors of the eukaryotes?

Actually, two ways have been proposed in which the original eukaryotic cell could have evolved from prokaryote precursors. The older view hypothesized that membranous organelles arose by multiple inpocketing and infolding of the cell membrane in an ancestral prokaryote. More recently set forth, the **endosymbiotic theory** suggests that mitochondria, chloroplasts, and perhaps even centrioles and flagella may have originated from a cooperative union among prokaryotes. Thus, mitochondria are seen as former bacteria, and chloroplasts as former cyanobacteria (Fig. 34–2). The theory further stipulates that each of these partners brought to the union something the others lacked. For example,

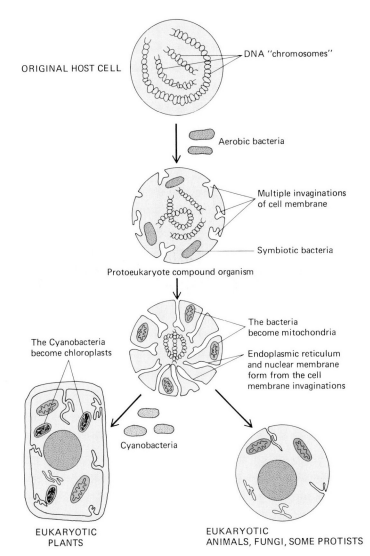

ORIGINAL HOST CELL

DNA "chromosomes"

Aerobic bacteria

Multiple invaginations of cell membrane

Symbiotic bacteria

Protoeukaryote compound organism

The Cyanobacteria become chloroplasts

The bacteria become mitochondria

Endoplasmic reticulum and nuclear membrane form from the cell membrane invaginations

Cyanobacteria

EUKARYOTIC PLANTS

EUKARYOTIC ANIMALS, FUNGI, SOME PROTISTS

Figure 34–2 Endosymbiotic theory of the origin of the eukaryotes. According to this view, the primitive ancestral eukaryote (called the *Urkaryote*) is invaded by bacteria or cyanobacteria, which confer on it the ability to respire aerobically or perform photosynthesis or both. The nuclear membrane originates from the endoplasmic reticulum.

mitochondria provided the ability to employ oxidative metabolism, which was lacking in the original host cell, and writhing spiral bacteria provided the ability to swim, eventually becoming flagella.

The principal evidence in favor of the endosymbiotic theory is that mitochondria and chloroplasts possess *some* (but not all) of their own genetic apparatus distinct from that of the cell's nucleus. Thus, they have both their own DNA and their own ribosomes. Chloroplasts have as much DNA as the average virus particle. Also, the DNA and ribosomes of these organelles are similar to those found in prokaryotes. Moreover, a group of anaerobic and presumably primitive bacteria, the **archaebacteria,** possess sequences of nucleic acid bases that could be similar to those of the ancestors of both the "host" and the endosymbiont prokaryotes; that is, their base sequence can be seen as intermediate between that of nuclear and mitochondrial DNA.

Biologists agree that the two most momentous events in the evolutionary history of life on earth were the origin of cells an estimated 3.5 billion years ago and the origin of complex multicellular animals and plants about 700 million years ago. Geological evidence seems to indicate that about 70% of the history of life (and half the duration of planet Earth) unfolded between these two events. For most of this vast span of time, little can be found in the fossil record but the most primitive fossils of prokaryotes resembling bacteria or cyanobacteria. Only during comparatively recent time is there evidence for eukaryotic cells with their characteristic makeup and properties: mitosis, meiosis, mitochondria, chloroplasts, and multicellularity. Presumably, the majority of evolutionary experiments would have been discarded in favor of the familiar varieties that survived.

It is difficult to summarize the geological record of the history of life (see Table 14–3, Life and the Geological Timetable). It is a fossil record, which means that what we are able to study today is not the total community of things living at a given past time, but only those in which accidents, often very freakish accidents, have allowed specimens of organisms to be preserved. In some cases the organism died in an aquatic environment, or was washed into one, but did not decay because of special conditions, such as a lack of oxygen in the muddy bottom sediments. Eventually, perhaps, its remains were entombed in the sediments, which became sedimentary or metamorphic rocks under subsequent heat and pressure.

In other instances the organism decayed but left a mold or impression of its body, which was subsequently filled with other material. In still others its remains were replaced by other kinds of mineral entirely, by a slow process of mineralization. In a few instances extinct organisms were frozen in Arctic permafrost. Usually the internal details of fossils are lost, but sometimes the preservation is so complete that even cellular structure can be studied.

One assumes that the upper layers of fossil-bearing rocks are younger than those beneath them (the **law of superposition**). That permits scientists to establish the relative ages of the organisms entombed in those sediments. Not only may organisms fail to be fossilized, but also subsequent events (such as erosion) may remove the rocks in which they occur. Even if this does not happen, most fossil-bearing rocks are out of our reach. We tend to discover fossils only as they are exposed by weathering of the rocks that contain them. It is not surprising that there are great gaps in our knowledge, but the biological conclusions of historic geology are, in general, as follows:

1. The very oldest formations, formerly thought to be essentially devoid of life, contain microfossils of prokaryote and (in their

Figure 34–3 Blue-green algae form colonies called stromatolites. Stromatolites like this from the Draken formation, Spitsbergen, Norway, appear to date from before the origin of eukaryotic life (about 800 million years old). (A.H. Knoll.)

younger parts) possible unicellular eukaryote organisms. They also contain the formations called **stromatolites,** cyanobacterial aggregations found today only in certain restricted localities (Fig. 34–3).

2. The formations next in age contain the simplest plants and very simple invertebrate animals, which may lack such internal structures as gut cavities (most or all such organisms are unknown today).

3. The next series of fossil communities includes jawless fish (of which the modern hagfish and lampreys are survivors); more complex invertebrates, such as trilobites; and ammonite mollusks and the like. But not all of these are necessarily in the same sediments. There is also some evidence of the presence of primitive land plants.

4. The next group of organisms includes the jawed vertebrates (including amphibians) and the insects. Vascular plants also appear in the fossil record. Many or most of the earlier forms persist.

5. The reptiles, and very soon thereafter, early mammals appear next in the fossil record. Seed plants, first gymnosperms and then angiosperms, occur in the same geological formations.

6. With the demise of the larger reptiles (dinosaurs), mammals and birds become dominant among vertebrates and very diverse.

Human Evolution

In Darwin's day there was little fossil evidence to apply to the question of human evolution. Darwin and his forerunners based their thinking principally on what we today would call comparative anatomy, discovering the closest relationship between humans and anthropoid apes. Some cited orangutans as the nearest living human relatives, others, chimpanzees—a debate that continues to this day. Still it seemed there was no fossil that could be classified as a clear link between human and apelike ancestor. The succeeding century has seen many such apparent missing links unearthed, together with some that, though human-like, do not seem to be links at all.

In 1855 the British anatomist Richard Owen proposed an argument that he believed would show that humans and the great apes were *not* related. With the exception of the orangutan, all apes possess a heavy ridge of bone above the eyes, the **supraorbital torus.** This occurs rarely in humans, and then only feebly. There is no obvious reason why such a ridge would be a disadvantage to humans, he reasoned. If apes were indeed the ancestors of humans, the torus should still be present in either living or dead varieties of humanity.

Unfortunately for Owen's choice of argument, quarry workers discovered the bones of a human being in a cave in the Neander valley of

Figure 34–4 Neanderthal skull. Note the very heavy ridge of bone, called the supraorbital torus, above the eyes and the protruding face. The brain size was, if anything, greater than that of modern humans.

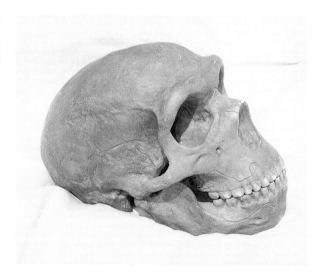

Figure 34–5 *Homo erectus.* (a) A replica of the skull. Note the massive bony ridges over the eyes, even greater than those of Neanderthal people. Also note the receding forehead, protruding jaw and absence of a chin. (b) A view of the femur of *Homo erectus* (*Pithecanthropus*), discovered by Eugène Dubois. The well-developed linea aspera (a long ridge that serves as a point of attachment for many hip muscles) indicates an erect posture for this hominid, as is reflected in its modern scientific name. The advanced bony tumor on the femur is possible evidence that sick *Homo erectus* hominids were aided by well ones. (a, photograph of Wenner-Gren foundation replica by David G. Gantt.)

Germany. The word for valley is *thal* in German, so this ancient gentleman became known to the world as **Neanderthal man** (Fig. 34–4). Among other distinctive features, Neanderthal man had a very heavy supraorbital torus. He also suffered from skeletal deformities, perhaps of arthritic origin, that would in life have produced a stooped, shambling, and somewhat ape-type gait. Although the discovery of numerous other Neanderthal skeletons has shown that these people stood as erect as we, the supraorbital torus has stood the test of time as a distinctive Neanderthal trait, along with a virtual absence of a chin and an extremely heavy skeleton generally. The size of Neanderthal man's brain was fully equal to ours, and probably ran a little larger on the average. However, the forehead was much lower than ours. Nevertheless, properly manicured and attired, Neanderthal man would probably not stand out in a shopping mall, although he would attract attention (not necessarily admiring) on a beach.

Neanderthal man was the first in a series of discoveries of fossil humans and others not clearly human that are now collectively known as **hominids.** The next putative human ancestor to be discovered was what we now know as **Homo erectus** (Fig. 34–5). That was not, however, its

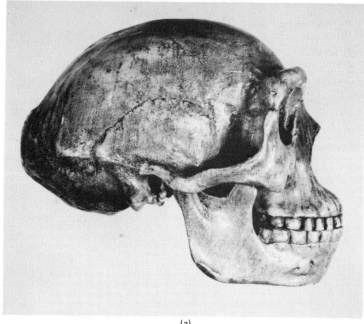

(a)

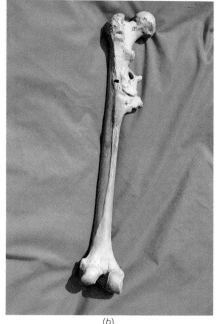

(b)

Figure 34–6 *Orangutan.* Dubois thought these asiatic apes to be more humanlike than any others, which impelled him to seek *Pithecanthropus* in Indonesia. (Tom McHugh, Photo Researchers, Inc.)

original name. (Unless the sex of a fossil hominid is relevant to the discussion, from here on it is referred to with a neuter pronoun.) Even before any actual bones had come to light, the German evolutionist Ernst Heinrich Haeckel in 1866 had proposed a scientific name for the yet-unknown hominid linking humanity and the apes. In Greek, ape is rendered *pithecos,* and human being, *anthropos.* An ape-man, then, would be *Pithecanthropus.* Haeckel, nevertheless, was not destined to discover the bones of his ape-man. This privilege was reserved by history to Eugène Dubois.

Although Dubois trained as a physician, his ambition was to discover *Pithecanthropus.* He reasoned that since, in the then-majority view, the East Indian ape known as the orangutan (Fig. 34–6) resembles humanity more closely than do the African apes, it was likely that *Pithecanthropus* would be found in the East Indies where orangutans occur. Dubois, a Dutchman despite his French name, joined the Dutch army and obtained an assignment to the East Indies (then a Dutch possession) as a military surgeon. Once there, he persuaded both the army and the colonial government to support his research, and in 1892 he unearthed the remains of what he was pleased to call *Pithecanthropus erectus.* There wasn't much left of it. *Pithecanthropus* consisted of a skullcap, a femur, and two teeth. The femur, except for what appeared to be a developing bone tumor, was very similar to a modern femur. The skull, as far as anyone could tell, would have had a cranial capacity of 855 cubic centimeters—very low indeed for an adult human. Yet the brain was far too large for any known ape.

Since Dubois's time, many fossils have been unearthed that seem to belong to the same general group as *Pithecanthropus.* Like *Pithecanthropus,* these occur in the East Indies and also in China, Africa, and even Europe. The skeletons of *Homo erectus* (as *Pithecanthropus* is now known) discovered since are much the same as those of modern humans—at least below the neck. However, the brain cavity is small and the jaw chinless. On the whole, *Homo erectus* was smaller than *Homo sapiens* but, like the Neanderthal, quite powerfully built. The intermediate traits of *Homo erectus* are almost entirely confined to the head, particularly the cranial cavity.

The next major discovery made in the investigation of humanity's possible origins was **Australopithecus,** a hominid whose investigation is the most actively pursued area of human paleontology even today.

Darwin and his colleagues had proposed that the origin of the human race was probably on the African continent, a suggestion that did not sit well with Dubois but did find favor with the Australian physician

Raymond Dart. Dart secured a position as a professor of anatomy at the University of Witwatersrand in South Africa, where the presence of a rich store of mammalian fossils was brought to his attention in 1924. Although the early fossil discoveries at the site, the Taung quarry, were in the tradition of Neanderthal man, quite a different creature was eventually found, a child with a dentition comparable to that of a modern human of perhaps 6 years of age. In due course, the skull was named *Australopithecus africanus.* Since there were no other specimens from which a more complete estimation of its skeletal anatomy could be gleaned, *Australopithecus* was initially of unclear significance and highly controversial. It was years before a better understanding of this hominid could be had; indeed, a great deal about it is still unclear.

The next chapter in the hominid story was written by Robert Broom, a South African physician and paleontologist who was much impressed by Dart's *Australopithecus* child. Broom's attention was drawn to another quarry, located near the village of Sterkfontein, especially two sites known as Kromdraai and Swartkrans. Here he found more *Australopithecus* material, including a fair amount of the postcranial skeleton, such as the pelvis and the upper half of one femur. The femur alone was enough to establish that *Australopithecus* could walk upright. Further work led to the conclusion that two, not one, species of *Australopithecus* were represented in Broom's collection. The second one is known today as *Australopithecus robustus* (Fig. 34–7). It has a generally heavier skeleton, an immense jaw, and dental evidence of a very coarse diet. It is considered to be a later comer than *A. africanus,* although the two species apparently coexisted for a considerable time. *A. robustus* is universally agreed as having died without evolutionary issue, a specialized animal more intelligent, perhaps, than modern chimpanzees, but of little more biological significance. *A. africanus,* whose brain size range also overlaps that of modern apes, is thought by many anthropologists to be directly ancestral to the genus *Homo,* including *H. erectus* and *H. sapiens.*

Louis Leakey, the child of missionary parents stationed near Nairobi, Kenya, was educated in England and received a degree in anthropology from Cambridge University. Leakey organized expeditions to Kenya and Tanzania initially in the 1920s. He was rewarded in a locale now famous throughout the world—the Olduvai Gorge in Tanzania. Expeditions by Leakey family members have unearthed a wealth of fossil hominid material, including, but not limited to, *Australopithecus.* Some of these were *A. africanus,* some were *A. robustus* (or, as Leakey called it, *Zinjanthropus*), and some were *Homo.* Of the *Homo* specimens, some

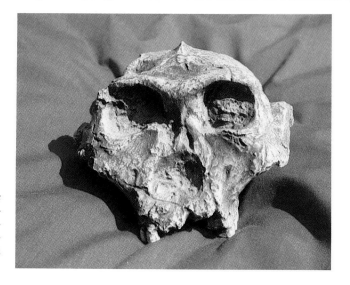

Figure 34–7 *Australopithecus robustus,* one of the first australopithecines to be discovered. This is evidently a specialized form, not considered ancestral to other known hominids.

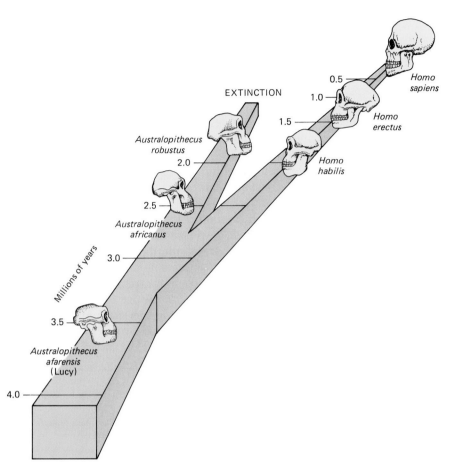

Figure 34-8 A hypothetical human family tree. Note the overlapping time periods of various species. It has been suggested that in many cases one hominid species was the direct cause of the extinction of others, but the reasons that *Australopithecus* and Neanderthal people are not still with us are unknown. (After Johanson, D.C., and T.D. White, *Science* 203: 321-330, 1979.)

had a brain size between those of *A. africanus* and *Homo erectus;* others were clearly *Homo erectus.* Leakey named the small-brained specimen *Homo habilis.* He felt that it was distinct from *Australopithecus,* and in this way he established, or attempted to establish, a human lineage that did not include *Australopithecus* at all. Had *Homo habilis* been discovered by another group of paleontologists, it probably would have been placed in the *Australopithecus* genus. In fact, something very like this was about to occur.

After a series of expeditions to the Afar triangle region of Ethiopia, Tim White and Donald Johanson announced in 1978 a new species of *Australopithecus* that, because of its age and to some degree its anatomy, they considered ancestral not only to other species of *Australopithecus* but to *Homo habilis* as well. The Leakey family agrees that this new hominid was probably ancestral to *Homo habilis* but does not see it as an australopithecine. They believe that the *Homo* and *Australopithecus* lineages have been distinct since perhaps the Miocene. As discoverers, however, Johanson and White have had the privilege of naming their fossil and have thus registered their opinion.[1] They called her **Australopithecus afarensis,** or more familiarly, just **Lucy.**[2]

The majority of archaeologists, anthropologists, and paleontologists knowledgeable in the field of hominid fossils would probably support a summary view of human evolution something like the following (Fig. 34-8):

Australopithecus afarensis was ancestral to all later hominids. *A. afarensis* itself was derived from some apelike ancestor, such as *Dryo-*

---

[1] The first to publish a scientific description of an organism is entitled to name it.
[2] The story goes that the name came from a popular Beatles song, "Lucy in the Sky with Diamonds."

pithecus (not described here), which also gave rise to the modern African and possibly Asiatic apes. This view is bolstered by the great similarity between chimpanzee and human proteins. *A. afarensis* in turn gave rise to the australopithecines on one hand and to the *Homo* lineage on the other. The australopithecines coexisted with *Homo* for as long as 2 million years, eventually becoming extinct. In their early history, there were at least two species of *Australopithecus: A. africanus* and *A. robustus.* It is possible that these hominids made a few simple tools and were eliminated by *Homo,* particularly *H. erectus,* who was known to have used a variety of tools. The australopithecines and early *Homo* were fully bipedal, with feet almost indistinguishable from those of modern human beings.

Homo erectus was the first hominid to spread beyond the African continent into Asia and Europe. It gave rise to *Homo sapiens.* Possibly the earliest *Homo sapiens* resembled Neanderthal people, but the most extreme Neanderthals were isolated in glacial cul-de-sacs in Europe, where they became specialized for an extremely strenuous life-style, perhaps involving close-encounter techniques of killing big game. The Neanderthals were eventually replaced by modern versions of *Homo sapiens,* such as the Cro-Magnon people, who left paintings of high artistic quality in caves in France and Spain.

EVOLUTIONARY EVIDENCE

The theory of evolution states that all organisms have gradually developed from a common, simple ancestral type. The theory is based on data from such areas as morphology, biochemistry, the fossil record, and observed microevolutionary processes. Some of the evidence that supports the theory of evolution follows.

Microevolution

Many cases of naturally occurring microevolution have been observed. For instance, bacterial populations have become resistant to antibiotics, flies and mosquitoes have become resistant to pesticides, and industrial melanism has occurred in moths. The domestic rabbit and the house mouse are subspecies that have evolved during the time of modern humans. A few plant species (such as maize) have evolved during recent years. If microevolution has been witnessed in a relatively short span of years, it seems quite plausible that over millions of years many minor microevolutionary changes have accumulated to produce significant evolution, that is, macroevolution. Furthermore, where can one draw the line between micro- and macroevolution? If a species can evolve, why not a whole genus, order, class, or phylum, given enough time?

Morphological Resemblances

Comparative anatomy has shown interesting patterns of structural similarity between organisms that appear to go beyond what could be expected of coincidence. These are most often explained as the result of genetic kinship among the similar organisms. For example, all vertebrates have the same pattern of circulation, nerves, muscles, bones, and other organs, and structural complexity gradually increases as one moves from agnathans to mammals. Even more striking is that these organs develop in the embryo in much the same manner in all vertebrates. Such *homologous* organs (Fig. 34–9), possessing as they do the same underlying anatomical plan and having the same embryonic ori-

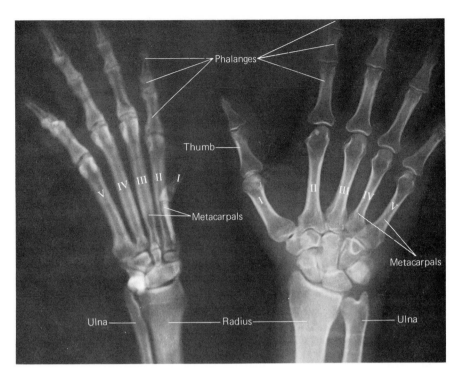

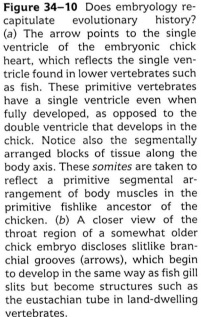

Figure 34-9 Homology in vertebrate forelimbs. Though differing in shape and proportion, the bones of the hand of a human being and those of the enlarged forefoot of a dog are fundamentally similar; the differences exist, as it were, within the similarities. This deep underlying similarity is taken to reflect a common evolutionary origin.

gin, are held to have a common ancestral and genetic origin, even though this can be very difficult to demonstrate conclusively.

Vestigial (meaning "trace"), apparently useless **structures** also provide evidence for evolution. All vertebrate embryos develop pharyngeal pouches (Fig. 34–10), which give rise to gills in fish and amphibians but during further development either disappear or are modified to form different structures in birds, reptiles, and mammals. Our wisdom teeth are thought to be vestigial structures, holdovers from our more vegetarian past. Each of us has a complete set of muscles for wiggling our ears, a useful ability in many animals but of little utility in human beings. The

Figure 34–10 Does embryology recapitulate evolutionary history? (a) The arrow points to the single ventricle of the embryonic chick heart, which reflects the single ventricle found in lower vertebrates such as fish. These primitive vertebrates have a single ventricle even when fully developed, as opposed to the double ventricle that develops in the chick. Notice also the segmentally arranged blocks of tissue along the body axis. These *somites* are taken to reflect a primitive segmental arrangement of body muscles in the primitive fishlike ancestor of the chicken. (b) A closer view of the throat region of a somewhat older chick embryo discloses slitlike branchial grooves (arrows), which begin to develop in the same way as fish gill slits but become structures such as the eustachian tube in land-dwelling vertebrates.

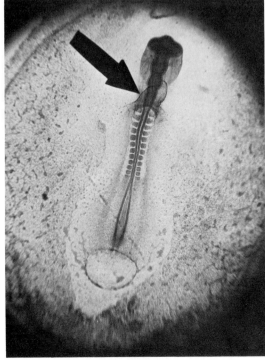

(a)

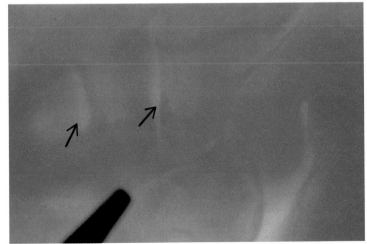

(b)

presence of the same anatomical, embryological, and vestigial structures suggests that all vertebrates have a great many genes in common. This, in turn, suggests a common ancestry.

Biochemistry

All organisms have the same fundamental biochemical mechanisms. All employ DNA, the citric acid cycle, cytochromes, and many other complex compounds. It seems inconceivable to most scientists that the biochemistry of living things would be so similar if all life did not develop from a common group of ancestors. Furthermore, the amino acid sequence of proteins is very similar in organisms thought on other grounds to be related. For example, the sequence of the 300 amino acids in hemoglobin is identical in humans and in chimpanzees. In the gorilla, two of the amino acids are different. Monkey hemoglobin differs in the sequence of 12 amino acids. Since DNA codes protein synthesis, protein similarity is a strong indicator of genetic similarity. (Actual DNA sequences have also been studied.) Further biochemical evidence (Fig. 34–11) is accumulating that supports most of the taxonomic relationships that have been previously proposed.

Genetics

In Chapter 33 we discussed the genetic theory of natural selection, along with the genetic mechanisms that are able to produce evolution. The very existence of such mechanisms is evidence of a sort, but more direct evidence comes from comparison of genes themselves by DNA hybridization techniques. The degree to which DNA from two organisms will hybridize is an indication of how much, in terms of base sequences, they have in common.

It is often possible to show that organisms of similar anatomy are similar genetically as well, which implies that the former is due to the latter. It is hard to think of any way in which extensive genetic similarities could exist except as a consequence of a common genealogy. Further-

Figure 34–11 One method of comparing proteins is by subjecting them to a high voltage electric field. Different proteins migrate through this field at differing rates, allowing them to be separated and compared. (a) In this apparatus the proteins being studied migrate in tubes of gel which are then specially stained to reveal the various proteins. (b) Similar patterns of proteins are assumed to indicate a close evolutionary relationship between the organisms being compared. (NASCO.)

(a)

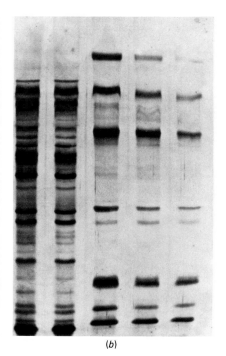

(b)

more, many organisms that seem to be related on the basis of other similarities also show close similarities in the shape, size, or markings of their chromosomes.

Biogeography and Distribution

Any child knows that polar bears do not live on the equator, since the tropics are obviously an unsuitable habitat for them. Less easy to explain is that these bears are confined to North Polar areas. Why don't they occur in Antarctica? And for that matter, why are there no penguins in Alaska? Clearly, climate and topography are not the only factors governing distribution.

The modern science of **biogeography** studies the distributions of plants and animals and seeks to draw evolutionary conclusions from this data. By comparing not only the organisms but also their distribution, biogeographers seek to infer the **center of origin** of each species, that is, the range of the population when the species evolved. They also attempt to learn something of the history of the migrations and other events that have produced the modern world of life.

Every species has a characteristic range (Fig. 34–12), which can be very small (as in the case of the Florida "panther") or very large (as in the case of the common cockroach or *Homo sapiens*). Organisms that are ecologically similar never or rarely occupy the same range, perhaps because one would certainly be better able to prosper and would displace its closely similar competitor. If organisms are *dis*similar in their adaptations, though, they might not directly compete and therefore could coexist. This view, *Gause's law,* is sometimes summed up in the phrase "complete competitors cannot coexist," which is sometimes further abbreviated as the *4-C law.*

Figure 34–12 Biogeographical realms. Large parts of the earth have characteristic animal species and genera, animals different from those in other large realms. Kangaroos, for example, occur naturally in Australia but not in the adjacent Oriental realm.

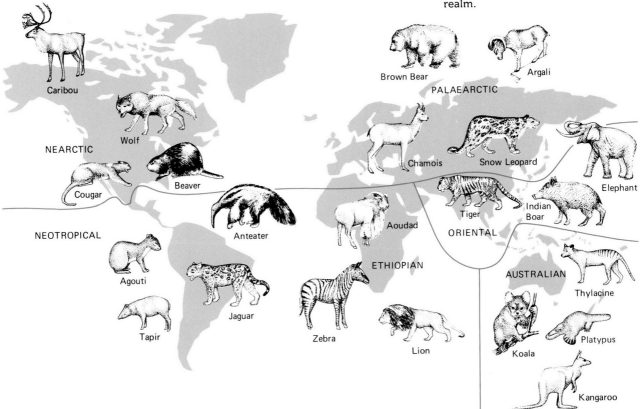

PLACENTALS

MARSUPIALS

Lemur

Cuscus

Anteater (*Myrmecophaga*)

Anteater (*Myrmecabius*)

Mouse (*Mus*)

Mouse (*Dasycerus*)

Flying squirrel

Flying phalanger

Wolf

Tasmanian wolf

Mole

Mole (*Notoryctes*)

Cat (*Felis*)

Cat (*Dasyurus*)

Figure 34–13 Parallel adaptations among Australian marsupials and placental mammals found in the rest of the world occurred as a result of convergent evolution. Other extinct marsupials from Australia and Latin America resembled the saber-toothed tiger, rhinoceros, and other large placental mammals.

The real problem, however, is in explaining how organisms very similar to one another yet seemingly unrelated can occupy different geographical areas. The classic example is Australia.

Think of the animals originally native to Europe and North America—wolves, moles, mice, cattle, woodchucks, and much else. These animals, though found throughout the Northern Hemisphere and even over much of the Southern, are absent from Australia. Or are they? There are, or were, animals in Australia that were closely similar to moles, mice, cattle, woodchucks, and wolves (Fig. 34–13). Yet the resemblance, in some cases astonishingly close, can be seen as superficial because the Northern Hemisphere animals possess placentas. The Australian equivalents—the marsupials—do not. They have external pouches in which their fetuses complete their late development.

If Australia originally contained primitive pouched mammals and most of the rest of the world contained primitive placental mammals, then natural selection might well have adapted a placental wolf to a wolf's life-style in Europe, and a marsupial "wolf" to a wolf's life-style in Australia. Each would have retained the fundamental structures, however, of its remote ancestors. Since deep-sea barriers (resulting, it is now thought, from continental drift; see Fig. 34–14) prevented the spread of Australian and European organisms into one another's habitat from perhaps the Mesozoic until before modern times, there was no way that European wolves could render Australian "wolves" extinct by competition. (It is interesting to note that native Australian people evidently brought dogs with them when they settled the continent in prehistoric times. By colonial times the marsupial "wolf" persisted only on the island of Tasmania, where there were no dogs.)

Actually, marsupials are found elsewhere than in Australia. One place is South America, which today has several species of opossums but used to have numerous other marsupial species. The modern isthmus of Panama evidently did not always exist, and it is believed that South America was once an isolated continent. When the isthmus became established, according to this view, South America was invaded by placental mammals, which put all the marsupials out of business except for the opossum. This doughty marsupial even invaded the North American continent, where it continues to raid chicken coops to this day.

We have already discussed the finches of the Galapagos Islands (near South America). Darwin did not confine his study to these birds but also catalogued all the other animals he could collect. He did the same for the Cape Verde Islands off Africa. The Galapagos species were similar to—almost a small sample of—South American ones, and the Cape Verde species were similar to African ones. Darwin concluded that the ancestors of each oceanic group of species were brought to their island by chance events, and in each case had originated on the adjacent continent. This strongly implied also that the modern island species had changed, and probably so had the continental ones, so that the island species and their continental cousins had become different from one another by evolution.

Structural Parallelism

Unrelated or distantly related organisms sometimes possess very similar adaptations. This is known as **structural parallelism.** Structural parallelism might at first be taken as evidence that such organisms are, in fact, closely related, except that the similar organs, when closely examined, may turn out to have vastly different embryonic origins.

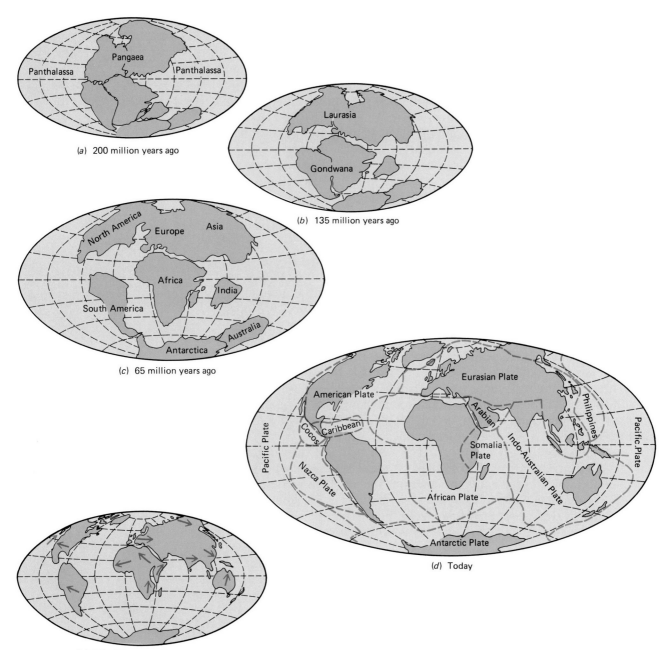

(a) 200 million years ago

(b) 135 million years ago

(c) 65 million years ago

(d) Today

(e) 50 million years from now

Figure 34–14 Continental drift. As currently understood, the continents of the world move at a rate of a few inches per year. At present their movement is drawing them mostly apart, but continental collisions are thought to have occurred in the past. The motion is made possible by the earth's hot and plastic interior, over which the continents float. Convection currents in the underlying material produce the actual movements of the continents.

Close study of the annelid eye shown in Figure 34–15 discloses that when compared with the vertebrate eye, there are substantial differences of detail (not all annelids have eyes like this, or even any eyes at all). The lens is not adjustable, there is no equivalent of an iris, there are accessory retinas that have no vertebrate equivalent, and the cellular arrangement of the retina is far different. Functionally, though, they are much the same. That is, they are **analogous** but not **homologous.**

How can this similarity be taken as evidence for evolution? It is argued that since the embryonic origins of the two structures are different, they must have had different ancestral origins. Yet the mechanical similarity shows that the demands of similar life-styles tend to produce the same or similar adaptations by natural selection. But since unrelated animals are genetically different, the forces of natural selection must operate on different sets of genes, so that even though the end results may be similar, the genetic basis of the two sets of similar adaptations is different. This is reflected in the differences of detail and also the differences of embryonic origin.

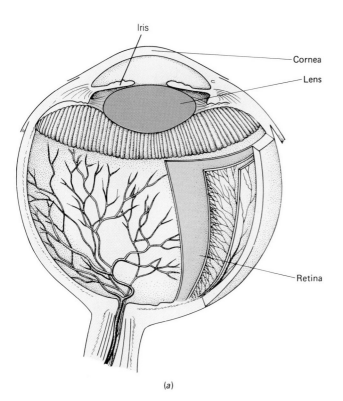

(a)

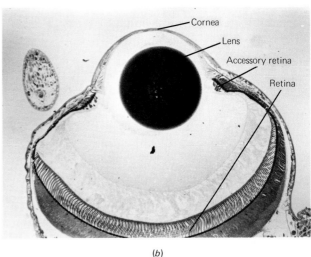

(b)

Figure 34–15 Structural parallelism in humans and an annelid worm. (a) Diagram of a human eye, similar to that of most vertebrates. (b) Photomicrograph of the eye of a polychaete annelid. The two are clearly not identical—the annelid has no iris, the vertebrate has no accessory retinas—but they are closely similar. Since the two seem to have no close evolutionary relationship, their similar organs are considered analogous but not homologous. (b, courtesy of Dr. George Wald and *Science*. From Wald, G. "Vision in Annelid Worms," *Science*, 24 July 1970, 1434–1439. Copyright 1970 by the American Association for the Advancement of Science.)

A nonevolutionary view of the origins of parallel adaptations would have difficulty accounting for these differences. Moreover, it is possible in this way to account in evolutionary terms for the similarities between the marsupial and placental mammals we have just discussed. A marsupial with a life-style like that of a wolf, for example, would have to possess adaptations similar to those of a wolf, so natural selection would have produced a wolflike body in both cases. Since different evolutionary pathways lead to much the same result in such instances, the development of similar adaptations in unrelated groups of organisms is known as **convergent evolution.**

Adaptive Radiation

How can evolution explain the *diversity* of species? Every potential habitat or ecological niche represents a source of resources. If the habitat or niche could be exploited, it would afford an advantage to any organism (and to its genes) that could become adapted to its demands. Because of the constant competition for food and living space, each group of organisms does tend to spread out and occupy as many different habitats and ecological niches[1] as possible. Within each habitat, those genotypes that produced superior phenotypic adaptations would propagate themselves better than those less well suited to the demands of the environment or the life-styles appropriate to it. Since the habitats and the ecological niches differ, the appropriate adaptations would differ as well, resulting in a variety of physical and behavioral specializations. This process of evolution from a single ancestral species to a variety of forms that occupy somewhat different habitats and ecological niches is termed **adaptive radiation.**

[1] As will be discussed in greater detail in Chapter 36, an ecological niche is a potential life-style. Organisms are adapted to their life-styles as well as to their habitat. Human beings and cockroaches are both well-adapted to live inside houses (but not, of course, in the same way!). Thus, they share the same habitat but not the same ecological niche, and their adaptations are quite different.

Figure 34–16 Adaptive radiation in mammals. A primitive, insectivorous, shrewlike cohabitant with the dinosaurs is believed to have been the ancestor of all the highly specialized orders of modern mammals.

One of the classic examples of adaptive radiation is the evolution of placental mammals (Fig. 34–16). Apparently, the earliest known mammal was an insect-eating, five-toed, short-legged creature that walked with the soles of its feet flat on the ground. Today we see a great variety of mammalian types. These include dogs and deer adapted for a terrestrial life in which running rapidly is important for survival; squirrels and primates adapted for life in the trees; bats equipped for flying; beavers and seals, which maintain an amphibious existence; the completely aquatic whales, porpoises, and sea cows; and the burrowing animals— moles, gophers, and shrews. In each of these, the number and shape of the teeth, the length and number of leg bones, the number and attachment sites of muscles, the thickness and color of fur, the length and shape of the tail, and so on are specifically adapted to the animal's lifestyle and environment.

Adaptive radiation that gives rise to several different types of descendants, adapted in different ways to different environments, is a result of **divergent evolution.** It is the opposite, in a way, of the convergent evolution that results in structural parallelism.

Let's return to the example of the Galapagos Islands. Darwin took them to be a microcosm of adaptive radiation, represented by the variety of ground finches present there today. Some of these birds live on the ground and feed on seeds, others feed mainly on cactus, and still others have taken to living in trees and eating insects. These variations in feeding have been accompanied by changes in the size and structure of the beak. This suggested to Darwin that the essence of adaptive radiation is the evolution from a single ancestral form to a variety of different forms, each of which is adapted and specialized in some unique way to survive in a particular habitat.

Dating Methods and the Fossil Record

When geologists explore fossils in sedimentary rocks, they often find that the fossils were deposited in a striking sequence, with the deepest and oldest strata containing the most primitive fossils. Buried in successive layers from the bottom to the top is a progression of fossils from simplest to most complex.

Careful studies have indicated that under the most typical present-day circumstances, on the average, approximately 1 foot of sediment is deposited every 5000 years. By measuring the depth at which fossils are buried in a deposit, scientists can estimate their age.

A more accurate method for determining the age of fossils is radioactive dating. Radioactive elements decay into stable products at specific, constant rates. This decay rate is not altered by such factors as temperature or pressure, as far as is known, and appears to proceed at the same rate at all times. One of the most useful systems of dating is the uranium-lead method. When igneous rocks were first formed, some contained uranium-238. The uranium immediately began to decay at a constant rate to form lead-206. The half-life of uranium-238 is 4.5 billion years, which means that half the atoms in a particular sample will be converted to lead during that time. By measuring the ratio of uranium-238 to lead in a rock sample, scientists can compute the age of the rock.

The potassium-argon method is based upon the decay of radioactive potassium-40 into argon and calcium. Potassium-40 has a half-life of 1.3 billion years. For archaeological artifacts or fossils less than 30,000 years old, a radiocarbon method is used. The half-life of carbon-14 is 5568 years. Since the amount of radioactive carbon generated in the earth's atmosphere by the sun varies, radioactive carbon dating has not been as easy to use as was initially thought, and has had to be revised in the light of such objective methods of dating as tree-ring counts. Unfortunately, it is harder to check the accuracy of most other dating methods. However, the rocks dated as oldest do seem to contain the fossils which appear on other grounds to be the most primitive (Fig. 34–17).

Figure 34–17 Fossils are the "hard" evidence we have of the history of life on earth. (*a*) Crinoids (ancient echinoderms) embedded in rock from the Mississippian period, more than 300 million years ago. (*b*) *Gallinuloides wyomingensis*, an Eocene bird, reconstructed from fossils found in the Green River, Wyoming. (E.R. Degginger.)

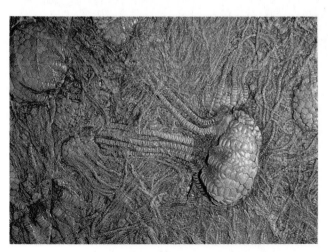

(a)

(b)

To sum up, the vast majority of biologists consider the evidence to indicate that the diversity of organisms is best and most simply explained in terms of evolution. Most scientists, while readily conceding that some of the hypotheses about particular events or mechanisms may have to be modified as new evidence is found, accept the concept of evolution as being one of the most fundamental theories in biology. But from the time of its birth evolutionary theory has been steeped in controversy. Even among those who feel sure that evolution is a fact, there has been a certain amount of disagreement as to its mechanisms.

CHAPTER SUMMARY

I. According to current theories, life originated on earth from naturally occurring macromolecules by a kind of spontaneous generation.

II. Eukaryotic cells may have originated from a symbiotic union of several prokaryotes.

III. The current consensus on human evolution is that an ape-like ancestor gave rise to both the hominids and apes. The earliest hominids resembled *A. afarensis,* which gave rise both to the australopithecines and to the genus *Homo.*

    A. The earliest known member of genus *Homo* was *H. habilis,* which gave rise to *H. erectus.*

    B. *H. erectus* gave rise to both the Neanderthal and modern versions of *H. sapiens.*

IV. Evidence for evolution is drawn from a wide area of scientific disciplines, including geology, biochemistry, and genetics.

    A. Microevolutionary evidence points to the establishment within recent years of new species that have developed such traits as pesticide resistance. It is argued that if species can evolve, so can larger categories, such as order, class, or phylum.

    B. Morphological evidence points to the anatomical similarities between some organisms, which suggests their evolutionary relationship, along with the presence of vestigial structures, for which there are no known current functions.

    C. Biochemical evidence is based upon the chemical similarities between related groups, suggesting a common ancestry.

    D. Genetic evidence is based on DNA hybridization techniques and on chromosome similarities among related organisms.

    E. Narrowly defined geographical occurrence of apparently related organisms suggests their evolution from a common ancestor living in that region.

    F. The fossil record suggests the systematic change of extinct organisms during geological times.

Post-Test

1. Modern experiments have indicated that complex _____ compounds, such as occur in living things today, could have originated _____ _____ on the primitive earth.

2. Most biologists think that life might have originated in the _____ of the earth but others have suggested that life began in _____ or _____ chambers or even in extraterrestrial environments.

3. Most biologists agree that the earliest living things were nutritionally _____ .

4. It is held that simple extensions of the kind of microevolution that is observable today could, given enough time, have produced major changes, or _____ .

5. Amino acid similarities in proteins reflect DNA _____ sequence, similarities that are assumed to indicate genetic relationships among the organisms being compared.

6. A fossil might be dated by using the known rate of decay or half-life of a radioactive _____ , such as potassium-40.

7. _____ organs have the same embryonic origin and underlying anatomical plan. This is taken to imply that organisms that possess homologous organs have a common _____ .

8. Functionally, the annelid eye and the vertebrate eye are _____ , but they are not _____ .

9. Similar adaptations in unrelated groups of organisms are referred to as examples of _____ _____ ; such similarities result from _____ evolution.

10. Adaptive radiation that gives rise to different types of organisms adapted in different ways to different environments is termed _____ evolution.

Review Questions

1. What is meant by homology? How does the existence of homologous structures in organisms appear to support the theory of evolution?

2. Is the fossil record an objective account of evolutionary events? Is it complete?

3. How might embryology be used as evidence for evolution?

4. The Australian continent contains many living marsupial mammals but very few original placentals, whereas everywhere else, placental mammals are dominant. Australia is also isolated geographically and appears to have been so for an immense span of time. Why might there be so few native Australian placentals? How do the many specialized forms of marsupial life in Australia argue for evolution?

5. How can the age of a rock be estimated on the basis of its radioactive elements? Explain the uranium-lead method of radioactive dating.

6. How is structural parallelism used as evidence for evolution?

Readings

Ambrose, E.J. *The Nature and Origin of the Biological World.* New York, John Wiley, 1982. An up-to-date evolution textbook that discusses some of the current controversial views, and treats them fairly.

Bambach, R.K. "Responses to Creationism," *Science,* 20 May 1983. Five anticreationist books and one creationist book are reviewed in this article. We note that the creationist book is perhaps not representative.

Cairns-Smith, A.G. *Genetic Takeover and the Mineral Origins of Life.* Cambridge, Cambridge University Press, 1982. The chicken-and-egg riddle of the origin of life has always been one of replication: How can a cell without information reproduce? How can information reproduce without a cell? This author presents an ingenious and eccentric hypothesis.

Cloud, P., and M.F. Glaessner. "The Ediacarian Period and System: Metazoa Inherit the Earth," *Science,* 27 August 1982. According to the authors, the naked metazoa of the Upper Precambrian now deserve a new name and status.

DeCamp, L. *The Great Monkey Trial.* New York, Doubleday, 1968. An entertaining, scholarly, and not always fair account of the Scopes "monkey trial" of Dayton, Tennessee, in 1925.

Faul, H. "A History of Geologic Time," *American Scientist,* 66: 159–165 (1978). How the concepts of historical geology originated and developed.

Greene, J.C. *Science, Ideology and World View.* Berkeley, University of California Press, 1982. How evolution influences the way we view the universe and our place in it.

Lewin, R. *Thread of Life: The Smithsonian Looks at Evolution.* Washington, D.C., Smithsonian Books (distributed by W.W. Norton, New York), 1982. A handsome and beautifully illustrated coffee-table volume.

McLoughlin, J.C. *Archosauria: A New Look at the Old Dinosaur.* New York, Viking Press, 1979. *Synapsida: A New Look into the Origin of Mammals.* New York, Viking Press, 1980. These volumes (two of a series) are exceptionally well illustrated and a lot of fun to read.

Mossman, D.J., and W.A.S. Sarjeant. "The Footprints of Extinct Animals," *Scientific American,* Vol. 24, No. 1, January 1983. An account of vertebrate evolution with emphasis on information gained from animal tracks.

Officer, C.B., and C.L. Drake. "The Cretaceous-Tertiary Transition," *Science,* 25 March 1983. These authors do not think that the dinosaurs and others were extraterrestrially expunged (see D.A. Russell's article, listed below).

Rukang, W., and L. Shenlong. "Peking Man," *Scientific American,* June 1983. *Homo erectus* was no dolt but possessed a respectable culture.

Russell, D.A. "The Mass Extinctions of the Late Mesozoic," *Scientific American,* January 1982. The new paleontological orthodoxy: The dinosaurs were extinguished by an asteroid.

Stebbins, G.L. *Darwin to DNA, Molecules to Humanity.* San Francisco, W.H. Freeman & Company, 1982.

Wilson, A.C. "The Molecular Basis of Evolution." *Scientific American,* October 1985. Molecular biologists are gaining new insights into evolution by tracking mutations in DNA.

Chapter 35
BEHAVIOR

Outline

Learning Objectives

After you have studied this chapter you should be able to:

1. Support the theses that behavior is (1) adaptive, (2) homeostatic, (3) flexible.
2. Define tropisms and taxes and give examples of each.
3. Cite examples of biological rhythms and suggest some of the mechanisms known or thought to be responsible for them.
4. Using appropriate examples, summarize the role of sign stimuli (releasers) in the expression of simple and complex programmed behavior.
5. Summarize the contributions of heredity, environment, and maturation to behavior.
6. Compare learning ability with innate behavior as adaptational systems and give at least one example of their interaction.
7. Discuss the adaptive significance of imprinting.
8. Postulate biological advantages for migration.
9. Given a description of an animal society, identify the cooperative result of actions of the organisms, the suppression of aggression, and the modes of communication the animals employ.
10. Present the concept of a dominance hierarchy, giving at least one example, and speculate its possible general adaptive significance and social function.
11. Distinguish between home range and territory and give three theories about the adaptive significance of territoriality.
12. Discuss the adaptive value of courtship behavior and describe a pair bond.
13. Compare the society of a social insect with human society.
14. Define kin selection and summarize its proposed role in the maintenance of insect and other animal societies.
15. Summarize the emphases of sociobiology.

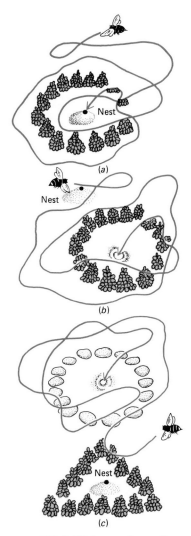

S uppose that your instructor were to arm you with a hypodermic syringe full of poison and demand that you find a particular type of insect (which you have never seen and which can fight back) and that you inject the ganglia of its nervous system (about which you have been taught nothing) with just enough poison to paralyze your victim, but not enough poison to kill it. You would be hard put to accomplish these tasks, but a solitary wasp no longer than the first joint of your thumb does it all with elegance and surgical precision, and without instruction.

The bee-killer wasp *Philanthus* captures bees, stings them, and places the paralyzed insects in burrows excavated in the sand. She then lays an egg on her victims, which are devoured alive by the larva that hatches from that egg. From time to time the *Philanthus* returns to her hidden nest to reprovision it until the larva becomes a hibernating pupa in the fall. Her offspring will repeat this, doing it to perfection without ever having seen it done.

When a *Philanthus* covers a nest with sand, she takes precise bearings on the location of the burrow before flying off again to hunt. There is no way in which knowledge of the location of the burrow could be genetically programmed in the wasp. How to dig it, how to cover it, how to kill the bees—these behaviors appear to be genetically programmed. But since a burrow can be dug only in a suitable spot, its location must be learned after it is dug. That this is so was determined by the Dutch investigator Nikko Tinbergen.

Tinbergen surrounded the wasp's burrow with a circle of pine cones, on which the wasp took her bearings (Fig. 35–1). Before she returned with another moribund bee, Tinbergen moved the circle of pine cones. The wasp could not find her burrow—the cones no longer surrounded it. Only when the experimenter restored the cones to their original location could the wasp find her burrow.

WHAT IS BEHAVIOR?

Behavior refers to the responses of an organism to signals from its environment. Notice how efficiently the wasp carried out a complex, though largely genetically programmed, sequence of behaviors. Very little of her behavior had to be learned. In contrast, the existence of programmed behavior is hard to demonstrate in human beings. We owe the complexity of our behavior to a *generalized* ability to learn. The *Philanthus* wasp's intelligence is as narrowly specialized as her stinger.

Much of what organisms do can be analyzed in terms of specific behavior patterns that occur in response to stimuli (changes) in the environment. A dog may wag its tail, a bird may sing, or a butterfly may release a volatile sex attractant. Behavior is just as diverse as biological structure and is just as characteristic of a given species as its structure and biochemistry.

BEHAVIOR AS ADAPTATION

Animal behavior used to be studied in isolation from the physical characteristics of animals. Perhaps few people stopped to think that behavioral patterns are as much adaptations as an animal's wings, legs, shell, or stinger. In fact, the physical traits of an animal make little sense without reference to its behavior.

Animal behavior also used to be studied almost exclusively under artificial laboratory conditions. This was partly due to a movement in psychology to make the behavioral sciences more objective by studying behavior in a simplified environment where controls were easier to main-

Figure 35–1 Tinbergen's sand wasp experiment. When the ring of pine cones is moved from position (*a*) to position (*b*), the *Philanthus* wasp behaves as if her nest were still located at the center. She has therefore learned its position in relation to the cones. That it is the arrangement of the cones rather than the cones themselves that the wasp responds to is shown by the substitution of a ring of stones for cones in (*c*). The learning ability of *Philanthus* is quite limited but is adequate for situations that normally arise in nature. (After Tinbergen.)

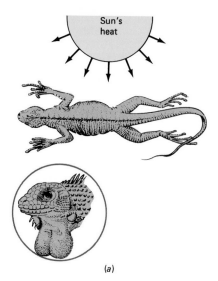

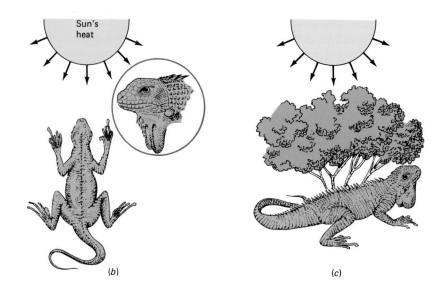

Figure 35–2 Behavioral thermoregulation in a lizard. The cold lizard (*a*) lies at right angles to the sunlight, puffing up its body to increase the surface area available for heat absorption. When too warm (*b*), the lizard orients itself parallel to the sun's rays and deflates its body. Eventually (*c*), it seeks shade. (Illustration concept, courtesy of Pam Godfrey.)

tain. The restoration of behavior studies to their proper place in natural history was the contribution of such scholars as Nikko Tinbergen and Konrad Lorenz, who with others founded the science of ethology in the 1920s and 1930s. **Ethology** is the study of behavior in natural environments from the point of view of adaptation. Since behavior is *adaptive*, ethologists believe it is best studied in the field or under conditions reflecting the natural life of the organism that exhibits it.

A particular behavior may help an organism obtain food or water, acquire and maintain territory in which to live, protect itself, or reproduce. Certain behavioral responses may lead to the death of the individual but increase the chance of survival of the population or species through the survival of the offspring.

Behavior tends to be *homeostatic* as well as adaptive. The body of a homeothermic organism has a collection of physiological responses that help to keep body temperature constant. For example, a human may shiver to generate more heat or perspire when too hot. A dog may pant, cooling the blood in the blood vessels of its respiratory tract.

Many poikilotherms (cold-blooded animals) can regulate their body temperature by behavioral adaptations. Lizards, for example, may warm their bodies by basking in the sunlight. To absorb the maximum amount of heat, the lizard places its body at right angles to the sun's rays, puffs itself up and spreads out all body membranes (Fig. 35–2; see also Fig. 1–5). If the lizard becomes too warm, it may first orient the body parallel to the rays of sunlight, decreasing the area exposed directly to sunlight. It may also retract its body membranes and shrink its body as much as possible. If that proves insufficient, the lizard will seek shade, spreading out all body membranes and the body itself to the maximum to radiate excess heat. This behavioral mechanism of thermoregulation is surprisingly effective. It has been shown that lizards infected with dangerous bacteria can maintain their temperatures several degrees above normal. In effect, these cold-blooded creatures run a fever. The fever and behavior responsible for it can even be abolished with aspirin.

SIMPLE BEHAVIOR

Even bacteria "make decisions": whether to move toward food (Fig. 35–3) or away from a toxic substance; toward a place with a certain temperature, or away from it if it is too hot or too cold. Some bacteria are sensitive to other stimuli, such as the earth's magnetic field. But no bacterium has a nervous system, specialized sense organs, or muscles. How does it sense stimuli and make an appropriate response? Evidence is accumulating that some proteins, such as those responsible for transporting food materials into a bacterium through its cell membrane, also function as receptors capable of detecting food substances.

Bacteria respond to many stimuli by moving toward or away from them. When the flagella rotate counterclockwise, they rotate together and the bacterium travels in a fairly straight line. Clockwise rotation pulls the bundle of flagella apart and makes the bacterium dance in place. Such microscopic dances are known as **twiddles** (Fig. 35–4). Resumption of the counterclockwise movement sends the bacterium in a straight line again, but not necessarily in the original direction. The bacterium is able to respond to gradients in a stimulating substance in the surrounding water. In the presence of a stimulus to which it responds positively, the bacterium employs less twiddling (random motion), so on the whole the bacterium tends to approach the source of the stimulus. Negative stimuli cause the bacterium to reverse this, so the organism tends to move away from the source of the noxious material or situation.

Such simple behavior appears to be basically a matter of physics and chemistry, little more complicated, perhaps, than the guidance systems of a military missile. But the physics and chemistry are so organized as to adapt the bacterium actively to the changes that are constantly occurring in its environment. Comparable mechanisms probably govern the phagocytosis of food by amebas or of bacteria by white blood cells. And even multicellular organisms display simple behavior, some of which can be simply explained.

Tropisms

Plants have neither muscles nor a nervous system, so how can they be said to behave? Yet plants certainly do grow, and time-lapse motion pictures demonstrate stimulus-oriented growth. Growth responses toward

(a)

(b)

Figure 35–3 Chemotaxis in bacteria. (a) In this highly magnified view, bacteria cluster about the opening of a capillary tube containing a chemical that is attractive to them. (b) This bacterium contains magnetic granules (line of dots) whose attraction by the earth's magnetic field tells the bacterium which direction is "down" (approximately ×62,000). (a, courtesy of Dr. Julius Adler and *Science.* Copyright 1972 by the American Association for the Advancement of Science. b, courtesy of Denise Maratea.)

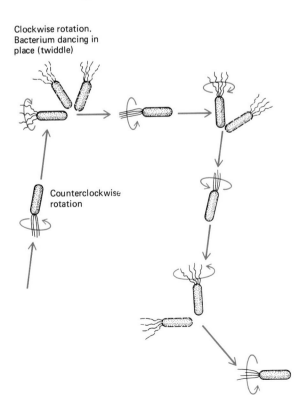

Clockwise rotation. Bacterium dancing in place (twiddle)

Counterclockwise rotation

Figure 35–4 Bacteria swim in straight lines when their flagella rotate counterclockwise but twiddle briefly when the rotation is reversed, for flagellar rotation cannot be coordinated when it is clockwise.

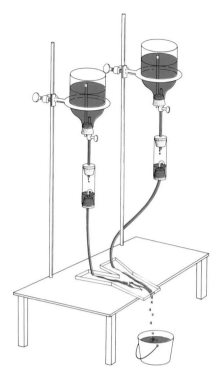

Figure 35–5 A simple maze for studying chemotaxis in a flatworm. The worm will turn left or right depending on the location of a source of an attractive chemical.

or away from a stimulus are known as **tropisms.** As discussed in Chapter 16, plant tropisms depend upon hormones; **phototropisms** occur in response to light, **geotropisms** in response to gravity, and **thigmotropisms** in response to a solid object. Plant behavior is not limited to tropisms. We have already seen in previous chapters that a fair number of plants employ effectors—such as traplike leaves with which they can even capture insect prey.

Taxes

Many animals far more complex than bacteria or plants respond to stimuli in much the same way, with simple orientational behaviors known as **taxes.** A taxis generally involves the reception of a stimulus and a movement toward or away from that stimulus. Thus, a positive geotaxis is a movement downward in response to gravity, and a negative phototaxis is a movement away from a light source. Flatworms, exhibiting positive chemotaxis, congregate on a piece of raw meat left overnight in a stream. They need no other stimulus than the chemical cues given off by the meat. If a flatworm is placed in the apparatus shown in Figure 35–5 and meat extract is placed in one of the bottles, the flatworm will swim into the arm of the trough that receives water from the bottle containing meat extract. By adjusting the lighting and the rate of flow from both bottles so that all other stimuli to which the worms might possibly respond are equal on both sides of the animal, it is possible to show that it is responding to the meat extract alone. Even such complex animals as insects and mammals have a large collection of simple orientational behaviors in their repertoires.

BIOLOGICAL RHYTHMS AND CLOCKS

It is to an organism's advantage that its metabolic processes and behavior be synchronized with the cyclic changes in the external environment. Cyclic control mechanisms change these processes at repetitive intervals, giving rise to daily rhythms, monthly cycles, or annual rhythms. Even some plant behavior, such as "sleep" movements in which the leaves fold to conserve water and heat at night, lacks such obvious triggers as physiological imbalance or external cues and follows a regular daily or longer cycle. In human beings, also, physiological processes seem to follow an intrinsic rhythm. Human body temperature, for example, follows a typical daily curve. The activities of many marine animals that live along the shore are linked to the cycle of the tides. Fiddler crabs on the eastern coast of the United States emerge from their burrows to feed at each low tide (twice every 24 hours).

Lunar Cycles

Some biological rhythms of animals, and perhaps plants, reflect the **lunar** (moon) **cycle.** The most striking ones are those in marine organisms that are tuned to the changes in the tides due to the phases of the moon. For instance, the swarming of the Pacific palolo worm at a particular time of year is governed by a combination of tidal, lunar, and annual rhythms. The Atlantic fireworm swarms in the surface waters surrounding Bermuda for 55 minutes after sunset on days of the full moon during the three summer months.

The grunion, a small fish of the Pacific coast of the United States, swarms from April through June on those three or four nights when the spring tide occurs. At precisely the high point of the tide the fish squirm onto the beach, deposit eggs and sperm in the sand and return to the sea in the next wave. By the time the next tide reaches that portion of the beach 15 days later, the young fish have hatched and are ready to enter the sea.

Circadian Rhythms

Periods of activity and sleep, feeding and drinking, body temperature, and many other processes have a cycle approximately 24 hours long. Hence they are called **circadian rhythms** (from the Latin words *circa,* approximately, and *dies,* day). Some animals are **diurnal,** exhibiting their greatest activity during the day, whereas others are **nocturnal** and most active during the hours of darkness. Still others are **crepuscular,** having their greatest activity during the twilight hours, at dawn, or both. If an animal's food is most plentiful in the early morning, for example, its cycle of activity must be regulated so that it becomes active shortly before dawn, even though dawn changes slightly from day to day. As the adage goes, "The early bird catches the worm."

What Controls the Biological Clock?

Current evidence seems to indicate that there is no single biological clock in most organisms. Instead, the interaction of a number of biochemical processes may produce the timed accumulation of certain substances to critical levels. These substances, whatever they may be, might be responsible for governing behavioral and physiological rhythms. The pineal gland is thought to play a role in the timing system of rats, birds, and some other vertebrates. Regions of the hypothalamus have been shown to be a part of the biological clock in mammals.

In many organisms the biological clock appears to have a genetic basis. Normal fruit flies, *Drosophila,* have a clock that has a running period of 24.2 hours. The running period is the clock's repetitive cycle when the animals are isolated from environmental cycles and kept under constant conditions. Mutant fruit flies have been discovered with free running periods of 19 and 28 hours. Each mutation has been traced to the same locus on the X chromosome.

Some investigators hold that biological rhythms are **endogenous;** that is, they are regulated internally by a biological clock capable of detecting the passage of time. According to this theory, no regular environmental stimulus is needed to keep the clock running. Snails and some other marine organisms whose activities vary with the tide continue to show the cyclic variations in activity when removed to an aquarium and protected from changes in light, temperature, and other factors. This persistence of rhythmic changes in activity, coordinated with the cyclic changes in the environment from which the animal was removed, is strong evidence for the endogenous explanation of biological clocks.

Other investigators argue that biological rhythms are **exogenous;** that is, they are controlled by environmental stimuli. It has been shown that biological clocks often interact to some extent with external and internal stimuli and often can be reset by such environmental cues. If animals that breed in the spring are transported from the Northern to the Southern Hemisphere, their cycle eventually shifts to coincide with the occurrence of spring in their new home.

THE GENETIC BASIS OF BEHAVIOR

An insect such as a bee can maintain an elaborate society because the instructions for that society are genetically inherited and programmed. A bee is capable of only the most limited learning—that which is required

Figure 35–6 A releaser stimulus. (a) The moth, when threatened by a predatory bird, abruptly exposes two "eyespots" on its lower wings (b). These resemble the eyes of an owl or other carnivore and apparently trigger an innate avoidance response in the insectivorous bird. Notice that the entire owl need not be present to trigger this behavior—just the specific eyespot stimulus.

(a)

(b)

by the immediate demands of its environment. The complexity of some genetically programmed behavior is wondrous, but no more so than the genetically determined complexity of the anatomy and physiology of any organism.

Although the distinction is not always clear, ethologists recognize two sorts of behavior, innate and learned. **Instinctive,** or **innate, behavior** is genetic; genes control the development of the programmed neural and motor patterns. In contrast, learned behaviors develop as a result of experience. Some innate behavior appears to be functional from the moment that the neural circuitry is in place, and does not seem to be modified by environmental factors. The first web of the orb-weaving spider, for example, is complete in all detail and repeatedly built in the same manner throughout the life of the spider.

Sign stimuli, also called **releasers,** often serve as triggers for fixed-action patterns of behavior. When quick action is essential, as in escaping from a predator, a danger sign is more useful than a detailed description of the danger. Alarm signs, whether they are sights or sounds, are usually simple and contrast sharply to the environment. Small birds typically show an immediate flight reaction to animals with large eyes—understandably, considering that their major predators, owls and hawks, have large eyes (Fig. 35–6).

In the spring male stickleback fish establish territories from which they drive other males. Investigators found that very simple model fish with red undersides were more effective in releasing defense behavior than more realistic models lacking the red underside. Apparently, the red belly of the male is the important sign stimulus responsible for releasing territorial defense behavior in sticklebacks (Fig. 35–7).

In many species innate behavior is capable of modification as a result of interaction with the environment. Herring gull chicks peck the beaks of the parents, which regurgitate partially digested food for them. The chicks are attracted by two sign stimuli: a red spot on the beak and the beak's shape and downward movement. This "begging behavior" is sufficiently functional to get the chicks their first meal, but much energy is wasted in pecking. Some pecks are off target and fail to reach the parent's beak. However, the begging behavior becomes more efficient over time, and in experiments with models of the parent's beak, the chicks become increasingly selective of the shape necessary to evoke the begging response (Fig. 35–8). Thus, the initial functional instinct is perfected by environmental interaction.

Behavior is essentially a property of the coordinating mechanisms of the body (that is, of the nervous and endocrine systems). The capacity for behavior is therefore subject to whatever genetic characteristics govern the development and *range of function* of these systems. One may think of a continuous scale of behaviors, ranging from the most rigidly programmed, genetically inherited types, through those that are somewhat modifiable, to those that, though containing a genetic component, are extensively developed through experience.

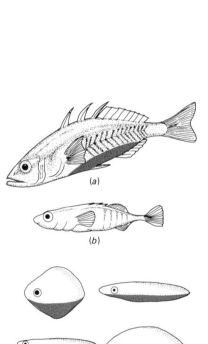

Figure 35–7 A sign stimulus is a particular feature that triggers an innate response. A male stickleback fish (a) will not attack a realistic model of another male stickleback if it lacks a red belly (b), but it will attack another model, however unrealistic, that has a red "belly" (c). Therefore, it is the specific red stimulus, rather than the recognition based on a combination of features, that triggers the aggressive behavior.

Figure 35–8 A young albatross begging food from a parent. (Paul Ehrlich, Stanford University/BPS.)

LEARNING

Learned behavior can be defined as behavior that is modified as a result of interaction with the environment. The simplest form of learning is **habituation,** learning to ignore repeated stimuli that are not followed by either benefit or obvious cost. Learning capabilities reflect the specialized mode of life of an animal. The same rat that has difficulty learning the artificial task of pushing a lever to get an immediate reward learns in a *single* trial to avoid a food that has made it ill as long as six hours after the food was eaten. Those who poison rats to get rid of them can readily appreciate the adaptive value of this learning talent to the rat. Such quick aversive learning forms the basis of warning coloration (Chapter 33), found in many poisonous insects and brilliantly colored but distasteful bird eggs. Once made ill by such an egg, the predators learn to avoid them.

The most complex learning is **insight learning,** the ability to remember past experiences that may involve different stimuli and to adapt these recalled events to solve a new problem. Insight learning is most easily demonstrated in primates (Fig. 35–9). A dog can be placed in a blind alley that it must *circumvent* in order to reach a reward. The difficulty of the problem appears to be that the animal must move *away* from the reward in order to get *to* it. At first the dog typically flings itself at the barrier nearest the food. Eventually, by trial and error, the frustrated dog may find its way around the barrier and reach the reward. A baboon placed in the same kind of situation is likely to see the solution immediately.

Learning is widely believed to depend upon changes in the readiness of individual neurons to form circuit relationships with one another and to transmit impulses in specific directions. In order to learn, an organism's neurons must have a large number of potential interactions with one another, and hence there must be many, not just the few required by programmed behavior. Since innate behavior is really a consequence of the biophysical properties of individual neurons and of their interconnections, the more narrowly stereotyped a system of behavior is, the more obvious is its genetic control. Yet without the necessary preexistence of the proper neural circuitry, even learned behavior would be impossible. Moreover, the kind of learned behavior that the organism typically and most easily develops also depends upon the layout of that circuitry.

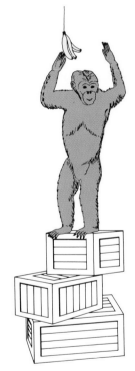

Figure 35–9 Insight learning, and in this case, simple tool use. Confronted with the problem of reaching food hanging from the ceiling, the chimpanzee stacks boxes until it can climb and reach the food. Many other examples of apparent insight are known from the behavior of these animals.

PHYSIOLOGICAL READINESS

Before any pattern of behavior can be exhibited, an organism must be physiologically ready to produce it. Breeding behavior does not ordinarily occur among birds or most mammals unless steroid sex hormones are

Figure 35–10 The formation of parent–offspring bonds. (*a*) Through imprinting, some young animals follow the first moving object they encounter. Usually, the object is their mother, although it is possible experimentally to imprint many such infants upon unnatural objects. Note that in this instance the ducklings are with one, not both, of the two hens. Even though the two hens would probably be indistinguishable to us, the ducklings know which one is their mother. (*b*) In large populations, such as this colony of seals at Cape Cross, Namibia, an offspring that is separated from its parent will easily remain lost and will starve to death. The bond, therefore between offspring and parent must be quickly established and must be maintained by a complex series of behavioral interactions. (*b*, Mitchell L. Osborne, The Image Bank.)

(*a*) (*b*)

present in their blood at certain concentrations. A human baby cannot walk unless its reflex and muscular development permits it. Yet these states of physiological readiness are themselves produced by a continuous interaction with the environment. The level of sex hormones in a bird's blood may be determined by seasonal variations in day length. The baby's muscles develop in response to exercise. Without the trial and error involved in learning how to walk, walking would be retarded.

Perhaps the best example of such interaction between readiness and environment is afforded by the white-crowned sparrow, which exhibits considerable regional variation in its song.[1] This bird, even if kept in isolation, eventually will sing a very poorly developed but recognizable white-crowned sparrow song. However, if it is allowed to grow up under the care of its parents for the first three months of life, when it matures, it will sing in the local "dialect" characteristic of its parents or foster parents. If such learning does not take place in those three months, it never will, and if the sparrow consorts with birds of other species, it will not learn their songs.

IMPRINTING

Anyone who has watched a mother duck with a swarm of ducklings must have wondered how she can keep track of such a horde of almost identical little creatures, tumbling about in the weeds and grass, let alone tell them from those belonging to another hen (Fig. 35–10). Though she is capable of recognizing her offspring to an extent, basically it is they that have the responsibility of keeping track of her—a far simpler chore. Each duckling follows her about, like a nail attracted to a magnet. Even if it gets into trouble, the duckling usually must take the initiative. It emits distress cries, and in response the mother will rescue it if she can. Clearly, the survival of the duckling requires an extremely rapid establishment of the behavioral bond between it and its parent. In fact, such a behavioral bond is usual between parent and offspring, especially if the offspring are able to follow the parent about. This form of learning—in

---

[1] Different species of birds vary greatly in the extent to which their songs are learned or innate. Among cowbirds, for example, which rely on birds of other species to raise their young, the song is *entirely innate*. Some song birds, on the other hand, are able to *learn* hundreds of songs in a single breeding season.

which a young animal forms a strong attachment to an individual, usually one of the parents, within a few hours of birth (or hatching)— is known as **imprinting.** An early investigator of imprinting, ethologist Konrad Lorenz, discovered that a newly hatched bird may imprint on a human, or even an inanimate object, if its parent is not present. Though the behavior itself is genetically determined, the bird learns the object.

Among many kinds of birds, especially ducks and geese, imprinting is established even before hatching. The older embryos of these birds are able to exchange calls with their nest mates and parents right through the porous eggshell. When they hatch, at least one parent is normally on hand, emitting the characteristic vocalizations that the hatchlings have already learned. During a brief critical period after hatching, the chicks learn to associate these vocalizations with the appearance of the parent.

Imprinting establishes the bond between mother and offspring among many mammals, as well as among birds. In many species the mother establishes a bond with her offspring while the offspring is imprinting upon her. The mother in some species of hoofed mammals will accept her offspring for only a few hours after its birth. If they are kept apart past that time, the young are thereafter rejected. Normally, this behavior enables the mother to distinguish her own offspring from those of others, evidently by olfactory cues.

MIGRATION

The dramatic seasonal migrations of birds have long excited human curiosity, but it is only recently that even their rudiments have been properly understood. Until Renaissance times, for instance, the winter disappearance of many migratory birds from Europe was attributed to hibernation in such unlikely places as the muddy bottoms of ponds. Widespread human travel and communication over long distances has led to our modern view that even without any obvious immediate motivation (such as hunger), many animals regularly travel long distances to breed or just to spend certain seasons of the year. Some migrations involve astonishing feats of endurance and navigation. Ruby-throated hummingbirds cross the vast reaches of the Gulf of Mexico twice each year, and the sooty tern travels across the entire South Atlantic from Africa to reach its tiny island breeding grounds south of Florida.

Though there is no reason to think that migration is purposeful, it can seem carefully planned (Fig. 35–11). Birds may feed heavily weeks before those food reserves will be needed, and often fly south even *before* the weather turns cold or food becomes scarce. Salmon swim into fresh water toward the end of their life cycles. Monarch butterflies fly southward, and the *next generation* flies north in the spring. The propensity for this behavior must be inherited and maintained by natural selection. Migration, in other words, must be a specific adaptation in the life-styles of many organisms. Yet the adaptive significance of migratory behavior is not always clear. It is obviously to the advantage of birds to fly south in the winter, but why certain eels migrate to the Sargasso Sea to spawn is a mystery.

The behavioral trigger that sets off migratory behavior varies. Some animals migrate upon maturation. In others explicit environmental cues trigger the process. In migratory birds, for example, changes in day length are sensed by the pineal gland. These trigger characteristic restless behavior called **Zugunruhe**—migratory restlessness. The bird evinces increased readiness to fly and flies for longer periods of time.

The *direction* of travel is also important, and this raises the general problem of animal navigation. Birds appear to navigate by a combina-

Figure 35–11 Seasonal changes in the physiology and behavior of the white-crowned sparrow. Note the increased rate of feeding and then restlessness (Zugunruhe) that precedes each period of migration. (From Alcock, J. *Animal Behavior: An Evolutionary Approach,* 2nd ed. Sunderland, Mass., Sinauer Associates, 1979.)

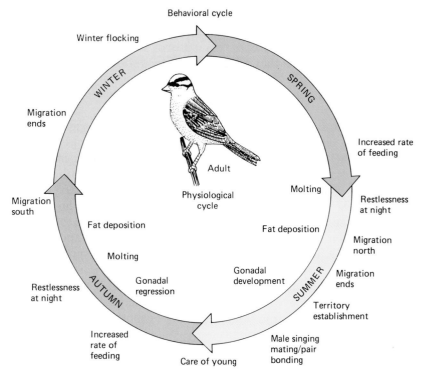

tion of celestial (sun and star) plus geographic and climatic cues (Fig. 35–12). Honeybees (perhaps) and some birds are sensitive to the earth's magnetic field. However, the cues employed by many animals to negotiate their migratory journeys are not understood.

One of the organisms in which migration *is* fairly well understood is the whitethroat, a small European warbler. Working in the 1950s, Franz and Eleonore Sauer hand-reared a number of these birds (presumably to rule out any possibility of transmission of information from parent to offspring). When (and only when) the birds could see the star patterns of

Figure 35–12 The northward migration of the Canada goose keeps pace with the arrival of spring in different parts of the North American continent. The geese are shown following lines that connect different points of the map according to a mean temperature of 2°C, or 35°F. (Modified after Lincoln.)

(a)

(b)

the night sky, they attempted to fly in the normal direction of migration for this species, a direction that they had no opportunity to learn. When the birds were brought into a planetarium and the night sky of a different locale was simulated on the planetarium dome, they attempted to fly in a direction that would have taken them to their normal wintering grounds from that locality. The conclusion seemed inescapable: Though the direction of migration was unlearned, the birds were able to find it by means of celestial navigation.

The "need" to migrate and the direction of migration are unlearned. The star patterns that make navigation possible *are* learned. But how to learn them is innate. The entire mechanism is so constructed, though, that the learned behavior is dependent upon the unlearned, so under normal circumstances all birds of the same population learn the same thing and behave, for the most part, identically during migration. This intricate interaction of learned and unlearned behavior does have one common theme: that of adaptation. If any component were to fail, the birds would not reach their destination.

WHAT IS SOCIAL BEHAVIOR?

In a pioneering work, *The Social Life of Animals,* W.C. Allee[1] showed that many animals are far more resistant to noxious environments in groups than alone. Schools of fish are less vulnerable to predators than single fish because large numbers tend to confuse their predators. Many fish have elaborate evasive maneuvers that are workable only when large numbers of individuals perform them together. Flocks of birds may be able to find food better than single individuals. Insects are able to construct elaborate nests and raise young by mass-production methods when they cooperate. A pack of wolves and a pride of lions have greater success in hunting than the individual animals hunting alone. Animals that are hunted may be better able to detect or discourage predators when some individuals in the group are always on watch. It seems clear that social behavior offers definite benefits (Fig. 35–13).

The mere presence of more than one individual does not mean that the behavior is social. Many factors of the physical environment bring animals together in **aggregations,** but whatever interaction they experience is circumstantial. A light shining in the dark is a stimulus that causes large numbers of moths to aggregate around it. The high humidity under a log may attract aggregations of wood lice. Although it may be adaptive for these organisms to aggregate, their behavior is not social.

(c)

Figure 35–13 The advantages in numbers. (*a*) This large school of grunts makes it difficult for a predator to focus on and chase any one individual at a time. (*b*) This group of elephants is clustered in what is called a "circle of protection." Large elephants, which few predators would dare attack, stand on the perimeter of the circle, while the young elephants are protected inside. If you look closely, you can see an infant standing at the feet of its protectors. (*c*) In many herds of mammals one or several animals watch for predators while others carry out other activities, such as grazing or drinking. These giraffes were photographed at a waterhole in Namibia. (*a*, Charles Seaborn. *b* and *c*, E.R. Degginger.)

---

[1] Allee, W.C. *The Social Life of Animals.* New York, W.W. Norton, 1938.

Figure 35–14 A male hylid frog of Costa Rica calling to locate a mate. (L.E. Gilbert, University of Texas at Austin/BPS.)

Ethologists generally define **social behavior** as adaptive **conspecific** (among members of the same species) interactions. Many species that engage in social behavior form societies. A **society** is a group of individuals belonging to the same species that cooperate in an adaptive manner. A hive of bees, a flock of birds, a pack of wolves, and a school of fish are examples of societies. Some societies are loosely organized, whereas others have a complex structure. In a well-organized society there is cooperation and a division of labor among animals of different sexes, age groups, or castes. A complex system of communication reinforces the organization of the society. The members of a society tend to remain together and to resist attempts by outsiders to enter the group.

COMMUNICATION

The ability to communicate is an essential ingredient of social behavior, for only by exchanging mutually recognizable signals can one animal influence the behavior of another (Fig. 35–14). **Communication** occurs when an animal performs an act that changes the behavior of another organism. Communication may facilitate finding food, as in the elaborate dances of the bees. It may hold a group together, warn a group of danger, indicate social status, solicit or indicate willingness to provide care, identify members of the same species, or indicate sexual maturity.

Modes of Communication

Animal communication differs significantly from most human communication in that it is not symbolic. As you read, information is conveyed to your mind by words; yet the words themselves are not the information. They stand for it. The relationship between the word "cat" and the animal itself is a learned one; a person who could read only Japanese would not recognize it. This is not to say that **signals** in some sense are not employed by animals. In a way, all releasers are signals. However, releasers are not necessarily learned, whereas true symbols are.

Although in humans some body communication is culturally determined learned behavior, a large part of it (such as smiling) is truly universal and appears to be physiologically determined. The pupil of the human eye dilates in certain emotional situations, such as sexual interest or excitement. Without realizing it, people respond to such subtle cues. In experiments, a photograph of a woman's face with the pupils retouched to appear greatly dilated was far more attractive to male subjects than one in which they were shown as pinpoints.

Signals are often transmitted involuntarily as an accompaniment of the physiological state of the organism. Information about an animal's emotional or mental state may be transmitted even if no other members of the species are nearby. For example, a bird automatically gives an alarm call when it sights a predator. Certainly there are times when human beings would rather not communicate their true feelings—yet there may be instances in which we do not really have any choice, as with pupil dilation. And who has not blushed at a time when he would have given almost anything not to have done so?

Do animals ever employ symbols, or are animal signals totally restricted to the equivalent of gasps of alarm? The matter is controversial. Many ethologists think that even dogs do not respond to spoken language as we do, but deduce the behavior we command from an astute reading of human facial expression, vocal intonation, and bodily attitude. On the other hand, chimpanzees have been taught to speak a very few words meaningfully, and to a limited extent can use sign language appropriately. Whether apes employ symbolic language in nature is unlikely, although the potential seems to exist.

The singing of birds is an obvious example of auditory communication, serving to announce the presence of a territorial male. Some animals communicate by scent rather than sound. Antelopes rub the secretions of facial glands on conspicuous objects in their vicinity (Fig. 35–15). Dogs mark territory by frequent urination. Certain fish, the gymnotids, use electric pulses for navigation and communication (Fig. 35–16), including territorial threat, in a fashion similar to bird vocalization. As E.O. Wilson has said, "The fish, in effect, sing electrical songs." Who would have guessed it?

Pheromones

Pheromones are chemical signals that convey information between members of a species. They are a simple, widespread means of communication. Secretion of pheromones is the only communication mechanism available to unicellular organisms and to many simple invertebrates because other communication channels require rather complex sending and receiving mechanisms. Pheromone communication has been discovered in nearly all organisms studied, including plants. Many types of signals can be conveyed by pheromones. Most act as releasers eliciting a very specific, immediate, but transitory type of behavior. Others act as primers triggering hormonal activities that may result in slow but longlasting responses. Some may act in both ways.

An advantage to pheromone communication is that little energy must be expended to synthesize the simple but distinctive organic compounds involved. Conspecific individuals have receptors attuned to the molecular configuration of the pheromone; other species ignore it. Pheromones are effective in the dark, they can pass around obstacles, and they last for several hours or longer. Major disadvantages of pheromone communication are slow transmission and limited information content. Some animals compensate for the latter disadvantage by secreting different pheromones with different meanings.

Pheromones are important in attracting the opposite sex and in sex recognition in many species. Many female insects produce pheromones

Figure 35–15 Pronghorn antelope marking territory by applying scent from its facial glands on any convenient object. (Harry Engels, Bruce Coleman Inc.)

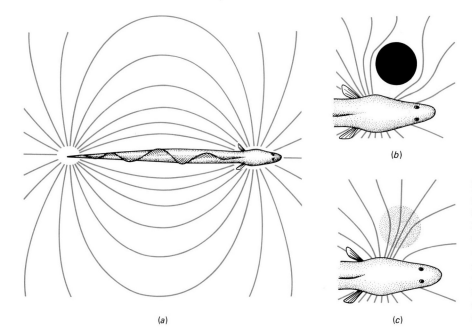

(a)

(b)

(c)

Figure 35–16 Certain fish navigate and communicate through electric signals. In *Gymnarchus*, an electrical field is radiated from its head to its tail (a). Objects that conduct less (b) or more electricity (c) than the surrounding water distort the electric field. Special sense organs along the sides of the fish help it use this information to navigate, intimidate territorial rivals, and locate mates.

that attract males of the appropriate species. We have taken advantage of some sex attractant pheromones to help control such pests as gypsy moths by luring the males to traps baited with synthetic analogs of the female pheromone.

Some aspects of the sexual cycle of vertebrates are affected by pheromones. The odor of a male mouse introduced among a group of females causes their estrous cycles to become synchronized. In some species of mice the odor of a strange male, a sign of high population density, causes a newly impregnated female to abort.

DOMINANCE HIERARCHIES

In the spring a paper-wasp nest may be founded cooperatively by females that have survived their winter hibernation. During the early course of construction a series of squabbles among the females takes place, in which the combatants bite one another's bodies or legs and (rarely) sting. In the end one of the young potential queens gets the upper hand over all the rest, and thereafter she is hardly ever challenged. The queen spends more and more time on the nest, and less and less time out foraging for herself. She takes the food she needs from the others as they return, and if they do not like it, they can leave; some do.

The queen then begins to take an interest in raising a family—her family. Since she is almost always at hand, she is able to prevent other wasps from laying eggs in the brood cells by rushing at them, jaws agape. At the same time, she cannot be stopped from laying all the eggs she wants, since she has already demonstrated that she cannot be successfully challenged. Furthermore, if any of the other females *do* manage to lay, she eats their eggs. *Her* eggs are undisturbed. Those subordinates that stay experience a definite regression of the ovaries, and eventually become sterile workers. The young that have been raised in the nest grow up as sterile workers from the outset, although they retain the potential of becoming reproductively competent in the event that the queen dies.

A careful analysis of this aggressive behavior discloses that the queen can bite any other wasp without serious fear of retaliation. There is usually another wasp, however, that can bite any wasp she chooses (other than the queen) without fear of retaliation. Thus, though the queen can bite any wasp in the nest, the other wasps are not equal in their relationships with one another. One can arrange the wasps into a definite **dominance hierarchy**, an arrangement of status that regulates aggressive behavior within the society:

$$\text{Queen} \longrightarrow \text{Wasp A} \longrightarrow \text{Wasp B} \longrightarrow \ldots \text{Wasp J} \longrightarrow \text{Wasp K}$$

Suppression of Aggression

Once a dominance hierarchy is established, little or no time is wasted in fighting. Subordinate wasps, upon challenge, generally exhibit submissive poses that inhibit the aggressive behavior of the queen toward them. Consequently, few or no colony members are lost through wounds sustained in fighting (Fig. 35–17).

Physiological Determinants

In some animals dominance is a simple function of aggressiveness, which is itself often influenced directly by sex hormones. Among chickens the cock is the most dominant. If a hen receives testosterone injections, her place in the dominance hierarchy shifts upward. If she is

Figure 35–17 Social animals use many signals to convey messages relating to social dominance. In the case of baboons, one signal serves two functions. Females turn their buttocks toward males to signify readiness to mate. Subordinate males use the same gesture to assure dominant males that they do not intend to challenge the higher-ranking animals. Here a subordinate male presents to a more dominant one. The dominant male in turn reassures the subordinate with a peaceful pat on the back. (T.W. Ransom/BPS.)

spayed, the reverse takes place. Recent tests on rhesus monkeys have shown that when males are dominant, their testosterone levels are much higher than when they have been defeated. Not only can estrogen reduce dominance and testosterone increase dominance, but dominance may even increase testosterone. It is not always easy to unscramble the situation!

In many species males and females have separate dominance systems, but in many monogamous animals, especially birds, the female takes on the dominance status of her mate by virtue of their relationship. However, this is not always the case. Like many fishes, labrid coral reef fishes are capable of sex reversal. What is odd is that the most dominant individual is always male, and the remaining fishes within his territory are always female. If the male dies or is removed, the most dominant female will become the new male. Should anything new happen to "him," the next ranking female will become the new sultan of the harem. Still other fishes exhibit the reverse behavior. In them the most dominant fish becomes a female.

TERRITORIALITY

Virtually all animals, and even some plants, maintain a minimum personal distance from their neighbors, as one can observe in the even spacing among the members of a flock of birds resting on a telephone line. Most animals have a geographical area that they seldom or never leave. Such an area is called a **home range** (Fig. 35–18). Since the animal has the opportunity to become familiar with everything in that range, it has an advantage over both its predators and its prey in negotiating cover and finding food. Some, but not all, animals defend a portion of the home range against other individuals of the same species and even against individuals of other species. Such a defended area is called a **territory.** The tendency to defend such a territory is known as **territoriality.**

Territoriality is easily studied in birds. Typically, the male chooses a territory at the beginning of the breeding season. This behavior results from high concentrations of sex hormones in the blood. The males of adjacent territories fight until territorial boundaries become fixed. Generally, the dominance of a cock varies directly with his nearness to the center of his territory. Thus, close to "home" he is a lion. When invading some other bird's territory, he is likely to be a lamb. The interplay of dominance values among territorial cocks eventually produces a neutral line at which neither is dominant. That line is the territorial boundary. Bird songs announce the existence of a territory and often serve as a substitute for violence. Furthermore, they announce to eligible females that a propertied male resides in the territory. Typically, male birds take

Figure 35–18 A coral reef has many secluded areas in which a territorial animal can establish a home range. Among the most territorial of coral reef fishes is the moray eel, pictured here, which will attack any animal (including a human diver) that comes too close to its shelter. (Charles Seaborn.)

Figure 35–19 Courtship displays.
(*a*) A male waved albatross from the
Galapagos Islands. (*b*) Courtship signals by male fiddler crabs are specific
to each species. This particular sequence of the motion of the large
right claw is characteristic of the species *Uca lactea*. (*a*, D.R. Paulson, University of Washington/BPS. *b*, after
Crane, J. *Zoologica* 42: 69–82,
1957.)

(*a*)

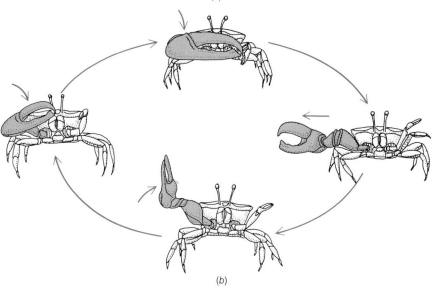

(*b*)

up a conspicuous station, sing, and sometimes display striking patterns
of coloration to their neighbors and rivals (Fig. 35–19).

Territoriality among animals may be adaptive in that it tends to
reduce conflict, control population growth, and ensure the most efficient
use of environmental resources by encouraging dispersion and spacing
organisms more or less evenly throughout a habitat. Usually, territorial
behavior is related to the specific life-style of the organism that displays
it, and to whatever aspect of its ecology is most critical to its reproductive
success. For instance, sea birds may range over hundreds of square
miles of open water but exhibit territorial behavior that is restricted to
nesting sites on a rock or island, their resource that is in the shortest
supply and for which competition is keenest.

SEXUAL BEHAVIOR AND REPRODUCTION

The minimum social contact, and for some species of animals (for example, many species of spiders) the only social contact, is the sex act. Fertilization and perhaps the rearing of young are some animals' only forms of
social behavior (Fig. 35–20). Let us consider the sex act as a basic example of social behavior, for the elements to which it can be reduced are
also the least common denominators of most social behavior.

Like other social relationships, the sex act is adaptive in that it
promotes the welfare of the species. It requires *cooperation,* the *temporary suppression of aggressive behavior,* and a *system of communication.* Among some jumping spiders, for example, mating is preceded by a

Figure 35–20 Male and female
jumping spiders, *Phidippus audax,* in
the courtship behavior that precedes
mating. The male performs an elaborate dance that inhibits the female's
natural aggression toward him, allowing him to get close enough to
inseminate her. (James H. Carmichael, Bruce Coleman, Inc.)

ritual courtship on the part of the male, the effect of which is to produce temporary paralysis in the female. While she is thus enthralled, the male inseminates her. Should she recover before he makes his escape, he becomes the main course at his own wedding feast. Whether he appreciates the opportunity or not, he is thus able to make the ultimate in material contributions to the eggs the female will presently produce. She would otherwise have to bear the metabolic burden of their production all by herself.

Since an individual that reproduces perpetuates its genes, it is not surprising that natural selection has favored mechanisms, including behavior, that promote successful reproduction. To fertilize as many females as possible, the males often compete intensely with one another. Sexual competition among males of the same species often has contributed to the evolution of large male size, brilliant breeding colors, ornaments, antlers, and other features that give a male an advantage in establishing dominance among his peers and attracting females.

Since the female usually chooses the mate, selection has favored those male characteristics that make a male most attractive. Selection has also favored those female attributes that enable her to determine that the male is worthy of her investment. Success of a male in dominance encounters with other males indicates his fitness to the female. Although the female of some species accepts the first male that attempts to court her, in other species the female tests the males by provoking encounters. Female baboons and chimpanzees in estrus have enlarged, brilliantly colored genital swellings that attract all males and incite competition among them (Fig. 35–21).

The victorious male courts the female. An important function of courtship is to ensure that the male is a member of the same species, but it also provides the female further opportunity to evaluate the quality of the male. Courtship may also be necessary as a signal to trigger nest building or ovulation. Courtship rituals may be long and complex. The first display of the male releases a counter behavior of a conspecific female. This, in turn, releases additional male behavior, and so on until the pair are ready for copulation. Certain male spiders make an offering of food to the female during courtship. This inhibits any aggressive tendencies that the female may have on being approached and also provides the female with some of the food needed for egg production.

Sexual selection has also led to strategies whereby a successful male protects an inseminated female from copulation with other males. After copulation a male damselfly continues to grasp and fly with the female until she has deposited her eggs. A successful drone honeybee discharges much of his genital apparatus into the virgin queen's genital passages, thereby blocking them against insemination by another male.

Figure 35–21 This female baboon signals periodic sexual readiness with her bright red posterior. (Clem Haagner, Bruce Coleman, Inc.)

Pair Bonds

A **pair bond** is a stable relationship between animals of the opposite sex that ensures cooperative behavior in mating and the rearing of the young (Fig. 35–22). In some species a newly arrived female is initially treated as a rival male. Then, through the use of instinctive appeasement postures and gestures by both male and female, the initial hostility is dissipated and mating takes place. Such sexual appeasement behavior may be very elaborate and gives rise to mating dances in some birds.

The releaser mechanisms involved in the establishment and maintenance of the pair bond are often remarkably detailed. A male flicker possesses a black, mustache-like marking under the beak. This is lacking in the female. If a "happily married" female flicker is captured and such a mustache is painted on her, her mate will vigorously attack her as if she were a rival male. He will accept her again if it is removed. Such cues enable courtship rituals to function as behavioral genetic isolating mechanisms among species.

Figure 35–22 A pair of nesting albatrosses. In many species pair bonds are maintained by grooming or other displays of affection. (E.R. Degginger.)

Care of the Young

Care of the young is an additional component of successful reproduction in many species, and it, too, requires a parental investment (Fig. 35–23). The benefit of parental care is the increased likelihood of the survival of the offspring, but the cost is a reduction in the number of offspring that

(a)

Figure 35–23 Examples of parental investment. (a) Cougars and black bears are normally mortal enemies and actively avoid each other. This confrontation was initiated when the cougar intruded in the area where a female bear was raising her cubs. (b) A chinstrap penguin regurgitating food for her young. Such an investment of time and energy on the part of the parent does not benefit the parent directly. It does help ensure the transmission of the parent's genes into succeeding generations. (c) A baby baboon rides on its mother's back during early infancy and comes to inherit some of her social status. (a, E.R. Degginger. b, P.R. Ehrlich, Stanford University/BPS. c, courtesy of Busch Gardens.)

(b)

(c)

can be produced. Because of the time spent carrying the developing embryo, the female has more to lose than the male if the young do not develop. Thus, females are more likely to brood eggs and young than males, and usually the females invest more in parental care.

Investing time and effort in care of the young is usually less advantageous to a male, for time spent in parenting is time lost in inseminating other females. Even worse, it may not be certain who fathered the offspring. Raising some other male's offspring is a definite genetic disadvantage. In some situations, however, it may be to the male's advantage to help rear his own young, or those of a genetic relative. Receptive females may be scarce. And gathering sufficient food may require more effort than one parent can provide. In some habitats the young may need protection against predators.

PLAY

Play is an important aspect of the development of behavior in many species, especially young mammals. It serves as a means of practicing adult patterns of behavior and perfecting means of escape, prey killing, and even sexual conduct. In true play the behavior may not be actually consummated. Thus a kitten pounces upon a dead leaf but of course does not kill it, even though the kitten administers a typical carnivore neck bite. When playing with a littermate, the same kitten may practice the disemboweling stroke with its hind claws (Fig. 35–24), but the littermate is not intentionally injured in the process.

HIGHLY ORGANIZED SOCIETIES

Some animal societies exhibit elaborate and complex patterns of social interactions. In these societies there is considerable division of labor not directly connected with the care of the young.

Insect Societies

Although many insects cooperate socially, such as tent caterpillars, which spin a communal nest, the most elaborate insect societies are found among the ants, wasps, and termites. Insect societies are held together by an elaborate system of releasers, and as a result they tend to be quite rigid. The social insects secrete pheromones that accomplish such tasks as suppressing the ovaries of worker honeybees or alerting an ant hill to the presence of an enemy ("alarm substance" is given off from a special abdominal gland of an excited worker). Virtually all the intercommunication within insect societies is comparable to the hormonal

Figure 35–24 Young lions playing in southern Africa. Play is behavior that is not consummated and often serves as a means of practicing behavior that will be used in earnest in later life, possibly in hunting, fighting for territory, or competing for mates. (Susan McCartney, Photo Researchers, Inc.)

communication within a single organism. The social coordination achieved approaches the behavior of the cells of a loosely organized individual of about the level of a sponge.

The social organization of honeybees has been studied more extensively than that of any social insect. A honeybee society generally consists of a single adult queen, up to 80,000 worker bees (all female), and at certain times, a few males called drones that fertilize newly developed queens. The queen deposits about 1000 fertilized eggs per day in the wax cells of a comb. Each fertilized egg develops into a larva that is fed by worker bees. After about 6 days the workers seal the cell off with wax and the larva develops into a pupa. About 12 days later an adult worker bee emerges from the wax cell.

Division of labor in the bee society is mostly determined by age. The youngest hive bees serve as nurse bees. They have special glands on the head that secrete **royal jelly** essential for the nutrition of all larval bees, and which, in larger quantities, produces queens. After about a week as nurse bees, workers begin to produce wax and build and maintain the

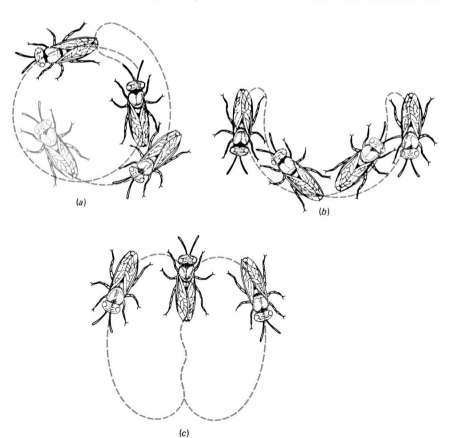

Figure 35–26 Three kinds of communication dances performed by honeybees, as observed by Karl von Frisch. (a) The round dance. (b) The sickle dance. (c) The waggle dance. (Modified from K. von Frisch and M. Lindauer.)

wax cells. Older workers are foragers, bringing home the vital honey (a half-teaspoonful per bee lifetime) and pollen. Most worker bees die at the ripe old age of 42—days, that is.

Behavioral cues tip off the bees when there is a labor shortage in any category. If there are too many larvae for the nurse bees, foragers will pitch in for the duration of the emergency, redeveloping royal-jelly glands, if need be.

The composition of a bee society is controlled by an antiqueen pheromone secreted by the queen. It acts as a releaser, inhibiting the workers from raising a new queen, and it also has a primer effect, for it inhibits the development of the ovaries in the workers (Fig. 35–25). If the queen dies, or if the colony becomes so large that the inhibiting effect of the pheromone is dissipated, the workers begin to feed some larvae the special food that promotes their development into new queens.

The most sophisticated known mode of communication among bees is a stereotyped series of body movements known as a **dance.** If a worker runs across a rich source of nectar, it can communicate this fact to the other bees within the hive by dancing on a vertical comb surface. If the food supply is nearby, the bee performs a **round dance** (Fig. 35–26), which generally excites the other bees and causes them to fly about in all directions till they have found the nectar. But if the source is distant, the bee performs a **waggle dance.** This "step" has a figure-eight configuration. As the bee treads the long axis of the figure-eight, she emits a series of distinctive sounds and waggles her abdomen from side to side. The angle between the long axis and the force of gravity is the same as the angle between the sun's rays and the shortest flight direction to the food source (Fig. 35–27).

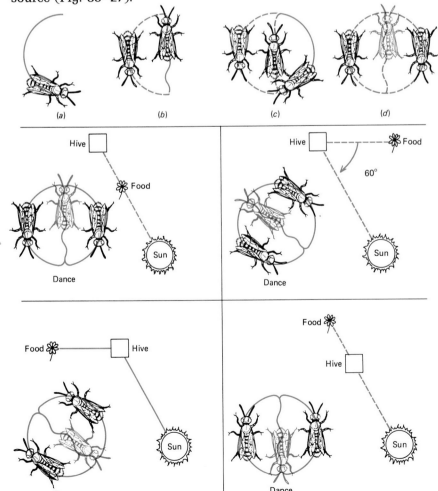

(a)   (b)   (c)   (d)

Hive   Food   Sun   Dance

Hive   Food   60°   Sun   Dance

Food   Hive   Sun   Dance

Food   Hive   Sun   Dance

Figure 35–27 Indication of direction by the waggle dance.

Focus on . . .

HYGIENIC BEHAVIOR IN BEES

Bees of the Van Scoy strain are very susceptible to a serious epidemic disease known as American foulbrood, which kills immature bees as they develop in brood cells of the honeycomb. Bees of the Brown strain are less susceptible because workers remove dead larvae from their waxen cells and discard them. This **hygienic behavior** has two components: (1) the removal of the wax cap of the cell, and (2) the removal of the dead larva. The Van Scoy bees leave the dead larvae to rot. Appropriately, this contrasting behavior is called **unhygienic.** The existence of these two variety-specific traits implied that hygienic behavior in honeybees was under genetic control.

In 1964 W.C. Rothenbuhler investigated the genetic basis of the behavior. Rothenbuhler crossed hygienic and unhygienic bees. The F_1 generation was unhygienic, so the hygienic trait was evidently recessive (see figure). Backcrossing these hybrid unhygienic bees with hygienic ones, he obtained four behavioral phenotypes in approximately equal proportions:

1. Worker bees that would neither uncap the cells of dead larvae nor remove the corpses even if the experimenter opened the cells for them.
2. Workers that both uncapped the cells and removed larvae.

3. Workers that opened the caps of cells but left the larvae untouched.
4. Workers that did not uncap cells but removed larvae if the cells were uncapped by the experimenter.

These results can be explained easily by the hypothesis that the ability to uncap is controlled by a single pair of allelic genes, and the ability to remove is controlled by an independently assorted pair. Notice that since the total behavior is sequential, the first part, the uncapping, serves as a releaser to trigger the second component, removal. But the appropriate neural pathways can develop in the nervous systems of the insects only if functional copies of the genes are present and unsuppressed by abnormal alleles.

Rothenbuhler did not demonstrate that the total uncapping behavior was completely specified by a single gene, nor was all of the removal behavior encoded in another single gene. Rather, he showed that there was at least one key gene necessary for the presence of each of these behaviors. Removing that gene was like removing a necessary cog in a machine. Even though the machine becomes nonfunctional, most of it is still there.

UURR X *uurr*

Uu Rr F$_1$

(phenotype: no uncapping, no removal)

UuRr X *uurr* (backcross)

Let *U* = no uncapping
 u = uncapping
R = no removal of dead larvae
 r = removal
U, R dominant
 u, r recessive

	UR	Ur	uR	ur
ur	*UuRr*	*Uurr*	*uuRr*	*uurr*
	no uncapping no removal	no uncapping removal	uncapping no removal	uncapping removal
	1	1	1	1

Inheritance of hygienic behavior in honeybees. The genes u and r represent the hygienic traits (see key); U and R represent unhygienic traits.

Vertebrate Societies

Among vertebrate societies we find a far greater range and plasticity of potential behavior than among insect societies. Vertebrate societies are far less rigid and much more adaptable to changing needs. In some ways they are also less complex. Vertebrate societies usually contain nothing comparable to the physically and behaviorally specialized castes of termites or ants. What is more, except for human beings, individual members of vertebrate societies are not as specialized in their tasks as are the social insects. A beehive may be considered collectively a kind of superorganism, but it is hard to interpret a wolf pack or a herd of red deer in that fashion.

Whereas the elaborate societies of social insects result from genetic programming of their behavior, the complexity of human society results from **culture,** behavior that is symbolically transmitted from generation to generation. It is true that human behavior is not wholly determined by culture. We even have a genetically determined capacity for acquiring and transmitting culture itself. However, the medium for cultural transmission is not the information contained in DNA but the information contained in language. Because of language and the culture it makes possible, the principal distinguishing characteristic of human societies is their great reliance upon *use of tools,* by which we have transformed the face of our planet.

KIN SELECTION

Altruistic behavior, in which one individual appears to act in such a way as to benefit others rather than itself, is frequently observed in the more complex social groups (Fig. 35–28). A particularly clear case of altruistic behavior has been observed by biologists Watts and Stokes in the mating of wild turkeys. Several groups of males, in each of which there is a dominance hierarchy, gather in a special mating territory and go through their displays of tail spreading, wing dragging, and gobbling in front of females who come to the area to copulate. One group attains dominance over other groups as a result of cooperation among the males within the group. The dominant member of the dominant group is the one to copulate most frequently with the females. Seemingly, the males that helped establish the dominant group but have low status within it gain nothing. Close analysis has shown that members of a group are brothers from the same brood. Since they share many genes with the successful male, they are indirectly perpetuating many of their genes. In this case altruism is closely related to **kin selection,** vicarious gene propagation among closely related individuals.

Kin selection may account for the evolution of the complex societies of social insects, in which some individuals specialize for reproduction and other close relatives do the chores of the colony. In the bee society the workers are sterile females, and the queen functions vicariously as their reproductive organ. If the queen produces offspring, at least a large portion of the genes shared by the queen and worker will have been passed on to the next generation, even though the worker herself has not reproduced.

This is particularly true of bees, wasps, and ants. Since the males are haploid, each male passes on his *entire* complement of genes to his offspring without the intervention of meiosis, and all sperm are genetically identical. Thus, each female bee has three-fourths of her genes in common with her sisters, instead of the more usual half. In fact, since the queen bee mates infrequently during her lifetime (she can store sperm for

Figure 35–28 Prairie dog. Low-ranking members of this social rodent group act as sentries. Sentries place their own lives in danger by exposing themselves outside their burrows. However, in this way they protect their siblings and by so doing ensure that the genes they share in common will be perpetuated in the population. This is a classic example of kin selection. (Tina Waisman.)

long periods between copulations), every bee alive in a hive at the same time may well have the same father. Expressed more exactly, any given worker bee will probably have half of her genes derived meiotically from the same father and therefore identical to half of the genes possessed by any other worker in the hive—plus half of the *remaining* half in common with any other worker in the hive that had, as it probably did, the same mother. That gives all workers a **coefficient of relatedness** of 75%, that is, 0.75.

Organisms in which both sexes are diploid, such as ourselves, have only a 0.50 coefficient of relatedness with our own children. To be sure, it is not important how closely related worker bees may be to one another, since they never have offspring in the normal course of events. However, sooner or later some of those sibling larvae being reared so assiduously by the worker bees will become drones or a new queen and *will* pass on those genes to the next generation. Apparently, therefore, it is more efficient for the worker to propagate her genes by proxy and the mass production of larvae than to lay her own eggs and go through the complete process of raising them to maturity herself.

Kin selection is not limited to social insects. Among some birds (Florida jays and others), nonreproducing individuals aid in the rearing of the young. Nests tended by these additional helpers as well as parents produce more young than nests with the same number of eggs overseen only by parents. The nonreproducing helpers, siblings of the parents, are apparently increasing their own biological success by ensuring the successful propagation of their genes through their siblings' offspring.

SOCIOBIOLOGY

Sociobiology is the school of ethology that focuses upon the evolution of behavior through natural selection. It represents a synthesis of population genetics, evolution, and ethology. Like many biologists of the past (such as Darwin), Edward O. Wilson and other sociobiologists emphasize the animal roots of human behavior, but they have attempted to inform their discipline with population genetics, with particular emphasis on the effect of kin selection on patterns of inheritance. Many of the concepts discussed in this chapter, such as altruism and paternal investment in care of the young, are based on contributions made by sociobiologists.

For the sociobiologist, the organism and its adaptations—including its behavior—are ways its genes have of making more copies of themselves. The cells and tissues of the body support the functions of the reproductive system. The reproductive system's job is the transmission of genetic information to succeeding generations.

There are unique pitfalls in attempting to reconstruct the evolution of behavior, since behavior rarely leaves an explicit fossil record. Additionally, by applying human social terms to behavior in animals that may be only superficially similar, we create the perhaps entirely false impression that it is the *same* behavior. It is an easy step from that to the assumption that the causes and utility of these behaviors are the same as those of corresponding human behavior. Consider, for instance, the question of whether humans are territorial. We do tend to preserve space between us as individuals, to defend our homes, and as groups to defend larger, political areas. However, do these behaviors have the same genetic and adaptive value in humans as in animals? And is human territoriality homologous with that of other animals, or is it merely analogous?

Also, problems of objectivity can exist. Any assumptions we may have about our own territoriality can cause us to look at the behavior of animals as a mirror of our own. Among closely related species of primates, social organization and the degree of territoriality and aggressive behavior vary widely. Which of these species should we choose as models for studying human behavior?

Most of the controversy that has been triggered by sociobiology seems related to its possible ethical implications. Sociobiology is often taken as denying that human behavior is flexible enough to permit substantial improvements in the quality of our social lives. Yet sociobiologists do not disagree with their critics that human behavior is flexible. The debate therefore seems to rest on the *degree* to which human behavior is genetic and the *extent* to which it can be modified.

As sociobiologists acknowledge, people through culture possess the ability to change their way of life far more profoundly in a few years than a hive of bees or a troop of baboons could accomplish in hundreds of generations of genetic evolution. This ability is indeed genetically determined, and that is a very great gift. How we use it and what we accomplish with it is not a gift but a responsibility upon which our own well-being and the well-being of other species depend.

CHAPTER SUMMARY

    I. Behavior consists of the responses of an organism to signals from its environment.

    II. Ethology is the scientific study of behavior under natural conditions from the point of view of adaptation.
       A. Behavior tends to be adaptive.
       B. Behavior tends to be homeostatic.

    III. Tropisms and taxes are examples of simple behavior.
       A. Plant behavior consists largely of tropisms—growth responses toward or away from a stimulus.
       B. Taxes are simple orientational behaviors, such as moving toward or away from a bright light.

    IV. It is adaptive for an organism's metabolic processes and behavior to be synchronized with the cyclical changes in the environment.
       A. Some biological rhythms reflect the lunar cycle or the changes in tides due to phases of the moon.
       B. In many species physiological processes and activity follow circadian rhythms.
       C. No single biological clock has been found. Biological rhythms are thought to be regulated by both endogenous and exogenous factors.

    V. Innate behavior is unlearned and may be triggered by a specific unlearned sign stimulus, or releaser.

    VI. Innate behavior is genetic; learned behaviors develop as a result of experience. The actual development of behavior is generally a product of a complex interaction between heredity and environment. Virtually all behavior is modifiable to some extent, and virtually all behavior possesses some genetic component or predisposition.

    VII. Before any pattern of behavior can be exhibited, an organism must be physiologically ready to produce it.

    VIII. In imprinting, a bond is established between the offspring and its mother during a critical period of development.

    IX. In some birds the need to migrate and the direction of migration appear to be genetically programmed, but how to navigate may be learned.

    X. Social behavior is adaptive conspecific interaction. A society is a group of individuals of the same species that cooperate in an adaptive manner.

A. In a society there is a means of communication, cooperation, division of labor, and a tendency to stay together.

B. Animals form societies because it is adaptive for them to do so.

XI. Animal communication involves the transmission of signals but does not utilize (as far as is known) symbolic language in the human sense.

A. Animals often transmit signals involuntarily as a result of their physiological state.

B. Pheromones are chemical signals that convey information between members of a species.

XII. Dominance hierarchies result in the suppression of aggressive behavior.

XIII. Organisms often inhabit a home range, from which they seldom or never depart. This range, or some portion of it, may be defended from members of the same (or occasionally different) species.

A. Defended areas are called *territories,* and the defensive behavior is *territoriality.*

B. Often, territorial defense is carried out by display behavior rather than actual fighting.

XIV. Courtship behavior ensures that the male is a member of the same species, and it permits the female to assess the quality of the male.

A. A pair bond is a stable relationship between a male and a female that ensures cooperative behavior in mating and rearing the young.

B. Parental care increases the probability that the offspring will survive. A high investment in parenting is less advantageous to the male than to the female.

XV. Play gives the young animal a chance to practice adult patterns of behavior.

XVI. Insect societies depend upon releasers and so tend to be rigid, with the role of the individual narrowly defined.

XVII. Vertebrate societies are far less rigid than insect ones. Though innate behavior is important in them, in general the role of the individual is learned.

A. Human society is by far the most complex of all vertebrate societies.

B. Human society is almost uniquely characterized by the possession of language, culture, and tool use.

XVIII. In altruistic behavior one individual appears to behave in such a way as to benefit others rather than itself.

A. Altruism may be closely related to kin selection.

B. Kin selection may account for the evolution of complex societies of social insects in which only a few members reproduce.

XIX. Sociobiology is a school of ethology that focuses on the evolution of behavior through natural selection.

Post-Test

1. Behavior may be defined as responses of an organism to _____ .

2. _____ is the study of behavior in natural environments from the point of view of adaptation.

3. A positive growth response of a plant to gravity is a _____ .

4. The movement of flatworms toward a piece of raw meat is a _____ _____ .

5. A biological rhythm with approximately a 24-hour cycle is a _____ _____ .

6. Animals that are most active at dawn or twilight are described as _____ .

7. Another name for a sign stimulus is a _____ _____ .

8. _____ behavior is mainly genetic; _____ behavior develops as a result of experience.

9. _____ is a form of learning in which a young animal forms a strong attachment to an individual (usually its parent) within a few hours of birth.

10. The term Zugunruhe refers to _____ _____ .

11. Adaptive interactions among members of a population are referred to as _____ _____ _____ .

12. A _____ is a group of individuals belonging to the same species that cooperate in an adaptive manner and have a means of communicating with one another.

13. A swarm of flies in a cow pasture is an example of an _____ .

14. An important difference between human and animal communication is that animal communication is not generally _____ .

15. _____ are chemical signals that convey information between members of a species.

16. An arrangement of members of a population by status is called a _____ _____ .

17. The geographical area that members of a population seldom leave is the _____ _____ _____ .

18. Territoriality tends to reduce _____ and control _____ growth.

19. A _____ _____ is a stable relationship between animals of the opposite sex that ensures cooperative behavior in mating and rearing the young.

20. In a beehive the youngest bees serve as _____ _____ bees; they secrete _____ _____ .

21. The extensive behavioral repertoire of the bee is almost entirely _____ (innate or learned).

22. Human society differs from other animal societies in that it depends mainly on the transmission of _____ _____ .

23. In _____ behavior one individual appears to act to benefit others rather than itself.

24. _____ selection favors the indirect perpetuation of an animal's genes by a relative.

25. According to sociobiology, an organism and its adaptations are ways that its genes have of _____ _____ _____ .

Review Questions

1. In what ways are the behaviors of *Philanthus,* the bee-killer wasp, adaptive?

2. What is the difference between a tropism and a taxis? Could plants be capable of taxes, or animals of tropisms?

3. What is imprinting and why is it considered a form of learning? What is its significance as an adaptation?

4. How does the response of a gull chick to feeding develop? Which components of the parent-chick interaction are learned and which innate? How do the innate components ensure uniformity of the learned responses?

5. Sensitivity to the earth's magnetic field has been observed in birds, bees, and bacteria. What possible reason would bacteria have to be sensitive to this stimulus?

6. Why is it adaptive for some species to be diurnal but others nocturnal or crepuscular?

7. How does physiological readiness affect instinctual behavior? learned behavior?

8. When Konrad Lorenz kept a greylag goose isolated from other geese for the first week of its life, the goose persisted in following human beings about in preference to other geese. How could this behavior be explained?

9. What distinguishes an organized society from an aggregation? Cite an example of an organized society, and describe characteristics that qualify the society as organized.

10. How many similarities between the transmission of information by symbolic language and by heredity can you think of? How many differences?

11. Contrast the "language" of bees with human language.

12. How does an organism learn its place in a dominance hierarchy? What determines this place? What are the advantages of a dominance hierarchy?

13. What is territoriality? What functions does it seem to serve?

14. What is sociobiology?

15. What is kin selection? How is kin selection used by sociobiologists to explain the evolution of altruistic behavior?

16. How is play behavior adaptive? Give specific examples.

17. How are pair bonds adaptive? Give specific examples.

18. What are some advantages of courtship rituals?

Readings

Alkon, D.L. "Learning in a Marine Snail," *Scientific American,* July 1983. Evidently for the first time, a cellular mechanism of learning has been defined.

Bonner, J.T. "Chemical Signals of Social Amoebae," *Scientific American,* April 1983. A discussion of the chemicals that cellular slime molds emit and respond to;

these chemicals permit different species to coexist and yet maintain their own identity.

Bonner, J.T. *The Evolution of Culture in Animals.* Princeton, Princeton University Press, 1980. The title tells the contents but fails to convey the charm of this scholarly little book.

Burgess, J.W. "Social Spiders," *Scientific American,* March 1976. A few species of spiders interact socially and build large communal webs.

Camhi, J.M. "The Escape System of the Cockroach," *Scientific American,* December 1980. A study of the mechanisms by which a roach rapidly escapes from predators.

Crews, D., and W.R. Garstka. "The Ecological Physiology of a Garter Snake," *Scientific American,* November 1982. An account of the physiological and behavioral adaptations of the red-sided garter snake to its harsh environment.

Gallup, G.G., Jr. "Self-Awareness in Primates," *American Scientist* July–August 1979. Chimpanzees and the like may, in a sense, be persons.

Gould, J.L. *Ethology.* New York, W.W. Norton, 1982. The best and most up-to-date general text so far.

Gould, S.J. "The Guano Ring," *Natural History,* January 1982. An easily studied and interpreted example of territoriality.

Heinrich, B. "The Regulation of Temperature in the Honeybee Swarm," *Scientific American,* June 1981. A discussion of thermoregulation in a swarm of bees.

Ligon, J.D., and S.H. Ligon. "The Cooperative Breeding Behavior of the Green Woodhoopoe," *Scientific American,* July 1982. Kin selection and altruism in a vertebrate society.

Matthews, G.V.T. Orientation and Position-Finding by Birds (booklet). Burlington, N.C., Carolina Biological Supply Company, 1974. Though written before the magnetic senses of animals were at all understood, this is a good summary of research on the use of visual cues by birds in navigation.

Partridge, B.L. "The Structure and Function of Fish Schools," *Scientific American,* June 1982. Schooling confuses predators and is coordinated by lateral line sensory input.

Von Frisch, K. *Animal Architecture.* New York, Harcourt Brace Jovanovitch, 1974. The construction projects of animals, from spider webs to anthills, not forgetting birds' nests.

Wicksten, M.K. "Decorator Crabs," *Scientific American,* February 1980. A discussion of a species of spider crabs that camouflage themselves by attaching materials to their exoskeletons.

Wilson, E.O. *Sociobiology, the New Synthesis.* Cambridge, Mass., Belknap Press, 1975. An important (to say the least) summary of Wilson's proposed mechanisms of the evolution of altruism and his views on kin selection.

Wursig, B. "Dolphins," *Scientific American,* March 1979. An interesting description of dolphin behavior and learning ability.

Chapter 36

ECOLOGY: THE BASIC PRINCIPLES

Outline

Learning Objectives

After you have read this chapter you should be able to:

1. Define, distinguish among, and give examples of populations, communities, and ecosystems.
2. Distinguish between habitat and ecological niche and relate them to the concept of adaptation.
3. Describe the effects of biotic potential and environmental resistance in keeping a population in balance.
4. Summarize the effects of competition and predation on population size.
5. Given an array of organisms, draw a food chain, identifying the mode of nutrition of organisms at each trophic level. Identify levels of greatest and least biomass and energy.
6. Summarize the energy relationships of an ecosystem (with particular attention to the second law of thermodynamics).
7. Describe the flow of nutrients through an ecosystem, using phosphorus as an example.
8. Describe the general effect of stress on communities and give an example.
9. Summarize the concept of ecological succession. Give a definition and an example of primary succession and secondary succession.

A decade or two ago one might have thought that ecology was invented by politicians. The concept seemed new to many people who heard for the first time (often in political speeches) about such issues as strip-mining, undrinkable water, all-but-unbreathable air, and vanishing wildlife. Today, despite the continued importance of environmental issues, many politicians have taken up new issues, and some look upon ecology as a passing fad, no longer a major problem of the 1980s. Unfortunately, the problems have not gone away, and our global life support system continues to deteriorate.

How could this have happened? The facts have not changed. Perhaps the explanation is that they were not perceived as facts but as doctrines of a social and political movement. Somehow in the heady excitement of the ecology movement we failed to communicate that ecology itself is not a movement. Ecology is a science. If that can be made clear, perhaps the facts can be approached once again in a new spirit. Dismantled programs can be reassembled, obscured goals rediscovered, jaded minds re-educated, and perhaps this time, priorities permanently reordered.

The first formal statement of the concept of ecology is attributed to the 19th-century German biologist, Ernst Haeckel. As proposed by him, the term embodies two Greek roots, *oikos*, house or habitat, and *logos*, study of. **Ecology** is, then, the branch of biology that examines the interactions between organisms and their environment, as well as with one another.

ECOLOGICAL ORGANIZATION

Groups of organisms may be associated in three main levels of ecological organization—populations, communities, and ecosystems. A **population** is a group of individuals of the same species that occupy a particular area at the same time. All the populations that occupy a particular area at a given time make up a **community** (Fig. 36–2). A community and its physical, nonliving environment constitute an **ecosystem.** As discussed in Chapter 1, an ecosystem may be as small as an aquarium containing fish, plants, and decomposers or as large as an ocean or forest. The largest, most self-sufficient ecosystem known is the **ecosphere—** our planet Earth and all of its inhabitants.

HABITAT AND ECOLOGICAL NICHE

No organism is distributed evenly over the face of the earth. Rather, each population occupies a particular habitat, a specific type of environment— a burrow in a prairie, the muddy bottom of a lake, or a treetop. Even the

Figure 36–1 The ecosphere. No significant loss or gain of matter occurs on the earth as a whole. All materials utilized by living things must be continuously recycled within this vast, closed system. (NASA.)

Figure 36–2 Ecology is the study of relationships. Organisms in a community, such as this coral reef community, are adapted to the demands of their environment and to one another and are often mutually interdependent. (Robin Lewis, Coastal Creations.)

water trapped in the leaves of a pitcher plant may contain protozoa and mosquito larvae that somehow escape digestion.

One thinks of habitat as horizontal—an area on a map. Yet communities generally show a variety of different vertical habitats that vary in physical factors, such as temperature, light, and oxygen. In a forest, for example, there is a vertical stratification of plant life, from mosses and herbs on the ground to shrubs, low trees, and tall trees. Each of these strata has a distinctive animal population. Even such highly mobile animals as birds are more or less restricted to certain layers. Some species of birds are found only in shrubs, others only in the tops of tall trees. There may be daily and seasonal changes in the populations found in each stratum, and some animals are found first in one, then in another habitat as they pass through their life histories.

An **ecological niche** is the role of a species in an ecosystem. To describe an organism's ecological niche, we must know what it eats, when it is active, where it goes, and how it affects other organisms and its nonliving environment. Ecological niche helps determine habitat. One would not expect to find a polar bear in an Amazon rain forest. It may be helpful to think of an organism's habitat as its address (where it lives) and of its ecological niche as its profession (how it functions in its community).

What determines where an organism can live and its ecological niche? Among the most important determinants are the organism's structural, functional, and behavioral adaptations. Also important are the competition of other organisms and the stage of development of the community.

POPULATION DYNAMICS

In a stable population there is a balance between biotic potential and environmental resistance (Fig. 36–3). **Biotic potential** is the capacity of a population to increase its numbers. This capacity is kept in check by **environmental resistance,** a combination of factors that limit the survival of the individuals of the population and prevent the population from increasing indefinitely. Environmental resistance includes such obvious limiting factors as lack of nutrients or water, adverse weather conditions, and lack of appropriate habitats. Less obvious factors, such as competition, predation, and disease, also limit the growth of a population.

If the environmental resistance for a population is less than its biotic potential, its numbers will increase. However, if environmental resistance is greater than biotic potential, its numbers will decrease. As

Figure 36–3 In a stable population there is a balance between the factors that increase numbers and those that limit population size.

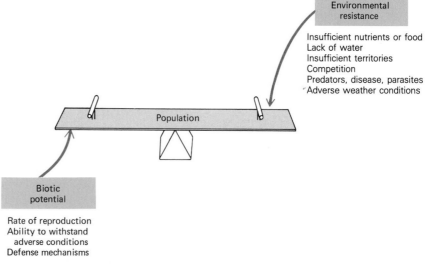

population size approaches the **carrying capacity,** the maximum population that the area can support without being damaged, the environmental resistance increases. (Population growth is discussed in further detail in Chapter 38.)

Adaptations and Biotic Potential

The environment includes **nonbiotic factors,** such as climate and the availability of light, water, and oxygen. Equally important are such **biotic factors** as competitors, predators, and parasites. To survive and to propagate its genes, an organism must be adapted to both biotic and nonbiotic factors. It must be able to withstand harmful factors and take advantage of life-sustaining factors. The better adapted an organism is to its environment, the greater its biotic potential—its ability to reproduce and increase its population.

A cactus, for instance, must be able to withstand drought and so has no leaves (Fig. 36–4) but does have a very extensive root system and much tissue in which water may be stored. It must also be able to withstand predation from animals that might take advantage of its water stores or accumulated food in an area where few plants of any kind are able to grow. Its spines certainly do discourage most grazers. As a primary producer, it is also adapted to the abundance of sunlight it receives and so has a green, often flattened stem provided with chlorenchyma and stomata for photosynthesis. Its biotic potential depends upon these adaptations and the extent to which they are appropriate for its environment and life-style.

Limiting Factors

Certain requirements must be met for an organism to survive and reproduce—an equable temperature range, adequate oxygen, an appropriate mixture of inorganic nutrients in the soil, sufficient food, adequate nesting sites, and so on. If some factor is essential, then it has the potential of posing a limit for the organism, or for growth of the population. A **limiting factor,** then, is whatever essential factor is in the shortest (or most excessive) supply. Were one to provide the optimum amounts of a former limiting factor, then something else would become limiting in turn (Fig. 36–5).

If a population of rats were supplied with a superabundance of food, for instance, the availability of water might nevertheless limit population growth. Were *that* supplied in large amounts, the availability of nesting sites might become limiting. If *those* were supplied, the stress engendered by excessive interaction among the animals might produce disease, or the crowding might permit disease organisms and parasites to pass from one to another so efficiently that epizootics (animal epidem-

Figure 36–4 Flowering barrel cactus. A cactus has an extensive root system for absorbing water and a great deal of tissue for storing water. However, it has no leaves, an adaptation that minimizes water loss. Its spines are an adaptation for discouraging would-be predators. (Noble Proctor, Photo Researchers, Inc.)

ics) would result. Indeed, we often do not realize what really is essential either to ourselves or to other organisms until it gets to be in short supply! (See Focus on Life Without the Sun.)

Limiting factors do more than limit potential populations. They also govern distribution and occurrence of populations within a community. As we will see in the next chapter, climatic limiting factors are primarily responsible for the differential distribution of major biotic communities on the earth. What is or is not an effective limiting factor depends on the adaptations of each type of organism. That is why one environment will give a particular species a competitive advantage over another, while another environment may see this reversed. It is also why organisms tend to remain in their ecological niches.

All known organisms produce more offspring than the number of individuals in the parental generation. One does not usually think of cactus plants as being capable of overrunning the earth, but ordinary North American prickly pear almost overran the continent of Australia earlier in this century. In the absence of natural enemies, the cactus spread without seeming limit, crowding out other vegetation and rendering vast tracts of land useless for grazing or agriculture. Only when one of the cactus's natural enemies, the *Cactoblastis* insect, was deliberately introduced did the prickly pear menace subside. In this case, the major component of the environmental resistance to the growth of the cactus population was a natural enemy. It was largely a **population-independent limiting factor** because the insects were able to search out and destroy most specimens of the cactus, even those isolated from one another. (Other examples of population-independent limiting factors are drought and frost.) But there are other components of the cactus's environmental resistance.

Considering the desert habitat of the cactus, one necessity that is bound to be in short supply is water. Water is a potential limiting factor for all desert organisms. Since the cactus requires a vast root system to collect whatever water does become available, there is a limit to how closely spaced cacti and other desert plants can be before their root systems compete to their mutual disadvantage. Water is therefore also a **population-dependent limiting factor** for them.

In actual practice many potential limiting factors may interact with one another, especially when one or more of them is near its tolerable extreme. Trees, for example, are less likely to be killed by frost when they are also abundantly watered. Thus, it has become fashionable to view the growth, population density, and geographic distribution of an organism as determined by a complex of factors, which *together* constitute the environmental resistance that a given species must tolerate.

The environmental resistance of a habitat, then, determines whether a particular organism can be a member of a specific community. A habitat's environmental resistance for one organism is far different from its resistance for another. The desert has much less environmental resistance for a plant like a cactus, which is adapted to it, than it has for one like a water lily, which is not.

Figure 36–5 How agricultural chemist Justus von Liebig, a 19th-century pioneer, viewed limiting factors. The capacity of the bucket depends on the length of the *shortest* stave. So, too, the welfare of an organism depends on the specific environmental factor that is least optimal for it. (Redrawn after Liebig's original version.)

Competition

One of the more important potential limiting factors in the life of any organism is competition. Competition can take many forms. Since two organisms of the same species have essentially identical ecological niches and much the same set of adaptations, **intraspecific competition** for the always limited resources of the environment can be expected to be acute. For instance, in the spring common North American gray squirrels

Focus on . . .

LIFE WITHOUT THE SUN

In the mid 1970s an oceanographic expedition studied the Galapagos Rift, a deep cleft in the ocean floor off the coast of Ecuador. The expedition revealed a series of extremely hot springs on the floor of the abyss in which seawater apparently had penetrated to be heated by the hot rocks below. During its sojourn within the earth the water had also been charged with mineral compounds, including hydrogen sulfide, H_2S. Certain bacteria, called chemosynthetic autotrophs, extract the energy they need for building carbon dioxide into carbohydrates not from light by photosynthesis but from the oxidation of inorganic compounds. Some of these bacteria can cause hydrogen sulfide to react with oxygen, producing water and sulfur or even sulfate. These reactions are exergonic and provide the energy to operate a version of the light-independent reactions of photosynthesis.

At the tremendous depths of the Galapagos Rift there is no light for photosynthesis and therefore little plankton except what rains down dead from the lighted surface layers. But the hot springs support a rich and bizarre life in great contrast to the surrounding lightless desert of the abyssal floor. Many of the species in these oases of life were new to science; some could not even be assigned with confidence, at that time, to known phyla. Some had astonishing adaptations. One species of clam, for example, is unique among invertebrates in possessing red blood cells that contain hemoglobin. The mystery is, what do these species live on?

The basis of the food chains in these aquatic oases is the chemically autotrophic bacteria we have mentioned, which are remarkable in being able to survive and multiply in water so hot (exceeding 300°) that it would not even remain in liquid form were it not under such extreme pressure. These bacteria function as primary producers. The energy that activates these living communities originated not in the thermonuclear reactions of the sun but in the heat released by slow radioactive decay in the depths of the earth.

The local ocean floor is dotted with the remains of extinct communities of this sort, which died out when the hot springs ceased to flow. Where new hot springs occur, however, swimming larval forms are apparently able to re-establish the communities in new locations. One would hardly think that brine, far hotter than the normal boiling point of water and charged with noxious gases and other chemicals, could be a necessity to any living thing. There is hardly a more dramatic example of a limiting factor.

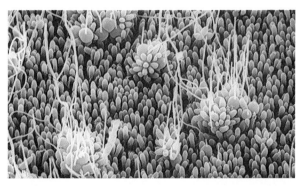

(a)

(b)

The inhabitants of the Galapagos Rift. (a) Scanning electron micrograph (×5200) of chemically autotrophic bacteria, which are the base of the food chain in hydrothermal vent systems. (b) Chemically autotrophic bacteria living in the tissues of these beard worms extract hydrogen sulfide and carbon dioxide from water to manufacture organic compounds. Because beard worms lack digestive systems, they depend on the organic compounds provided by the endosymbiotic bacteria, along with materials filtered from the surrounding water and digested extracellularly. Also visible in the photograph are some filter-feeding mollusks. (a, Carl D. Wirson, Woods Hole Oceanographic Institution. b, J. Fredrick Grassle, Woods Hole Oceanographic Institution.)

are very much in evidence. Often they attract attention with risky antics as they pursue one another through the treetops, over lawns, and across roadways. One is observing here, in most cases, young male squirrels being driven away from the nest by older, established territorial animals. Even in a densely urban habitat young squirrels have been shown to travel miles before they are either killed (as most of them are) by such hazards as cats or automobiles or are able to occupy territories vacated by the death or advancing age of previous tenants.

Interspecific competition can also be sharp among members of different but ecologically similar species (Fig. 36–6). A classic instance involves two species of anole (sometimes incorrectly called "chameleon") lizards in Florida. The native American anole used to occur almost everywhere in the lower southeastern states, but when Florida was colonized by the imported and larger Cuban anole, the American anole largely disappeared from urban habitats. Interspecific competition among songbirds affords another example, more easily studied in most parts of the country. One need only observe events at a winter bird feeder to see intense interspecific competition and interspecific dominance relationships among grackles, blue jays and song sparrows. Yet almost no competition is observable among birds of radically differing ecological niches. Song sparrows do not normally compete with buzzards, for instance.

According to Gause's law, often referred to as the principle of competitive exclusion, two species that compete in every way cannot coexist, at least not indefinitely. One of them is bound to have a superior set of adaptations to its life-style and sooner or later will force its competitor into local extinction.

Competition certainly helps determine whether an organism can live in a given area. In Florida, for example, many native coastal mangrove forests have been displaced by imported Brazilian pepper trees. The mangroves have grown there for millennia, and although other environmental disturbances (to which pepper trees are more resistant) have certainly played a part, it is the competitive superiority of the pepper trees in this habitat that has produced the demise of the mangroves. A well-tended lawn is a different kind of case. As long as it is properly fertilized and watered, it will have few weeds, since under those circumstances the grass has a competitive advantage. If it is neglected, however, the advantage goes to the weeds, which promptly overgrow the lawn.

When organisms possess specialized adaptations, they must usually give up something in return. Take the cactus we have been discussing. Most grass cannot grow where the cactus will survive, but the cactus *could* grow almost anywhere grass can grow. Why don't cacti take over the world? The answer probably is that where grass and other less specialized plants can grow, water is not normally an important limiting factor. Their stomata can remain open all day, and they can have well-developed leaves or blades with much surface area for gas exchange and light absorption. They can tolerate the accompanying water loss because water is available in relative abundance. The cactus, however, has no leaves and must rely on the small surface area of its stem for photosynthesis. Moreover, it stores carbon dioxide for photosynthesis during the *night,* keeping its stomata closed during the day (see Focus on Photosynthesis in Desert Plants, Chapter 17). So the sacrifice that the cactus makes to the demands of its harsh environment is photosynthetic inefficiency. This puts it at such a competitive disadvantage in areas of water abundance that it usually occurs only where the soil will not retain water.

Predation

Predation is nothing more than one living organism killing and eating another. By this definition, even herbivores can be predators upon plants, as in the case of tent caterpillars (Fig. 36–7) feeding upon the leaves of trees, possibly killing the trees by complete defoliation. However, we tend to think of predation as having an animal victim.

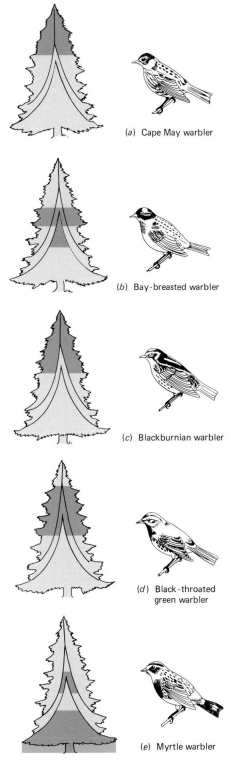

(a) Cape May warbler

(b) Bay-breasted warbler

(c) Blackburnian warbler

(d) Black-throated green warbler

(e) Myrtle warbler

Figure 36–6 Interspecific competition is reduced among warblers in the genus *Dendroica* as a result of vertical stratification. Each species forages for insects in different parts of the same spruce trees. The colored regions indicate where each species spends at least half its feeding time. (After MacArthur.)

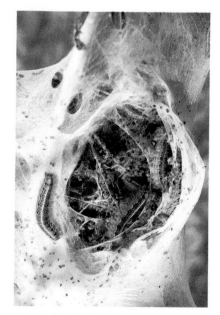

Figure 36–7 The larvae of the tent caterpillar live in a tentlike nest of silk, emerging from it daily to feed. Several large nests of these caterpillars can easily defoliate a small to medium-size deciduous tree. In this photo the nest has been opened to show the caterpillars within.

Unlike plants, animals can usually respond actively to attack. Large ruminants in vigorous health are formidable antagonists for wolves, and if they are alert, difficult to approach as well. Among smaller organisms defenses against predation rely less upon brute force. **Aposematic,** or warning, behavior or coloration warns predators of some nasty defense possessed by a potential victim. Doesn't the loud buzz and striking yellow-and-black color scheme of a hornet convey a clear message to you? It evidently makes a lasting impression on birds who have tried to eat such insects. In some forms of mimicry an animal presents the aposematic behavior without anything to back it up; a predator with experience with (or in some cases an innate aversion to) the model will avoid the mimic also (Fig. 36–8).

Other strategies of predator avoidance include rapid escape or withdrawal to inaccessible location, as one can observe in a cockroach under hot pursuit. A nocturnal pattern of activity keeps some organisms, again like cockroaches, out of the way of diurnal organisms such as ourselves, although there are many predators adapted for night hunting. Camouflage and protective coloration also tend to conceal potential prey organisms from possible predators.

Predators also have many adaptational strategies for overcoming the defenses of their prey: Wolves hunt in packs, thereby saturating the prey's defenses; lions ambush their prey; African hunting dogs tire the prey by chasing in relays; and coyotes attack vulnerable parts of the prey, like the soft bellies of armadillos. The prey-predator relationship is sometimes viewed as a kind of evolutionary arms race, in which predators must counter increasingly efficient defensive adaptations on the part of their prey with increasingly efficient methods of predation. On the other hand, a too-efficient predator might render its prey—and therefore itself—extinct.

Why this does not happen more often probably results from: (1) habitat complexity and (2) prey population reduction. When simplified prey-predator systems have been studied in the laboratory, the extinction of prey (and ultimately, predator) populations was much less likely to occur when habitat diversity was great. The more nooks and crannies where prey organisms can hide, or the more out-of-the-way places where they can live, the less likely it is that even the most efficient predator can find them all. Prey population reduction is related to this. If the predator is indeed very efficient, it will succeed in reducing the population density of the prey. The rarer the prey organisms become, the harder they are to find, especially in a diverse habitat. The predators starve, or at least raise fewer young successfully, and the predator population then also declines. The reduction in predators may permit the prey

Figure 36–8 American hognose snake. This snake has adopted a threatening posture reminiscent of a rattlesnake. It is all a bluff, however, or rather, mimicry. The harmless snake cannot be induced to bite defensively.

Figure 36–9 Robber fly consuming its insect prey. These predaceous flies consume a wide variety of species, especially houseflies. However, since the reproductive ability of most of the prey species far outstrips that of this predator, robber flies probably exert no significant effect on the population densities of their victims. (Rod Planck, Tom Stack & Associates.)

population to recover to some extent, which in turn produces an upswing in the density of predators, and so on. This negative feedback mechanism therefore tends to regulate the populations of both predator and prey, often in cyclic fashion.

To be sure, nature is rarely so simple. When predator populations are subsidized from some other food source, the predators can become numerous enough to wipe out local populations of prey completely. Rats living in a garbage dump can become so numerous that one cannot find nestling birds within a quarter of a mile—yet normally, rats would never be so numerous as to have a substantial impact on the bird population.

Predators are only one of the several components of environmental resistance, and not necessarily the most important. The impact of predators upon prey populations seems to vary with the particular prey-predator relationship. In some cases, such as the wolf-moose interaction, the predators seem able to take only old and sick prey. Since these probably do not have the potential to reproduce anyway, the wolves could influence moose population density only if they were able to take many young moose. An even better example is afforded by the interaction of houseflies and robber flies (Fig. 36–9). It is unlikely that robber flies reduce the housefly population even as well as a flyswatter does. Acting together, however, the *many* housefly predators may have an important influence on housefly population density.

Actually, the effect of prey population density upon that of predators is better understood. In a classic instance, the Hudson Bay Company has kept careful records of all pelts brought to their Canadian agents in trade since colonial times. Their records seem to indicate that as the populations of hares vary, so, too, do the much smaller populations of the fur-bearing predators, like the lynx, that live largely on the hares (Fig. 36–10).

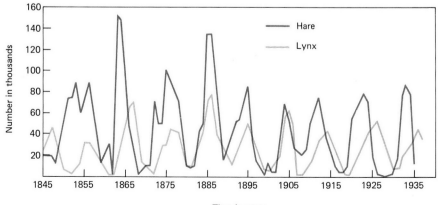

Figure 36–10 Changes in the abundance of the lynx and the snowshoe hare, as indicated by the number of pelts received by the Hudson's Bay Company. This is a classic case of cyclic oscillation in population density and of dependence of predator upon prey.

The most efficient predator of all is doubtless *Homo sapiens.* Few strategies can protect prey organisms from the attacks of this predator. Its ability to seek out and destroy every last specimen of a prey species or to destroy habitat has rendered many species extinct. The island of Mauritius once harbored a kind of giant flightless pigeon that became extinct in the 1700s because European sailors hunted and killed every one of them. Yet the dodo was neither the first nor the last. When the human population has encountered the combination of limiting factors that must ultimately stabilize it, one wonders how many other species of organisms will remain in what is bound then to be a biotically impoverished world.

ECOSYSTEM DYNAMICS

An ecosystem consists of producers, consumers, and decomposers, and their nonliving environment (Fig. 36–11). Producers, principally plants and algae, capture solar energy and with this energy make complex organic compounds from carbon dioxide and water. This food is both an energy-rich fuel and a building material for tissues. Producers typically make more food than they immediately need and can themselves be partially or totally consumed by consumers. The consumers—animals or protists—then use the food for their own energy and tissue growth. They consume this food in their own tissues and store the excess. Consumers both eat the food and breathe the oxygen that producers have made. Decomposers are microbes, principally fungi and bacteria, which metabolize dead material and body wastes.

Nutritional Life-Styles

There are really only two ways to obtain food. Producers are **autotrophs** (self-nourishers) and make their own food from simple inorganic materials. All other organisms are **heterotrophs,** organisms unable to synthesize their own foodstuffs from inorganic materials. Heterotrophs *must* live at the expense of autotrophs or upon decaying organic matter. All consumers and decomposers are heterotrophs.

PRODUCERS

The producers, or autotrophs, include plants, algal protists, and autotrophic bacteria. Autotrophs require only water, carbon dioxide, inor-

Figure 36–11 A small freshwater pond provides an example of an ecosystem. The great tropical Victoria lilies support a rich array of invertebrates, which are consumed by larger invertebrates, and vertebrates, which may consume one another. In the end all are decomposed by bacteria and fungi. (Joe Hill.)

ganic salts, and a source of energy. It is the source of energy that varies. Plants and certain bacteria are **photosynthetic autotrophs,** deriving the energy needed for biosynthetic processes from sunlight. A few bacteria are **chemosynthetic autotrophs** and obtain energy by oxidizing certain inorganic substances, such as ammonia or hydrogen sulfide. These bacteria have special enzyme systems that catalyze the oxidation of these substances and couple the oxidation with the generation of energy-rich phosphates. For example, nitrite bacteria *(Nitrosomonas)* oxidize ammonia to nitrites; nitrate bacteria *(Nitrobacter)* oxidize nitrites to nitrates; iron bacteria oxidize ferrous to ferric iron; and still other bacteria oxidize hydrogen sulfide to sulfates. The energy derived from these oxidations is used to synthesize all the organic materials necessary to maintain life and growth. The nitrite and nitrate bacteria are important in the cyclic use of nitrogen, for together they convert ammonia to nitrate, a form more readily used by many plants.

CONSUMERS

All consumers are heterotrophs. Most obtain food as solid particles that must be eaten, digested, and absorbed (holozoic nutrition). They must constantly locate, catch, and eat other organisms. To do this, animals possess a variety of sensory, nervous, and muscular structures to find and catch food, and digestive systems that convert food into molecules small enough to be absorbed.

Herbivores are animals that eat plants and obtain energy-rich compounds from the contents of the plant cells, compounds made by the plant using energy derived from sunlight. **Carnivores** eat other animals; **omnivores** eat either plants or animals.

Many **symbionts** (discussed in Chapter 27; see also Focus on Symbiosis), such as parasites, commensals, and mutualistic partners, are consumers. Parasites may obtain their nutrients by ingesting and digesting solid particles or by absorbing organic molecules through their cell walls from the body fluids or tissues of the host.

Detritus feeders, usually known as **scavengers,** have some of the characteristics of decomposers (Fig. 36–12). Many detritus feeders, like earthworms, prepare material for attack by decomposers. Others, like fly maggots, eat the decomposers themselves.

DECOMPOSERS

Fungi and most bacteria can neither make their nutrients by autotrophic processes nor ingest solid food. They absorb their required organic nutrients directly through a cell membrane. This type of heterotrophic nutrition is known as **saprobic** nutrition. Saprobes grow wherever there are decomposing bodies of animals or plants, or masses of plant and animal by-products. The humus of the uppermost layers of the forest floor, for instance, or the bottom ooze of some aquatic habitats is largely composed of decaying matter.

Yeasts are good examples of saprobes. They need only inorganic salts, oxygen, and some kind of sugar. From sugar they derive the energy to make all the other substances needed for life—proteins, fats, nucleic acids, vitamins, and so on. When plenty of oxygen is available, yeasts obtain energy by oxidizing glucose completely to carbon dioxide and water via the citric acid cycle. When the supply of oxygen is limited, they ferment glucose and form alcohol and carbon dioxide. This, as we have seen, yields only about one-twentieth as much energy as the complete oxidation of glucose, and therefore yeasts and other organisms that are capable of both aerobic and anaerobic metabolism grow very slowly in the absence of oxygen.

Figure 36–12 Detritus feeders. The greatest part of these fungi consists of microscopic hyphae dispersed throughout the decaying organic matter on which they are feeding. Only the fruiting bodies are visible. (Courtesy of Leo Frandzel.)

Focus on . . .

SYMBIOSIS

The most delicate and precise adaptations organisms must have are probably adaptations to one another. This is evident in ecological relationships between predator and prey, but the kind of mutual relationship called symbiosis is probably the most exacting of them all.

Symbiosis, which means "living together," is a term that encompasses a spectrum of relationships. At one extreme are the parasitic relationships, where one member of the association benefits at the expense of the other. At the other extreme is mutualism, in which *both* parties seem to benefit equally. Somewhere in the middle lies commensalism, in which the benefits of the relationship are one-sided, yet there is little harm to the other party (Fig. A).

How would you describe the relationship of the pilot fish to the shark (Fig. B)? These fish cling to the body of the shark host, feeding on scraps that it probably would not eat anyway. They might do a little harm to the shark in slowing it down, or causing it to expend more

B. A shark with several pilot fish attached. (Charles Seaborn.)

energy in swimming, yet it is possible that they aid it in subtle ways as well, perhaps by sensing potential enemies.

Algae live inside the body cells of *Hydra viridissima*. The hydra subsists on the organic products of the alga's photosynthesis. Is it a parasite? Perhaps the nitrogenous wastes of the hydra's metabolism make an important contribution to the nutrition of the algae (Fig. C).

Cleaning symbiosis is an especially interesting case of mutual adaptation. A tiny, brilliantly colored wrasse, which could be swallowed at a gulp by the much larger fish (such as groupers) it services, swims about them with impunity (Fig. D). The host may even open its mouth so that the little cleaner may pick decaying food from its teeth or devour parasites deep in the pharynx. The bright color of the wrasse serves as a negative releaser (see Chapter 35), which inhibits predation from its fish hosts. Indeed, so strong is the host's inhibition that the relationship serves as the basis of a kind of mimicry in which a brightly colored imitator of the wrasse swims up to the host fish. When the host presents itself for grooming, the impostor may steal into its mouth. Quickly grasping a mouthful of succulent flesh, the mimic darts off before the startled host can react.

A. This Caribbean shrimp (genus Pereclemines) *might receive some leftover food from the host anemone, but, primarily, it is given protection from predators. The anemone probably neither benefits nor is harmed by the relationship. (Charles Seaborn.)*

Food Chains and Food Webs

A series of organisms that eat and in turn are eaten by others is called a **food chain.** The position of an organism in a food chain is called its **trophic** (feeding) **level.** Producers are found at the first trophic level, herbivores at the second, and carnivores at the third and higher levels.

A very simple food chain might consist of grass eaten by a cow, which in turn is eaten by a human being (Fig. 36–13a). As in all food chains, the producer, in this case the grass, occupies the first trophic level. When the cow, a **primary consumer,** eats the grass, some of the chemical energy stored in the grass's organic molecules is transferred to the second trophic level. The cow now enjoys the benefits of the photo-

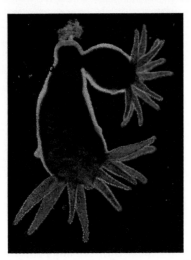

C. A budding Hydra with symbiotic algae. (Walker England, Photo Researchers, Inc.)

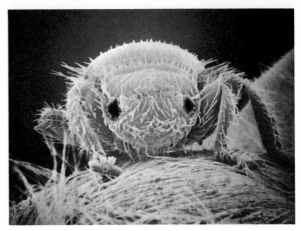

E. A scanning electron micrograph (falsely colored to show contrasts) of a wingless fly on the back of a bee. (Dr. Brad Adams, Photo Researchers, Inc.)

Perhaps the most interesting relationships, at least from an evolutionary point of view, are the frankly parasitic ones. Often parasites lack body parts that their nonparasitic relatives possess. An example is the wingless fly *(Braula)* shown riding on the back of a bee host (Fig. E). What use are wings to the fly? The bee does all the flying; why carry excess baggage?

Such adaptations severely limit many parasites in their choice of a host. Bird lice, for example, are best transmitted from generation to generation of their hosts by nest contact, but this tends to minimize the chance that they will be transmitted to other species of bird. Through this and many other adaptations, parasites become not only host-dependent but host-specific as well. If hosts evolve a resistance to parasites, then the parasites may well have to coevolve, keeping up with the new "design features" of their hosts. In fact, evolutionary taxonomists have used the more easily interpreted features of the related species of parasites to give them clues as to the possible evolution of their hosts. If two species of parasites seem closely related, the reasoning goes, so must be their hosts.

D. A wrasse with a grouper. (Charles Seaborn.)

synthetic activities of the grass. The human being in this case is a **secondary consumer,** who by eating the cow gains some of the chemical energy originally captured by the producer but now stored in the bonds of the cow's own organic compounds. Thus, in a food chain, energy is transferred from the eaten to the eater.

An animal may be a member of several food chains and may occupy different trophic levels in those food chains (Fig. 36–13b). For example, a human may be a primary consumer in one food chain, eating plant foods, but a secondary or tertiary consumer in other chains, eating herbivores or carnivores. In most communities there are many food chains that interconnect at various levels, forming a **food web** (Fig. 36–14). A food web may be extremely complex, involving thousands of types of

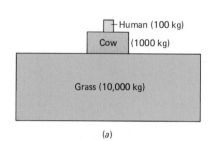

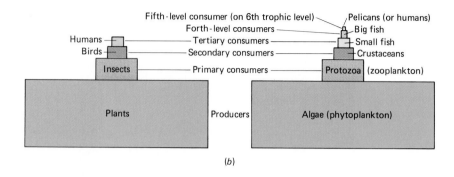

(a) *(b)*

Figure 36–13 *(a)* A simple food chain. Note that over a given period 10,000 kilograms of grass can support only 1000 kilograms of cow, which in turn can support only 100 kilograms of human being (if that person ate only beef). *(b)* Comparison of a terrestrial *(left)* with an aquatic *(right)* food chain. Draw several food chains of your own design.

organisms. This produces many alternative pathways of energy flow, which contribute to the stability of the community.

Organic waste products and corpses of producers and consumers support decomposers. In turn, decomposers serve as the basis of subsidiary food chains that support such organisms as detritus feeders. For example, a food chain might begin with plants whose dead leaves are decomposed by fungi and bacteria (Fig. 36–15). The microorganisms are eaten by detritus feeders, perhaps maggots. Then, chickens might consume the maggots. Finally, the chicken or its eggs might be eaten by human beings.

If we were to count the number of organisms in each trophic level, we would typically find that there are more producers than herbivores, and more herbivores than carnivores. This can depicted by a **pyramid of numbers** (Fig. 36–16). Sometimes, the pyramid of numbers can be inverted. For example, a few large trees can support many insects and small birds.

Energy Transfer Through Food Chains

As we saw in Chapter 6, energy is transferred from autotrophs, principally plants, to heterotrophs by means of organic compounds. One of the functions of these food compounds is to provide energy that is liberated

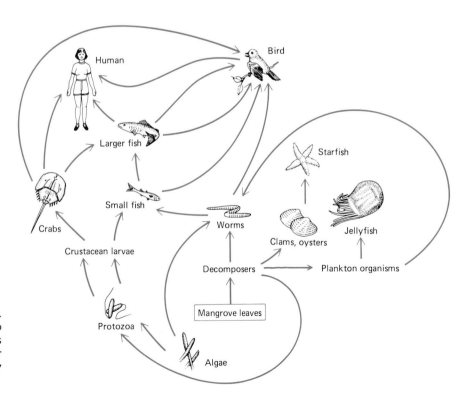

Figure 36–14 A marine food web. Arrows point from each organism to its consumer. Note that even this web is simplified. Human beings, for example, could eat algae, thereby acting as primary consumers.

(a)

(b)

Figure 36–15 The beginning of a detritus food chain. Coastal mangrove trees (a) shed their leaves, which decay under water. (b) The decaying leaves and the decay organisms themselves are eaten by tiny crustaceans, which in turn are eaten by fish.

by such processes as the citric acid cycle. Another function is to provide the building materials for bodies. Virtually any chemical element that is necessary for life is first absorbed from the environment by some plant, which initially incorporates it into food molecules. These then pass into the metabolisms of herbivores and carnivores and ultimately organisms of decay. Eventually, those organisms of decay liberate all the elements put into food by plants in the form of simple, inorganic substances.

Only a small portion of the energy captured by a producer can be transferred to the herbivore that eats it. This is because the producer uses a great deal of energy for its own biological activities. Energy must be used for synthesis of needed compounds, growth, transport of materials, and many other processes within the plant. Then, also, for every energy process that the plant carries on, a great deal of energy is lost as heat, as explained by the second law of thermodynamics (Chapter 6). Similar energy consumption occurs within the herbivore. Additional energy is lost when undigested food passes out of the body in feces or when the organism dies. Some of that energy is captured by decomposers and detritus feeders. Decomposing organisms also produce heat. A manure pile can be seen to steam on a cold day, and curing hay produces enough heat to be a fire hazard.

On average, the energy transfers from one trophic level to the next are only about 10% efficient. This means that only about 10% of the energy captured by the producers can be passed on to the primary consumers. Then, only about 10% of the energy within the primary consumers can be passed on to the secondary consumers, and so on. In studying energy transfer it is convenient to think of a food chain as a food pyramid that shows the relative size decrease of each level (Fig. 36–13).

Figure 36–16 Pyramids of numbers. (a) A pyramid depicting the approximate numbers of organisms at each trophic level in a given area (perhaps 1000 square meters) of a temperate grassland ecosystem. (b) The relationships in a forest community are illustrated by an inverted pyramid. Most of the producers are large trees and shrubs, and each can support many consumers.

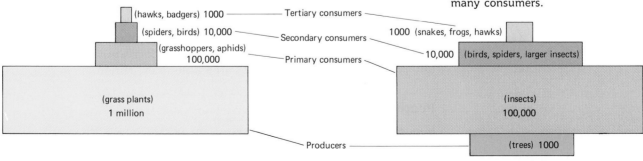

(hawks, badgers) 1000 — Tertiary consumers
(spiders, birds) 10,000 — Secondary consumers — 1000 (snakes, frogs, hawks)
(grasshoppers, aphids) 100,000 — Primary consumers — 10,000 (birds, spiders, larger insects)
(grass plants) 1 million — (insects) 100,000
Producers — (trees) 1000

(a)   (b)

Figure 36–17 A pyramid of biomass for a given area of a temperate grassland. Compare with the pyramid of numbers shown in Figure 36–16 (*a*).

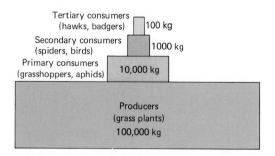

Tertiary consumers (hawks, badgers) 100 kg

Secondary consumers (spiders, birds) 1000 kg

Primary consumers (grasshoppers, aphids) 10,000 kg

Producers (grass plants) 100,000 kg

Biomass and Trophic Levels

A number of studies of communities have indicated that the primary producers are typically the most numerous and, collectively, weigh the most (Fig. 36–17). The **biomass,** weight of living material, decreases with increase in trophic level, and there are usually also fewer carnivores in any ecosystem than there are herbivores. In many cases the herbivorous animals have a biomass about 10% that of the plants. In turn, highly specialized carnivores may together weigh only 10% of the collective weight of the herbivores. Thus, there could be a 1000-fold reduction of biomass from primary producers, such as grass, to ultimate predators (those that have no natural predators), such as hawks. Predators, especially large ones, are never common, for it takes many square miles of habitat to support each one.

The biomass ratio varies considerably among communities and even within the same community at different times. The reason is that producers and primary consumers may be eaten at different rates so that, at least temporarily, predators might eat so many herbivores that they outnumber their prey. However, over a period of time the rate of production of primary producer biomass cannot be outpaced by the rate of production of any of the consumers dependent upon it. Productivity studies over a period of time always yield progressive reductions with each trophic level.

One might wonder why it is advantageous for an organism to be a consumer at all, let alone a carnivore. But consider how little energy is obtainable from most kinds of plant food relative to its bulk; it is no accident that many reducing diets emphasize lettuce. Indeed, most herbivores spend most of their time just eating, as one can readily observe by watching a cow spend her day chewing cud. The consumer of the cow, however, eats energy that is far more concentrated. Carnivores can often go for days between meals. Their population densities may be low, but individually, they live well. Herbivores may be thought of as rather inefficient transformers of energy, taking energy from a large source of low concentration and transforming it to a small product of high concentration. This suggests a possible reason why there are seldom four links in a typical food chain. The ultimate predators of such a chain would be specialized for eating carnivores, but there would be nothing to be gained by always eating carnivores rather than herbivores (the energy concentration of their tissues would be about the same). In a long food chain the total biomass of the ultimate predators would be very small indeed; they would be rare and vulnerable to extinction.

Productivity

In most ecosystems all energy input is via the producers. The rate of total energy storage by producers in an ecosystem over a certain period of time is referred to as **gross productivity.** However, much of the energy stored

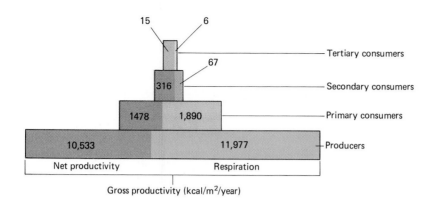

Figure 36–18 A pyramid of energy for a river ecosystem illustrating gross productivity, net productivity, and net community productivity. Measurements are in kilocalories per square meter per year. (Redrawn from Turk, J., and A. Turk. *Environmental Science*, 3rd ed. Philadelphia, Saunders College Publishing, 1984.)

by producers is used in cellular respiration and other activities. The rate of energy storage in the biomass of the producers, after their rate of energy use has been subtracted, is **net productivity** (Fig. 36–18). This energy may be stored in the form of new biomass (growth and new producers), stored organic compounds, or even in seeds and fruit. **Net community productivity** is the rate at which energy is stored in the community as a whole. To compute it, respiration in producers, consumers, and decomposers is subtracted from gross productivity.

Different kinds of ecosystems have different characteristic productivities (Fig. 36–19). Over most temperate regions annual net production is about 500 to 2000 grams per square meter. Productivity depends upon environmental conditions favorable to growth of the producers. Fertile soil, sufficient sunlight and moisture, and a long growing season contribute to high productivity. This is why estuaries, marshes, and many tropical ecosystems are very productive, whereas desert areas have low productivity. Except near the coasts, oceans are not very productive, owing mainly to a lack of sufficient nutrients. Dead plants and animals sink to the ocean bottom, where they are decomposed. However, the nutrients released are not readily used because there is not enough light to support photosynthesis. About 60% of the earth's surface is covered by open ocean, where the annual net productivity is only about 500 grams per square meter or less. On agricultural land measurements of productivity reflect crop yields.

Nutrient Cycling

Simple inorganic substances can be recycled indefinitely, being built up into organic materials and broken down again into simple components any number of times (at least in theory). By their action decomposers usually prevent the permanent accumulation of corpses and wastes, and

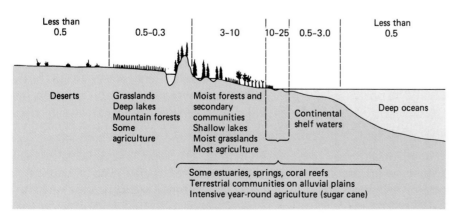

Figure 36–19 The world distribution of primary production in grams of dry matter per square meter per day, as indicated by average daily rates of gross production in major ecosystems. (Redrawn from Odum, E.P. *Fundamentals of Ecology*, 3rd ed. Philadelphia, Saunders College Publishing, 1971.)

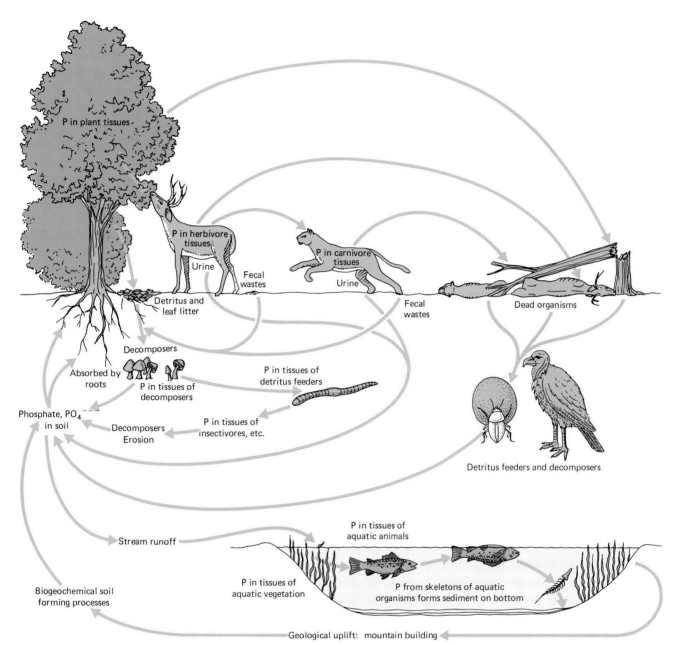

Figure 36–20 The phosphorus cycle is an example of a biogeochemical cycle.

more important, liberate recyclable materials not liberated by the metabolism of consumers. These include not only carbon, hydrogen, and oxygen, but also nitrogen, sulfur, potassium, phosphorus, iron, and other elements. Some may be lost by runoff from precipitation, but in undisturbed ecosystems most find their way back to the roots of the primary producers. Detritus feeders, such as maggots, lobsters, and earthworms, play an auxiliary role to decomposers.

Both decomposers and detritus feeders are themselves eaten by consumers and are also decomposed by other decomposers. Ultimately, the entire biomass produced by the primary producers is reconverted to inorganic materials by the action of decomposers. Both biological and geological activities are involved in nutrient cycling. Appropriately, such pathways are referred to as **biogeochemical cycles.** The phosphorus cycle is shown in Figure 36–20. (The nitrogen cycle is illustrated in Chapter 19.)

STRESS AND ECOSYSTEMS

Stress may prove a limiting factor for some organisms; it may also become a limiting factor sooner for some organisms than for others. If stress is applied to a community, it will reduce species diversity and complexity by eliminating or reducing the population of those organisms least well adapted to it. On the other hand, those organisms well adapted to the stress (and thereby freed from competition) may prosper greatly as long as the stress persists.

The earliest studies of community stress were performed at Brookhaven National Laboratories on Long Island to determine what effect large doses of ionizing radiation might have on natural communities. The researchers discovered that the sparse oak woodland exposed to the radiation was completely destroyed at the highest intensities, with progressively less destruction in areas farther from the radiation source, until at a considerable distance there was no obvious effect. The important thing was that the number of species, or more exactly, the species diversity increased with the distance from the radiation source. In other words, small amounts of radiation killed rather few species, larger amounts killed more, and the very largest amounts killed all.

Though this may seem obvious in retrospect, a less than obvious implication of the study, and one that has been repeatedly confirmed with many stressing agents, is that stress simplifies communities (Fig. 36–21). This is because the stressing agent leaves only those organisms that are resistant to it. These may, however, attain great numbers of biomass, since they are freed from competition with the stress-sensitive species. The nature of the stressing agent makes a difference in the exact species composition of the community; two such agents might simplify the same community comparably but in different ways, for each would leave a different assemblage of organisms unscathed.

An excellent example is a lawn. As long as it is regularly mowed, all the plants in it (not only grass but also chickweed, plantain, dandelion, portulaca, and other pests) must be resistant to the mowing. The grass is resistant by virtue of its basal meristem (Chapter 19). Most of the others are resistant as a result of their low, spreading growth habit, which enables them to miss the blades. Should the mowing stop, a much greater variety of plants could and would live there—shrubs and tall weeds, for instance. In time the diversity of the lawn would become very great and by then it would, in fact, have ceased to be a lawn and would have begun the successional process (soon to be discussed) that would perhaps ultimately produce woodland.

Figure 36–21 The Dolly Sods area of West Virginia, a community subject to considerable natural stress. Branches are missing from one side of many coniferous trees, mostly because of the prevailing winds and the development of rime ice at low temperatures. Very few plant species are able to grow in this area of sparse soil and harsh climate. The stress is not entirely natural, however, for the area was denuded of much of its original topsoil by disastrous fires in past years.

Stress renders a community more susceptible to further stress because it reduces its diversity. A disease that kills a particular species of grass would be of little consequence in a woodland or even a prairie but could be a disaster in your front lawn. Since the pollution and other stress that we place on natural communities around us tends to simplify them in just this way, we approach ever nearer the point at which they can absorb no further punishment and some random factor causes their collapse.

SUCCESSION

Communities of organisms do not spring into existence in final form but typically pass through a series of stages, beginning with a **pioneer community** and ending with a **climax community,** which as far as one can tell is ultimately stable and undergoes little further change. The whole sequence of communities leading to the climax is known as a **sere.** The progressive series of changes in the kinds and numbers of organisms leading to a climax community is known as **ecological succession** (Fig. 36–22).

Generally, the earliest stages of ecological successions are the most highly stressed, since habitats unmodified by living things are usually harsh. Therefore, early stages of successions tend to have low species

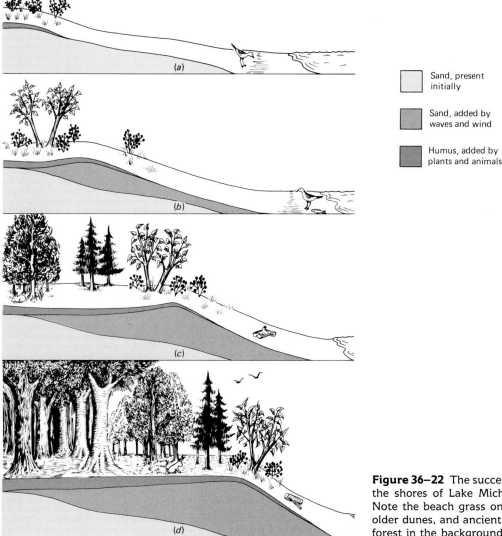

Sand, present initially

Sand, added by waves and wind

Humus, added by plants and animals

Figure 36–22 The succession of communities along the shores of Lake Michigan in northern Indiana. Note the beach grass on the foredunes, shrubs on older dunes, and ancient dunes with an established forest in the background. (After Buchsbaum.)

diversity. As they proceed, however, they tend to ameliorate the environment, which produces new ecological niches that can be occupied by other, more diverse organisms. Stress may even produce conditions to which newcomers are better adapted than are the members of the pioneer community, so the pioneers are succeeded by interlopers that could not have existed under the original conditions. A climax is reached when a community is established that contains no potential ecological niches that could be filled by, or better filled by, any organisms not already present in the community. Stress tends to delay succession, or rather to produce a kind of climax community that will persist as long as the stress is applied. In the absence of the stress, succession will continue.

A Typical Succession

In theory, soil nutrients originate from the parent rocky material that, when weathered, gave rise to the soil. This is such a slow process that the mineral needs of mature plant communities cannot be sustained solely by continued weathering of the rock. The very earliest plant communities to colonize a habitat, however, may indeed have obtained their mineral nutrients from the rocky materials upon which they grew. A succession that begins in such an initially lifeless habitat is said to be a **primary succession** (Fig. 36–23). This is different from a **secondary succession** (as might take place in a neglected lawn) which begins with a pre-existing community (Fig. 36–24).

Imagine a bare rock surface on a moderately tall mountain in a warm climate. Few organisms could do more than rest briefly on its surface, for it would provide almost none of the conditions needed for life. One form of life could exist and even thrive there, however: lichens.

The very first lichens might form a layer growing among the grains of a rock surface, or even a millimeter or two inside the rock itself, but soon they would be visible on the rock surface. As has already been discussed, a lichen is not a plant; it is a compound organism consisting of one of several species of a fungus in association with one or another species of algae or cyanobacteria. Lichens are amazingly hardy, tolerating extremes of dehydration that would kill any true plant. This obviously suits them to the harsh, bare-rock habitat. Equally important, lichens have an incredible ability to digest or solubilize that rock and to concentrate nutrients from it and from the surrounding air, probably via rainfall.

Lichens are not immortal, however, and when they die and decay, the minerals they have accumulated in their bodies are released by decomposers and taken up by living lichens, in addition to whatever the new generation of lichens can liberate afresh from the rock. Thus, the lichen community gradually accumulates an ever-increasing capital of minerals. These minerals are stored in the bodies of living lichens, in the organisms that are now living on them, and to a lesser extent, in the organic detritus that is beginning to accumulate under and around them. By now, fine particles of rock have weathered away from the surface but have been trapped by the lichen community and cannot be washed away as before. Eventually, the mineral particles and organic detritus form a thin layer of true soil capable of briefly holding some moisture. This soil contains a substantial part of the accumulating mineral capital of the developing community of organisms.

As its thickness increases and more minerals accumulate in it, the soil becomes suited to the requirements of some real plants—mosses. They now take hold among the lichens and grow vigorously. In fact, these mosses are better adapted to the less rigorous conditions the lichens

Figure 36–23 Primary succession. This view shows small plants growing on recently cooled volcanic lava in Hawaii. (Charles Seaborn)

Figure 36–24 Secondary succession in the taiga. The taiga is a coniferous community, but when it is disturbed by logging or fire, the mid-succession trees are usually hardwoods, such as the brilliant yellow aspens shown here. In time, conifers will displace the aspens. (Tina Waisman.)

have created than are the lichens themselves. Gradually they crowd the lichens out, going through many cycles of life and decay themselves in the process.

Minerals continue to weather out of the parent rock, now buried, after perhaps a thousand years, beneath as much as an inch of soil. These minerals are added over the course of years to the accumulating capital of the now radically changed community. Few or no minerals leave the community, for as soon as they are released from the dead organisms by decay bacteria and fungi, they are either reincorporated into young organisms or held in the soil temporarily in absorbed form.

Eventually, the moss may produce enough soil, especially in the deeper crevices, so that grass and even small trees can take root. Even later these successor forms may so dominate the community that the lichens and moss that preceded them may be almost impossible to find.

Why Successions Occur

Why does ecological succession take place at all? Why must a sere occur to prepare the habitat for the climax community? Each community modifies the habitat, so it becomes less harsh and stressful. That permits the establishment of another community that is more complex and more exacting in its requirements. The reason for this becomes evident when one considers the effects of stress upon communities. What environment could be more stressful than that of a pioneer community? The extremes of temperature often are almost outside the range that most living organisms can tolerate. Water is usually in very short supply, and mineral nutrients can be almost unavailable. Only a few of the very toughest organisms can withstand a pioneer environment in a primary succession.

In succession, the actions of pioneer organisms on the habitat may make it more tolerable, but the very pioneer organisms that have made the habitat less harsh are bound to lose in the competition that will likely ensue with less specialized organisms. These, in turn, modify the habitat further and are replaced by still other organisms until, finally, no more efficiently productive organisms are to be found and a climax community results. Since a stressed community is also a simple community, as conditions moderate in the course of succession, the community becomes steadily more diverse until in the near-climax* state diversity and complexity are at a maximum.

CHAPTER SUMMARY

I. Ecology is the study of interactions among living things and between living things and their environment.
II. Groups of organisms form populations, communities, and ecosystems.
   A. A population is a group of individuals of the same species occupying a particular area at the same time.
   B. A community consists of all the populations that occupy a given area at a given time.
   C. An ecosystem is a fairly self-sufficient community consisting of producers, consumers, and decomposers, plus their nonliving environment.
III. Each population occupies a particular habitat and a specific ecolog-

---

* Sometimes climax communities are less diverse than their immediate predecessors, owing to dominance by a few very well adapted species.

ical niche. An ecological niche is the role an organism plays in the ecosystem as defined by its adaptations.

IV. In a stable population there is a balance between biotic potential and environmental resistance.

  A. A limiting factor is an essential factor that is in the shortest supply. Potential limiting factors tend to interact to determine population density and size and the distribution of individual organisms and communities.

  B. Intraspecific and interspecific competition are important limiting factors. According to the principle of competitive exclusion, complete competitors cannot coexist.

  C. Predation does not render the prey extinct because of habitat complexity and because a reduction of the prey population results in a corresponding decrease in the predator population.

V. An ecosystem usually contains producers, which practice autotrophic nutrition (for the most part plants and protists), and consumers and decomposers, which are heterotrophs. The decomposers recycle waste products and corpses in the form of simple inorganic compounds for reuse by the producers.

  A. Biomass and energy are lost on each trophic level of a food chain.

  B. Food chains are usually not simple in nature. Rather, energy passes through most communities in a network of paths, a food web, which tends to render the community more stable than if all food chains were mutually independent.

  C. Net community productivity is the rate at which energy is stored in the organic compounds of a community after energy used in respiration has been subtracted.

  D. Matter travels cyclically through ecosystems, in an essentially closed cycle, so it is recycled as well as reused.

VI. Any kind of stress tends to simplify ecosystems.

VII. In ecological succession communities progress through a series of stages and finally form a climax community.

Post-Test

1. All the populations that occupy a particular area at a given time make up a _____ .
2. The role an organism plays in an ecosystem is its ecological _____ ; the part of the environment in which it lives is its _____ .
3. The capacity of a population to increase its numbers is its _____ _____ .
4. Nutrients, water, and predation are examples of _____ factors.
5. The displacement of the American anole lizard in Florida by the Cuban anole is an example of _____ _____ .
6. The principle of competitive exclusion explains that complete competitors cannot _____ for very long.
7. When the hare population increases, its predator population _____ (increases or decreases).
8. In a food chain herbivores are most likely to be found on the _____ _____ level; producers would be on the _____ _____ _____ level.
9. In saprobic nutrition organisms _____ nutrients through the _____ _____ _____ .
10. In a food chain the greatest biomass is found in the _____ _____ level; the least amount of energy is concentrated in the _____ _____ _____ level.
11. A growing calf converts about _____ % of its food into edible beef.
12. The term biomass refers to the _____ .
13. The rate at which energy is stored in a community as a whole is its _____ _____ _____ .
14. In general, stress tends to _____ complex communities.
15. The last stage in a sere is a _____ _____ .
16. A _____ succession begins in a pre-existing community.
17. Lichens are often members of _____ communities.

Review Questions

1. In a sense, people and cockroaches occupy the same habitat, but not the same ecological niche. Describe the differences in niche between the two as completely as you can. Can the two organisms be said to be competitors? If so, how is it that they manage to coexist?

2. What is a limiting factor? What actual or potential limiting factors might govern the population of houseflies? of starlings? of lions? of people?

3. The biotic potential of a tapeworm is enormous. Why isn't the world overrun by tapeworms?

4. When a bounty was placed on coyotes and wolves, effectively eradicating them, the deer population on Arizona's Kaibab Plateau increased rapidly. However, after several years the herd population declined sharply. What factors may have contributed to that decline?

5. Identify each of the following as population-dependent or population-independent and justify your answer: (1) competition for food, (2) competition for territories, (3) a drought.

6. It has been said that predators live on capital whereas parasites live on interest. Explain, and postulate which life-style has more biological advantages.

7. Why is biomass lost on each successive trophic level of a food chain? Why is it ecologically as well as economically more expensive to eat steak than rice?

8. What types of consumers might be found in a typical ecosystem? List them and give their typical food sources.

9. How does a detritus feeder differ from other consumers? Should it be considered a decomposer?

10. Draw a food chain for pelicans, algae, microscopic crustacean larvae, small fish, and large fish. Indicate the trophic levels with the greatest and the least biomass and energy.

11. Describe or diagram the phosphorus cycle.

12. What is the difference between a sere and a climax community? between primary and secondary succession?

13. In parts of the American South, hardwoods are the usual climax vegetational type, but they are economically much less valuable than the slash and yellow pine trees that characterize an earlier successional stage. Often farmers burn their woodlands periodically in a controlled fashion that, if properly done, does not harm the pines but eliminates the hardwoods. What ecological principle or principles does this practice illustrate?

Readings

Brownell, P.H. "Prey Detection by the Sand Scorpion," *Scientific American,* December 1984, 86–97. The sand scorpion is a nocturnal hunter of the Mojave Desert that does not see or hear the insects it feeds on. Instead it has receptors on its legs that are sensitive to any disturbance in the sand.

Cloud, P. "Evolution of Ecosystems," *American Scientist,* 62:54–66, 1974. Summary of the geological timetable from an ecological point of view.

Clutton-Brock, T.H. "Reproductive Success in Red Deer," *Scientific American,* February 1985, 86–92. This report on the determinants of the lifetime breeding success in red deer discusses interactions that are characteristic of many mammal species.

Deyrup, M. "Deadwood Decomposers," *Natural History,* March 1981. Decaying Douglas fir trunks support a community of organisms whose insects alone number more than 300 species.

Edmond, J., and K. von Damm. "Hot Springs on the Ocean Floor," *Scientific American,* April 1983. A unique and extreme environment that supports the only known nonsolar ecosystems.

Goreau, T.F., N.I. Goreau, and T.J. Goreau. "Corals and Coral Reefs," *Scientific American,* August 1979. An account of tiny coral polyps that live in symbiosis with photosynthetic algae and build large limestone reefs.

Gosz, J.R., et al. "The Flow of Energy in a Forest Ecosystem," *Scientific American,* March 1978.

Horn, H.S. "Forest Succession," *Scientific American,* May 1975. Soil moisture retention influences succession, but changes in tree species composition are also related to leaf mosaic geometry of the involved trees.

Krebs, J.R., and N.B. Davies, eds. *Behavioral Ecology: An Evolutionary Approach.* New York, Sinauer, 1978. This book could just as well have been listed under behavior, but behavior is a part of ecology, after all.

Kroodsma, D.E. "The Ecology of Avian Vocal Learning," *Bioscience,* March 1983. Learned and innate acquisition of birdsong, related to ecological niche.

Likens, G.E., et al. "Recovery of a Deforested Ecosystem," *Science,* 3 February 1978. The replacement of biomass and nutrients lost in cutting down forests may require a human generation.

Moore, J. "Parasites That Change the Behavior of Their Host," *Scientific American,* May 1984. Certain parasites, such as thorny-headed worms that infect pill bugs, make the host more vulnerable to predation by their next host.

Schaller, G.B. *Golden Shadows, Flying Hooves.* New York, Knopf, 1973. A fascinating personal glimpse of behavioral ecology in a Tanzanian ecosystem. (You might also want to consult Schaller's *Year of the Gorilla.*)

Chapter 37

ECOLOGY: THE MAJOR COMMUNITIES

Learning Objectives

After you have studied this chapter you should be able to:

1. Briefly describe the principal biomes of the earth.
2. Describe at least three ways in which aquatic habitats differ significantly from terrestrial habitats.
3. Describe such marine communities as the estuarine, intertidal, continental shelf, planktonic, and abyssal communities, summarizing (where applicable) their food-web relationships, their zones of greatest productivity, their principal environmental constraints, and their general ecological significance.
4. Describe thermal and nutrient stratification in an aquatic habitat.
5. Describe the processes of eutrophication and of ecological succession in a typical freshwater pond or lake.

The properties of air and water determine the distinctive adaptations of the organisms that live in each habitat—sometimes in surprising ways. Thermoregulation, for instance, is easier for an animal living in air because air does not conduct heat away from the body as efficiently as water does. When the surroundings are hot, the evaporation of water can be used to cool a terrestrial organism. Yet temperatures are usually more extreme in air than in water, for water tends to resist temperature change to such an extent that it even modifies the climate of adjacent land masses.

TERRESTRIAL LIFE ZONES: BIOMES

Biogeographical realms are regions made up of entire continents or large parts of a continent separated by major geographical barriers and characterized by the presence of certain types of animals and plants. Within these biogeographical realms are large, distinct, easily differentiated community units called **biomes,** (Fig. 37–1), whose boundaries are established by a complex interaction of climate, physical factors, and biotic factors. A biome is a large community unit characterized by the kinds of plants and animals present.

Unfortunately, many communities do not fit into any one biome type. Some terrestrial communities are produced by very special circumstances and do not closely resemble even the most general description of a biome. Moreover, intergrading areas between biomes, called **ecotones,** and subclimax successional stages within them may produce local assemblages of organisms quite different from those only a few miles away. With these caveats in mind, let us consider some of the major terrestrial communities.

Tundra

No fully terrestrial organisms live at the North Pole for the simple reason that it is permanently covered with seawater and floating ice. South of that, many areas are permanently covered with glacial ice caps, as in Greenland. However, in the extreme north, wherever the snow melts seasonally, there exists a distinctive circumpolar **tundra** community (Fig. 37–2). (The Southern Hemisphere has no equivalent because it has no land in the proper latitudes.) For the most part, the land of the tundra is of low relief and drainage is slow. That, combined with the low rates of evaporation that result from cold annual temperatures, produces a swampy landscape of broad shallow lakes, sluggish streams, bogs, and fens. In addition, tundra has a layer (varying in thickness) of permanently frozen ground, the **permafrost,** which interferes with subterranean drainage. Frozen mammals, especially extinct mammoths, have been unearthed from the permafrost, with the meat still edible after tens of thousands of years—at least edible by dogs.

The very short growing season of the tundra affects the life of everything in it. The great natural stress to which this community is subjected results in low species diversity dominated by reindeer moss (actually a lichen), grasses, sedges, and annuals. There are no real trees or shrubs except in very sheltered localities, though dwarf willows and other dwarfed trees do grow widely. The need to complete an entire life cycle in a span of weeks produces frantic rates of growth and development in annuals, made possible in large part by the great length of each summer day. In many places the sun does not set at all for a considerable number of days in midsummer. Many arctic flowering plants have parabolic flowers designed like solar collectors to focus the sun's heat on developing ovules and to provide a warm haven for pollinating insects.

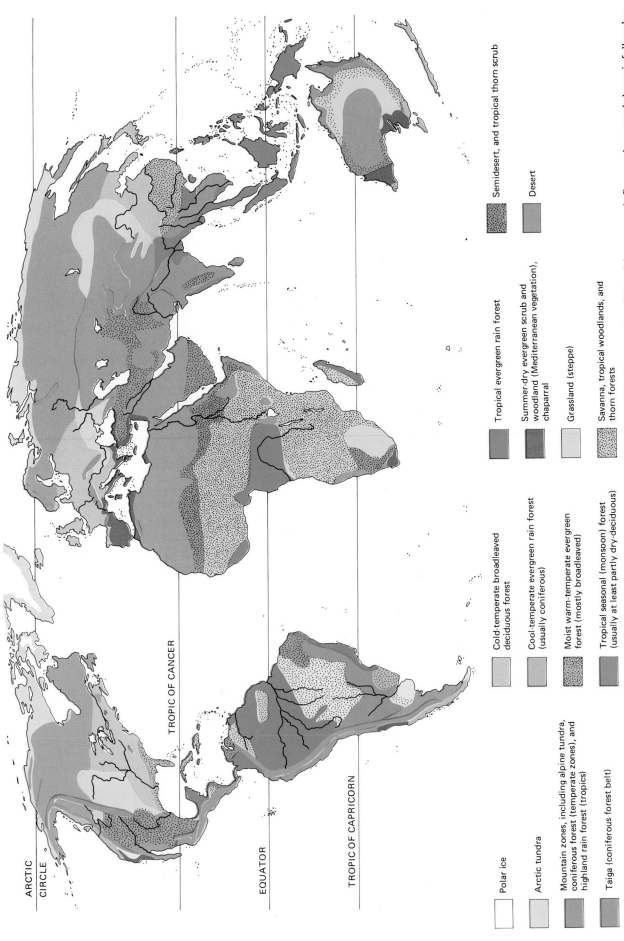

Figure 37–1 A map of the biomes of the world. Note that only the tundra and the taiga are roughly circumpolar. Other biomes are influenced as much by rainfall as by growing season and therefore have a much less continuous distribution.

ARCTIC CIRCLE

TROPIC OF CANCER

EQUATOR

TROPIC OF CAPRICORN

Polar ice

Arctic tundra

Mountain zones, including alpine tundra, coniferous forest (temperate zones), and highland rain forest (tropics)

Taiga (coniferous forest belt)

Cold-temperate broadleaved deciduous forest

Cool-temperate evergreen rain forest (usually coniferous)

Moist warm-temperate evergreen forest (mostly broadleaved)

Tropical seasonal (monsoon) forest (usually at least partly dry-deciduous)

Tropical evergreen rain forest

Summer-dry evergreen scrub and woodland (Mediterranean vegetation), chaparral

Grassland (steppe)

Savanna, tropical woodlands, and thorn forests

Semidesert, and tropical thorn scrub

Desert

(a)

(b)

(c)

(d)

Figure 37–2 The tundra biome. (a) Alaskan tundra in the fall. Very like the trees of temperate deciduous forests, many of the low plants of the tundra, such as bearberry (b), change their foliage colors. The brilliant hues probably attract birds and mammals, which by eating the fruit, disperse the seeds. Wildlife of the tundra includes (c) the snowy owl, which, like most owls, is a nocturnal predator, and (d) the arctic fox. The color of this animal varies according to the season. During the long winter, its coat is almost entirely white. (a, b, and d, Tina Waisman. c, Louis W. Campbell.)

Most modern animal life (Fig. 37–2) of the tundra is small, consisting of the ratlike lemmings, weasels, arctic foxes, snowshoe hares, ptarmigan, snowy owls, hawks, and the like. The immense musk-ox is a modern exception, saved from extinction in the nick of time by human agency. The great woolly mammoth elephants, known not only from their remains but also from the cave paintings of our ancestors, were probably rendered extinct by prehistoric human predation. Caribou and reindeer have survived. There are no reptiles or amphibians. One might also expect few or no insects, but insects are able to survive over winter as eggs or pupae and occur in great numbers. Mosquitos are particularly numerous.

Some equivalent of the tundra community occurs on a few mountains, but the thin dry air and usual absence of a permafrost layer make these mountain "tundras" not fully comparable to their prototype. All tundra regenerates very slowly after disturbance, and successional processes are notably slow.

Taiga

Below the tundra a second circumpolar community, the **taiga,** stretches across the world. Dominated by coniferous evergreens, this community is characterized by a great depth of partly decomposed pine and spruce needles, an acidic soil, and a somewhat longer but still short growing season. Deciduous trees may form striking subclimax communities but otherwise are rare. The probable reason is that the growing season is so short that photosynthesis is needed year-round. The conifers, with their needlelike leaves, are ideally suited to withstand the physiological drought of the winter months and yet can photosynthesize to some extent

(a)

(b)

even then (Fig. 37–3). Though most of the taiga is not well suited to agriculture—because of its short growing season and mineral-poor soil—the taiga yields vast quantities of lumber and pulpwood, plus furs and other forest products.

The taiga extends quite far south on the western coast of North America, reaching as far as northern California. Here, the short growing season of the subarctic has an equivalent, produced by the heavy seasonal rains characteristic of the West Coast. The extensive drought that follows the rains every year is equivalent to the tundra's physiological drought of winter. As with the tundra, the taiga has a mountain equivalent often found at the middle altitudes.

Taiga animal life consists mainly of caribou (which migrate into the area from the tundra in winter) and a few other large animals, such as wolves, bears, and moose (Fig. 37–3), but most animal life is medium-sized to small: many species of rodents, rabbits, and fur-bearing predators, such as lynx, sable, and fisher. Insects are abundant, but amphibians and reptiles are sparsely represented except in the southern extensions.

Figure 37–3 (a) An example of taiga, or northern coniferous forest: Spruce Valley, McKinley Park, Alaska. (b) A common inhabitant of the coniferous forest, a cow moose. Also seen in this photograph is the limited development of the underbrush typical of a dense coniferous forest. (a, E.R. Degginger. b, J.N.A. Lott, McMaster University/BPS.)

The Temperate Communities

Below the taiga, communities cease to be circumpolar. Precipitation plays as great a part in their distribution as does growing season, and in temperate latitudes rainfall varies greatly with longitude. In general, continental interiors tend to be dry, but this results from a variety of causes. Permanent high-pressure areas, such as those over the Sahara Desert, may nudge moist air masses aside. Air passing over a large land mass also may dry out without having the opportunity to be recharged with fresh moisture. The climate of the North American continent, however, is dominated by **rain shadows** cast by mountain ranges, especially in the West (Fig. 37–4).

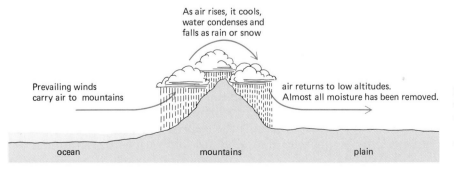

As air rises, it cools, water condenses and falls as rain or snow

Prevailing winds carry air to mountains

air returns to low altitudes. Almost all moisture has been removed.

ocean   mountains   plain

Figure 37–4 Rain shadow. As prevailing winds pass over mountains, the air cools and moisture condenses. Little is left for areas downwind of the mountains, so deserts tend to develop in such locales.

(a)

(b)

Figure 37–5 Seasonal changes in a temperate deciduous forest. (a) Dense, green foliage of hardwoods during summer. (b) Color changes in foliage during fall. (b, E.R. Degginger.)

As prevailing easterly winds push against the bases of the Rocky Mountains, masses of moist air from the Pacific Ocean are forced upward. As they rise to altitudes of low pressure, they become rarefied, and in accordance with simple physical principles, they cool. When the dew point is reached, precipitation occurs. Thus, the westerly slopes of the mountains are so well watered that a kind of rain forest develops. Considerable rain falls in the upper reaches of the eastern slopes also, but by the time the air has sunk back to lower altitudes, most of the available moisture has been wrung from it and its relative humidity is very low indeed.

DECIDUOUS FORESTS

Where precipitation is greatest, **temperate deciduous forests** tend to develop (Fig. 37–5). Those of the northeastern and middle eastern United States might serve as an example. Typically, the soil of a deciduous forest consists of a deep clay-rich **horizon** layer and a surface horizon that is rich in humus. As organic materials in the humus decay, mineral ions are released. If they are not immediately absorbed by the roots of the living trees, these ions are washed into the clay, where they may be retained. Deep plowing that mixes these two horizons produces a fair agricultural soil, since it mixes the mineral-bearing clay with humus and makes all of it accessible to crop roots. Erosion resulting from the lack of a proper root structure in the soil can be serious, as can both the loss of minerals by runoff and the consumption of crop plants at some geographically distant point.

Deciduous forests were among the first communities to be converted to agricultural use, yet it is surprising how well they have survived into the 20th century. Although in Europe and Asia many original forest habitats have been cultivated for thousands of years by traditional methods without substantial loss in fertility, the American frontier mentality often perceived land as a limitless resource. American farmers of the 18th and 19th centuries often abandoned the wise soil conservation practices of their forebears and allowed erosion and other forms of soil depletion to destroy their land. In many areas, especially New England, the highly glaciated mountainous terrain had only a thin layer of soil, which was swiftly exhausted. Ironically, the abandonment of the land that resulted from poor agricultural productivity permitted succession to restore much of the original forest, though doubtless in stunted form. Nevertheless there is probably more wildlife habitat in the eastern United States today than in the continuously cultivated countryside of Western Europe.

The deciduous woodland originally contained the familiar (to Americans) range of large mammalian fauna: puma, wolves, deer, bison, bears, and others now extinct, plus many small mammals and birds (Fig. 37–6). Both reptiles and amphibians abounded, together with much denser and more varied insect life than exists today.

TEMPERATE GRASSLANDS

It is difficult to say where the eastern deciduous forest originally left off and where the grassland began because the North American Indians often burned forests and grasslands by accident, for agriculture, or to clear land for traveling and hunting. Repeated burning encourages grassland and discourages the development or continuance of forests. It is likely, however, that originally the eastern deciduous forest extended through Illinois and well into Missouri and westward. As rainfall de-

Figure 37–6 A family of brown bears. The largest land predators, bears are common animals in relatively unsettled regions of northern forests. (Art Wolfe, The Image Bank.)

creased with longitude, there was a greater tendency for minerals to accumulate in a marked horizon just below the topsoil, instead of washing out of the soil. The soil had considerable humus content, for prairie grasses are similar to deciduous plants in that the above-ground portions of the plants die off each winter, while the roots survive underground. Certain species of grass grew as tall as a man on horseback, and the land was covered with herds of grazing animals, mainly bison. The principal predators were wolves, though in sparser, drier areas their place was taken by coyotes. Smaller fauna included great villages of prairie dogs and their predators (foxes, black-footed ferrets, and various raptorial birds), grouse, reptiles (such as rattlesnakes), and great numbers of insects.

This biome was so well suited to agriculture that little of it persists except in such places as abandoned railroad rights-of-way. Almost nowhere can we see even an approximation of what our ancestors saw as they settled the Midwest (Fig. 37–7). It is not surprising that the American Midwest, the Ukraine, and other such areas became the breadbaskets of the world. These habitats are ideal for such grasses as the cereal grains.

DESERTS

Not all deserts are temperate in climate; indeed, by strict definition, none of them are. The low water content of the desert atmosphere leads to temperature extremes of heat and cold, so that a 100°F change of temperature in a single day is sometimes possible. Deserts vary greatly, depending upon the amount of precipitation received. A **chaparral** biome (Fig. 37–8), though dry, is almost like a forest of shrubs and a few widely

Figure 37–7 Western grassland of the North American plains. (E.R. Degginger.)

(a)

(b)

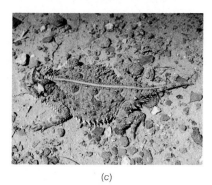

(c)

Figure 37–8 The chaparral biome. (a) A scrub juniper and oak community in Arizona. (b) The yucca is a plant that grows in both chaparral and moist desert environments. It spends a considerable part of its energy budget to create the large flowering stalk that you see; since vegetation is widely spaced, the stalk is necessary for it to be visible to its pollinator, the yucca moth. (c) A horned lizard, showing protective coloration, an important adaptation to life in a region where ground cover is scarce. (a, J. Robert Waaland, University of Washington/BPS. b, L. Egede-Nissen/BPS. c, L.E. Gilbert, University of Texas at Austin/BPS.)

scattered trees of drought-resistant species. Chaparral develops in warm areas with relatively abundant but sharply seasonal rainfall. If the rainfall is less seasonally distributed, a sagebrush or sparse grassland is more likely to develop. Less rainfall produces the desert of the Hollywood western—picturesque large and small cacti, yucca plants, contorted Joshua trees, and widely scattered bunchgrass—or its ecological equivalent. Still less rainfall produces a Sahara-like desert with widely scattered plant life and, in places, virtually no life of any sort.

Desert soil tends to be rich in minerals and hence fertile when irrigated. However, as irrigation water evaporates, minerals accumulate, often producing inarable saline soil that must be washed with large amounts of water to be restored to productivity. As one might predict, desert soils have little humus content and are often sandy.

Desert animals tend to be small and of cryptic habits, remaining undercover or returning to shelter periodically during the heat of the day (Fig. 37–9). At night they come out to forage. In addition to specialized insects, there are many specialized desert reptiles: lizards, tortoises, and snakes, especially venomous snakes, such as the American sidewinder rattlesnake and similarly specialized Old World vipers. Mammals include such rodents as the American kangaroo rat, which does not have to drink water but can subsist solely on the water content of its food plus metabolically generated water. In American deserts there are also jackrabbits, and in Australian deserts, ecologically equivalent specialized kangaroos. Carnivores like the African fennec fox and some raptorial birds, especially owls, live on the rodents and rabbits. A few larger herbivores, such as antelopes, may also be found.

The Tropics

Tropical and subtropical habitats are at least as varied as temperate ones, and like temperate life zones they are determined mainly by precipitation. Thus, there are tropical forests, grasslands, and deserts. In the tropics the seasonal distribution of rainfall is especially important. There are grasslands that would be rain forests except that almost all their rainfall occurs during 2 months of the year. One can hardly expect lush vegetation to persist for 10 months unwatered, especially where evaporation is encouraged by high temperatures. Of the tropical habitats we will consider two: rain forests and savannah grasslands.

TROPICAL RAIN FORESTS

Of all the life zones in the world, the coral reef and the tropical rain forest (Fig. 37–10) are unexcelled in species diversity and variety. No one species dominates the tropical rain forest; one can sometimes travel for a quarter mile without encountering two members of the same species of tree, especially in subclimax communities.

Despite what you may have seen in Tarzan movies, the vegetation is not dense at ground level except near stream banks and in areas undergoing early and middle succession. The continuous canopy of leaves overhead produces a dark habitat with an extremely humid microclimate. A large part of the precipitation received by the tropical rain forest is locally recycled water that comes from the transpiration of the forest's own trees. If trees are removed over an extensive area, the rainfall may be so reduced that re-establishment of the original forest becomes unlikely.

A fully developed rain forest has three distinct stories of vegetation. The topmost story consists of the crowns of occasional very tall trees. It is entirely exposed to the sunlight. The middle story forms a continuous canopy of leaves that lets in very little sunlight for the support of the sparse understory. The understory itself consists not only of plants specialized for life there, but also the seedlings of taller trees.

All stories of vegetation support extensive **epiphyte** communities of smaller plants that grow in crotches, on bark, or even on the leaves of their hosts (Fig. 37–11). The epiphytes are not parasitic in the usual sense, but their numbers can become so great as to break branches or interfere with photosynthesis. Since so little light penetrates to the understory, many of its plants are adapted to climb upon already established host trees rather than invest their meager photosynthetic resources in the dead cellulose tissues of their own trunks. Tropical vines as thick as a man's thigh abound. This adaptation can be seen in several species of **strangler tree.** The strangler fig (Fig. 37–11), for instance,

(a)

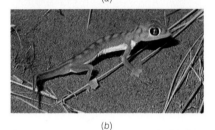

(b)

Figure 37–9 Two desert inhabitants. (a) The giant saguaro cactus of the southwestern United States. (b) A gecko, from the Central Namib Desert (Africa). Its webbed feet act something like snowshoes so that the lizard can run rapidly on the surface of the sand. (a, E.R. Degginger. b, F.J. Odendaal, Stanford University/BPS.)

Figure 37–10 The tropical rain forest. (a) A broad view of a rain forest on one of the Hawaiian Islands. It is possible that some trees in this photograph are species introduced by the settlers and traders who began to arrive in the mid-19th century. (b) Tropical rain forest trees typically possess elaborate systems of buttress roots to hold their ground in the wet soil. (a, Charles Seaborn.)

(a)

(b)

(b)

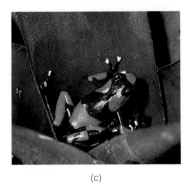

(c)

(a)

Figure 37–11 Life in the rain forest. (a) Strangler fig growing on the trunk of a palm tree. (b) Large epiphytes grow on tree trunks and branches. The dense epiphytes produce a unique aerial community with many distinctive inhabitants, such as the brilliantly colored arrow-poison frog (c), living inside the cup of a bromeliad. (c, Edmund D. Brodie, Jr., Adelphi University/BPS.)

overgrows the trunk of the host, eventually killing it after it has served the fig's purpose.

Certain large epiphytes known as bromeliads store as much as a gallon of rain water in their leaf cups and absorb the water thus stored between rains by means of tiny scales on the bases of the leaves; these scales function somewhat like root hairs. Mosquito larvae, other insects, and even specialized species of crabs and frogs live in this odd aerial swamp. Recent evidence suggests that epiphyte communities are penetrated extensively by adventitious roots of their host trees, and probably contribute to the host trees' nutrition. The trees of the tropical rain forest are usually evergreen (though not coniferous), but there are exceptions. Their roots are often shallow, forming a mat as thick as 2 or 3 feet on the surface of the soil. Swollen bases or flying buttresses hold them upright despite their poor anchorage.

Since the temperature is high year-round, decay organisms decompose organic litter before it can become humus. At the same time, the usually high rate of rainfall leaches nutrients rapidly from the soil. Highly developed mycorrhizae, however, extract nutrients from decomposing material and transfer them to the roots of living plants before they have a chance to enter the soil (Chapter 14). The absorptive mechanisms of the community are so efficient that the runoff water often has a lower mineral content than the rain that falls there.

Rain forest animals include the most abundant and varied insect, reptile, and amphibian fauna on the face of the earth. Birds, too, are varied and often brilliantly colored. Much mammalian life is arboreal (sloths, monkeys), although some larger ground-dwelling forms, including even elephants, are to be found there.

Explosive population growth in tropical countries may spell the end of all rain forests by the end of the century. It is believed that many rain forest organisms will be rendered extinct in this way before they have even been scientifically described. The total ecological impact of rain forest destruction is unknown at present, but it is likely that the burning or decay of felled trees will contribute substantially to the world atmospheric carbon dioxide content while removing one of the largest carbon dioxide fixation areas of active photosynthesis.

The local ecological impact is likely to be more immediate and substantial. Once the protective mat of roots has been removed from the soil, erosion is likely to become an even more severe problem than in temperate climates, and will often be accompanied by **laterization,** the permanent chemical hardening of clay-rich soils. Since the mineral content of the rain forest community resides mostly in its vegetation, rain forest soils tend to be poor. Even if left alone, areas of destruction may not succeed to mature forest again if the soil is excessively depleted. Moreover, if the area destroyed is very extensive, the rather inefficiently dispersed seeds of the forest species may not be able to make their way to much of the area involved.

(a)

(b)

SAVANNAH

The **savannah** (or **veld**) life zone is a tropical grassland or very open woodland, depending on one's viewpoint (Fig. 37–12). Widely dispersed trees, such as acacia, usually bristling with thorns for protection against herbivores, grow amid long grasses. The greatest herbivore biomass in the modern world occurs in the African savannah. Here live the great herds of antelope, giraffe, zebra and the like. Large predators, such as lions and hyenas, kill and scavenge the herbivores. Savannah is produced naturally either by low rainfall or by sharply seasonal rainfall. In areas of seasonally varying rainfall, the herds and their predators may migrate annually, much as caribou migrate between tundra and taiga in the north.

The tropical grasslands are being rapidly converted to range for cattle and other animals. Severe overgrazing in places has converted marginal savannah to actual desert. On the other hand, lumbering and grazing has converted much bordering rain forest to savannah. In both cases large amounts of original wildlife habitat are being subjected to probably irreversible destruction.

Figure 37–12 The savannah and its inhabitants. (a) Cape buffalo grazing in Uganda. (b) The cheetah is uniquely adapted as a pursuit predator. Anatomically more like a dog than a cat in many ways, the cheetah is the fastest known mammal, often able to attack and kill antelopes. Its numerous adaptations, such as sight-hunting and long legs, would render it unable to exist except in the open savannah habitat. (a, E.R. Degginger. b, Warren and Genny Garst, Tom Stack Associates.)

AQUATIC HABITATS

Aquatic habitats vary greatly, but all are governed by the special properties of water. Oxygen is usually far less available in water than it is in air. As water warms, it holds less oxygen, and that is just when the body temperatures of aquatic organisms, typically poikilothermic, demand the most oxygen. Water is far denser than air. An aquatic organism requires little in the way of a skeleton and only a small expenditure of energy to float. Water brings dissolved substances to all parts of any plant suspended in it, so water plants do not need much of a vascular system. Finally, water greatly interferes with the penetration of light, so floating photosynthesizers must remain near the surface, and vegetation attached to the bottom can grow only in the shallowest zones.

Probably the most fundamental division in aquatic ecology is that of fresh versus salt water. The physiological adaptations required to permit osmoregulation in freshwater animals must be a major strain on the organism's resources. Some animals, such as echinoderms, are never found in fresh water at all. Such organisms, adapted to a very narrow range of salinity, are known as **stenohaline;** those that can live in a variety of salinities are **euryhaline.** Some fish and crustaceans are euryhaline as adults but stenohaline as larvae. Others, such as salmon and eels, live in fresh water during part of their life cycles and salt during another part, migrating between the two.

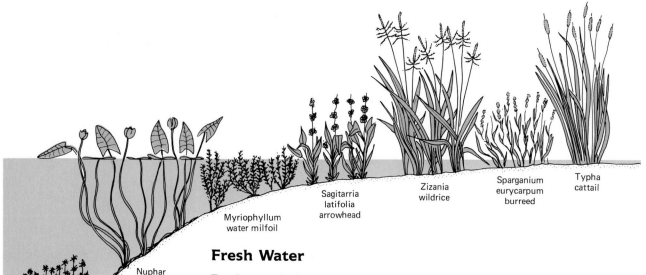

Zizania
wildrice

Sparganium
eurycarpum
burreed

Typha
cattail

Sagitarria
latifolia
arrowhead

Myriophyllum
water milfoil

Nuphar
spatterdock

Chara
muskgrass

Figure 37–13 Zones of vegetation about lakes and along riverbanks. Note the changes in vegetation with water depth. (After Dansereau.)

Fresh Water

Freshwater habitats include streams, lakes, and ponds. Even temporary ponds have a typical assemblage of inhabitants: mosquito larvae, tadpoles, fairy shrimp, which hatch from eggs that lie dormant during dry spells in the bottom, and water strider insects (see Fig. 2–7), which live atop the surface film while the water lasts (they can fly away when it dries up). The kinds of organisms found in streams vary greatly, mostly in accordance with the strength of the current. In fast streams the water is often cold and highly oxygenated, and the inhabitants may have adaptations, such as suckers, that keep them from being swept away. Large, slow-moving streams resemble lakes ecologically, and it is on lakes that we will focus our emphasis.

LAKES AND FRESHWATER LIFE ZONES

A typical lakeshore is inhabited by emergent air-breathing vegetation, such as cattails and water lilies (Fig. 37–13). The lake shore plus several other concentric communities of deeper-dwelling, entirely aquatic plants constitute the **littoral zone.** It is the most highly productive zone of the lake. A shallow lake may consist entirely of a littoral zone. Algae, particularly filamentous algae and diatoms, may exceed the biomass of the higher plants in the littoral zone. The littoral zone contains frogs and their tadpoles, turtles, annelid worms, crayfish and other crustaceans, insect larvae, and many non-game fish, such as perch or carp, plus some sport fish, such as bass. Here, too, at least in the quieter areas, one finds surface film dwellers, sometimes collectively called the **neuston,** such as water striders and whirligig beetles.

The deeper **limnetic zone** is sparser in all life, but it is here that larger fish spend most of their time, although they may visit the littoral zone to feed and breed. Due to its depth, less vegetation grows here.

The deepest, **profundal zone** is out of reach of effective quantities of light for photosynthesis. Primary producers could not exist in this zone because it is below their **compensation point,** at which the products of photosynthesis balance the rate of catabolism. Nevertheless, much food drifts into this zone from adjacent zones or falls into it from the lighted waters above. When dead plants and animals reach the profundal zone, decay bacteria liberate the minerals their bodies contain, but these cannot be effectively recycled because primary producers do not exist there. In consequence, the profundal habitat tends to be both mineral rich and anaerobic, with few forms of higher life occupying it.

These conditions can develop because of the marked **thermal stratification** characteristic of large lakes. Cool water sinks to the bottom in the summer, being separated from the warmer water above by a marked and abrupt temperature transition, the **thermocline.** In temperate zone lakes in the fall, high winds and falling temperatures cause a mixing of the lake waters. This is called the **fall turnover.** The presence of large amounts of mineral ions in the surface waters encourages the development of high algal populations, which form temporary **blooms** then and again in the spring. Such turnover, and the thermocline responsible for it, is a temperate zone phenomenon largely absent in the tropics.

LAKE SUCCESSION

Lake or pond succession largely depends on the deposition of silt and is therefore more a geologically based than a biologically based phenomenon, in contrast to usual terrestrial successions. Nevertheless, it is just as relentless, if not more so. The woodland pond has been described by one ecologist as a temporary wet spot on the floor of the forest. In the absence of intervention, a freshwater lake or pond will become progressively shallower until it becomes a part of the surrounding terrain. (This can also happen in some estuaries and bays of the marine habitat.) The pond's aquatic vegetation will change at the same time. Bladderwort and tape grass may be among the earliest stages in this succession. Eventually, water lilies or similar plants may appear. With the establishment of a marsh, a cattail and bulrush community may develop. The climax of this aquatic succession is not aquatic at all, but woodland, prairie, or bog.

Figure 37–14 Cultural eutrophication, a result of pollution of water by sewage or fertilizers.

The associated fauna undergoes dramatic changes in keeping with the plant successions. The more "desirable" species of fish are usually associated with the **oligotrophic** (early) and intermediate stages of succession. Successional changes in fish fauna are chiefly associated with rising water temperatures, decreasing oxygen content, and bottom changes affecting egg laying and breeding. These **eutrophic** changes, as they are known, are even more directly related to progressive nutritional enrichment of the water and increase in the total biomass of the community, especially the growth of aquatic plants.

The excessive fertilization of aquatic communities that results from human practices, such as fertilizer-rich runoff from farms, septic tank drainage, and the like, is known as **cultural eutrophication** (Fig. 37–14). Cultural eutrophication can be so marked that the plants compete with the animals for oxygen, especially at night or on cloudy days, and can form such a tangle that there is virtually nowhere to swim. In addition, cyanobacteria prosper in cultural eutrophication, displacing the algae over which they have a competitive advantage in such situations. In extreme cases rotting vegetation can produce extensive fish kills.

Cultural eutrophication is often reversible, but only if the abnormal nutrient input can be stopped. Attempts to kill the excessive vegetation with herbicides or by importing herbivorous fish or mammals (such as water buffalo) are not usually successful in the long run because the dead or defecated vegetation simply decays and liberates the minerals it contains. This in turn produces further eutrophication. The actual physical removal of the vegetation and its transport to a distant locale is the only after-the-fact approach that appears to work at all.

Life in the Sea

Most of the world's biomass probably resides in the oceans, though the oceans are by no means uniformly productive. This oceanic biomass can be classified in three categories: plankton, nekton, and benthos.

Figure 37–15 Tiny crustaceans form part of the zooplankton, the animal component of the floating basis of most food chains. Not all are adults; some are the larval forms of large crustaceans such as lobsters and crabs. A planktonic stage is part of the life cycle of many large aquatic animals. During this stage the young are dispersed by water currents to new habitats perhaps hundreds of miles from where they were hatched. (H.W. Pratt/BPS.)

PLANKTON

Plankton (Fig. 37–15) are mostly small or microscopic forms: tiny crustaceans, diatoms and other protists, jellyfish, swimming tunicates, ocean sunfish, arrow worms, and the larval forms of many bottom-dwelling organisms. One can see from this that "plankton" is not a taxonomic classification. Indeed, one is hard put to define it with exactitude, but for our purposes plankton consists of organisms with feeble powers of locomotion. Though some of the members of this community are capable of surprising vertical migration, their horizontal movements are mostly passive and at the mercy of wind and water currents. Plankton are at the base of most marine food chains, but not all plankton eaters are small. The baleen whale, the largest organism ever known to have lived, eats plankton directly, straining it out of the water through the whalebone in its mouth (Fig. 37–16).

NEKTON

Nekton consists of organisms (mostly fish) whose powers of locomotion render them somewhat independent of water movements. These tend to occur, however, where plankton and benthos (below) do, because they ultimately depend on floating and fixed producer organisms.

BENTHOS

Benthos consists of all burrowing and bottom-dwelling organisms. It includes detritus feeders (such as many worms and bivalve mollusks), many crustaceans (such as crabs, lobsters, and some shrimp), sessile

Figure 37–16 A baleen whale filtering the surface of the ocean for plankton. The bird nearby is hoping to capture some fish that have been stirred near the surface of the water by the motion of the whale. (David W. Hamilton, The Image Bank.)

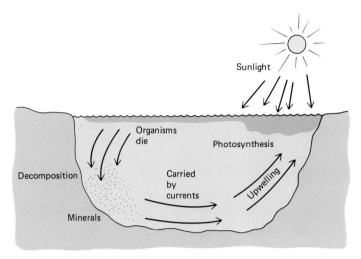

Figure 37–17 In the open sea and large lakes, organisms die and sink to the bottom, forming the basis of the food chain that terminates in the detritus feeders. Decomposition releases nutrients, but these are out of reach of the planktonic producers near the surface. Nutrients are, however, carried to an area of upwelling, usually off the coast of a continent. There they come to the surface and stimulate the growth of productive planktonic communities.

creatures (such as sea anemones, corals, and tunicates), and even some fish (such as flatfish, rays, and moray eels). Aquatic plants equipped with holdfasts are also part of the shallow-water benthos.

MARINE LIFE ZONES

Although marine and freshwater life zones are comparable in many ways, their shallowest and deepest areas differ substantially. The depths of even the deepest lakes do not approach those of the oceanic abysses, extremely deep areas that extend miles below the sunlit surface. Nutrient accumulation in the abyss is substantial, but usually there are no seasonal turnovers to remix the deep and surface waters. Wind and tidal action, however, produce deep currents, which in time bring the nutrients to the surface again, sometimes thousands of miles from their place of original deposit, by a process termed **upwelling** (Fig. 37–17).

In general, the shallow waters of the continental shelf are the most productive both because they *are* shallow and easily penetrated by light and, even more important, because they are close to the continental river drainage systems, which bring nutrient runoff, weathered from the soil and rocks, into the sea. Some of those nutrients, particularly compounds of phosphorus, are permanently or semipermanently lost to the world of life because they are insoluble. Once deposited on the bottom, they remain there—barring geological uplift, which may re-expose them in the remote future. This one-way travel of phosphate, essential for the production of ATP, creatine phosphate, and the nucleic acids, is one factor in rendering phosphate scarce and making it a limiting factor in many terrestrial ecosystems.

We may be thankful, however, for the oceanic sink that converts carbon dioxide into the calcium carbonate shells of microorganisms and mollusks. The vast accumulation of chalk and similar sediments on the sea bottom has removed much carbon from the atmosphere. Were it not for this, the greenhouse effect caused by the continual volcanic production of carbon dioxide would long ago have rendered the earth as uninhabitable as the planet Venus.

SALT MARSHES AND ESTUARIES. Where the sea meets the land there may be one of several kinds of ecosystems: a rocky shore, a sandy beach, an intertidal mud flat, or a tidal estuary containing salt marshes (Fig. 37–18). An **estuary** is a coastal body of water, partly surrounded by land, with access to the open sea and usually a supply of fresh water from a river. Its salinity is intermediate between sea water and fresh water. Many estuaries undergo marked variations in temperature, salinity, and

(a)

(b)

Figure 37–18 The estuarine environment. (a) The Brigantine salt marsh estuary in New Jersey. Estuaries are the most biologically productive communities and the basis for many aquatic food chains. However, they are subject to extensive destruction from land development and pollution from rivers that feed into them from populated areas. (b) The glasswort, *Salicornia virginia*, is an estuary plant that grows just above the high tide mark in marshy areas. It has adapted to salty soil conditions by its ability to accumulate comparable concentrations of salt in its own tissues. (a, E.R. Degginger. b, Roy R. Lewis, Coastal Creations.)

other physical properties in the course of a year. To survive there, organisms must have a wide range of tolerance to these changes.

The waters of estuaries are among the most fertile in the world, often having a much greater productivity than the adjacent sea or the fresh water up the river. This high productivity is brought about by (1) the action of the tides, which promote a rapid circulation of nutrients and help remove waste products, (2) the importation of nutrients into the estuarine ecosystem from the land drained by rivers and creeks that run into the estuary, and (3) the presence of many kinds of plants, which provide an extensive photosynthetic carpet, and whose roots and stems also mechanically trap much potential food material. As leaves and plants die, they decay, forming the basis of many detritus food chains. The majority of commercially important fin and shell fish spend their larval stages in estuaries among the protecting roots and tangle of decaying stems.

Estuaries and salt marshes have often impressed uninformed people as being worthless. In consequence they have been used as dumps for the castoffs of industrial civilization and have become severely polluted. More recently they have been "filled" with dredged bottom material to form artificial land for residential and industrial development. A large part of the total productivity of the marine environment has been lost in this way.

THE INTERTIDAL ZONE. The gravitational pulls of both sun and moon produce two tides a day throughout the ocean, but the height of those tides varies seasonally and by the influence of the local topography. The area between low and high tide is the **intertidal zone.** The high levels of light and nutrients together with an abundance of oxygen make the intertidal zone a potentially excellent habitat. Yet it is a very stressful one. If an intertidal beach is sandy, the inhabitants must contend with a constantly shifting environment, which both threatens to engulf them and gives them scant protection against wave action. Consequently, most sand-dwelling organisms are continuous and active burrowers. They are, however, able to follow the tides up and down the beach, and so do not usually have any notable adaptations to drying and exposure.

On the other hand, a rocky shore (Fig. 37–19) provides a fine anchorage but is exposed to wave action when submerged and to drying (to

Figure 37–19 Intertidal life zonation. Several clearly defined areas can be seen in this photograph of a rocky beach at low tide in the vicinity of Botany Beach, British Columbia. The barnacles covering the higher rocks show up as white, and the mussels, on lower rocks, as dark blue or black. Pink tones are contributed by coralline algae, with green and brown (kelp) algae in the very lowest areas. (J. Robert Waaland, University of Washington/BPS.)

say nothing of seasonal cooking and freezing) when exposed to the air. A typical rocky shore denizen will have some way of sealing in moisture (perhaps by closing its shell, if it has one), plus powerful means of anchorage to the rocks. Mussels, for example, have horny, threadlike anchors, and barnacles have a special cement gland. Rocky shore intertidal plants usually have thick, gummy polysaccharide coats, which dry out slowly when exposed, and flexible bodies not easily broken by wave action. Some members of this community hide in burrows or crevices at low tide, and some small semiterrestrial crustaceans run about the splash line, following it up and down the beach.

THE SUBTIDAL ZONE   The **subtidal zone** is below the lowest tide but still shallow enough for vigorous photosynthesis. Largely protected from wave action, this area supports a varied community of echinoderms, such as sea stars, fish, burrowing worms, eelgrass, and the like. Samples of this community can be seen in tidal pools, temporary marine ponds left behind on rocky shores by the receding tide. Shore birds find the subtidal and intertidal zones to be rich hunting and fishing grounds. (The subtidal zone can be considered to be the shallowest portion of the neritic zone [below].)

THE NERITIC (CONTINENTAL SHELF) ZONE.   Nekton and larger benthic organisms are mostly confined to the shallower **neritic** waters of less than 200 feet in depth because that is where their food is. Not only is there considerable vegetation on the bottom, but as you can see from Figure 37–20, this is also where the chlorophyll content of the water itself is high, though not as high as in an estuary. That means that very little commercial fishing is worthwhile more than a hundred miles or so offshore, a fact that is not without significance in determining a nation's "territorial waters."

DEEP WATERS.   Some 88% of the ocean is more than a mile deep, far below any algal compensation depth and therefore, for the most part, an aquatic desert. What we sometimes hear about feeding the world's swelling population from the "limitless resources of the sea" ignores the fact that they are anything but limitless. Indeed, our exploitation even now approaches their limits both because of increased fishing and because of inshore marine habitat destruction. Most of what life does exist under the tremendous pressures and eternal darkness of the abyss depends entirely upon whatever food filters into its habitat from the upper lighted regions. The principal exceptions are the hot spring oases that were discussed in Chapter 36, Focus on Life Without the Sun.

Figure 37-20 The coastal enrichment of the ocean is graphically illustrated by this photograph, which was prepared using a special instrument that is sensitive to the chlorophyll content of the water. Note that chlorophyll levels are highest at continental margins; most of the open sea is therefore little more than an aquatic desert. (NASA.)

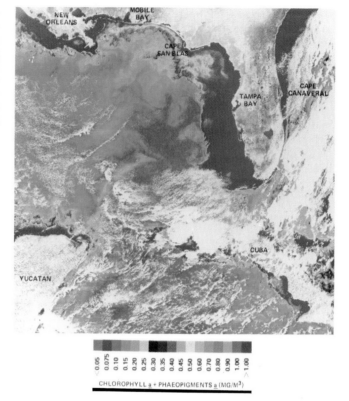

CHLOROPHYLL a + PHAEOPIGMENTS a (MG/M³)

Animals of the abyss are strikingly adapted to the darkness and scarcity of food (Fig. 37–21). Many have illuminated organs, enabling them to see one another for mating. Some abyssal squid even emit a luminous cloud of ink to distract predators. Many abyssal organisms are predators and live in dispersed populations. Some are remarkably adapted killing machines prepared to take quick and maximum advantage of any rare opportunity to feed. Their scattered population suggests that it might be hard for them to find mates, but studies to confirm this must await the development of means of observing the ocean bottom over long periods. One species of deep-sea anglerfish hangs onto her mate once she has found him, for he permanently attaches to her body and grows into it—the perfect male career parasite.

LIFE ZONE INTERACTION

Not one of the biomes or life zones we have discussed exists in isolation; all interact. It is profitable, for example, to plant bushes around a trout stream or pond, for some insects living on the bushes fall in the water, where the fish devour them. To take another example, when parts of the Amazon rain forest flood annually, the fish leave the stream beds and

Figure 37-21 A deep-sea fish. Most fish that live at great ocean depths have weak or vestigial eyes. Many have luminous organs for locating each other for mating. Inside the nearly transparent body of this particular fish you can see the remains of small crustaceans, which probably drifted down from the surface waters. (Peter David, Photo Researchers, Inc.)

range widely over the forest floor, where they have been shown to play a role in dispersing the seeds of many species of plants.

Inhabitants of various biomes may interact over wide distances, even global distances in the case of migratory birds and fish. Migratory birds commonly spend critical parts of their life cycles in entirely different countries, which can make their conservation difficult. It does little good, for instance, to protect a songbird in one country if the inhabitants of the next put it in the cooking pot as soon as it lands in their neighborhood. This concept of large-scale interaction makes ecology difficult for many people to grasp—and adds to the problems discussed in the next chapter.

CHAPTER SUMMARY

   I. The circumpolar biomes whose limits depend mainly on the growing season are the tundra and taiga.
   - A. The tundra usually has a layer of permafrost. It is treeless and wet.
   - B. The taiga is dominated by coniferous evergreens.
  II. The temperate zone biomes tend to be delimited by precipitation.
   - A. Deciduous woodland has a relatively good soil consisting of a layer of humus underlain with clay.
   - B. Prairie or grassland receives less rainfall than deciduous forest and retains more soil nutrients. It is dominated by grasses.
   - C. Deserts, which receive even less rainfall, have saline soils and specialized plant and animal communities.
 III. The tropical biomes are rain forests, savannah, and deserts.
   - A. Rain forests are produced by yearlong heavy precipitation. The soil is mineral-poor, and most of the minerals of the community are incorporated into its biomass.
   - B. Savannah consists of grassland mixed with open woodlands. Natural savannah supports great herds and their predators.
   - C. Deserts are sparsely vegetated areas of low rainfall.
  IV. Aquatic habitats are strongly influenced by the penetration of light and the mineral content of the water. There are three main categories of aquatic communities.
   - A. Plankton are passive, small drifters constituting a very large part of the biomass of most aquatic communities.
   - B. Nekton consists of active swimmers.
   - C. Benthos is the term for bottom dwellers.
   V. Freshwater and saltwater habitats are subdivided somewhat differently.
   - A. Freshwater lakes have a littoral inshore zone, highly productive with some emergent and some submerged vegetation. Larger fish are found in the limnetic zone, but the profundal zone is relatively sterile. Its mineral-rich water mixes with the surface waters during the fall and spring turnover.
   - B. Aquatic succession depends largely upon deposition of silt and mineral enrichment. Cultural eutrophication results from accelerated nutrient enrichment and is profoundly damaging to aquatic habitats.
   - C. Marine communities include the very productive estuarine community, the intertidal zone (in which organisms must withstand exposure and wave action), the subtidal zone, the neritic region of the continental shelves, and the relatively unproductive abyss.
  VI. Though called ecosystems, none of the major communities are totally self-sufficient, and all interact to some extent.

Post-Test

1. Within the biogeographical realms are _____ _____ , large differentiated communities containing certain characteristic organisms.
2. Areas of contact between biomes are known as _____ .
3. The major _____ biome is subarctic and arctic, and circumpolar. For the most part it is underlain by a permanently frozen layer of soil, the _____ .
4. South of this community there is another circumpolar one, the _____ , which is characterized by the dominance of coniferous evergreens.
5. The tropical rain forest is characterized by a very shallow layer of _____ and extremely effective mechanisms of _____ mineral nutrients.
6. High _____ of the soil is a potential problem faced by irrigation agriculture in desert areas.
7. In deep aquatic, especially oceanic habitats, mineral nutrients tend to accumulate in the _____ areas.
8. In lakes, there is often a seasonal mixture of water in the spring and fall known as the _____ ; this produces _____ of plankton.
9. Estuaries are noted for their _____ (high, low) salinity and _____ (high, low) biological productivity.
10. The _____ zone contains organisms that often must possess adaptations to resist wave action.
11. The portion of the ocean that overlies the continental shelf is known as the _____ region.
12. Freshwater lakes are usually most productive in their _____ zone.
13. An older, nutritionally enriched lake may be described as _____ .

Review Questions

1. What are the main biologically important differences between aquatic and terrestrial habitats?
2. Compare the tundra with the tropical rain forest.
3. Why is the species diversity of the tundra low?
4. What appears to give evergreens a competitive advantage over deciduous trees in the taiga?
5. Why does the soil of the prairie tend to be more mineral-rich than that of the deciduous forest?
6. What is a rain shadow? What effect does it have upon climate?
7. Why would it be difficult to farm in a tropical rain forest?
8. Compare the behavior of mineral nutrients in lakes and in the ocean.
9. What is cultural eutrophication?
10. What factors are responsible for the high productivity of estuaries?

Readings

Ayensu, E.S. *Jungles.* New York, Crown Publishers, 1980. A handsome coffee-table volume filled with fascinating information, this book is little short of a revelation.

Boerner, R.E.J. "Fire and Nutrient Cycling in Temperate Ecosystems," *Bioscience,* March 1982. By liberating nutrients stored in woody tissues, fire can increase ecosystem productivity.

Caufield, C. "The Rain Forests," *The New Yorker,* 14 January 1985. An especially noteworthy if depressing chronicle of the rapid disappearance of this immense and vital biotic resource.

Connell, J.H. "Diversity in Tropical Rain Forests and Coral Reefs," *Science,* 24 March 1978. Even in the tropics, climax communities may have a species equilibrium of relatively low diversity.

Koehl, M.A.R. "The Interaction of Moving Water and Sessile Organisms," *Scientific American,* December 1982. Adaptations of intertidal and subtidal community organisms.

Leigh, E.G., A.S. Rand, and D.M. Windsor. *The Ecology of a Tropical Rain Forest: Seasonal Rhythms and Long-Term Changes.* Washington, D.C., Smithsonian Institution Press, 1983. "By far the best study of a tropical ecosystem by a team of competent specialists," according to the noted biologist Ernst Mayr.

Marquis, R.E. "Microbial Barobiology," *Bioscience,* April 1982. The biosphere may extend for miles underground.

Perry, D.R. "The Canopy of the Tropical Rain Forest," *Scientific American,* November 1984. An interesting discussion of the ecology of the tropical rain forest and how it can be studied.

Van Cleve, K., et al. "Taiga Ecosystems in Interior Alaska," *Bioscience,* January 1983.

Waring, R.H. "Land of the Giant Conifers," *Natural History,* October 1982. Why does the taiga extend so far south on the Pacific coast of the North American continent? A peculiar local climate makes it so.

White, P.T. "Rain Forests: Nature's Dwindling Treasures," *National Geographic,* January 1983. Unexcelled photographs, often almost as rich a source of information as any text, and enjoyable as well.

Chapter 38
HUMAN ECOLOGY

Learning Objectives

After you have studied this chapter you should be able to:

1. Review the development and impact of modern human life-styles upon the ecosystems of the earth.
2. Contrast an agricultural community with the natural community it replaces, relating the differences to ecological instability.
3. Describe two problems associated with pesticide use and two problems specifically associated with the use of persistent pesticides (chlorinated hydrocarbons).
4. Summarize the sources of water pollution and describe the effects of dumping organic wastes into a stream.
5. Discuss the principal ecological effects and climatic implications of air pollution.
6. Describe the principal methods of solid-waste disposal and discuss the relevance of recycling to waste disposal.
7. Discuss nuclear power and solar power as energy options.
8. Summarize the process of extinction, listing factors that contribute to the decline and extinction of endangered species, and providing an example of each.
9. Relate overpopulation to specific environmental problems and explain how humans can temporarily expand the carrying capacity of their habitat.

Human beings have an ecology quite as much as do prairie dogs or ivy plants. And although one rarely speaks of a species as a geological or ecological disaster, that is the category in which *Homo sapiens* surely belongs. No organism in the history of life has had a greater impact on the planet than we have had. In a few short generations we have utterly transformed the face of the earth and greatly accelerated the rate of extinctions of species. One might be able to take a neutral position at this point; after all, who cares what happens to snail darter fish or the furbish lousewort plant, or even the chimpanzee? Yet what we face is not merely the absence of a few species of interest to bird watchers, or the destruction of forests of concern only to those who have time to enjoy them. What we face is the real possibility of the heedless disassembly of our planet's life support system, for as you must understand by now, the cycling of nutrients and gases, and the flow and storage of biologically useful energy, depend on the interaction of a variety of living things. In other words, the conditions required by living things, ourselves included, have been produced by other living things,[1] and the life support system of life is life itself.

THE ECOLOGY OF AGRICULTURE

In large measure the ecological impact of humanity varies widely with the kinds of human societies and their specific technologies. Of these, perhaps no technology has had more direct and indirect ecological impact than agriculture.

Such techniques as scientific crop rotation, use of chemical fertilizers and pesticides, mechanical tillage, and sophisticated irrigation systems did not exist 200 years ago. What has made most of them widespread and practical is the ready availability of energy. Energy of one kind or another chemically fixes nitrogen in fertilizer factories, turns the wheels of tractors, and mines water from the ground (Fig. 38–1).

Agricultural Communities

An agricultural community, such as a field of corn or cabbages, usually has characteristics that set it apart from the community it displaced. First, *an agricultural community is unstable* because it can be maintained only by human intervention and energy input. Maize (corn), for example, can reproduce only with human aid. Another way of looking at the matter is that humans are essential for the continuance of agricultural communities.

Second, *an agricultural community is simpler than any natural ecosystem.* The plants of an agricultural community may be of only a single species. This **monoculture** greatly simplifies the community, and since the plant that is being monocultured can be consumed by human beings, the community that results is highly and desirably productive from a human viewpoint (Fig. 38–2). However, that very monoculture renders the agricultural community unstable and highly susceptible to insect pests or diseases, which can spread from one host organism to another with little hindrance.

In the natural state host plants (those attacked by pests) are interspersed with others that a particular pest species will not eat. Moreover, wild plants are usually genetically diverse and often possess natural pest-control adaptations, such as poisonous alkaloids. Plant breeders

---

[1] For instance, the equable range of temperatures on the earth could not exist if not for the relative lack of carbon dioxide in the atmosphere. The oxygen we breathe and the ozone that protects us from the ultraviolet radiation of the sun are products of photosynthesis for which, like the consumption of the carbon dioxide, we may thank protists and plants.

(a)

(b)

usually develop strains without the alkaloids, but that makes the varieties they produce much more susceptible to insect attack. Furthermore, plant breeders try to produce varieties with increased yield, improved marketability, greater ease of shipping or harvesting, and resistance to particular diseases. At any one time agriculture tends to be dominated by the latest varieties, so any threat to that single variety is potentially able to wipe out an entire major crop for one or more years. (This actually happened several years ago with a dominant variety of corn, which was uniquely susceptible to a new strain of fungus disease.) In modern monocultural agriculture, therefore, unnatural pest-control methods and other practices have become necessary. Inevitably, these methods have

Figure 38–1 Mechanized versus traditional agriculture. (a) The extraordinary productivity of Western agriculture today is due to mechanization. However, modern farming methods have ecological side-effects, including accelerated soil erosion and pollution of groundwater and surface runoff with fertilizers and pesticides. (b) Though less productive per acre, traditional agriculture may better preserve those acres for the use of future generations. The future of agricultural science may lie in the improvement of such methods; indeed, we may yet see them used again in the developed world. (USDA photos.)

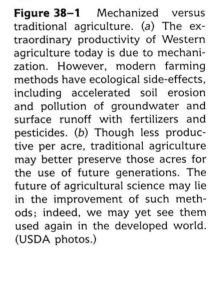

Figure 38–2 Monoculture. This vast field of strawberries consists not only of a single species but even of a single variety of that species of plant. (Ed Cooper Photos)

substantial ecological impact, often extending far outside the boundaries of the areas where they are applied.

Third, *the chemical cycles of an agricultural community are often incomplete.* The cultivated species are consumed at a spot remote from where they are grown, and their substance is not usually returned to the soil but escapes from the terrestrial ecosystem through a sewage disposal plant or garbage dump. This interruption of nutrient cycles causes the soil to deteriorate unless artificially fertilized—which brings with it problems of its own (such as increased eutrophication of adjacent bodies of water).

Fourth, *competition among species is greatly reduced.* Farmers go to great lengths to suppress species considered undesirable (called **weeds**) that compete with cultivated varieties or interfere with their harvest.

It should be obvious that an agricultural community is not and cannot be the same as the natural community it has replaced. The demands of agriculture simply do not permit this, and this realization has led to the development of game parks and preserves in many countries. Yet even on farmland itself some compromise is possible, and with enlightened management, agricultural lands can be made to yield a modest return of wild game, or merely to serve as homes for wild species. However, even in the better-educated nations of the West, a knowledge of ecology sufficient to induce most farmers and other citizens to sacrifice even a small part of the potential food production of their land for seemingly impractical purposes is lacking. In the heavily populated and economically less developed countries, the situation is even worse. Hungry peasants encroach continually on wild game preserves and expunge most surviving large wild animals from the land they have already acquired. (This is sometimes abetted by money supplied by Westerners for illegal trophies.)

Forestry

The deforestation of all countries has proceeded with alarming speed since the Industrial Revolution and, in some localities, since ancient times. One reads, for instance, of the Biblical cedars of Lebanon, and the modern Lebanese flag even depicts one. Yet there is practically nothing left of the great forests whose timber was used in the construction of the palace and temple of King Solomon. This deforestation resulted not only from simply cutting the trees down, but also from heavy grazing by sheep and goats that prevented the growth of seedlings.

More recently, firewood and wood charcoal fueled the early Industrial Revolution. To this day the greatest use of forest products worldwide is for fuel. In countries that have limited or no fossil fuel resources, this often leads to dramatic deforestation. So acute is deforestation in modern India, for example, that the major fuel used over vast areas is dried cattle dung. However, there are many other uses for timber as well. In the 19th century forests were extensively and wantonly cut to provide wood for construction timber and paper pulp, with fires often completing the destruction that the logging had begun. To some extent such practices continue today.

Many nations have embarked on extensive reforestation programs. However, the trees planted often are not native and do not fit in well with the local ecology. In any case they are usually managed as a crop, so ecologically, this **silviculture,** as it is called, resembles agriculture. Tree farms are not forest communities; usually they are single-species aggregations (Fig. 38–3) with little wildlife, and usually they do not contain climax species. Douglas fir seedlings, for instance, will not grow well in

Figure 38–3 A fir-tree seedling nursery. After the trees reach a certain height, they are transferred to another location where they are planted in rows like a crop, periodically thinned, and weed trees removed. Little wildlife inhabits such farms. (USDA)

(a)

(b)

the shade of mature trees and must be planted in clear-cut areas from which all mature trees have been removed.

PESTICIDES

The **first-generation pesticides**—inorganic chemicals, such as sulfur, lead, arsenic, and mercury—have been used to repel or kill pests for hundreds of years. Many of these substances are quite toxic, and they sometimes accumulated in the soil in amounts that inhibited plant growth. Pests also became resistant to them (a case of natural selection), so over time they lost their effectiveness.

The era of **second-generation pesticides** began about 1940 with the discovery of DDT. These pesticides are synthetic organic compounds that can be classified into three groups: (1) **DDT** and related **chlorinated hydrocarbons** (Chlordane, Dieldrin, Mirex), (2) **organophosphates** (malathion, parathion), and (3) **carbamates** (Sevin, Temik). These highly effective organic pesticides have been developed and put into widespread use by everyone from farmers and health departments to suburban homeowners.

The chlorinated hydrocarbons interfere with nerve action by antagonizing the sodium pumps in nerve cell membranes. Most other organic pesticides are anticholinesterases that render the enzyme cholinesterase incompetent (see Chapter 22). Anticholinesterases are, in general, more toxic than chlorinated hydrocarbons. Human beings are less susceptible to the action of both classes of pesticides than insects, partly because of their large body size and partly because of differences in physiology.

Problems With Using Pesticides

Most widely used pesticides are broad-spectrum poisons that kill nonpest species as well as pests. Another major problem with using pesticides is that many insects become resistant to them. Although initially the pesticide may cause a rapid decline in the pest population, resistant mutants gradually replace the susceptible insects. By 1980 more than 200 DDT-resistant species were known. Often, a pesticide-resistant population evolves, while the natural predators of these insects do not become resistant. With their natural enemies devastated, the pest population increases in numbers and may become a greater threat than before the pesticides were used.

Predators do not develop resistance as quickly as pests because they are often larger and reproduce at a slower rate. They are also fewer in number because they are at a higher level on the food chain. Thus, the

Figure 38–4 Biocides, necessary to control insect pests (*a*), become concentrated as they are passed along the food chain. When predators consume poisoned insects the concentration of pesticides increases. Ultimate predators such as hawks, bald eagles, and peregrine falcons that feed on insectivorous birds are prime victims. (*b*) Pesticides interfere with the embryonic development of these predatory birds, causing thin-shelled, fragile eggs that rarely hatch in the wild. (*a*, USDA. *b*, Michigan Department of Natural Resources.)

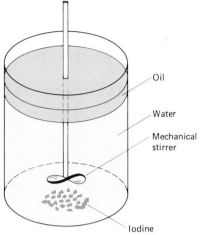

Figure 38–5 How a substance almost insoluble in water can become concentrated in another medium. Imagine a simple experiment with iodine, water, and mineral oil. The iodine is placed on the bottom of a vessel, and water is added. On top of the water is placed a layer of oil. The iodine is quite insoluble in water but very soluble in oil; however, it must pass through the water to get to the oil. Very soon, the water is saturated with iodine, which gives it a slight yellow tinge that never deepens. However, as hours pass, the iodine in the water enters the mineral oil, which absorbs it avidly. There is now less iodine in the water, and more dissolves to replace it. Still more is absorbed by the oil, so more dissolves in the water, and so on, until almost all of the iodine is dissolved in the oil even though little was dissolved in the intervening water at any time.

Like the iodine in this experiment, chlorinated hydrocarbons tend to be soluble in fat but not in water, and most are very stable compounds. Thus, although they are present in only a few parts per billion or even trillion in surface water, they accumulate rapidly in the fats and oils of living things. This represents the first stage of biological magnification.

probability that resistant individuals are present or that resistant strains will be selected is lower.

DDT and Other Chlorinated Hydrocarbons

The organophosphate and carbamate pesticides are **biodegradable,** which means they decompose several weeks or months after being sprayed. DDT and other chlorinated hydrocarbons are **persistent pesticides.** They are not biodegradable. Another problem with the chlorinated hydrocarbon pesticides is that they do not remain confined to the areas where they are sprayed. Food chains, air movements, and to a lesser extent, water currents distribute persistent pesticides globally, so there is now probably no organism on land or in the sea whose tissues are completely free from them. Traces of DDT have even been found in the fat of penguins in Antarctica, thousands of miles from where it was sprayed.

The chlorinated hydrocarbon pesticides are insoluble in water (Fig. 38–5). For this reason they are neither readily broken down metabolically nor easily excreted. When an animal eats food containing DDT, wastes from the food are eventually excreted, but a large portion of the DDT remains in the body and accumulates in body fat. This leads to the process known as **biological magnification,** in which chlorinated hydrocarbons (and some other substances, including certain radioactive elements) are concentrated in ecosystems (Fig. 38–6). Thus, in a sprayed lawn the cricket that eats a little grass each day eventually accumulates more pesticide than the grass did. The toad that eats these crickets soon has a greater DDT content than the crickets, and so on. In this way an ultimate predator, such as a hawk or human, concentrates in its body an appreciable portion of the pesticides absorbed by many acres of vegetation. Milk of nursing mothers has been found to contain concentrations of DDT higher than acceptable limits set by the Food and Drug Administration for cows' milk.

Reducing Pesticide Use

Because DDT poses a serious threat to ecosystems, its use in the United States was banned in 1973. However, in the early 1980s it was still being manufactured in the United States for export to the many other countries

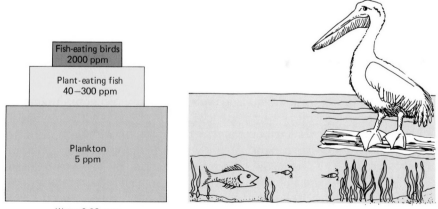

Figure 38–6 A toxicant pyramid, which provides further stages of concentration. Most of the biomass consumed on each level is excreted, but DDT and similar fat-soluble pesticides are not excreted to anywhere near the same extent and therefore tend to become progressively concentrated in the higher members of the food chain. Thus, they are most dangerous to ultimate predators. The numbers indicate concentration of DDT in parts per million.

that still use it. Several closely related chlorinated hydrocarbon pesticides have also been banned, but others continue to be used widely.

Although the problems connected with pesticide use are reduced to some extent when biodegradable compounds, such as the carbamates or organophosphates, are substituted, *there is no known pesticide that is free from undesirable ecological consequences.* The carbamate Sevin, for instance, is almost completely harmless to mammals but it is instant death for bees, and bees are vital for pollination of food and wild plants. Since all pesticides kill more than the pests against which they are directed, in the end our pesticide problems can be solved only if we can avoid them. Yet this seems less likely to happen than ever. In fact, we now use twice the amount of pesticides used in 1962, the year Rachel Carson's *Silent Spring* was published.[1]

Why are pesticides still used so widely? The reasons are not hard to find. The World Health Organization contends that an international ban of DDT would be a disaster to world health. The agricultural production of the United States might decrease by 30% if all pesticides were banned, and poorer nations would suffer even more from decreased food supplies and increased incidences of insect-borne diseases, especially malaria. Ironically, however, the mosquitoes that carry malaria have in recent years developed behavioral and other resistance to DDT, so although DDT remains damaging to everything else, it is steadily less effective against the pests for which it is intended.

Moreover, the consequences of population growth made possible by malaria control may well outweigh the good that has been accomplished. We can avoid using pesticides, to be realistic, only if we are content to produce less food. We can produce less food only if we have fewer mouths to feed. With fewer mouths to feed, we could, perhaps, confine human settlement to the parts of the globe ecologically suited to us—for example, by avoiding malarial swamps.

Figure 38–7 Biological control, an alternative to the use of biocides against screw worm flies, which produce tremendous damage in livestock. The male flies are sterilized and then released to mate with wild females. Since the females mate only once, they will never reproduce. (USDA)

Alternatives to Pesticides

There *are* alternatives to the use of pesticides short of complete abandonment of our crops to the insect world, although none permit as high a level of agricultural production as our overpopulated world demands. Some are effective by themselves; others help reduce the need for pesticides. Among these alternatives are the use of natural enemies, the release of sterile males, and such practices as careful crop rotation to break pest life cycles by depriving them regularly of their preferred food plant (Fig. 38–7).

WASTE DISPOSAL AND POLLUTION

Waste may be defined as any product of our civilization that is usually discarded rather than used, or a formerly useful product no longer used for its original purpose or for any other. **Pollution** is a reduction in the quality of the environment by the addition of materials (or heat) not normally found there. Pollution exists when wastes or other substances have a significantly damaging effect upon public health, property, ecosystems, or aesthetic values.

Polluting costs us a great deal. Why do we do it? Most of the reasons are economic.

---

[1] *Silent Spring* is often used to date the beginning of the modern environmental movement, although this movement certainly has much deeper roots. In *Silent Spring* Carson eloquently described the ecological damage done by pesticides, especially when carelessly applied without thought of the consequences.

1. By avoiding the costs of waste disposal, the polluter compels society to bear either the costs of his pollution or the costs of correcting it.
2. By passing on—as in the case of dumps of toxic chemicals—the cost of cleanup to future generations, we avoid having to pay for it ourselves. The unborn have no effective recourse against us for despoiling their heritage.
3. By providing convenience of use through disposable packaging, we increase sales of many goods, while compelling the public to bear the costs, ecological and otherwise, of that disposal. (What would your garbage collection bill amount to if the 60% to 80% of it that is packaging did not need to be thrown out?)
4. By deliberately designing goods to wear out or become unstylish, and by otherwise encouraging wasteful consumption, we increase the flow of goods and services through the economy.

Waste, of course, results in the need for waste disposal, plus the depletion of virgin resources that future generations will need.

WATER POLLUTION

At present, the two major sources of water pollution are industrial wastes and municipal sewage. How do they produce their effects?

Organic wastes provide a rich source of nutrients for decay bacteria and fungi. Hence feces, blood from slaughterhouses, oxygen-demanding wastes from paper mills, and peelings from vegetable-processing plants (among many other things) stimulate the growth of bacteria whose metabolism rapidly removes oxygen from the water. Industrial wastes may also contain large amounts of sediment, chemically combined nitrogen, phosphorus, carbon dioxide, methane, hydrogen sulfide, and smaller amounts of miscellaneous chemicals, heavy-metal ions, and even pesticides.

Industry accounts for most water pollution in the United States. Usually far more concentrated than municipal sewage, industrial waste produces up to 12 times the amount of pollution per gallon of effluent that municipal wastes do. But what they lack in concentration municipal wastes somewhat make up in volume, especially with street drainage and surface runoff after rain and snow.

A polluted environment is a demanding one. Not surprisingly, few organisms can tolerate it. Yet those able to exist in it often attain astronomical numbers and very large biomass. Most aquatic organisms, including, unfortunately, those we hold most desirable, are sensitive to the effects of pollution and are destroyed by it.

When organic wastes are dumped into a stream, a predictable sequence of events occurs (Fig. 38–8). Near the source of pollution, surprisingly, numbers of fish and other organisms may persist because the organic wastes have not had time to decay, and the dissolved oxygen level is high. In severe instances, somewhat farther away, conditions worsen, with bacteria and fungi that degrade organic sewage using up so much oxygen that anaerobic conditions exist. Still farther away, the polluted water has begun to purify itself: Organic materials start to disappear, oxygen diffuses into the water from the air, and except for cultural eutrophication, something like a normal ecology is established.

About half the population of the United States depends upon rivers for drinking water (the rest of us use groundwater). Although legislation requires industries and municipalities to treat their wastes before dumping them in the nearest stream or river, the purification process is never

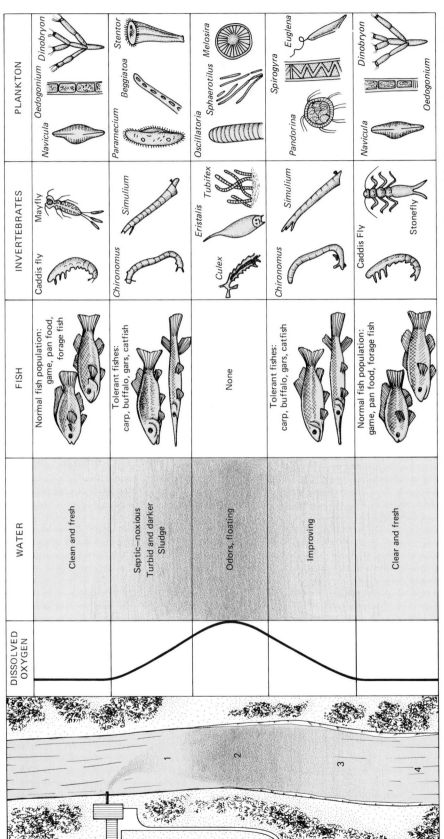

Figure 38-8 Zonation of pollution ecology in a stream receiving untreated sewage. As the amount of oxygen dissolved in the water decreases, fishes disappear. Only organisms able to obtain oxygen from the surface, tolerate low oxygen tensions, or respire anaerobically can survive. When the sewage has been all decomposed by bacteria, the species of plants and animals present in the stream return approximately to normal, although eutrophic changes may persist. (From Odum.)

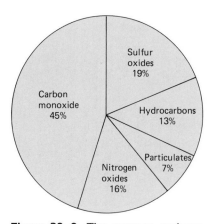

Figure 38–9 The average makeup of air pollution.

100% effective. Thus, many pollutants, including suspected carcinogens, are present in the drinking water of many communities. Eventually, many river pollutants (like air and land pollutants) find their way to the oceans—the ultimate **sinks** for most pollution.

AIR POLLUTION

Have you ever considered the plight of a fish in a polluted lake? It is exposed to a continuous dose of toxic substances every moment, every day, from which it usually has no escape. Air pollution puts us in the position of the fish, for each of us must breathe about 20,000 times every day. With each breath we inhale a wide variety of toxic gases and particles spewed into the air by automobiles, power plants, industries, and cigarette smokers. The United States alone dumps more than 220 million tons of pollutants into the air annually.

The pollutants that account for most air pollution are carbon monoxide, sulfur oxides, nitrogen oxides, hydrocarbons, and particulates. Their relative contributions to air pollution are indicated in Figure 38–9. These pollutants interact to form secondary pollutants and smog. Cigarette smoke is also an important indoor air pollutant. It contains several other pollutants and represents a highly concentrated source of pollution. Although the total volume of polluted air produced by smokers does not begin to compete with the amount spewed forth by industry, the health consequences are more severe, for the air polluted by the smoker is the air closest to us—the very air we must breathe. Industrial pollution is generally somewhat diluted by the time it wafts our way. Nevertheless, the effects of industrial and other gross pollution on public health and the ecology are substantial.

Ecological Effects of Air Pollution

Air pollution has many sources, ranging from the smokestacks of power plants to the exhaust pipe of the family car. Our consumptive life-style ensures a multitude of sources of such air pollution, especially in manufacturing areas and in the vicinity of cities. In Chicago, air pollution at times reduces the amount of sunlight about 40%, and in New York City, 25%. When sunlight is reduced, photosynthesis is also diminished. In addition, trees and other plants absorb great quantities of pollutants directly from the air. For the most part damage occurs in the photosynthetic tissue (mesophyll) of the leaf (Fig. 38–10). The surfaces of mesophyll cells, which are moist to facilitate gas exchange in photosynthesis, are vulnerable to attack by toxic substances in the air. Fluoride particu-

Figure 38–10 Effects of air pollution (sulfur dioxide) on white birch leaf (right). (USDA)

(a)

(b)

(c)

Figure 38–11 The meteorology of air pollution: three views of downtown Los Angeles. (a) In a normal pattern of air flow, warm air close to the ground rises, carrying with it most atmosphere pollutants. (b) A thermal inversion. At 75 meters a lid of warm air prevents the circulation of air from below. (c) Heavy pollution trapped under an inversion layer at 450 meters. (Courtesy of Los Angeles Air Pollution Control District.)

lates, photochemical smog (generated from automobile emissions by a chemically complex atmospheric process), sulfur dioxide, and perhaps even very fine soot can kill vegetation. Some species of plants are more susceptible to certain pollutants than other species. At a concentration of less than one part per million, the common pollutant nitrogen dioxide reduces the growth of tomato plants by 30%.

In Los Angeles County and areas up to 60 miles away, forests are being damaged by photochemical smog. Photosynthesis is reduced by 66% when smog concentrations are 0.25 part per million. Such decreases in photosynthesis slow the flow of resins under the bark, in turn rendering the trees susceptible to plant disease and insect pests. With the increase in air pollution that present trends are producing, vast damage to vegetation may occur in North America and even the world. This damage is caused not only by the direct means already mentioned, but also by acid rain.

Meteorology of Air Pollution

Atmospheric inversion is the cause of most air pollution episodes (Fig. 38–11). In such cases weather conditions form a lid of warm air above cooler, polluted air. Though inversions do not actually increase pollution, they seal the pollutants below and prevent them from being dispersed. In Los Angeles County, where conditions favor them, inversion layers form when warm air moves in from the deserts to the east and lies over the mountains that surround Los Angeles. Beneath this warm air is a layer of cooler air that has moved in from the sea. Such inversions also form, though less regularly, in virtually every area of the country. Simple air stagnation, which can occur anywhere, is almost as bad, for then pollution is not dispersed by normal wind action.

It is that very wind action, helpful in moving polluted air, that unfortunately results in acid rain perhaps hundreds of miles from where the pollutants originated. It is common for acid rain to be caused by air pollution in other nations, and this has caused international disputes among such countries as Great Britain and Sweden, and the United States and Canada. Acid rain can add sulfur to crops that need this element (not always present in fertilizer). But in comparison to the ecological harm that is done, this benefit is trivial. Acid rain damages trees and other vegetation directly (see Fig. 2–10), probably leaches far more nutrients from the soil than it could possibly add, and where the surface waters are soft (that is, where they lack natural buffering capacity) kills fish and other aquatic life over vast areas, reaching into wilderness lakes and ponds seldom visited by people but despoiled by them nonetheless.

Air pollution probably also affects climate on a global scale. Some scientists fear that if we continue to inject great quantities of particulates into the atmosphere, we may bring on another ice age. Particulate matter could serve as nuclei for the condensation of high clouds, which would reflect sunlight away from the earth. One scientist has calculated that the addition of only 50 million tons of pollutant particles to the atmosphere could reduce the average surface temperature of our planet from its present 60°F (15°C) to about 40°F (7°C). Most forms of plant life could not survive in such a cold climate. Other scientists fear that the vast amounts of carbon dioxide we are producing may have a **greenhouse effect,** holding heat in so that average temperatures will rise.

The really upsetting aspect of global air pollution is our ignorance of its probable effects and our inability to predict them with certainty. Whatever the long-term effects on climate, pollutants *are* likely to affect the delicate balance of the ecosphere. It seems foolish to await a practical demonstration of just which prediction turns out to be correct.

RADIOACTIVE POLLUTION

Radioactive strontium has been released into the atmosphere mostly by the atmospheric testing of strategic nuclear weapons, a practice forsworn in 1963 by those countries that signed the Nuclear Test Ban Treaty. In the event of even limited nuclear war, however, not only radioactive strontium but also radioactive iodine and many other such substances would find their way into the food chain, producing a harvest of disease for generations after the acute deaths from the nuclear exchange. Even during peace, widespread generation of nuclear wastes by power plants could have a similar effect, particularly in the event of accidents encountered in transporting those wastes, destruction of a nuclear power plant by sabotage or warfare, or diversion of nuclear fuel.

The radioactive isotope of strontium is chemically similar to calcium and so travels in ecosystems much as calcium does. Like calcium, it becomes incorporated into bones and teeth and can produce such damage as bone cancer. It has a half-life of 28 years—that is, in 28 years half of it will still remain, the other half having decayed into nonradioactive products. In another 28 years the remaining half will not have declined to zero but will itself have been halved, so one might say the quarter life of radioactive strontium is 56 years. That is a long time to carry such a substance in one's bones.

This is not idle speculation. Reindeer "moss," the arctic lichen mentioned in Chapter 37, has, like other lichens, the ability to greatly concentrate substances from precipitation and perhaps from atmospheric dust as well. It greatly concentrates radioactive strontium, which is further concentrated, somewhat like persistent pesticides, when reindeer eat the lichens. When Laplanders or Eskimos milk the reindeer, the radioactive strontium is still further concentrated and may end up in the body of a child. Less marked but similar biological magnification takes place in the grass-cow-human food chain.

The indirect ecological effects of widespread nuclear war are beyond calculation, but it is likely that a large nuclear exchange involving only two nations would ultimately destroy all or almost all the earth's major ecosystems as radioactive fallout was carried around the globe by atmospheric circulation. (See Focus on The Last Winter.)

SOLID WASTE DISPOSAL

Each of us accounts for about 3.6 kilograms (almost 10 pounds) of garbage, trash, and other solid wastes per day, and the amount is rising steadily. Until quite recently there was little thought of doing anything other than picking solid wastes up from one place and putting them down in another. But we are rapidly running out of rugs under which to sweep the debris of modern society.

Solid waste disposal is a relatively recent problem. Through colonial times the principal household waste other than excrement was organic garbage, and not much of that. Disposable containers were unknown. A bottle or a pot would ordinarily be discarded only if it were broken and unsalvageable. Bones were gnawed clean or boiled down for soup, or both. The little that found its way out into the yard was rapidly consumed by the traditional garbage disposal of primitive communities—pigs, dogs, and chickens. In turn, such animals were often slaughtered and eaten—a good example of the recycling of wastes.

Today's American consumer discards large quantities of paper (newspapers, paper bags, cups, plates, cartons, and other packaging materials) and substantial amounts of edible food scraps, as well as nonreturnable glass and plastic bottles, steel and aluminum cans, and wood and garden refuse, little of which is recycled (Fig. 38–12). In addition to these municipal wastes, agricultural activities generate more than 1.8 billion metric tons (a metric ton is about 2205 pounds) of wastes each year, mainly manure. Mining and industrial wastes add to the problem.

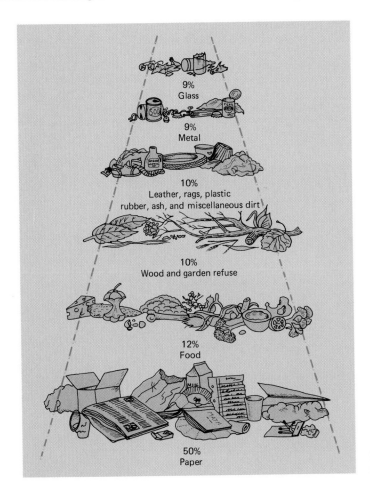

9%
Glass

9%
Metal

10%
Leather, rags, plastic
rubber, ash, and miscellaneous dirt

10%
Wood and garden refuse

12%
Food

50%
Paper

Figure 38–12 Composition of municipal trash in the U.S.

THE LAST WINTER

There has never been any doubt in the minds of informed people that nuclear war would be unimaginably disastrous, but until recently there had been surprisingly little attention paid to the probable ecological consequences of nuclear exchange. Scientists assembled at a two-day conference in the fall of 1983 agreed that of all possible ecological damage, the consequences of nuclear war would be by far the most serious, producing a horrendous but perhaps fitting climax to the environmental despoliations of our civilization.

The biological concentration of such isotopes as strontium-90 has been studied since the 1950s, and it is obvious that even a limited nuclear exchange would produce widespread environmental pollution by radioactive fallout. It has been clear for some time that the destruction of much of the protective atmospheric ozone layer by nuclear-generated nitrogen oxides would produce widespread crop damage, cancer, and disruption of the visual systems of animals whose sight is sensitive to light of very short wavelengths. What is new is an appreciation of what the smoke, soot, and pulverized earth produced by nuclear explosions would do to the terrestrial climate.

Such widespread dust has entered the earth's atmosphere before. There is much speculation that dust stirred up by asteroidal bombardment of the earth may have been responsible for several widespread extinctions of life that are observable in the fossil record. Even within historic times large volcanic eruptions have placed enough ash in the atmosphere to produce marked climatic changes. The 1815 eruption of the Indonesian volcano Tambora ejected some 25 cubic *miles* of debris, much of which did not fall to earth immediately. In 1816 it produced a disastrous year for agriculture, the "year without a summer," in which there were three killing frosts during the New England growing season and widespread hardship throughout the Northern Hemisphere.

It would appear that the dust and soot produced by even a "moderate" nuclear exchange could be expected to almost completely obscure sunlight over the entire Northern Hemisphere and probably the Southern Hemisphere as well. Even in summer, prevailing temperatures would immediately fall below freezing, dropping as low as perhaps $-15°$ to $-20°C$. The cold would persist, freezing natural bodies of water, in many cases to the bottom. Since the oceans would remain relatively warm for a considerable time, the resulting marked temperature difference between land and sea would produce storms of unprecedented violence. The darkness and cold lasting for months on end would cause most animals and plants to die, and probably most of them to become extinct, according to the conferees. This would be especially true of the tropical forms, but even temperate zone species would be decimated, particularly if the exchange occurred in the summer.

Although recovery of a sort could be expected within a few years, conventional agriculture would be impossible, not only because of climatic disruption but also because of the destruction of the industrial-based needs of agriculture, such as fertilizer plants, fuel delivery, availability of agricultural machinery and spare parts, and the like. Starving people would hunt down any surviving animals that they could, and eventually would starve themselves. Moreover, radiation sickness and other disease states, brought about by widespread chemical pollution resulting from the burning of synthetic material, could be expected to weaken even the survivors. The authors of a report summarizing the conference's conclusions emphasized "that survivors, at least in the Northern Hemisphere, would face extreme cold, water shortages, lack of food and fuel, heavy burdens of radiation and pollutants, disease and psychological stress—all in twilight or darkness."[1] The authors also predicted that most tropical plants and animals would be rendered extinct, as would most temperate zone vertebrates. The report concluded on a particularly chilling note: Under these circumstances the extinction of *Homo sapiens* cannot be excluded.

[1]"Long-term Biological Consequences of Nuclear War," by Paul R. Ehrlich, et al., *Science,* 23 December 1983.

The first atomic bomb detonated, at Trinity Site on White Sands Missile Range, July 16, 1945. This began the process that led to the unprecedented power of humans to render all life on earth extinct. (U.S. Army, White Sands Missile Range.)

Although some rural communities still dispose of their solid wastes in open dumps, and many coastal cities practice ocean dumping, the two principal means of solid waste disposal today are the **sanitary landfill** and incineration. After waste is dumped in a sanitary landfill, it may be further compacted by bulldozers. Each day a layer of soil is pushed over the garbage to discourage flies and rats. Using the sanitary landfill method, abandoned strip mines have been filled and eventually reclaimed, and artificial mountains have been constructed for skiing in the Midwestern plains. Not all land should be filled, however. Marshes and even lakes are favorite sites for landfills, despite the ecological value of these wetlands (see Chapter 37).

Landfills, like dumps, can pollute groundwater as contaminants leak into the ground. Another problem with sanitary landfills is that filled land has a limited number of suitable uses. If used as a building site, settling may cause walls and foundations to crack. Methane gas resulting from anaerobic decomposition in the depths of the fill may seep into buildings and constitute an explosion hazard. For these reasons filled land is best used for parks and other recreational purposes.

Many communities burn their garbage rather than dump it. In a modern incinerator the trash is burned in a carefully engineered furnace but some air pollution does result. The heat produced by the fire may be used to boil water and generate steam, which can be sold for industrial use.

RECYCLING

Many pollutants may be thought of as resources out of place. Solid wastes can be recycled and reused, and recycling can also be applied to sewage. After all, the minerals and organic matter that originate in Midwestern prairies end up flowing into New York Harbor, where they are only a dangerous nuisance. This sewage could be reclaimed, and so could solid wastes.

Technology is available for recycling most solid wastes. Yet less than 10% of consumer goods are recycled. About 20% of the paper used in the United States each year is recycled. If we increased this to 50%, the amount recycled in Japan, we could save about 100 million trees. Enough energy would be saved to supply 750,000 homes with electricity. Why don't we practice recycling to a greater extent?

Perhaps the main reason recycling has not become more popular is that it is generally thought to be expensive. Manufacturers consider it cheaper to consume the energy needed to produce goods from virgin materials than to hire the labor necessary to recycle. An aluminum can is worth about 1 cent in the United States. For that penny, many people would rather throw the can in the trash than worry about recycling it. However, it takes about 20 times more energy to manufacture a can from raw ore than from recycled scrap metal. If the costs of manufacturing and then disposing of the can were considered, the value of recycling would be more readily appreciated. Changes in tax laws, which now encourage use of virgin materials over recycling, and other legislative incentives might be useful in promoting recycling.

ENERGY OPTIONS

Just as there could be no life on earth without the energy from the sun, there could be no modern society without the energy harnessed by human beings. Yet as population has increased and technology has expanded, we have come to realize that we have an energy problem. This problem has resulted from too many people consuming too much energy, both directly and indirectly, by consuming too many goods that required energy input for manufacture (Fig. 38–13). During the past several years

Figure 38–13 Differences in energy consumption between the United States and Europe. At midnight, when most people are asleep, the United States is brightly lit, as can be seen in this satellite photograph (*a*). In contrast, most of Europe is dark, so that much less energy is wasted (*b*). (*a*, U.S. Air Force photo; *b*, U.S. Defense Meteorological Satellite Program photo.)

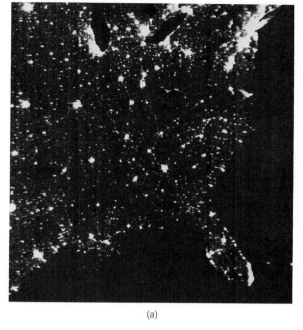

(a)

(b)

it has become increasingly evident that we are depleting oil and natural gas resources at an alarming rate, and that the amount of coal available is also limited. New technologies are being developed that will enable us to replace traditional energy sources with new ones, but at present we still depend on fossil fuels (mainly petroleum) for about 90% of our energy needs (Fig. 38–14).

Two important energy options being developed are nuclear power and solar power. Thirty years ago nuclear experts heralded the age of nuclear power with the claim that nuclear power would be "too cheap to meter." These words could not have been less true. With changes in the economy and massive cost overruns, nuclear power has become very expensive. Questions about environmental impact, availability of uranium ore, and safety have seriously threatened the nuclear power industry. (Only two nuclear plants ordered to be built in the United States during the past 9 years have not been later canceled, and these two are unlikely to ever be completed.) An average of 12 mishaps per day occur in nuclear plants, mainly due to equipment problems or failures, and the problem of safe disposal of nuclear wastes has never been solved.

The nuclear power plants currently in use are **fission** plants (Fig. 38–15). A controlled chain reaction is established in some suitable material, usually a uranium isotope of plutonium. The heat that is produced boils water or some other fluid, producing steam that drives a turbine and generator. **Fusion**-type nuclear power depends on the fusion of atoms of light elements, such as deuterium, tritium, and lithium, to produce the heat. Nuclear fusion may prove less harmful than fission, but the technology is still in its developmental stages and no one knows when this energy option will become a reality. Meanwhile, it is probable that fission-type nuclear power plants will continue to account for a small percentage (perhaps 5%) of the world energy budget.

Many experts consider solar power a safe and viable alternative to nuclear power. The energy is free and very abundant, it causes no air pollution and few environmental hazards of other kinds, and the technology promises to be less expensive than alternatives. Solar energy can be

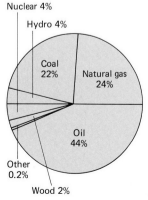

Figure 38–14 Composition of fossil fuels depended on for 90% of our energy needs.

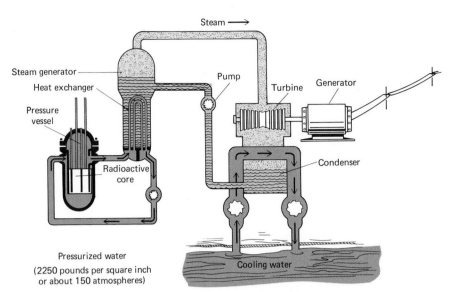

Figure 38–15 Schematic illustration of a nuclear fission plant powered by a pressurized water reactor.

used in buildings through passive designs maximizing the use of natural sunlight and through the installation of solar collectors, which trap and store heat (Fig. 38–16).

Technology is available for converting sunlight into direct-current electricity, but thus far it is expensive. Photovoltaic cells, tiny cells similar to the silicon semiconductor chips used in calculators, can convert up to 20% of the sunlight striking their surface into electricity. Another solar technology is the brine pond that traps heat at high temperatures; that heat can be converted to electricity.

Other energy options include hydropower and wind, tide, and geothermal and ocean thermal power. With the exception of hydropower, these sources are not economically feasible at present and are not expected to contribute more than about 1% of the annual energy budget by the year 2000. Energy experts project that we will still be tightly locked in to the fossil fuels for our energy needs as we enter the 21st century. In their 800-page report on the energy situation from 1985 to 2010, the Committee on Nuclear and Alternative Energy Systems concluded that the highest priority of the United States' national energy policy should be reducing the growth of energy demand.

Figure 38–16 Solar collectors on a house. (Courtesy of Energy Systems Division of the Grumman Corporation, manufacturers of Sunstream Solar Collectors.)

Figure 38–17 Some animal species faced with extinction. (*a*) The orangutan, once thought by some to be the living species closest to *Homo sapiens*. (*b*) A flightless bird from a small island north of the Hawaiians. (*c*) Many species of the tortoises that Darwin observed in the Galapagos are nearing extinction. The irony of this is rich, for it was this very abundance of species that helped Darwin formulate some of his concepts of natural selection. (*d*) Predators are often persecuted out of proportion to the amount of damage they actually cause ranchers. This photograph displays the attitude of whoever shot this coyote, left to hang on a ranch fence, presumably as a lesson to other coyotes. (*a*, Ira Block, The Image Bank. *b*, Charles Seaborn. *c*, J.N.A. Lott, McMaster University/ BPS. *d*, Bob Martin.)

(a)

(b)

(c)

(d)

EXTINCTION

Although extinctions have always occurred, human beings directly or indirectly have raised the rate at which they occur perhaps 1000-fold (Fig. 38–17). The species most vulnerable to extinction appear to be large, predatory, and migratory, such as the polar bear. However, extinction also pursues animals with other combinations of characteristics: those that require large tracts of wilderness or solitude, live in very specialized or restricted habitats, compete with human beings in any respect, or yield economically valuable products.

Extinctions also result from importing alien species against which an endangered organism has no effective defense. Recently, environmental pollution has also been involved in some extinctions. Although it is difficult to be sure, the **destruction of habitat** appears to account, at least in part, for the bulk of extinctions and endangerments of species. The extinction of the passenger pigeon probably resulted from the destruction of the beech forests in which it nested as much as from the market hunting that is usually blamed. Another extinct species, the ivory-billed woodpecker, required large virgin tracts of cypress forest. Today little of that habitat is left.

In large measure extinction—whether from hunting, habitat destruction, or other cause—results from the tendency of our species, like any other species, to multiply to the limits of environmental resistance. The resultant conversion of as much other life as possible to human biomass, plus our formidable technology-aided competition with them for resources or space assures the destruction of any species that is not directly and obviously valuable to us. Much of the destruction has resulted more from thoughtlessness than from any need. But how can we prevent the extinction of most species with which we come in contact if our population continues to increase? Surely we may yet discover that species long considered useless have potential for domestication or play some vital role in the ecosystem to which we ourselves belong. Let us hope that when that time comes, it will not be too late.

OVERPOPULATION

Fifty years ago many communities dumped their raw sewage into the nearest river or bay. They could depend on natural bacterial decomposition to break it down with little negative effect upon the ecology of the waterway. As communities grew, however, additional numbers of people produced more sewage than nature could handle. As water became more and more polluted, communities had to invest in expensive sewage treatment systems. Technology, in other words, had to take over nature's functions.

More people means more than just increased sewage, however. More people need more food, clothing, houses, schools, roads, automobiles, television sets, energy, and all the material goods our society holds so dear. Each of these can be translated into increased pressure on the environment. For example, the need for more food results in the use of more pesticides and more chemical fertilizers, which results in damage to the soil and increased land and water pollution. Production of chemical fertilizers requires the use of more petroleum, contributing to energy shortages. At the same time, more people need housing, stores, schools, and roads, so more land is taken out of agricultural production, leaving farmers to try to raise more food on less land.

The one-third of the world's population that lives in the developed countries consumes 85% of the earth's resources and is responsible for most of the stress placed on the environment. The other two-thirds of humanity strive to consume at our level. It has been estimated that the maximum world population that could be supported at the United States' level of affluence is less than *1 billion.* The environmental impact of enriching even this billion to current United States norms would involve increased industrial pollution, increased erosion of agricultural land, further depletion of natural resources, and much more. Such a goal is totally unrealistic, however, because at the current growth rate world population will reach 8 billion by the year 2100. And inevitably most human beings living then will subsist in unprecedented conditions of poverty.

Overpopulation is thus one of the most pressing problems of our time—a problem from which we in the United States are *not* insulated. Aside from the moral issues involved, the National Security Council has described population increases around the world as a threat to our national security. How much greater a threat it must be to the Third World nations in which population growth is now the greatest—and to their neighbors (Fig. 38–18).

Figure 38–18 In many heavily populated countries, people often lack what those in developed nations consider the basic necessities of life. Some people in Indian cities spend most of their lives without homes, sleeping on the streets. (United Nations photo.)

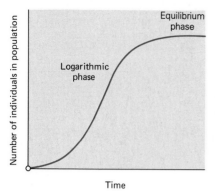

Figure 38–19 A typical sigmoid (S-shaped) population growth curve. The total number of individuals is plotted against elapsed time. The environment initially offers little resistance and the population grows exponentially. When the population becomes sufficiently dense, competition for food, habitat, and other necessities begins to limit growth. The population stabilizes at the carrying capacity.

How Population Grows

Early in this century the Russian ecologist G.F. Gause studied the growth of populations of *Paramecium,* the ciliate protist. Given optimal conditions, the population increased slowly at first, in what is sometimes called the **lag phase** of population growth. Soon, however, the population entered a rapid exponential growth phase, the **logarithmic phase** (Fig. 38–19). Ultimately, population growth slowed and stopped in the **equilibrium phase.** The overall curve is S-shaped.

Gause's observations have since been extended to cover a multitude of different organisms, including mammals. When the biotic potential of an organism initially greatly exceeds the resistance of its environment, its population growth closely resembles that of Gause's microorganisms. When the population becomes sufficiently dense, competition for food and other resources limits further growth. Population stabilizes at the **carrying capacity,** the maximum population that the environment can sustain indefinitely. If, in achieving population growth, the organisms have consumed essential *non*renewable resources, or have consumed renewable resources so excessively that they lose their ability for renewal, there is little or no equilibrium phase. In such a case the essentials for supporting even a small population are lacking and the population enters a *phase of decline,* itself often logarithmic. In this way even very dense populations can "crash" and may suffer almost instant extinction (Fig. 38–20).

It is instructive to view human population growth in this light. Beginning with 14,000 years ago up through the middle ages the human species was essentially in its lag phase of population growth. Disease and food shortages served as powerful environmental resistance. But around the time of the Industrial Revolution our population entered the logarithmic phase of its growth, and is now increasing by about 200,000 persons per day (Fig. 38–21). Today, in some nations, human population is *doubling* every 15 years, or even more rapidly.

At any one time population depends upon a balance between two factors, *birth rate* and *death rate.* Different organisms have different ways of balancing the two. A very high death rate could be compensated for by a high birth rate, for example, but if the death rate were low, high populations could be maintained by a much lower birth rate.

The populational, or demographic, events of the past 250 years are shown in Figure 38–22. Notice that among developed countries, birth rate has declined over the long run. Death rate has also declined, mostly due to the control of infectious disease. Since birth rate and death rate have mostly marched in step, the population of such countries as Swe-

Figure 38–20 Growth function for reindeer on St. Paul Island. Twenty-five reindeer were introduced on the island in 1910. Favorable environmental conditions, including a lack of predators, allowed the population to increase exponentially. By 1940, there were 2000 reindeer. At that point, however, there was no longer enough food to support such a large population, and mass starvation followed. (From Krebs: *Ecology,* New York, Harper and Row, 1972.)

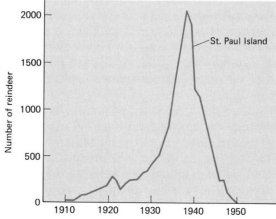

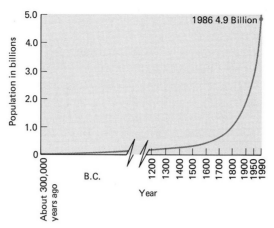

Figure 38–21 The growth of the human population.

den has increased only modestly, and in some, such as West Germany, it appears to be in a state of slight decline. The birth rate in less developed countries has also fallen dramatically, but modern medicine has brought mortality down even faster, so that their population has increased greatly despite the decline in birth rate.

Current and past population growth affects age distribution in a nation's population. In the graph of the United States population shown in Figure 38–23, notice that there are very few individuals in the uppermost age ranges. Mostly because of the post World War II baby boom, there is a bulge in the population profile at around the teen and young adult ages.

Sweden, shown in the second graph, has a stable population history. Due to widespread use of birth control methods and ready availability of excellent medical care for the entire population, about as many Swedes are born each year as die. Thus, Sweden has achieved *zero population growth.* However, the average Swede lives to quite an advanced age so that the numbers of people in *all* age ranges are about the same up to about age seventy.

Mexico, shown in the third graph, typifies heavy population growth in the third world. In such a nation, children have always been perceived as an economic asset—extra hands to work subsistence farms, and eventually, providers of care to aged parents. Since, until recently, many children died early from infectious disease, powerful economic motivation has caused widespread resistance to population control.

But here there is a conflict of economic interest between individuals and their society. In all three graphs we have stippled the age ranges in which people consume resources without, on the whole, making substantial economic contributions. A demand for schooling, pediatric care, nursery care, and much else is generated by the children of Mexico. Yet the resources of Mexican society are largely insufficient to meet this de-

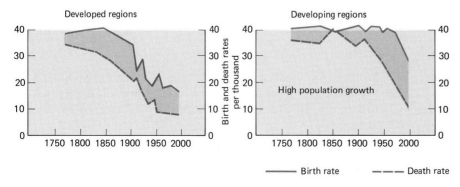

Figure 38–22 Birth and death rates in the developed and underdeveloped world for the past 250 years.

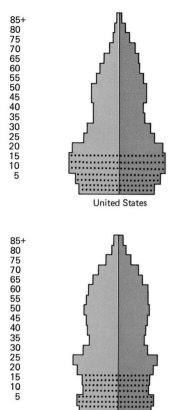

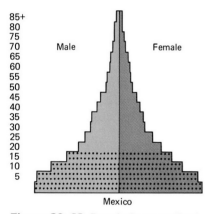

Figure 38–23 Population age distribution in three countries. In each graph, the stippled regions indicate the ages during which people consume resources without making economic contributions.

mand. Moreover, the growth rates of the Mexican population result in the doubling of the need for *all* of society's services every 15 years. Even an affluent country like Sweden or the United States could not readily double all its roads, schools, hospitals, sewage disposal plants, fire departments, apartment buildings, and more every 15 years. And if it did, think of the ecological consequences. However, the developing countries cannot even run fast enough to stay where they are, yet understandably, they aspire to the same standards of affluence that the developed nations enjoy.

What has produced the demographic differences between the developed and less-developed nations? Evidently, the citizens of the developed nations ceased some time ago to regard children as an economic asset. In the United States, for example, it costs in excess of $60,000 to raise a child to economic independence (not counting the cost of a college education), almost none of which is returned to the parent. Upwardly mobile people therefore tend to limit their family size. Most developed countries went through a difficult time of social transition in the 19th and early 20th centuries, when they accumulated the capital necessary to establish the industrial economies in which children are no longer an economic asset. Not so the less-developed nations: With all capital consumed as fast as it is generated by the urgent needs of their populations, they are not likely to achieve industrial economies or undergo the kind of demographic transition that has placed the developed countries in their present enviable position.

When Is a Nation Overpopulated?

It is commonly held in the United States that *other* countries, especially less developed nations, are overpopulated, but surely not the United States itself. However, as suburban sprawl stretches out, merging one city into another, as farms are sold and the cost of food rises, as resources dwindle and we experience fuel and other shortages, as water pollution and air pollution produce increasing blight, many individuals no longer need to be convinced that overpopulation is a reality, a present reality, in this country.

In the long run the ideal population for a nation, continent, or planet is one that can be sustained indefinitely. Higher population densities than this generally result from *temporary expansion of the carrying capacity of the habitat by technological means.* These expansions thus far have always depended on the consumption of nonrenewable resources and our passing some of the costs to the ecosystem. Since nonrenewable resources will dwindle and finally cease, and since the ability of the ecosystem to absorb pollution is limited, expansion cannot continue indefinitely. Populations dependent on present-day mechanical technology will eventually be drastically reduced. Since more resources will by that time have been consumed, the resulting population is likely to be both poorer materially and fewer in number.

THE FUTURE

The quality of life in our world depends on the ability of human beings to limit their numbers. It is desirable that this goal be accomplished through planning, not through violence or disease. Although many environmental problems are rooted in human overpopulation, their severity is linked to our choices regarding our life-style and how we go about supporting it. As living organisms, we share much in common with the fate of other life forms on the planet. We differ from all other organisms, however, in our unique capacity to reflect on and probe our biological identities. This talent is the key to ensuring our, and the planet's, continued survival.

CHAPTER SUMMARY

I. Modern agriculture requires substantial energy and technologic input and has great environmental impact.
   A. In agriculture, people replace diverse natural communities with those consisting of one or two specially bred species.
      1. These artificial communities are unstable, simple, and usually lacking in some important ecological mineral cycles.
      2. Some wildlife can coexist with agricultural communities.
   B. Monoculture of agricultural species promotes attack by pests and disease. This is typically combatted using pesticides.
   C. Silviculture resembles agriculture in that single-species tree farms are cultivated, rather than forest communities.
II. Most pesticides are broad-spectrum poisons that kill nonpest species as well as pests. Many pest populations become resistant to pests, whereas predator populations are often eradicated.
   A. Persistent pesticides are not biodegradable. They become widely distributed and become concentrated in predators by biological magnification.
   B. Alternatives to pesticide use exist and should be employed where possible.
III. Pollution exists when wastes degrade the environment's suitability for any of the organisms that would normally inhabit it, or when damage to property, aesthetic values, or public health results.
IV. Severe aquatic pollution by organic substances produces a characteristic zonation. Eventually, much of the pollution is rendered harmless by natural processes, but only after a portion of the habitat has been degraded.
V. With every breath we inhale various pollutants deposited in the air and elsewhere by motor vehicles, industries, power plants, and other sources.
   A. Trees and other plants are damaged by air pollution.
   B. An inversion layer may develop when a layer of warm air acts as a lid, sealing pollutants beneath so that they cannot be dispersed.
   C. Particulate pollution may cause a planetary cooling trend, whereas carbon dioxide accumulation in the atmosphere is causing a greenhouse effect, which may lead to a warmer climate.
VI. Radioactive materials cannot easily be disposed of, and they persist and are passed extensively through food chains. At present most such radioactive pollution has resulted from atmospheric tests of nuclear weapons. The ecological effects of nuclear war would be catastrophic.
VII. Sanitary landfills and incineration are two main ways of solid waste disposal.
VIII. Technology is available for recycling most solid wastes, yet less than 10% of consumer goods are recycled.
IX. The nuclear power industry has been threatened by serious questions regarding its safety and environmental impact. Solar power is considered a safe alternative for the future.

X. Species may become endangered or extinct by a variety of processes—overexploitation, persecution, environmental pollution, or the introduction of exotic species—but habitat destruction is the most important.

XI. Increases in population translate into increased needs for food, shelter, clothing, material goods, and educational, medical, and social services. These increases result in additional environmental stress.

A. We have temporarily extended the carrying capacity of the earth for human beings by our technology. This expansion depends on consumption of nonrenewable resources and the passage of some of the costs to the environment in the form of pollution.

B. A nation may be considered overpopulated if its total population is too great to be sustained permanently by its resources. By this criterion, virtually all areas of the modern world are overpopulated *at the present time.*

Review Questions

1. What are some of the reasons that agricultural communities are unstable compared with the natural communities they replace?

2. Why are pest populations more likely to become pesticide-resistant than predator populations?

3. Why was DDT banned? Explain.

4. What are some specific alternatives to pesticide use?

5. What are wastes? How are wastes related to pollution? Why do people pollute the environment?

6. What are two important sources of water pollution?

7. What happens when organic wastes are dumped into a stream or river?

8. What are some immediate ecological effects of air pollution?

9. What is an inversion layer? What is the greenhouse effect?

10. Why are persistent pesticides and radioactive pollutants both subject to biological magnification? Explain this process.

11. Suggest ways to reduce the problem of solid waste disposal.

12. What are some of the reasons for the current skepticism regarding the nuclear power industry? Why do many ecologists consider solar power a better energy alternative?

13. Identify some of the ways in which we hasten the extinction of many species.

14. When should a nation be considered overpopulated? Is the United States overpopulated? Explain.

15. Business costs that can be passed on to someone who does not profit from the business are called externalities. What externalities are connected with pollution?

16. Which contributes more, in your opinion, to our ecological crisis: overpopulation or wasteful consumption? Show how each contributes to the major ecological problems outlined in this chapter.

Post-Test

1. Agricultural communities are generally _____ _____ (simpler, more complex) than natural communities.

2. Predators do not develop resistance to pesticides as quickly as pests because they are often _____ _____ and _____ at a slower rate.

3. Pesticides that are _____ decompose within a few weeks or months after they have been sprayed.

4. DDT and other chlorinated hydrocarbons are _____ _____ (soluble, insoluble) in water and very _____ (soluble, insoluble) in body fat.

5. Due to _____ _____ , persistent pesticides, such as DDT, tend to accumulate in the _____ levels of food chains.

6. _____ is a reduction in the quality of the environment by the addition of materials not normally found there.

7. In the area of most recent pollution of a stream, the organic content is _____ (high, low) and the dissolved _____ level is still high.

8. In an atmospheric inversion a layer of _____ air forms a lid above one of _____ polluted air.

9. Carbon dioxide pollution may have a _____ effect that could cause the climate to become considerably _____ .

10. Radioactive strontium becomes incorporated in _____ and _____ .

11. Two main ways of solid disposal today are the _____ _____ and _____ _____ .

12. The nuclear power plants currently in use are _____ plants.
13. The bulk of extinctions now occurring appear to be due to _____ _____ .
14. At the present time the human population appears to be in the _____ phase of growth.

15. The ideal human population size for our planet is one that can be sustained _____ . Higher population densities result from temporary expansion of the _____ _____ by technological means.

Readings

Altieri, M., D. Letourneau, and J.R. Davis. "Developing Sustainable Agroecosystems," *Bioscience*, January 1983. Energy-intensive agricultural ecosystems are not practical in the long run and must be replaced with alternative, indefinitely stable technologies.

Bastian, R.K., and J. Benforado. "Waste Treatment: Doing What Comes Naturally," *Technology Review*, February–March 1983. Mother Nature's way of waste treatment is best, especially if one can make it happen in a sewage disposal facility.

Beck, M. et al. "The Bitter Politics of Acid Rain," *Newsweek*, 25 April 1983. A neat, brief summation of the political and economic reasons for acid rain, and much else besides.

Bosch, P.W. "A Neolithic Flint Mine," *Scientific American*, June 1979. Mining and the consumption of nonrenewable resources at the dawn of agriculture.

Carothers, S., and R. Dolan. "Dam Changes on the Colorado River," *Natural History*, January 1982. Ecological changes produced by the Glen Canyon Dam.

Eckholm, E. "Human Wants and Misused Lands," *Natural History*, June 1982. How does one reconcile human needs for land with the land's needs?

Ehrlich, P., and A. Ehrlich. *Extinction*. New York, Random House, 1981. A powerful and compelling compilation of arguments for the preservation of species. These authors have also written many other excellent books in the environmental sciences, for instance, *The End of Affluence* and *The Population Bomb*.

Elias, T.S., and H.S. Irwin. "Urban Trees," *Scientific American*, November 1976. The kinds of trees that can grow in Brooklyn—an extreme example of ecosystem simplification by stress.

Graham, F., Jr. *Since Silent Spring*, Boston, Houghton Mifflin, 1970. Rather than refer you to the original book, *Silent Spring*, we offer a somewhat updated version.

Gwatkin, D.R., and S.K. Brandel. "Life Expectancy and Population Growth in the Third World," *Scientific American*, May 1982. A decrease in death rate is the basic motor of population growth in less-developed countries.

Hornblower, M. "How Dangerous is Acid Rain?" *National Wildlife*, June–July 1983. Very dangerous.

Kinkead, E. "Tennessee Small Fry," *The New Yorker*, 8 January 1979. Portrait of an endangered species—the snail darter fish.

Lockeretz, W. "The Lessons of the Dust Bowl," *American Scientist*, September–October 1978. The principal lesson in this paper appears to be that we have not learned our lesson about "the most severe man-made environmental problem the United States has ever seen."

Moran, J.M., M.D. Morgan, and J.H. Wiersma. *Introduction to Environmental Science*. San Francisco, W.H. Freeman, 1980. A fine introductory textbook written as if the reader had some intelligence.

Radcliffe, B., and L.P. Gerlach. "The Ecology Movement After Ten Years," *Natural History*, January 1981. Among educated ecologically aware persons, environmental issues are still important, as to a lesser extent they are among the general population.

Revelle, R. "Carbon Dioxide and World Climate," *Scientific American*, August 1982. As a result of fossil fuel use and the clearing and burning of forests, the global atmospheric content of carbon dioxide is increasing. What will this do to us?

Stommel, H., and E. Stommel. "The Year Without a Summer," *Scientific American*, June 1979. A natural environmental mishap, produced by the explosion of the volcano Krakatoa, illustrates the fragility of the heat balance of the earth's atmosphere.

Ward, G.M., T.M. Sutherland, and J.M. Sutherland. "Animals as an Energy Source in Third World Agriculture," *Science*, 9 May 1980. A model of clear thinking and careful discussion whose implications reach far beyond the issues immediately addressed by the authors. Appropriate technology is often traditional technology. But even traditional technology could be improved.

Westoff, C.F. "Marriage and Fertility in the Developed Countries," *Scientific American*, December 1978. What are the consequences of demographic changes in the developed countries, whose populations may actually begin to decline before the year 2015?

POST-TEST ANSWERS

Chapter 1

1. the study of life 2. metabolism 3. homeostasis 4. mutation 5. adaptations 6. cells 7. organs 8. communities 9. Ecology 10. photosynthesis 11. water, energy 12. decomposers 13. Protista 14. genus 15. hypothesis 16. theory

Chapter 2

1. atom 2. neutrons 3. orbitals 4. It consists of 2 carbon atoms, 6 hydrogen atoms, and 1 oxygen atom. 5. electron donors 6. covalent 7. reduction 8. acid 9. basic (alkaline) 10. buffer 11. capillary action; adhesive, cohesive 12. hydrogen

Chapter 3

1. c 2. d 3. b 4. e 5. c 6. d 7. a 8. b 9. amino acids 10. amino acid sequence in its polypeptide chains 11. Cellulose 12. glycogen

Chapter 4

1. resolving power 2. ribosomes 3. smooth ER 4. Golgi complex 5. mitochondria 6. Chloroplasts 7. microtubules 8. Microfilaments 9. nucleus; nuclear envelope 10. chromatin; chromosomes 11. genes 12. cell wall 13. vacuole 14. nuclear membrane; membranous 15. function it performs

Chapter 5

1. selectively permeable 2. hydrophilic; hydrophobic 3. hydrophobic tails 4. Microvilli 5. cellulose 6. Desmosomes 7. gap junctions 8. diffusion 9. osmosis 10. hypertonic 11. isotonic 12. exocytosis 13. phagocytosis 14. dissolved materials (solutes) 15. active transport

Chapter 6

1. potential energy 2. thermodynamics 3. first law of thermodynamics 4. free 5. Endergonic 6. equilibrium 7. enzymes 8. activation energy 9. substrates 10. coenzymes 11. noncompetitive 12. ATP 13. electrons 14. three 15. cytochromes

Chapter 7

1. photons 2. thylakoids 3. split water 4. electron; electron acceptor 5. Energy; ADP; ATP 6. noncyclic 7. ATP; NADPH

8. back to photosystem I 9. protons; thylakoid 10. energy; ATP 11. CO_2, NADPH, ATP 12. carbon fixation 13. Six 14. 4-carbon

Chapter 8

1. glycolysis 2. cytoplasm; mitochondria 3. carbon dioxide, hydrogen 4. NAD; oxygen 5. ATP 6. lactate 7. ethyl alcohol; carbon dioxide 8. 2; 36–38 9. fermentation 10. oxidized

Chapter 9

1. Genetics 2. chromatin 3. diploid, 2n 4. haploid, n 5. homologous 6. cell cycle 7. mitosis 8. cytokinesis 9. interphase 10. synthesis (S) 11. centromeres 12. metaphase 13. anaphase 14. Chalones 15. two; four 16. tetrads 17. crossing over 18. genetic recombination 19. tubules; testis 20. Golgi complex 21. oocyte; polar body 22. conjugation

Chapter 10

1. nineteenth; Gregor Mendel 2. gene; protein (polypeptide) 3. phenotype 4. homologous 5. X; son 6. multiple; codominant; dominant 7. recessive; same 8. heterozygous; recessive 9. can 10. inactivate 11. trisomy; mothers; fathers

Chapter 11

1. DNA 2. double helix 3. sugar (deoxyribose), phosphate group, base 4. TTGCCAGT 5. thymine; guanine, cytosine 6. replication fork 7. DNA 8. semiconservative 9. DNA polymerase 10. nucleus

Chapter 12

1. messenger RNA, ribosomal RNA, transfer RNA 2. mRNA 3. rRNA, protein 4. triplet; codon 5. transcription 6. RNA polymerase; promoter 7. ribosomes 8. A-site 9. initiation; elongation; termination 10. Operons; regulatory 11. repressor; structural genes 12. induction

Chapter 13

1. genetic engineering 2. restriction endonucleases; viral 3. recognition sites 4. plasmids 5. antibiotics 6. hybridizing, DNA (or mRNA) 7. restriction endonuclease 8. ligase; cloned (or cultured) 9. crown gall; bacterium; plasmid 10. self-replicating; protein (or product)

Chapter 14

1. family 2. classes; kingdoms 3. genus; species 4. viroid
5. a 6. c 7. d 8. f 9. b 10. b 11. a 12. b 13. a 14. c
15. bacillus; coccus 16. cell walls; bacteria 17. syphilis
18. pseudopods 19. diatoms 20. fungus 21. spores;
basidia 22. Zygomycetes 23. coenocytic 24. dinoflagellate
25. Sporozoa

Chapter 15

1. green algae 2. conjugation tube; iso 3. accessory; cyanobacteria 4. generations; iso; hetero 5. bryophyta; vascular
6. sporophyte; gametophyte 7. phloem; xylem; spores; gametes 8. gymnosperms; angiosperms; seed 9. pollen grain
10. endosperm

Chapter 16

1. invertebrates 2. bilaterally symmetrical 3. body cavity
4. echinoderms, chordates 5. spicules 6. eggs, sperm
7. stinging cells 8. a 9. e 10. f 11. b 12. d 13. c 14. g
15. c 16. e 17. f 18. g 19. d 20. d 21. eating poorly
cooked pork; hookworms 22. birds, mammals 23. amphibians 24. terrestrial; keeps the embryo moist 25. lays eggs;
pouches

Chapter 17

1. anchorage; absorption 2. guard cells 3. stomata; water
4. palisade parenchyma 5. potassium 6. source; sink; leaves
7. water; minerals; cohesion 8. food; heart 9. root (hydrostatic) pressure; guttation

Chapter 18

1. scutellum 2. hypocotyl 3. germination inhibitor 4. pollen;
water 5. staminate 6. pistillate; pollen tube; embryo 7. petals 8. male; female (or stamens and pistils)

Chapter 19

1. maturation 2. primary meristems 3. leaves 4. water
5. primary thickening 6. darker 7. lowers; cellulose cross-
links 8. ethylene 9. phytochrome; red 10. nitrate
11. ammonia; nitrate 12. cyanobacteria 13. apoplast
14. Casparian 15. mycorrhizae

Chapter 20

1. tissue 2. gland 3. squamous; cuboidal; columnar 4. b
5. d 6. a 7. e 8. c 9. b 10. c 11. c 12. d 13. b 14. a,e
15. a 16. e 17. intercalated disks 18. skeletal, cardiac
19. axons, dendrites 20. endocrine 21. integumentary

Chapter 21

1. epidermis; dermis 2. corneum 3. keratin 4. hydrostatic
5. molt 6. endoskeleton 7. compact; cancellous 8. lubricant;
joints 9. actin, myosin 10. d. nerve impulse; c. acetylcholine
release; b. T-system depolarization; a. calcium release; e. uncovering of the binding sites of the actin filaments; f. cross-
bridges flex; g. cross-bridges release binding sites (actually the
last two steps alternate repeatedly) 11. store energy
12. glycogen; ATP

Chapter 22

1. reception 2. muscles, glands 3. synapse 4. reflex 5. sensory 6. glial; neuron 7. cell body 8. axon 9. cellular sheath
(neurilemma) 10. ganglion 11. sodium pumps 12. neural
impulse, action potential (or wave of depolarization)
13. threshold 14. refractory period 15. neurotransmitter
16. acetylcholine 17. convergence

Chapter 23

1. nerve net 2. central nervous system 3. cerebral ganglia
4. brain, spinal cord 5. retina; rods, cones 6. taste, smell
7. pressure 8. protect the brain and spinal cord from mechanical injury 9. cerebrum 10. motor areas 11. beta waves
12. maintaining consciousness 13. c 14. d 15. a 16. b
17. e 18. referred pain 19. sympathetic, parasympathetic
20. tolerance

Chapter 24

1. open 2. diffusion 3. plasma 4. c 5. d 6. b 7. a 8. c
9. b 10. anemia 11. thrombin 12. c 13. d 14. a 15. e
16. c 17. blood pressure 18. left atrium 19. liver 20. baroreceptors 21. mitral 22. vasoconstrictors 23. oxygen
24. atherosclerosis 25. interstitial (tissue) fluid; lymph

Chapter 25

1. antigens; antibodies 2. thymus 3. killer T cells 4. plasma
cells 5. passive immunity 6. c 7. d 8. e 9. a 10. b
11. phagocytosis 12. Systemic anaphylaxis 13. members of
the same species 14. graft rejection 15. cornea

Chapter 26

1. thin; diffusion; moist; blood vessels. 2. tracheal; spiracles
3. lungs. 4. air sacs 5. trachea 6. alveoli 7. diaphragm
8. e 9. a 10. b 11. d 12. c 13. Decompression sickness
14. inhaling dirty air 15. emphysema 16. cigarette smoking

Chapter 27

1. ingestion 2. Digestion 3. Elimination 4. herbivore
5. canine 6. symbiont 7. it has only one opening 8. two
openings 9. salivary amylase; carbohydrate (starch) 10. b
11. d 12. a 13. c 14. b 15. e 16. rugae 17. villi
18. duodenum 19. peptic ulcer 20. store bile

Chapter 28

1. minerals 2. coenzymes 3. c 4. a 5. d 6. e 7. b
8. triacylglycerols 9. amino acids; glucose 10. joined to form
glycogen 11. amino acids; urea 12. basal metabolic rate
13. a person gains weight 14. protein

Chapter 29

1. excretion 2. uric acid 3. urea 4. protonephridia; flame
5. metanephridia 6. antennal (green) glands 7. Malpighian
tubules 8. nephrons 9. a 10. c 11. e 12. d 13. b 14. c
15. capillaries; Bowman's capsule 16. afferent arteriole; efferent arteriole 17. glomerular filtrate 18. excreted in the urine
19. hypertonic 20. collecting ducts; reabsorbed; decreased

Chapter 30

1. ducts; hormones 2. hormone 3. cyclic AMP 4. hypothalamus 5. Juvenile 6. c 7. d 8. b 9. a 10. e 11. a
12. e 13. c 14. b 15. d 16. hypo; thyroxine 17. goiter
18. cortisol 19. thyroid 20. diabetes mellitus

Chapter 31

1. vas deferens 2. erectile; penis 3. b 4. c 5. d 6. a 7. b
8. b 9. puberty 10. estrogen, progesterone 11. b 12. e
13. d 14. a 15. c 16. e 17. ovulation 18. anterior pituitary;
follicles 19. 14th 20. ovulation

Chapter 32

1. cells arrange themselves to produce the form of the organism
2. cellular differentiation 3. b 4. e 5. c 6. d 7. a
8. embryo 9. corpus luteum; pregnancy has begun 10. ectoderm; endoderm 11. 2nd 12. an opening in the wall between
the atria 13. newborn infant 14. adolescence

Chapter 33

1. change; organic 2. acquired 3. naturalist; finches
4. Genetic drift 5. reduced; decline 6. species; reproductive
(genetic) 7. sympatric 8. prezygotic; courtship 9. punctuated
equilibrium

Chapter 34

1. organic; spontaneously 2. shallow waters; clouds; volcanic
3. heterotrophic 4. macroevolution 5. base (or nucleotide)
6. isotope 7. homologous; ancestry 8. analogous; homologous 9. structural parallelism; convergent 10. divergent

Chapter 35

1. signals from the environment 2. Ethology 3. geotropism
4. positive chemotaxis 5. circadian rhythm 6. crepuscular
7. releaser 8. Instinctive; learned 9. imprinting
10. migratory restlessness 11. social behavior 12. society
13. aggregation 14. symbolic 15. Pheromones 16. dominance hierarchy 17. home range 18. conflict; population
19. pair bond 20. nurse; royal jelly 21. innate 22. culture
23. altruistic 24. Kin 25. making more copies of themselves

Chapter 36

1. community 2. niche; habitat 3. biotic potential 4. limiting 5. interspecific competition 6. coexist 7. increases
8. 2nd trophic; 1st trophic 9. absorb; cell membrane 10. 1st
trophic; highest trophic 11. 10 12. weight of living material
13. net community productivity 14. simplify 15. climax community 16. secondary 17. pioneer

Chapter 37

1. biome 2. ecotones 3. tundra; permafrost 4. taiga
5. humus (topsoil); recycling 6. salinity 7. deepest (abyssal)
8. turnover; blooms 9. low; high 10. intertidal 11. neritic
12. littoral 13. eutrophic

Chapter 38

1. simpler 2. larger; reproduce 3. biodegradable 4. insoluble; soluble 5. biological magnification; higher 6. Pollution 7. high; oxygen 8. warm; cooler 9. greenhouse; warmer
10. bones, teeth 11. sanitary landfill, incineration 12. fission
13. habitat destruction 14. logarithmic 15. indefinitely; carrying capacity

APPENDIX A
DISSECTING TERMS

Your task of mastering new terms will be greatly simplified if you learn to dissect each new word. Many terms can be divided into a prefix, the part of the word that precedes the main root, the word root itself, and often a suffix, a word ending that may add to or modify the meaning of the root. As you progress in your study of biology, you will learn to recognize the more common prefixes, word roots, and suffixes. Such recognition will help you analyze new terms so that you can more readily determine their meaning, and will also help you remember them.

PREFIXES

a-, ab- from, away, apart (abduct, lead away, move away from the midline of the body)

a-, an- un-, -less, lack, not (asymmetrical, not symmetrical)

ad- (also **af-, ag-, an-, ap-**) to, toward (adduct, move toward the midline of the body)

ambi- both sides (ambidextrous, able to use either hand)

ante- forward, before (anteflexion, bending forward)

anti- against (anticoagulant, a substance that prevents coagulation of blood)

bi- two (biceps, a muscle with two heads of origin)

bio- life (biology, the study of life)

brady- slow (bradycardia, abnormally slow heart beat)

circum-, circ- around (circumcision, a cutting around)

co-, con- with, together (congenital, existing with or before birth)

contra- against (contraception, against conception)

crypt- hidden (cryptorchidism, undescended or hidden testes)

cyt- cell (cytology, the study of cells)

di- two (disaccharide, a compound made of two sugar molecules chemically combined)

dis- (also **di-** or **dif-**) apart, un-, not (dissect, cut apart)

dys- painful, difficult (dyspnea, difficult breathing)

end-, endo- within, inner (endoplasmic reticulum, a network of membranes found within the cytoplasm)

epi- on, upon (epidermis, upon the dermis)

eu- good, well (euphoria, a sense of well-being)

ex-, e-, ef- out from, out of (extension, a straightening out)

extra- outside, beyond (extraembryonic membrane, a membrane such as the amnion that protects the embryo)

hemi- half (cerebral hemisphere, lateral half of the cerebrum)

hetero- other, different (heterogeneous, made of different substances)

homo-, hom- same (homologous, corresponding in structure)

hyper- excessive, above normal (hypersecretion, excessive secretion)

hypo- under, below, deficient (hypodermic, below the skin; hypothyroidism, insufficiency of thyroid hormones)

in-, im- not (imbalance, condition in which there is no balance)

inter- between, among (interstitial, situated between parts)

intra- within (intracellular, within the cell)

iso- equal, like (isotonic, equal strength)

mal- bad, abnormal (malnutrition, poor nutrition)

mega- large, great (megakaryocyte, giant cell of bone marrow)

meta- after, beyond (metaphase, the stage of mitosis after prophase)

neo- new (neonatal, newborn during the first 4 weeks after birth)

oligo- small, deficient (oliguria, abnormally small volume of urine)

oo- egg (oocyte, developing egg cell)

orth-, ortho- straight (orthodontist, one who straightens teeth)

para- near, beside, beyond (paracentral, near the center)

peri- around (pericardial membrane, membrane that surrounds the heart)

poly- many, much, multiple, complex (polysaccharide, a carbohydrate composed of many simple sugars)

post- after, behind (postnatal, after birth)

pre- before (prenatal, before birth)

retro- backward (retroperitoneal, located behind the peritoneum)

semi- half (semilunar, half-moon)

sub- under (subcutaneous tissue, tissue immediately under the skin)

super-, supra- above (suprarenal, above the kidney)

syn- with, together (syndrome, a group of symptoms which occur together and characterize a disease)

trans- across, beyond (transport, carry across)

SUFFIXES

-able, -ible able (viable, able to live)

-ac pertaining to (cardiac, pertaining to the heart)

-ad used in anatomy to form adverbs of direction (cephalad, toward the head)

-asis, -asia, -esis condition or state of (hemostasis, stopping of bleeding)

-cide kill, destroy (biocide, substance that kills living things)

-ectomy surgical removal (appendectomy, surgical removal of the appendix)

-emia condition of blood (anemia, without enough blood)

-gen something produced or generated or something that produces or generates (pathogen, something that can cause disease)

-gram record, write (electrocardiogram, a record of the electrical activity of the heart)

-graph record, write (electrocardiograph, an instrument for recording the electrical activity of the heart)

-ic adjective-forming suffix which means *of* or *pertaining to* (ophthalmic, of or pertaining to the eye)

-ist one who practices, deals with, or does (biologist, one who studies biology)

-itis inflammation of (appendicitis, inflammation of the appendix)

-logy study or science of (cytology, study of cells)

-oid like, in the form of (thyroid, in the form of a shield)

-oma tumor (carcinoma, a malignant tumor)

-osis indicates disease (psychosis, a mental disease)

-ous, -ose full of (poisonous, full of poison)

-pathy disease (dermopathy, disease of the skin)

-plasty reconstruction (rhinoplasty, reconstruction of the nose)

-scope instrument for viewing or observing (microscope, instrument for viewing small objects)

-stomy refers to a surgical procedure in which an artificial opening is made (colostomy, surgical formation of an artificial anus)

-tomy cut, incision into (appendectomy, incision into the appendix)

-uria refers to urine (polyuria, excessive production of urine)

SOME COMMON WORD ROOTS

aden gland, glandular (adenosis, a glandular disease)

alg pain (neuralgia, nerve pain)

angi, angio vessel, vascular (lymphangiogram, an x-ray of lymphatic vessels following injection of a radiopaque contrast media)

arthr joint (arthritis, inflammation of the joints)

bi, bio life (biology, study of life)

blast a formative cell, germ layer (osteoblast, cell that gives rise to bone cells)

brachi arm (brachial artery, blood vessel that supplies the arm)

bronch branch of the trachea (bronchitis, inflammation of the bronchi)

bry grow, swell (embryo, an organism in the early stages of development)

carcin cancer (carcinogenic, cancer-producing)

cardi heart (cardiac, pertaining to the heart)

cephal head (cephalad, toward the head)

cerebr brain (cerebral, pertaining to the brain)

cervic, cervix neck (cervical, pertaining to the neck)

chol bile (cholecystogram, an x-ray of the gallbladder)

chondr cartilage (chondrocyte, a cartilage cell)

chrom color (chromosome, deeply staining body in nucleus)

col, coli, colo colon (colitis, inflammation of the colon)

cran skull (cranial, pertaining to the skull)

cyst, cysti, cysto urinary bladder (cystitis, inflammation of the urinary bladder; cystogram, an x-ray of the urinary bladder)

cyt cell (cytology, study of the cells)

derm skin (dermatology, study of the skin)

duct, duc lead (duct, passageway)

ecol dwelling, house (ecology, the study of organisms in relation to their environment)

encephal, encephalo brain (encephalitis, inflamation of the brain)

enter intestine (enteritis, inflammation of the intestine)

evol to unroll (evolution, descent of complex organisms from simpler ancestors)

gastr stomach (gastritis, inflammation of the stomach)

gen generate, produce (gene, a hereditary factor)

glyc, glyco sweet, sugar (glycogen, storage form of glucose)

gon semen, seed (gonad, an organ producing gametes)

hem, em blood (hematology, the study of blood)

hepat, hepar liver (hepatitis, inflammation of the liver)

hist tissue (histology, study of tissues)

hom, homeo same, unchanging, steady (homeostasis, reaching a steady state)

hydr water (hydrolysis, a breakdown reaction involving water)

hyster, hystero uterus (hysterectomy, surgical removal of all or part of the uterus; hysterogram, an x-ray of the uterus)

lapar, laparo abdomen (laparotomy, incision into the abdomen)

laryng, laryngo larynx (laryngitis, inflammation of the larynx)

leuk white (leukocyte, white blood cell)

lith, litho stone or calculus (lithonephritis, inflammation of the kidney due to the presence of calculi)

macro large (macrophage, large phagocytic cell)

mamm breast (mammary glands, the glands that produce milk to nourish the young)

mening, meningo meninges (meningitis, inflammation of the membranes of the brain or spinal cord)

micro small (microscope, instrument for viewing small objects)

morph form (morphogenesis, development of body form)

my, myo muscle (myocardium, muscle layer of the heart)

nephr kidney (nephron, microscopic unit of the kidney)

neur, nerv nerve (neuralgia, pain associated with a nerve)

occiput back part of the head (occipital, back region of the head)

odont, odonto tooth (odontotomy, incision into a tooth)

ophthal, ophthalmo eye (ophthalmopathy, disease of the eye)

orchi, orchido, orchid testis (orchitis, inflammation of the testes; orchiectomy, surgical removal of the testis)

ost, oss bone (osteology, study of bones)

ot, oto ear (otitis, inflammation of the ear; otoscope, an instrument for examination of the ear)

path disease (pathologist, one who studies disease processes)

ped child (pediatrics, branch of medicine specializing in children)

ped, pod foot (biped, organism with two feet)

phag eat (phagocytosis, process by which certain cells ingest particles and foreign matter)

phil love (hydrophilic, a substance that attracts water)

phleb, phlebo vein (phlebitis, inflammation of a vein)

phren, phreno diaphragm (phrenocolic, of or pertaining to the diaphragm and colon)

proct anus (proctoscope, instrument for examining rectum and anal canal)

psych mind (psychology, study of the mind)

pyel, pyelo pelvis or kidney (pyelitis, inflammation of the renal pelvis)

rect, recto rectum (rectocolitis, inflammation of the rectum and colon)

rhin nose (rhinalgia, pain in the nose)

salping, salpingo uterine tube (salpingectomy, surgical removal of the uterine tube)

scler hard (atherosclerosis, hardening of the arterial wall)

som body (chromosome, deeply staining body in the nucleus)

splen, spleno spleen (splenectomy, surgical removal of the spleen)

stas, stat stand (stasis, condition in which blood stands, as opposed to flowing)

thromb clot (thrombus, a clot within the body)

thym, thymo thymus (thymectomy, surgical removal of the thymus)

ur urea, urine (urologist, a physician specializing in the urinary tract)

visc pertaining to an internal organ or body cavity (viscera, internal organs)

THE METRIC SYSTEM

Standard metric units		Abbreviations
Standard unit of mass	gram	g
Standard unit of length	meter	m
Standard unit of volume	liter	l

Some common prefixes		Examples
kilo	1,000	a kilogram is 1,000 grams
centi	0.01	a centimeter is 0.01 meter
milli	0.001	a milliliter is 0.001 liter
micro (μ)	one-millionth	a micrometer is 0.000001 (one-millionth) of a meter
nano (n)	one-billionth	a nanogram is 10^{-9} (one-billionth) of a gram
pico (p)	one-trillionth	a picogram is 10^{-12} (one-trillionth) of a gram

SOME COMMON UNITS OF LENGTH

Unit	Abbreviation	Equivalent
meter	m	approximately 39 in
centimeter	cm	10^{-2} m
millimeter	mm	10^{-3} m
micrometer	μm	10^{-6} m
nanometer	nm	10^{-9} m
angstrom	Å	10^{-10} m

Length conversions

1 in	=	2.5 cm	1 mm	=	0.039 in
1 ft	=	30 cm	1 cm	=	0.39 in
1 yd	=	0.9 m	1 m	=	39 in
1 mi	=	1.6 km	1 m	=	1.094 yd
			1 km	=	0.6 mi

To convert	Multiply by	To obtain
inches	2.54	centimeters
feet	30	centimeters
centimeters	0.39	inches
millimeters	0.039	inches

Think Metric!

A 154-lb person weighs 70 kilograms (kg).

A 5′6″ person is 165 cm long.

You are driving down the highway at 85.8 km per hour. That is the same speed as 55 mph.

A 70-kg human male has 5.6 liters of blood. That is about 6 quarts.

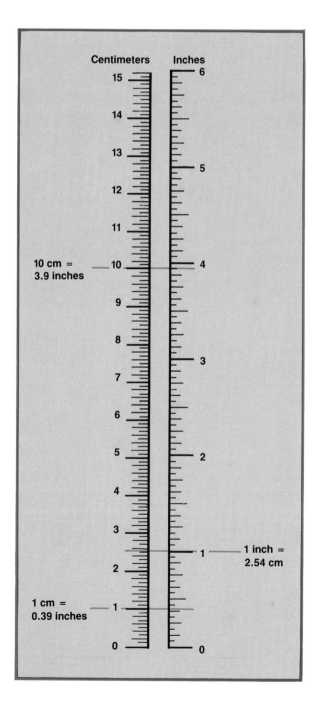

SOME COMMON UNITS OF VOLUME

Unit	Abbreviation	Equivalent
liter	l	approximately 1.06 qt
milliliter	ml	10^{-3} l (1 ml = 1 cm^3 = 1 cc)
microliter	μl	10^{-6} l

Volume conversions

1 tsp	= 5 ml	1 ml	= 0.03 fl oz
1 tbsp	= 15 ml	1 l	= 2.1 pt
1 fl oz	= 30 ml	1 l	= 1.06 qt
1 cup	= 0.241 l	1 l	= 0.26 gal
1 pt	= 0.47 l		
1 qt	= 9.95 l		
1 gal	= 3.8 l		

To convert	Multiply by	To obtain
fluid ounces	30	milliliters
quart	0.95	liters
milliliters	0.03	fluid ounces
liters	1.06	quarts

SOME COMMON UNITS OF WEIGHT

Unit	Abbreviation	Equivalent
kilogram	kg	10^3 g (approximately 2.2 lb)
gram	g	approximately 0.035 oz
milligram	mg	10^{-3} g
microgram	μg	10^{-6} g
nanogram	ng	10^{-9} g
picogram	pg	10^{-12} g

Weight conversions

1 oz	= 28.3 g	1 g	= 0.035 oz
1 lb	= 453.6 g	1 kg	= 2.2 lb
1 lb	= 0.45 kg		

To convert	Multiply by	To obtain
ounces	28.3	grams
pounds	453.6	grams
pounds	0.45	kilograms
grams	0.035	ounces
kilograms	2.2	pounds

APOTHECARY SYSTEM OF WEIGHT AND VOLUME*

Metric weight		Apothecary weight	Metric volume		Apothecary volume
30 g	=	1 ounce	1,000 ml	=	1 quart
15 g	=	4 drams	500 ml	=	1 pint
10 g	=	2.5 drams	250 ml	=	8 fl ounces
4 g	=	60 grains	90 ml	=	3 fl ounces
		(= 1 dram)	30 ml	=	1 fl ounce
2 g	=	30 grains			
1 g	=	15 grains			

*Used by pharmacists in preparing medications.

Temperature conversions	Some equivalents
$°C = \dfrac{(°F - 32) \times 5}{9}$	1°C = 1.8°F
	10°C = 18°F
$°F = \dfrac{°C \times 9}{5} + 32$	16°C = 61°F

Think Celsius!

When room temperature is 20°C, you probably will not feel cold. That is the same as 68°F.

When the temperature reaches 100°C, water boils.

At 0°C, water freezes.

Normal human body temperature is about 37°C.

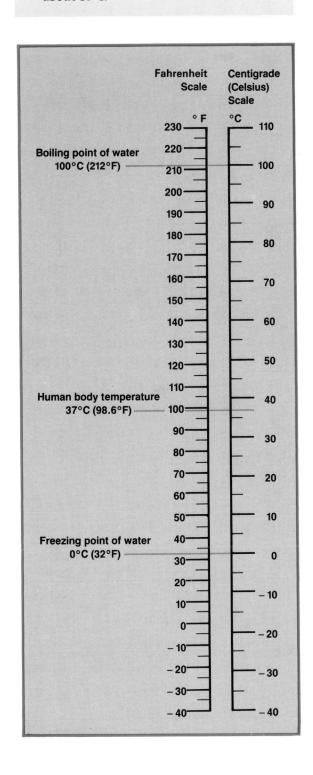

Boiling point of water
100°C (212°F)

Human body temperature
37°C (98.6°F)

Freezing point of water
0°C (32°F)

GLOSSARY/INDEX*

*Boldface page numbers indicate pages on which the index term is defined; "il" following the page number indicates an illustration, "t" a table, "f" a focus, and "n" a footnote.

Helix (**hee**-licks) Any spiral, but in biochemistry, usually one that does not vary in diameter. Also the name of an important genus of helix-shaped snail

Hemizygous genes Genes carried on a heterokaryotic sex chromosome, **185**

Hemocoel (**he**-mow-seel) A network of large blood-filled sinuses which makes up an open circulatory system in mollusks and arthropods, 320–497

Hemoglobin (**hee**-muh-**glo**-bin) The respiratory pigment of red blood cells that has the property of taking up oxygen or releasing it, 500
chemical structure of, 56il, 57

Hemoglobinopathic disease *See* Anemia

Hemolysis (he-**mol**-i-sis) The destruction of red blood cells and the resultant escape of hemoglobin, **189**

Hemophilia, 187f
Hemorrhoids, **511**
Heparin, 502

Hepatic (he-**pat**-ik) Pertaining to the liver

Hepatic portal system, 516–517

Herbivore (**herb**-i-vore) Animals which only consume plants or algae for nutritional requirements, 559–560, 763

Heredity *See* Genes; Genetic code; Genetics; Hardy-Weinberg law

Hermaphrodite (her-**maph**-ro-dite) An organism which can produce both male and female gametes, **633**

Heroin, 677t

Heterothallic (het-er-o-**thal**-ik) An organism having two mating types; only by combining a plus strain and a minus strain can sexual reproduction occur

Heterotroph(s), 15, **762**

Heterozygous alleles (het-ur-oh-**zye**-gus) Genetically mixed. Usually this term refers to a single pair of allelic genes that are unlike each other, **181**

Hexose, **45**
Hirudinea, 322, 322il
Histamine, **502**
Histones, 208, **231**
Holdfast, 290
Holothuroidea, **330**

Homeostasis (home-ee-oh-**stay**-sis) The balanced internal environment of the body; the automatic tendency of an organism to

Homeostasis (*Continued*)
maintain such a steady state, **6,** 726

Homeotherms (**home**-ee-o-therms) Organisms that maintain a constant body temperature metabolically rather than behaviorally, **726**

Hominids, evolution of, 707–712
Homo erectus, 708, 708il

Homologous chromosomes Chromosomes that have similar gene loci in the same sequence; in humans there are 23 pairs of homologous chromosomes; one member of each pair is derived from the mother and one from the father, 166il

Homologous structures (ho-**mol**-uh-gus) Corresponding in embryologic origin, often in structure, and often in presumed evolutionary origin, **718**, 713il, 719il

Homology Similarity in basic structural plan and development assumed to reflect a common genetic ancestry

Homothallic (home-o-**thal**-ik) A plant that employs only one physically recognizable mating type *See* Ulva

Homozygous alleles (hoh-moh-**zye**-gus) Genetically the same. Usually this term refers to a single pair of allelic genes that are alike, **181**

Honey bees, 51il, **744**
Hooke, Robert, 63

Hormone (**hor**-mone) A chemical messenger produced by an endocrine gland or by certain cells; a hormone is usually transported in the blood and regulates some aspect of metabolism, 611 *See also* individual hormones
invertebrate, 613–614
mechanisms of action, 611–613
vertebrate, 614–629

Horseshoe crabs, 326il
Horsetails, 293, 294il
Host and parasites, **562**
Host-mothering, 680t
Human ecology, 798–818
Human evolution, 707–712
Human growth hormone, 917t, 920–921
Human life cycle, 678, 679t
Humus, **372,** 782
Hydra, 310–311, 311il
Hydrogen,
atomic structure of, 30il, 33il
as energy carrier *See* Chemiosmosis
ions of, and pH scale, 38–40, 41
Hydrogen bond, 36il

Hydrolysis (high-**drol**-i-sis) A chemical reaction involving water in which a large molecule is usually broken down into smaller products with the chemical addition of water, 48, 48il

Hydrolytic enzymes Enzymes that split a larger chemical into smaller components by the chemical addition of water

Hydrophilic (high-droe-**fil**-ic) Attracted to water

Hydrophobic (high-droe-**fobe**-ic) Repelled by water

Hydrostatic skeleton, 435–436
Hydrothermal vent, 758f
Hydrozoa, **309**
Hygienic behavior (in bees), 746f
Hymen, **644**
Hyperglycemia, **625**

Hyperopia (high-per-**oh**-pee-ah) A refractive error of the eye in which only distant objects can be brought into focus

Hypersecretion (high-pur-se-**kree**-shun) Excessive secretion, e.g., when a gland releases too much hormone

Hypersensitivity, **535**–537

Hypertension (**high**-pur-**ten**-shn) High blood pressure, **510**

Hypertonic (**high**-pur-**ton**-ick) Having an osmotic pressure or solute concentration greater than that of some other solution which is taken as a standard, **94**, 95t

Hyperventilation (**high**-pur-ven-ti-**lay**-shun) Abnormally rapid, deep breathing

Hypha (Hyphae), 274
Hypocotyl, 383

Hypoglycemia (**high**-po-glye-**see**-mee-uh) Reduction of blood-glucose level below normal, 626

Hypostasis, genetic, **181**

Hypothalamus (high-poe-**thal**-uh-mus) A part of the brain that functions in regulating the pituitary gland, the autonomic system, emotional responses, body temperature, water balance, and appetite; located below the thalamus, **616**–617, 616t

Hypotheses, scientific, **21**

Hypotonic (**high**-poh-**ton**-ick) Referring to a solution whose osmotic pressure or solute content is less than that of some standard of comparison, **94**, 95t

Hypoxia (high-**pock**-see-uh) Oxygen deficiency

IG *See* Immunoglobulins

Prokaryotes (pro-**kar**-ee-oats) Simple unicellular organisms with no true nucleus such as bacteria or cyanobacteria, 66–69, **172**–173, 704–706

Prolactin (pro-**lak**-tin) A hormone secreted by the anterior lobe of the pituitary gland that stimulates lactation (milk production), 616t, 619, 645

Promoter A recognition signal encoded in DNA that functions to initiate transcription, 217

Pronuclei, 651
Propagules, **377**
Prophase, 160–163
Prop roots, **372**, 785il

Prostaglandins (pros-tuh-**glan**-dins) A group of local hormones which are released by many different tissues and which perform a wide range of physiological actions, **629**, 656–657

Prostate gland (**pros**-tate) In the male the gland that surrounds the beginning of the urethra and contributes to the semen, **637**

Protective coloration, **694**–695, 696il

Protein (**pro**-teen) A large, complex organic compound composed of chemically linked amino acid subunits; contains carbon, hydrogen, oxygen, nitrogen, and sulfur; the principal structural constituent of cells, 46t, **53**–57, 581–582
amino acid content of, **54**
biosynthesis of See Translation
classification of, 54t
deficiency of, **588**–589
digestion of, 566, 568t, 570
of plasma membrane, 88
quality of, 582
structure of, **55**–57, 56il

Prothoracic gland, **614**

Protists (Protista) Plant or animal-like, eukaryotes that occur as single-cell or which form colonies, 260t, 269–274

Protoderm, **395**

Proton (**pro**-ton) A positively charged subatomic particle located in the atomic nucleus, **27**

Proton gradients, and chemiosmosis See Mitochondria
Protonephridia, **312**, 596

Protoplasm Obsolete term for cellular contents, **69**

Protoplast Plant, fungal or bacterial cell without its cell wall
and cell fusion techniques, 249, 249il

Protostome An animal in which the mouth forms from the blastopore during embryonic development, 305–307

Protozoa (pro-toe-**zo**-uh) Single-celled or colonial heterotrophic eukaryotes, 260t, 269–273

Provascular cylinder, **394**
Proximal convoluted tubule, of nephron, 603il

Pseudocoelom (**soo**-dow-see-loam) Body cavity which develops between the mesoderm and endoderm, found principally in nematodes, **305**

Pseudopod (**soo**-dow-pod) A temporary protrusion of a portion of the cytoplasm of an ameboid cell used in locomotion or food engulfment, **270**

Peudostratified (soo-dow-**strat**-i-fide) columnar epithelium An epithelial tissue which appears layered but all of whose cells are attached to the same basement membrane, 416t

Puberty (**pew**-bur-tee) Period of sexual maturation; period during which the secondary sexual characteristics begin to develop and the ability to reproduce is attained, **637**

Pulmonary (**pul**-muh-ner-ee) Pertaining to the lungs

Pulmonary arteries, 512
Pulmonary circulation, **511**–512, 516il
Pulmonary emphysema, **554**, 555il
Pulmonary veins, 512

Pulse The rhythmic expansion of an artery which may be felt with the finger; it is due to blood ejected with each cardiac contraction and so is felt in time with the heartbeat, **509**–511

Pulvinus, **396**
Punctuated equilibrium, 699, 700f
Punnett square, 180il, 181, 182il, 183il, 185il, 187il, 192il

Purines Double-ring nucleotides, 57il

Pylorus, **566**

Pyrimidines Single-ring nucleotides, 57il See also DNA

Pyruvate, 143f, 145f, 146
Quantum, 31
Quaternary structure, of proteins, 55

Radial symmetry Body plan in which a multitude of planes will divide the organism into mirror-image halves, but in which one axis of the body is longer than the other; found principally in cnidarians, **303**, 304il

Radiata, **303**
Radicle See Root

Radioactive dating See Dating of fossils
Radioactive pollution, **808**–809, 810f
Rain forest, **785**, 785il
Rain shadow, **781**, 781il
Rapid eye movement (REM) sleep, 484
RAS (reticular activating system), 482–483
Reabsorption, in kidney, 605
Reaction centers, in photosynthesis, **129**
Reaction wood, **399**
Reagins, **536**
Receptacles, **380**

Receptor (re-**sep**-tur) 1. A specialized neural structure which is excited by a specific type of stimulus. 2. A site on the cell surface specialized to combine with a specific substance such as a hormone or transmitter substance

Recessive genes (re-**ses**-iv) Genes not expressed in the heterozygous state, **181**

Recombinant DNA, 237–243, 238il, 240il
ethical problems of, 250–251
Recombinant RNA, **250**, 251il
Rectum, 572
Recycling of wastes, 811
Red blood cells, 500, 501il

Reduction (re-**duk**-shun) In chemistry, the gain of electrons by a substance, or the chemical addition of hydrogen; the opposite of oxidation, **36**

Reflex action (**re**-flex) A predictable, automatic sequence of stimulus-response usually involving a sequence of at least three neurons: a sensory neuron, an association neuron, and a motor neuron, **458**–459, 459il

Refractory period, in neural transmission, **462**
Regulator genes, **226**–227
Releasers, **730**
Releasing hormones, 619, 620il

Renal (**re**-nal) Pertaining to the kidney

Renal tubule, 603il, 605

Renin (**re**-nin) A proteolytic enzyme released by the kidney in response to ischemia or lowered pulse pressure, which changes angiotensinogen into angiotensin, leading to increase in blood pressure, **511**

Replication fork, **210**
Repolarization, of neurons, **460**
Repression, of enzymatic end products, 227–228
Repressor, 228il
Reproduction
in animals, **633**–660
asexual, **9**

Stimulus (plural, stimuli) A physical or chemical change in the internal or external environment of an organism potentially capable of provoking a behavior response, **453**

Stipe, **290**

Stirrup, of ear *See* Stapes

Stomach, **565**–566, 565il

Stomata, 358, 358il, 359il
  role in photosynthesis, 359f

Stratified squamous epithelium, 416t
  See also Skin

Stratum basale (**stray**-tum ba-**say**-lee) The deepest layer of the epidermis that consists of cells that continuously divide, 433

Stratum corneum, **433**

Stress, **628**–629, 629il

Stroma
  of chloroplast, **126**
  of connective tissue, **416**

Stromatolites, **707**, 707il

Strontium-90, 808

Structural genes, **226**

Structural parallelism, 717–719

Structural proteins *See* Proteins

Style, of flower Long slender portion of the pistil above the ovulary which contains the pollen tube, **380**

Subatomic particles, 27–31

Subpharyngeal ganglia, **324**

Substrate-level phosphorylation, 118
  See also Glycolysis

Substrates, **113**

Subtidal zone, **793**

Successions, ecological, **772**–774, 772il, 773il
  lake, **789**

Sucrose, 46t, 47, 48il

Sugars, 45–48 *See also* Photosynthesis, Phloem, Glucose

Sulcus (**sul**-kus) (plural, sulci) A groove, trench, or depression, especially one of those on the surface of the brain separating the gyri, **478**

Sulfur, 577t

Sulfur bacteria *See* Photosynthetic bacteria

Surface tissues, of plants, 354, 355t

Sweat gland, 7il, 8, 433

Swim bladder, **546**

Symbiosis (sim-bee-**oh**-sis) An intimate relationship between two or more organisms of different species, **561**–562, 763, 764f *See also* Commensalism, Mutualism, Parasitism

Symmetry, 304f, 304il, 330

Sympathetic nervous system (**sim**-pah-**thet**-ik) The thoracolumbar portion of the autonomic nervous system; its general effect is to

Sympathetic nervous system (*Continued*) mobilize energy, especially during stress situations; prepares body for fight-or-flight response, **490**

Synapse (**sin**-aps) The junction between two neurons or between a neuron and an effector, 457

Synapsis, **166**

Synaptic cleft, **457**

Synaptic knob, **453**

Synaptic vesicles, 464il

Synovial fluid, 440

Syphilis, 660t

Systemic circulation, **511**–516, 513il, 516il

Systems, body *See* individual systems

Systole (**sis**-tuh-lee) The contraction phase of the cardiac cycle, **508**

Tadpole, **335**

Taiga, **780**–781, 781il

Tannins, 368

Tapeworms, 313–316, 315il

Target tissue, of hormone, **611**

Taste, sense of, 477

Taste buds, **477**

Taxes, **728**, 728il

Taxonomy, **258**–264

T-system of muscle cells, 443, 444il

Teeth, 560, 563

Telophase, **163**

Temperate communities, 781–783

Terminal bud, **389**

Territoriality, 739–740

Tertiary structure, of proteins, 55

Testis (**tes**-tis) (plural, testes) One of the paired male sex glands which produce both male sex hormone and sperm, 617t, **635**, 635il

Testosterone (tes-**tos**-tur-ohn) The principal male sex hormone; secreted by testes, 617t, 638

Tetanic contraction, **449**

Tetrad, 166, 166il

Thalamus (**thal**-uh-mus) The part of the brain that serves as a main relay center transmitting information between the spinal cord and the cerebrum, 479t, 481il

Thalidomide, 676il, 677t

Theory, scientific, **22**

Therapsids, **339**

Thermocline, **789**

Thermodynamics Principles governing heat or energy transfer, 107
  laws of, 107–112, 108il

Thermoregulation *See* Poikilotherm, Homeotherm

Thigmotropism, **396**

Thoracic (tho-**ras**-ik) Pertaining to the chest

Thoracic (lymphatic) duct, 518

Thrombin (**throm**-bin) One of several proteins in the blood important in the clotting process, 502

Thrombus (**throm**-bus) A blood clot formed within a blood vessel or within the heart, 514f

Thylakoids, **76**, 76il, 125

Thymine, 57il *See also* DNA

Thymosin *See* Thymus gland

Thymus gland (**thy**-mus) An endocrine gland that functions as part of the lymphatic system; necessary for the proper development of the immune response mechanism, **533**, 629

Thyroid gland (**thy**-roid) An endocrine gland that lies anterior to the trachea and releases hormones that regulate the rate of metabolism, 616t, 620–623
  disorders of, 622–623

Thyroid hormones, 616t, 618t

Thyroid stimulating hormone (TSH), 616t, **621**

Thyroxine (T_4) (thy-**rok**-sin) The principal hormone of the thyroid gland, 616t

Tight junctions, 91, 92il

Tinbergen, Nikko, 725

Tissue (**tish**-you) A group of closely associated, similar cells that work together to carry out specific functions
  animal, 413
  plant, 353–356

Tissue culture, 377il

Tissue fluid, 518

T lymphocyte A type of white blood cell responsible for cell-mediated immunity; also called T cell, **527**–528, 527il

Tonsils (**ton**-sils) Aggregate of lymph nodules in the throat region; located strategically to deal with pathogens that enter through the mouth or nose, 518

Topsoil, 783

Touch, sense of, 475

Trachea 1. In land-dwelling vertebrates, the main air tube extending from the lower end of the larynx to its division into the bronchi. 2. In land-dwelling arthropods and certain other invertebrates, tiny air tubes conveying oxygen directly to body cells
  of humans, **549**
  of insects, **325**, 544, 544il

Tracheids, 365

Tracheophytes, 292–296

Tranquilizers, 492t

Transcription The synthesis of RNA from a DNA template, **216**, 216il, 217, 218il